Statistical Mechanics of Liquids and Solutions

The statistical mechanical theory of liquids and solutions is a fundamental area of physical sciences with important implications in other fields of science and industrial applications. *Statistical Mechanics of Liquids and Solutions: Intermolecular Forces, Structure and Surface Interactions, Vol II* is the second in a series of two on this subject. While the first volume introduced equilibrium statistical mechanics in general and statistical mechanics of liquids and solutions at an introductory level, the present volume presents an advanced treatment of the subject and penetrates much deeper into liquid state theory.

A major theme in both books is the intimate relationship between forces in a fluid and the fluid structure – a relationship that is paramount for the understanding of the subject of interactions in dense fluids. Using a microscopic, molecular approach, the text emphasizes clarity of physical explanations for phenomena and mechanisms relevant to fluids, addressing the structure and behavior of liquids and solutions under various conditions. A notable feature is the author's treatment of intermolecular interactions in liquids and solutions that include interactions between nanoparticles, macroparticles and surfaces. The book provides an in-depth treatment of simple liquids, molecular fluids, particle dispersions, dense ionic fluids and electrolyte solutions with molecular solvents, both homogeneous bulk fluids and inhomogeneous fluids near surfaces and in confinement. It contains, for example, a unified exact treatment of electrolyte solutions, ionic liquids and polar fluids as well as approximate theories and applications. The book combines physical transparency, theoretical sharpness and a pedagogical and accessible style. It gives explicit and clear textual explanations and physical interpretations for the mathematical relationships and derivations.

Statistical Mechanics of Liquids and Solutions will be an invaluable resource for graduate and postgraduate students in physics, chemistry, soft matter science, surface and colloid science and related fields, as well as professionals and instructors in those areas of science.

Roland Kjellander earned a master's degree in chemical engineering, a Ph.D. in physical chemistry, and the title of docent in physical chemistry from the Royal Institute of Technology, Stockholm, Sweden. He is currently a professor emeritus of physical chemistry in the Department of Chemistry and Molecular Biology at the University of Gothenburg, Sweden. His previous appointments include roles in various academic and research capacities at the University of Gothenburg, Sweden; Australian National University, Canberra; Royal Institute of Technology, Stockholm, Sweden; Massachusetts Institute of Technology, Cambridge, USA; and Harvard Medical School, Boston, USA. He was awarded the 2004 Pedagogical Prize from the University of Gothenburg, Sweden, and the 2007 Norblad-Ekstrand Medal from the Swedish Chemical Society. Professor Kjellander's field of research is statistical mechanics, in particular liquid state theory.

Statistical Mechanics of Liquids and Solutions

Intermolecular Forces, Structure and Surface Interactions

Volume II

Roland Kjellander

CRC Press
Taylor & Francis Group
Boca Raton London New York

CRC Press is an imprint of the
Taylor & Francis Group, an **informa** business

Designed cover image: Shutterstock_1323284378

First edition published 2025
by CRC Press
2385 NW Executive Center Drive, Suite 320, Boca Raton FL 33431

and by CRC Press
4 Park Square, Milton Park, Abingdon, Oxon, OX14 4RN

CRC Press is an imprint of Taylor & Francis Group, LLC

ISBN: 978-1-032-25728-0 (hbk)
ISBN: 978-1-032-26162-1 (pbk)
ISBN: 978-1-003-28688-2 (ebk)

DOI: 10.1201/9781003286882

Typeset in Latin Modern font
by KnowledgeWorks Global Ltd.

Contents

Contents from Volume I

Chapter 7 ▪ Inhomogeneous and Confined Simple Fluids

Chapter 8 ▪ Surface Forces

Preface

This book is the second volume of two on equilibrium statistical mechanics of liquids and solutions with focus on intermolecular forces, surface forces and the structure of fluids. A major theme is the intimate relationship between forces in a fluid and the fluid structure – a relationship that is paramount for the understanding of the subject of interactions in dense fluids.

Interparticle interactions in liquids and solutions are immensely important in a large number of fields like solution chemistry, electrochemistry, soft matter physics, surface and colloid science, molecular biosciences and in many applications. Such interactions have traditionally been treated by using rather simple theories, but with the development of more sophisticated theoretical methods and, in particular, with the increasing power of computers, the realism of the theoretical treatments has been greatly improved during the last several decades.

A major tool in the theoretical investigation of liquids and solutions is computer simulations. There also exist various numerical methods based on approximations of considerable sophistication and accuracy. The use of such methods in parallel to simulations is complementary to the latter and their further development is important for future advances in the study of interactions in dense fluid media. This kind of methods rely more heavily on the understanding of fundamental statistical mechanics of fluids than simulations, while the latter are in general more simple to perform and more commonly used.

With the increased reliance on computer simulations for the study of molecular phenomena in liquids and solutions, there is, however, a risk of using such techniques as "black boxes" where an understanding of the statistical mechanical foundations is not prioritized. The results from simulations are numbers and without a sufficient theoretical framework, the relevance and reliability of these numbers can be hard to evaluate and may be overestimated. Furthermore, the links between various entities and their connection with the underlying phenomena may be out of sight. A deep statistical mechanical understanding of these matters can be of great advantage in the decisions of what calculations to do and how they can be interpreted. Thus, education in fundamental statistical mechanics for liquids and solutions is well motivated for all workers in the field.

For experimentalists, a knowledge of statistical mechanical theories of interparticle interactions beyond the traditional, simple ones is of great value for the design and interpretation of the increasingly more sophisticated methods available, like direct measurements of interparticle interactions and many techniques where such interactions are paramount for the phenomena studied. In many systems with strong coupling, for example concentrated electrolyte solutions and ionic liquids, and in a lot of other cases where many-body correlations are important, the traditional theories are clearly insufficient. The correct interpretation of the experimental results then must rely on more complete liquid state theory.

The first volume of this treatise contained a comprehensive introduction to general statistical mechanics for equilibrium systems and the statistical mechanics of simple fluids and electrolytes at an introductory level. The treatment included homogeneous fluids, fluids in confinement, inhomogeneous fluids in general and interactions between particles of any size and shape immersed in simple fluids, including forces between macroscopic particles and surface forces. Some simple, common approximations for fluids and electrolytes were treated as well as formally exact theory. Distribution function theory for spherical particles in homogeneous and inhomogeneous systems was the basis for the study. For electrolytes the Poisson-Boltzmann (PB) approximation was studied in some detail including rather recent findings for interparticle interactions in that approximation. Such systems were also treated in an exact general formalism (Dressed ion theory) that makes it possible to understand quite complex properties of electrolytes in a manner that is similar to that of the PB approximation despite that the theory is exact.

The current book contains liquid state theory for fluids consisting of spherical and/or nonspherical particles and uses more advanced statistical mechanical concepts and techniques. Thereby it greatly expands and deepens the study of liquids and solutions compared to the first volume, but it nevertheless shares the same basic philosophy in its presentation of the subject. Like the previous book it is limited to classical statistical mechanics of fluids at equilibrium. It presents and uses the major parts of the tool box of liquid-state theory and contains both formally exact theory and important approximate methods for homogeneous and inhomogeneous fluids. A special feature is a unified exact treatment of electrolyte solutions, ionic liquids and polar fluids. The contents of the book is summarized in the section *Overview of contents* and each chapter starts with a summary.

Insights in the underlying mechanisms for interactions in liquids constitute an essential ingredient in the analysis of many aspects of nature. Statistical mechanics of fluids should therefore be a part of the curriculum for a wider range of students than often is the case today. For these and many other reasons, text books like the present treatise have an important role to play in the education of chemists, physicists and students in neighboring fields. The books may also serve as an inspiration to a wider audience in the scientific community.

The inspiration to write these books came from my experiences from courses in statistical thermodynamics and liquid state theory at University of Gothenburg. The material that I wrote for these courses is integrated to a large extent in the two books. Major parts of the books are, however, newly written.

I am very much indebted to our students for the feedback that has helped in the development of a large part of the contents of the books. Both the questions and comments of the students during the lectures and after the courses, including ideas of what to improve and clarify, have been absolutely essential for me in the evolution of the material. Therefore, I would like to thank the students for their input in the process. I am also indebted to my present and former colleagues at the University of Gothenburg.

Special thanks go to Jean-Marc Bomont and Lloyd L. Lee for kindly providing me with original data used in Figures 10.2 and 10.3 and to Ruslan L. Davidchack for doing the same for Figures 10.8 and 10.9. Andreas Härtel is warmly thanked for the original data to Figures 10.10, 12.9, 12.11 and 12.10 and for sending me the originals for 10.11 and 10.12. I also want to cordially thank Susan Perkin and Timothy S. Groves for the original data for Figure 13.2 and Susan Perkin and Alexander Smith for the unpublished data for Figure 13.4. Luc Belloni is warmly thanked for providing me with the unpublished data for Figure 13.6. Finally, I would like to express my gratitude to Alexei Kornyshev, Godhard Sutmann and Hélène Berthoumieux for the original data for Figure 14.1. Everyone is also acknowledged for giving me the right to use the data and/or figures in the book and, in most cases, for reading the corresponding parts of the book. This is very much appreciated.

Overview of Contents

Volume I of this treatise contains Part I (Chapters 1–4) and Part II (Chapters 5–8).[1] The current volume, which can be read independently of the first one,[2] starts with **Part III** that treats *liquid state theory for homogeneous and inhomogeneous fluids.*

Chapter 9 presents the theory of simple fluids, that is, fluids consisting of spherically symmetric particles. Several types of correlation functions of various orders, like distribution function and direct correlation functions, are treated and their relationships are explored. A special emphasis is laid on physical interpretations of the mathematical relationships in the theory. We study linear response theory and derive several exact integral-equation relationships for correlation functions, density distributions and other entities. Thereafter nonlinear response is treated and the associated bridge functions on the singlet and pair levels are defined. Furthermore, higher order direct-correlation functions are given a physical interpretation. Various important closure approximations in integral equation theory are introduced. The concept of functionals and the use of functional derivatives, expansions and integrations, are established and extensively utilized to formulate and derive the theory. The foundation of density functional theory is laid out.

Chapter 10 deals with important approximate theories. Several classical integral-equation approximations are studied and some modern, accurate integral-equation theories are explored. Special attention is placed on the application of several types of consistency criteria in such theories. Various ways to assess the accuracy of the approximations for bulk fluids are explored with the objective to do systematic improvements. Integral equation theories for inhomogeneous fluids on both singlet and pair distribution levels are also treated. Approximate Density Functional Theories are introduced and, in particular, Fundamental Measure Theory (FMT) for hard sphere mixture is explored in some depth. Special emphasis is placed on the introduction of the concepts used in FMT in a gradual and logical manner. Various refinements of the original version of FMT are presented as well as a more consistent approach.

[1]Part I. *Basis of Equilibrium Statistical Mechanics* and Part II. *Fluid Structure and Interparticle Interactions.* A table of contents of Volume I is included in the present book.

[2]Throughout the book we will recapitulate essential parts of the theory from Volume I along the way. References to the previous volume will be given when appropriate, so the reader should be able to easily find the corresponding material in Volume I.

The general statistical mechanics for fluids consisting of nonspherical particles is treated in **Chapter 11**. Distribution function theory in the canonical and the grand canonical ensembles is thereby explored in considerable detail, including probability densities, particle density distributions and different types of correlation functions of various orders. Functional derivative relationships between the various correlation functions, density distributions and free energy entities are established. Many-body chemical potentials are introduced and utilized in the formulation of the theory. Some classical theories for effective solute-solute interactions in solutions are also presented, where the solvent is treated implicitly. This includes the McMillan-Mayer theory for solutions, which have important conceptual features, and the Adelman theory for effective potentials and direct correlation functions.

Part IV contains *a unified exact treatment of electrolyte solutions, ionic liquids and polar fluids*.

Chapter 12 deals with simple electrolytes. It presents an exact theory for inhomogeneous and homogeneous electrolytes consisting of spherically symmetric ions, including the primitive model of electrolyte solutions (where the solvent is modeled as a dielectric continuum), classical plasmas and molten salts. Particularly important subjects are dielectric properties of electrolytes, nonlocal electrostatics, a general screened Coulomb potential and the simultaneous existence of multiple decay modes for electrostatic interactions and correlations. The modes have different decay lengths and have either monotonic or oscillatory exponential decay with distance. A key element is the fact that each decay length in bulk electrolytes is not simply a parameter that is explicitly expressed in terms of the system parameters, like the Debye length in the Debye-Hückel approximation. Instead it is a solution of an equation with multiple solutions derived from the dielectric response function of the electrolyte; each solution gives a decay mode. This exact equation can be written in a form that is very similar to the usual expression for the Debye length. Many different aspects of electrolyte systems are treated, including various types of interparticle correlations, surface interactions and the phenomena of underscreening and overscreening.

Electrolytes consisting of nonspherical ions, molecules and/or other particles are treated in **Chapter 13**. It contains a general, exact theory for ionic liquids and electrolyte solutions with molecular solvent that is a generalization of the corresponding theory for spherical-ion electrolytes in Chapter 12. The key features are the same in both cases. In particular, the analysis of dielectric properties of electrolytes, nonlocal electrostatics, the screened Coulomb potential and the existence of multiple decay modes is essentially the same. The main difference is the fact that the interactions between ions/particles are dependent on particle orientations, which makes the formalism to differ in detail but not in its over-all features. An important ingredient is the concept of "multipolar effective charges," that is, effective charges with values that depend on particle orientations. They describe the strength and orientation dependence of the interparticle interactions for each decay mode, whereby each mode has its own value of these entities. Concrete applications of the theory include the interpretation of experimental results for surface forces in room-temperature ionic liquids and other fluids. This includes a theoretical analysis of underscreening and the nature of hydration/solvation forces.

Chapter 14 deals with pure polar fluids. One important topic is the dielectric properties of such fluids, especially the static dielectric function and the dielectric constant. The interparticle interactions and correlations are dominated for large distances by the Coulombic contributions, but for short separations oscillatory contributions dominate in an interval up to several molecular diameters. They have very much larger amplitude than the Coulombic forces. These oscillatory, exponentially decaying contributions constitute multiple decay modes of the same kinds as for electrolytes and satisfy the same equation that is derived from the dielectric response function. They originate from the structure of the fluid and cause a kind of structural force. For surface forces they give rise to a "solvation force" (often called "hydration force" in aqueous systems).

References to equations in Volume I of this treatise are written in the format "Equation I:x.x," where x.x is the equation number in Volume I. The table of contents of Volume I is included.

Text marked with ★ contains material that is complementary or more advanced than the rest and can be skipped in the first reading.

A ***list of symbols*** used in the book is included.

Exercises are integrated in the text throughout the book and are marked with **{Exercise: ... ...}**.

Author

Roland Kjellander earned a master's degree in chemical engineering, a Ph.D. in physical chemistry, and the title of docent in physical chemistry from the Royal Institute of Technology, Stockholm, Sweden. He is currently a professor emeritus of physical chemistry in the Department of Chemistry and Molecular Biology at the University of Gothenburg, Sweden. His previous appointments include roles in various academic and research capacities at the University of Gothenburg, Sweden; Australian National University, Canberra; Royal Institute of Technology, Stockholm, Sweden; Massachusetts Institute of Technology, Cambridge, USA; and Harvard Medical School, Boston, USA. He was awarded the 2004 Pedagogical Prize from the University of Gothenburg, Sweden, and the 2007 Norblad-Ekstrand Medal from the Swedish Chemical Society. Professor Kjellander's field of research is statistical mechanics, in particular liquid state theory. (http://orcid.org/0000-0003-2651-8775)

List of Symbols

Symbol	*Explanation*	*SI unit*
A	Helmholtz energy	J
$\mathcal{A}$	intrinsic Helmholtz energy = A minus the interaction energy with external potential	J
A	area	m^2
A^{h}	surface area of a hard sphere	m^2
A^{ex}	excess Helmholtz energy	J
A^{ideal}	ideal Helmholtz energy	J
$\mathcal{A}^{ex}$	excess intrinsic Helmholtz energy	J
$b^{(1)}(\mathbf{r})$	singlet bridge function	unitless
$b^{(2)}(r)$, $b^{(2)}(\mathbf{r},\mathbf{r}')$	pair bridge function	unitless
$c^{(1)}(\mathbf{r})$, $c^{(1)}(\mathbf{R})$	singlet direct-correlation function	unitless
$c^{(1)}(\mathbf{R}\|\mathbf{R}')$	singlet direct-correlation function for a particle with coordinates $\mathbf{R} = (\mathbf{r}, \boldsymbol{\omega})$ when a particle is located at $\mathbf{r}'$ and has orientation $\boldsymbol{\omega}'$. Analogously for $c^{(1)}(\mathbf{r}\|\mathbf{r}')$.	unitless
$c^{(1)\mathrm{R}}(\mathbf{r})$	residual part of singlet direct-correlation function (defined in Equations 9.202 and 9.206)	unitless
$c^{(2)}(r)$, $c^{(2)}(\mathbf{R},\mathbf{R}')$	pair direct-correlation function	unitless
$c^{\star}_{ij}(r)$, $c^{\star}_{ij}(\mathbf{R},\mathbf{R}')$	short-range part of $c^{(2)}$, defined as $c^{\star}_{ij} \equiv c^{(2)}_{ij} + \beta u^{el}_{ij}$	unitless
$C^{(2)}_{ij}(\mathbf{r},\mathbf{r}')$	$\equiv \delta^{(3)}_{ij}(\mathbf{r}-\mathbf{r}')/n_i(\mathbf{r}) - c^{(2)}_{ij}(\mathbf{r},\mathbf{r}')$ (spherical particles)	unitless

Symbol	*Explanation*	*SI unit*
$C_{ij}^{(2)}(\mathbf{R},\mathbf{R}')$	analogous to $C_{ij}^{(2)}(\mathbf{r},\mathbf{r}')$ (Equation 11.141)	unitless
$C_{ij}^{\star}(\mathbf{r},\mathbf{r}')$	$\equiv \delta_{ij}^{(3)}(\mathbf{r}-\mathbf{r}')/n_i(\mathbf{r}) - c_{ij}^{\star}(\mathbf{r},\mathbf{r}')$ (spherical particles)	unitless
$C_{ij}^{\star}(\mathbf{R},\mathbf{R}')$	analogous to $C_{ij}^{\star}(\mathbf{r},\mathbf{r}')$ (Equation 13.31)	unitless
$c^{(l)}(\mathbf{r}_1,\ldots,\mathbf{r}_l)$	direct correlation function of order l	unitless
d^{h}	diameter of hard core	m
d^{LJ}	particle size parameter in Lennard-Jones potential	m
$D^{\mathcal{S}}$	surface separation	m
$d^{\mathcal{S}}$	distance of closest approach of particle centers to a surface	m
$\mathbf{E}$	electrostatic field	$\mathrm{V\,m^{-1}}$
$\mathcal{E}_r^{\mathrm{eff}}$	effective relative dielectric permittivity	unitless
$\mathcal{E}_r^{\star}$	dielectric factor	unitless
$\mathbf{F}$	force vector, $\mathbf{F} = (F_x, F_y, F_z)$	N
F	force	N
$\mathbf{F}^{\mathrm{intr}}$	intrinsic force = force from particles of the fluid (not from external field)	N
G	Gibbs energy	J
$g^{(1)}(\mathbf{r})$	singlet distribution function $g^{(1)}(\mathbf{r}) \equiv n^{(1)}(\mathbf{r})/n^{\mathrm{b}}$	unitless
$g^{(2)}(r)$, $g_{ij}^{(2)}(r)$	pair distribution function for spherical particles in bulk fluid (radial distribution function)	unitless
$g_{ij}^{(2)}(\mathbf{R},\mathbf{R}')$	pair distribution function (general case)	unitless
$g^{(l)}$	l-point (l-particle) distribution function	unitless

Symbol	*Explanation*	*SI unit*
$g_N^{(l)}$	l-point (l-particle) distribution function (canonical ensemble with N particles)	unitless
H	enthalpy	J
h	Planck's constant ($h = 6.62607 \cdot 10^{-34}$ J s)	J s
$\mathcal{H}(x)$	Heaviside step function ($\mathcal{H}(x) = 0$ for $x < 0$ and 1 for $x \geq 0$)	unitless
$h^{(2)}(r)$, $h_{ij}^{(2)}(r)$	pair correlation function for spherical particles in bulk fluid, $h^{(2)} = g^{(2)} - 1$	unitless
$h_{ij}^{(2)}(\mathbf{R}, \mathbf{R}')$	pair correlation function (general case), $h_{ij}^{(2)} = g_{ij}^{(2)} - 1$	unitless
$h_N^{(2)}(\mathbf{R}, \mathbf{R}')$	pair correlation function, $h_N^{(2)} = g_N^{(2)} - 1$ (canonical ensemble with N particles)	unitless
$h_{ij}^{\star}(r)$, $h_{ij}^{\star}(\mathbf{R}, \mathbf{R}')$	pair correlation function for the dress of an ion or particle (defined in Equations 12.33 and 13.25)	unitless
$h_{\mathrm{S}}^{(2)}(r)$	sum correlation function, $h_{\mathrm{S}}^{(2)} = \frac{1}{2}[h_{++}^{(2)} + h_{+-}^{(2)}]$	unitless
$h_{\mathrm{D}}^{(2)}(r)$	difference correlation function, $h_{\mathrm{D}}^{(2)} = \frac{1}{2}[h_{++}^{(2)} - h_{+-}^{(2)}]$	unitless
$H_{ij}^{(2)}(\mathbf{r}, \mathbf{r}')$	$\equiv n_i(\mathbf{r})\delta_{ij}^{(3)}(\mathbf{r} - \mathbf{r}') + n_i(\mathbf{r})n_j(\mathbf{r}')h_{ij}^{(2)}(\mathbf{r}, \mathbf{r}')$ (spherical particles)	m^{-6}
$H_{ij}^{(2)}(\mathbf{R}, \mathbf{R}')$	analogous to $H_{ij}^{(2)}(\mathbf{r}, \mathbf{r}')$ (Equation 11.140)	m^{-6}
$H_{ij}^{\star}(\mathbf{r}, \mathbf{r}')$	$\equiv n_i(\mathbf{r})\delta_{ij}^{(3)}(\mathbf{r} - \mathbf{r}') + n_i(\mathbf{r})n_j(\mathbf{r}')h_{ij}^{\star}(\mathbf{r}, \mathbf{r}')$ (spherical particles)	m^{-6}
$H_{ij}^{\star}(\mathbf{R}, \mathbf{R}')$	analogous to $H_{ij}^{\star}(\mathbf{r}, \mathbf{r}')$ (Equation 13.29)	m^{-6}

Symbol	*Explanation*	*SI unit*
$H_{\mathrm{NN}}, H_{\mathrm{NQ}}, H_{\mathrm{QQ}}$	density-density, density-charge and charge-charge correlation functions (defined in Equations 12.204 and 13.98)	m^{-6}
$H^{\star}_{\mathrm{NN}}, H^{\star}_{\mathrm{NQ}}, H^{\star}_{\mathrm{QQ}}$	defined as H_{NN}, H_{NQ} and H_{QQ} but with $H^{\star}_{ij}$ instead of $H^{(2)}_{ij}$	m^{-6}
i	imaginary unit	unitless
$\mathbb{I}$	identity matrix (diagonal elements equal to 1 and the rest equal to 0)	unitless
$\Im(\cdot)$	imaginary part (of a complex number)	
$\mathbf{j}$	angular momentum vector, $\mathbf{j} = (j_x, j_y, j_z)$	$\mathrm{kg\,m^2\,s^{-1}}$
$\mathbf{k}$	wave vector, $\mathbf{k} = k\hat{\mathbf{k}}$	m^{-1}
k	wave number, $k = \lvert\mathbf{k}\rvert = 2\pi/\lambda$	m^{-1}
k_B	Boltzmann's constant ($k_B = 1.38065 \cdot 10^{-23}\,\mathrm{J\,K^{-1}}$)	$\mathrm{J\,K^{-1}}$
M	concentration unit "molar" = $\mathrm{mol\,dm^{-3}}$ ($1\,\mathrm{M} = 10^3\,\mathrm{mol\,m^{-3}}$)	$1\,\mathrm{mol\,m^{-3}} = 10^{-3}\,\mathrm{M}$
m	mass	kg
m	in some obvious cases m is an integer	
m^{d}	dipole moment	$\mathrm{C\,m}$
$\mathbf{m}^{\mathrm{d}}$	dipole moment vector	$\mathrm{C\,m}$
$\mathbf{m}^{\mathrm{d}\star}$	dipole moment vector of a dressed particle	$\mathrm{C\,m}$
$\mathbf{m}^{\mathrm{d,eff}}$	effective dipole moment vector of a particle	$\mathrm{C\,m}$
$\mathbb{M}^{\mathrm{Q}}$	quadrupole moment tensor	$\mathrm{C\,m^2}$
$\mathbb{M}^{\mathrm{Q}\star}$	quadrupole moment tensor of a dressed particle	$\mathrm{C\,m^2}$
$\mathbb{M}^{\mathrm{Q,eff}}_i$	effective quadrupole moment tensor of a particle	$\mathrm{C\,m^2}$

Symbol	*Explanation*	*SI unit*
N	number of particles	unitless
N_i	number of particles of species i	unitless
N_{Av}	Avogadro's constant ($N_{Av} = 6.02214 \cdot 10^{23}$ mol^{-1})	mol^{-1}
n, $n^{(1)}$	number density (number of particle centers per unit volume)	m^{-3}
n_i, $n_i^{(1)}$	number density of species i (number of particle centers per unit volume)	m^{-3}
n_i^{b}	number density of species i in bulk phase	m^{-3}
n_{tot}	total number density, $n_{\mathrm{tot}} = \sum_i n_i$	m^{-3}
$n_{\mathrm{tot}}^{\mathrm{b}}$	total number density in bulk phase	m^{-3}
$\mathfrak{n}(\mathbf{R}) \equiv \mathfrak{n}(\mathbf{r}, \boldsymbol{\omega})$	number density at $\mathbf{r}$ of particles with orientation $\boldsymbol{\omega}$	m^{-3}
$n_i^{(1)}(\mathbf{r}\vert\mathbf{r}'; j)$	number density of species i at $\mathbf{r}$ when there is a fixed particle of species j at $\mathbf{r}'$	m^{-3}
$\mathfrak{n}_i^{(1)}(\mathbf{R}\vert\mathbf{R}'; j)$	number density at $\mathbf{r}$ of i-particles with orientation $\boldsymbol{\omega}$ [where $(\mathbf{r}, \boldsymbol{\omega}) = \mathbf{R}$] when there is a fixed j-particle with orientation $\boldsymbol{\omega}'$ at $\mathbf{r}'$	m^{-3}
$\mathfrak{n}_{i;j}^{(1)}(\mathbf{R}\vert\mathbf{R}')$	$\equiv \mathfrak{n}_i^{(1)}(\mathbf{R}\vert\mathbf{R}'; j)$ (an alternative notation). Analogously for $n_{i;j}^{(1)}(\mathbf{r}\vert\mathbf{r}')$.	m^{-3}
$n(\mathbf{r}\vert\mathbf{r}_1, \ldots, \mathbf{r}_l)$	number density at $\mathbf{r}$ when l particles are located at positions $\mathbf{r}_1, \ldots, \mathbf{r}_l$	m^{-3}
$\mathfrak{n}(\mathbf{R}\vert\mathbf{R}_1, \ldots, \mathbf{R}_l)$	number density at $\mathbf{r}$ for particles with orientation $\boldsymbol{\omega}$ [where $(\mathbf{r}, \boldsymbol{\omega}) = \mathbf{R}$] when l fixed particles have coordinates $\mathbf{R}_1, \ldots, \mathbf{R}_l$	m^{-3}

Symbol	*Explanation*	*SI unit*
$\breve{n}^{(1)}_{\{\mathbf{r}^N\}}(\mathbf{r})$	microscopic number density distribution at $\mathbf{r}$ for particle configuration $\mathbf{r}^N$ (spherical particles)	m^{-3}
$\breve{\mathfrak{n}}^{(1)}_{\{\mathbf{R}^N\}}(\mathbf{R})$	microscopic number density distribution for particles with coordinates $\mathbf{R} = (\mathbf{r}, \boldsymbol{\omega})$ for particle configuration $\mathbf{R}^N$.	m^{-3}
$n^{(l)}$, $\mathfrak{n}^{(l)}$	l-point (l-particle) density distribution function ($n^{(1)} \equiv n$, $\mathfrak{n}^{(1)} \equiv \mathfrak{n}$)	m^{-3l}
$n^{(l)}_N$, $\mathfrak{n}^{(l)}_N$	l-point (l-particle) density distribution function (canonical ensemble with N particles)	m^{-3l}
$\mathfrak{n}_0$, $\mathfrak{n}_R$, $\mathfrak{n}_A$, $\mathfrak{n}_V$	weighted densities used in FMT (defined in Equations 10.152 and 10.154)	various
$\vec{\mathfrak{n}}_{R^{\hat{r}}}$, $\vec{\mathfrak{n}}_{A^{\hat{r}}}$	vector-valued weighted densities used in FMT (defined in Equation 10.154)	various
P	pressure = force per unit area	$\mathrm{N\,m^{-2}}$
$\mathcal{P}$	probability	unitless
$\mathbf{p}$	momentum vector, $\mathbf{p} = (p_x, p_y, p_z) = m\mathbf{v}$	$\mathrm{kg\,m\,s^{-1}} = \mathrm{N\,s}$
p	length of momentum vector, $p = \lvert\mathbf{p}\rvert$	$\mathrm{kg\,m\,s^{-1}} = \mathrm{N\,s}$
$\mathfrak{p}$	probability	unitless
P^{b}	pressure in bulk phase	$\mathrm{N\,m^{-2}}$
$\mathscr{P}(N)$	probability that an open system contains N particles	unitless
$\mathcal{P}^{(l)}$	l-point (l-particle) configurational probability density	m^{-3l}
$\mathcal{P}^{(l)}_N$	l-point (l-particle) configurational probability density (canonical ensemble with N particles)	m^{-3l}

Symbol	*Explanation*	*SI unit*
$\mathbf{p}^N$	momenta in $3N$-dimensional space, $\mathbf{p}^N \equiv \mathbf{p}_1, \mathbf{p}_2, \cdots, \mathbf{p}_N$	$\mathrm{kg\,m\,s^{-1}} = \mathrm{N\,s}$
Q_N	canonical partition function for N particles	unitless
q	charge	C
q	heat (added to the system)	J
q_e	elementary charge (charge of a proton) ($q_\mathrm{e} = 1.60218 \cdot 10^{-19}\,\mathrm{C}$)	C
q_i	charge of a particle of species i	C
q_i^eff	effective charge of a particle of species i	C
$q_i^\star$	dressed particle charge of species i, $q_i^\star = \int d\mathbf{r}\, \rho_i^\star(\mathbf{r})$, $q_i^\star = \int d\mathbf{r}\, \rho_i^\star(\mathbf{r}, \boldsymbol{\omega})$	C
$\overline{q}_i^\star$	average dressed particle charge, defined in Equation 13.42 (for anisotropic bulk phase)	C
$q_\pm$	$q_\pm \equiv \mathrm{x}_+^\mathrm{b} q_+ + \mathrm{x}_-^\mathrm{b} \lvert q_- \rvert$ (for binary electrolyte)	C
$\mathsf{Q}^\mathrm{eff}(\hat{\mathbf{r}}, \boldsymbol{\omega})$	multipolar effective charge (direction dependent) of particle with orientation $\boldsymbol{\omega}$	C
$\mathbf{r}$	cartesian coordinate vector, $\mathbf{r} = (x, y, z)$	m
r	length of vector $\mathbf{r}$, $r = \lvert \mathbf{r} \rvert$	m
$\hat{\mathbf{r}}$	unit vector $\hat{\mathbf{r}} \equiv \mathbf{r}/r$	unitless
$\mathbf{r}^N$	coordinates in $3N$-dimensional space, $\mathbf{r}^N \equiv \mathbf{r}_1, \mathbf{r}_2, \cdots, \mathbf{r}_N$ (particle configuration)	m
$\mathbf{R}$	$\mathbf{R} \equiv (\mathbf{r}, \boldsymbol{\omega})$ a composite variable for location and orientation of a particle	(m, unitless)
$\mathbf{R}^N$	$\mathbf{R}^N \equiv \mathbf{R}_1, \mathbf{R}_2, \cdots, \mathbf{R}_N$ (particle configuration)	(m, unitless)

Symbol	*Explanation*	*SI unit*
$\Re(\cdot)$	real part (of a complex number)	
R^{h}	radius of a hard sphere	m
S	entropy	$\mathrm{J\,K^{-1}}$
T	absolute temperature	K
t	time	s
U	energy	J
$\langle U\rangle$, $\bar{U}$	internal energy ($\langle U\rangle \equiv \bar{U}$)	J
$\mathfrak{u}$	energy of single particle	J
$\mathscr{U}_i$	energy of quantum state i of a system	J
$u(r)$, $u_{ij}(r)$	pair interaction potential (spherical particles)	J
$u_{ij}(\mathbf{R},\mathbf{R}')$	pair interaction potential (general case)	J
u^{el}	electrostatic part of the pair interaction potential	J
u^{sh}	short-range (non-electrostatic) part of the pair interaction potential	J
$u^{\ell}(r)$, $u^{\mathrm{L}}(r)$	a long-range part of the pair potential as used when $u(r)$ is split up in practical calculations (cf. Figure 9.1)	J
$u^{s}(r)$, $u^{\mathrm{S}}(r)$	a short-range part of the pair potential as used when $u(r)$ is split up in practical calculations (cf. Figure 9.1)	J
$\breve{U}_N^{\mathrm{pot}}(\mathbf{r}^N)$, $\breve{U}_N^{\mathrm{pot}}(\mathbf{R}^N)$	instantaneous potential energy for N particles	J
V	volume	$\mathrm{m^3}$
$\mathbf{v}$	velocity vector, $\mathbf{v} = (v_x, v_y, v_z)$	$\mathrm{m\,s^{-1}}$
$v(\mathbf{r})$, $v(\mathbf{R})$	external potential	J

Symbol	*Explanation*	*SI unit*
v^{el}	electrostatic part of the external potential (the interaction energy between a particle and Ψ^{ext})	J
v^{sh}	short-range (non-electrostatic) part of the external potential	J
$v(\mathbf{R}\|\mathbf{R}')$	external potential in the presence of a fixed external particle with coordinates $\mathbf{R}' = (\mathbf{r}', \boldsymbol{\omega}')$, i.e., $v(\mathbf{R}\|\mathbf{R}') = v(\mathbf{R}) + u(\mathbf{R}, \mathbf{R}')$. Analogously for $v(\mathbf{r}\|\mathbf{r}')$.	J
V^{h}	volume of a hard sphere	m^3
$\mathtt{W}$	work (done on the system)	J
w	potential of mean force	J
$w^{(1)}(\mathbf{r})$, $w^{(1)}(\mathbf{R})$	potential of mean force for a particle, alternative notations: $w(\mathbf{r})$ and $w(\mathbf{R})$	J
$w_{ij}^{(2)}$	pair potential of mean force for particles of species i and j	J
$w_{ij}^{(2)\text{el}}$	the linear electrostatic part of $w_{ij}^{(2)}$ (see Table 12.2 on page 303 and Equation 13.28)	J
w^{intr}	intrinsic potential of mean force, i.e., due to particles of the fluid (not from external potential)	J
$w(\mathbf{R}\|\mathbf{R}')$	potential of mean force for a particle with coordinates $\mathbf{R} = (\mathbf{r}, \boldsymbol{\omega})$ when a particle is located at $\mathbf{r}'$ and has orientation $\boldsymbol{\omega}'$. Analogously for $w(\mathbf{r}\|\mathbf{r}')$.	J
$w^{(l)}(\mathbf{R}_1, \ldots, \mathbf{R}_l)$	l-point (l-particle) potential of mean force	J
$\mathscr{W}^{S}_{\mathtt{w}_1,\mathtt{w}_2}$	free energy of interaction per unit area between two planar walls ($\mathtt{w}_1$ and $\mathtt{w}_2$)	$\text{J}\,\text{m}^{-2}$

Symbol	*Explanation*	*SI unit*
x_i	mole fraction of species i; for a mixture of substances 1, 2 and 3: $\mathsf{x}_i = N_i/(N_1 + N_2 + N_3) = n_i/(n_1 + n_2 + n_3)$	unitless
$\mathsf{x}_+^{\mathrm{b}}$, $\mathsf{x}_-^{\mathrm{b}}$	$\mathsf{x}_i^{\mathrm{b}} = n_i^{\mathrm{b}}/(n_+^{\mathrm{b}} + n_-^{\mathrm{b}})$ for $i = +, -$ in binary electrolyte bulk phase	unitless
$y^{(2)}(\mathbf{r}, \mathbf{r}')$	cavity function, $y^{(2)} \equiv e^{\gamma^{(2)} + b^{(2)}}$. It satisfies $g^{(2)}(\mathbf{r}, \mathbf{r}') = y^{(2)}(\mathbf{r}, \mathbf{r}')e^{-\beta u(\mathbf{r}, \mathbf{r}')}$.	unitless
Z_N	configurational partition function for N particles	m^{3N}
z_i	valency of an ion of species i, $\mathsf{z}_i = q_i/q_{\mathrm{e}}$	unitless
$\mathsf{Z}_{N+l}^{(l)}(\mathbf{R}_1, \ldots, \mathbf{R}_l)$	configurational partition function for a system with N mobile particles when l fixed particles have coordinates $\mathbf{R}_1, \ldots, \mathbf{R}_l$	m^{3N}
β	$\beta = (k_B T)^{-1}$	J^{-1}
γ	activity coefficient (activity factor)	unitless
γ^{b}	activity coefficient (activity factor) in bulk phase	unitless
$\gamma^{(1)}(\mathbf{r})$	$\gamma^{(1)}(\mathbf{r}_1) \equiv \int d\mathbf{r}_2\, c^{(2)}(r_{12})[n(\mathbf{r}_2) - n^{\mathrm{b}}]$	unitless
$\gamma^{(2)}(\mathbf{r}, \mathbf{r}')$	$\gamma^{(2)}(\mathbf{r}, \mathbf{r}') \equiv h^{(2)}(\mathbf{r}, \mathbf{r}') - c^{(2)}(\mathbf{r}, \mathbf{r}')$	unitless
δ_{ij}	Kronecker delta (1 if $i = j$ and 0 otherwise)	unitless
$\delta(x)$	Dirac's delta function	m^{-1}
$\delta^{(2)}(\mathbf{r})$	two-dimensional Dirac delta function, $\delta^{(2)}(\mathbf{r}) \equiv \delta(x)\delta(y)$	m^{-2}
$\delta^{(3)}(\mathbf{r})$	three-dimensional Dirac delta function, $\delta^{(3)}(\mathbf{r}) \equiv \delta(x)\delta(y)\delta(z)$	m^{-3}
$\delta_{ij}^{(3)}(\mathbf{r})$	$\delta_{ij}^{(3)}(\mathbf{r}) \equiv \delta_{ij}\delta^{(3)}(\mathbf{r})$	m^{-3}
$\delta^{\omega}(\boldsymbol{\omega})$	Dirac function for orientational variable	unitless

Symbol	*Explanation*	*SI unit*
$\delta^{(3)}(\mathbf{R})$	$\delta^{(3)}(\mathbf{R}) \equiv \delta^{(3)}(\mathbf{r})\delta^{\omega}(\boldsymbol{\omega})$	m^{-3}
$\delta^{(3)}_{ij}(\mathbf{R})$	$\delta^{(3)}_{ij}(\mathbf{R}) \equiv \delta^{(3)}(\mathbf{r})\delta^{\omega}(\boldsymbol{\omega})\delta_{ij}$	m^{-3}
$\varepsilon^{\mathrm{LJ}}$	energy parameter in Lennard-Jones potential	J
ε_r	relative dielectric permittivity (dielectric constant)	unitless
$\varepsilon_r^{\mathrm{cont}}$	dielectric constant for a dielectric continuum	unitless
ε_0	permittivity of vacuum ($\varepsilon_0 = 8.85419{\cdot}10^{-12}\ C^2\mathrm{J}^{-1}\mathrm{m}^{-1}$)	$C^2\mathrm{J}^{-1}\mathrm{m}^{-1}$
$\tilde{\epsilon}(\mathbf{k})$	dielectric function (longitudinal static), $\tilde{\epsilon}(\mathbf{k}) = \tilde{\epsilon}_{\mathrm{sing}}(k) + \tilde{\epsilon}_{\mathrm{reg}}(\mathbf{k})$	unitless
$\tilde{\epsilon}_{\mathrm{sing}}(k)$	singular part of $\tilde{\epsilon}(\mathbf{k})$, $\tilde{\epsilon}_{\mathrm{sing}}(k) = \beta \sum_i n_i^{\mathrm{b}} q_i \bar{q}_i^{\star}/[\varepsilon_0 k^2]$	unitless
$\tilde{\epsilon}_{\mathrm{reg}}(\mathbf{k})$	regular part of $\tilde{\epsilon}(\mathbf{k})$ (is finite for $k = 0$)	unitless
$\tilde{\epsilon}^{\mathrm{R}}(\mathbf{k})$	reciprocal dielectric function, $\tilde{\epsilon}^{\mathrm{R}}(\mathbf{k}) = 1/\tilde{\epsilon}(\mathbf{k})$	unitless
$\epsilon(\mathbf{r})$, $\epsilon(\mathbf{r};\mathbf{r}')$	(spatial) dielectric permittivity function	m^{-3}
$\epsilon^{\mathrm{R}}(\mathbf{r})$, $\epsilon^{\mathrm{R}}(\mathbf{r};\mathbf{r}')$	(spatial) reciprocal dielectric-permittivity function	m^{3}
ζ	activity, $\zeta = \eta e^{\beta\mu}$; for spherical particles $\zeta = e^{\beta\mu}/\Lambda^3$	m^{-3}
η	$\eta = \mathfrak{q}_{\mathrm{r}}/\Lambda^3$ ($\mathfrak{q}_{\mathrm{r}}$ is defined in Equation 11.6); for spherical particles $\mathfrak{q}_{\mathrm{r}} = 1$	m^{-3}
η^{hs}	hard sphere packing fraction, $\eta^{\mathrm{hs}} = \pi(d^{\mathrm{h}})^3 n^{\mathrm{b}}/6$	unitless
Θ	grand potential	J
κ	decay parameter ($1/\kappa$ = decay length)	m^{-1}
κ_{D}	Debye parameter ($1/\kappa_{\mathrm{D}}$= Debye length)	m^{-1}

Symbol	*Explanation*	*SI unit*
$\kappa_{\Re}$, $\kappa_{\Im}$	$\kappa_{\Re} = \Re(\kappa)$, $\kappa_{\Im} = \Im(\kappa)$ and $\kappa = \kappa_{\Re} + \mathrm{i}\kappa_{\Im}$	m^{-1}
Λ	thermal de Broglie wave length	m
λ	wave length	m
μ	chemical potential	J
μ_i	chemical potential of species i	J
μ^{ex}	excess chemical potential	J
$\mu^{\mathrm{ex,b}}$	excess chemical potential in bulk phase	J
$\mu^{\mathrm{ex}}(\mathbf{R}\|\mathbf{R}')$	μ^{ex} for a particle with coordinates $\mathbf{R} = (\mathbf{r}, \boldsymbol{\omega})$ when a particle is located at $\mathbf{r}'$ and has orientation $\boldsymbol{\omega}'$. Analogously for $\mu^{\mathrm{ex}}(\mathbf{r}\|\mathbf{r}')$.	J
$\mu^{(2)\mathrm{ex}}$	two-particle excess chemical potential	J
μ^{ideal}	ideal chemical potential	J
$\mu^{\mathrm{intr}}(\mathbf{r})$	intrinsic chemical potential, $\mu^{\mathrm{intr}}(\mathbf{r}) \equiv \mu - v(\mathbf{r})$	J
$\mu_{\pm}$	$\mu_{\pm} = \mathsf{x}_{+}^{\mathrm{b}}\mu_{+} + \mathsf{x}_{-}^{\mathrm{b}}\mu_{-}$ (for binary electrolyte)	J
Ξ	grand canonical partition function	unitless
$\Xi^{(l)}(\mathbf{R}_1, \ldots, \mathbf{R}_l)$	grand canonical partition function for a system with fixed particles at coordinates $\mathbf{R}_1, \ldots, \mathbf{R}_l$	unitless
ξ	coupling parameter, $0 \leq \xi \leq 1$	unitless
ξ_0, ξ_{R}, ξ_{A}, ξ_{V}	weighted-density parameters in FMT for bulk mixtures (defined in Equation 10.82)	various
ρ	charge density	$\mathrm{C\,m}^{-3}$
ρ_i	charge density around a particle of species i. For bulk electrolytes it is the same as ρ_i^{cloud}	$\mathrm{C\,m}^{-3}$

Symbol	*Explanation*	*SI unit*
$\rho_i^\star$	dressed charge density of a particle of species i, $\rho_i^\star = \sigma_i + \rho_i^{\text{dress}} \equiv \rho_i^{\text{tot}} - \rho_i^{\text{lin}}$	$\text{C}\,\text{m}^{-3}$
ρ_i^{cloud}	the charge density of the ion cloud around a particle of species i. For bulk electrolytes it is the same as ρ_i.	$\text{C}\,\text{m}^{-3}$
ρ_i^{dress}	charge density of the dress of a particle of species i, $\rho_i^{\text{dress}} \equiv \rho_i^{\text{cloud}} - \rho_i^{\text{lin}}$	$\text{C}\,\text{m}^{-3}$
ρ_i^{lin}	linear part of electrostatic polarization charge density due to an ion of species i	$\text{C}\,\text{m}^{-3}$
ρ_i^{tot}	total charge density associated with a particle of species i, $\rho_i^{\text{tot}} = \sigma_i + \rho_i^{\text{cloud}}$	$\text{C}\,\text{m}^{-3}$
σ_i	internal charge density of a particle of species i; for spherically symmetric ions $\sigma_i(\mathbf{r}) = q_i\delta^{(3)}(\mathbf{r})$	$\text{C}\,\text{m}^{-3}$
$\sigma^{\mathcal{S}}$	surface charge density	$\text{C}\,\text{m}^{-2}$
$\sigma^{\mathcal{S}\text{eff}}$	effective surface charge density	$\text{C}\,\text{m}^{-2}$
σ_X	root mean square deviation of quantity X, $\sigma_X^2 = \left\langle (X - \langle X\rangle)^2 \right\rangle$	same as X
$\Phi^{\text{ex}}(\mathbf{r})$	$\Phi^{\text{ex}}(\mathbf{r})/\beta$ is the intrinsic excess free-energy density; for a bulk fluid $\Phi^{\text{ex,b}} = \beta\mathcal{A}^{\text{ex,b}}/V$	m^{-3}
$\phi_{\text{Coul}}(r)$	unit Coulomb potential, $\phi_{\text{Coul}}(r) = (4\pi\varepsilon_0 r)^{-1}$	$\text{J}\,\text{C}^{-2}$
$\phi^\star_{\text{Coul}}(\mathbf{r})$, $\phi^\star_{\text{Coul}}(\mathbf{r},\mathbf{r}')$	unit screened Coulomb potential	$\text{J}\,\text{C}^{-2}$
$\varphi^{\text{R}}(r)$	a function used in the "renormalization" of integral equation closures (see Sections 9.1.3.2, 9.1.3.3 and 10.1)	unitless
$\varphi^{\text{ex,b}}$	$\varphi^{\text{ex,b}}/\beta$ is the excess free energy per particle in bulk	unitless

Symbol	*Explanation*	*SI unit*
$\chi^{\rho}(\mathbf{r})$, $\chi^{\rho}(\mathbf{r},\mathbf{r}')$	polarization response function for external electrostatic potential	$\mathrm{C^2\,J^{-1}m^{-6}}$
$\chi^{\star}(\mathbf{r})$, $\chi^{\star}(\mathbf{r},\mathbf{r}')$	polarization response function for total electrostatic potential	$\mathrm{C^2\,J^{-1}m^{-6}}$
$\chi^{\mathrm{el}}(\mathbf{r})$, $\chi^{\mathrm{el}}(\mathbf{r};\mathbf{r}')$	electric susceptibility function	$\mathrm{m^{-3}}$
χ_T	isothermal compressibility	$\mathrm{m^2\,N^{-1}}$
Ψ	total electrostatic potential	$\mathrm{V = J\,C^{-1}}$
Ψ^{ext}	external electrostatic potential (from a charge distribution external to the system)	$\mathrm{V = J\,C^{-1}}$
Ψ^{pol}	electrostatic potential from polarization charge density	$\mathrm{V = J\,C^{-1}}$
ψ_i	mean electrostatic potential due to a particle of species i	$\mathrm{V = J\,C^{-1}}$
$\boldsymbol{\omega}$	normalized orientational variable, $\boldsymbol{\omega} = (\frac{\phi}{2\pi}, \frac{\cos\theta}{2}, \frac{\chi}{2\pi})$ with Euler angles (ϕ, θ, χ)	unitless
$\omega(r)$	weight function	$\mathrm{m^{-3}}$
ω_i^0, ω_i^{R}, ω_i^{A}, ω_i^{V}	weight functions in FMT, defined in Equation 10.140	various
$\vec{\omega}_i^{\mathrm{A}^{\hat{\mathbf{r}}}}(\mathbf{r})$, $\vec{\omega}_i^{\mathrm{R}^{\hat{\mathbf{r}}}}(\mathbf{r})$	vector-valued weight functions in FMT, defined in Equation 10.142	various
∇	gradient operator, $\nabla = \left(\frac{\partial}{\partial x}, \frac{\partial}{\partial y}, \frac{\partial}{\partial z}\right)$	$\mathrm{m^{-1}}$
∇^2	Laplace operator, $\nabla^2 = \frac{\partial^2}{\partial x^2} + \frac{\partial^2}{\partial y^2} + \frac{\partial^2}{\partial z^2}$	$\mathrm{m^{-2}}$
$\otimes$	convolution operator	

III

Liquid state theory for homogeneous and inhomogeneous fluids

CHAPTER 9

Basics of Liquid State Theory

In this chapter we will build upon what we learned in Volume I of this treatise about general statistical mechanics for equilibrium systems and, in particular, the statistical mechanics of homogeneous and inhomogeneous fluids and solutions at an introductory level. However, we will recapitulate essential parts of the theory along the way, so it should be possible to read the current text independently from the former (references to the previous volume will nevertheless be given). The treatment in Volume I was restricted to fluids consisting of spherical particles and dealt with, for example, the structure of such fluids and interparticle interactions in them, including interactions between immersed bodies of any size and shape as well as surface forces. The study included confined and inhomogeneous simple fluids and electrolytes treated on two different levels: the density distribution (singlet) level and the pair distribution level with anisotropic pair-correlation functions. Many explicit examples of the results from the theory were given.

Here, we will initially restrict ourselves to spherical particles and treat fluids consisting of nonspherical particles later, in Chapter 11. While the previous treatment was based on various kinds of density distributions, we will here utilize a larger part of the toolbox of liquid state theory. To start with, we will introduce the concept of direct correlation functions of various orders and study linear response theory, whereafter nonlinear response and the associated bridge functions will be included. Several exact integral-equation relationships will be derived and integral equation approximations will be introduced. Essential tools in the theory, both conceptually and technically, are the concept of functionals and the use of functional derivatives, expansions and integrations, which will be established and extensively utilized to formulate and derive the theory. In the last section of the present chapter we will introduce density functional theory (DFT), which is an exact formulation of

DOI: 10.1201/9781003286882-9

statistical mechanics, while DFT approximations will be treated in the next chapter.

9.1 DIRECT CORRELATION FUNCTIONS AND INTEGRAL EQUATIONS

In Part II of this treatise (in the first volume) we built statistical mechanics of fluids on various distribution functions. On the **singlet level** for a one-component system of spherical particles, we have, for example, the

- *singlet density distribution* $n^{(1)}(\mathbf{r}) \equiv n(\mathbf{r})$ (or simply the *density distribution*),

where $\mathbf{r} = (x, y, x)$, and on the **pair level**, we have, for example, the

- *pair distribution function*[1] $g^{(2)}(\mathbf{r}, \mathbf{r}')$,
- *pair correlation function* $h^{(2)}(\mathbf{r}, \mathbf{r}') = g^{(2)}(\mathbf{r}, \mathbf{r}') - 1$ (also called the "*total correlation function*") and the
- *pair density distribution function* $n^{(2)}(\mathbf{r}, \mathbf{r}') = n(\mathbf{r})n(\mathbf{r}')g^{(2)}(\mathbf{r}, \mathbf{r}')$.

The density $n(\mathbf{r})$ is the number density of particle centers at the point $\mathbf{r}$, that is, $n(\mathbf{r})d\mathbf{r}$ is the average number of particle centers in the infinitesimal volume element $d\mathbf{r} = dx\,dy\,dz$ that surrounds the point $\mathbf{r}$. The pair density can be written as

$$n^{(2)}(\mathbf{r}, \mathbf{r}') = n(\mathbf{r})n^{(1)}(\mathbf{r}'|\mathbf{r}), \qquad \text{where } n^{(1)}(\mathbf{r}'|\mathbf{r}) \equiv n(\mathbf{r}')g^{(2)}(\mathbf{r}, \mathbf{r}').$$

The density $n^{(1)}(\mathbf{r}'|\mathbf{r})$ is the number density at $\mathbf{r}'$ when a particle is located at $\mathbf{r}$ (a conditional density) and it can also be written as $n(\mathbf{r}'|\mathbf{r})$ without superscript (1). Of course, we also have $n^{(2)}(\mathbf{r}, \mathbf{r}') = n(\mathbf{r}')n^{(1)}(\mathbf{r}|\mathbf{r}')$ when there is a particle located at $\mathbf{r}'$. In order to develop the theory further, we need to introduce a new type of correlation functions, the so-called **direct correlations functions**. They exist as singlet, pair, triplet, quartet etc. functions, and we will start with the singlet one. Singlet functions are also called *one-point functions,* since they depend on one coordinate point $\mathbf{r}$, or, alternatively, *one-particle functions,* since they are usually associated with the coordinates of a particle in the system. Likewise, pair functions are also called *two-point functions* or *two-particle functions.*

9.1.1 The singlet direct-correlation function $c^{(1)}$

Consider an inhomogeneous fluid with density distribution $n(\mathbf{r})$ in equilibrium with a bulk fluid of number density n^{b}, which constitutes an infinitely large

[1]The function $g^{(2)}(\mathbf{r}, \mathbf{r}')$ is sometimes called the "pair correlation function," but in this treatise this name is reserved for $h^{(2)} = g^{(2)} - 1$.

reservoir of particles (superscript b denotes "bulk"). We will treat the system in the grand canonical ensemble where the chemical potential μ and the absolute temperature T are constant. The volume V, which is also constant, is defined by the boundaries of the system. These boundaries do not need to be physical boundaries like wall surfaces; they can also be mathematical surfaces inside the fluid that are located wherever we choose to place them. We assume, however, that V is macroscopically large at least in one of the three dimensions.

The fluid is exposed to an **external potential** $v(\mathbf{r})$, which describes the interaction between a particle at $\mathbf{r}$ and something that is not part of the system, for example an external body that may have a surface in contact with the fluid or be fully immersed in it. Remember that it is always up to us to decide what comprises the system and what is not part of it.

A **configuration** of the particles is the set of points $\{\mathbf{r}_\nu\}_{\nu\geq 1}$ that gives the locations of all particles at a certain moment in time, whereby $\mathbf{r}_\nu$ is the center of particle ν. When the particles have a certain configuration, the interaction energy with the external potential is equal to $\sum_\nu v(\mathbf{r}_\nu)$, where the sum is taken over all particles in the system. In most cases we will assume that the interactions between the particles in the fluid are pairwise additive, so the total interactional energy of the system, the potential energy $\breve{U}^{\mathrm{pot}}$ for each instantaneous configuration, equals[2]

$$\breve{U}^{\mathrm{pot}} = \sum_{\nu} v(\mathbf{r}_\nu) + \sum_{\nu,\nu'>\nu} u(\mathbf{r}_\nu, \mathbf{r}_{\nu'}), \tag{9.1}$$

where $u(\mathbf{r}_\nu, \mathbf{r}_{\nu'})$ is the **pair interaction potential** and the second term is over all pairs of particles, $\sum_{\nu,\nu'>\nu} \equiv \sum_\nu \sum_{\nu';\,\nu'>\nu}$. We will neglect the internal degrees of freedom of the particles.

The entire system can be inhomogeneous, for example a fluid between two surfaces that are near each other. It is also possible that a part of the system can be in the bulk state, for example for a fluid in contact with a surface, where a part of the fluid far away from the surface but inside the volume V can be in the bulk state. In both cases the fluid is in equilibrium with the bulk fluid of density n^{b} (superscript b stands for "bulk") and chemical potential μ. We may also select a volume V inside the bulk fluid, so the entire system is homogeneous. The potential $v(\mathbf{r})$ is normally set equal to zero in the bulk (this is possible since we may add a constant to a potential without changing the physics).

To start with, we will recapitulate some matters from Chapter 5 in Volume I of this treatise (equation numbers from the first volume are written in the format "Equation I:x.x," where x.x is the equation number in Volume I). The density distribution is expressed in terms of the **potential of mean force**, $w(\mathbf{r})$, as $n(\mathbf{r}) = n^{\mathrm{b}} e^{-\beta w(\mathbf{r})}$, where $\beta = 1/(k_B T)$, k_B is Boltzmann's constant

[2]The symbol "ˇ" over a quantity, like in $\breve{U}$, is used throughout this treatise to indicate that the value is the instantaneous one, that is, when the particles are located at $\{\mathbf{r}_\nu\}_{\nu\geq 1}$.

and T is the absolute temperature. In order to distinguish this potential from the pair potential of mean force $w^{(2)}$ that we will encounter later, it is also called the ***singlet potential of mean force*** and denoted $w^{(1)}(\mathbf{r}) \equiv w(\mathbf{r})$, but the superscript (1) is usually omitted (like it is for $n^{(1)}(\mathbf{r})$). By convention we always set the value of w in the bulk phase equal to zero. The mean force $\mathbf{F}(\mathbf{r})$ that acts on a particle at $\mathbf{r}$ is equal to $\mathbf{F}(\mathbf{r}) = -\nabla w(\mathbf{r})$. It is given by the sum of the force from the external potential $-\nabla v(\mathbf{r})$ and the mean force $\mathbf{F}^{\text{intr}}(\mathbf{r})$ from all other particles in the system, the *intrinsic mean force*. The latter force is the average taken over all possible configurations of these other particles. The potential of mean force $w(\mathbf{r})$ can therefore be divided into two parts, the external potential $v(\mathbf{r})$ and the **intrinsic potential of mean force** $w^{\text{intr}}(\mathbf{r})$, which accordingly is the contribution from the interactions between the particles of the system and we have $\mathbf{F}^{\text{intr}}(\mathbf{r}) = -\nabla w^{\text{intr}}(\mathbf{r})$. Accordingly,

$$n(\mathbf{r}) = n^{\text{b}} e^{-\beta w(\mathbf{r})} = n^{\text{b}} e^{-\beta[v(\mathbf{r}) + w^{\text{intr}}(\mathbf{r})]} \tag{9.2}$$

(cf. Equations I:5.19 and I:5.25), which we can write as

$$n(\mathbf{r}) = n^{\text{b}} g^{(1)}(\mathbf{r}), \tag{9.3}$$

where $g^{(1)}(\mathbf{r}) = e^{-\beta[v(\mathbf{r}) + w^{\text{intr}}(\mathbf{r})]}$ is the **singlet distribution function** (in the bulk phase $g^{(1)}(\mathbf{r}) = 1$).

When one moves a particle very slowly from one position, $\mathbf{r}_1$, to another, $\mathbf{r}_2$, the reversible work performed on the system against the interactions is given by the line integral

$$-\int_{\mathbf{r}_1}^{\mathbf{r}_2} \mathbf{F}(\mathbf{r}) \cdot d\mathbf{r} = \int_{\mathbf{r}_1}^{\mathbf{r}_2} \nabla w(\mathbf{r}) \cdot d\mathbf{r} = w(\mathbf{r}_2) - w(\mathbf{r}_1). \tag{9.4}$$

The movement is thereby done so slowly that the distribution of the other particles around the particle is fully relaxed all the time (a quasistatic change in the position of the particle). The reversible work is equal to the change in free energy during the process[3] and since the free energy is a state function, the difference $w(\mathbf{r}_2) - w(\mathbf{r}_1)$ is independent of the path taken for the particle. Hence $w(\mathbf{r})$ is a free-energy quantity.

The potential of mean force w is closely related to the chemical potential μ. For an inhomogeneous fluid, the latter is given by

$$\mu = \mu^{\text{ideal}}(\mathbf{r}) + \mu^{\text{ex}}(\mathbf{r}) = k_B T \ln(\Lambda^3 n(\mathbf{r})) + \mu^{\text{ex}}(\mathbf{r}) \tag{9.5}$$

(Equation I:5.24), where $\mu^{\text{ideal}}(\mathbf{r}) = k_B T \ln(\Lambda^3 n(\mathbf{r}))$ is the **ideal chemical potential** (the non-interactional part of μ, cf. Equation I:3.34), $\Lambda \equiv h/(2\pi m k_B T)^{1/2}$ is the thermal de Broglie wave length, m is the particle mass and h is Planck's constant. The **excess chemical potential** $\mu^{\text{ex}}(\mathbf{r})$ is equal

[3]This is a standard thermodynamical result that is discussed in more detail in footnote 11 on page 15.

to the reversible work done against interactions when one inserts a new particle at position $\mathbf{r}$ in the system (and keeps it there). The insertion can, for example, be done by moving the particle to $\mathbf{r}$ from a place where it does not interact with anything ("infinitely far away").[4] Note that $\boldsymbol{\mu}^{\text{ex}}(\mathbf{r})$ ***contains all interactional contributions to the chemical potential***, both from the interactions between the particles in the system (the intrinsic part) and from the interaction with the external potential $v(\mathbf{r})$.[5]

The difference $\mu^{\text{ex}}(\mathbf{r}_2) - \mu^{\text{ex}}(\mathbf{r}_1)$ must be equal to the reversible work $w(\mathbf{r}_2) - w(\mathbf{r}_1)$ in Equation 9.4 because of the path independence. In the present case, the path is taken via the place at infinity, whereby $-\mu^{\text{ex}}(\mathbf{r}_1)$ is the reversible work when the particle is moved from $\mathbf{r}_1$ to infinity, and $\mu^{\text{ex}}(\mathbf{r}_2)$ is the work when it moved from there to $\mathbf{r}_2$. It follows that $w(\mathbf{r})$ and $\mu^{\text{ex}}(\mathbf{r})$ differ only by an additive constant

$$\mu^{\text{ex}}(\mathbf{r}) = w(\mathbf{r}) + \text{const} = v(\mathbf{r}) + w^{\text{intr}}(\mathbf{r}) + \text{const}, \tag{9.6}$$

where the constant equals $\mu^{\text{ex,b}}$, that is, the excess chemical potential in the bulk phase. This applies when $w = 0$ in the bulk, which is the normal choice used here (otherwise the constant is $\mu^{\text{ex,b}} - w^{\text{b}}$, where w^{b} is the value of w in bulk).

The total chemical potential μ is the change in free energy[6] when a new particle is added to the system and is free to move like any other particle. Its excess part, $\mu^{\text{ex}}(\mathbf{r})$, is the change when a particle is added and kept fixed at $\mathbf{r}$. Thus, the term $\mu^{\text{ideal}}(\mathbf{r})$ in Equation 9.5 is equal to the change in free energy when the particle placed at $\mathbf{r}$ is released and allowed to move around freely as all other particles (the entropy changes when it is released). At equilibrium μ has the same value everywhere but the two parts $\mu^{\text{ideal}}(\mathbf{r})$ and $\mu^{\text{ex}}(\mathbf{r})$ depend on the coordinate point $\mathbf{r}$; only their sum is constant. Thus Equation 9.5 constitutes the equilibrium condition of equal μ at all points.

Equation 9.5 implies that

$$n(\mathbf{r}) = \zeta e^{-\beta\mu^{\text{ex}}(\mathbf{r})}, \tag{9.7}$$

---

[4]Alternatively, we may create the new particle at $\mathbf{r}$ and keep it there, whereby $\mu^{\text{ex}}(\mathbf{r})$ is equal to the work against the interparticle interactions during this process. The self-energy of the particle during its creation is thereby *not* included in $\mu^{\text{ex}}(\mathbf{r})$. It could be included, but if it is, it does not influence differences like $\mu^{\text{ex}}(\mathbf{r}_2) - \mu^{\text{ex}}(\mathbf{r}_1)$.

[5]In a different convention that sometimes can be found in the literature, μ^{ex} does not include the external potential and then μ^{ex} is equal to the intrinsic part of the chemical potential.

[6]The free energy considered here is the total free energy of the system (which is open) and the reservoir of particles that gives the constant μ and T provided that the reservoir is huge (in principle in the thermodynamic limit). The combined system (our system and the reservoir) can be envisaged as being closed and treated in the canonical ensemble, so the total free energy is the Helmholtz free energy A_{tot}. Thereby μ is the change in A_{tot} when a particle is added. The free energy change $\mu^{\text{ex}}(\mathbf{r})$ when one inserts a new particle at $\mathbf{r}$ also refers to the total free energy A_{tot} of the combined system, but we can alternatively express it in terms of the free energy of the open system itself, that is, the grand potential Θ. This will be dealt with in footnote 11.

where $\zeta = e^{\beta\mu}/\Lambda^3$ is the **activity**. This relationship is equivalent to Equation 9.2. Note that we do not need to specify a bulk density in Equation 9.7 since ζ is determined solely by μ and T. The **activity coefficient** γ^{b} (also called the **activity factor**) for the bulk phase is defined as $\gamma^{\mathrm{b}} = \exp(\beta\mu^{\mathrm{ex,b}})$ and is related to the activity as $\zeta = n^{\mathrm{b}}\gamma^{\mathrm{b}}$. For a given density n^{b} we can obtain the activity if γ^{b} is known.

From Equation 9.6 it is obvious that $\mu^{\mathrm{ex}}(\mathbf{r})$, like $w(\mathbf{r})$, can be divided into two parts: the work against the external potential and the work against the interactions with the other particles. The latter part, the intrinsic excess chemical potential, differs from $w^{\mathrm{intr}}(\mathbf{r})$ by an additive constant according to Equation 9.6. Instead of introducing a special notation for this intrinsic entity we define the **singlet direct-correlation function** $c^{(1)}(\mathbf{r})$ from[7]

$$\mu^{\mathrm{ex}}(\mathbf{r}) = v(\mathbf{r}) - k_B T\, c^{(1)}(\mathbf{r}), \tag{9.8}$$

so $-k_B T\, c^{(1)}(\mathbf{r})$ is the intrinsic part of μ^{ex} and is equal to the reversible work against interactions with the particles of the system when one inserts a new particle at position $\mathbf{r}$. The factor $-k_B T$ in the definition of $c^{(1)}$ is included for historic reasons (it is the same factor as in the definitions of free energies in statistical mechanics) and we see that $c^{(1)}$ is a dimensionless quantity, just like the functions $g^{(1)}$ and $g^{(2)}$.

We have now introduced several potential functions that are closely related to each other. For clarity, these entities, their relationships, and values in the bulk phase are summarized in Table 9.1. It is practical to use both $w(\mathbf{r})$ and $\mu^{\mathrm{ex}}(\mathbf{r})$ despite that they differ only by a constant; the former can be set to zero in the bulk phase while the latter always has the value $\mu^{\mathrm{ex,b}}$ there, i.e., the reversible work against interactions to insert a new particle in the bulk phase. It is likewise practical to use both $w^{\mathrm{intr}}(\mathbf{r})$ and $-k_B T c^{(1)}(\mathbf{r})$.

The gradient $k_B T\,\nabla c^{(1)}(\mathbf{r})$ is the average force from interparticle interactions on a particle located at $\mathbf{r}$, $k_B T\,\nabla c^{(1)}(\mathbf{r}) = \mathbf{F}^{\mathrm{intr}}(\mathbf{r}) = -\nabla w^{\mathrm{intr}}(\mathbf{r})$, and it is explicitly given in terms of the pair distribution function as (cf. Equation I:5.118)

$$k_B T\,\nabla c^{(1)}(\mathbf{r}) = \mathbf{F}^{\mathrm{intr}}(\mathbf{r}) = -\int d\mathbf{r}'\, g^{(2)}(\mathbf{r},\mathbf{r}')n(\mathbf{r}')\nabla u(\mathbf{r}',\mathbf{r}), \tag{9.9}$$

where the integral is taken over the whole space. This integrand can be interpreted as follows: $-\nabla u(\mathbf{r}',\mathbf{r})$ is the force acting on a particle at $\mathbf{r}$ from the interaction with a particle at $\mathbf{r}'$ and $g^{(2)}(\mathbf{r},\mathbf{r}')n(\mathbf{r}')$ is the density of particles at $\mathbf{r}'$ when a particle is located at $\mathbf{r}$. The force on the particle at $\mathbf{r}$ from all other particles is obtained by summing over the latter, that is, by integration over $\mathbf{r}'$.

---

[7] This function is a one-point (singlet) function and we include the superscript "(1)" in order to distinguish it from higher order direct correlation functions $c^{(2)}$, $c^{(3)}$ etc. that we will introduce later.

Table 9.1 Some singlet (one-point) potential functions, their relationships and values in bulk.

Entity	Description	Relationships	Bulk value
$v(\mathbf{r})$	External potential		$0^{\dagger}$
$w(\mathbf{r})$	Potential of mean force	$w(\mathbf{r}) = \mu^{\mathrm{ex}}(\mathbf{r}) - \mu^{\mathrm{ex,b}}$	$0^{\dagger}$
$w^{\mathrm{intr}}(\mathbf{r})$	Intrinsic part of w	$w^{\mathrm{intr}}(\mathbf{r}) = w(\mathbf{r}) - v(\mathbf{r})$	$0^{\dagger}$
$\mu^{\mathrm{ex}}(\mathbf{r})$	Excess chemical potential*	$\mu^{\mathrm{ex}}(\mathbf{r}) = w(\mathbf{r}) + \mu^{\mathrm{ex,b}}$	$\mu^{\mathrm{ex,b}}$
$c^{(1)}(\mathbf{r})$	Singlet direct-correlation function $-k_BTc^{(1)}$ = intrinsic part of μ^{ex}	$c^{(1)}(\mathbf{r}) = -\beta[\mu^{\mathrm{ex}}(\mathbf{r}) - v(\mathbf{r})]$	$-\beta\mu^{\mathrm{ex,b}}$

*$\mu^{\mathrm{ex}}(\mathbf{r}) = \mu - \mu^{\mathrm{ideal}}(\mathbf{r})$ = reversible work to move a particle from infinity to $\mathbf{r}$.

†By convention.

From the fact that $\mathbf{F}(\mathbf{r}) = -\nabla w(\mathbf{r}) = k_BT\,\nabla \ln n(\mathbf{r})$, it follows that

$$\begin{aligned}\nabla \ln n(\mathbf{r}) &= \beta\mathbf{F}(\mathbf{r}) = \beta\left[-\nabla v(\mathbf{r}) + \mathbf{F}^{\mathrm{intr}}(\mathbf{r})\right] \\ &= -\beta\left[\nabla v(\mathbf{r}) + \int d\mathbf{r}'\, g^{(2)}(\mathbf{r},\mathbf{r}')n(\mathbf{r}')\nabla u(\mathbf{r}',\mathbf{r})\right], \quad (9.10)\end{aligned}$$

where we have used Equation 9.9 to obtain the last equality. This is the **first Born-Green-Yvon (BGY) equation** (Equation I:5.119).

> Whenever an integral appears without any explicit limits like in Equations 9.9 and 9.10, it will be assumed throughout this treatise that the integration is taken over all relevant values of the integration variables (like over the whole available space in the system). Likewise, whenever a sum is written without explicit limits, one sums over all relevant possibilities (like over all particles in the system as in Equation 9.1 or, for multicomponent systems, over all species present).

The excess chemical potential $\mu^{\mathrm{ex}}(\mathbf{r})$ is expressed in terms of $c^{(1)}(\mathbf{r})$ in Equation 9.8 and by inserting this into Equation 9.5 for μ, we can write the latter as

$$\mu = k_BT \ln(\Lambda^3 n(\mathbf{r})) + v(\mathbf{r}) - k_BT\, c^{(1)}(\mathbf{r}). \quad (9.11)$$

This is another way to write the equilibrium condition that the chemical potential is the same for all $\mathbf{r}$. Since $-k_BT\, c^{(1)}(\mathbf{r})$ is the intrinsic part of the excess chemical potential, the singlet direct-correlation function can be calculated by using any method for chemical potential calculations described in the first volume of this treatise.[8]

We can write Equation 9.11 as

$$k_BT\ln\zeta = k_BT\ln n(\mathbf{r}) + v(\mathbf{r}) - k_BT\,c^{(1)}(\mathbf{r}) \tag{9.12}$$

and Equation 9.7 can be written as

$$n(\mathbf{r}) = \zeta e^{-\beta v(\mathbf{r})+c^{(1)}(\mathbf{r})}. \tag{9.13}$$

It is often useful to express the density at one point $\mathbf{r}$ in terms of the density at another point, say, $\mathbf{r}_0$, and it follows from Equation 9.13 that

$$n(\mathbf{r}) = n(\mathbf{r}_0)e^{-\beta[v(\mathbf{r})-v(\mathbf{r}_0)]+c^{(1)}(\mathbf{r})-c^{(1)}(\mathbf{r}_0)}, \tag{9.14}$$

where we have eliminated ζ. If we select $\mathbf{r}_0$ to be in the bulk phase where $v = 0$ and $n = n^{\mathrm{b}}$, this equation yields

$$n(\mathbf{r}) = n^{\mathrm{b}}e^{-\beta v(\mathbf{r})+c^{(1)}(\mathbf{r})-c^{(1)\mathrm{b}}}, \tag{9.15}$$

where $c^{(1)\mathrm{b}}$ is the value of $c^{(1)}$ in the bulk, which according to Equation 9.8 is given by $c^{(1)\mathrm{b}} = -\beta\mu^{\mathrm{ex,b}}$ (cf. Table 9.1). By comparing with Equations 9.2 we see that $w^{\mathrm{intr}}(\mathbf{r}) = -k_BT\left[c^{(1)}(\mathbf{r}) - c^{(1)\mathrm{b}}\right]$.

We have thus obtained ***the following important equation*** that constitutes a starting point for further developments

$$n(\mathbf{r}) = n^{\mathrm{b}}e^{-\beta v(\mathbf{r})+\Delta c^{(1)}(\mathbf{r})}, \tag{9.16}$$

where $\Delta c^{(1)}(\mathbf{r}) = c^{(1)}(\mathbf{r}) - c^{(1)\mathrm{b}}$. From Equation 9.3 we get

$$g^{(1)}(\mathbf{r}) = e^{-\beta v(\mathbf{r})+\Delta c^{(1)}(\mathbf{r})}. \tag{9.17}$$

This means that in order to obtain $n(\mathbf{r})$ and $g^{(1)}(\mathbf{r})$, one needs to determine the difference $\Delta c^{(1)}(\mathbf{r})$. This is one of the major topics in the following sections.

---

[8] One can, for example, use Widom's formula (I:5.144), which can be written as

$$c^{(1)}(\mathbf{r}) = \ln\left\langle e^{-\beta\sum_\nu u(\mathbf{r},\mathbf{r}_\nu)}\right\rangle^{\circ}\Big|_{\mathbf{r}\text{ fix}},$$

where a so-called test particle is placed fixed at $\mathbf{r}$ and the other particles do not feel the interaction with this particle (the test particle is treated like a "ghost"). In the exponent, $u(\mathbf{r},\mathbf{r}_\nu)$ is the pair interaction between the test particle and a particle at $\mathbf{r}_\nu$. The average is taken over all configurations of all particles with locations $\{\mathbf{r}_\nu\}_{\nu\geq 1}$, whereby the configurations are not affected by the presence of the test particle (this is indicated by the superscript "o" on the average).

Alternatively, one can interpret the average in the equation simply as a mathematical average over $\{\mathbf{r}_\nu\}_{\nu\geq 1}$ of the exponential function with an argument containing $u(\mathbf{r},\mathbf{r}_\nu)$ without any test particle actually being placed at $\mathbf{r}$. In any way, the equation shows that $c^{(1)}(\mathbf{r})$ can be directly expressed in terms of the interparticle interactions $u(\mathbf{r},\mathbf{r}_\nu)$ in the system. Widom's formula will be treated in more detail in Chapter 11 (see for example Equation 11.102).

It is also possible to express the pair distribution function $g^{(2)}$ in terms of $c^{(1)}$. The number density at $\mathbf{r}$ in a system with a particle placed fixed at $\mathbf{r}'$ is equal to

$$n(\mathbf{r}|\mathbf{r}') \equiv g^{(2)}(\mathbf{r},\mathbf{r}')n(\mathbf{r}). \tag{9.18}$$

The fluid is here exposed to the potential $v(\mathbf{r}) + u(\mathbf{r},\mathbf{r}')$, which we will write as

$$v(\mathbf{r}|\mathbf{r}') \equiv v(\mathbf{r}) + u(\mathbf{r},\mathbf{r}'). \tag{9.19}$$

By formally regarding the particle at $\mathbf{r}'$ as being external to the system, we may describe the situation as if the fluid with chemical potential μ is exposed to an external potential $v(\mathbf{r}|\mathbf{r}')$. All expressions above can be adapted to this situation, for instance Equation 9.13 becomes

$$n(\mathbf{r}|\mathbf{r}') = \zeta e^{-\beta v(\mathbf{r}|\mathbf{r}')+c^{(1)}(\mathbf{r}|\mathbf{r}')}, \tag{9.20}$$

where $c^{(1)}(\mathbf{r}|\mathbf{r}')$ is the singlet direct-correlation function at $\mathbf{r}$ for this system, that is, in the presence of an external particle located at $\mathbf{r}'$. Since the interaction $u(\mathbf{r},\mathbf{r}')$ with the fixed particle at $\mathbf{r}'$ is considered to be a part of the external potential, it is not contained in $c^{(1)}(\mathbf{r}|\mathbf{r}')$, which is an intrinsic entity for the system consisting of all other particles (remember that the intrinsic excess chemical potential is equal to $-k_BTc^{(1)}$). Note that the external particle and the particles in the system belong to the same species in the present case.

From Equation 9.18 we see that

$$g^{(2)}(\mathbf{r},\mathbf{r}') = \frac{n(\mathbf{r}|\mathbf{r}')}{n(\mathbf{r})} = \frac{\zeta e^{-\beta v(\mathbf{r}|\mathbf{r}')+c^{(1)}(\mathbf{r}|\mathbf{r}')}}{\zeta e^{-\beta v(\mathbf{r})+c^{(1)}(\mathbf{r})}},$$

where we have inserted 9.13 and 9.20. It follows that

$$g^{(2)}(\mathbf{r},\mathbf{r}') = e^{-\beta u(\mathbf{r},\mathbf{r}')+c^{(1)}(\mathbf{r}|\mathbf{r}')-c^{(1)}(\mathbf{r})}, \tag{9.21}$$

which can be written as

$$g^{(2)}(\mathbf{r},\mathbf{r}') = e^{-\beta u(\mathbf{r},\mathbf{r}')+\Delta c^{(1)}(\mathbf{r}|\mathbf{r}')}, \tag{9.22}$$

where $\Delta c^{(1)}(\mathbf{r}|\mathbf{r}') = c^{(1)}(\mathbf{r}|\mathbf{r}') - c^{(1)}(\mathbf{r})$. This is ***another important equation*** because it says that the pair correlation function can be obtained when we have determined $\Delta c^{(1)}(\mathbf{r}|\mathbf{r}')$. This is another major topic in the following sections.

It is customary to introduce the **pair potential of mean force** $w^{(2)}$ from

$$g^{(2)}(\mathbf{r},\mathbf{r}') = e^{-\beta w^{(2)}(\mathbf{r},\mathbf{r}')} = e^{-\beta[u(\mathbf{r},\mathbf{r}')+w^{(2)\mathrm{intr}}(\mathbf{r},\mathbf{r}')]}, \tag{9.23}$$

where $w^{(2)\mathrm{intr}}$ is the **intrinsic part** of $w^{(2)}$.[9] The designation "intrinsic" is used because the fixed particle at $\mathbf{r}'$ is here considered to be external to the

[9]Sometimes $-\beta w^{(2)\mathrm{intr}}$ is called the "**thermal pair potential**," see the definition of $\omega_{\mathrm{M}}^{(2)}$ on page 78.

system. (Nevertheless, for simplicity we will use the notation $w^{(2)\text{intr}}$ even in cases when the particle at $\mathbf{r}'$ belongs to the system.) It follows that

$$w^{(2)\text{intr}}(\mathbf{r}, \mathbf{r}') = -k_B T \, \Delta c^{(1)}(\mathbf{r}|\mathbf{r}') = -k_B T \left[c^{(1)}(\mathbf{r}|\mathbf{r}') - c^{(1)}(\mathbf{r}) \right]. \qquad (9.24)$$

Once we know the difference $\Delta c^{(1)}(\mathbf{r}|\mathbf{r}')$, we know $w^{(2)\text{intr}}(\mathbf{r}, \mathbf{r}')$ and hence $g^{(2)}(\mathbf{r}, \mathbf{r}')$.

We can say that the distribution function $g^{(1)}$ describes how the bulk density n^{b} is influenced by the exposure to the potential v from an external source and thereby becomes $n(\mathbf{r})$; we have $g^{(1)}(\mathbf{r}) = n(\mathbf{r})/n^{\text{b}}$. Likewise, the function $g^{(2)}$ describes how the inhomogeneous density $n(\mathbf{r})$ is affected by the exposure to the potential u from a fixed particle at $\mathbf{r}'$ and hence becomes $n(\mathbf{r}|\mathbf{r}')$, whereby $g^{(2)}(\mathbf{r}, \mathbf{r}') = n(\mathbf{r}|\mathbf{r}')/n(\mathbf{r})$. Note that

$$\frac{n(\mathbf{r}|\mathbf{r}')}{n(\mathbf{r})} = g^{(2)}(\mathbf{r}, \mathbf{r}') = g^{(2)}(\mathbf{r}', \mathbf{r}) = \frac{n(\mathbf{r}'|\mathbf{r})}{n(\mathbf{r}')}$$

as a consequence of the symmetry between $\mathbf{r}$ and $\mathbf{r}'$.

The common feature of the two important results, Equations 9.17 and 9.22, is that $\Delta c^{(1)}$ is the change in $c^{(1)}$ when the field that causes the deviation in density is turned on. In both cases, $\Delta c^{(1)}$ ***expresses the complicated part of the response of the fluid*** to this field, that is, the indirect effect from the interactions with and between the mobile particles in the system.

As we have seen, $c^{(1)}$ and w^{intr} are very closely related – they express essentially the same thing. In the mathematical development of the theory we will mostly make use of the direct correlation function, but we will remind ourselves about the links to the potential of mean force and the excess chemical potential whenever it is physically motivated to do so.

It is useful to introduce the **pair cavity function** $y^{(2)}$ defined from

$$y^{(2)}(\mathbf{r}, \mathbf{r}') = e^{\Delta c^{(1)}(\mathbf{r}|\mathbf{r}')} = e^{-\beta w^{(2)\text{intr}}(\mathbf{r}, \mathbf{r}')}, \qquad (9.25)$$

so Equation 9.23 can be written

$$g^{(2)}(\mathbf{r}, \mathbf{r}') = e^{-\beta u(\mathbf{r}, \mathbf{r}')} y^{(2)}(\mathbf{r}, \mathbf{r}'), \qquad (9.26)$$

where the trivial factor $e^{-\beta u}$ is separated out in $g^{(2)}$. We have already encountered the function $y^{(2)}$ in, for example, Section 5.7 of Volume I (Equation I:5.78 ff.) where we expressed the pressure in terms of the pair distribution function of a hard core bulk fluid.

Let us see what these equations mean. Say that there is a particle located at $\mathbf{r}'$ (the "central" particle). The factor $e^{-\beta u}$ in $g^{(2)}$ is small or zero when $\mathbf{r}$ is located where the pair potential is strongly repulsive, for example in the core region where u is infinitely large and the factor is zero, which makes $g^{(2)}$ to be zero there. Such a region around the central particle can be described as a cavity that contains the particle itself, but where the other particles are not allowed to enter. This motivates the name "cavity function." In fact, $y^{(2)}(\mathbf{r},\mathbf{r}')$ is defined everywhere, including inside any hard core region. This is due to the fact that $c^{(1)}(\mathbf{r}|\mathbf{r}')$ and $c^{(1)}(\mathbf{r})$ that occur in $\Delta c^{(1)}(\mathbf{r}|\mathbf{r}')$ are intrinsic functions that do not contain any contribution from the external potential, that is, $v(\mathbf{r})$ and $v(\mathbf{r}) + u(\mathbf{r},\mathbf{r}')$, respectively. We will return to this issue later.

It is illustrative to express the pair potential of mean force $w^{(2)}$ in a different manner. By applying Equation 9.8 to the situation with an external particle placed at $\mathbf{r}'$, it becomes

$$\mu^{\text{ex}}(\mathbf{r}|\mathbf{r}') = v(\mathbf{r}|\mathbf{r}') - k_B T\, c^{(1)}(\mathbf{r}|\mathbf{r}')$$

and we have

$$\begin{aligned}\mu^{\text{ex}}(\mathbf{r}|\mathbf{r}') - \mu^{\text{ex}}(\mathbf{r}) &= v(\mathbf{r}|\mathbf{r}') - v(\mathbf{r}) - k_B T\left[c^{(1)}(\mathbf{r}|\mathbf{r}') - c^{(1)}(\mathbf{r})\right] \\ &= v(\mathbf{r}|\mathbf{r}') - v(\mathbf{r}) - k_B T\,\Delta c^{(1)}(\mathbf{r}|\mathbf{r}'). \qquad (9.27)\end{aligned}$$

In Chapter 5 (in Volume I) we introduced the **two-particle excess chemical potential**, $\mu^{(2)\text{ex}}(\mathbf{r},\mathbf{r}')$, as the change in free energy when we simultaneously insert two particles, one at $\mathbf{r}$ and one at $\mathbf{r}'$ and keep them there (Equation I:5.163). Before the insertion, none of these two particles interact with anything else; each them is infinitely far away from the system and from each other. Let us compare $\mu^{(2)\text{ex}}$ with the entity $\mu^{\text{ex}}(\mathbf{r}|\mathbf{r}')$, which is the change in free energy for the situation where the particle at $\mathbf{r}'$ is already present when we insert the particle at $\mathbf{r}$. The difference between these entities must therefore be the free energy change when the particle at $\mathbf{r}'$ was inserted, that is, $\mu^{\text{ex}}(\mathbf{r}')$, so we have

$$\mu^{(2)\text{ex}}(\mathbf{r},\mathbf{r}') - \mu^{\text{ex}}(\mathbf{r}|\mathbf{r}') = \mu^{\text{ex}}(\mathbf{r}'). \qquad (9.28)$$

By taking $\mu^{\text{ex}}(\mathbf{r}|\mathbf{r}')$ from this equation and using $v(\mathbf{r}|\mathbf{r}') = v(\mathbf{r}) + u(\mathbf{r},\mathbf{r}')$, the relationship 9.27 can be written

$$\mu^{(2)\text{ex}}(\mathbf{r},\mathbf{r}') - \mu^{\text{ex}}(\mathbf{r}') - \mu^{\text{ex}}(\mathbf{r}) = u(\mathbf{r},\mathbf{r}') - k_B T\,\Delta c^{(1)}(\mathbf{r}|\mathbf{r}'). \qquad (9.29)$$

The rhs is equal to $-k_B T\, w^{(2)}(\mathbf{r},\mathbf{r}')$ so we can conclude that

$$w^{(2)}(\mathbf{r},\mathbf{r}') = \mu^{(2)\text{ex}}(\mathbf{r},\mathbf{r}') - \mu^{\text{ex}}(\mathbf{r}') - \mu^{\text{ex}}(\mathbf{r}). \qquad (9.30)$$

Here, the terms $\mu^{\text{ex}}(\mathbf{r})$ and $\mu^{\text{ex}}(\mathbf{r}')$ correspond to separate and independent single-particle insertions into the system. This means that $w^{(2)}(\mathbf{r},\mathbf{r}')$ is the difference in excess chemical potentials when we insert the two particles at the same time and when we insert only one particle either at $\mathbf{r}$ or at $\mathbf{r}'$ and sum the results. In Equation 9.30 we clearly see the symmetry $w^{(2)}(\mathbf{r},\mathbf{r}') = w^{(2)}(\mathbf{r}',\mathbf{r})$, so we have $w^{(2)\text{intr}}(\mathbf{r},\mathbf{r}') = w^{(2)\text{intr}}(\mathbf{r}',\mathbf{r})$, which is not immediately obvious from Equation 9.24.

Let us return to the cavity function $y^{(2)} = e^{-\beta w^{(2)\text{intr}}}$. We have $w^{(2)\text{intr}} = w^{(2)} - u$ and from Equation 9.30 we therefore obtain

$$w^{(2)\text{intr}}(\mathbf{r},\mathbf{r}') = \mu^{(2)\text{ex}}(\mathbf{r},\mathbf{r}') - u(\mathbf{r},\mathbf{r}') - \mu^{\text{ex}}(\mathbf{r}') - \mu^{\text{ex}}(\mathbf{r}).$$

Now, $\mu^{(2)\text{ex}}(\mathbf{r},\mathbf{r}')$ contains a term $u(\mathbf{r},\mathbf{r}')$ that gives the interaction between the two particles placed at $\mathbf{r}$ and $\mathbf{r}'$, respectively. Therefore, this pair potential is absent from $\mu^{(2)\text{ex}}(\mathbf{r},\mathbf{r}') - u(\mathbf{r},\mathbf{r}')$. Let us define

$$\mu^{(2)\text{ex-}u}(\mathbf{r},\mathbf{r}') \equiv \mu^{(2)\text{ex}}(\mathbf{r},\mathbf{r}') - u(\mathbf{r},\mathbf{r}'), \tag{9.31}$$

which contains all contributions to the free energy change for an insertion of the two particles apart from the pair interaction u between them. Thus $\mu^{(2)\text{ex-}u}$ can be interpreted as the change in free energy when we insert two (imagined) particles that *do not interact with each other*, but interact fully with all other particles and with the external potential. We have

$$w^{(2)\text{intr}}(\mathbf{r},\mathbf{r}') = \mu^{(2)\text{ex-}u}(\mathbf{r},\mathbf{r}') - \mu^{\text{ex}}(\mathbf{r}) - \mu^{\text{ex}}(\mathbf{r}'), \tag{9.32}$$

so we can calculate $w^{(2)\text{intr}}$ by inserting a pair of particles that do not interact with each other. The construct with an insertion of such a pair is quite useful because it allows us to calculate $w^{(2)\text{intr}}(\mathbf{r},\mathbf{r}')$ and hence $y^{(2)}(\mathbf{r},\mathbf{r}')$ for all $\mathbf{r}$ and $\mathbf{r}'$, including cases where $u(\mathbf{r},\mathbf{r}')$ is infinite.

We finish this section with some comments on pair functions like the pair distribution function. For spherically symmetric particles, the pair potential is isotropic $u(\mathbf{r}_1,\mathbf{r}_2) = u(r_{12})$, where $r_{12} = |\mathbf{r}_{12}|$ and $\mathbf{r}_{12} = \mathbf{r}_2 - \mathbf{r}_1$. In bulk fluids of such particles, the pair distribution function and the other pair functions are also isotropic, $g^{(2)}(\mathbf{r}_1,\mathbf{r}_2) = g^{(2)}(r_{12})$, For inhomogeneous fluids of spherically symmetric particles, the pair functions like $g^{(2)}(\mathbf{r}_1,\mathbf{r}_2)$ are, however, anisotropic, that is, they depend on the locations of the two points $\mathbf{r}_1$ and $\mathbf{r}_2$ and not only on the distance r_{12} between them. In the general case they therefore depend on six variables, x_1, y_1, z_1, x_2, y_2 and z_2, but when there is some symmetry in the system, this number can usually be reduced.

Let us, for example, consider a fluid outside a planar wall with a uniform smooth surface, where the external potential depends only on the coordinate z in the perpendicular direction from the surface, $v(\mathbf{r}) = v(z)$. Therefore, all singlet functions depend solely on z, like $n(\mathbf{r}) = n(z)$, in this case. Due to the translational symmetry in the lateral direction along the surface, the pair functions depend only on three independent variables, like the distances of the two points $\mathbf{r}_1$ and $\mathbf{r}_2$ from the surface and the separation between these points $r_{12} = |\mathbf{r}_{12}|$, where $\mathbf{r}_{12} = \mathbf{r}_2 - \mathbf{r}_1$. Any three independent variables related to $\mathbf{r}_1$ and $\mathbf{r}_2$ are thereby possible; a preferred choice is to use z_1, z_2 and $s_{12} = |\mathbf{s}_2 - \mathbf{s}_1|$, where $\mathbf{s}_\nu = (x_\nu, y_\nu)$, so s_{12} is the length of the vector $\mathbf{r}_{12}$ projected on a plane parallel to the surface. We then have $g^{(2)}(\mathbf{r}_1,\mathbf{r}_2) = g^{(2)}(s_{12}, z_1, z_2)$ and likewise for the other pair functions apart from $u(r_{12})$.

9.1.2 Linear response theory and the pair direct-correlation function $c^{(2)}$

In the previous section we saw that the change in the singlet direct-correlation function, $\Delta c^{(1)}$, due to certain variations in external potential is the key quantity needed to obtain the singlet distribution function $g^{(1)}(\mathbf{r})$ and the pair distribution function $g^{(2)}(\mathbf{r}, \mathbf{r}')$, Equations 9.17 and 9.22. From $g^{(1)}$ one can calculate the density distribution $n(\mathbf{r})$ via Equation 9.3.

To obtain $g^{(1)}$ and $g^{(2)}$ we need to determine how $c^{(1)}$ changes when we do the following variations: in Equation 9.17 the external potential goes from zero to $v(\mathbf{r})$ and in Equation 9.22 it goes from $v(\mathbf{r})$ to $v(\mathbf{r}|\mathbf{r}')$, that is, to $v(\mathbf{r})+u(\mathbf{r}, \mathbf{r}')$. To make the arguments physically transparent, we will first investigate how $\mu^{\text{ex}}(\mathbf{r})$ changes and then obtain the change in $c^{(1)}(\mathbf{r})$ via $k_B T\, c^{(1)}(\mathbf{r}) = v(\mathbf{r}) - \mu^{\text{ex}}(\mathbf{r})$ (from Equation 9.8).

As a first step we will investigate how μ^{ex} changes when we do small variations in external potential: $v(\mathbf{r}) \to v(\mathbf{r}) + \delta v(\mathbf{r})$, where $\delta v(\mathbf{r})$ is close to zero for all $\mathbf{r}$. The variation $\delta v(\mathbf{r})$ expresses how much we change the external potential at each point $\mathbf{r}$. When we do this variation in v, the function μ^{ex} at an arbitrary point $\mathbf{r}'$ changes from $\mu^{\text{ex}}(\mathbf{r}')$ to $\mu^{\text{ex}}(\mathbf{r}') + \delta\mu^{\text{ex}}(\mathbf{r}')$, where $\delta\mu^{\text{ex}}$ is, as yet, unknown. Provided that $\delta v(\mathbf{r})$ is infinitesimally small everywhere, $\delta\mu^{\text{ex}}(\mathbf{r}')$ is linearly related to $\delta v(\mathbf{r})$.[10]

Remember that $\mu^{\text{ex}}(\mathbf{r}')$ is the change in free energy when one inserts a new particle at $\mathbf{r}'$ and keeps it fixed there. When we do this insertion the density distribution at other points $\mathbf{r}$ changes from $n(\mathbf{r})$ to $n(\mathbf{r}|\mathbf{r}')$, that is, by

$$\Delta n(\mathbf{r}|\mathbf{r}') \equiv n(\mathbf{r}|\mathbf{r}') - n(\mathbf{r}) = g^{(2)}(\mathbf{r}, \mathbf{r}')n(\mathbf{r}) - n(\mathbf{r}) = h^{(2)}(\mathbf{r}, \mathbf{r}')n(\mathbf{r}). \quad (9.33)$$

As we saw earlier, this variation in density can be regarded as the result of a change in external potential from $v(\mathbf{r})$ (which we will call case A) to $v(\mathbf{r}|\mathbf{r}') = v(\mathbf{r})+u(\mathbf{r}, \mathbf{r}')$ (case B, where the new particle is present). Thus, if one goes from A to B, the difference in free energy between these two cases equals $\mu^{\text{ex}}(\mathbf{r}')$ and provided the particle insertion is done slowly (reversibly) this is the reversible work done. The relevant free energy is the **grand potential** $\Theta = A - \mu\bar{N}$ of the system, where A is the Helmholtz free energy and $\bar{N}$ the average number of particles,[11] so we have $\mu^{\text{ex}} = \Delta\Theta = \Theta_{\text{B}} - \Theta_{\text{A}}$.

---

[10]The linear relationship between the two entities means that if $\delta\mu^{\text{ex}}_{\text{I}}$ is an change in μ^{ex} due to the variation δv_{I} and $\delta\mu^{\text{ex}}_{\text{II}}$ is the change due to δv_{II}, then $\delta\mu^{\text{ex}}_{\text{I}} + \delta\mu^{\text{ex}}_{\text{II}}$ is the change due to the variation $\delta v_{\text{I}} + \delta v_{\text{II}}$ where δv_{I} and δv_{II} are small but otherwise arbitrary.

We can compare this with the change in the value of a function $f(x)$ when we vary the argument x by a small amount dx. When $x \to x + dx_{\text{I}}$ the function f changes by $df(x)_{\text{I}} = (df(x)/dx)dx_{\text{I}}$ and when $x \to x + dx_{\text{II}}$ the change is $df(x)_{\text{II}} = (df(x)/dx)dx_{\text{II}}$. Obviously, when $x \to x + dx_{\text{I}} + dx_{\text{II}}$ the change is $df(x)_{\text{I}} + df(x)_{\text{II}} = (df(x)/dx)[dx_{\text{I}} + dx_{\text{II}}]$ provided that dx_{I} and dx_{II} are infinitesimally small.

[11]The fact that the grand potential Θ is the relevant free energy for the reversible work here can be realized as follows. For a reversible process with work $\mathtt{w}^{\text{rev}}$ and heat $\mathtt{q}^{\text{rev}} = T\Delta S$ at constant temperature, we have the standard thermodynamic result that the change in Helmholtz free energy is $\Delta A = \Delta\bar{U} - T\Delta S = \mathtt{q}^{\text{rev}} + \mathtt{w}^{\text{rev}} - T\Delta S = \mathtt{w}^{\text{rev}}$, where $\Delta\bar{U} = \mathtt{q}^{\text{rev}} + \mathtt{w}^{\text{rev}}$ is the change in internal energy. In the grand canonical ensemble, where the chemical potential is constant, $\bar{N}$ will in general vary when we change the external potential. Then the relevant free energy is the grand potential $\Theta = A - \mu\bar{N}$, where Θ, A and $\bar{N}$ are properties of the open system with volume V and we have $\Delta\Theta = \Delta A - \mu\Delta\bar{N} = \mathtt{w}^{\text{rev}} - \mu\Delta\bar{N}$ for a reversible process at constant T. The "controllable" reversible work, i.e., the work *apart from* the work associated with the "automatic" change in number of mobile particles, is equal to $\mathtt{w}^{\text{rev}} - \mu\Delta\bar{N}$, which is equal to $\Delta\Theta$. It is this controllable work that we discuss

Let us return to the task to find out how much μ^{ex} changes when the external potential varies by δv. Since $d\mu^{\text{ex}} = d\Theta_{\text{B}} - d\Theta_{\text{A}}$, we can do this by determining $d\Theta_{\text{A}}$ and $d\Theta_{\text{B}}$ individually and then take the difference. Thus, we will first determine how much Θ_{A} changes when the external potential varies by δv and then do the same for Θ_{B}. Recall that by definition, work is the change in energy due to the change in an external variable (in the current case the external potential is such a variable). For case A, where the external potential goes from $v(\mathbf{r})$ to $v(\mathbf{r}) + \delta v(\mathbf{r})$, the work is $\delta v(\mathbf{r}) \times n(\mathbf{r})d\mathbf{r}$ because the change in energy for a particle located at $\mathbf{r}$ is $\delta v(\mathbf{r})$ and since there are on average $n(\mathbf{r})d\mathbf{r}$ particles in the volume element $d\mathbf{r}$ around the point $\mathbf{r}$, the differential work on them from the variation in external potential is $\delta v(\mathbf{r}) \times n(\mathbf{r})d\mathbf{r}$. The entire work due to $\delta v(\mathbf{r})$ on all mobile particles is therefore

$$d\Theta = \int d\mathbf{r}\, n(\mathbf{r})\delta v(\mathbf{r}). \tag{9.34}$$

We will later give a more strict proof of Equation 9.34 in Section 9.2.4.1.

We can obtain the corresponding result for case B, where we have inserted the new particle at $\mathbf{r}'$, by replacing the density $n(\mathbf{r})$ by $n(\mathbf{r}|\mathbf{r}')$ in Equation 9.34. To obtain $d\Theta_{\text{B}} - d\Theta_{\text{A}}$ we therefore take the difference of the right hand side of this equation for the respective cases

$$\int d\mathbf{r}\, n(\mathbf{r}|\mathbf{r}')\delta v(\mathbf{r}) - \int d\mathbf{r}\, n(\mathbf{r})\delta v(\mathbf{r}) = \int d\mathbf{r}\, \Delta n(\mathbf{r}|\mathbf{r}')\delta v(\mathbf{r}).$$

This is the intrinsic part of $\delta\mu^{\text{ex}}(\mathbf{r}')$ for the new particle at $\mathbf{r}'$, that is, from the interactions with the mobile particles in the system. In addition we have the contribution from the interaction of the new particle itself with the external potential. The latter gives a contribution $\delta v(\mathbf{r}')$, i.e., the change in interaction with v that the particle experiences. The total change in μ^{ex} from the variation δv is therefore

$$\begin{aligned}\delta\mu^{\text{ex}}(\mathbf{r}') &= \delta v(\mathbf{r}') + \int d\mathbf{r}\, \Delta n(\mathbf{r}|\mathbf{r}')\delta v(\mathbf{r}) \\ &= \delta v(\mathbf{r}') + \int d\mathbf{r}\, h^{(2)}(\mathbf{r},\mathbf{r}')n(\mathbf{r})\delta v(\mathbf{r}),\end{aligned} \tag{9.35}$$

where we have inserted Equation 9.33. This is the relationship that we have been looking for.[12]

---

here. Incidentally, in the isobaric-isothermal ensemble the controllable reversible work is, instead, equal to the change in Gibbs free energy, $\Delta G = \Delta A + P\Delta\bar{V} = \mathtt{w}^{\text{rev}} + P\Delta\bar{V}$, where the volume change $\Delta\bar{V}$ occurs "automatically" when P is constant. This latter result for ΔG at constant N, P and T is more commonly used in standard textbooks of thermodynamics than the former result for $\Delta\Theta$ at constant μ, V and T.

[12] The second term in Equation 9.35 can alternatively be rationalized as follows. The number of particles in volume element $d\mathbf{r}$ at the point $\mathbf{r}$ is changed by $\Delta n(\mathbf{r}|\mathbf{r}')d\mathbf{r}$ when the new particle is inserted at $\mathbf{r}'$, i.e., particles are moved in to or out from this volume element (the probability that they are there increases or decreases). The integrand equals the work

Equation 9.35 implies that the excess chemical potential at $\mathbf{r}'$ is not only affected by the change in external potential in the same point (the first term in the rhs), but also the change in the neighborhood of this point (the integral). The integrand says that there is a contribution from the variation in the external potential at other points, $\mathbf{r}$, whereby the change in $\mu^{\text{ex}}(\mathbf{r}')$ is proportional to $\delta v(\mathbf{r})$ with a proportionality factor $h^{(2)}(\mathbf{r},\mathbf{r}')n(\mathbf{r})$. The range of influence from $\mathbf{r}$ to $\mathbf{r}'$ is therefore determined by decay of $h^{(2)}(\mathbf{r},\mathbf{r}')n(\mathbf{r})$ as function of the distance $|\mathbf{r}-\mathbf{r}'|$. The total change in $\mu^{\text{ex}}(\mathbf{r}')$ is obtained by summation of the contributions from all points $\mathbf{r}$ as expressed by the integral in the equation and by adding $\delta v(\mathbf{r}')$.

Since $\delta\mu^{\text{ex}}(\mathbf{r}') = \delta v(\mathbf{r}') - k_BT\,\delta c^{(1)}(\mathbf{r}')$ (Equation 9.8), we see by comparison with Equation 9.35 that

$$\delta c^{(1)}(\mathbf{r}') = -\beta \int d\mathbf{r}\, h^{(2)}(\mathbf{r},\mathbf{r}')n(\mathbf{r})\delta v(\mathbf{r}). \tag{9.36}$$

Note that the factors $n(\mathbf{r})$ and $h^{(2)}(\mathbf{r},\mathbf{r}')$ in the integrand are the functions *before* the variation δv in external potential, that is, for the unperturbed system with external potential $v(\mathbf{r})$.

Equation 9.36 is exact when $\delta v(\mathbf{r})$ is infinitesimally small for all $\mathbf{r}$. This is the relationship between variations in $c^{(1)}$ and v that we set out to find as a first step in the determination of $\Delta c^{(1)}$. Since Equation 9.36 concerns small variations, it is, however, not sufficient for use in Equations 9.17 and 9.22 that require $\Delta c^{(1)}$ due to finite variations Δv. For such variations one must add nonlinear terms to the right hand side of Equation 9.36, which, as it stands, gives the linear contributions only. Rather than complementing Equation 9.36 with such nonlinear terms, we will seek the required change in $c^{(1)}$ in a different, but related manner that focuses on the relationship between variations in $c^{(1)}$ and n rather than v. This leads to new important concepts and when they are introduced, we will return to the question of determining $\Delta c^{(1)}$ in the next section.

From the results above we can, in fact, determine how much the density distribution $n(\mathbf{r}')$ in the system varies when the external potential is changed by $\delta v(\mathbf{r})$. Say that the density changes from $n(\mathbf{r}')$ to $n(\mathbf{r}') + \delta n(\mathbf{r}')$. Since we hold μ constant (equilibrium is maintained with the same bulk fluid), the activity ζ is also constant and Equation 9.12 implies that $k_BT\delta\ln(n(\mathbf{r}')) + \delta v(\mathbf{r}') - k_BT\,\delta c^{(1)}(\mathbf{r}') = 0$. To linear order we have[13] $\delta\ln(n(\mathbf{r}')) = \delta n(\mathbf{r}')/n(\mathbf{r}')$

due to the additional potential $\delta v(\mathbf{r})$ when the number of particles is changed in $d\mathbf{r}$ and the integral is the reversible work from the change in density in the entire volume. We here assume that the volume V of the system encloses the region where $\delta v(\mathbf{r})$ is nonzero.

[13]Normal differentiation rules applies to small variations in functions, whereby the differentiation $d\ln f(\mathbf{r}) = df(\mathbf{r})/f(\mathbf{r})$ corresponds to the variation $\delta\ln f(\mathbf{r}) = \delta f(\mathbf{r})/f(\mathbf{r})$. This can be realizes as follows. When we change $f(\mathbf{r})$ to $f(\mathbf{r})+\delta f(\mathbf{r})$, the function $\ln f(\mathbf{r})$ changes by

$$\begin{aligned}\delta\ln f(\mathbf{r}) &= \ln[f(\mathbf{r})+\delta f(\mathbf{r})] - \ln f(\mathbf{r}) = \ln\{[f(\mathbf{r})+\delta f(\mathbf{r})]/f(\mathbf{r})\}\\ &= \ln\{1+\delta f(\mathbf{r})/f(\mathbf{r})\} \to \delta f(\mathbf{r})/f(\mathbf{r})\end{aligned}$$

when $\delta f(\mathbf{r}) \to 0$ for all $\mathbf{r}$, so it follows that $\delta\ln f(\mathbf{r}) = \delta f(\mathbf{r})/f(\mathbf{r})$.

and hence

$$\frac{\delta n(\mathbf{r}')}{n(\mathbf{r}')} = -\beta\delta v(\mathbf{r}') + \delta c^{(1)}(\mathbf{r}'). \tag{9.37}$$

By inserting Equation 9.36 we obtain the change in density

$$\delta n(\mathbf{r}') = -\beta n(\mathbf{r}')\left[\delta v(\mathbf{r}') + \int d\mathbf{r}\, h^{(2)}(\mathbf{r},\mathbf{r}')n(\mathbf{r})\delta v(\mathbf{r})\right] \tag{9.38}$$

This equation is called **the first Yvon equation.** It gives the ***linear response*** of the density distribution when the system is exposed to a small change in external potential. We have encountered this equation earlier in Appendix 6B of Volume I, Equation I:6.205, where we derived it in the canonical ensemble. Here, we have obtained it in the grand canonical ensemble in a completely different manner.

Equation 9.36 gives the change of $c^{(1)}(\mathbf{r}')$ due to the variation $\delta v(\mathbf{r})$ in external potential at various points $\mathbf{r}$, whereby the proportionality factor in the integrand is $-\beta h^{(2)}(\mathbf{r},\mathbf{r}')n(\mathbf{r})$. As indicated earlier, it is also relevant to know how much $c^{(1)}(\mathbf{r}')$ (and thereby $\mu^{\text{ex}}(\mathbf{r}')$) is influenced by a change in *density*, $\delta n(\mathbf{r})$, at various points $\mathbf{r}$.[14] We therefore ask:

> When the density is varied by adding $\delta n(\mathbf{r})$, the average number of particles with centers in the volume element $d\mathbf{r}$ around the point $\mathbf{r}$ is changed by $\delta n(\mathbf{r})d\mathbf{r}$. How much does $c^{(1)}(\mathbf{r}')$ at various points $\mathbf{r}'$ then change via the interparticle interactions?

Since small changes in $c^{(1)}$, n and v are related by linear relationships as explored above, it should be possible to express $\delta c^{(1)}$ linearly in terms of δn; in other words, it should be possible to eliminate δv from Equations 9.36 and 9.38. To this end we define the **pair direct-correlation function** $c^{(2)}(\mathbf{r},\mathbf{r}')$ from

$$\delta c^{(1)}(\mathbf{r}') = \int d\mathbf{r}\, c^{(2)}(\mathbf{r},\mathbf{r}')\delta n(\mathbf{r}), \tag{9.39}$$

so the contribution to the change $\delta c^{(1)}(\mathbf{r}')$ from point $\mathbf{r}$ is proportional to $\delta n(\mathbf{r})$ with a proportionality factor $c^{(2)}(\mathbf{r},\mathbf{r}')$, as yet unknown. It is possible to express $c^{(2)}$ in terms of n and h from Equations 9.36, 9.38 and 9.39. If one combines these three equations and uses the fact that $\delta v(\mathbf{r})$ is an infinitesimally small but otherwise arbitrary variation, one finds after some algebra (see Exercise 9.1) that the pair direct-correlation function must satisfy the equation

$$h^{(2)}(\mathbf{r},\mathbf{r}') = c^{(2)}(\mathbf{r},\mathbf{r}') + \int d\mathbf{r}''\, c^{(2)}(\mathbf{r},\mathbf{r}'')n(\mathbf{r}'')h^{(2)}(\mathbf{r}'',\mathbf{r}'), \tag{9.40}$$

which is called the **Ornstein-Zernike (OZ) equation.**[15] For a given density

[14]This density change can in turn be caused by a change in external potential according to Equation 9.38.

[15]We will also prove the OZ equation in a different, but related manner in Section 11.4.

distribution $n(\mathbf{r})$, this equation can be used to determine the function $c^{(2)}$ when $h^{(2)}$ is known and, vice versa, to determine $h^{(2)}$ when $c^{(2)}$ is known.

{Exercise 9.1:

Show the OZ equation 9.40 from the previous results, whereby you can proceed in the following manner:

Eliminate $\delta c^{(1)}$ by combining Equations 9.36 and 9.39. Then, insert δn from Equation 9.38 and obtain an equation where δv is the only remaining variational quantity. Finally, obtain the OZ equation by setting the arbitrary $\delta v(\mathbf{r})$ equal to a constant times the Dirac delta function $\delta^{(3)}(\mathbf{r}-\mathbf{r}')$ for an arbitrary $\mathbf{r}'$ and then performing the integrations that involve delta functions. Thereby you can use the property of the delta function $\int d\mathbf{r}\, f(\mathbf{r})\delta^{(3)}(\mathbf{r}-\mathbf{r}') = f(\mathbf{r}')$ for any reasonable function $f(\mathbf{r})$. (The Dirac delta function is described in Appendix 5A of Volume I.)}

The OZ equation is often used as the definition of $c^{(2)}$ from given functions n and $h^{(2)}$ and then Equation 9.39 follows as a consequence of this definition. Here, we have instead used Equation 9.39 as the definition and then the Ornstein-Zernike equation follows. The two points of view are equivalent. The function $c^{(2)}$ is a symmetric function, $c^{(2)}(\mathbf{r},\mathbf{r}') = c^{(2)}(\mathbf{r}',\mathbf{r})$, which is not immediately apparent from its definition 9.39. In the integrand of the OZ equation the functions $h^{(2)}$ and $c^{(2)}$ can change places with each other (which follows from a swapping of $\mathbf{r}$ and $\mathbf{r}'$ in this equation).

The OZ equation can be used to express $h^{(2)}$ entirely in term of $c^{(2)}$. By substituting $\mathbf{r}'' \to \mathbf{r}_4$ and then $\mathbf{r} \to \mathbf{r}''$ in Equation 9.40 we obtain the equation

$$h^{(2)}(\mathbf{r}'',\mathbf{r}') = c^{(2)}(\mathbf{r}'',\mathbf{r}') + \int d\mathbf{r}_4\, c^{(2)}(\mathbf{r}'',\mathbf{r}_4)n(\mathbf{r}_4)h^{(2)}(\mathbf{r}_4,\mathbf{r}').$$

If we insert this in the right hand side (rhs) of Equation 9.40 and write $\mathbf{r}_3$ instead of $\mathbf{r}''$, we obtain

$$\begin{aligned} h^{(2)}(\mathbf{r},\mathbf{r}') &= c^{(2)}(\mathbf{r},\mathbf{r}') + \int d\mathbf{r}_3\, c^{(2)}(\mathbf{r},\mathbf{r}_3)n(\mathbf{r}_3) \\ &\times \left[c^{(2)}(\mathbf{r}_3,\mathbf{r}') + \int d\mathbf{r}_4\, c^{(2)}(\mathbf{r}_3,\mathbf{r}_4)n(\mathbf{r}_4)h^{(2)}(\mathbf{r}_4,\mathbf{r}')\right] \\ &= c^{(2)}(\mathbf{r},\mathbf{r}') + \int d\mathbf{r}_3\, c^{(2)}(\mathbf{r},\mathbf{r}_3)n(\mathbf{r}_3)c^{(2)}(\mathbf{r}_3,\mathbf{r}') \\ &+ \int d\mathbf{r}_3 d\mathbf{r}_4\, c^{(2)}(\mathbf{r},\mathbf{r}_3)n(\mathbf{r}_3)c^{(2)}(\mathbf{r}_3,\mathbf{r}_4)n(\mathbf{r}_4)h^{(2)}(\mathbf{r}_4,\mathbf{r}'). \end{aligned}$$

By continuing in the same way, it is easy to realize that if one continues to successively substitute the $h^{(2)}$ function in the rhs in this fashion, one obtains an infinite sum of integrals which only contain $c^{(2)}$ and n. Accordingly, if we write $\mathbf{r}_1, \mathbf{r}_2$ instead of $\mathbf{r}, \mathbf{r}'$, we obtain

$$h^{(2)}(\mathbf{r}_1, \mathbf{r}_2) = c^{(2)}(\mathbf{r}_1, \mathbf{r}_2) + \int d\mathbf{r}_3 \, c^{(2)}(\mathbf{r}_1, \mathbf{r}_3) n(\mathbf{r}_3) c^{(2)}(\mathbf{r}_3, \mathbf{r}_2) \qquad (9.41)$$

$$+ \int d\mathbf{r}_3 d\mathbf{r}_4 \, c^{(2)}(\mathbf{r}_1, \mathbf{r}_3) n(\mathbf{r}_3) c^{(2)}(\mathbf{r}_3, \mathbf{r}_4) n(\mathbf{r}_4) c^{(2)}(\mathbf{r}_4, \mathbf{r}_2) + \dots,$$

where the next term is an integral containing the product $c^{(2)} n c^{(2)} n c^{(2)} n c^{(2)}$ and the following term contains $c^{(2)} n c^{(2)} n c^{(2)} n c^{(2)} n c^{(2)}$ etc., with functions evaluated at the respective coordinates. This can be illustrated pictorially as

$$h^{(2)}(\mathbf{r}_1, \mathbf{r}_2) = \; \text{○——○} \; + \; \text{○——●——○} \; + \; \text{○——●——●——○}$$

$$+ \; \text{○——●——●——●——○} \; + \; \text{○——●——●——●——●——○} \; + \; \dots$$

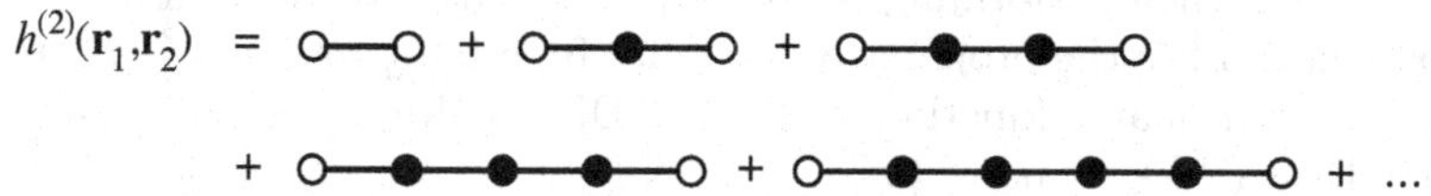

where the open circles represent the points $\mathbf{r}_1$ and $\mathbf{r}_2$, each filled circle represents a point $\mathbf{r}_\nu$ ($\nu \neq 1$ and 2) including a factor $n(\mathbf{r}_\nu)$, the lines represent $c^{(2)}$ and integration is performed over the whole space for the filled points. Thus the pair correlation function (total correlation function) $h^{(2)}(\mathbf{r}_1, \mathbf{r}_2)$ can be expressed in terms of an infinite sum of "chains" of direct correlation functions. This is a reason for the name "total" correlation function since $h^{(2)}$ constitutes the total sum of chains of the direct correlations. Each separate illustration of the chains is an example of what is called a "diagram."

The pair direct-correlation function has a very important role in the theory of fluids. From Equation 9.8 we have $\delta \mu^{\text{ex}}(\mathbf{r}') = \delta v(\mathbf{r}') - k_B T \, \delta c^{(1)}(\mathbf{r}')$ and by inserting Equation 9.39 it follows that

$$\delta \mu^{\text{ex}}(\mathbf{r}') = \delta v(\mathbf{r}') + \int d\mathbf{r} \left[-k_B T c^{(2)}(\mathbf{r}, \mathbf{r}') \right] \delta n(\mathbf{r}).$$

This relationship can be interpreted as follows: since $\mu^{\text{ex}}(\mathbf{r}')$ is the work against interactions when a new particle is inserted at $\mathbf{r}'$, the equation expresses how this work changes due to a variation δv in the external potential and the consequent change δn in density. The integral is the change in the intrinsic part of the work, i.e., the interparticle interaction part, while $\delta v(\mathbf{r}')$ is the change in work against the external potential. In the integrand, the product $d\mathbf{r} \, \delta n(\mathbf{r})$ is the change in number of particle with centers in the volume element $d\mathbf{r}$ and $-k_B T c^{(2)}(\mathbf{r}, \mathbf{r}')$ gives the contribution per particle to the change in work from interactions with particles in $d\mathbf{r}$.

Thus, $-k_B T c^{(2)}$ acts as a kind of "effective" interaction per particle with the new particle at $\mathbf{r}'$. In fact, we generally have[16]

$$-k_B T c^{(2)}(\mathbf{r}, \mathbf{r}') \sim u(\mathbf{r}, \mathbf{r}') \quad \text{when } |\mathbf{r} - \mathbf{r}'| \to \infty, \qquad (9.42)$$

---

[16]This relationship, which holds asymptotically for large separations $|\mathbf{r} - \mathbf{r}'|$, will be discussed in more detail later.

so this effective interaction approaches the ordinary pair interaction for large separations.[17] For small separations the function $c^{(2)}$ is less simple, but it is in general appreciably simpler than $h^{(2)}$. For this reason it is often easier to approximate $c^{(2)}$ than $h^{(2)}$ and once this is done one can determine $h^{(2)}$ via the OZ equation. This will be discussed in the next section.

The first Yvon equation, Equation 9.38, gives the change in density $\delta n(\mathbf{r}')$ when we have exposed our system to a given small change in external potential $\delta v(\mathbf{r})$. We are now in a position where we can answer the following "reverse" question: What change in potential, $\delta v(\mathbf{r})$, is necessary in order to induce a given small change in density, $\delta n(\mathbf{r}'')$? Equation 9.37 can be written

$$\delta v(\mathbf{r}) = -\frac{1}{\beta}\left[\frac{1}{n(\mathbf{r})}\delta n(\mathbf{r}) - \delta c^{(1)}(\mathbf{r})\right]$$

and by using the definition (9.39) of $c^{(2)}$ we obtain **the second Yvon equation**

$$\delta v(\mathbf{r}) = -\frac{1}{\beta}\left[\frac{1}{n(\mathbf{r})}\delta n(\mathbf{r}) - \int d\mathbf{r}''\, c^{(2)}(\mathbf{r}'',\mathbf{r})\delta n(\mathbf{r}'')\right], \tag{9.43}$$

which is the answer to our question, that is, it gives δv in terms of δn. One can use the Ornstein-Zernike equation to show that Equation 9.43 follows from Equation 9.38 and vice versa. Both Yvon equations are exact for infinitesimally small variations δv and δn; they express the *linear part of the response* due to changes in v and n, respectively.

One can derive some particularly useful formulas by applying the results above to special kinds of variations of the functions. Let $\hat{\mathbf{a}}$ be a unit vector in an arbitrary direction. We now shift the source of the external potential by a small arbitrary distance da in the direction $\hat{\mathbf{a}}$, that is, the position is shifted by the vector $d\mathbf{a} \equiv da\,\hat{\mathbf{a}}$. Then $v(\mathbf{r})$ changes by $\delta v(\mathbf{r}) = -\nabla v(\mathbf{r})\cdot d\mathbf{a}$ and since it is $v(\mathbf{r})$ that causes the inhomogeneity in the density distribution, the latter is also shifted by an equal distance in the same direction. This means that $n(\mathbf{r})$ changes by $\delta n(\mathbf{r}) = -\nabla n(\mathbf{r})\cdot d\mathbf{a}$. Likewise, $c^{(1)}(\mathbf{r})$ changes by $\delta c^{(1)}(\mathbf{r}) = -\nabla c^{(1)}(\mathbf{r})\cdot d\mathbf{a}$. If we insert this variation in Equation 9.36 we obtain[18]

$$-\nabla_1 c^{(1)}(\mathbf{r}_1)\cdot d\mathbf{a} = -\beta\int d\mathbf{r}_2\, h^{(2)}(\mathbf{r}_1,\mathbf{r}_2)n(\mathbf{r}_2)\left[-\nabla_2 v(\mathbf{r}_2)\cdot d\mathbf{a}\right].$$

[17]The role of $-k_BTc^{(2)}$ as a kind of effective interaction can also be seen by considering the gradient $k_BT\nabla_1 c^{(2)}(\mathbf{r},\mathbf{r}_1)$, which is a contribution to the force that acts on a new particle at $\mathbf{r}_1$. This can be understood from the following. The gradient $k_BT\,\nabla_1 c^{(1)}(\mathbf{r}_1)$, which equals $-\nabla_1 w^{\text{intr}}(\mathbf{r}_1)$, is the total force from interparticle interactions on the new particle *before* the change δn in density (cf. Section 9.1.1). Since $c^{(2)}$ gives the change in $c^{(1)}$ from variations in particle density according to Equation 9.39, the force $k_BT\nabla_1 c^{(2)}(\mathbf{r},\mathbf{r}_1)$ is the contribution per particle from the *change* in number of particles with centers in the volume element $d\mathbf{r}$ located at $\mathbf{r}$.

[18]This result applies provided the shift in the source of the inhomogeneity has no other effect on the interactions than a change in $v(\mathbf{r})$. If the interparticle interactions are also changed, for instance for charged particles near a surface that induce polarization charges on the surface (see Section 7.1.4 in Volume I), then there are terms that involve higher order correlation functions too.

Since $d\mathbf{a}$ is arbitrary but small, it follows that

$$\nabla_1 c^{(1)}(\mathbf{r}_1) = -\beta \int d\mathbf{r}_2\, h^{(2)}(\mathbf{r}_1, \mathbf{r}_2) n(\mathbf{r}_2) \nabla_2 v(\mathbf{r}_2). \tag{9.44}$$

This is an expression for $\nabla_1 c^{(1)}$ that can be compared with Equation 9.9, which can be written

$$\nabla_1 c^{(1)}(\mathbf{r}_1) = \beta \mathbf{F}^{\text{intr}}(\mathbf{r}_1) = -\beta \int d\mathbf{r}_2\, g^{(2)}(\mathbf{r}_1, \mathbf{r}_2) n(\mathbf{r}_2) \nabla_1 u(\mathbf{r}_2, \mathbf{r}_1). \tag{9.45}$$

Equation 9.44 involves the gradient of the external potential and is valid even if the interparticle interactions are not pairwise additive, while Equation 9.45 involves the pair potential and requires pairwise additivity.

The mean force acting on a particle located at $\mathbf{r}_1$ is given by $\mathbf{F}(\mathbf{r}_1) = -\nabla_1 v(\mathbf{r}_1) + \mathbf{F}^{\text{intr}}(\mathbf{r}_1) = -\nabla_1 v(\mathbf{r}_1) + k_B T\, \nabla_1 c^{(1)}(\mathbf{r}_1)$ and is, as we have seen, related to the density gradient by $\mathbf{F}(\mathbf{r}_1) = k_B T\, \nabla_1 \ln n(\mathbf{r}_1)$. By combining this with (9.44) we obtain

$$\begin{aligned} \nabla_1 \ln n(\mathbf{r}_1) &= \beta \mathbf{F}(\mathbf{r}_1) \\ &= -\beta \left[\nabla_1 v(\mathbf{r}_1) + \int d\mathbf{r}_2\, h^{(2)}(\mathbf{r}_1, \mathbf{r}_2) n(\mathbf{r}_2) \nabla_2 v(\mathbf{r}_2) \right], \end{aligned} \tag{9.46}$$

which can be compared with the first Born-Green-Yvon equation 9.10.

The analogous arguments applied to Equation 9.39 imply that

$$\nabla_1 c^{(1)}(\mathbf{r}_1) = \int d\mathbf{r}_2\, c^{(2)}(\mathbf{r}_1, \mathbf{r}_2) \nabla_2 n(\mathbf{r}_2)$$

and

$$\begin{aligned} \nabla_1 \ln n(\mathbf{r}_1) &= \beta \mathbf{F}(\mathbf{r}_1) \\ &= -\beta \nabla_1 v(\mathbf{r}_1) + \int d\mathbf{r}_2\, c^{(2)}(\mathbf{r}_1, \mathbf{r}_2) \nabla_2 n(\mathbf{r}_2). \end{aligned} \tag{9.47}$$

Here, the gradient of the density appears in the integrand instead of an interaction potential. Both Equations 9.46 and 9.47 are integro-differential equation for the density distribution $n(\mathbf{r})$, which, like the BGY equation 9.10, can be used to determine $n(\mathbf{r})$ when the pair correlation function $h^{(2)}$ or the pair direct-correlation function $c^{(2)}$ is known (or at least approximately known). The two equations 9.46 and 9.47 are known in the literature under the name **Triezenberg-Zwanzig-Lovett-Mou-Buff-Wertheim (TZLMBW) equations**[19] or some other permutation of all or some of these names(!). They

[19] D. G. Triezenberg and R. Zwanzig, Phys. Rev. Lett. B, **28** (1972) 1183 (https://doi.org/10.1103/PhysRevLett.28.1183); D. G. Triezenberg, PhD Thesis, University of Maryland, 1973; R. Lovett, C. Y. Mou, and F. P. Buff, J. Chem. Phys., **65** (1976) 570 (https://doi.org/10.1063/1.433110); M. S. Wertheim, J. Chem. Phys., **65** (1976) 2377 (https://doi.org/10.1063/1.433352).

are equivalent and one can transform them into one another by utilizing the Ornstein-Zernike equation 9.40.

Next, we consider a case where we have a body B immersed in an inhomogeneous fluid with its center of mass located at $\mathbf{r}_\mathrm{B}$. The body can have any shape and size and it is considered to be external to the fluid. Say that the interactions of the fluid particles with it is given by the $v_\mathrm{B}(\mathbf{s})$, where $\mathbf{s}$ is counted from the center of mass of the body. The total external potential for the fluid particles in the presence of the body is

$$v(\mathbf{r};\mathbf{r}_\mathrm{B}) = v(\mathbf{r}) + v_\mathrm{B}(\mathbf{r}-\mathbf{r}_\mathrm{B}), \tag{9.48}$$

where $v(\mathbf{r})$ is the external potential in the absence of B and we have explicitly shown in $v(\mathbf{r};\mathbf{r}_\mathrm{B})$ that the total v depends on the location $\mathbf{r}_\mathrm{B}$ of B. We assume that the body has a fixed orientation relative to the laboratory frame of reference and we have suppressed the orientational variable for it in Equation 9.48. Let us move the location of body B by the infinitesimal distance $d\mathbf{r}_\mathrm{B}$, whereby the external potential at any point $\mathbf{r}_2$ changes by

$$\delta v(\mathbf{r}_2) = \nabla_\mathrm{B} v(\mathbf{r}_2;\mathbf{r}_\mathrm{B})\cdot d\mathbf{r}_\mathrm{B} = \nabla_\mathrm{B} v_\mathrm{B}(\mathbf{r}_2-\mathbf{r}_\mathrm{B})\cdot d\mathbf{r}_\mathrm{B}.$$

The density at point $\mathbf{r}_1$ in the presence of B is denoted $n(\mathbf{r}_1;\mathbf{r}_\mathrm{B})$ and this density changes by $\delta n(\mathbf{r}_1) = \nabla_\mathrm{B} n(\mathbf{r}_1;\mathbf{r}_\mathrm{B})\cdot d\mathbf{r}_\mathrm{B}$ when B is moved by $d\mathbf{r}_\mathrm{B}$. By inserting δv and δn into the first Yvon equation 9.38 and using the fact that $d\mathbf{r}_\mathrm{B}$ can be arbitrary selected, we obtain[20]

$$\begin{aligned}\nabla_\mathrm{B} n(\mathbf{r}_1;\mathbf{r}_\mathrm{B}) \;&=\; -\beta n(\mathbf{r}_1;\mathbf{r}_\mathrm{B})\Big[\nabla_\mathrm{B} v_\mathrm{B}(\mathbf{r}_1-\mathbf{r}_\mathrm{B}) \\ &+ \int d\mathbf{r}_2\, h^{(2)}(\mathbf{r}_1,\mathbf{r}_2;\mathbf{r}_\mathrm{B}) n(\mathbf{r}_2;\mathbf{r}_\mathrm{B}) \nabla_\mathrm{B} v_\mathrm{B}(\mathbf{r}_2-\mathbf{r}_\mathrm{B})\Big]\end{aligned} \tag{9.49}$$

where we have explicitly shown that $h^{(2)}$ depend on the location of the body. The quantity $\nabla_\mathrm{B} n(\mathbf{r}_1;\mathbf{r}_\mathrm{B})$ gives the "rate of change" of density at $\mathbf{r}_1$ when body B is moved under the condition of constant chemical potential of the fluid (a quasistatic move so equilibrium is maintained). We will now apply this equation to a special case.

Equation 9.49 is particularly useful when the external potential $v(\mathbf{r})$ is caused by a planar wall I and the body B is another planar wall II parallel to the former and located at distance D^S from it. The fluid in the slit between the two walls is assumed to be in equilibrium with a bulk fluid with a given density and thereby it has a given chemical potential. We assume that both walls have infinite lateral extents and that they are smooth so the external potentials depend solely on the distance ℓ from each wall surface: $v_\mathrm{I}(\ell)$ and $v_\mathrm{II}(\ell)$. The coordinate system is placed with the origin at surface I and the z

[20]R. Kjellander and S. Sarman, Chem. Phys. Letters **149** (1988) 102 (https://doi.org/10.1016/0009-2614(88)80357-9). Note that there is a misprint in sign in the rhs of Equation 4 of this paper.

axis in the perpendicular direction out from the surface. The total external potential of the fluid in Equation 9.48 is then given by

$$v(\mathbf{r};\mathbf{r}_\mathrm{B}) = v(z;D^S) \equiv v_\mathrm{I}(z) + v_\mathrm{II}(D^S - z),$$

where we use the notation $v(z;D^S)$ to show that the external potential depends on the surface separation. The fluid density in the slit also depends on D^S and is $n(z;D^S)$. It is assumed to be zero for $z < 0$.

A pertinent question is how the average number of particles $\bar{\mathscr{N}}$ per unit area in the slit varies as a function of wall separation D^S *under the condition of constant chemical potential.* Since $\bar{\mathscr{N}}(D^S) = \int dz\, n(z;D^S)$, where the integral is taken across the slit, we can obtain the derivative $d\bar{\mathscr{N}}/dD^S$ from Equation 9.49 as in shown Appendix 9A, where the equation that correspond to this equation is derived (Equation 9.234). The results are as follows.

When wall II is a hard, $v_\mathrm{II}(\ell)$ is zero outside the wall but infinite inside. For this case all contributions to $d\bar{\mathscr{N}}/dD^S$ arise from particles in contact with wall II (they arise from particle collisions with the wall). This is analogous to the hard core contribution, the so-called contact-density term, to entities like the pressure, cf. Section 5.7 in Volume I. We obtain in the Appendix (Equations 9.236 and 9.237)

$$\frac{d\bar{\mathscr{N}}(D^S)}{dD^S} = \bar{\mathscr{N}}'_\mathrm{core,II}(D^S) \qquad \text{(hard wall)}, \tag{9.50}$$

where the rhs is the contact contribution given by

$$\begin{aligned}\bar{\mathscr{N}}'_\mathrm{core,II}(D^S) &= n_\mathrm{II}^\mathrm{cont}(D^S)\left\{1 + \int_0^{D^S - d_\mathrm{II}^\mathrm{h}} dz_1\, n(z_1;D^S)\right.\\ &\quad \times \left.\int d\mathbf{s}_{12}\, h^{(2)}(s_{12}, z_1, D^S - d_\mathrm{II}^\mathrm{h}; D^S)\right\},\end{aligned} \tag{9.51}$$

where $\mathbf{s}_{12} = (x_2 - x_1, y_2 - y_1)$, d_II^h is the distance of closest approach to wall II for the centers of the fluid particles, $n_\mathrm{II}^\mathrm{cont}(D^S)$ is the contact density at wall II for surface separation D^S ($n_\mathrm{II}^\mathrm{cont}$ is defined in the Appendix) and $h^{(2)}(s_{12}, z_1, z_2; D^S) = h^{(2)}(\mathbf{r}_1, \mathbf{r}_2; D^S)$. In the integral over $\mathbf{s}_{12}$, a particle at $z_2 = D^S - d_\mathrm{II}^\mathrm{h}$ is in contact with wall II, while particles at z_1 correlate with it as described by $h^{(2)}$. This describes the collisional contributions from the particles in contact with the wall, while they correlate with the other particles.

For the case of a *hard core wall* II, where $v_\mathrm{II}(\ell)$ is nonzero outside the hard core of the wall, there are contributions from the interactions between the wall and all particles in the slit. In the Appendix it is shown that[21]

$$\frac{d\bar{\mathcal{N}}(D^S)}{dD^S}$$
$$= \bar{\mathcal{N}}'_{\text{core,II}}(D^S) - \beta \int_0^{D^S - d^{\text{h}}_{\text{II}}} dz_1\, n(z_1; D^S) \Big[v'_{\text{II}}(D^S - z_1) \tag{9.52}$$
$$+ \int_0^{D^S - d^{\text{h}}_{\text{II}}} dz_2 \int d\mathbf{s}_{12}\, h^{(2)}(s_{12}, z_1, z_2; D^S) n(z_2; D^S) v'_{\text{II}}(D^S - z_2) \Big]$$

(Equation 9.238), where $v'_{\text{II}}(\ell) = dv_{\text{II}}(\ell)/d\ell$. For a *soft wall* (no hard core) the contact contribution is zero and only the integral remains (with $d^{\text{h}}_{\text{II}} = 0$ and the z integrals taken across the entire slit). Analogous results can be obtained for wall I.

9.1.3 Aspects of nonlinear response and integral equation approximations

Important tasks in liquid state theory are to determine the singlet distribution function $g^{(1)}(\mathbf{r})$, the density distribution $n(\mathbf{r}) = n^{\text{b}} g^{(1)}(\mathbf{r})$ and the pair distribution function $g^{(2)}(\mathbf{r}, \mathbf{r}')$. In Section 9.1.1 we found that

$$g^{(1)}(\mathbf{r}) = e^{-\beta v(\mathbf{r}) + \Delta c^{(1)}(\mathbf{r})} \tag{9.53}$$

(Equation 9.17) and

$$g^{(2)}(\mathbf{r}, \mathbf{r}') = e^{-\beta u(\mathbf{r}, \mathbf{r}') + \Delta c^{(1)}(\mathbf{r}|\mathbf{r}')}, \tag{9.54}$$

(Equation 9.22) so we need to determine $\Delta c^{(1)}$. Recall that $-k_B T\, c^{(1)}(\mathbf{r})$ is the intrinsic part of the excess chemical potential $\mu^{\text{ex}}(\mathbf{r})$, so the task is to investigate properties of the chemical potential. In the case of $g^{(1)}$, the change $\Delta c^{(1)}(\mathbf{r})$ is caused by a variation in external potential from zero to $v(\mathbf{r})$. For $g^{(2)}$, the change $\Delta c^{(1)}(\mathbf{r}|\mathbf{r}')$ is caused by a change in potential from $v(\mathbf{r})$ to $v(\mathbf{r}|\mathbf{r}') \equiv v(\mathbf{r}) + u(\mathbf{r}, \mathbf{r}')$ due to the insertion of a particle at location $\mathbf{r}'$. As we have seen, the density distribution changes from n^{b} to $n(\mathbf{r})$ and from $n(\mathbf{r})$ to $n(\mathbf{r}|\mathbf{r}')$, respectively, in these two cases.

In the previous section 9.1.2 we determined how $c^{(1)}$ changes due to infinitesimally small variations in external potential $\delta v(\mathbf{r})$

$$\delta c^{(1)}(\mathbf{r}') = -\beta \int d\mathbf{r}\, h^{(2)}(\mathbf{r}, \mathbf{r}') n(\mathbf{r}) \delta v(\mathbf{r}) \tag{9.55}$$

[21] R. Kjellander and S. Sarman, Mol. Phys. **70** (1990) 215 (https://doi.org/10.1080/00268979000100961).

(Equation 9.36) and in density $\delta n(\mathbf{r})$

$$\delta c^{(1)}(\mathbf{r}') = \int d\mathbf{r}\, c^{(2)}(\mathbf{r},\mathbf{r}')\delta n(\mathbf{r}), \tag{9.56}$$

(Equation 9.39), which constitutes the definition of $c^{(2)}$. The variation $\delta c^{(1)}(\mathbf{r}')$ is the linear response due to the change $\delta v(\mathbf{r})$ and the relationship between $\delta c^{(1)}(\mathbf{r}')$ and $\delta n(\mathbf{r})$ is also linear. When the changes in external potential and in density are not infinitesimally small, like in the present case, we must complement these linear response results with nonlinear terms. This is the task in front of us.

9.1.3.1 The bridge function

Let $\Delta n(\mathbf{r}_3)$ be the change in density at $\mathbf{r}_3$ and $\Delta c^{(1)}(\mathbf{r}_1)$ the change in singlet direct-correlation function at $\mathbf{r}_1$ when the external potential is varied by $\Delta v(\mathbf{r})$. The variations are not small, so the response is nonlinear and $\Delta c^{(1)}$ depends nonlinearly on Δn. If we apply Equation 9.56 to this case, we have

$$\Delta c^{(1)}(\mathbf{r}_1) = \int d\mathbf{r}_3\, c^{(2)}(\mathbf{r}_1,\mathbf{r}_3)\Delta n(\mathbf{r}_3) + \text{terms nonlinear in } \Delta n, \tag{9.57}$$

where the first term on the rhs gives the *linear part* of $\Delta c^{(1)}$ due to the change Δn and the rest is the contributions to $\Delta c^{(1)}$ that depends nonlinearly on Δn. In Equation 9.57 $c^{(2)}$ is evaluated for the system *before* we change the external potential.

In the case of $g^{(1)}(\mathbf{r})$ in Equation 9.53, we have $\Delta n(\mathbf{r}_3) = n(\mathbf{r}_3) - n^{\mathrm{b}}$ and we can write Equation 9.57 as

$$\Delta c^{(1)}(\mathbf{r}_1) = \int d\mathbf{r}_3\, c^{(2)\mathrm{b}}(\mathbf{r}_1,\mathbf{r}_3)\Delta n(\mathbf{r}_3) + b^{(1)}(\mathbf{r}_1), \tag{9.58}$$

where $c^{(2)\mathrm{b}}$ is the pair direct-correlation function for the bulk phase and $b^{(1)}(\mathbf{r}_1)$, which is called the **singlet bridge function**, is the sum of all terms that depends nonlinearly on Δn.[22] One can, in fact, express $b^{(1)}(\mathbf{r}_1)$ explicitly in terms of $\Delta n(\mathbf{r}_3)$, n^{b} and $h^{(2)\mathrm{b}}(\mathbf{r}_3,\mathbf{r}_4)$ as an infinite sum of complicated multicenter integrals (called "*cluster integrals*"),[23] but this is in practice not an accessible route to calculate $b^{(1)}(\mathbf{r}_1)$. Instead one approximates the bridge

[22] The nonlinear terms will be treated in more detail an Section 9.2.3.

[23] A simple example of an infinite sum of simple cluster integrals that we have already seen is given in Equation 9.41. A more complicated cluster integral appears in the forthcoming Equation 9.161. We have also encountered such integrals earlier in Section 3.4 in Volume I when dealing with virial expansions for real gases, where the virial coefficients can be expressed as cluster integrals, like in Equations I:3.54 and I:3.56 (see also Equation 10.10). In Volume I after Equation I:3.56 there is a misprint in the definition of the Mayer f-function f_{M} that lacks a minus sign in the exponent; it should be $f_{\mathrm{M}}(r_{ij}) \equiv \exp[-\beta u(r_{ij})] - 1$. The same error occurs in the preceding shaded text box.

function in various ways as will be explored later. The density distribution satisfies the equation

$$n(\mathbf{r}_1) = n^{\mathrm{b}} e^{-\beta v(\mathbf{r}_1) + \int d\mathbf{r}_3\, c^{(2)\mathrm{b}}(r_{13})[n(\mathbf{r}_3) - n^{\mathrm{b}}] + b^{(1)}(\mathbf{r}_1)}, \tag{9.59}$$

which follows from Equations 9.3, 9.53 and 9.58, where we have inserted $c^{(2)\mathrm{b}}(\mathbf{r}_1, \mathbf{r}_3) = c^{(2)\mathrm{b}}(r_{13})$ for the bulk fluid. By introducing the notation

$$\gamma^{(1)}(\mathbf{r}_1) = \int d\mathbf{r}_3\, c^{(2)\mathrm{b}}(r_{13})[n(\mathbf{r}_3) - n^{\mathrm{b}}] \tag{9.60}$$

we can write Equation 9.59 as

$$n(\mathbf{r}_1) = n^{\mathrm{b}} e^{-\beta v(\mathbf{r}_1) + \gamma^{(1)}(\mathbf{r}_1) + b^{(1)}(\mathbf{r}_1)}, \tag{9.61}$$

which will be useful later. Distinguish the function $\gamma^{(1)} = \gamma^{(1)}(\mathbf{r}_1)$ from the activity coefficient γ^{b} for a bulk phase.

Turning to the pair distribution $g^{(2)}$, we consider $\Delta c^{(1)}$ for an insertion of a particle at $\mathbf{r}_2$, so the change Δn in density at $\mathbf{r}_3$ is equal to $\Delta n(\mathbf{r}_3|\mathbf{r}_2) = h^{(2)}(\mathbf{r}_2, \mathbf{r}_3)n(\mathbf{r}_3)$, as shown in Equation 9.33. If we apply Equation 9.57 to calculate $\Delta c^{(1)}$ to be used in Equation 9.54, we obtain

$$\Delta c^{(1)}(\mathbf{r}_1|\mathbf{r}_2) = \int d\mathbf{r}_3\, c^{(2)}(\mathbf{r}_1, \mathbf{r}_3)\Delta n(\mathbf{r}_3|\mathbf{r}_2) + b^{(2)}(\mathbf{r}_1, \mathbf{r}_2), \tag{9.62}$$

where $b^{(2)}(\mathbf{r}_1, \mathbf{r}_2)$ is the sum of all terms that depends nonlinearly on Δn for this case and is called the **pair bridge function**. One can, in fact, express $b^{(2)}(\mathbf{r}_1, \mathbf{r}_2)$ explicitly in terms of $n(\mathbf{r}_3)$ and $h^{(2)}(\mathbf{r}_3, \mathbf{r}_4)$ as an infinite sum of cluster integrals,[24] but again one has to resort to approximations in practice, as will be seen later. In this case we define

$$\begin{aligned} \gamma^{(2)}(\mathbf{r}_1, \mathbf{r}_2) &= \int d\mathbf{r}_3\, c^{(2)}(\mathbf{r}_1, \mathbf{r}_3)\Delta n(\mathbf{r}_3|\mathbf{r}_2) \\ &= \int d\mathbf{r}_3\, c^{(2)}(\mathbf{r}_1, \mathbf{r}_3)h^{(2)}(\mathbf{r}_2, \mathbf{r}_3)n(\mathbf{r}_3), \end{aligned} \tag{9.63}$$

so we have

$$\Delta c^{(1)}(\mathbf{r}_1|\mathbf{r}_2) = \gamma^{(2)}(\mathbf{r}_1, \mathbf{r}_2) + b^{(2)}(\mathbf{r}_1, \mathbf{r}_2) \tag{9.64}$$

and Equation 9.54 can be written

$$g^{(2)}(\mathbf{r}_1, \mathbf{r}_2) = e^{-\beta u(\mathbf{r}_1, \mathbf{r}_2) + \gamma^{(2)}(\mathbf{r}_1, \mathbf{r}_2) + b^{(2)}(\mathbf{r}_1, \mathbf{r}_2)} \tag{9.65}$$

(cf. Equation 9.61).

---

[24]The terms in the infinite sum for $b^{(2)}$ are multicenter integrals of products of $n(\mathbf{r})$ and $h^{(2)}(\mathbf{r}, \mathbf{r}')$; so-called "bridge diagrams" or "elementary diagrams," where the term "diagram" is used since the cluster integrals can be represented pictorially by a diagram (like the set of diagrams that represents Equation 9.41, so-called chain diagrams). The first bridge diagram is given in Equation 9.161 and the figure accompanying this equation.

From the OZ equation 9.40 it follows that Equation 9.63 is the same as

$$\gamma^{(2)}(\mathbf{r}_1, \mathbf{r}_2) = h^{(2)}(\mathbf{r}_1, \mathbf{r}_2) - c^{(2)}(\mathbf{r}_1, \mathbf{r}_2). \tag{9.66}$$

Thus, Equation 9.65 can be written

$$\begin{aligned} g^{(2)}(\mathbf{r}_1, \mathbf{r}_2) &= h^{(2)}(\mathbf{r}_1, \mathbf{r}_2) + 1 \\ &= e^{-\beta u(\mathbf{r}_1,\mathbf{r}_2)+h^{(2)}(\mathbf{r}_1,\mathbf{r}_2)-c^{(2)}(\mathbf{r}_1,\mathbf{r}_2)+b^{(2)}(\mathbf{r}_1,\mathbf{r}_2)}, \end{aligned} \tag{9.67}$$

which is the most commonly used form of the relationship between $h^{(2)}$, u, $c^{(2)}$ and $b^{(2)}$. All these functions are symmetric with respect to the interchange of the indices $1 \leftrightarrow 2$, for example $b^{(2)}(\mathbf{r}_1, \mathbf{r}_2) = b^{(2)}(\mathbf{r}_2, \mathbf{r}_1)$. Note that Equations 9.62 and 9.67 are exact relationships provided that the exact bridge functions are used. For spherically symmetric particles the pair potential is isotropic $u(\mathbf{r}_1, \mathbf{r}_2) = u(r_{12})$, but in inhomogeneous fluids, the pair functions $h^{(2)}$, $c^{(2)}$ and $b^{(2)}$ are, as we have seen, anisotropic and depend on the locations of the two points $\mathbf{r}_1$ and $\mathbf{r}_2$.

It is useful to write $g^{(2)}$ in Equation 9.65 in terms of the cavity function $y^{(2)}$

$$g^{(2)}(\mathbf{r}_1, \mathbf{r}_2) = e^{-\beta u(\mathbf{r}_1,\mathbf{r}_2)} y^{(2)}(\mathbf{r}_1, \mathbf{r}_2)$$

(Equation 9.26) and we see from Equation 9.66 and 9.67 that

$$y^{(2)}(\mathbf{r}_1, \mathbf{r}_2) = e^{\gamma^{(2)}(\mathbf{r}_1,\mathbf{r}_2)+b^{(2)}(\mathbf{r}_1,\mathbf{r}_2)} = e^{h^{(2)}(\mathbf{r}_1,\mathbf{r}_2)-c^{(2)}(\mathbf{r}_1,\mathbf{r}_2)+b^{(2)}(\mathbf{r}_1,\mathbf{r}_2)}. \tag{9.68}$$

Alternatively this follows from the definition of $y^{(2)}$ in Equation 9.25 and the fact that

$$w^{(2)\text{intr}}(\mathbf{r}_1, \mathbf{r}_2) = -k_B T \left[\gamma^{(2)}(\mathbf{r}_1, \mathbf{r}_2) + b^{(2)}(\mathbf{r}_1, \mathbf{r}_2)\right] \tag{9.69}$$

as can be seen from Equations 9.24 and 9.64. For spherical particles with a hard core of diameter d^{h}, the factor $\exp[-\beta u(\mathbf{r}_1, \mathbf{r}_2)]$ in $g^{(2)}(\mathbf{r}_1, \mathbf{r}_2)$ is zero inside the hard core, $r_{12} < d^{\text{h}}$, since u is infinite there. This makes $g^{(2)}(\mathbf{r}_1, \mathbf{r}_2) = 0$ for $r_{12} < d^{\text{h}}$, while $y^{(2)}$ is nonzero there. The function $g^{(2)}$ has a stepwise discontinuity at $r_{12} = d^{\text{h}}$ caused by the factor $e^{-\beta u}$, but the functions $\gamma^{(2)}$, $w^{(2)\text{intr}}$ and $y^{(2)}$ are continuous and have also a continuous derivative.[25]

All equations obtained so far in the current chapter can be applied to both homogeneous bulk fluids and inhomogeneous fluids consisting of spherically symmetric particles. In the inhomogeneous case, singlet functions like $n(\mathbf{r})$ depend in general on three independent variables, x, y and z, and pair functions like $g^{(2)}(\mathbf{r}_1, \mathbf{r}_2)$ depend on six variables, x_1, y_1, z_1, x_2, y_2 and z_2. The number of independent variables can, however, often be reduced when there is some symmetry, like for a fluid outside a planar smooth wall as we saw at the end of Section 9.1.1.

[25]The function $\gamma^{(2)} = h^{(2)} - c^{(2)}$ is equal to the integral in the OZ equation 9.40, which is continuous. In the difference $h^{(2)} - c^{(2)}$ the discontinuity of $h^{(2)} = g^{(2)} - 1$ is therefore cancelled by an equal discontinuity of $c^{(2)}$.

For bulk fluids of spherical particles with density $n(\mathbf{r}) = n^{\mathrm{b}}$, the pair functions are isotropic, for instance $h^{(2)}(\mathbf{r}_1, \mathbf{r}_2) = h^{(2)}(r_{12})$, where we drop the superscript b of $h^{(2)\mathrm{b}}$ and other pair functions in the bulk phase for the time being. The Ornstein-Zernike equation 9.40 can in this case be written as

$$h^{(2)}(r_{12}) = c^{(2)}(r_{12}) + n^{\mathrm{b}} \int d\mathbf{r}_3\, c^{(2)}(r_{13})h^{(2)}(r_{32}) \quad \text{(bulk fluid)} \qquad (9.70)$$

and Equation 9.67 becomes

$$h^{(2)}(r_{12}) + 1 = e^{-\beta u(r_{12}) + h^{(2)}(r_{12}) - c^{(2)}(r_{12}) + b^{(2)}(r_{12})} \quad \text{(bulk fluid).} \qquad (9.71)$$

These equations are exact. The OZ equation 9.70 can alternatively be written as

$$h^{(2)}(r) = c^{(2)}(r) + n^{\mathrm{b}} \int d\mathbf{r}'\, c^{(2)}(r')h^{(2)}(|\mathbf{r} - \mathbf{r}'|) \quad \text{(bulk fluid),} \qquad (9.72)$$

where we have set $\mathbf{r} = \mathbf{r}_{12}$, $\mathbf{r}' = \mathbf{r}_{13}$ and hence $\mathbf{r} - \mathbf{r}' = \mathbf{r}_{12} - \mathbf{r}_{13} = \mathbf{r}_{32}$. Note that $d\mathbf{r}' = d\mathbf{r}_{13} = d(\mathbf{r}_3 - \mathbf{r}_1) = d\mathbf{r}_3$ because we can keep $\mathbf{r}_1$ fixed.

Assume that we know $u(r_{12})$, that is, we use some model of the system with a given pair potential. Say that we are interested in this system for a given density n^{b}. Equations 9.70 and 9.71 constitute a set of two equations with three unknown functions, $h^{(2)}$, $c^{(2)}$ and $b^{(2)}$. In order to be able to determine these functions, we need one more equation. This could be an equation that relates $b^{(2)}$ to the other functions, $h^{(2)}$ and $c^{(2)}$, or to one of them. As mentioned earlier one can express $b^{(2)}$ explicitly in terms of n^{b} and $h^{(2)}$ as an infinite sum of cluster integrals, so in principle one has a third equation that is also exact, but in practice one cannot use it because these integrals are very complicated to evaluate. Thus one needs some approximation for $b^{(2)}$.

In order to see how this can be done, we will introduce some approximations that are commonly used in liquid state theory and can be applied to both homogeneous and inhomogeneous fluids. We will consider both types of fluids in the next section, but the main treatment of the inhomogeneous case is postponed to Sections 10.1.3 and 10.1.4. For a bulk fluid we will use Equation 9.71 with an approximate $b^{(2)}$ (expressed in terms of, for example, $h^{(2)}$ and/or $c^{(2)}$) together with the OZ equation 9.70, whereby there are only two unknowns, $h^{(2)}$ and $c^{(2)}$. These two equations can therefore be solved (at least numerically). The approximation for $b^{(2)}$ is a **closure approximation**[26] because it results in a closed set of equations that can be solved. This is the basis of the approximate **integral equation theories** of fluids.[27] The term "integral equation theory" is used since we obtain an integral equation when the closure and Equation 9.71 are inserted in the OZ equation. Many

[26] See Section 5.9 in Volume I for an introduction to closure approximations.

[27] In Section 5.12 in Volume I, we have already encountered simple integral equations: the mean field approximations obtained by neglecting triplet correlation functions in the second Born-Green-Yvon equation.

closure approximations have been suggested and used in the literature; in the next section we are going to give a survey of some of the most common ones (without any detailed motivations for the time being). We will also introduce a couple of approximations that we will investigate later on.

In statistical mechanics there usually exist alternative formulas that can be used to calculate the value of a thermodynamical quantity or some other entity. When one makes approximations, it often happens that the value depends on *which* formula is used, that is, *how* the value is calculated. This means that there are inconsistencies in approximate theories. In general there exist several exact relationships that express a thermodynamical quantity in terms of various correlation functions. Some them are given in Section 5.7 of Volume I and more will be derived in the present book. Provided that the correlation functions are known exactly, these relationships give the same value of the quantity, but when approximations of these functions are used, the relationships may give different results. The requirement that a pair of such relationships give the same result constitute an example of a **consistency criterium**. The corresponding is true for the calculation of other entities and there exists several criteria that an exact theory must fulfill but that approximate theories do not fulfill in general. It is, in fact, unavoidable that inconsistencies of one type or another appear in approximate theories and a goal is to have approximations that satisfy as many consistency criteria as possible, at least to a good approximation.

A technique to improve approximations is to introduce one or a few parameters in the closure approximation and to select the value(s) of the parameter(s) such that one or some consistency criteria are fulfilled (one needs an equal number of independent criteria as there are parameters). The idea is that the best possible approximation is obtained when the criteria are fulfilled. Some closures of this kind are introduced in the following sections. More examples of consistency criteria and their use will be discussed in Section 10.1.2.

9.1.3.2 *Some common closure approximations for integral equations*

The simplest approximation for the pair bridge function is simply to set $b^{(2)} = 0$, which is the so-called **Hypernetted Chain (HNC) approximation**. We have from Equations 9.67 and 9.68

$$g^{(2)}(\mathbf{r}_1, \mathbf{r}_2) = e^{-\beta u(\mathbf{r}_1,\mathbf{r}_2)+h^{(2)}(\mathbf{r}_1,\mathbf{r}_2)-c^{(2)}(\mathbf{r}_1,\mathbf{r}_2)} \qquad \text{(HNC)} \qquad (9.73)$$

and

$$y^{(2)}(\mathbf{r}_1, \mathbf{r}_2) = e^{h^{(2)}(\mathbf{r}_1,\mathbf{r}_2)-c^{(2)}(\mathbf{r}_1,\mathbf{r}_2)}. \qquad \text{(HNC)} \qquad (9.74)$$

To set $b^{(2)} = 0$ means that one has neglected the nonlinear contributions to $\Delta c^{(1)}$ in terms of the variation Δn in Equation 9.62. Despite this, the HNC approximation is a nonlinear approximation (in a different sense) since there is a nonlinear relationship between $y^{(2)}$ and the exponent in the expression

9.74. Experience has shown that the HNC approximation is often quite good for particles that interact with long-ranged potentials. This will be discussed further in Sections 9.2.3 and 9.2.4.

Equation 9.73 can be written as

$$c^{(2)} = -\beta u + h^{(2)} - \ln(1 + h^{(2)}). \qquad \text{(HNC)} \tag{9.75}$$

For a bulk fluid, one can insert this into the OZ equation 9.70, whereby one obtains an integral equation that only contains $h^{(2)}(r_{12}) = h^{(2)}(\mathbf{r}_1, \mathbf{r}_2)$ as unknown. To solve this equation one must resort to numerical methods. In practice, one usually formulates the equation in such a way that one can solve it iteratively. This is described in the shaded text box below, which can be skipped in the first reading.

★[28] In order to obtain the pair correlation functions one usually resorts to Fourier transform techniques, which are described in Appendix 9B. The integral in rhs of the OZ equation 9.70 is a convolution integral (cf. Equation 9.250 in the Appendix), which simply becomes a product in Fourier space. For a bulk fluid of density n^{b}, the OZ equation therefore becomes

$$\tilde{h}^{(2)}(k) = \tilde{c}^{(2)}(k) + n^{\mathrm{b}}\tilde{c}^{(2)}(k)\tilde{h}^{(2)}(k) \quad \text{(bulk fluid)}, \tag{9.76}$$

where $\tilde{h}^{(2)}(k)$ and $\tilde{c}^{(2)}(k)$ are the Fourier transforms of $h^{(2)}(r)$ and $c^{(2)}(r)$, respectively, and k is the wave number. We can write this as

$$\tilde{h}^{(2)}(k) = \frac{\tilde{c}^{(2)}(k)}{1 - n^{\mathrm{b}}\tilde{c}^{(2)}(k)} \quad \text{(bulk fluid)} \tag{9.77}$$

so we have $\tilde{h}^{(2)}$ expressed in terms of $\tilde{c}^{(2)}$. Since Equation 9.75 gives $c^{(2)}$ in terms of $h^{(2)}$, it would be suitable to eliminate $c^{(2)}$ from the two equations, but the logarithm in Equation 9.75 complicates the matter since the Fourier transform of $\ln(1 + h^{(2)}(r))$ is not simple.

One way to proceed is to solve the problem in an iterative manner. Starting from an approximate $h^{(2)}(r)$ given numerically on a grid on the r axis, one can insert this in Equation 9.75 and obtain an approximate $c^{(2)}(r)$ on the same grid. Next, one takes the Fourier transform of $c^{(2)}(r)$ in a numerical manner,[29] whereby the values of $\tilde{c}^{(2)}(k)$ is obtained on a grid of k values. By inserting $\tilde{c}^{(2)}(k)$ into Equation 9.77, one obtains $\tilde{h}^{(2)}(k)$ which yields a

[28] Text marked with ★ contains material that is complementary or more advanced than the rest and can be skipped in the first reading.

[29] It is suitable to use so-called *fast Fourier transform* techniques here.

new $h^{(2)}(r)$ by taking the inverse Fourier transform. One then repeats the cycle until convergence is reached; in practice one must use various numerical techniques in order to facilitate convergence.[30]

A closure with a linear relationship between $h^{(2)}$ and $c^{(2)}$ is obtained if one approximates $y^{(2)} \approx 1 + h^{(2)} - c^{(2)}$, which is obtained by taking the linear terms in a power expansion of the exponential function in Equation 9.74. This is the **Percus-Yevick (PY) approximation**, which hence says that

$$g^{(2)}(\mathbf{r}_1, \mathbf{r}_2) = e^{-\beta u(\mathbf{r}_1, \mathbf{r}_2)}[1 + h^{(2)}(\mathbf{r}_1, \mathbf{r}_2) - c^{(2)}(\mathbf{r}_1, \mathbf{r}_2)] \qquad \text{(PY)} \qquad (9.78)$$

(see Example 9:B on page 51 for a justification of this approximation). It works well for hard sphere fluids provided that the density is not too high; at the highest densities, the accuracy is only semiquantitative (see Section 10.1.1). The PY closure can equivalently be expressed as the following approximation for the bridge function

$$b^{(2)} = \ln(1 + h^{(2)} - c^{(2)}) - [h^{(2)} - c^{(2)}]. \qquad \text{(PY)} \qquad (9.79)$$

We can alternatively write this as

$$b^{(2)} = \ln(1 + \gamma^{(2)}) - \gamma^{(2)} \qquad \text{(PY)} \qquad (9.80)$$

since $h^{(2)} - c^{(2)} = \gamma^{(2)}$.

Another linear approximation is the **Mean Spherical Approximation (MSA)**, which is used for particles with a hard core. In this approximation one sets $g^{(2)}(\mathbf{r}_1, \mathbf{r}_2) = 0$ inside the hard core, $r_{12} < d^{\mathrm{h}}$, as required, and outside the core one makes the approximation

$$c^{(2)}(\mathbf{r}_1, \mathbf{r}_2) = -\beta u(\mathbf{r}_1, \mathbf{r}_2) \quad \text{for } r_{12} \geq d^{\mathrm{h}}. \qquad \text{(MSA)} \qquad (9.81)$$

The latter Ansatz arises from the general property of $c^{(2)}$ mentioned earlier, namely that $c^{(2)}(\mathbf{r}_1, \mathbf{r}_2) \sim -\beta u(\mathbf{r}_1, \mathbf{r}_2)$ when $r_{12} \to \infty$ (Equation 9.42) and one assumes that this limiting property holds as an equality all the way down to

[30] In practice one uses both the new $h^{(2)}$ and the old $h^{(2)}$ in each iteration step in order to predict the function $h^{(2)}$ to be inserted into Equation 9.75. There exists several more or less sophisticated numerical schemes to use in order to improve the convergence of the iterative procedure. A very simple option is to perform a so-called Picard iteration. Then one usually takes a fraction α of $h^{(2)}_{\mathrm{new}}(r)$ and mixes it with the fraction $1-\alpha$ of $h^{(2)}_{\mathrm{old}}(r)$ in order to obtain the function $h^{(2)}(r) = \alpha h^{(2)}_{\mathrm{new}}(r) + (1-\alpha) h^{(2)}_{\mathrm{old}}(r)$ that one inserts into Equation 9.75. This is done in each iteration cycle, whereby one selects α sufficiently small so that convergence is reached (whether convergence can be reached or not in a simple Picard iteration may depend on the system). In addition, there are many other details that has to be attended to in the numerical procedure, for example how to handle discontinuities that occur in the correlations functions for hard core particles and the handling of long-range tails of the functions for large r and/or large k values. We will, however, not treat these matters here.

contact between the particles. Alternatively and equivalently, one can express MSA as an approximation for the bridge function

$$b^{(2)} = \ln(1 + h^{(2)}) - h^{(2)}. \quad \text{for } r_{12} \geq d^{\text{h}}. \qquad \text{(MSA)} \qquad (9.82)$$

For some fluids with simple pair interaction potentials, for instance the hard sphere fluid, one can obtain analytic solutions for the pair correlation functions in both the PY approximation and MSA, see Section 10.1. For hard spheres, these two approximations are identical because the PY approximation then becomes $g^{(2)} = 1 + h^{(2)} - c^{(2)}$ for $r_{12} \geq d^{\text{h}}$, so $c^{(2)} = 0$ there (since $g^{(2)} = 1 + h^{(2)}$). For particles with various other pair potentials, one can solve the OZ equation together with the closure in numerical manner (cf. the HNC approximation above).

In the **Generalized Mean Spherical Approximation (GMSA)**, one improves MSA by adding some empirical function to $b^{(2)}$ in Equation 9.82 (equivalently, one can subtract this empirical function from the MSA Ansatz 9.81 for $c^{(2)}$). This empirical function can, for example, be an exponentially decaying function like $\mathcal{K}\exp(-\alpha r)/r$ with fitting parameters $\mathcal{K}$ and α. These parameters are selected such that some consistency criteria hold, for example the criterium that a thermodynamical property (often the pressure) of the system should have the same value irrespectively in which manner it is calculated, cf. the discussion at the end of the previous section 9.1.3.1.

For particles with a soft core, that is, with a pair potential that goes continuously to infinity when $r \to 0$, one can use the **Soft-core Mean Spherical Approximation (SMSA)**, which is given by

$$g^{(2)}(\mathbf{r}_1, \mathbf{r}_2) = e^{-\beta u^s(\mathbf{r}_1,\mathbf{r}_2)}[1 + h^{(2)}(\mathbf{r}_1, \mathbf{r}_2) - c^{(2)}(\mathbf{r}_1, \mathbf{r}_2) - \beta u^\ell(\mathbf{r}_1, \mathbf{r}_2)] \quad \text{(SMSA)}, \qquad (9.83)$$

where one has split the pair potential in two parts $u = u^s + u^\ell$ such that $u^\ell(\mathbf{r}_1, \mathbf{r}_2) = u^\ell(r_{12})$ is the long-range part (usually attractive)

$$u^\ell(r) = \begin{cases} u(r), & r \geq r_0 \\ u(r_0), & r < r_0 \end{cases} \qquad (9.84)$$

and $u^s(\mathbf{r}_1, \mathbf{r}_2) = u^s(r_{12})$ is the short-range part (usually repulsive)

$$u^s(r) = \begin{cases} 0, & r \geq r_0 \\ u(r) - u(r_0), & r < r_0, \end{cases} \qquad (9.85)$$

where r_0 is suitably chosen. As an example we show how this can be done for the Lennard-Jones (LJ) potential

$$u(r) = u^{\text{LJ}}(r) \equiv 4\varepsilon^{\text{LJ}} \left[\left(\frac{d^{\text{LJ}}}{r} \right)^{12} - \left(\frac{d^{\text{LJ}}}{r} \right)^{6} \right], \qquad (9.86)$$

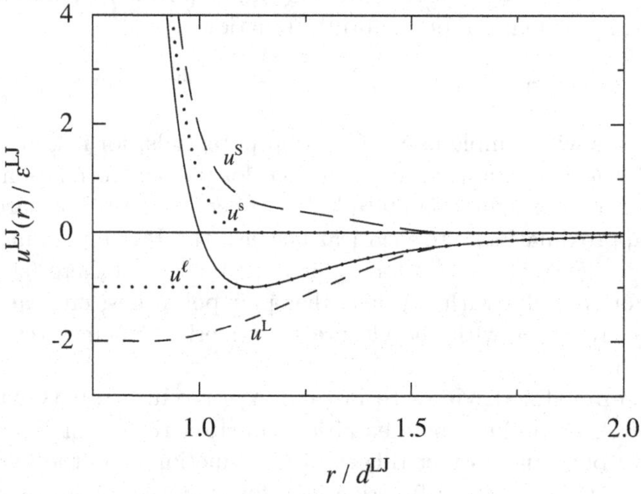

Figure 9.1 A plot of the Lennard-Jones potential $u^{\mathrm{LJ}}(r)$ (full curve) and its partitioning into long-range and short-range parts. The dotted curves show the Weeks-Chandler-Anderson partitioning, where $u^{\ell}(r)$ and $u^{s}(r)$ are defined in Equations 9.84 and 9.85, respectively, and r_0 is placed at the minimum of $u^{\mathrm{LJ}}(r)$. The dashed curves show the long-range and the short-range parts, $u^{\mathrm{L}}(r)$ and $u^{\mathrm{S}}(r)$, according to the Bomont-Bretonnet partitioning defined in Section 10.1 (u^{L} is given by Equation 10.30).

where $\varepsilon^{\mathrm{LJ}}$ and d^{LJ} are parameters (see Equation I:5.6). In this case, r_0 is usually selected at the minimum of the potential, see Figure 9.1 where $u^{\mathrm{LJ}}(r)$, $u^{s}(r)$ and $u^{\ell}(r)$ are plotted. This manner to partition the LJ potential into a repulsive short-range and an attractive long-range part is often called the Weeks-Chandler-Anderson (WCA) partitioning after the authors who first used it. We will discuss other ways to partition u^{LJ} later.

Note that SMSA reduces to the PY approximation when $r_0 \to \infty$. Furthermore, MSA is a special case of SMSA because for a hard-core fluid with $r_0 = d^{\mathrm{h}}$, the short-range part $u^{s}(r)$ is ∞ for $r < r_0$ and zero otherwise, whereby Equation 9.83 for $r_{12} \geq r_0 = d^{\mathrm{h}}$ implies Equation 9.81 because $u^{s} = 0$, $\exp(-\beta u^{s}) = 1$ and $u^{\ell} = u$ there.

We can alternatively write SMSA as

$$b^{(2)} = \ln(1 + \gamma^{(2)} - \beta u^{\ell}) - \gamma^{(2)} + \beta u^{\ell} \qquad \text{(SMSA)}, \tag{9.87}$$

that is, as an approximation for the bridge function. For hard core fluids, Equation 9.83 implies that $\gamma^{(2)} - \beta u^{\ell} = h^{(2)}$ for $r \geq r_0 = d^{\mathrm{h}}$, so MSA expressed as in Equation 9.82 is included in Equation 9.87 for this case.

There exist a whole class of closure approximations that uses the Ansatz

$$b^{(2)} \approx f(h^{(2)} - c^{(2)}) = f(\gamma^{(2)}), \tag{9.88}$$

where f is some function. The PY approximation, where we have $f(\gamma^{(2)}) = \ln(1+\gamma^{(2)}) - \gamma^{(2)}$ (see Equation 9.80), belongs to this class. Another example is the **Martynov-Sarkisov (M-S) approximation**[31] where one takes

$$b^{(2)} = \left[1 + 2\gamma^{(2)}\right]^{1/2} - \gamma^{(2)} - 1 \qquad \text{(M-S)}. \tag{9.89}$$

As we will see in Section 10.1, this closure is much better than the PY approximation for hard sphere fluids at high densities. The M-S closure can alternatively be written[32]

$$b^{(2)} = -\frac{[\beta w^{(2)\mathrm{intr}}]^2}{2} \qquad \text{(M-S)}, \tag{9.90}$$

which was used in its original justification by Martynov and Sarkisov as will be discussed in Section 10.1.1.

Another closure in the class given by the Ansatz 9.88 is the **Verlet approximation**[33] that uses the approximation

$$b^{(2)} = -\frac{\tau_1[\gamma^{(2)}]^2}{2} \cdot \frac{1 + \tau_2\gamma^{(2)}}{1 + \tau_3\gamma^{(2)}} \qquad \text{(Verlet)}, \tag{9.91}$$

where the values of the parameters τ_1, τ_2 and τ_3 are selected such that one achieves a good over-all thermodynamical consistency for the system studied. The functional form of the rhs is an example of a so-called *Padé approximant* for $b^{(2)} = b^{(2)}(\gamma^{(2)})$ as function of $\gamma^{(2)}$.[34] The great advantage with this kind of approach is that a Padé approximant closure is quite flexible and can therefore be adopted to several consistency criteria, which allow improvements of the quality of the approximation in a systematic manner.

In all these examples of closures, the pair bridge function $b^{(2)}$ is approximated by an expression that involves some other pair functions. Another approach is to approximate $b^{(2)} \approx b^{(2)}_{\mathrm{Ref}}$, where $b^{(2)}_{\mathrm{Ref}}$ is the bridge function from another system, called *the reference system* (usually a hard sphere fluid) where $b^{(2)}_{\mathrm{Ref}}$ is known with reasonably high accuracy or can be accurately calculated. This is the so-called **Reference HNC (RHNC) approximation.** It is a very useful approach that will be discussed further in Section 9.2.4. One can select the diameter of the hard spheres of the reference system such that some

[31] G. A. Martynov and G. N. Sarkisov, Mol. Phys. **49** (1983) 1495 (https://doi.org/10.1080/00268978300102111).

[32] This follows from the fact that Equation 9.69 can be written $\beta w^{(2)\mathrm{intr}} = -\gamma^{(2)} - b^{(2)}$, which means that Equation 9.90 says that $b^{(2)} = -[\gamma^{(2)} + b^{(2)}]^2/2$. This is a quadratic equation in $b^{(2)}$ and Equation 9.89 is one of its solutions (the other solution can be discarded for physical reasons).

[33] L. Verlet, Mol. Phys. **41** (1980) 183 (https://doi.org/10.1080/00268978000102671) and *ibid* **42** (1981) 1291 (https://doi.org/10.1080/00268978100100971).

[34] A Padé approximant for a function $f(x)$ is a rational function of given order in x, where the coefficients of the polynomials in the numerator and denominator are selected such that $f(x)$ is approximated in the best possible manner.

consistency criterion holds (cf. the discussion at the end of the previous section 9.1.3.1). An alternative name of this approximation is the **Modified HNC (MHNC) approximation**, but the latter can also include other approximations for the bridge function (the nomenclature in the literature is not fully consistent).

All of these closures can be used for both homogeneous an inhomogeneous fluids. For bulk fluids it is sufficient to solve the OZ equation together with the closure approximation for $b^{(2)}$, but for inhomogeneous fluids one must also determine the density distribution $n(\mathbf{r})$ given by

$$n(\mathbf{r}_1) = n^{\mathrm{b}} e^{-\beta v(\mathbf{r}_1) + \int d\mathbf{r}_3\, c^{(2)\mathrm{b}}(r_{13})[n(\mathbf{r}_3) - n^{\mathrm{b}}] + b^{(1)}(\mathbf{r}_1)}$$

(Equation 9.59). The most common type of approach for the inhomogeneous case is the use of *singlet-level integral equation theories*, where approximate $c^{(2)\mathrm{b}}(r)$ and $b^{(1)}(\mathbf{r})$ are inserted into this equation. Such theories utilize two closure approximations, one for the bulk fluid in order to calculate $c^{(2)\mathrm{b}}(r)$ and another closure to obtain an expression for the singlet bridge function $b^{(1)}(\mathbf{r})$. To calculate $c^{(2)\mathrm{b}}(r)$ one thereby uses a closure for $b^{(2)\mathrm{b}}$ and solves the OZ equation 9.70 for the bulk fluid. An example is the **HNC/PY approximation** that uses the singlet HNC approximation $b^{(1)}(\mathbf{r}) = 0$ and takes $c^{(2)\mathrm{b}}(r)$ from the PY approximation for the bulk phase. This kind of theories are treated in more detail in Section 10.1.3.

The *pair-level integral equation theories* for inhomogeneous fluids are in general considerably more accurate than the singlet level theories. In these theories the pair correlation functions for the inhomogeneous fluid are calculated by solving the inhomogeneous version of the OZ equation 9.40 together with any of the closure approximation for $b^{(2)}$. This can be compared to the singlet level theories, which use pair correlation functions from the bulk obtained from the solution of the bulk version of the OZ equation (Equation 9.70) together with a closure for $b^{(2)\mathrm{b}}$. As mentioned earlier, the pair functions for spherical particles are isotropic in the bulk fluid, while they are anisotropic for inhomogeneous fluids: The latter kind of pair functions are explicitly calculated in the present case.

The density distribution, which also can be written as

$$n(\mathbf{r}) = n^{\mathrm{b}} e^{-\beta v(\mathbf{r}) + c^{(1)}(\mathbf{r}) - c^{(1)\mathrm{b}}} = \zeta e^{-\beta v(\mathbf{r}) + c^{(1)}(\mathbf{r})}$$

(Equations 9.13 and 9.15), can be determined in the pair level theories by solving any of the exact equations for $n(\mathbf{r})$ like the BGY equation 9.10 or any of the TZLMBW equations 9.46 and 9.47. Alternatively and much simpler, for several closures (including all that belong to the kind given in Equation 9.88) one can calculate $c^{(1)}(\mathbf{r})$ directly from the pair correlations functions as described in Section 10.1.5. Examples of pair level theories include the **Anisotropic HNC (AHNC) approximation** and the **Anisotropic**

PY (APY) approximation (Anisotropic Percus-Yevick approximation),[35] which use the HNC closure 9.73 and PY closure 9.78, respectively. They and other theories on the pair level are described in more detail in Section 10.1.4.

9.1.3.3 *More refined integral equation closures*

An important kind of improvement of the closures with Ansatz 9.88 uses instead

$$b^{(2)} \approx f(\gamma^{(2)} - \varphi^{\mathrm{R}}) = f(\gamma^{*(2)}), \tag{9.92}$$

where $\gamma^{*(2)} \equiv \gamma^{(2)} - \varphi^{\mathrm{R}}$ and φ^{R} is some suitable function. As will be justified below, one can, for example, select $\varphi^{\mathrm{R}} = \beta u^{\ell}$ and we will see how one can judge whether this kind of approach is suitable (via so-called Duh-Haymet plots). In fact, this selection is made in the SMSA expression 9.87, which can be obtained from the PY approximation, Equation 9.80, by replacing $\gamma^{(2)}$ by $\gamma^{(2)} - \beta u^{\ell}$. The splitting of the pair potential in a short-range and a long-range part $u = u^s + u^{\ell}$ as in Equations 9.84 and 9.85 is, however, quite arbitrary and the choice made in these equations is not the optimal one for use in $b^{(2)}$. One can instead select another long-range part u^{L} that is defined in a different manner than in Equation 9.84 and select $\varphi^{\mathrm{R}} = \beta u^{\mathrm{L}}$. The short-range part is then $u^{\mathrm{S}}(r) = u(r) - u^{\mathrm{L}}(r)$. The various choices of $u^{\mathrm{L}}(r)$ differ for small r and, as we will see, the values there matter for the quality of the closure. This topic will be treated in Section 10.1 and Figure 9.1 shows one example of u^{L} and u^{S} that will be discussed later.

The subtraction of φ^{R} from $\gamma^{(2)}$ when $\gamma^{*(2)}$ is obtained has been called a "renormalization" of $\gamma^{(2)}$. The choice of φ^{R} is not restricted to βu^{ℓ} and βu^{L}; other suitable functions have also been used for φ^{R}.[36] Some of these renormalizations have been quite successful, but the best way to do this in general is still an open question. For each closure that uses the Ansatz 9.92, the definition of φ^{R} will be explicitly specified in what follows.

When one selects $\varphi^{\mathrm{R}} = \beta u^{\ell}$, the closure is

$$b^{(2)} \approx f(\gamma^{(2)} - \beta u^{\ell}). \tag{9.93}$$

An example of this kind of closure is the **Vompe-Martynov (V-M) approximation**[37] that modifies the M-S approximation in Equation 9.89 by replacing $\gamma^{(2)}$ by $\gamma^{(2)} - \beta u^{\ell}$, so one obtains

$$b^{(2)} = \left[1 + 2(\gamma^{(2)} - \beta u^{\ell})\right]^{1/2} - (\gamma^{(2)} - \beta u^{\ell}) - 1 \quad \text{(V-M)}. \tag{9.94}$$

[35] Alternative names are the **Inhomogeneous PY approximation** and the **PY2 approximation**.

[36] Various other ways to choose φ^{R} include $\beta n^{\mathrm{b}}(d^{\mathrm{h}})^3 u^{\mathrm{L}}$ and other functions that depend on density. Another choice is a φ^{R} that is proportional to a Mayer f-function $\exp[-\beta^* u^*(r)] - 1$, where $u^*(r)$ is some kind of pair potential and β^* a parameter. The choice of φ^{R} will be given when each approximation is presented.

[37] A. G. Vompe and G. A. Martynov J. Chem. Phys., **100** (1994) 5249 (https://doi.org/10.1063/1.467189).

or, equivalently,

$$b^{(2)} = -\frac{[\beta(w^{(2)\mathrm{intr}} + u^{\ell})]^2}{2} \qquad \text{(V-M)} \tag{9.95}$$

instead of Equation 9.90. Let us now see why the replacements of $\gamma^{(2)}$ by $\gamma^{(2)} - \beta u^{\ell}$ and $w^{(2)\mathrm{intr}}$ by $w^{(2)\mathrm{intr}} + u^{\ell}$ are reasonable. The potential of mean force $w^{(2)}$ can always be expressed as

$$\begin{aligned} \beta w^{(2)} &= \beta[u + w^{(2)\mathrm{intr}}] = \beta u^{s} + \beta[u^{\ell} + w^{(2)\mathrm{intr}}] \\ &= \beta u^{s} - [\gamma^{(2)} - \beta u^{\ell}] - b^{(2)} \end{aligned}$$

(from Equations 9.23 and 9.69). Here, we see that in the expression for $w^{(2)}$, we have the combinations $u^{\ell} + w^{(2)\mathrm{intr}}$ (in the first line) and $\gamma^{(2)} - \beta u^{\ell}$ (in the second line). As it turns out in practice, the use of $u^{\ell} + w^{(2)\mathrm{intr}}$ (as in Equation 9.95) is better as a basis of the approximation for $b^{(2)}$ than the use of $w^{(2)\mathrm{intr}}$ by itself in the approximate closure (as in Equation 9.90).[38] Likewise, the use of $\gamma^{(2)} - \beta u^{\ell}$ is, as we will see, better than the use of $\gamma^{(2)}$ alone, which is the basis for the use of Ansatz 9.93 instead of 9.88.

However, a similar argument can be used for other selections of the long-range part, u^{L}, of the pair potential, which is then partitioned into a long-range and a short range part, $u = u^{\mathrm{S}} + u^{\mathrm{L}}$, in a different manner than the WCA partitioning given in Equation 9.84 and 9.85. In the **Bomont-Bretonnet (B-B) approximation,**[39] which is an extension of the V-M approximation, one uses $\varphi^{\mathrm{R}} = \beta u^{\mathrm{L}}$ with a choice of $u^{\mathrm{L}}(r)$ that is plotted together with $u^{\mathrm{S}}(r)$ in Figure 9.1 for LJ fluids (u^{L} is defined in Equation 10.30 of Section 10.1.2, where the justification of this choice is given). The B-B closure is

$$b^{(2)} = -\frac{[\beta(w^{(2)\mathrm{intr}} + u^{\mathrm{L}})]^2 + \tau(\gamma^{*(2)})^2}{2} \qquad \text{(B-B).} \tag{9.96}$$

or, equivalently,

$$b^{(2)} = \left[1 + 2\gamma^{*(2)} + \tau(\gamma^{*(2)})^2\right]^{1/2} - \gamma^{*(2)} - 1 \qquad \text{(B-B),} \tag{9.97}$$

where $\gamma^{*(2)} = \gamma^{(2)} - \beta u^{\mathrm{L}}$ and τ is a parameter that is chosen such that a thermodynamic consistency criterium is fulfilled (details will be given in

[38] As mentioned in connection to Equation 9.62, $b^{(2)}$ can be explicitly expressed as a sum of cluster integrals containing the pair correlation function

$$h^{(2)} = e^{-\beta u} e^{-\beta w^{(2)\mathrm{intr}}} - 1 = e^{-\beta u^{s}} e^{-\beta[u^{\ell} + w^{(2)\mathrm{intr}}]} - 1$$

and the density n (cf. Section 9.2.3). In an analysis of the cluster integrals, Vompe and Martynov[37] found that the approximation for $b^{(2)}$ in terms of $u^{\ell} + w^{(2)\mathrm{intr}}$ (as in the rhs of the expression for $h^{(2)}$) is better than in terms of $w^{(2)\mathrm{intr}}$ alone.

[39] J. M. Bomont and J. L. Bretonnet, J. Chem. Phys. **119** (2003) 2188 (https://doi.org/10.1063/1.1583675);
Mol. Phys. **101** (2003) 3249 (https://doi.org/10.1080/00268970310001619313).

Section 10.1.2). The essential difference from the V-M approximation, apart from the selection of u^{L}, is the appearance of the term $\tau(\gamma^{*(2)})^2$ in Equations 9.96 and 9.97. These differences make the B-B approximation more accurate than the V-M approximation, which in turn is better in general than the M-S approximation. Note that if $\tau = 1$, Equation 9.97 becomes the HNC closure $b^{(2)} = 0$.

In Section 10.1.2 we will also treat a refinement of the Verlet closure, the **Zero-separation (ZSEP) approximation** by Lloyd L. Lee,[40] where $\gamma^{(2)}$ in Equation 9.91 is replaced by $\gamma^{*(2)}$, so this approximation is also of the kind given by the Ansatz 9.92. In this approach, the best values of the parameters τ_1, τ_2 and τ_3 in the Padé approximant are determined from a set of consistency criteria, which will be discussed in Section 10.1.2. As will be described there, another function than βu^{L} is used for φ^{R}. The ZSEP and B-B approximations are very accurate for the calculation of a wide variety of quantities in the systems where they have been tested.

Before proceeding, we will illustrate why closures of the form $b^{(2)} \approx f(\gamma^{(2)})$ can work and why a renormalization of $\gamma^{(2)}$, where it is replaced by $\gamma^{(2)} - \beta u^{\ell}$ or $\gamma^{(2)} - \beta u^{\mathrm{L}}$ in the closure, can give better results. As an example we will investigate Monte Carlo (MC) simulation results[41] for LJ fluids. A relationship $b^{(2)}(r) = f(\gamma^{(2)}(r))$ implies that if one makes an x, y plot of $y = b^{(2)}(r)$ and $x = \gamma^{(2)}(r)$ for a sequence of r values (that is, the trajectory of the point (x, y) in the x, y plane that is obtained when one varies r), one obtains a plot of the function $y = f(x)$.

In order to see if $b^{(2)}$ of a system satisfies this kind of relationship, one can take $b^{(2)}(r)$ and $\gamma^{(2)}(r)$ calculated by computer simulation of the system and make such an x, y plot, which is called a **Duh-Haymet plot.**[42] A necessary condition for the $b^{(2)}(r)$ and $\gamma^{(2)}(r)$ data to be represented by a function $b^{(2)} = f(\gamma^{(2)})$ is that there is only one value of $b^{(2)}$ for each value of $\gamma^{(2)}$, that is, it is *not* allowed to have two r values, r_1 and r_2, such that $b^{(2)}(r_1) \neq b^{(2)}(r_2)$ when $\gamma^{(2)}(r_1) = \gamma^{(2)}(r_2)$. A function $f(\gamma^{(2)})$ has only one value for each $\gamma^{(2)}$, not multiple values. Since we consider approximations where $b^{(2)} \approx f(\gamma^{(2)})$ one can be a little less strict and allow that $b^{(2)}(r_1) \neq b^{(2)}(r_2)$ provided that $b^{(2)}(r_1) \approx b^{(2)}(r_2)$ for all instances when $\gamma^{(2)}(r_1) = \gamma^{(2)}(r_2)$. One can then say that the data is approximately represented by the function $b^{(2)} = f(\gamma^{(2)})$.

---

[40] L. L. Lee, J. Chem. Phys. **103** (1995) 9388 (https://doi.org/10.1063/1.469998); *ibid* **107** (1997) 7360 (https://doi.org/10.1063/1.474974); *ibid* **110** (1999) 7589 (https://doi.org/10.1063/1.478661); L. L. Lee, D. Ghonasgi, and E. Lomba, J. Chem. Phys. **104** (1996) 8058 (https://doi.org/10.1063/1.471522).

[41] M. Llano-Restrepo and W. G. Chapman, J. Chem. Phys. **97** (1992) 2046 (https://doi.org/10.1063/1.463142).

[42] This type of plot was first suggested and used for electrolytes, D.-M. Duh and A. D. J. Haymet, J. Chem. Phys. **97** (1992) 7716 (https://doi.org/10.1063/1.463491), and later also applied to LJ fluids by the same authors, *ibid* **103** (1995) 2625 (https://doi.org/10.1063/1.470724), and by M. Llano-Restrepo and W. G. Chapman, *ibid* **100** (1994) 5139 (https://doi.org/10.1063/1.467241).

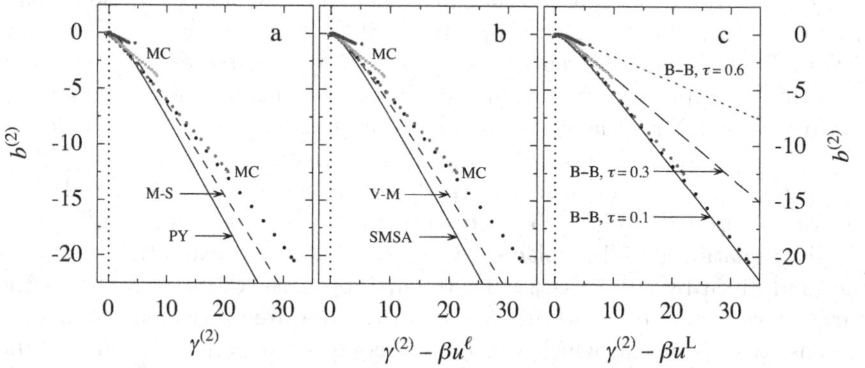

Figure 9.2 **(See color insert).** (a) A so-called Duh-Haymet plot that shows the values of the bridge function $b^{(2)}$ as ordinate and the corresponding values of the function $\gamma^{(2)} \equiv h^{(2)} - c^{(2)}$ as abscissa. The symbols show the MC simulation results for a Lennard-Jones bulk fluid at a reduced temperature $T^* = k_B T/\varepsilon^{\mathrm{LJ}} = 1.5$ and for various reduced densities $n^{\mathrm{b}*} = n^{\mathrm{b}}(d^{\mathrm{LJ}})^3$: from top to bottom $n^{\mathrm{b}*} = 0.4$ (violet), 0.6 (green), 0.8 (red) and 0.9 (black). The full curve shows the PY closure with $b^{(2)}$ as function of $\gamma^{(2)}$ according to Equation 9.80 and the dashed curve shows the M-S closure, Equation 9.89, plotted in the same manner. (b) The same plot of the MC data, but with $\gamma^{(2)} - \beta u^{\ell}$ as abscissa. Here, the full curve shows the SMSA closure with $b^{(2)}$ as function of $\gamma^{(2)} - \beta u^{\ell}$ according to Equation 9.87 and the dashed curve likewise shows the V-M closure, Equation 9.94. (c) The same plot of the MC data, but with $\gamma^{(2)} - \beta u^{\mathrm{L}}$ as abscissa with u^{L} from the B-B partitioning (see Figure 9.1). The curves show the B-B closure in Equation 9.97 with $b^{(2)}$ as function of $\gamma^{*(2)} = \gamma^{(2)} - \beta u^{\mathrm{L}}$ for various values of the parameter τ. The MC data are taken from the work of Llano-Restrepo and Chapman.[41]

Figure 9.2a shows a Duh-Haymet plot of the MC data for the LJ fluid at various densities, whereby the data points obtained in the simulation for each density are represented by a set of symbols of the same color. Each set shows the trajectory in the x, y plane of $y = b^{(2)}(r)$ and $x = \gamma^{(2)}(r)$ for the set of r values used in the simulation. The rightmost symbol in each set shows $\gamma^{(2)}(r)$ and $b^{(2)}(r)$ at $r \approx 0$ and consecutive symbols going to the left correspond to larger r values. The total range of the data is smaller for low densities than for higher ones. In the figure the MC data points form nice curves, so it seems as if the trajectories of the MC data can be well represented by functions $b^{(2)} \approx f(\gamma^{(2)})$ with different f for different densities, at least on the scale displayed. The curves drawn in frame (a) of the figure show the PY closure $b^{(2)} = \ln(1 + \gamma^{(2)}) - \gamma^{(2)}$ and the M-S closure $b^{(2)} = \left[1 + 2\gamma^{(2)}\right]^{1/2} - \gamma^{(2)} - 1$, which do not agree with the data, so these approximations will not yield good results for the LJ fluid.

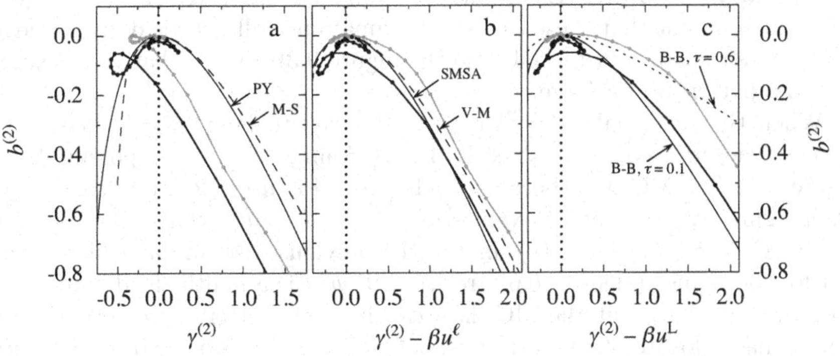

Figure 9.3 **(See color insert).** Magnified views of Figures 9.2 a–c near the origin, showing Duh-Haymet plots of the MC results for the LJ fluid at $T^* = 1.5$ with reduced densities $n^{\mathrm{b}*} = 0.6$ and 0.9 (green and black symbols, respectively, connected by straight lines). The same closures as in Figure 9.2 are also plotted, but in frame (c) the B-B curves for only $\tau = 0.1$ and 0.6 are shown.

Let us now investigate the leftmost part of each set of data points (near the origin) that correspond to larger r values where the functions are appreciably smaller. That part of the data is shown in Figure 9.3a in a greatly magnified view and only two of the densities are displayed for reasons of clarity. We can see that each trajectory near the origin forms loops, so for some values of $\gamma^{(2)}$ there are more than one value of $b^{(2)}$ giving a multivalued "function," which means that our condition above is not fulfilled. The part with loops correspond to the functions $\gamma^{(2)}(r)$ and $b^{(2)}(r)$ for $r \gtrsim 1.2\,d^{\mathrm{LJ}}$ in ordinary space, where both functions are oscillatory. Note that if we have $b^{(2)} = f(\gamma^{(2)})$ for oscillatory functions, the point $y = b^{(2)}(r)$ and $x = \gamma^{(2)}(r)$ for increasing r will move back and forth along the curve of the function $f(\gamma^{(2)})$ without any loops. Thus, this part of the MC data are not represented by a function $b^{(2)} = f(\gamma^{(2)})$ to a good approximation. This shortcoming could not be seen in Figure 9.2a because of the difference in scale.

Let us now go to the cases of renormalized $\gamma^{(2)}$ and we start with $b^{(2)} \approx f(\gamma^{(2)} - \beta u^{\ell})$ as in Equation 9.93, i.e., the type of closure used in the SMSA and V-M approximations. Figure 9.2b shows the same simulation data is in frame (a), but plotted with $x = \gamma^{(2)}(r) - \beta u^{\ell}(r)$. On the scale of the figure there is not much difference between the plots in frames (a) and (b), but in Figure 9.3b, which shows the region around the origin in a greatly magnified view, we see that the loops are less prominent in this case. Thus, $b^{(2)}$ is a still a multivalued "function" of $\gamma^{(2)} - \beta u^{\ell}$ there, but the values of $y = b^{(2)}(r)$ are closer together for equal $x = \gamma^{(2)}(r) - \beta u^{\ell}(r)$ than in the previous case with $x = \gamma^{(2)}(r)$. We can therefore conclude that the data can be *better* represented by a closure of the form $b^{(2)} \approx f(\gamma^{(2)} - \beta u^{\ell})$ than $b^{(2)} \approx f(\gamma^{(2)})$, despite the remaining shortcomings. In the figures there are also plotted curves for the

SMSA and V-M closures, Equations 9.83 and 9.94, respectively. From Figure 9.2b it is apparent that these these approximations will not yield good bridge functions for small r in the LJ fluid. In Figure 9.3b we see that the closures will do a better job for large r.

When the renormalized $\gamma^{(2)}$ is instead taken to be $\gamma^{*(2)} = \gamma^{(2)} - \beta u^{\mathrm{L}}$, with u^{L} chosen according to the B-B partitioning of the pair potential (cf. Figure 9.1), the MC data become as shown in Figures 9.2c and 9.3c. In the latter figure we see that the values of $y = b^{(2)}(r)$ are even closer together for equal $x = \gamma^{(2)}(r) - \beta u^{\mathrm{L}}(r)$ than in the previous case, in particular in the plot for the reduced density 0.6 (green). Still, $b^{(2)}$ is a multivalued "function" close to the origin, but the MC data can be even better represented by an approximate closure of the form $b^{(2)} \approx f(\gamma^{(2)} - \beta u^{\mathrm{L}})$ than before. In Figure 9.2c the B-B closure is plotted for various values of the parameter τ and we see that this approximation has the potential to work well for the LJ fluid, in particular since it has the possibility to describe the variation of the MC data with density by changing τ. This is confirmed in Section 10.1.2, where we will see that the B-B approximation yields very good results for distribution functions and other properties of LJ fluids. The value of τ, which is determined by means of a thermodynamic consistency criterium in this approximation, in effect makes a compromise between a good description of the behavior for small r, corresponding to Figure 9.2c, and that for large r, corresponding to figure 9.3c. In fact, this approximation works very well also for hard sphere fluids as we will see in Section 10.1.2. It will also be shown there that the ZSEP approximation too is very accurate for both LJ and hard sphere fluids.

Finally, we will make some further remarks regarding integral equation approximations. In the development of such theories throughout many decades, there have appeared a huge number of closures that we will not deal with in this treatise. We will, however, mention the following one that belongs to the class that uses the Ansatz 9.93. In the **Zerah-Hansen (Z-H) closure,**[43] also called the **HMSA closure**, one makes an interpolation between the SMSA and HNC approximations by setting

$$\begin{aligned} b^{(2)}(r) &= \ln\left(1 + \frac{\exp\left[\mathfrak{f}(r)\{\gamma^{(2)}(r) - \beta u^{\ell}(r)\}\right] - 1}{\mathfrak{f}(r)}\right) \\ &- \gamma^{(2)}(r) + \beta u^{\ell}(r) \qquad \text{(Z-H)}, \end{aligned} \tag{9.98}$$

where $\mathfrak{f}(r) = 1 - e^{-\alpha r}$ and α is a parameter that is selected such that a thermodynamic consistency criterium is fulfilled (see Section 10.1). The function $\mathfrak{f}(r)$ has been called a "switching function" because when $r \to \infty$, where $\mathfrak{f}(r) \to 1$, the Z-H closure reduces to the HNC closure and when $r \to 0$, where $\mathfrak{f}(r) \to 0$, it reduces to the SMSA closure. The closure is based the fact that the HNC approximation is good for systems where long-range interactions

[43] G. Zerah and J. P. Hansen, J. Chem. Phys. **84** (1986) 2336 (https://doi.org/10.1063/1.450397).

dominate and the SMSA approximation treats the effects of rapidly changing short range interactions well. It is therefore useful for various kinds of fluids. For a hard sphere fluid, where $u^{\ell}(r) = 0$, the Z-H closure becomes equal to the **Rogers-Young (R-Y) closure**[44] that preceded it

$$b^{(2)}(r) = \ln\left(1 + \frac{\exp\left[\mathfrak{f}(r)\gamma^{(2)}(r)\right] - 1}{\mathfrak{f}(r)}\right) - \gamma^{(2)}(r) \quad \text{(R-Y)}. \tag{9.99}$$

It belongs to the class that uses the Ansatz 9.88. The R-Y closure makes an interpolation between the PY and HNC approximations and was designed to treat systems with repulsive pair potentials, not only hard spheres. These two closures have played important roles in the history of integral equations.

9.2 SOME USES OF FUNCTIONALS IN STATISTICAL MECHANICS

9.2.1 Functionals and functional derivative relationships

The mathematical concept of a functional is very important in statistical mechanics. We recall that a (one-dimensional) function is a mapping from a set of numbers to another (or the same) set of numbers $x \mapsto f(x)$. One may say that a function is the "rule" that takes *a number* as input and gives *a number* as output. A **functional**, on the other hand, is a mapping from a set of functions to a set of numbers, for example $f \mapsto f(x_0)$ where a function is mapped to the value of the function at a certain point x_0. Thus, one may say that a functional is a rule that takes *a function* as input and gives *a number* as output; for each function the functional returns a number that depends on the function. Other simple examples of functionals are $f \mapsto df(x)/dx|_{x=x_0}$ (i.e., the value of the derivative of f at x_0), $f \mapsto \int_{x_1}^{x_2} dx\, f(x)$ and $f \mapsto \int_{x_1}^{x_2} dx\, \{f(x)\}^2$. We will use the notation $F[f]$ (with square brackets) to denote a functional F, for instance $F[f] = \int_{x_1}^{x_2} dx\, f(x)$, so the value of $F[f]$ is the number that the functional associates with the function f. A *linear* functional F satisfies $F[a_1 f_1 + a_2 f_2] = aF[f_1] + bF[f_2]$ for arbitrary numbers a_1 and a_2 and functions f_1 and f_2 in the set. The functional $f \mapsto \int_{x_1}^{x_2} dx\, \{f(x)\}^2$ is an example of a *nonlinear* functional.

The functionals $f \mapsto f(x_0)$ and $f \mapsto f(\mathbf{r}_0)$ for functions in the one-dimensional and three-dimensional spaces, respectively, are called the **Dirac delta functionals**. Let us denote this functional $\delta_{\mathbf{r}_0}[f]$, so we have $\delta_{\mathbf{r}_0}[f] = f(\mathbf{r}_0)$ in 3D space (distinguish this from the variation $\delta f(\mathbf{r})$ of the function $f(\mathbf{r})$). In physics it is very common to express this functional in terms of the **Dirac delta function** $\delta^{(3)}(\mathbf{r})$ that is described in Appendix 5A in Volume I. This function has the property that for any (well-behaved) function $f(\mathbf{r})$ one

[44] F. J. Rogers and D. A. Young, Phys. Rev. A **30** (1984) 999 (https://doi.org/10.1103/PhysRevA.30.999).

has

$$\int d\mathbf{r}\, \delta^{(3)}(\mathbf{r} - \mathbf{r}_0) f(\mathbf{r}) = f(\mathbf{r}_0), \tag{9.100}$$

so the functional $\delta_{\mathbf{r}_0}[f] \equiv \int d\mathbf{r}\, \delta^{(3)}(\mathbf{r}-\mathbf{r}_0) f(\mathbf{r})$. The Dirac delta function $\delta^{(3)}(\mathbf{r})$ itself is accordingly associated with the functional $f \mapsto f(\mathbf{0})$.[45]

In statistical mechanics for fluids, a simple example of a functional is the fact that the internal energy depends on the pair interaction potential $u(r)$ between the particles. For each pair interaction u, the internal energy has some value and if we change the potential $u(r)$ from one function of r to another, the energy changes. Hence, the internal energy is a *functional* of u, $U = U[u]$. The internal energy is also a *function*, for example, it is a function of the temperature, volume and number of particles. This means that an entity can simultaneously be a functional of some entities (functions) and a function of some other entities (numbers): $U = U(T, V, N; [u])$, where the square brackets around u indicates that U is a functional of u. For the time being we shall, however, focus on the functional aspects of various entities.

For an inhomogeneous fluid, the internal energy is a functional of both the pair interaction $u(r)$ and the external potential $v(\mathbf{r})$, so we have $U = U[u, v]$ and the value of U changes when we vary $u(r)$ and/or $v(\mathbf{r})$. When we focus on the dependence of, say, v but keep u the same, we will explicitly write out only the v dependence, like $U = U[v]$. This is relevant, for instance, when we expose a fluid of a particular kind of particles to various external potentials. Then, u is unchanged while v is varied and we will only show the functional dependence of U on v explicitly. An especially important functional is the grand potential as a functional of the external potential, $\Theta = \Theta[v]$, which we will investigate later.

Another example of a functional is how the density $n(\mathbf{r}')$ at one point $\mathbf{r}'$ in an inhomogeneous fluid depends on the external potential $v(\mathbf{r})$. As we have seen, when the function v is changed, the density at $\mathbf{r}'$ changes so $n(\mathbf{r}')$ is a functional of v. The density changes at other points too when v is varied (in principle everywhere else), so $n(\mathbf{r}')$ for all $\mathbf{r}'$ is a functional of v and we write $n = n(\mathbf{r}'; [v])$. The density is accordingly a function of $\mathbf{r}'$ and a functional of v. The values of $v(\mathbf{r})$ for *all* $\mathbf{r}$ influence the value of n for *each* $\mathbf{r}'$ (at least in principle) and this is implicit in this notation. The influence of $v(\mathbf{r})$ is of course different for different points $\mathbf{r}'$. If $v(\mathbf{r})$ changes only in a certain region of space, the density far away from this region changes just a little and the change is negligible in locations very far away.

In Section 9.1.2 we dealt with the change in density obtained when one varies the external potential, which means that we, in fact, considered the density distribution as a functional of the external potential $n = n(\mathbf{r}'; [v])$ (without saying this). We also dealt with the reverse question: what variation in potential is needed to obtain a given change in density, that is, we considered

[45] In mathematics, the linear mapping $f \mapsto f(\mathbf{0})$ of a so-called test function $f(\mathbf{r})$ is called a *Dirac distribution*.

v as a functional of n. This is the inverse functional $v = v(\mathbf{r}; [n])$ of the former one. In fact, in the grand canonical ensemble there is a one-to-one relationship between the density distribution $n(\mathbf{r}')$ and the external potential $v(\mathbf{r})$ of a system: for a given $v(\mathbf{r})$ there exists a *unique* density distribution $n(\mathbf{r}')$ that no other external potential gives rise to. This is shown in Appendix 9D. Therefore, one can obtain a unique answer to the question: what $v(\mathbf{r})$ gives rise to a given density distribution $n(\mathbf{r}')$, which means that the inverse functional to $n = n(\mathbf{r}'; [v])$ exists. This also implies that one can consider, for example, the grand potential as a functional not only of v, but also of n, that is, $\Theta = \Theta[n]$. To see this, assume that we know $\Theta = \Theta[v]$. Since we (in principle) can figure out which v gives rise to a given n, that is, $v = v[n]$, we can define $\Theta[n] \equiv \Theta[v[n]]$.

The linear response results in Section 9.1.2 can be formulated in a powerful way by introducing derivatives of functionals. As a first example we consider variations in free energy when the external potential is varied. In the grand canonical ensemble we thereby study the functional $\Theta = \Theta[v]$ when we make the variation $v(\mathbf{r}) \to v(\mathbf{r}) + \delta v(\mathbf{r})$. As we have seen, we have (Equation 9.34)

$$d\Theta = \int d\mathbf{r}\, n(\mathbf{r})\delta v(\mathbf{r}) \tag{9.101}$$

for infinitesimally small variations $\delta v(\mathbf{r})$. We shall now introduce a compact notation for this relationship between variations in Θ and v. Let us write Equation 9.101 as

$$d\Theta = \int d\mathbf{r}\, \frac{\delta\Theta}{\delta v(\mathbf{r})}\delta v(\mathbf{r})$$

where we have introduced the notation

$$\frac{\delta\Theta}{\delta v(\mathbf{r})} = n(\mathbf{r}) \tag{9.102}$$

The entity $\delta\Theta/\delta v(\mathbf{r})$ is called the **functional derivative** of Θ with respect to $v(\mathbf{r})$ and the expression 9.102 simply means *by definition* that Equation 9.101 is fulfilled for infinitesimally small variations $\delta v(\mathbf{r})$. Thus, the density distribution $n(\mathbf{r})$ is the functional derivative of the grand potential with respect to the external potential when μ is constant (T and V are also constant).

In general, when an entity $\mathcal{F}$ is a functional of a function $s(\mathbf{r})$, so we have $\mathcal{F} = \mathcal{F}[s]$, we say that the function $t(\mathbf{r})$ is the functional derivative

$$t(\mathbf{r}) = \frac{\delta\mathcal{F}}{\delta s(\mathbf{r})}$$

provided that

$$d\mathcal{F} = \int d\mathbf{r}\, t(\mathbf{r})\delta s(\mathbf{r})$$

for infinitesimally small variations $s(\mathbf{r}) \to s(\mathbf{r}) + \delta s(\mathbf{r})$. The two equations are simply equivalent to each other.[46]

Likewise, if an entity f is a functional of $s(\mathbf{r})$ as well as a function of $\mathbf{r}'$, so we have $f = f(\mathbf{r}'; [s])$, we say that the function $\tau(\mathbf{r}', \mathbf{r})$ is the functional derivative

$$\tau(\mathbf{r}', \mathbf{r}) = \frac{\delta f(\mathbf{r}')}{\delta s(\mathbf{r})}$$

provided that the variation $\delta f(\mathbf{r}')$ of f satisfies

$$\delta f(\mathbf{r}') = \int d\mathbf{r}\, \tau(\mathbf{r}', \mathbf{r}) \delta s(\mathbf{r})$$

for infinitesimally small variations $s(\mathbf{r}) \to s(\mathbf{r}) + \delta s(\mathbf{r})$; that is

$$\delta f(\mathbf{r}') = \int d\mathbf{r}\, \frac{\delta f(\mathbf{r}')}{\delta s(\mathbf{r})} \delta s(\mathbf{r}).$$

We found earlier that $\delta c^{(1)}(\mathbf{r}') = -\beta \int d\mathbf{r}\, h^{(2)}(\mathbf{r}, \mathbf{r}') n(\mathbf{r}) \delta v(\mathbf{r})$ (Equation 9.55) and it follows that this can be expressed as

$$\frac{\delta c^{(1)}(\mathbf{r}')}{\delta v(\mathbf{r})} = -\beta h^{(2)}(\mathbf{r}', \mathbf{r}) n(\mathbf{r}). \tag{9.103}$$

Likewise the definition 9.56 of $c^{(2)}$ can be expressed as

$$\frac{\delta c^{(1)}(\mathbf{r}')}{\delta n(\mathbf{r})} = c^{(2)}(\mathbf{r}', \mathbf{r}), \tag{9.104}$$

which accordingly can be adopted as an *alternative definition* of $c^{(2)}$. These relationships are important examples of functional derivatives.

A mathematically important relationship can be concluded from Equation 9.100 for Dirac's delta function applied to a variation $\delta f(\mathbf{r})$

$$\int d\mathbf{r}\, \delta^{(3)}(\mathbf{r} - \mathbf{r}') \delta f(\mathbf{r}) = \delta f(\mathbf{r}').$$

Interpreted in terms of a functional derivative relationship this implies that

$$\frac{\delta f(\mathbf{r}')}{\delta f(\mathbf{r})} = \delta^{(3)}(\mathbf{r} - \mathbf{r}') \tag{9.105}$$

for any function f. This is a very useful relationship.

Another mathematically important relationship is the "**chain rule**" for functional differentiation

$$\frac{\delta f(\mathbf{r}')}{\delta s(\mathbf{r})} = \int \frac{\delta f(\mathbf{r}')}{\delta t(\mathbf{r}'')} \frac{\delta t(\mathbf{r}'')}{\delta s(\mathbf{r})} d\mathbf{r}'' \tag{9.106}$$

[46]Note that the unit of the functional derivative $\delta \mathcal{F}/\delta s(\mathbf{r})$ is not equal to the unit of $\mathcal{F}/s(\mathbf{r})$. The unit of $\delta \mathcal{F}/\delta s(\mathbf{r})$ is that of $\mathcal{F}/s(\mathbf{r})$ times an inverse volume, whereby the latter cancels the unit of volume from $d\mathbf{r}$ in the integral.

for three functions $f(\mathbf{r}')$, $s(\mathbf{r})$ and $t(\mathbf{r}'')$ that are functionally dependent of each other.

{**Exercise 9.2:** Prove Equation 9.106 by first writing $\delta f(\mathbf{r}')/\delta t(\mathbf{r}'')$ and $\delta t(\mathbf{r}'')/\delta s(\mathbf{r})$ in terms of their definitions as integrals with the appropriate variations included. Then one eliminates the variation $\delta t(\mathbf{r}'')$ from the resulting equations and identifies $\delta f(\mathbf{r}')/\delta s(\mathbf{r})$.}

A special case of the chain rule is

$$\int \frac{\delta f(\mathbf{r}')}{\delta t(\mathbf{r}'')}\frac{\delta t(\mathbf{r}'')}{\delta f(\mathbf{r})} d\mathbf{r}'' = \delta^{(3)}(\mathbf{r}-\mathbf{r}'), \tag{9.107}$$

which we obtain by setting $s(\mathbf{r}) = f(\mathbf{r})$ in Equation 9.106 and using Equation 9.105. This relationship expresses that $\delta f(\mathbf{r}')/\delta t(\mathbf{r}'')$ and $\delta t(\mathbf{r}'')/\delta f(\mathbf{r})$ are **functional inverses** of each other.

Example 9.A: The Yvon equations revisited

To take a functional derivative of an expression is very similar to take an ordinary derivative. As an example, let us take the functional derivative of

$$k_BT \ln\zeta = k_BT \ln n(\mathbf{r}) + v(\mathbf{r}) - k_BT\, c^{(1)}(\mathbf{r}) \tag{9.108}$$

(which is Equation 9.12) with respect to $v(\mathbf{r}')$ when μ, T and therefore the activity ζ are constant. We obtain

$$0 = k_BT\frac{1}{n(\mathbf{r})}\frac{\delta n(\mathbf{r})}{\delta v(\mathbf{r}')} + \frac{\delta v(\mathbf{r})}{\delta v(\mathbf{r}')} - k_BT\frac{\delta c^{(1)}(\mathbf{r})}{\delta v(\mathbf{r}')}$$

(cf. Equation 9.37 and footnote 13), which yields

$$\begin{aligned}\frac{1}{n(\mathbf{r})}\frac{\delta n(\mathbf{r})}{\delta v(\mathbf{r}')} &= -\beta\frac{\delta v(\mathbf{r})}{\delta v(\mathbf{r}')} + \frac{\delta c^{(1)}(\mathbf{r})}{\delta v(\mathbf{r}')} = \\ &= -\beta\delta^{(3)}(\mathbf{r}-\mathbf{r}') - \beta h^{(2)}(\mathbf{r},\mathbf{r}')n(\mathbf{r}'),\end{aligned}$$

where we have inserted Equation 9.103 and used Equation 9.105. We can write this as

$$\frac{\delta n(\mathbf{r})}{\delta v(\mathbf{r}')} = -\beta n(\mathbf{r})\left[\delta^{(3)}(\mathbf{r}-\mathbf{r}') + h^{(2)}(\mathbf{r},\mathbf{r}')n(\mathbf{r}')\right]. \tag{9.109}$$

This expression is the **first Yvon equation** written in functional derivative form. It is equivalent to Equation 9.38.

Likewise, from the functional derivative of Equation 9.108 with respect to $n(\mathbf{r}')$ at constant ζ and T we obtain

$$\frac{1}{n(\mathbf{r})}\frac{\delta n(\mathbf{r})}{\delta n(\mathbf{r}')} = -\beta\frac{\delta v(\mathbf{r})}{\delta n(\mathbf{r}')} + \frac{\delta c^{(1)}(\mathbf{r})}{\delta n(\mathbf{r}')}$$

and together with Equation 9.104 this yields

$$\frac{\delta v(\mathbf{r})}{\delta n(\mathbf{r}')} = -\frac{1}{\beta}\left[\frac{1}{n(\mathbf{r})}\delta^{(3)}(\mathbf{r}-\mathbf{r}') - c^{(2)}(\mathbf{r},\mathbf{r}')\right], \tag{9.110}$$

which is the **second Yvon equation** written in functional derivative form and is equivalent to Equation 9.43.

If we define

$$H^{(2)}(\mathbf{r},\mathbf{r}') \equiv n(\mathbf{r})\delta^{(3)}(\mathbf{r}-\mathbf{r}') + n(\mathbf{r})n(\mathbf{r}')h^{(2)}(\mathbf{r},\mathbf{r}'), \tag{9.111}$$

the first Yvon equation 9.109 can be written as

$$\frac{\delta n(\mathbf{r})}{\delta v(\mathbf{r}')} = -\beta H^{(2)}(\mathbf{r},\mathbf{r}') \tag{9.112}$$

and, likewise, if we define

$$C^{(2)}(\mathbf{r},\mathbf{r}') \equiv \frac{1}{n(\mathbf{r})}\delta^{(3)}(\mathbf{r}-\mathbf{r}') - c^{(2)}(\mathbf{r},\mathbf{r}'), \tag{9.113}$$

the second Yvon equation 9.110 can be written

$$\frac{\delta v(\mathbf{r})}{\delta n(\mathbf{r}')} = -\frac{1}{\beta}C^{(2)}(\mathbf{r},\mathbf{r}'). \tag{9.114}$$

Since $\delta n(\mathbf{r})/\delta v(\mathbf{r}')$ and $\delta v(\mathbf{r})/\delta n(\mathbf{r}')$ are functional inverses of each other we see from Equations 9.112 and 9.114 that the functional inverse of $H^{(2)}$ is

$$[H^{(2)}]^{-1}(\mathbf{r},\mathbf{r}') = C^{(2)}(\mathbf{r},\mathbf{r}'). \tag{9.115}$$

This means that we have (cf. Equation 9.107)

$$\int d\mathbf{r}''\, H^{(2)}(\mathbf{r}',\mathbf{r}'')C^{(2)}(\mathbf{r}'',\mathbf{r}) = \delta^{(3)}(\mathbf{r}-\mathbf{r}'). \tag{9.116}$$

The latter equation with Equations 9.111 and 9.113 inserted is equivalent to the OZ equation 9.40, which can be derived by simplifying the resulting expression. The OZ equation accordingly expresses this functional inverse relationship.

{**Exercise 9.3:** Obtain the OZ equation 9.40 by inserting the definitions of $H^{(2)}$ and $C^{(2)}$ into Equation 9.116 and simplifying.}

From Equations 9.112 and 9.102 it follows that

$$-\beta H^{(2)}(\mathbf{r},\mathbf{r}') = \frac{\delta n(\mathbf{r})}{\delta v(\mathbf{r}')} = \frac{\delta}{\delta v(\mathbf{r}')}\left[\frac{\delta \Theta}{\delta v(\mathbf{r})}\right] = \frac{\delta^2 \Theta}{\delta v(\mathbf{r}')\delta v(\mathbf{r})}, \tag{9.117}$$

that is, a second order functional derivative. This can be used as an alternative definition of $H^{(2)}(\mathbf{r},\mathbf{r}')$.

We have now expressed the important linear response results of Section 9.1.2 as functional derivative relationships. Let us proceed to show some further important relationships. In the grand canonical ensemble (constant T, V and μ) let us make a reversible change[47] of the external potential by $\delta v(\mathbf{r})$, whereby the average number of particles, $\bar{N}$, changes by, say, $d\bar{N}$. The **Helmholtz free energy** $A = \Theta + \mu\bar{N}$ then changes by $dA = d\Theta + \mu d\bar{N} = \int d\mathbf{r}\, n(\mathbf{r})\delta v(\mathbf{r}) + \mu d\bar{N}$, where we have used Equation 9.101. Since $d\bar{N} = \int d\mathbf{r}\, \delta n(\mathbf{r})$, we have

$$dA = \int d\mathbf{r}\, n(\mathbf{r})\delta v(\mathbf{r}) + \mu \int d\mathbf{r}\, \delta n(\mathbf{r}). \tag{9.118}$$

For an ideal gas exposed to an external potential $v(\mathbf{r})$ we obtained in Section 3.3 of Volume I the result (Equation I:3.47)

$$A = k_B T \int d\mathbf{r}\, n(\mathbf{r}) \left[\ln(\Lambda^3 n(\mathbf{r})) - 1\right] + \int d\mathbf{r}\, n(\mathbf{r})v(\mathbf{r}) \quad \text{(ideal gas)}, \tag{9.119}$$

where $\int d\mathbf{r}\, n(\mathbf{r})v(\mathbf{r})$ is the interaction energy between the particles in gas and the external potential. The first term is the **intrinsic ideal free energy**

$$\mathcal{A}^{\text{ideal}} = k_B T \int d\mathbf{r}\, n(\mathbf{r}) \left[\ln(\Lambda^3 n(\mathbf{r})) - 1\right], \tag{9.120}$$

where we follow a common notation and use script $\mathcal{A}$ to denote the intrinsic part of A (that is, instead of writing it as A^{intr}).

Let now turn to a system with interacting particles. In the presence of interparticle interactions we have a free energy contribution due to them, namely the *intrinsic excess free energy* $\mathcal{A}^{ex}$. The sum $\mathcal{A} = \mathcal{A}^{\text{ideal}} + \mathcal{A}^{ex}$ constitute the **intrinsic Helmholtz free energy**, while the total Helmholtz free energy is

$$A = \mathcal{A}^{\text{ideal}} + \mathcal{A}^{\text{ex}} + \int d\mathbf{r}\, n(\mathbf{r})v(\mathbf{r}). \tag{9.121}$$

Thus, we have

$$\mathcal{A} \equiv A - \int d\mathbf{r}\, n(\mathbf{r})v(\mathbf{r}), \tag{9.122}$$

[47] For reversible work in the grand canonical ensemble, see footnote 11 .

that is, $\mathcal{A}$ is the part of A that does not *explicitly* depends on the external potential.

For small changes $\delta v(\mathbf{r})$ and $\delta n(\mathbf{r})$, we have from Equation 9.122

$$d\mathcal{A} = dA - \int d\mathbf{r}\, v(\mathbf{r})\delta n(\mathbf{r}) - \int d\mathbf{r}\, n(\mathbf{r})\delta v(\mathbf{r})$$

and by inserting Equation 9.118, we obtain

$$d\mathcal{A} = \int d\mathbf{r}\, [\mu - v(\mathbf{r})]\delta n(\mathbf{r}).$$

Thus, $\mu - v(\mathbf{r})$ is the functional derivative of $\mathcal{A}$ with respect to $n(\mathbf{r})$, so we have

$$\frac{\delta \mathcal{A}}{\delta n(\mathbf{r})} = \mu - v(\mathbf{r}) = k_B T \ln(\Lambda^3 n(\mathbf{r})) - k_B T\, c^{(1)}(\mathbf{r}), \tag{9.123}$$

where the last equality follows from Equation 9.11. The quantity $\mu - v(\mathbf{r})$ is the so-called **intrinsic chemical potential**

$$\mu^{\mathrm{intr}}(\mathbf{r}) = \mu - v(\mathbf{r})$$

and we see from Equation 9.123 that it is the sum of the ideal chemical potential $k_B T \ln(\Lambda^3 n(\mathbf{r}))$ (cf. Equation 9.5) and the intrinsic excess chemical potential $-k_B T\, c^{(1)}(\mathbf{r})$ (cf. Equations 9.8 and 9.11). In Exercise 9.4 it is shown that functional differentiation of Equation 9.120 with respect to $n(\mathbf{r})$ gives

$$\frac{\delta \mathcal{A}^{\mathrm{ideal}}}{\delta n(\mathbf{r})} = k_B T \ln(\Lambda^3 n(\mathbf{r})). \tag{9.124}$$

{**Exercise 9.4:** Show Equation 9.124 by performing the functional differentiation of Equation 9.120.}

It follows from Equations 9.123, 9.124 and $\mathcal{A} = \mathcal{A}^{\mathrm{ideal}} + \mathcal{A}^{ex}$ that

$$\frac{\delta \mathcal{A}^{\mathrm{ex}}}{\delta n(\mathbf{r})} = -k_B T\, c^{(1)}(\mathbf{r}). \tag{9.125}$$

This can be used as an *alternative definition* of $c^{(1)}(\mathbf{r})$

$$c^{(1)}(\mathbf{r}) \equiv -\beta \frac{\delta \mathcal{A}^{\mathrm{ex}}}{\delta n(\mathbf{r})} \tag{9.126}$$

that is equivalent to the previous one in Equation 9.8. From Equation 9.104 we see that

$$c^{(2)}(\mathbf{r}', \mathbf{r}) = \frac{\delta c^{(1)}(\mathbf{r}')}{\delta n(\mathbf{r})} = -\beta \frac{\delta^2 \mathcal{A}^{\mathrm{ex}}}{\delta n(\mathbf{r})\delta n(\mathbf{r}')}. \tag{9.127}$$

The order of the differentiation with respect to $n(\mathbf{r})$ and $n(\mathbf{r}')$ does not matter, so $c^{(2)}(\mathbf{r}',\mathbf{r}) = c^{(2)}(\mathbf{r},\mathbf{r}')$ as mentioned earlier. These two relationships are fundamental for the direct correlation functions.

The grand potential $\Theta = A - \mu\bar{N} = A - \mu \int d\mathbf{r}'\, n(\mathbf{r}')$ can be expressed as

$$\Theta = \mathcal{A}^{\text{ideal}} + \mathcal{A}^{\text{ex}} + \int d\mathbf{r}'\, n(\mathbf{r}')[v(\mathbf{r}') - \mu], \tag{9.128}$$

where Equation 9.121 has been utilized. By taking the functional derivative with respect to $n(\mathbf{r})$ we obtain

$$\begin{aligned} \frac{\delta\Theta}{\delta n(\mathbf{r})} &= \frac{\delta \mathcal{A}^{\text{ideal}}}{\delta n(\mathbf{r})} + \frac{\delta \mathcal{A}^{\text{ex}}}{\delta n(\mathbf{r})} + \int d\mathbf{r}'\, \delta^{(3)}(\mathbf{r}-\mathbf{r}')[v(\mathbf{r}') - \mu] \\ &= k_B T \ln(\Lambda^3 n(\mathbf{r})) - k_B T\, c^{(1)}(\mathbf{r}) + v(\mathbf{r}) - \mu, \end{aligned}$$

where we have used $\delta n(\mathbf{r}')/\delta n(\mathbf{r}) = \delta^{(3)}(\mathbf{r}-\mathbf{r}')$ (i.e., Equation 9.105 with $f = n$). Equation 9.11 implies that the rhs is zero, so we have

$$\frac{\delta\Theta}{\delta n(\mathbf{r})} = 0. \tag{9.129}$$

This is a very important equation. From the derivation here we can see that Equation 9.129 and the equilibrium condition 9.11 are equivalent to each other. In Section 9.3 these matters will be discussed in more detail from a fundamental point of view.

Example 9.B: Justification for the Percus-Yevick approximation

As a further example of the use of functional differentiation, we will give a justification for the PY approximation. Consider the function

$$y^{(1)}(\mathbf{r}_1) = e^{c^{(1)}(\mathbf{r}_1)}, \tag{9.130}$$

which according to Equation 9.13 satisfies

$$n(\mathbf{r}_1) = \zeta y^{(1)}(\mathbf{r}_1) e^{-\beta v(\mathbf{r}_1)}. \tag{9.131}$$

It is the analogue of the cavity function $y^{(2)}$ that satisfies

$$g^{(2)}(\mathbf{r}_1,\mathbf{r}_2) = y^{(2)}(\mathbf{r}_1,\mathbf{r}_2) e^{-\beta u(\mathbf{r}_1,\mathbf{r}_2)} \tag{9.132}$$

(Equation 9.26) and is given by $y^{(2)}(\mathbf{r}_1,\mathbf{r}_2) = e^{\Delta c^{(1)}(\mathbf{r}_1|\mathbf{r}_2)}$, where $\Delta c^{(1)}(\mathbf{r}_1|\mathbf{r}_2) = c^{(1)}(\mathbf{r}_1|\mathbf{r}_2) - c^{(1)}(\mathbf{r}_1)$ (cf. Equation 9.22).

In the presence of a fixed particle at $\mathbf{r}_2$, which is considered to be external to the fluid but belongs to the same species, Equation 9.131 becomes

$$n(\mathbf{r}_1|\mathbf{r}_2) = \zeta y^{(1)}(\mathbf{r}_1|\mathbf{r}_2)e^{-\beta[u(\mathbf{r}_1,\mathbf{r}_2)+v(\mathbf{r}_1)]}, \tag{9.133}$$

where $n(\mathbf{r}_1|\mathbf{r}_2) = n(\mathbf{r}_1)g^{(2)}(\mathbf{r}_1,\mathbf{r}_2)$ as before. We have $y^{(1)}(\mathbf{r}_1|\mathbf{r}_2) = e^{c^{(1)}(\mathbf{r}_1|\mathbf{r}_2)}$, which can be written as

$$y^{(1)}(\mathbf{r}_1|\mathbf{r}_2) = e^{\Delta c^{(1)}(\mathbf{r}_1|\mathbf{r}_2)+c^{(1)}(\mathbf{r}_1)} = y^{(2)}(\mathbf{r}_1,\mathbf{r}_2)y^{(1)}(\mathbf{r}_1). \tag{9.134}$$

The functional derivative of Equation 9.130 with respect to n is

$$\frac{\delta y^{(1)}(\mathbf{r}_1)}{\delta n(\mathbf{r}_3)} = \frac{\delta e^{c^{(1)}(\mathbf{r}_1)}}{\delta n(\mathbf{r}_3)} = e^{c^{(1)}(\mathbf{r}_1)}\frac{\delta c^{(1)}(\mathbf{r}_1)}{\delta n(\mathbf{r}_3)} = y^{(1)}(\mathbf{r}_1)c^{(2)}(\mathbf{r}_1,\mathbf{r}_3)$$

and hence for small variations $\delta y^{(1)}(\mathbf{r}_1)$ and $\delta n(\mathbf{r}_3)$, we have

$$\delta y^{(1)}(\mathbf{r}_1) = y^{(1)}(\mathbf{r}_1)\int d\mathbf{r}_3\, c^{(2)}(\mathbf{r}_1,\mathbf{r}_3)\delta n(\mathbf{r}_3). \tag{9.135}$$

This result is valid if δn is small, but we may use it for large Δn to obtain an approximation for $\Delta y^{(1)}$, that is,

$$\Delta y^{(1)}(\mathbf{r}_1) \approx y^{(1)}(\mathbf{r}_1)\int d\mathbf{r}_3\, c^{(2)}(\mathbf{r}_1,\mathbf{r}_3)\Delta n(\mathbf{r}_3). \tag{9.136}$$

In particular, we may use it for the changes $\Delta y^{(1)}$ and Δn when we insert a new particle at $\mathbf{r}_2$, namely

$$\begin{aligned}\Delta y^{(1)}(\mathbf{r}_1|\mathbf{r}_2) &\approx y^{(1)}(\mathbf{r}_1)\int d\mathbf{r}_3\, c^{(2)}(\mathbf{r}_1,\mathbf{r}_3)\Delta n(\mathbf{r}_3|\mathbf{r}_2)\\ &= y^{(1)}(\mathbf{r}_1)\int d\mathbf{r}_3\, c^{(2)}(\mathbf{r}_1,\mathbf{r}_3)n(\mathbf{r}_3)h^{(2)}(\mathbf{r}_3,\mathbf{r}_2)\\ &= y^{(1)}(\mathbf{r}_1)\left[h^{(2)}(\mathbf{r}_1,\mathbf{r}_2) - c^{(2)}(\mathbf{r}_1,\mathbf{r}_2)\right] \end{aligned} \tag{9.137}$$

where we have used Equation 9.33 and the OZ equation 9.40. From Equation 9.134 it follows that

$$\Delta y^{(1)}(\mathbf{r}_1|\mathbf{r}_2) \equiv y^{(1)}(\mathbf{r}_1|\mathbf{r}_2) - y^{(1)}(\mathbf{r}_1) = \left[y^{(2)}(\mathbf{r}_1,\mathbf{r}_2) - 1\right]y^{(1)}(\mathbf{r}_1), \tag{9.138}$$

so by dividing Equation 9.137 by $y^{(1)}(\mathbf{r}_1)$, we obtain

$$y^{(2)}(\mathbf{r}_1,\mathbf{r}_2) - 1 \approx h^{(2)}(\mathbf{r}_1,\mathbf{r}_2) - c^{(2)}(\mathbf{r}_1,\mathbf{r}_2). \quad \text{(PY)} \tag{9.139}$$

This is the **Percus-Yevick approximation** because when this expression is inserted into Equation 9.132 we obtain Equation 9.78.

The key assumption here is that $\Delta y^{(1)}$ and Δn to a good approximation are linearly related to each other, which means that the approximate, linear relationship 9.136 can be used for the finite changes $\Delta y^{(1)}$ and Δn, as we did in the first line of Equation 9.137. For a hard sphere fluid this is reasonable as can be realized in the following manner.

Outside the hard core, $r_{12} \geq d^{\mathrm{h}}$ (where d^{h} is the hard core diameter), we have $u(\mathbf{r}_1, \mathbf{r}_2) = 0$ for a hard sphere fluid. Hence, from Equations 9.131 and 9.133 we obtain the exact relationship for hard spheres

$$\Delta n(\mathbf{r}_1|\mathbf{r}_2) = \zeta \Delta y^{(1)}(\mathbf{r}_1|\mathbf{r}_2) e^{-\beta v(\mathbf{r}_1)} \quad \text{when } r_{12} \geq d^{\mathrm{h}}. \tag{9.140}$$

Thus, for a given external potential $v(\mathbf{r}_1)$, the finite variations $\Delta y^{(1)}$ and Δn are linearly related to each other at each point $\mathbf{r}_1$ outside the hard core. This means that the linearity assumed in Equation 9.137 is satisfied and *is not an approximation* for $r_{12} \geq d^{\mathrm{h}}$ in a hard sphere fluid.[48] The linearity is, however, an approximation for $r_{12} < d^{\mathrm{h}}$, but since the functions $y^{(2)}$ and $\Delta y^{(1)}$ are continuous and have continuous derivatives across the hard core boundary, the linearity inside the core is a reasonable approximation. There are nevertheless considerable deviations from the accurate values for r_{12} near the origin as we will see in Section 10.1.1.

Equation 9.101 says that $d\Theta = \int d\mathbf{r}\, n(\mathbf{r})\delta v(\mathbf{r})$, which is valid when μ, T and V are constant. Let us generalize this slightly so we include cases when both $v(\mathbf{r})$ and μ can vary while T and V remain constant. This equation for the differential $d\Theta$ is then replaced by $d\Theta = \int d\mathbf{r}\, n(\mathbf{r})\delta v(\mathbf{r}) - \bar{N} d\mu$ because $(\partial\Theta/\partial\mu)_{T,V} = -\bar{N}$. Since $\bar{N} = \int d\mathbf{r}\, n(\mathbf{r})$ we can write this as

$$d\Theta = \int d\mathbf{r}\, n(\mathbf{r})[\delta v(\mathbf{r}) - d\mu] = -\int d\mathbf{r}\, n(\mathbf{r})\delta\mu^{\mathrm{intr}}(\mathbf{r}),$$

which means that

$$\frac{\delta\Theta}{\delta[\mu - v(\mathbf{r})]} = \frac{\delta\Theta}{\delta\mu^{\mathrm{intr}}(\mathbf{r})} = -n(\mathbf{r}). \tag{9.141}$$

This expression replaces Equation 9.102 for cases where μ is not constant. All previous equations in the present section have been derived for applications where μ is constant. When μ is allowed to vary, the difference is essentially that $v(\mathbf{r})$ should be replaced by $v(\mathbf{r}) - \mu$, that is by $-\mu^{\mathrm{intr}}(\mathbf{r})$, in the functional

[48] For any fluid with a pair potential of finite range, the same is true for distances r_{12} where the potential $u(r_{12})$ is zero.

derivative relationships. Equation 9.103 is then replaced by

$$\frac{\delta c^{(1)}(\mathbf{r}')}{\delta \mu^{\text{intr}}(\mathbf{r})} = \beta h^{(2)}(\mathbf{r}', \mathbf{r}) n(\mathbf{r}) \tag{9.142}$$

and Equation 9.117) by

$$-\beta H^{(2)}(\mathbf{r}, \mathbf{r}') = \frac{\delta^2 \Theta}{\delta \mu^{\text{intr}}(\mathbf{r}') \delta \mu^{\text{intr}}(\mathbf{r})}. \tag{9.143}$$

In the grand canonical ensemble μ is constant, so the previous formulas with derivatives with respect to $v(\mathbf{r})$ are valid. Whenever we work in this ensemble, which we normally will do, we will use the formulas with v-derivatives because they are more easy to interpret physically. It is, however, common in the literature to use Equations 9.141 - 9.143 with derivatives with respect to μ^{intr}, that is, $\mu - v(\mathbf{r})$, even when one works in the grand canonical ensemble.

9.2.2 Functional expansions and integrations

Let us consider a functional $\mathcal{F}$ of a one-variable function $f(x)$, so we have $\mathcal{F} = \mathcal{F}[f]$. Its functional derivative $\delta\mathcal{F}/\delta f(x)$ satisfies, as we have seen,

$$d\mathcal{F} = \int dx \, \frac{\delta \mathcal{F}}{\delta f(x)} \delta f(x), \tag{9.144}$$

where $d\mathcal{F} = \mathcal{F}[f + \delta f] - \mathcal{F}[f]$ is the change in $\mathcal{F}$ for an infinitesimally small variation $f(x) \to f(x) + \delta f(x)$. In order to get another perspective on this expression, let us consider the total differential of a multidimensional function $F = F(y_1, y_2, \ldots, y_l)$ when $y_\nu \to y_\nu + dy_\nu$ for $1 \leq \nu \leq l$ and we have

$$dF = \sum_\nu \frac{\partial F}{\partial y_\nu} dy_\nu.$$

By introducing $y_\nu = f(x_\nu)$ and $dy_\nu = \delta f(x_\nu)$ for a set of x values $\{x_\nu\}_{1 \leq \nu \leq l}$, we can write this as

$$dF = \sum_\nu \frac{\partial F}{\partial f(x_\nu)} \delta f(x_\nu), \tag{9.145}$$

which is similar to Equation 9.144. While the value of $F = F(f(x_1), f(x_2), \ldots, f(x_l))$ depends on the values of $f(x)$ only at the discrete set of x values, $\mathcal{F}[f]$ depends in the general case on the values of $f(x)$ for all x within some range (the range depends on how $\mathcal{F}[f]$ is defined). The functional $\mathcal{F}[f]$ is in a way a generalization of $F(f(x_1), f(x_2), \ldots, f(x_l))$ since one goes from a discrete number of points to a continuum of points, whereby the sum in Equation 9.145 is replaced by an integral in Equation 9.144.[49]

[49] As pointed out in Footnote 46, the unit of $\delta\mathcal{F}/\delta f(x)$ is not equal to the unit of $\mathcal{F}/f(x)$; here there is a factor of inverse length in the unit of the former. In contrast, $\partial F/\partial f(x_\nu)$ and $F/f(x_\nu)$ have the same units.

When we make a finite variation $y_\nu \to y_\nu + \Delta y_\nu$ for the function F, we have the Taylor expansion

$$\begin{aligned}\Delta F &= F(y_1 + \Delta y_1, \ldots, y_l + \Delta y_l) - F(y_1, \ldots, y_l) \\ &= \sum_\nu \frac{\partial F}{\partial y_\nu}\Delta y_\nu + \frac{1}{2!}\sum_{\nu,\nu'} \frac{\partial^2 F}{\partial y_\nu \partial y_{\nu'}}\Delta y_\nu \Delta y_{\nu'} + \ldots.\end{aligned}$$

By inserting $y_\nu = f(x_\nu)$ we can write this as

$$\begin{aligned}\Delta F &= F\left(f(x_1) + \Delta f(x_1), \ldots, f(x_l) + \Delta f(x_l)\right) - F\left(f(x_1), \ldots, f(x_l)\right) \\ &= \sum_\nu \frac{\partial F}{\partial f(x_\nu)}\Delta f(x_\nu) + \frac{1}{2!}\sum_{\nu,\nu'} \frac{\partial^2 F}{\partial f(x_\nu)\partial f(x_{\nu'})}\Delta f(x_\nu)\Delta f(x_{\nu'}) + \ldots,\end{aligned}$$

where the values for the derivatives are evaluated at $y_\nu = f(x_\nu)$, that is, before the change by Δf.

Analogously, for the functional $\mathcal{F}[f]$ with a finite variation $f(x) \to f(x) + \Delta f(x)$, we have the **functional Taylor expansion**

$$\begin{aligned}\Delta\mathcal{F} &= \mathcal{F}[f + \Delta f] - \mathcal{F}[f] \qquad (9.146)\\ &= \int dx \frac{\delta\mathcal{F}}{\delta f(x)}\Delta f(x) + \frac{1}{2!}\int dx\, dx' \frac{\delta^2\mathcal{F}}{\delta f(x)\delta f(x')}\Delta f(x)\Delta f(x') + \ldots,\end{aligned}$$

where the values of the functional derivatives are evaluated using the function $f(x)$ before the change by $\Delta f(x)$.

Formally, functional derivatives and functional Taylor expansions can be defined as follows. Consider the functional $\mathcal{F}[f + \xi s]$, where $s = s(x)$ is an arbitrary function and ξ is a number. Since $\mathcal{F}[f + \xi s]$ is an ordinary function of ξ we can take the derivative

$$\left.\frac{d\mathcal{F}[f + \xi s]}{d\xi}\right|_{\xi=0} = \lim_{\xi\to 0}\frac{\mathcal{F}[f + \xi s] - \mathcal{F}[f]}{\xi}$$

and likewise the second and higher derivatives. We thereby have the ordinary Taylor expansion for finite values of ξ

$$\Delta\mathcal{F} = \mathcal{F}[f + \xi s] - \mathcal{F}[f] = \left.\frac{d\mathcal{F}[f + \xi' s]}{d\xi'}\right|_{\xi'=0}\xi + \frac{1}{2!}\left.\frac{d^2\mathcal{F}[f + \xi' s]}{d\xi'^2}\right|_{\xi'=0}\xi^2 + \ldots.$$

We can define the functional derivatives from these derivatives in the following manner[50]

$$\left.\frac{d\mathcal{F}[f+\xi' s]}{d\xi'}\right|_{\xi'=0} = \int dx\, \frac{\delta\mathcal{F}[f]}{\delta f(x)} s(x),$$
$$\left.\frac{d^2\mathcal{F}[f+\xi' s]}{d\xi'^2}\right|_{\xi'=0} = \int dx\, dx'\, \frac{\delta^2\mathcal{F}[f]}{\delta f(x')\delta f(x)} s(x)s(x')$$

and likewise for higher derivatives

$$\left.\frac{d^l\mathcal{F}[f+\xi' s]}{d\xi'^l}\right|_{\xi'=0} = \int dx_1 \dots dx_l\, \frac{\delta^l\mathcal{F}[f]}{\delta f(x_1)\dots\delta f(x_l)} s(x_1)\dots s(x_l).$$

The functional Taylor expansion (9.146) is obtained by applying these formulas by inserting $\xi s(x) = \Delta f(x)$ with finite ξ values.

The relationships between functional derivatives of various orders are, for instance,

$$\frac{\delta^2\mathcal{F}}{\delta f(x_1)\delta f(x_2)} = \frac{\delta\left(\frac{\delta\mathcal{F}}{\delta f(x_1)}\right)}{\delta f(x_2)},$$

(compare with Equations 9.117 and 9.127) and likewise

$$\frac{\delta^l\mathcal{F}}{\delta f(x_1)\dots\delta f(x_l)} = \frac{\delta\left(\frac{\delta^{l-1}\mathcal{F}}{\delta f(x_1)\dots\delta f(x_{l-1})}\right)}{\delta f(x_l)} \tag{9.147}$$

for any $l \geq 2$.

It is straightforward to generalize these formulas to functionals defined for functions in three-dimensional space. For example, the functional Taylor expansion of a functional $\mathcal{F}[f]$ when the function $f = f(\mathbf{r})$ can vary between two "states" $f_{\mathrm{I}}(\mathbf{r})$ and $f_{\mathrm{II}}(\mathbf{r})$ is given by

$$\begin{aligned}\Delta\mathcal{F}[f] &= \mathcal{F}[f_{\mathrm{II}}] - \mathcal{F}[f_{\mathrm{I}}] \qquad (9.148)\\ &= \sum_{l=1}^{\infty}\frac{1}{l!}\int d\mathbf{r}_1\dots d\mathbf{r}_l \left.\frac{\delta^l\mathcal{F}}{\delta f(\mathbf{r}_1)\dots\delta f(\mathbf{r}_l)}\right|_{f=f_{\mathrm{I}}} \Delta f(\mathbf{r}_1)\dots\Delta f(\mathbf{r}_l),\end{aligned}$$

where $\Delta f(\mathbf{r}) = f_{\mathrm{II}}(\mathbf{r}) - f_{\mathrm{I}}(\mathbf{r})$.

A functional Taylor expansion, like in Equation 9.148, constitute one way to obtain differences of functionals between two states like in

[50]The definition can be understood as follows. The derivative $d\mathcal{F}[f+\xi s]/d\xi$ is a linear functional of $s(x)$ for infinitesimally small ξ values. Rietz' representation theorem in functional analysis says (under quite general conditions that are fulfilled in all cases we will consider) that for each linear functional of $s(x)$ there exists a unique (generalized) function $D(x)$ such that the functional can be written in the form $\int dx\, D(x)s(x)$ (see, for example, I. Stakgold, 1967, *Boundary Value Problems of Mathematical Physics*, Vol. 1, New York: Macmillan). Therefore, $d\mathcal{F}[f+\xi s]/d\xi = \int dx\, D(x)s(x)$ in the limit $\xi \to 0$ and the function $D(x)$ is, by definition, the functional derivative $\delta\mathcal{F}/\delta f(x)$.

$\Delta\mathcal{F}[f] = \mathcal{F}[f_{\rm II}] - \mathcal{F}[f_{\rm I}]$. Another way to obtain $\Delta\mathcal{F}[f]$ is by integration of $d\mathcal{F}/d\xi$, a so-called **functional integration**. To illustrate this concept we start with a practical example and then we will give the general formulas.

Example 9.C: An example of a functional integration

Our task is to obtain the change in grand potential Θ when the external potential $v(\mathbf{r})$ is changed from $v_{\rm I}(\mathbf{r})$ to $v_{\rm II}(\mathbf{r})$, that is, $\Delta\Theta = \Theta_{\rm II} - \Theta_{\rm I}$. The change $d\Theta$ when we do a small variation δv in potential is given by

$$d\Theta = \int d\mathbf{r}\, \frac{\delta\Theta}{\delta v(\mathbf{r})}\delta v(\mathbf{r}) = \int d\mathbf{r}\, n(\mathbf{r})\delta v(\mathbf{r})$$

(from Equations 9.101 and 9.102). Since a finite change $\Delta v(\mathbf{r})$ in potential can be obtained from a series of steps with small changes $\delta v(\mathbf{r})$, we can add the small changes $d\Theta$ for every step to obtain $\Delta\Theta$. The equation above is valid only for the first step since $n(\mathbf{r})$ is the density *before* the change in potential. To obtain $\Delta\Theta$, the density $n(\mathbf{r})$ to be used in each step must be for the system at the beginning of the step; we have to let $n(\mathbf{r})$ to change along the path we follow. This is done by means of a functional integration (coupling-parameter integration) where we use the parameter ξ that varies between 0 and 1.

We have $\Delta v(\mathbf{r}) = v_{\rm II}(\mathbf{r}) - v_{\rm I}(\mathbf{r})$ and we let the potential for $0 \le \xi \le 1$ be equal to

$$v(\mathbf{r};\xi) = v_{\rm I}(\mathbf{r}) + \xi\Delta v(\mathbf{r}) = \xi v_{\rm II}(\mathbf{r}) + (1-\xi)v_{\rm I}(\mathbf{r}) \qquad (9.149)$$

so $v(\mathbf{r};0) = v_{\rm I}(\mathbf{r})$ and $v(\mathbf{r};1) = v_{\rm II}(\mathbf{r})$. We likewise let $\Theta(\xi)$ and $n(\mathbf{r};\xi)$ denote the grand potential and density, respectively, for the intermediate stages, for instance, $n(\mathbf{r};\xi)$ is the density of the system in the presence of the external potential $v(\mathbf{r};\xi)$.

When the potential is $v(\mathbf{r};\xi)$ we have

$$d\Theta(\xi) = \int d\mathbf{r}\, n(\mathbf{r};\xi)\delta v(\mathbf{r};\xi),$$

where $d\Theta(\xi)$ and $\delta v(\mathbf{r};\xi)$ are the variations when $\xi \to \xi + d\xi$, that is,

$$d\Theta(\xi) = \frac{d\Theta(\xi)}{d\xi}d\xi \quad \text{and} \quad \delta v(\mathbf{r};\xi) = \frac{\partial v(\mathbf{r};\xi)}{\partial\xi}d\xi.$$

We can now obtain $\Delta\Theta$ simply by an integration over ξ from 0 to 1, which yields

$$\Delta\Theta = \int_0^1 d\xi \int d\mathbf{r}\, n(\mathbf{r};\xi)\frac{\partial v(\mathbf{r};\xi)}{\partial\xi} \qquad (9.150)$$

and since $\partial v(\mathbf{r};\xi)/\partial\xi = \Delta v(\mathbf{r})$ we obtain

$$\Delta\Theta = \int_0^1 d\xi \int d\mathbf{r}\, n(\mathbf{r};\xi)\Delta v(\mathbf{r}). \tag{9.151}$$

It is, however, not necessary to vary $v(\mathbf{r})$ is a linear manner with ξ like in Equation 9.149. In fact $v(\mathbf{r};\xi)$ can be any continuous differentiable function of ξ that satisfies

$$v(\mathbf{r};\xi) = \begin{cases} v_{\mathrm{I}}(\mathbf{r}), & \text{when } \xi = 0 \\ v_{\mathrm{II}}(\mathbf{r}) & \text{when } \xi = 1 \end{cases} \tag{9.152}$$

and $\Delta\Theta$ is then given by Equation 9.150. The result in Equation 9.151 is valid only for the linear variation of $v(\mathbf{r};\xi)$.

Let us now obtain $\Delta\mathcal{F}[f] = \mathcal{F}[f_{\mathrm{II}}] - \mathcal{F}[f_{\mathrm{I}}]$ in the general case and we first set $f(\mathbf{r};\xi) = f_{\mathrm{I}}(\mathbf{r}) + \xi\Delta f(\mathbf{r})$ where $\Delta f(\mathbf{r}) = f_{\mathrm{II}}(\mathbf{r}) - f_{\mathrm{I}}(\mathbf{r})$. We obtain

$$\begin{aligned} \Delta\mathcal{F} &= \mathcal{F}[f_{\mathrm{I}} + \Delta f] - \mathcal{F}[f_{\mathrm{I}}] = \int_0^1 d\xi \frac{d\mathcal{F}[f_{\mathrm{I}} + \xi\Delta f]}{d\xi} \\ &= \int_0^1 d\xi \int d\mathbf{r}\, \left.\frac{\delta\mathcal{F}[f]}{\delta f(\mathbf{r})}\right|_{f=f_{\mathrm{I}}+\xi\Delta f} \Delta f(\mathbf{r}). \end{aligned} \tag{9.153}$$

Incidentally, we may note that this is the same kind of integration as the one we encountered when we performed coupling-parameter integrations in Volume I, for example in Section 5.7 when we calculated the chemical potential (Equations I:5.83 and I:5.84).

As in the example, we can alternatively use $f = f(\mathbf{r};\xi)$ for any continuous and differentiable function of ξ that satisfies

$$f(\mathbf{r};\xi) = \begin{cases} f_{\mathrm{I}}(\mathbf{r}), & \text{when } \xi = 0 \\ f_{\mathrm{II}}(\mathbf{r}) & \text{when } \xi = 1, \end{cases} \tag{9.154}$$

whereby we have

$$\Delta\mathcal{F} = \int_0^1 d\xi \int d\mathbf{r}\, \left.\frac{\delta\mathcal{F}[f]}{\delta f(\mathbf{r})}\right|_{f=f(\mathbf{r};\xi)} \frac{\partial f(\mathbf{r};\xi)}{\partial\xi}. \tag{9.155}$$

This is a very useful formula.

9.2.3 Higher order direct-correlation functions $c^{(m)}$ and their physical meanings

In Section 9.2.1 we saw that the direct-correlation functions $c^{(1)}$ and $c^{(2)}$ can be obtained by functional differentiation of the excess part $\mathcal{A}^{\text{ex}}$ of the intrinsic Helmholtz free energy with respect to the density distribution $n(\mathbf{r})$ (Equations 9.126 and 9.127)

$$\begin{aligned} c^{(1)}(\mathbf{r}_1) &= -\beta\frac{\delta\mathcal{A}^{\text{ex}}}{\delta n(\mathbf{r}_1)} \\ c^{(2)}(\mathbf{r}_1,\mathbf{r}_2) &= -\beta\frac{\delta^2\mathcal{A}^{\text{ex}}}{\delta n(\mathbf{r}_1)\delta n(\mathbf{r}_2)}. \end{aligned}$$

In the same manner we can define higher order direct-correlation functions $c^{(m)}$ from

$$c^{(m)}(\mathbf{r}_1,\mathbf{r}_2,\ldots,\mathbf{r}_m) = -\beta\frac{\delta^m\mathcal{A}^{\text{ex}}}{\delta n(\mathbf{r}_1)\delta n(\mathbf{r}_2)\ldots\delta n(\mathbf{r}_m)}, \tag{9.156}$$

which are related to each other via [cf. Equation 9.147]

$$c^{(m)}(\mathbf{r}_1,\mathbf{r}_2,\ldots,\mathbf{r}_m) = \frac{\delta c^{(m-1)}(\mathbf{r}_1,\mathbf{r}_2,\ldots,\mathbf{r}_{m-1})}{\delta n(\mathbf{r}_m)} = \frac{\delta^{m-1}c^{(1)}(\mathbf{r}_1)}{\delta n(\mathbf{r}_2)\ldots\delta n(\mathbf{r}_m)} \tag{9.157}$$

The order of differentiation does not matter and $c^{(m)}(\mathbf{r}_1,\mathbf{r}_2,\ldots,\mathbf{r}_m)$ is symmetric with respect to permutations of its arguments $\mathbf{r}_1,\mathbf{r}_2,\ldots,\mathbf{r}_m$. We will now show how these functions can be used in liquid state theory and what physical meanings they have.

In Section 9.1.3.1 we considered nonlinear contributions to $\Delta c^{(1)}$ from a density change Δn as induced by a variation in external potential, Δv. We saw that we need to obtain $\Delta c^{(1)}$ in order to calculate density distributions (singlet distribution functions) and pair distributions functions, Equations 9.53 and 9.54 respectively, where $\Delta c^{(1)}(\mathbf{r}) = c^{(1)}(\mathbf{r}) - c^{(1)\text{bulk}}$ and $\Delta c^{(1)}(\mathbf{r}|\mathbf{r}') = c^{(1)}(\mathbf{r}|\mathbf{r}') - c^{(1)}(\mathbf{r})$ in the two cases. In the first case the density change is induced by exposing a bulk fluid to an external potential and in the second case it is induced by placing a particle at $\mathbf{r}'$ in an inhomogeneous fluid. Here, we are going to investigate the nonlinear response in more detail.

We saw in Equation 9.57, that in general

$$\Delta c^{(1)}(\mathbf{r}) = \int d\mathbf{s}_1\, c^{(2)}(\mathbf{r},\mathbf{s}_1)\Delta n(\mathbf{s}_1) + B(\mathbf{r}), \tag{9.158}$$

where $\mathbf{s}_1 = (x_1, y_1, z_1)$ and $B(\mathbf{r})$ denotes "terms that depends nonlinearly on Δn" in Equation 9.57, that is, terms that are nonlinear functionals of Δn (we saw particular examples of $B(\mathbf{r})$ in Equations 9.58 and 9.62). Remember that $c^{(2)}$ is the pair direct-correlation function for the fluid *before* the change Δn in density. Let us first interpret the first term in the rhs of Equation 9.158.

The product $d\mathbf{s}_1\Delta n(\mathbf{s}_1)$ in the integrand is the change in average number of particles in the volume element $d\mathbf{s}_1$ around point $\mathbf{s}_1$ due to the variation Δn in density and $c^{(2)}(\mathbf{r},\mathbf{s}_1) \equiv \delta c^{(1)}(\mathbf{r})/\delta n(\mathbf{s}_1)$ describes how strongly this variation affects $c^{(1)}$ at point $\mathbf{r}$ (i.e., the influence on the intrinsic excess chemical potential). Mathematically this is expressed by

$$\left\{\text{contribution from } d\mathbf{s}_1\right\} = \frac{\delta c^{(1)}(\mathbf{r})}{\delta n(\mathbf{s}_1)}\Delta n(\mathbf{s}_1)d\mathbf{s}_1.$$

We know that $\delta c^{(1)}(\mathbf{r}) = \int d\mathbf{r}\, c^{(2)}(\mathbf{r},\mathbf{s}_1)\delta n(\mathbf{s}_1)$ (Equation 9.56), so this expression is correct for infinitesimally small variations δn in density (then the response is linear), but it is not correct for finite variations Δn. The problem in the latter case is that when we sum over all points $\mathbf{s}_1$, it is assumed that the contribution from each volume element is independent of that from the other volume elements. When there is a finite change in density in $d\mathbf{s}_1$ and also in another volume element $d\mathbf{s}_2$, the two contributions are simply added to each other as if the density change at one place does not affect the contribution from the other place. This is not correct and there is a contribution to $\Delta c^{(1)}$ from $d\mathbf{s}_1$ and $d\mathbf{s}_2$ that is not included. We can account for this by making the correction that $\delta c^{(1)}(\mathbf{r})/\delta n(\mathbf{s}_1) \equiv c^{(2)}(\mathbf{r},\mathbf{s}_1)$ varies when the density is changed at $\mathbf{s}_2$, that is, by considering

$$\frac{\delta}{\delta n(\mathbf{s}_2)}\left[c^{(2)}(\mathbf{r},\mathbf{s}_1)\right] = \frac{\delta}{\delta n(\mathbf{s}_2)}\left[\frac{\delta c^{(1)}(\mathbf{r})}{\delta n(\mathbf{s}_1)}\right] = \frac{\delta^2 c^{(1)}(\mathbf{r})}{\delta n(\mathbf{s}_2)\delta n(\mathbf{s}_1)} \equiv c^{(3)}(\mathbf{r},\mathbf{s}_1,\mathbf{s}_2).$$

We can thereby add the following to $\Delta c^{(1)}$

$$\left\{\text{additional contribution from } d\mathbf{s}_1 \text{ and } d\mathbf{s}_2\right\}$$
$$= \frac{\delta^2 c^{(1)}(\mathbf{r})}{\delta n(\mathbf{s}_2)\delta n(\mathbf{s}_1)}\left(\Delta n(\mathbf{s}_1)d\mathbf{s}_1\right)\left(\Delta n(\mathbf{s}_2)d\mathbf{s}_2\right).$$

Thus we have a new term that depends nonlinearly on Δn and we obtain

$$\Delta c^{(1)}(\mathbf{r}) \approx \int d\mathbf{s}_1\, c^{(2)}(\mathbf{r},\mathbf{s}_1)\Delta n(\mathbf{s}_1) + \frac{1}{2}\int d\mathbf{s}_1 d\mathbf{s}_2\, c^{(3)}(\mathbf{r},\mathbf{s}_1,\mathbf{s}_2)\Delta n(\mathbf{s}_1)\Delta n(\mathbf{s}_2)$$

where the factor 1/2 is included to correct for double counting.

However, this does not consider that there is a finite change in density at third volume element $d\mathbf{s}_3$ that affects the contributions from $d\mathbf{s}_1$ and $d\mathbf{s}_2$. This is taken into account by considering

$$\frac{\delta}{\delta n(\mathbf{s}_3)}\left[c^{(3)}(\mathbf{r},\mathbf{s}_1,\mathbf{s}_2)\right] = \frac{\delta^3 c^{(1)}(\mathbf{r})}{\delta n(\mathbf{s}_3)\delta n(\mathbf{s}_2)\delta n(\mathbf{s}_1)} \equiv c^{(4)}(\mathbf{r},\mathbf{s}_1,\mathbf{s}_2,\mathbf{s}_3)$$

and we have

$$\left\{\text{additional contribution from } d\mathbf{s}_1, d\mathbf{s}_2 \text{ and } d\mathbf{s}_3\right\}$$
$$= \frac{\delta^3 c^{(1)}(\mathbf{r})}{\delta n(\mathbf{s}_3)\delta n(\mathbf{s}_2)\delta n(\mathbf{s}_1)}\left(\Delta n(\mathbf{s}_1)d\mathbf{s}_1\right)\left(\Delta n(\mathbf{s}_2)d\mathbf{s}_2\right)\left(\Delta n(\mathbf{s}_3)d\mathbf{s}_3\right).$$

This obviously continues for ever and in the end, we have

$$\Delta c^{(1)}(\mathbf{r}) = \int d\mathbf{s}_1\, c^{(2)}(\mathbf{r},\mathbf{s}_1)\Delta n(\mathbf{s}_1) + \sum_{m=3}^{\infty}\frac{1}{(m-1)!} \times \int d\mathbf{s}_1 \dots d\mathbf{s}_{m-1}\, c^{(m)}(\mathbf{r},\mathbf{s}_1,\dots,\mathbf{s}_{m-1})\Delta n(\mathbf{s}_1)\dots\Delta n(\mathbf{s}_{m-1}), \tag{9.159}$$

where $1/(m-1)!$ corrects for multiple counting and $c^{(m)}(\mathbf{r},\mathbf{s}_1,\dots,\mathbf{s}_{m-1})$ is the direct correlation function given by Equations 9.156 and 9.157. In the expansion, $c^{(m)}$ for $m \geq 2$ are all evaluated for the system *before* the change in density. In most cases, the high order direct-correlation functions, $m \geq 3$, have a short range; increasingly shorter when m increases. They have in general a shorter range than the pair correlations functions when the coordinate points of the function argument are moved away from each other. We can recognize the expansion 9.159 as the functional Taylor expansion of $c^{(1)}(\mathbf{r})$ in $\Delta n(\mathbf{s})$, that is, Equation 9.148 with $\mathcal{F}[f] = c^{(1)}(\mathbf{r};[n])$.

The sum $\sum_{m=3}^{\infty}$ in Equation 9.159 equals $B(\mathbf{r})$ of Equation 9.158. This is the **bridge function** in

$$\Delta c^{(1)}(\mathbf{r}_1) = \int d\mathbf{r}_3\, c^{(2)\mathrm{b}}(\mathbf{r}_1,\mathbf{r}_3)\Delta n(\mathbf{r}_3) + b^{(1)}(\mathbf{r}_1),$$

$$\Delta c^{(1)}(\mathbf{r}_1|\mathbf{r}_2) = \int d\mathbf{r}_3\, c^{(2)}(\mathbf{r}_1,\mathbf{r}_3)\Delta n(\mathbf{r}_3|\mathbf{r}_2) + b^{(2)}(\mathbf{r}_1,\mathbf{r}_2),$$

(Equations 9.58 and 9.62), where in the former case all $c^{(m)}$ are evaluated for the bulk fluid and in the latter case they are evaluated for the density distribution before a new particle that causes the change Δn in density is inserted. From the discussion here and in Section 9.1.3 we can understand that the bridge function expresses how much the change in chemical potential at $\mathbf{r}$ is affected by the *interdependence* of influence from density changes at the various points in the neighborhood of $\mathbf{r}$. It is this interdependence that is ignored in the HNC approximation where one sets $B = 0$. In other approximations it is modeled in various manners by using some approximate B.

Explicitly, the bridge function in Equation 9.62 and hence in Equation 9.67 is given by

$$b^{(2)}(\mathbf{r},\mathbf{r}') = \sum_{m=3}^{\infty}\frac{1}{(m-1)!}\int d\mathbf{s}_1 \dots d\mathbf{s}_{m-1}\, c^{(m)}(\mathbf{r},\mathbf{s}_1,\dots,\mathbf{s}_{m-1}) \times n(\mathbf{s}_1)h^{(2)}(\mathbf{s}_1,\mathbf{r}')\dots n(\mathbf{s}_{m-1})h^{(2)}(\mathbf{s}_{m-1},\mathbf{r}'), \tag{9.160}$$

as obtained from Equation 9.159 where we have inserted

$$\Delta n(\mathbf{s}_\nu|\mathbf{r}') = n(\mathbf{s}_\nu)h^{(2)}(\mathbf{s}_\nu,\mathbf{r}')$$

(cf. Equation 9.33). As we know, the bridge function is symmetric, $b^{(2)}(\mathbf{r},\mathbf{r}') = b^{(2)}(\mathbf{r}',\mathbf{r})$, although this cannot easily be seen from this expression. Equation 9.160 is of interest theoretically, but is not very useful as it stands for practical numerical calculations.

It is possible to write each function $c^{(m)}$ as an infinite sum of terms involving n and $h^{(2)}$ (virtually all terms are cluster integrals). For instance, the function $c^{(3)}$ can be written

$$c^{(3)}(\mathbf{r},\mathbf{s}_1,\mathbf{s}_2) = h^{(2)}(\mathbf{r},\mathbf{s}_1)h^{(2)}(\mathbf{r},\mathbf{s}_2)h^{(2)}(\mathbf{s}_1,\mathbf{s}_2) + \text{ more complicated terms}$$

and inserted in Equation 9.160 this gives rise to the simplest term in $b^{(2)}$

$$\begin{aligned} b^{(2)}(\mathbf{r},\mathbf{r}') &= \frac{1}{2}\int d\mathbf{s}_1 d\mathbf{s}_2\, h^{(2)}(\mathbf{r},\mathbf{s}_1)h^{(2)}(\mathbf{r},\mathbf{s}_2)n(\mathbf{s}_1)h^{(2)}(\mathbf{s}_1,\mathbf{s}_2) \\ &\times\ n(\mathbf{s}_2)h^{(2)}(\mathbf{s}_1,\mathbf{r}')h^{(2)}(\mathbf{s}_2,\mathbf{r}') + \text{ more complicated terms.} \end{aligned} \quad (9.161)$$

This integral is an example of a cluster integral and it can pictorially be represented by a diagram

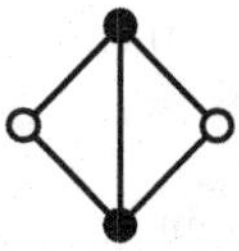

where the open circles represent the points $\mathbf{r}$ and $\mathbf{r}'$, each filled circles represents a point $\mathbf{s}_\nu$ including a factor $n(\mathbf{s}_\nu)$, the lines represent $h^{(2)}$ and integration is performed over the whole space for the filled points. The full expansion of the bridge function in terms of cluster integrals involving n and $h^{(2)}$ was derived by Morita and Hiroike in 1960 and will not be given here.[51]

{**Exercise 9.5:** Use a functional Taylor expansion according to Equation 9.148 to write the difference in excess intrinsic free energy $\mathcal{A}^{\text{ex}}$ for a bulk fluid with density n^{b} and an inhomogeneous fluid with density $n(\mathbf{r})$ in terms of the difference in density $\Delta n(\mathbf{r}) = n(\mathbf{r}) - n^{\text{b}}$. Thereby, use $\Delta f(\mathbf{r}_\nu) = \Delta n(\mathbf{r}_\nu)$ in Equation 9.148 and the functional derivatives of $\mathcal{A}^{\text{ex}}$ according to Equation 9.156.}

9.2.4 Functional integrations involving distribution functions and direct correlation functions

9.2.4.1 Some free energy relationships for distribution functions

In the grand canonical ensemble, the number density distribution $n^{(1)}(\mathbf{r}) \equiv n(\mathbf{r})$ for an open system is given by

$$n^{(1)}(\mathbf{r}) = \sum_{N=1}^{\infty} \mathscr{P}(N) n_N^{(1)}(\mathbf{r}) \quad (9.162)$$

[51]T. Morita and K. Hiroike, Progr. Theor. Phys. **23** (1960) 1003 (https://doi.org/10.1143/PTP.23.1003).

(Equation I:5.104), where $\mathscr{P}(N)$ is the probability that the system contains N particles and $n_N^{(1)}(\mathbf{r})$ is the canonical density distribution function for a system with N particles. The latter is equal to the canonical ensemble average

$$n_N^{(1)}(\mathbf{r}) = \left\langle \breve{n}^{(1)}_{\{\mathbf{r}^N\}}(\mathbf{r}) \right\rangle \tag{9.163}$$

of the **microscopic density distribution**

$$\breve{n}^{(1)}_{\{\mathbf{r}^N\}}(\mathbf{r}) \equiv \sum_{\nu=1}^{N} \delta^{(3)}(\mathbf{r}-\mathbf{r}_\nu) \tag{9.164}$$

(microscopic *number density* distribution), where $\mathbf{r}^N \equiv \mathbf{r}_1, \mathbf{r}_2, \ldots, \mathbf{r}_N$ (see Section 5.8 in Volume I, Equations I:5.89-90).

Let us recapitulate from Volume I that the canonical ensemble average of a function $f(\mathbf{r}^N)$ is

$$\langle f(\mathbf{r}^N) \rangle = \int d\mathbf{r}^N f(\mathbf{r}^N) \mathcal{P}_N^{(N)}(\mathbf{r}^N) \tag{9.165}$$

(Equation I:3.14), where $d\mathbf{r}^N \equiv d\mathbf{r}_1 d\mathbf{r}_2 \ldots d\mathbf{r}_N$, $\mathcal{P}_N^{(N)}$ is the configurational probability density

$$\mathcal{P}_N^{(N)}(\mathbf{r}^N) = \frac{e^{-\beta \breve{U}_N^{\mathrm{pot}}(\mathbf{r}^N)}}{Z_N}, \tag{9.166}$$

Z_N is the configurational partition function $Z_N = \int d\mathbf{r}^N \, e^{-\beta \breve{U}_N^{\mathrm{pot}}(\mathbf{r}^N)}$ (Equation I:3.9 - 10) and the potential energy for the N particles is

$$\breve{U}_N^{\mathrm{pot}}(\mathbf{r}^N) = \breve{U}_N^{\mathrm{intr}}(\mathbf{r}^N) + \sum_{\nu=1}^{N} v(\mathbf{r}_\nu), \tag{9.167}$$

where $\breve{U}_N^{\mathrm{intr}}$ is the intrinsic part of $\breve{U}_N^{\mathrm{pot}}$, that is, the interaction energy between the particles with configuration $\mathbf{r}^N$. Furthermore, we have from Volume I

$$\mathscr{P}(N) = \frac{\zeta^N Z_N}{N!\,\Xi} \tag{9.168}$$

(Equation I:3.38), where Ξ is the grand canonical partition function

$$\Xi = \sum_{N=0}^{\infty} \frac{\zeta^N Z_N}{N!} = \sum_{N=0}^{\infty} \frac{\zeta^N}{N!} \int d\mathbf{r}^N \, e^{-\beta \breve{U}_N^{\mathrm{pot}}(\mathbf{r}^N)} \tag{9.169}$$

(Equation I:3.37). We have $Z_0 = 1$ and the first term ($N = 0$) in the rhs of this equation does not include any integration.

We will now evaluate some functional derivatives of the grand potential $\Theta = -k_B T \ln \Xi$ and we start with

$$\frac{\delta \Theta}{\delta v(\mathbf{r})} = -\frac{k_B T}{\Xi} \frac{\delta \Xi}{\delta v(\mathbf{r})} = \frac{1}{\Xi} \sum_{N=1}^{\infty} \frac{\zeta^N}{N!} \int d\mathbf{r}^N \, e^{-\beta \breve{U}_N^{\mathrm{pot}}(\mathbf{r}^N)} \frac{\delta \breve{U}_N^{\mathrm{pot}}(\mathbf{r}^N)}{\delta v(\mathbf{r})}$$

(the sum starts with $N = 1$ because the $N = 0$ term in Ξ is constant so its derivative is zero). It is convenient to write the potential energy in Equation 9.167 as

$$\breve{U}_N^{\text{pot}}(\mathbf{r}^N) = \int d\mathbf{r}\, v(\mathbf{r})\breve{n}^{(1)}_{\{\mathbf{r}^N\}}(\mathbf{r}) + \breve{U}_N^{\text{intr}}(\mathbf{r}^N)$$

and it follows that

$$\frac{\delta \breve{U}_N^{\text{pot}}(\mathbf{r}^N)}{\delta v(\mathbf{r})} = \int d\mathbf{r}'\, \delta^{(3)}(\mathbf{r}' - \mathbf{r})\breve{n}^{(1)}_{\{\mathbf{r}^N\}}(\mathbf{r}') = \breve{n}^{(1)}_{\{\mathbf{r}^N\}}(\mathbf{r}).$$

We therefore obtain using Equations 9.165 - 9.168

$$\frac{\delta\Theta}{\delta v(\mathbf{r})} = \sum_{N=1}^{\infty} \mathscr{P}(N) \int d\mathbf{r}^N\, \mathcal{P}_N^{(N)}(\mathbf{r}^N)\breve{n}^{(1)}_{\{\mathbf{r}^N\}}(\mathbf{r}) = \sum_{N=1}^{\infty} \mathscr{P}(N) \left\langle \breve{n}^{(1)}_{\{\mathbf{r}^N\}}(\mathbf{r}) \right\rangle,$$

so we have from Equations 9.162 and 9.163

$$\frac{\delta\Theta}{\delta v(\mathbf{r})} = n^{(1)}(\mathbf{r}). \tag{9.170}$$

We obtained this relationship earlier in a different manner, see Equations 9.34, 9.101 and 9.102, but the present derivation is more strict than the previous one. Note that Equation 9.170 is valid irrespectively of the values of $\breve{U}_N^{\text{intr}}$, so we have, for example, not assumed that the interactions between the particles in the system are pair-wise additive.

Equation 9.170 can be used to obtain the change in grand potential for finite variations in external potential by means of functional integration (see also Example 9.C). Consider a system exposed to an external potential that is changed from $v_{\text{I}}(\mathbf{r})$ to $v_{\text{II}}(\mathbf{r})$ whereby the grand potential changes by $\Delta\Theta = \Theta_{\text{II}} - \Theta_{\text{I}}$. From Equation 9.155 with $f = v$ and $\mathcal{F}[f] = \Theta[v]$ it follows that

$$\Delta\Theta = \int_0^1 d\xi \int d\mathbf{r}\, \left.\frac{\delta\Theta}{\delta v(\mathbf{r})}\right|_{v=v(\mathbf{r};\xi)} \frac{\partial v(\mathbf{r};\xi)}{\partial \xi} = \int_0^1 d\xi \int d\mathbf{r}\, n^{(1)}(\mathbf{r};\xi)\frac{\partial v(\mathbf{r};\xi)}{\partial \xi}$$

where $n^{(1)}(\mathbf{r};\xi) = n^{(1)}(\mathbf{r})\big|_{v=v(\mathbf{r};\xi)}$ and we have used Equation 9.170. For the special case $v(\mathbf{r};\xi) = v_{\text{I}}(\mathbf{r}) + \xi[v_{\text{II}}(\mathbf{r}) - v_{\text{I}}(\mathbf{r})]$, where v depends linearly on ξ, this becomes

$$\Theta_{\text{II}} = \Theta_{\text{I}} + \int d\mathbf{r}\, [v_{\text{II}}(\mathbf{r}) - v_{\text{I}}(\mathbf{r})] \int_0^1 d\xi\, n^{(1)}(\mathbf{r};\xi). \tag{9.171}$$

In order to use these results one must be able to calculate the density distribution $n^{(1)}(\mathbf{r};\xi)$ for all ξ when $0 \le \xi \le 1$, at least approximately.

Next, we will consider the pair density-distribution function, which in the grand canonical ensemble is given by

$$n^{(2)}(\mathbf{r},\mathbf{r}') = \sum_{N=2}^{\infty} \mathscr{P}(N) n_N^{(2)}(\mathbf{r},\mathbf{r}') \tag{9.172}$$

(Equation I:5.103), where $n_N^{(2)}$ is the canonical pair density distribution

$$n_N^{(2)}(\mathbf{r},\mathbf{r}') = \left\langle \breve{n}^{(2)}_{\{\mathbf{r}^N\}}(\mathbf{r},\mathbf{r}')\right\rangle \tag{9.173}$$

(Equation I:5.92) and

$$\breve{n}^{(2)}_{\{\mathbf{r}^N\}}(\mathbf{r},\mathbf{r}') \equiv \sum_{\nu=1}^{N} \sum_{\substack{\nu'=1\\(\nu'\neq\nu)}}^{N} \delta^{(3)}(\mathbf{r}-\mathbf{r}_\nu)\delta^{(3)}(\mathbf{r}'-\mathbf{r}_{\nu'}), \tag{9.174}$$

is the **microscopic pair density-distribution** function (Equation I:5.91).

Let us restrict ourselves to systems with pair-wise additive interaction potentials between the particles, so we have

$$\breve{U}_N^{\mathrm{pot}}(\mathbf{r}^N) = \sum_{\nu=1}^{N} v(\mathbf{r}_\nu) + \sum_{\nu=1}^{N} \sum_{\nu'=\nu+1}^{N} u(\mathbf{r}_\nu,\mathbf{r}_{\nu'})$$

(cf. Equation 9.1), which we can write as

$$\breve{U}_N^{\mathrm{pot}}(\mathbf{r}^N) = \int d\mathbf{r}\, v(\mathbf{r})\breve{n}^{(1)}_{\{\mathbf{r}^N\}}(\mathbf{r}) + \frac{1}{2}\int d\mathbf{r}d\mathbf{r}'\, \breve{n}^{(2)}_{\{\mathbf{r}^N\}}(\mathbf{r},\mathbf{r}')u(\mathbf{r},\mathbf{r}'). \tag{9.175}$$

The factor of 1/2 compensates for the double counting due to the second sum in Equation 9.174. When the pair potential is varied while $v(\mathbf{r})$ is unchanged it follows from Equation 9.175 that[52]

$$\frac{\delta \breve{U}_N^{\mathrm{pot}}(\mathbf{r}^N)}{\delta u(\mathbf{r},\mathbf{r}')} = \frac{1}{2}\breve{n}^{(2)}_{\{\mathbf{r}^N\}}(\mathbf{r},\mathbf{r}').$$

From the definition 9.169 of the grand canonical partition function we see that Ξ is a functional of both v and u via $\breve{U}_N^{\mathrm{pot}}$, which hence also applies to Θ, so we have $\Theta = \Theta[u,v]$.

Completely analogously to the derivation of Equation 9.170 we obtain[53]

$$\frac{\delta\Theta}{\delta u(\mathbf{r},\mathbf{r}')} = \frac{1}{2}\sum_{N=2}^{\infty} \mathscr{P}(N)\left\langle \breve{n}^{(2)}_{\{\mathbf{r}^N\}}(\mathbf{r},\mathbf{r}')\right\rangle$$

(the $N = 0$ and 1 terms in Ξ do not contain u so they give zero in $\delta\Theta/\delta u$) and we conclude using Equations 9.172 and 9.173 that

$$\frac{\delta\Theta}{\delta u(\mathbf{r},\mathbf{r}')} = \frac{n^{(2)}(\mathbf{r},\mathbf{r}')}{2}. \tag{9.176}$$

[52] We have used $\delta f(\mathbf{r},\mathbf{r}')/\delta f(\mathbf{r}'',\mathbf{r}''') = \delta^{(3)}(\mathbf{r}-\mathbf{r}'')\delta^{(3)}(\mathbf{r}'-\mathbf{r}''')$, which follows from the fact that $\delta f(\mathbf{r},\mathbf{r}') = \int d\mathbf{r}''d\mathbf{r}'''\delta^{(3)}(\mathbf{r}-\mathbf{r}'')\delta^{(3)}(\mathbf{r}'-\mathbf{r}''')\delta f(\mathbf{r}'',\mathbf{r}''')$.

[53] Note that ζ is fixed during the differentiation because μ and T are constant. In general, μ changes when the pair interactions are changed, for instance for a bulk phase with density n^{b}. In the present case with constant μ, a variation in the pair interactions implies a change in the density n^{b} for the bulk phase in equilibrium with the system. This is not the case when solely the external potential varies.

Thus, functional differentiations of Θ with respect to v and u give the singlet and pair density distribution functions, respectively. The pair function can also be obtained by taking the second derivative

$$\begin{aligned}\frac{\delta^2\Theta}{\delta v(\mathbf{r})\delta v(\mathbf{r}')} &= -\beta H^{(2)}(\mathbf{r},\mathbf{r}') \\ &= -\beta\left[n^{(1)}(\mathbf{r})\delta^{(3)}(\mathbf{r}'-\mathbf{r})+n^{(2)}(\mathbf{r},\mathbf{r}')-n^{(1)}(\mathbf{r})n^{(1)}(\mathbf{r}')\right],\end{aligned}$$

see Equations 9.111 and 9.117.

{**Exercise 9.6:** Use Equation 9.176 and functional integration to derive an expression for $\Delta\Theta = \Theta_{\text{II}} - \Theta_{\text{I}}$ when the pair potential varies from $u_{\text{I}}(\mathbf{r},\mathbf{r}')$ to $u_{\text{II}}(\mathbf{r},\mathbf{r}')$ while the external potential $v(\mathbf{r})$ is kept unchanged. The density distributions $n^{(1)}(\mathbf{r})$ and $n^{(2)}(\mathbf{r},\mathbf{r}')$ change when u is varied.}

The grand potential is given by

$$\Theta = \mathcal{A}^{ex} + \int d\mathbf{r}\, n(\mathbf{r})\left(k_BT\left[\ln(\Lambda^3 n^{(1)}(\mathbf{r}))-1\right]+v(\mathbf{r})-\mu\right). \tag{9.177}$$

(Equations 9.120 and 9.128). Let us investigate what happens with Θ and $\mathcal{A}^{ex}$ when u is varied. The pair potential u occurs explicitly in $\mathcal{A}^{ex}$, but not in the other terms. The latter depend only indirectly on u via the density distribution $n^{(1)}(\mathbf{r})$, which changes when u is varied. The u-dependence of the entire Θ via $n^{(1)}(\mathbf{r})$ is given by

$$\int d\mathbf{r}''\,\frac{\delta\Theta}{\delta n^{(1)}(\mathbf{r}'')}\cdot\frac{\delta n^{(1)}(\mathbf{r}'')}{\delta u(\mathbf{r},\mathbf{r}')} = 0,$$

where the integral is zero because $\delta\Theta/\delta n^{(1)}(\mathbf{r}'') = 0$ (from Equation 9.129). This means that the variation in $n^{(1)}(\mathbf{r})$ does not contribute to the variation $\delta\Theta$, so the only contributions to $\delta\Theta$ when u is varied arise from $\mathcal{A}^{ex}$ and can be evaluated as if $n^{(1)}(\mathbf{r})$ does not change. It therefore follows from Equation 9.176 that

$$\left.\frac{\delta\mathcal{A}^{ex}}{\delta u(\mathbf{r},\mathbf{r}')}\right|_{n^{(1)}\text{ fixed}} = \frac{n^{(2)}(\mathbf{r},\mathbf{r}')}{2}, \tag{9.178}$$

where the density distribution $n^{(1)}(\mathbf{r})$ is kept fixed when u varies.

Let us now use these relationships to calculate differences in excess free energy $\Delta\mathcal{A}^{ex}$ when the pair potential is changed from $u_{\text{I}}(\mathbf{r},\mathbf{r}')$ to $u_{\text{II}}(\mathbf{r},\mathbf{r}')$ for all particles in the system. We take $u(\mathbf{r},\mathbf{r}';\xi) = u_{\text{I}}(\mathbf{r},\mathbf{r}')+\xi[u_{\text{II}}(\mathbf{r},\mathbf{r}')-u_{\text{I}}(\mathbf{r},\mathbf{r}')]$ and calculate $\Delta\mathcal{A}^{ex}$ by means of a functional integration, whereby we can utilize Equation 9.178 in the following manner. We will maintain $n^{(1)}(\mathbf{r})$ fixed when going from I to II and select $n^{(1)}(\mathbf{r}) = n^{(1)}_{\text{II}}(\mathbf{r})$, which is the equilibrium density distribution when $u = u_{\text{II}}(\mathbf{r},\mathbf{r}')$, that is, when $\xi = 1$. Obviously, for a system where the pair potential is $u(\mathbf{r},\mathbf{r}';\xi)$ with $\xi < 1$, the equilibrium density distribution is not equal to $n^{(1)}_{\text{II}}(\mathbf{r})$, so by setting $n^{(1)}(\mathbf{r}) = n^{(1)}_{\text{II}}(\mathbf{r})$ when $\xi < 1$ we are treating a system that is artificially removed from equilibrium. Nevertheless, one can calculate the pair distribution $g^{(2)}(\mathbf{r},\mathbf{r}';\xi)$ for such a system[54] and we have $n^{(2)}(\mathbf{r},\mathbf{r}';\xi) = n^{(1)}(\mathbf{r})n^{(1)}(\mathbf{r}')g^{(2)}(\mathbf{r},\mathbf{r}';\xi)$. Analogously to the derivation of Equation 9.171 we obtain

$$\begin{aligned} \mathcal{A}^{ex}_{\text{II}} &= \mathcal{A}^{ex}_{\text{I}} + \frac{1}{2}\int d\mathbf{r}d\mathbf{r}'\,[u_{\text{II}}(\mathbf{r},\mathbf{r}') - u_{\text{I}}(\mathbf{r},\mathbf{r}')] \int_0^1 d\xi\, n^{(2)}(\mathbf{r},\mathbf{r}';\xi) \\ &= \mathcal{A}^{ex}_{\text{I}} + \frac{1}{2}\int d\mathbf{r}d\mathbf{r}'\,[u_{\text{II}}(\mathbf{r},\mathbf{r}') - u_{\text{I}}(\mathbf{r},\mathbf{r}')] \\ &\times\; n^{(1)}(\mathbf{r})n^{(1)}(\mathbf{r}') \int_0^1 d\xi\, g^{(2)}(\mathbf{r},\mathbf{r}';\xi) \qquad (n^{(1)}\text{ fixed}). \end{aligned} \tag{9.179}$$

Note that $\mathcal{A}^{\text{ideal}} = k_BT \int d\mathbf{r}\, n^{(1)}(\mathbf{r}) \left[\ln(\Lambda^3 n^{(1)}(\mathbf{r})) - 1\right]$ irrespectively of the value of ξ since $n^{(1)}(\mathbf{r})$ remains the same. We have $\mathcal{A}_L = \mathcal{A}^{\text{ideal}} + \mathcal{A}^{ex}_L$ for $L = \text{I}$ or II. For a bulk phase this formula can always be used because $n^{(1)} = n^{\text{b}}$ is unchanged.

An interesting application of this result is to obtain the excess free energy $\mathcal{A}^{ex}$ for a system with pair potential $u(\mathbf{r},\mathbf{r}')$ and density distribution $n^{(1)}(\mathbf{r})$. In Equation 9.179 we take $u_{\text{I}} = 0$ and $u_{\text{II}} = u$, whereby $u(\mathbf{r},\mathbf{r}';\xi) = \xi u(\mathbf{r},\mathbf{r}')$ and $\mathcal{A}^{ex}_{\text{I}} = 0$. The initial system for $\xi = 0$ is an ideal gas with density distribution $n^{(1)}(\mathbf{r})$. The Helmholtz free energy $\mathcal{A}^{ex}$ for the fully coupled system with pair potential u (i.e., for $\xi = 1$) is $\mathcal{A}^{ex} = \mathcal{A}^{ex}_{\text{II}}$, so we obtain

$$\begin{aligned} \mathcal{A}^{ex} &= \frac{1}{2}\int d\mathbf{r}d\mathbf{r}'\, u(\mathbf{r},\mathbf{r}')n^{(1)}(\mathbf{r})n^{(1)}(\mathbf{r}') \\ &\times \int_0^1 d\xi\, g^{(2)}(\mathbf{r},\mathbf{r}';\xi) \qquad (n^{(1)}\text{ fixed}) \end{aligned} \tag{9.180}$$

and $\mathcal{A} = \mathcal{A}^{ex} + \mathcal{A}^{\text{ideal}}$. For the initial system with $\xi = 0$, we have $\mathcal{A} = \mathcal{A}^{\text{ideal}}$.

9.2.4.2 *Expressions for $\Delta c^{(1)}$, the free energy and the bridge function*

In Section 9.2.3 we determined $\Delta c^{(1)}$ when the density is changed by Δn, which in turn was induced by a variation in external potential, Δv. To obtain $\Delta c^{(1)}$ we made a functional Taylor expansion to infinite order in the direct correlation functions. Here, we will use the functional integration strategy to

---

[54] For example, in integral equation approximations for the pair distribution functions of inhomogeneous systems this can be done as follows. One solves the OZ equation 9.40 for the given density distribution $n^{(1)}(\mathbf{r})$ using any closure of Sections 9.1.3.2 or 9.1.3.3 with the pair potential $u(\mathbf{r},\mathbf{r}';\xi)$ inserted in the closure. This yields $g^{(2)}(\mathbf{r},\mathbf{r}';\xi)$. Such integral equation approximations are treated in Section 10.1.4.

obtain $\Delta c^{(1)}$. Recall that $\Delta c^{(1)}$ is, for example, needed in order to obtain the density distribution functions from Equations 9.53 or 9.54.

The change in $c^{(1)}$ for small variations δn in density is given by

$$\delta c^{(1)}(\mathbf{r}) = \int d\mathbf{s}\, \frac{\delta c^{(1)}(\mathbf{r})}{\delta n(\mathbf{s})} \delta n(\mathbf{s}) = \int d\mathbf{s}\, c^{(2)}(\mathbf{r}, \mathbf{s}) \delta n(\mathbf{s})$$

(Equations 9.56 and 9.127). We will use this to obtain $\Delta c^{(1)}$ by functional integration (using the general formula 9.155). Let us denote the initial value of the external potential as $v_{\mathrm{I}}(\mathbf{r})$ and the final $v_{\mathrm{II}}(\mathbf{r})$, so $\Delta v(\mathbf{r}) = v_{\mathrm{II}}(\mathbf{r}) - v_{\mathrm{I}}(\mathbf{r})$. Likewise $\Delta n(\mathbf{r}) = n_{\mathrm{II}}(\mathbf{r}) - n_{\mathrm{I}}(\mathbf{r})$ and $\Delta c^{(1)}(\mathbf{r}) = c_{\mathrm{II}}^{(1)}(\mathbf{r}) - c_{\mathrm{I}}^{(1)}(\mathbf{r})$. We now let $v(\mathbf{r}; \xi)$ and $n(\mathbf{r}; \xi)$ with $0 \leq \xi \leq 1$ denote the potential and the density, respectively, for the intermediate stages between I and II, where $n(\mathbf{r}; \xi)$ is the density of the system in the presence of the external potential $v(\mathbf{r}; \xi)$. The *only* requirements we have are that n and v must be continuous differentiable functions of ξ and that

$$v(\mathbf{r}; \xi) = \begin{cases} v_{\mathrm{I}}(\mathbf{r}) & \text{for } \xi = 0 \\ v_{\mathrm{II}}(\mathbf{r}) & \text{for } \xi = 1 \end{cases} \quad \text{and} \quad n(\mathbf{r}; \xi) = \begin{cases} n_{\mathrm{I}}(\mathbf{r}) & \text{for } \xi = 0 \\ n_{\mathrm{II}}(\mathbf{r}) & \text{for } \xi = 1 \end{cases} \qquad (9.181)$$

(cf. Equations 9.154 and 9.152). Otherwise the ξ dependence is completely arbitrary. Since $c^{(1)}$ is a free energy quantity, it is a state function so $\Delta c^{(1)}$ is independent of the path between $\xi = 0$ and 1 (here the term "path" implies the selection of how the functions vary with ξ). A particularly simple choice is to take $v(\mathbf{r}; \xi) = (1 - \xi) v_{\mathrm{I}}(\mathbf{r}) + \xi v_{\mathrm{II}}(\mathbf{r})$ so the variation is linear in ξ. Another simple choice is to use the one-to-one correspondence between the density distribution and the external potential (utilized in Section 9.2.1) and to set $n(\mathbf{r}; \xi) = (1 - \xi) n_{\mathrm{I}}(\mathbf{r}) + \xi n_{\mathrm{II}}(\mathbf{r})$. The potential $v(\mathbf{r}; \xi)$ is then defined as the potential that gives the density $n(\mathbf{r}; \xi)$. Then, $v(\mathbf{r}; \xi)$ is in general not known beforehand for intermediate ξ values and does not vary linearly with ξ, but, on the other hand, for the first choice, $n(\mathbf{r}; \xi)$ is in general not known instead and does not vary with ξ in a linear manner. When one of the functions has a simple variation, the other one has a complicated variation and vice versa. Sometimes it may be motivated to have some complicated ξ dependence for both, which is all right provided the conditions above are fulfilled.

We now let $c^{(2)}(\mathbf{r}, \mathbf{s}; \xi)$ denote the pair direct-correlation function for the system with external potential $v(\mathbf{r}; \xi)$ and density $n(\mathbf{r}; \xi)$, so we have

$$\delta c^{(1)}(\mathbf{r}; \xi) = \int d\mathbf{s}\, c^{(2)}(\mathbf{r}, \mathbf{s}; \xi) \delta n(\mathbf{s}; \xi),$$

where $\delta c^{(1)}(\mathbf{r}; \xi)$ and $\delta n(\mathbf{s}; \xi)$ are the variations when $\xi \to \xi + d\xi$, that is

$$\delta n(\mathbf{s}; \xi) = \frac{\partial n(\mathbf{s}; \xi)}{\partial \xi} d\xi$$

and the corresponding relationship for $c^{(1)}$. We obtain $\Delta c^{(1)}$ by integration over $0 \leq \xi \leq 1$, which yields

$$\Delta c^{(1)}(\mathbf{r}) = \int_0^1 d\xi \int d\mathbf{s}\, c^{(2)}(\mathbf{r},\mathbf{s};\xi) \frac{\partial n(\mathbf{s};\xi)}{\partial \xi} \tag{9.182}$$

(cf. Equation 9.155). In order to write this result in the same form as in

$$\Delta c^{(1)}(\mathbf{r}) = \int d\mathbf{s}\, c^{(2)}(\mathbf{r},\mathbf{s}) \Delta n(\mathbf{s}) + B(\mathbf{r})$$

(Equation 9.158), we add and subtract $\int d\mathbf{s}\, c^{(2)}(\mathbf{r},\mathbf{s})\Delta n(\mathbf{s})$ (with $c^{(2)}$ for the initial state $\xi = 0$) to obtain

$$\begin{aligned} \Delta c^{(1)}(\mathbf{r}) &= \int d\mathbf{s}\, c^{(2)}(\mathbf{r},\mathbf{s}) \Delta n(\mathbf{s}) \\ &+ \int_0^1 d\xi \int d\mathbf{s} \left[c^{(2)}(\mathbf{r},\mathbf{s};\xi) - c^{(2)}(\mathbf{r},\mathbf{s};0) \right] \frac{\partial n(\mathbf{s};\xi)}{\partial \xi}, \end{aligned} \tag{9.183}$$

where the last term can be identified as the bridge function $B(\mathbf{r})$ (for clarity have shown that the second $c^{(2)}$ in the integrand is evaluated for $\xi = 0$). Thus we have

$$B(\mathbf{r}) = \int_0^1 d\xi \int d\mathbf{s} \left[c^{(2)}(\mathbf{r},\mathbf{s};\xi) - c^{(2)}(\mathbf{r},\mathbf{s};0) \right] \frac{\partial n(\mathbf{s};\xi)}{\partial \xi}. \tag{9.184}$$

From this formula we see that the HNC approximation, where $B = 0$, can be described as the neglect of the ξ dependence of $c^{(2)}(\mathbf{r},\mathbf{s};\xi)$, which is thereby approximated by $c^{(2)}(\mathbf{r},\mathbf{s};0)$, so B is set to zero. This neglect implies that one as an approximation has set $c^{(m)} = 0$ for $m \geq 3$.[55] It is, in fact, sufficient to make the approximation $c^{(3)} \equiv 0$ to obtain the HNC approximation. Since $c^{(m)}$ for $m > 3$ are functional derivatives of $c^{(3)}$ it follows that they must also be zero.

Let us return to the general case. As mentioned earlier, Equation 9.42, we have $c^{(2)}(\mathbf{r},\mathbf{s}) \sim -\beta u(\mathbf{r},\mathbf{s})$ for large distances $|\mathbf{r}-\mathbf{s}|$. Since the pair interaction among the particles in the fluid is independent of ξ, we have $c^{(2)}(\mathbf{r},\mathbf{s};\xi) \sim -\beta u(\mathbf{r},\mathbf{s})$ for all ξ. Therefore, the long-range tail of $c^{(2)}$ only contributes to the first term of $\Delta c^{(1)}(\mathbf{r})$ in Equation 9.183 and not to $B(\mathbf{r})$ in Equation 9.184, where the tail is cancelled in the square bracket of the integrand. The influence of the long-range part of u on B is therefore only indirect. In cases where long-range effects of the pair potential are important, like systems where electrostatic effects dominate, it is indeed found that the *HNC approximation often works well*, at least partly for this reason.

[55] From the properties of the functional Taylor series for $c^{(2)}$ follow that the functional derivatives of this function, that is, $c^{(m)}$ for $m \geq 3$, describe how $c^{(2)}$ varies when the density of the system is changed. To ignore the variation of $c^{(2)}$ due to a change in density is therefore equivalent to the statement that all $c^{(m)}$ for $m \geq 3$ are set equal to zero.

To proceed, let us select one of the simple paths that was mentioned above, namely $n(\mathbf{r};\xi) = (1-\xi)n_{\mathrm{I}}(\mathbf{r}) + \xi n_{\mathrm{II}}(\mathbf{r}) = n_{\mathrm{I}}(\mathbf{r}) + \xi\Delta n(\mathbf{r})$, which implies that $\partial n(\mathbf{r};\xi)/\partial\xi = \Delta n(\mathbf{r})$ is independent of ξ, so we have from Equation 9.182

$$\Delta c^{(1)}(\mathbf{r}) = \int_0^1 d\xi \int d\mathbf{s}\, c^{(2)}(\mathbf{r},\mathbf{s};\xi)\Delta n(\mathbf{s}). \tag{9.185}$$

In order to emphasize that $c^{(2)}$ in the integrand is the pair direct-correlation function for the fluid with density $n(\mathbf{r};\xi)$, we can write $c^{(2)}(\mathbf{r},\mathbf{s};\xi) = c^{(2)}(\mathbf{r},\mathbf{s};\{n(\mathbf{r};\xi)\})$ for clarity. This will be done in what follows.

In the special case of a bulk fluid, we can use the expression 9.185 for $\Delta c^{(1)}$ to obtain very useful results for $c^{(1)}$ itself. To do this let us select $n_{\mathrm{I}}(\mathbf{r}) = n_{\mathrm{I}}^{\mathrm{b}} = 0$ and $n_{\mathrm{II}}(\mathbf{r}) = n_{\mathrm{II}}^{\mathrm{b}}$ whereby $n(\mathbf{r};\xi) = \xi n_{\mathrm{II}}^{\mathrm{b}}$ and $c^{(2)}(\mathbf{r},\mathbf{s};\xi) = c^{(2)\mathrm{b}}(|\mathbf{r}-\mathbf{s}|;\{\xi n_{\mathrm{II}}^{\mathrm{b}}\})$. In the bulk, $\Delta c^{(1)}(\mathbf{r})$ is a constant independent of $\mathbf{r}$ so we can set $\mathbf{r} = 0$ and we have $\Delta c^{(1)} = c^{(1)\mathrm{b}}\{n_{\mathrm{II}}^{\mathrm{b}}\}$ since $c^{(1)\mathrm{b}}\{n_{\mathrm{I}}^{\mathrm{b}}\} = 0$ when $n_{\mathrm{I}}^{\mathrm{b}} = 0$. Equation 9.185 then becomes

$$c^{(1)\mathrm{b}}\{n_{\mathrm{II}}^{\mathrm{b}}\} = \int_0^1 d\xi \int d\mathbf{s}\, c^{(2)\mathrm{b}}(s;\{\xi n_{\mathrm{II}}^{\mathrm{b}}\})n_{\mathrm{II}}^{\mathrm{b}} = \int_0^{n_{\mathrm{II}}^{\mathrm{b}}} dn'^{\mathrm{b}} \int d\mathbf{s}\, c^{(2)\mathrm{b}}(s;\{n'^{\mathrm{b}}\}), \tag{9.186}$$

where we have set $n'^{\mathrm{b}} = \xi n_{\mathrm{II}}^{\mathrm{b}}$ and made a variable substitution from ξ to n'^{b} in the ξ integral. Note that $c^{(2)\mathrm{b}}(s;\{n'^{\mathrm{b}}\})$ is the pair direct-correlation function for a bulk fluid of density n'^{b}. The quantity $-k_BTc^{(1)\mathrm{b}}$ is the excess chemical potential for a bulk fluid and we have

$$\mu = k_BT\ln(\Lambda^3 n^{\mathrm{b}}) - k_BT\int_0^{n^{\mathrm{b}}} dn'^{\mathrm{b}} \int d\mathbf{s}\, c^{(2)\mathrm{b}}(s;\{n'^{\mathrm{b}}\}) \quad \text{(bulk fluid)} \tag{9.187}$$

(from Equation 9.11 with $v = 0$), where we have written n^{b} instead of $n_{\mathrm{II}}^{\mathrm{b}}$. This equation can be used to calculate the chemical potential provided that the fluid remains as a single phase when the density varies from 0 to n^{b} (no phase transitions).

By taking the derivative of Equation 9.187 with respect of n^{b} we obtain

$$\frac{\partial\mu}{\partial n^{\mathrm{b}}} = \frac{k_BT}{n^{\mathrm{b}}} - k_BT\int d\mathbf{s}\, c^{(2)\mathrm{b}}(s;\{n^{\mathrm{b}}\}) \quad \text{(bulk fluid).} \tag{9.188}$$

Since the isothermal compressibility χ_T is given by

$$\chi_T = \frac{1}{n^{\mathrm{b}}}\left(\frac{\partial n^{\mathrm{b}}}{\partial P}\right)_T = \frac{1}{[n^{\mathrm{b}}]^2}\left(\frac{\partial n^{\mathrm{b}}}{\partial\mu}\right)_T \tag{9.189}$$

(see Equation I:2.130), we can write Equation 9.188 as

$$\frac{1}{k_BTn^{\mathrm{b}}\chi_T} = 1 - n^{\mathrm{b}}\int d\mathbf{s}\, c^{(2)\mathrm{b}}(s;\{n^{\mathrm{b}}\}) \quad \text{(bulk fluid).} \tag{9.190}$$

The rhs can alternatively be expressed in terms of the pair correlation function $h^{(2)\mathrm{b}}$. By integrating the OZ equation 9.70 with respect to $\mathbf{r}_2$ over the whole space we obtain

$$\int d\mathbf{r}_2 h^{(2)\mathrm{b}}(r_{12}) = \int d\mathbf{r}_2 c^{(2)\mathrm{b}}(r_{12}) + n^{\mathrm{b}} \int d\mathbf{r}_3\, c^{(2)\mathrm{b}}(r_{13}) \left[\int d\mathbf{r}_2 h^{(2)\mathrm{b}}(r_{32}) \right]$$

and by realizing that the square bracket is a constant independent of $\mathbf{r}_3$ we can deduce that[56]

$$1 - n^{\mathrm{b}} \int d\mathbf{s}\, c^{(2)\mathrm{b}}(r) = \frac{1}{1 + n^{\mathrm{b}} \int d\mathbf{s}\, h^{(2)\mathrm{b}}(r)} \quad \text{(bulk fluid).} \tag{9.191}$$

By inserting this in Equation 9.190 we see that

$$k_B T n^{\mathrm{b}} \chi_T = 1 + n^{\mathrm{b}} \int d\mathbf{s}\, h^{(2)\mathrm{b}}(s; \{n^{\mathrm{b}}\}) \quad \text{(bulk fluid).} \tag{9.192}$$

Equations 9.190 and 9.192 are different forms of the so-called **compressibility equation**. One can use them to obtain the pressure P by integrating $1/(n^{\mathrm{b}}\chi_T)$ from 0 to n^{b}, as can be seen from the first equality of Equation 9.189.

Let us now return to inhomogeneous fluids. Since $\delta\mathcal{A}^{\mathrm{ex}}/\delta n(\mathbf{r}) = -k_B T c^{(1)}(\mathbf{r})$ (Equation 9.126), we can obtain changes in excess intrinsic free energy, $\Delta\mathcal{A}^{\mathrm{ex}} = \mathcal{A}^{\mathrm{ex}}_{\mathrm{II}} - \mathcal{A}^{\mathrm{ex}}_{\mathrm{I}}$, by functional integration of $c^{(1)}$ over $n(\mathbf{r};\xi) = (1-\xi)n_{\mathrm{I}}(\mathbf{r}) + \xi n_{\mathrm{II}}(\mathbf{r})$. We obtain

$$\Delta\mathcal{A}^{\mathrm{ex}} = \int_0^1 d\xi \int d\mathbf{r}\, \left.\frac{\delta\mathcal{A}^{\mathrm{ex}}}{\delta n(\mathbf{r})}\right|_{n=n(\mathbf{r};\xi)} \Delta n(\mathbf{r}) = -k_B T \int_0^1 d\xi \int d\mathbf{r}\, c^{(1)}(\mathbf{r};\xi)\Delta n(\mathbf{r})$$

(cf. the general integration formula 9.155). The function $c^{(1)}(\mathbf{r};\xi)$ can be obtained from a slight generalization of Equation 9.185, namely

$$\Delta c^{(1)}(\mathbf{r};\xi) = c^{(1)}(\mathbf{r};\xi) - c^{(1)}(\mathbf{r};0) = \int d\mathbf{s} \left[\int_0^{\xi} d\xi'\, c^{(2)}(\mathbf{r},\mathbf{s};\xi') \right] \Delta n(\mathbf{s}),$$

where the integration over ξ' stops at ξ. By combining these two equations and using the fact that for any function $F(x)$, we have[57] $\int_0^1 d\xi \int_0^{\xi} d\xi'\, F(\xi') = \int_0^1 d\xi\,(1-\xi)F(\xi)$, we obtain

$$\begin{aligned} \Delta\mathcal{A}^{\mathrm{ex}} &= -k_B T \int d\mathbf{r}\, c^{(1)}(\mathbf{r};0)\Delta n(\mathbf{r}) \\ &\quad - k_B T \int_0^1 d\xi\,(1-\xi) \int d\mathbf{r}\, d\mathbf{s}\, c^{(2)}(\mathbf{r},\mathbf{s};\xi)\Delta n(\mathbf{r})\Delta n(\mathbf{s}). \end{aligned}$$

[56]Equation 9.191 can alternatively be obtained from the Fourier transform of the OZ equation. From Equation 9.77 it follows that $1 - n^{\mathrm{b}}\tilde{c}^{(2)\mathrm{b}}(k) = 1/[1 - n^{\mathrm{b}}\tilde{h}^{(2)\mathrm{b}}(k)]$ and we obtain Equation 9.191 by using Equation 9.245 applied to $\tilde{c}^{(2)\mathrm{b}}(k)$ and $\tilde{h}^{(2)\mathrm{b}}(k)$ at $k=0$.

[57]This relationship follows by partial integration $\int_0^1 d\xi \int_0^{\xi} d\xi'\, F(\xi') = [\xi \int_0^{\xi} d\xi'\, F(\xi')]_{\xi=0}^{\xi=1} - \int_0^1 d\xi\, \xi F(\xi) = \int_0^1 d\xi'\, F(\xi') - \int_0^1 d\xi\, \xi F(\xi)$, which can be written as $\int_0^1 d\xi\,(1-\xi)F(\xi)$.

The entire change in Helmholtz free energy ΔA is obtained by adding ΔA^{ideal} and the contribution from the external potential.

To finish this section we will obtain some further results for the bridge function $B(\mathbf{r})$ given by Equation 9.184. In Equation 9.185, which we can write as

$$\Delta c^{(1)}(\mathbf{r}) = \int d\mathbf{s} \left[\int_0^1 d\xi \, c^{(2)}(\mathbf{r}, \mathbf{s}; \xi) \right] \Delta n(\mathbf{s}),$$

we introduce the average

$$\bar{c}^{(2)}(\mathbf{r}, \mathbf{s}) \equiv \int_0^1 d\xi \, c^{(2)}(\mathbf{r}, \mathbf{s}; \xi) = c^{(2)}(\mathbf{r}, \mathbf{s}; \bar{\xi}(\mathbf{r}, \mathbf{s})), \qquad (9.193)$$

where $\bar{\xi}(\mathbf{r}, \mathbf{s})$ is a function with values $0 < \bar{\xi} < 1$. The last equality follows from the mean value theorem of integral calculus[58] applied to the ξ integration for each pair of $\mathbf{r}$ and $\mathbf{s}$ values. Note that $\bar{c}^{(2)}(\mathbf{r}, \mathbf{s}) \sim -\beta u(\mathbf{r}, \mathbf{s})$ for large $|\mathbf{r} - \mathbf{s}|$ independently of $\bar{\xi}$.

It follows that we can write

$$\Delta c^{(1)}(\mathbf{r}) = \int d\mathbf{s} \, \bar{c}^{(2)}(\mathbf{r}, \mathbf{s}) \Delta n(\mathbf{s}) \qquad (9.194)$$

and we have from Equation 9.184

$$B(\mathbf{r}) = \int d\mathbf{s} \left[\bar{c}^{(2)}(\mathbf{r}, \mathbf{s}) - c^{(2)}(\mathbf{r}, \mathbf{s}; 0) \right] \Delta n(\mathbf{s}). \qquad (9.195)$$

Although the function $\bar{\xi}$ is not known, this result is useful since it constrains $\bar{c}^{(2)}(\mathbf{r}, \mathbf{s})$ as follows: for each $(\mathbf{r}, \mathbf{s})$ the function $\bar{c}^{(2)}$ has a value that $c^{(2)}(\mathbf{r}, \mathbf{s}; \xi)$ *actually assumes* in the system for some intermediate density $n(\mathbf{r}; \xi)$ between $n_{\text{I}}(\mathbf{r})$ and $n_{\text{II}}(\mathbf{r})$ (i.e. for ξ between 0 and 1). This indicates that a suitable approximation for $\bar{c}^{(2)}$ may be sought by constructing $c^{(2)}$ for an intermediate density used locally for each pair of $\mathbf{r}$ and $\mathbf{s}$. *This kind of strategy is often used in density functional theories for fluids*, which are introduced in Section 9.3.

Let us specialize to the calculation of pair distributions given by

$$\begin{aligned} g^{(2)}(\mathbf{r}, \mathbf{r}') &= e^{-\beta u(\mathbf{r}, \mathbf{r}') + \Delta c^{(1)}(\mathbf{r}|\mathbf{r}')} \\ &= e^{-\beta u(\mathbf{r}, \mathbf{r}') + \gamma^{(2)}(\mathbf{r}, \mathbf{r}') + b^{(2)}(\mathbf{r}, \mathbf{r}')} \end{aligned} \qquad (9.196)$$

(Equations 9.54 and 9.65), where $\gamma^{(2)} = h^{(2)} - c^{(2)}$. In this case, the density change Δn is, as we have seen, induced by the insertion of a particle at $\mathbf{r}'$, $\Delta n(\mathbf{s}) = n(\mathbf{s}) h^{(2)}(\mathbf{s}, \mathbf{r}')$. We then have a special version of the requirements

[58]The mean value theorem states that for a continuous function $f(x)$ we have $\int_a^b dx \, f(x) = (b - a) f(\bar{x})$, where $a < \bar{x} < b$.

in (9.181)

$$
\begin{aligned}
v(\mathbf{r}|\mathbf{r}';\xi) &= \begin{cases} v(\mathbf{r}) & \text{for } \xi = 0 \\ v(\mathbf{r}) + u(\mathbf{r},\mathbf{r}') & \text{for } \xi = 1 \end{cases} \\
&\text{and} \\
n(\mathbf{r}|\mathbf{r}';\xi) &= \begin{cases} n(\mathbf{r}) & \text{for } \xi = 0 \\ n(\mathbf{r})g^{(2)}(\mathbf{r},\mathbf{r}') & \text{for } \xi = 1. \end{cases}
\end{aligned}
\tag{9.197}
$$

When $0 < \xi < 1$, the particle at $\mathbf{r}'$ interacts with the other particles with a pair potential $u(\mathbf{r},\mathbf{r}';\xi)$, so it is partially coupled to them (the other particles are fully coupled with each other all the time). For this case the bridge function is given by

$$
b^{(2)}(\mathbf{r},\mathbf{r}') = \int d\mathbf{s}\, \left[\bar{c}^{(2)}(\mathbf{r},\mathbf{s}|\mathbf{r}') - c^{(2)}(\mathbf{r},\mathbf{s};0)\right] n(\mathbf{s})h^{(2)}(\mathbf{s},\mathbf{r}'), \tag{9.198}
$$

where we have applied Equation 9.195 to the present case and selected the path $n(\mathbf{r}|\mathbf{r}';\xi) = n(\mathbf{r})[1+\xi h^{(2)}(\mathbf{s},\mathbf{r}')]$ to obtain the expression for $b^{(2)}$. This equation looks deceptively simple, but $\bar{c}^{(2)}(\mathbf{r},\mathbf{s}|\mathbf{r}')$ contains at lot of complexities as apparent from the discussion of $b^{(2)}$ in the previous section.

In $b^{(2)}(\mathbf{r},\mathbf{r}')$ only the difference in short-range contributions to $c^{(2)}$ matters (since the tail part $-\beta u$ of $c^{(2)}$ cancels as pointed out earlier). In fact, $b^{(2)}$ is in many cases not overly sensitive to the details of the pair interaction. It is found in practice that a major contribution to $b^{(2)}$ in dense fluids comes from the "packing" of the particles, a short-range effect. This lies behind the philosophy of the **RHNC approximation** where a bridge function for one system (often a hard sphere fluid) is used as an approximation to $b^{(2)}$ for another system.

9.2.4.3 *The determination of $c^{(1)}(\mathbf{r})$ and μ for inhomogeneous fluids and for bulk, general results*

In the previous section we determined changes $\Delta c^{(1)}$ of the direct correlation function by functional integration. Here, we will use this technique to determine $c^{(1)}$ itself. We thus have to find the reversible work $-k_BTc^{(1)}(\mathbf{r}')$ against interparticle interactions when we insert (or create) a new particle at $\mathbf{r}'$. As we have seen many times, the insertion of a fixed particle at $\mathbf{r}'$ gives the same effect on the other particles as a change in the external potential at $\mathbf{r}$ by an amount equal to the pair potential $u(\mathbf{r},\mathbf{r}')$. We will gradually increase the interaction from $v(\mathbf{r})$ to $v(\mathbf{r})+u(\mathbf{r},\mathbf{r}')$ and for each step, when the interaction is increased by $\delta v(\mathbf{r}) = \delta u(\mathbf{r},\mathbf{r}')$, we calculate the differential of the reversible work, which is equal to the change $d\Theta$ in free energy.[59] By summing $d\Theta$ for every step we obtain the total work $-k_BTc^{(1)}(\mathbf{r}')$.

[59] See the discussion in footnote 11 on page 15.

The differential $d\Theta$ for each step can be calculated from $d\Theta = \int d\mathbf{r}\, n(\mathbf{r})\delta v(\mathbf{r})$ (Equation 9.34), where $n(\mathbf{r})$ is the density before the change $\delta v(\mathbf{r}) = \delta u(\mathbf{r}, \mathbf{r}')$ in potential. To follow the same strategy as in the previous section, we let the external potential and hence the density change gradually as we slowly increase the interaction with the new particle and we have $v = v(\mathbf{r}|\mathbf{r}'; \xi)$ and $n = n(\mathbf{r}|\mathbf{r}'; \xi)$ when $0 \le \xi \le 1$. The initial and final states of the system are those specified in Equation 9.197 and the interaction with the new, partially coupled particle is given by the pair potential $u(\mathbf{r}, \mathbf{r}'; \xi)$ for $0 < \xi < 1$. When $\xi \to \xi + d\xi$ we have $\delta v(\mathbf{r}) = \delta u(\mathbf{r}, \mathbf{r}') = [\partial u(\mathbf{r}, \mathbf{r}'; \xi)/\partial\xi]d\xi$. By inserting the relevant quantities in $d\Theta = \int d\mathbf{r}\, n(\mathbf{r})\delta v(\mathbf{r})$, we have in this case for each value of ξ

$$-k_BT\delta c^{(1)}(\mathbf{r}'; \xi) = d\Theta(\xi) = \int d\mathbf{r}\, n(\mathbf{r}|\mathbf{r}'; \xi)\frac{\partial u(\mathbf{r}, \mathbf{r}'; \xi)}{\partial \xi}d\xi$$

and by integrating over ξ we obtain

$$\begin{aligned} c^{(1)}(\mathbf{r}') &= -\beta\int_0^1 d\xi \int d\mathbf{r}\, n(\mathbf{r}|\mathbf{r}'; \xi)\frac{\partial u(\mathbf{r}, \mathbf{r}'; \xi)}{\partial \xi} \\ &= -\beta\int_0^1 d\xi \int d\mathbf{r}\, n(\mathbf{r})g^{(2)}(\mathbf{r}, \mathbf{r}'; \xi)\frac{\partial u(\mathbf{r}, \mathbf{r}'; \xi)}{\partial \xi}, \end{aligned} \tag{9.199}$$

where we have set $n(\mathbf{r}|\mathbf{r}'; \xi) = n(\mathbf{r})g^{(2)}(\mathbf{r}, \mathbf{r}'; \xi)$ in the last line to show that the density around the partially coupled new particle is described by a pair distribution $g^{(2)}(\mathbf{r}, \mathbf{r}'; \xi)$.

By using Equation 9.199 we obtain the following expression for the calculation of the chemical potential μ of inhomogeneous fluids

$$\begin{aligned} \mu &= k_BT\ln(\Lambda^3 n(\mathbf{r}')) + v(\mathbf{r}') - k_BT\, c^{(1)}(\mathbf{r}') \qquad (9.200) \\ &= k_BT\ln(\Lambda^3 n(\mathbf{r}')) + v(\mathbf{r}') + \int_0^1 d\xi \int d\mathbf{r}\, n(\mathbf{r})g^{(2)}(\mathbf{r}, \mathbf{r}'; \xi)\frac{\partial u(\mathbf{r}, \mathbf{r}'; \xi)}{\partial \xi}, \end{aligned}$$

where the sum of the last two terms equals the excess part $\mu^{\text{ex}}(\mathbf{r}')$. This relationship constitutes a generalization of the equation for μ of a bulk fluid given by Equation I:5.84 of Volume I. It gives μ in terms of the pair distribution function $g^{(2)}$ and the density distribution n, which is useful in many cases.

In computer simulations, however, it is not necessary to calculate the distribution functions explicitly in order to obtain the chemical potential. Instead, one can calculate $\mu^{\text{ex}}(\mathbf{r}')$ directly from the sequences of particle configurations of the system generated in the simulations. This is described in Section 5.13 of Volume I, see for example Equations I:5.161 and I:5.162 which gives expressions for $\mu^{\text{ex}}(\mathbf{r}')$ in inhomogeneous fluids.

Here, we will continue the investigation of the links between $c^{(1)}(\mathbf{r}')$, μ and various pair functions. The ξ integration in Equation 9.199 for $c^{(1)}$ can, in fact, be performed analytically to a considerable extent and in the shaded text box below it is shown that this integration yields

$$c^{(1)}(\mathbf{r}') = -\int d\mathbf{r}\, n(\mathbf{r}) \left(\frac{1}{2} \left[h^{(2)}(\mathbf{r},\mathbf{r}') \right]^2 - c^{(2)}(\mathbf{r},\mathbf{r}') + g^{(2)}(\mathbf{r},\mathbf{r}') b^{(2)}(\mathbf{r},\mathbf{r}') \right) + \frac{\gamma^{(2)}(\mathbf{r}',\mathbf{r}')}{2} + c^{(1)\mathrm{R}}(\mathbf{r}'), \tag{9.201}$$

where

$$c^{(1)\mathrm{R}}(\mathbf{r}') = \int_0^1 d\xi \int d\mathbf{r}\, n(\mathbf{r}) \frac{\partial h^{(2)}(\mathbf{r},\mathbf{r}';\xi)}{\partial \xi} b^{(2)}(\mathbf{r},\mathbf{r}';\xi) \tag{9.202}$$

is a **"residual" part of $\mathbf{c}^{(1)}$**, where the coupling-parameter integration remains. All other terms contain only the pair and singlet functions of the fluid with fully coupled particles.

Equation 9.201 can be used in Equation 9.200 to calculate μ. For a bulk phase we obtain

$$\beta\mu^{\mathrm{ex,b}} = -c^{(1)\mathrm{b}} = n^{\mathrm{b}} \int d\mathbf{r} \left(\frac{1}{2} \left[h^{(2)b}(r) \right]^2 - c^{(2)\mathrm{b}}(r) + g^{(2)\mathrm{b}}(r) b^{(2)\mathrm{b}}(r) \right) - \frac{\gamma^{(2)\mathrm{b}}(0)}{2} - c^{(1)\mathrm{R,b}}, \tag{9.203}$$

which hence gives the excess chemical potential in bulk.

Let us now prove Equation 9.201. The function $g^{(2)}(\mathbf{r},\mathbf{r}';\xi)$ in Equation 9.199 is given by the following version of Equation 9.196

$$g^{(2)}(\mathbf{r},\mathbf{r}';\xi) = e^{-\beta u(\mathbf{r},\mathbf{r}';\xi) + \gamma^{(2)}(\mathbf{r},\mathbf{r}';\xi) + b^{(2)}(\mathbf{r},\mathbf{r}';\xi)},$$

where all functions are applicable for the situation with the partially coupled particle surrounded by the other particles; the latter are fully coupled to each other all the time. By taking the derivative of $g^{(2)}$ with respect to ξ we obtain

$$\frac{\partial g^{(2)}}{\partial \xi} = g^{(2)} \left[-\beta \frac{\partial u}{\partial \xi} + \frac{\partial \gamma^{(2)}}{\partial \xi} + \frac{\partial b^{(2)}}{\partial \xi} \right].$$

After rearrangement, this can be written as

$$\begin{aligned} \beta g^{(2)} \frac{\partial u}{\partial \xi} &= -\frac{\partial [h^{(2)}+1]}{\partial \xi} + [h^{(2)}+1] \frac{\partial \gamma^{(2)}}{\partial \xi} + [h^{(2)}+1] \frac{\partial b^{(2)}}{\partial \xi} \\ &= -\frac{\partial c^{(2)}}{\partial \xi} + \frac{\partial b^{(2)}}{\partial \xi} + h^{(2)} \frac{\partial \gamma^{(2)}}{\partial \xi} + h^{(2)} \frac{\partial b^{(2)}}{\partial \xi}, \end{aligned} \tag{9.204}$$

where we have inserted $\gamma^{(2)} - h^{(2)} = -c^{(2)}$. The product $g^{(2)}[\partial u / \partial \xi]$ in the lhs appears in the integrand of Equation 9.199.

The first two term on the rhs of Equation 9.204 can be directly integrated over $0 \leq \xi \leq 1$ and yield $-c^{(2)} + b^{(2)}$ for $\xi = 1$ (the lower limit $\xi = 0$ gives zero). This gives a contribution $-\int d\mathbf{r}\, n\left\{-c^{(2)} + b^{(2)}\right\}$ in the rhs of Equation 9.199. The coupling-parameter integration of the second last term can also be done analytically and in Appendix 9C it is shown that

$$\begin{aligned}\int_0^1 d\xi \int d\mathbf{r}\, n(\mathbf{r}) h^{(2)}(\mathbf{r},\mathbf{r}';\xi) \frac{\partial \gamma^{(2)}(\mathbf{r},\mathbf{r}';\xi)}{\partial \xi} \\ = \frac{1}{2}\int d\mathbf{r}\, n(\mathbf{r}) \left[h^{(2)}(\mathbf{r},\mathbf{r}')\right]^2 - \frac{\gamma^{(2)}(\mathbf{r}',\mathbf{r}')}{2}.\end{aligned} \tag{9.205}$$

The ξ integration of the last term in Equation 9.204 yields by means of partial integration

$$\int_0^1 d\xi\, h^{(2)} \frac{\partial b^{(2)}}{\partial \xi} = \left[h^{(2)} b^{(2)}\right]_{\xi=0}^{\xi=1} - \int_0^1 d\xi \frac{\partial h^{(2)}}{\partial \xi} b^{(2)},$$

where the first term on the rhs equals $h^{(2)}b^{(2)}$ for the fully coupled particle with $\xi = 1$ (the lower limit $\xi = 0$ gives zero). Gathering these results and inserting them into Equation 9.199 we obtain Equation 9.201.

One can, in fact, explicitly perform the coupling-parameter integration of the rest-term $c^{(1)\mathrm{R}}(\mathbf{r}')$ in Equation 9.202 by making use of the functional expansion 9.160 for $b^{(2)}(\mathbf{r},\mathbf{r}')$. Thereby, one obtains $c^{(1)\mathrm{R}}$ as an infinite series involving the higher order direct correlation functions $c^{(m)}$ for $m \geq 3$, namely,

$$\begin{aligned} c^{(1)\mathrm{R}}(\mathbf{r}') &= \sum_{m=3}^{\infty} \frac{1}{m!} \int d\mathbf{s}_1 \ldots d\mathbf{s}_m\, c^{(m)}(\mathbf{s}_1, \ldots, \mathbf{s}_m) \\ &\times\; n(\mathbf{s}_1) h^{(2)}(\mathbf{s}_1,\mathbf{r}') \ldots n(\mathbf{s}_m) h^{(2)}(\mathbf{s}_m,\mathbf{r}'), \end{aligned} \tag{9.206}$$

as shown in Appendix 9C. In this equation all functions are evaluated for the fully coupled fluid, but, like in the functional expansion 9.160 for $b^{(2)}(\mathbf{r},\mathbf{r}')$, Equation 9.206 is not useful as it stands for practical numerical calculations. Instead one makes approximations for $c^{(1)\mathrm{R}}$ as described later in Section 10.1.5. Thereby, one can calculate $c^{(1)}$ from Equation 9.201 and hence μ.

When $c^{(1)}(\mathbf{r}')$ has been calculated, one can obtain the density distribution $n(\mathbf{r}')$ of an inhomogeneous fluid from

$$n(\mathbf{r}') = n^{\mathrm{b}} e^{-\beta v(\mathbf{r}') + c^{(1)}(\mathbf{r}') - c^{(1)\mathrm{b}}} = \zeta e^{-\beta v(\mathbf{r}') + c^{(1)}(\mathbf{r}')} \tag{9.207}$$

(Equation 9.15), where $c^{(1)\mathrm{b}} = -\beta\mu^{\mathrm{ex,b}}$ is the value of $c^{(1)}$ in the bulk (cf. Table 9.1) and $\zeta = e^{\beta\mu}/\Lambda^3$ is the activity. Recall that ζ is given in terms of the activity coefficient γ^{b} of the bulk phase as $\zeta = n^{\mathrm{b}}\gamma^{\mathrm{b}} = n^{\mathrm{b}}\exp(\beta\mu^{\mathrm{ex,b}})$ (do not confuse γ^{b} with the correlation function $\gamma^{(2)}$). This expression is particularly useful when the activity of the system is known, for example when the inhomogeneous fluid is in equilibrium with a bulk fluid with a given density and known $\gamma^{\rm b}$.

The density distribution can alternatively be written in terms of the singlet bridge function $b^{(1)}$ as

$$n(\mathbf{r}') = n^{\rm b} e^{-\beta v(\mathbf{r}')+\gamma^{(1)}(\mathbf{r}')+b^{(1)}(\mathbf{r}')}$$

(Equation 9.61), where $\gamma^{(1)}(\mathbf{r}') = \int d\mathbf{r}\, c^{(2)\rm b}(|\mathbf{r}-\mathbf{r}'|)\,[n(\mathbf{r})-n^{\rm b}]$. By comparing the exponent with that in Equation 9.207 we see that

$$b^{(1)}(\mathbf{r}') = c^{(1)}(\mathbf{r}') - c^{(1)\rm b} - \int d\mathbf{r}\, c^{(2)\rm b}(|\mathbf{r}-\mathbf{r}'|)\left[n(\mathbf{r}) - n^{\rm b}\right].$$

By inserting the expression 9.201 for $c^{(1)}$ and Equation 9.203 for $c^{(1)\rm b}$ we obtain an expression for $b^{(1)}$ that corresponds to Equation 9.201. After simplification we get

$$\begin{aligned} b^{(1)}(\mathbf{r}') &= -\int d\mathbf{r}\,\Delta\left\{n(\mathbf{r})\left(\frac{1}{2}\left[h^{(2)}(\mathbf{r},\mathbf{r}')\right]^2 + g^{(2)}(\mathbf{r},\mathbf{r}')b^{(2)}(\mathbf{r},\mathbf{r}')\right)\right\} \\ &+ \int d\mathbf{r}\, n(\mathbf{r})\Delta c^{(2)}(\mathbf{r},\mathbf{r}') + \frac{\Delta\gamma^{(2)}(\mathbf{r}',\mathbf{r}')}{2} + \Delta c^{(1)\rm R}(\mathbf{r}'), \quad (9.208) \end{aligned}$$

where Δ in front of any function F means $\Delta F(\mathbf{r},\mathbf{r}') = F(\mathbf{r},\mathbf{r}') - F^{\rm b}(|\mathbf{r}-\mathbf{r}'|)$ with $F^{\rm b}$ equal to F evaluated in the bulk phase. This meaning of Δ applies also for the first term where F is given by an expression.

★ *Different kinds of one-particle bridge and direct correlation functions.*

The terminology for bridge and direct correlation functions is unfortunately not universal in the literature. The definitions adopted for the singlet bridge and pair direct-correlation functions in the current treatise are the most common ones, but there exist other functions that have been called *one-particle bridge functions* and *one-particle direct correlation functions* in the literature. In contrast, the nomenclature for the pair bridge function $b^{(2)}(\mathbf{r},\mathbf{r}')$ and the pair direct-correlation function $c^{(2)}(\mathbf{r},\mathbf{r}')$ is virtually universal, although the notation "$b^{(2)}$" may vary and sometimes the sign of the pair bridge function can be the opposite. In order to help the reader to understand the literature, we will here explain the most important one-particle functions that were initially introduced in the work by Georgy A. Martynov and subsequently used in some publications by various authors.[60]

[60]G. A. Martynov, Mol. Phys. **42** (1981) 329 (https://doi.org/10.1080/00268978100100291); G. A. Martynov, 1992, *Fundamental Theory of Liquids, Method of Distribution Functions*, Bristol: Adam Hilger. See also the review by J.-M. Bomont, 2008, *Recent Advances in the Field of Integral Equation Theories: Bridge Functions and Applications to Classical Fluids*, in Advances in Chemical Physics, Vol. 139, p. 1, edited by Stuart A. Rice, Hoboken: Wiley (https://doi.org/10.1002/9780470259498.ch1). This review article uses the Martynov nomenclature.

Following Martynov, one uses the singlet and pair **thermal potentials** $\omega_{\mathrm{M}}^{(1)}$ and $\omega_{\mathrm{M}}^{(2)}$ that can be defined from

$$\begin{aligned}\omega_{\mathrm{M}}^{(1)}(\mathbf{r}) &\equiv \Delta c^{(1)}(\mathbf{r}) = -\beta w_i^{\mathrm{intr}}(\mathbf{r}) \\ \omega_{\mathrm{M}}^{(2)}(\mathbf{r},\mathbf{r}') &\equiv \Delta c^{(1)}(\mathbf{r}|\mathbf{r}') = -\beta w_{ij}^{(2)\mathrm{intr}}(\mathbf{r},\mathbf{r}')\end{aligned}$$

in our notation, so we have

$$\begin{aligned}\omega_{\mathrm{M}}^{(1)}(\mathbf{r}) &= c^{(1)}(\mathbf{r}) - c^{(1)b} = c^{(1)}(\mathbf{r}) + \beta\mu^{\mathrm{ex,b}} \\ \omega_{\mathrm{M}}^{(2)}(\mathbf{r},\mathbf{r}') &= \gamma^{(2)}(\mathbf{r},\mathbf{r}') + b^{(2)}(\mathbf{r},\mathbf{r}').\end{aligned}$$

We will use the ω_{M} notation in the current text box only (subscript M stands for "Martynov"). Distinguish ω_{M} from other entities that are denoted by ω in the current treatise.

In this nomenclature one uses a kind of singlet "bridge function," here denoted $B_{\mathrm{M}}^{(1)}$, that is defined such that $\omega_{\mathrm{M}}^{(1)}$ is given by the following relationship that equivalent to our Equation 9.201 for $c^{(1)}$

$$\begin{aligned}\omega_{\mathrm{M}}^{(1)}(\mathbf{r}') &= \int d\mathbf{r}\, n(\mathbf{r}) \left[h^{(2)}(\mathbf{r},\mathbf{r}') - \omega_{\mathrm{M}}^{(2)}(\mathbf{r},\mathbf{r}') - \frac{1}{2}h^{(2)}(\mathbf{r},\mathbf{r}')\omega_{\mathrm{M}}^{(2)}(\mathbf{r},\mathbf{r}')\right] + \\ &+ B_{\mathrm{M}}^{(1)}(\mathbf{r}') + \beta\mu^{\mathrm{ex,b}}. \qquad (9.209)\end{aligned}$$

This implies that

$$B_{\mathrm{M}}^{(1)}(\mathbf{r}') = c^{(1)\mathrm{R}}(\mathbf{r}') - \frac{1}{2}\int d\mathbf{r}\, n(\mathbf{r}) h^{(2)}(\mathbf{r},\mathbf{r}') b^{(2)}(\mathbf{r},\mathbf{r}') \qquad (9.210)$$

in our notation. Note that $B_{\mathrm{M}}^{(1)}$ is different from the singlet bridge function $b^{(1)}$. Using Equations 9.160 and 9.206, one readily obtains from Equation 9.210 the functional expansion

$$\begin{aligned}B_{\mathrm{M}}^{(1)}(\mathbf{r}') &= -\frac{1}{2}\sum_{m=3}^{\infty}\frac{m-2}{m!}\int d\mathbf{s}_1 \ldots d\mathbf{s}_m\, c^{(m)}(\mathbf{s}_1, \ldots, \mathbf{s}_m) \\ &\times\; n(\mathbf{s}_1)h^{(2)}(\mathbf{s}_1,\mathbf{r}')\ldots n(\mathbf{s}_m)h^{(2)}(\mathbf{s}_m,\mathbf{r}'), \qquad (9.211)\end{aligned}$$

where we have made the variable substitution $\mathbf{s}_m = \mathbf{r}$ and used the symmetry of $c^{(m)}$.

Equation 9.209 can be written as

$$\begin{aligned}\omega_{\mathrm{M}}^{(1)}(\mathbf{r}') &= \int d\mathbf{r}\, n(\mathbf{r}) \Big[h^{(2)}(\mathbf{r},\mathbf{r}') - \omega_{\mathrm{M}}^{(2)}(\mathbf{r},\mathbf{r}') \\ &- \frac{1}{2}h^{(2)}(\mathbf{r},\mathbf{r}')\left(\omega_{\mathrm{M}}^{(2)}(\mathbf{r},\mathbf{r}') + B_{\mathrm{M}}^{(2)}(\mathbf{r},\mathbf{r}')\right)\Big] + \beta\mu^{\mathrm{ex,b}}, \qquad (9.212)\end{aligned}$$

where $B^{(2)}_{\rm M}$ is yet another kind of "bridge function" that satisfies

$$B^{(1)}_{\rm M}(\mathbf{r}') = -\frac{1}{2}\int d\mathbf{r}\, n(\mathbf{r})h^{(2)}(\mathbf{r},\mathbf{r}')B^{(2)}_{\rm M}(\mathbf{r},\mathbf{r}'). \tag{9.213}$$

and can be expressed as

$$\begin{aligned} B^{(2)}_{\rm M}(\mathbf{r},\mathbf{r}') &= \sum_{m=3}^{\infty}\frac{m-2}{m!}\int d\mathbf{s}_1 \ldots d\mathbf{s}_{m-1}\, c^{(m)}(\mathbf{r},\mathbf{s}_1,\ldots,\mathbf{s}_{m-1}) \\ &\times\; n(\mathbf{s}_1)h^{(2)}(\mathbf{s}_1,\mathbf{r}')\ldots n(\mathbf{s}_{m-1})h^{(2)}(\mathbf{s}_{m-1},\mathbf{r}'). \end{aligned} \tag{9.214}$$

By inserting this expansion for $B^{(2)}_{\rm M}$ into Equation 9.213 we obtain Equation 9.211, whereby we again have made the variable substitution $\mathbf{s}_m = \mathbf{r}$ and used the symmetry of $c^{(m)}$. The functional expansion 9.214 for $B^{(2)}_{\rm M}$ differs from that for $b^{(2)}$ in Equation 9.160 only by the coefficient in front of the integral.

Despite that $B^{(2)}_{\rm M}$ is a pair function, it has been called "one-particle bridge function" and denoted $B^{(1)}$, where, again, the superscript (1) does *not* mean that it is a singlet function. It has been denoted as a "*bridge function of the first kind*" while the usual pair bridge function $b^{(2)}$ then is a "*bridge function of the second kind.*" This terminology is not the common one and will not be used here.

The function $c^{(1)\rm R}(\mathbf{r}')$ in our notation can be obtained from Equations 9.210 and 9.213 as

$$c^{(1)\rm R}(\mathbf{r}') = \frac{1}{2}\int d\mathbf{r}\, n(\mathbf{r})h^{(2)}(\mathbf{r},\mathbf{r}')\left[b^{(2)}(\mathbf{r},\mathbf{r}') - B^{(2)}_{\rm M}(\mathbf{r},\mathbf{r}')\right]. \tag{9.215}$$

By inserting the functional expansions for $b^{(2)}$ and $B^{(2)}_{\rm M}$ in Equations 9.160 and 9.214, respectively, we obtain the expansion 9.206 for $c^{(1)\rm R}$.

In the Martynov nomenclature one uses a kind of "direct correlation function" $C^{(2)}_{\rm M}$ that satisfies

$$\omega^{(1)}_{\rm M}(\mathbf{r}') = \int d\mathbf{r}\, n(\mathbf{r})C^{(2)}_{\rm M}(\mathbf{r},\mathbf{r}') + \beta\mu^{\rm ex,b}, \tag{9.216}$$

which means that

$$c^{(1)}(\mathbf{r}') = \int d\mathbf{r}\, n(\mathbf{r})C^{(2)}_{\rm M}(\mathbf{r},\mathbf{r}')$$

in our notation. $C^{(2)}_{\rm M}$ is given by

$$C^{(2)}_{\rm M}(\mathbf{r},\mathbf{r}') = h^{(2)}(\mathbf{r},\mathbf{r}') - \omega^{(2)}_{\rm M}(\mathbf{r},\mathbf{r}') - \frac{1}{2}h^{(2)}(\mathbf{r},\mathbf{r}')\left(\omega^{(2)}_{\rm M}(\mathbf{r},\mathbf{r}') + B^{(2)}_{\rm M}(\mathbf{r},\mathbf{r}')\right)$$

and constitute the terms within the square brackets in Equation 9.212, which is obtained when this expression is inserted into Equation 9.216. Despite that

$C_{\mathrm{M}}^{(2)}$ is a pair function, it has been called "one-particle direct-correlation function" and denoted $C^{(1)}$ or $c^{(1)}$, where superscript (1) does *not* imply that it is a singlet function. The superscript "(1)" instead indicates that it is a "direct correlation function of the first order" in Martynov's nomenclature, which we do not use here. It is, of course, different from the commonly used singlet (one-particle) direct-correlation function $c^{(1)}$ and also different from the usual pair direct-correlation function $c^{(2)}$.

For a bulk fluid $\omega_{\mathrm{M}}^{(1)} = 0$ so Equation 9.212 can be written as

$$\begin{aligned}\beta\mu^{\mathrm{ex,b}} &= -n^{\mathrm{b}} \int d\mathbf{r} \left[h^{(2)\mathrm{b}}(r) - \omega_{\mathrm{M}}^{(2)\mathrm{b}}(r) \right. \\ &\quad \left. - \frac{1}{2} h^{(2)\mathrm{b}}(r) \left(\omega_{\mathrm{M}}^{(2)\mathrm{b}}(r) + B_{\mathrm{M}}^{(2)\mathrm{b}}(r) \right) \right] \end{aligned} \tag{9.217}$$

and can hence be used to calculate the excess chemical potential in the bulk. This expression is equivalent to Equation 9.203, only the notations differ.

9.3 THE BASIS OF DENSITY FUNCTIONAL THEORY (DFT)

Density functional theory (DFT) is in principle a fundamental and exact manner to formulate statistical mechanics. There exist a closely related density functional theory for quantum mechanics, but in this treatise we will solely consider DFT in classical statistical mechanics. The main practical use of the theory is to formulate approximations of various degrees of sophistication. Such approximations are also usually called "density functional theory."

9.3.1 Basic formalism

From the basic results of classical statistical mechanics presented in Chapter 3 in the Volume I it follows that distribution probabilities for the particles in an open system at equilibrium can be written in terms of the product of a canonical ensemble probability for N particles, $\mathcal{P}_N^{(N)\mathrm{tot}}$ defined below, and the probability $\mathscr{P}(N)$ that there are N particles in the system. Therefore, in the grand canonical ensemble the probability density $\mathcal{P}^{(N)\mathrm{tot}}(\mathbf{r}^N, \mathbf{p}^N)$ that the system contains N particles and that these particles have coordinates $\mathbf{r}^N \equiv \mathbf{r}_1, \mathbf{r}_2, \ldots, \mathbf{r}_N$ and momenta $\mathbf{p}^N \equiv \mathbf{p}_1, \mathbf{p}_2, \ldots, \mathbf{p}_N$ is

$$\mathcal{P}^{(N)\mathrm{tot}}(\mathbf{r}^N, \mathbf{p}^N) = \mathcal{P}_N^{(N)\mathrm{tot}}(\mathbf{r}^N, \mathbf{p}^N)\mathscr{P}(N),$$

where

$$\mathscr{P}(N) = \frac{Q_N e^{\beta\mu N}}{\Xi} \tag{9.218}$$

(Equation I:2.112), and Q_N and Ξ are the canonical and grand canonical partition functions, respectively. Subscript "N" on $\mathcal{P}_N^{(N)\mathrm{tot}}$ and Q_N means that there are N particles in the system (in the canonical ensemble). Superscript "tot" indicates that the probabilities are based on the total energy $\breve{U}_N^{\text{tot}}(\mathbf{r}^N,\mathbf{p}^N)$ so they concern both the spatial coordinates and momenta of the particles, $\mathbf{r}^N$ and $\mathbf{p}^N$. However, for the grand canonical probability we will henceforth use the notation $\mathcal{P}_{\text{tot}}^{(N)} \equiv \mathcal{P}^{(N)\text{tot}}$ for compactness.

As we saw in Section 3.2 (in Volume I), we have

$$\mathcal{P}_N^{(N)\text{tot}}(\mathbf{r}^N,\mathbf{p}^N) = \frac{e^{-\beta \breve{U}_N^{\text{tot}}(\mathbf{r}^N,\mathbf{p}^N)}}{Q_N h^{3N} N!} = \frac{e^{-\beta \breve{U}_N^{\text{tot}}(\mathbf{r}^N,\mathbf{p}^N)}}{\int d\mathbf{r}'^N \int d\mathbf{p}'^N e^{-\beta \breve{U}_N^{\text{tot}}(\mathbf{r}'^N,\mathbf{p}'^N)}},$$

where

$$\begin{aligned} \breve{U}_N^{\text{tot}}(\mathbf{r}^N,\mathbf{p}^N) &= \breve{U}_N^{\text{kin}}(\mathbf{p}^N) + \breve{U}_N^{\text{pot}}(\mathbf{r}^N) \\ &= \breve{U}_N^{\text{kin}}(\mathbf{p}^N) + \breve{U}_N^{\text{intr}}(\mathbf{r}^N) + \sum_{\nu=1}^{N} v(\mathbf{r}_\nu) \end{aligned}$$

is the sum of total kinetic and potential energies of the N particles (with $\breve{U}_N^{\text{intr}}$ being the interparticle interaction energy) and where Q_N is given by

$$Q_N = \frac{1}{h^{3N} N!} \int d\mathbf{r}^N \int d\mathbf{p}^N e^{-\beta \breve{U}_N^{\text{tot}}(\mathbf{r}^N,\mathbf{p}^N)} \tag{9.219}$$

(Equation I:3.5). The complete expression for the probability density in the grand canonical ensemble is accordingly

$$\mathcal{P}_{\text{tot}}^{(N)}(\mathbf{r}^N,\mathbf{p}^N) \equiv \mathcal{P}^{(N)\text{tot}}(\mathbf{r}^N,\mathbf{p}^N) = \frac{e^{-\beta[\breve{U}_N^{\text{tot}}(\mathbf{r}^N,\mathbf{p}^N) - \mu N]}}{\Xi\, h^{3N} N!} \tag{9.220}$$

and we have $\Xi = \sum_{N=0}^{\infty} Q_N \exp(\beta\mu N)$ (Equation I:3.35).

The average of a quantity X, which has $\mathbf{r}$ and $\mathbf{p}$ dependence $\breve{X}_N(\mathbf{r}^N,\mathbf{p}^N)$, is given by

$$\langle X \rangle = \sum_{N=0}^{\infty} \int d\mathbf{r}^N \int d\mathbf{p}^N \mathcal{P}_{\text{tot}}^{(N)}(\mathbf{r}^N,\mathbf{p}^N) \breve{X}_N(\mathbf{r}^N,\mathbf{p}^N) \equiv \sum_N \iint_{\{\mathbf{r},\mathbf{p}\}} \left(\mathcal{P}_{\text{tot}}^{(N)} \breve{X}_N\right),$$

where we have introduced a simplified notation for the average expression in the rhs.

The grand potential Θ can be written as

$$\Theta = \bar{U} - \mu \bar{N} - TS \tag{9.221}$$

(Equation I:2.118). In quantum statistical mechanics the entropy can be expressed in the form

$$S = -k_B \sum_{N,i} \mathfrak{p}(\mathscr{U}_i(N),N) \ln \mathfrak{p}(\mathscr{U}_i(N),N) = -k_B \left\langle \ln \mathfrak{p}(\mathscr{U}_i(N),N) \right\rangle$$

(Equation I:2.120), where i is the quantum state index, $\mathscr{U}_i$ is the energy of state i and

$$\mathfrak{p}(\mathscr{U}_i, N) = \mathscr{P}(N)\frac{e^{-\beta\mathscr{U}_i(N)}}{Q_N} = \frac{e^{-\beta[\mathscr{U}_i(N)-\mu N]}}{\Xi}. \tag{9.222}$$

The corresponding expression for S in classical statistical mechanics can be obtained by comparing Equations 9.220 and 9.222, whereby it is apparent that the probability that corresponds to $\mathfrak{p}$ is equal to $\mathcal{P}^{(N)}_{\text{tot}} h^{3N} N!$ (see also the discussion regarding Equation I:3.4 in Section 3.2 in Volume I). This means that

$$S = -k_B \left\langle \ln(\mathcal{P}^{(N)}_{\text{tot}} h^{3N} N!) \right\rangle$$

and hence Equation 9.221 can be written

$$\Theta = \left\langle \breve{U}^{\text{tot}}_N - \mu N + k_B T \ln(\mathcal{P}^{(N)}_{\text{tot}} h^{3N} N!) \right\rangle .$$

Explicitly, we have

$$\begin{aligned}\Theta &= \sum_{N=0}^{\infty} \int d\mathbf{r}^N \int d\mathbf{p}^N \mathcal{P}^{(N)}_{\text{tot}}(\mathbf{r}^N, \mathbf{p}^N) \left\{ \breve{U}^{\text{tot}}_N(\mathbf{r}^N, \mathbf{p}^N) - \mu N \right. \\ &+ \left. k_B T \ln\left(\mathcal{P}^{(N)}_{\text{tot}}(\mathbf{r}^N, \mathbf{p}^N) h^{3N} N!\right)\right\}\end{aligned}$$

or in the simplified notation[61]

$$\Theta = \Theta\left[\{\mathcal{P}^{(N)}_{\text{tot}}\}\right] = \sum_{N} \iint_{\{\mathbf{r},\mathbf{p}\}} \left(\mathcal{P}^{(N)}_{\text{tot}} \left\{\breve{U}^{\text{tot}}_N - \mu N + k_B T \ln\left(\mathcal{P}^{(N)}_{\text{tot}} h^{3N} N!\right)\right\}\right), \tag{9.223}$$

where we have written $\Theta = \Theta\left[\{\mathcal{P}^{(N)}_{\text{tot}}\}\right]$ in order to emphasize that Θ is a functional of the set of probabilities $\mathcal{P}^{(N)}_{\text{tot}}$ for $N \geq 0$.

9.3.2 Variational principles

$\mathcal{P}^{(N)}_{\text{tot}}$ in the expression 9.223 is the equilibrium probability density for N particles, but we can use the same expression as a *definition* of a functional

[61] In a commonly used alternative notation one writes the probability density as $\mathfrak{p}^{(N)}(\mathbf{r}^N, \mathbf{p}^N) \equiv \mathcal{P}^{(N)}_{\text{tot}}(\mathbf{r}^N, \mathbf{p}^N) h^{3N} N!$ and then the grand canonical average is expressed as

$$\langle X \rangle = \sum_{N=0}^{\infty} (h^{3N} N!)^{-1} \int d\mathbf{r}^N \int d\mathbf{p}^N \mathfrak{p}^{(N)}(\mathbf{r}^N, \mathbf{p}^N) \breve{X}_N(\mathbf{r}^N, \mathbf{p}^N) = Tr_{\text{cl}}\left(\mathfrak{p}^{(N)} \breve{X}_N\right),$$

where Tr_{cl} is the "classical trace" defined as $Tr_{\text{cl}} = \sum_{N=0}^{\infty} (h^{3N} N!)^{-1} \int d\mathbf{r}^N \int d\mathbf{p}^N$. Thereby $\Theta = \Theta[\mathfrak{p}^{(N)}] = Tr_{\text{cl}}\left(\mathfrak{p}^{(N)}\{\breve{U}^{\text{tot}}_N - \mu N + k_B T \ln \mathfrak{p}^{(N)}\}\right)$. We do not use this notation here. Instead we adhere to the notation that is used in the rest of this treatise. Note that in our simplified notation the symbol $\sum_{N} \iint_{\{r,\,p\}}$ corresponds to the classical trace.

$\acute{\Theta}\left[\{\mathscr{P}_{\text{tot}}^{(N)}\}\right]$ for *any* set of probability densities $\mathscr{P}_{\text{tot}}^{(N)}(\mathbf{r}^N, \mathbf{p}^N)$

$$\acute{\Theta}\left[\{\mathscr{P}_{\text{tot}}^{(N)}\}\right] = \sum_N \iint_{\{\mathbf{r},\mathbf{p}\}} \left(\mathscr{P}_{\text{tot}}^{(N)}\left\{\breve{U}_N^{\text{tot}} - \mu N + k_B T \ln(\mathscr{P}_{\text{tot}}^{(N)} h^{3N} N!)\right\}\right). \tag{9.224}$$

The diacritic mark above $\acute{\Theta}$ is used to distinguish it from the equilibrium Θ. When $\mathscr{P}_{\text{tot}}^{(N)} = \mathcal{P}_{\text{tot}}^{(N)}$ for all N the functional $\acute{\Theta}$ assumes the value Θ, that is, the grand potential for the equilibrium system. We are now going to see that for *any other* probability densities $\mathscr{P}_{\text{tot}}^{(N)}$, we have $\acute{\Theta}\left[\{\mathscr{P}_{\text{tot}}^{(N)}\}\right] > \acute{\Theta}\left[\{\mathcal{P}_{\text{tot}}^{(N)}\}\right] = \Theta$.

From the logarithm of Equation 9.220 it follows that

$$\breve{U}_N^{\text{tot}} - \mu N = -k_B T \ln(\Xi h^{3N} N! \mathcal{P}_{\text{tot}}^{(N)}) = \Theta - k_B T \ln(h^{3N} N!) - k_B T \ln \mathcal{P}_{\text{tot}}^{(N)},$$

where we have used $\Theta = -k_B T \ln \Xi$. By inserting this into Equation 9.224 we obtain

$$\begin{aligned} \acute{\Theta}\left[\{\mathscr{P}_{\text{tot}}^{(N)}\}\right] &= \sum_N \iint_{\{\mathbf{r},\mathbf{p}\}} \left(\mathscr{P}_{\text{tot}}^{(N)}\left\{\Theta - k_B T \ln \mathcal{P}_{\text{tot}}^{(N)} + k_B T \ln \mathscr{P}_{\text{tot}}^{(N)}\right\}\right) \\ &= \Theta - k_B T\left[\sum_N \iint_{\{\mathbf{r},\mathbf{p}\}} \left(\mathscr{P}_{\text{tot}}^{(N)} \ln \mathcal{P}_{\text{tot}}^{(N)}\right) - \sum_N \iint_{\{\mathbf{r},\mathbf{p}\}} \left(\mathscr{P}_{\text{tot}}^{(N)} \ln \mathscr{P}_{\text{tot}}^{(N)}\right)\right]. \end{aligned} \tag{9.225}$$

Now, for any two probability densities $\mathscr{P}_{\text{A}}^{(N)}$ and $\mathscr{P}_{\text{B}}^{(N)}$, we have from Gibbs' inequality 9.267 derived in Appendix 9D

$$\sum_N \iint_{\{\mathbf{r},\mathbf{p}\}} \left(\mathscr{P}_{\text{B}}^{(N)} \ln \mathscr{P}_{\text{A}}^{(N)}\right) - \sum_N \iint_{\{\mathbf{r},\mathbf{p}\}} \left(\mathscr{P}_{\text{B}}^{(N)} \ln \mathscr{P}_{\text{B}}^{(N)}\right) \leq 0.$$

This inequality is valid for any kind of probability density. By applying this to $\mathscr{P}_{\text{A}}^{(N)} = \mathcal{P}_{\text{tot}}^{(N)}$ and $\mathscr{P}_{\text{B}}^{(N)} = \mathscr{P}_{\text{tot}}^{(N)}$ we can conclude from Equation 9.225 that

$$\acute{\Theta}\left[\{\mathscr{P}_{\text{tot}}^{(N)}\}\right] \geq \Theta = \acute{\Theta}\left[\{\mathcal{P}_{\text{tot}}^{(N)}\}\right], \tag{9.226}$$

where equality applies only when $\mathscr{P}_{\text{tot}}^{(N)} = \mathcal{P}_{\text{tot}}^{(N)}$ for all N. This means that *the equilibrium probability density minimizes* $\acute{\Theta}\left[\{\mathscr{P}_{\text{tot}}^{(N)}\}\right]$. This is, as we will see, a very important result. It is closely related to the fact that the equilibrium state of a system in the grand canonical ensemble corresponds to the minimal value of the grand potential.

In Section 9.2.1 we noted that the grand potential is a functional of the equilibrium density distribution $n(\mathbf{r})$. This is a consequence of the fact (proven in Appendix 9D) that there is a one-to-one relationship between the external potential $v(\mathbf{r}')$ and $n(\mathbf{r})$, so v is a functional of n: we have $v = v(\mathbf{r}'; [n])$. Since Q_N is a functional of v (via $\breve{U}_N^{\text{tot}}$ in Equation 9.219), it is also a functional of

n. The same applies therefore to Ξ, $\Theta = -k_BT \ln \Xi$ and $\mathcal{P}^{(N)}_{\text{tot}}$ via their dependence of Q_N and we write $\Xi = \Xi[n]$, $\Theta = \Theta[n]$ and $\mathcal{P}^{(N)}_{\text{tot}} = \mathcal{P}^{(N)}_{\text{tot}}(\mathbf{r}^N, \mathbf{p}^N; [n])$.

The grand potential can be written $\Theta = A - \mu\bar{N} = A - \mu \int d\mathbf{r}\, n(\mathbf{r})$, where $A = \bar{U} - TS$ is Helmholtz free energy. In Section 9.2.1 we introduced the intrinsic Helmholtz free energy $\mathcal{A}$ in Equation 9.122 and since $A = \mathcal{A} + \int d\mathbf{r}\, n(\mathbf{r})v(\mathbf{r})$ it follows that

$$\Theta[n] = \mathcal{A}[n] + \int d\mathbf{r}\, n(\mathbf{r})v(\mathbf{r}) - \mu \int d\mathbf{r}\, n(\mathbf{r}), \qquad (9.227)$$

where we have explicitly shown that both Θ and $\mathcal{A}$ are functionals of $n(\mathbf{r})$. The functional $\mathcal{A}[n]$ is known if $\Theta[n]$ is known and vice versa. Since we have

$$\langle U^{\text{pot}} \rangle = \langle U^{\text{intr}} \rangle + \int d\mathbf{r}\, n(\mathbf{r})v(\mathbf{r})$$

(Equation I:5.112) it follows from Equation 9.223 that

$$\mathcal{A}\left[\{\mathcal{P}^{(N)}_{\text{tot}}\}\right] = \sum_N \iint_{\{\mathbf{r},\mathbf{p}\}} \left(\mathcal{P}^{(N)}_{\text{tot}} \left[\breve{U}^{\text{kin}}_N + \breve{U}^{\text{intr}}_N + k_BT \ln(\mathcal{P}^{(N)}_{\text{tot}} h^{3N} N!)\right]\right).$$

For any probability density $\mathscr{P}^{(N)}_{\text{tot}}(\mathbf{r}^N, \mathbf{p}^N)$ that is not the equilibrium one can calculate a value of $\mathcal{A}$ from this equation by inserting $\mathscr{P}^{(N)}_{\text{tot}}$ instead of $\mathcal{P}^{(N)}_{\text{tot}}$. The resulting functional is denoted $\acute{\mathcal{A}}\left[\{\mathscr{P}^{(N)}_{\text{tot}}\}\right]$. The value of $\acute{\mathcal{A}}$ is independent of the external potential since $\breve{U}^{\text{kin}}_N$ and $\breve{U}^{\text{intr}}_N$ do not depend on $v(\mathbf{r})$, which emphasizes that $\acute{\mathcal{A}}$ is truly an *intrinsic property of the fluid itself.* Given a set of probability densities $\mathscr{P}^{(N)}_{\text{tot}}$ one can, of course, calculate the corresponding average density distribution, which we denote $\acute{n}(\mathbf{r})$. In general it is different from the equilibrium distribution $n(\mathbf{r})$ for the system in presence of the external potential $v(\mathbf{r}')$. The equilibrium density distribution $n(\mathbf{r})$ corresponds to the minimal value of $\acute{\Theta}\left[\{\mathscr{P}^{(N)}_{\text{tot}}\}\right]$ according to Equation 9.226.

Let us summarize the two most important results we have obtained so far in this section:

(i) The intrinsic Helmholtz free energy $\mathcal{A}$ and the grand potential Θ are unique functionals of the equilibrium density distribution $n(\mathbf{r})$ for given T, μ and functions $\breve{U}^{\text{kin}}_N$ and $\breve{U}^{\text{intr}}_N$ of the particle momenta and coordinates, respectively.

(ii) Consider some singlet density distribution $\acute{n}(\mathbf{r})$. For a given external potential $v(\mathbf{r})$, the functional

$$\acute{\Theta}[\acute{n}] = \acute{\mathcal{A}}[\acute{n}] + \int d\mathbf{r}\, \acute{n}(\mathbf{r})v(\mathbf{r}) - \mu \int d\mathbf{r}\, \acute{n}(\mathbf{r})$$

assumes its minimal value when $\acute{n}(\mathbf{r})$ is equal to the equilibrium distribution $n(\mathbf{r})$; the minimal value is $\acute{\Theta}[n] = \Theta =$ the grand potential of the system. These results constitute the so-called **Hohenberg-Kohn-Mermin theorems**.

Since $\acute{\Theta}$ assumes its minimum when $\acute{n}(\mathbf{r}) = n(\mathbf{r})$, the latter result implies the variational principle

$$\left.\frac{\delta\acute{\Theta}[\acute{n}]}{\delta\acute{n}(\mathbf{r}')}\right|_{\acute{n}(\mathbf{r}')=n(\mathbf{r}')} = 0, \tag{9.228}$$

which yields

$$\left.\frac{\delta\acute{\mathcal{A}}[\acute{n}]}{\delta\acute{n}(\mathbf{r}')}\right|_{\acute{n}(\mathbf{r}')=n(\mathbf{r}')} + v(\mathbf{r}) - \mu = 0.$$

We thus have

$$\frac{\delta\mathcal{A}[n]}{\delta n(\mathbf{r})} + v(\mathbf{r}) - \mu = 0, \tag{9.229}$$

which is the so-called **Euler-Lagrange equation** that results from the variational principle above. The functional $\mathcal{A}[n]$ can be split in an ideal and an excess part and we have

$$\begin{aligned} \mathcal{A}[n] &= \mathcal{A}^{\text{ideal}}[n] + \mathcal{A}^{\text{ex}}[n] \\ &= k_B T \int d\mathbf{r}\, n(\mathbf{r}) \left[\ln(\Lambda^3 n(\mathbf{r})) - 1\right] + \mathcal{A}^{\text{ex}}[n] \end{aligned} \tag{9.230}$$

($\mathcal{A}^{\text{ideal}}$ is given in Equation 9.120). By performing the functional derivative in Equation 9.229 we can write the Euler-Lagrange equation as

$$k_B T \ln(\Lambda^3 n(\mathbf{r})) - k_B T\, c^{(1)}(\mathbf{r}) + v(\mathbf{r}) - \mu = 0, \tag{9.231}$$

where we have used our previous results from Section 9.2.1

$$\frac{\delta\mathcal{A}^{\text{ideal}}}{\delta n(\mathbf{r})} = k_B T \ln(\Lambda^3 n(\mathbf{r}))$$

and

$$\frac{\delta\mathcal{A}^{\text{ex}}}{\delta n(\mathbf{r})} = -k_B T\, c^{(1)}(\mathbf{r}).$$

Equation 9.231 is the usual equilibrium condition (Equation 9.11) and it implies that

$$n(\mathbf{r}) = \zeta e^{-\beta v(\mathbf{r}) + c^{(1)}(\mathbf{r})}, \tag{9.232}$$

which is the same as Equation 9.13 with $\zeta = e^{\beta\mu}/\Lambda^3$. Since $\mathcal{A}[n]$ is a unique functional and $\mathcal{A}^{\text{ideal}}[n]$ is explicitly known, $\mathcal{A}^{ex}[n]$ is also a unique functional of $n(\mathbf{r})$. The basic equation 9.232 for the density distribution is accordingly *equivalent to the Euler-Lagrange equation obtained from the variational principle of minimization of the grand potential as functional of the density distribution.*

Since we are dealing with equilibrium systems, we can, in fact, assume that the momentum distribution for the particles are at thermal equilibrium for the given temperature, which means that the momenta are Maxwell-Boltzmann distributed (see Section 3.2 in Volume I). This simplifies the formalism a bit because it implies that we can write

$$\begin{aligned}\mathscr{P}^{(N)}_{\text{tot}}(\mathbf{r}^N,\mathbf{p}^N) &= \mathscr{P}^{(N)}(\mathbf{r}^N)\prod_{\nu=1}^{N}\frac{e^{-\frac{\beta p_\nu^2}{2m_\nu}}}{(2\pi m k_B T)^{3/2}}\\ &= \mathscr{P}^{(N)}(\mathbf{r}^N)\left(\frac{\Lambda}{h}\right)^{3N}\prod_{\nu=1}^{N}e^{-\frac{\beta p_\nu^2}{2m_\nu}},\end{aligned}$$

where $\mathscr{P}^{(N)}(\mathbf{r}^N)$ is a configurational probability density for the particle coordinates and Λ is the thermal de Broglie wave length. Then, since $\breve{U}_N^{\text{kin}} = \sum_{\nu=1}^{N}\frac{p_\nu^2}{2m_\nu}$, we have

$$\acute{\mathcal{A}}\left[\{\mathscr{P}^{(N)}\}\right] = \sum_N \int_{\{\mathbf{r}\}}\left(\mathscr{P}^{(N)}\left[\breve{U}_N^{\text{intr}} + k_B T\ln(\mathscr{P}^{(N)}\Lambda^{3N}N!)\right]\right), \qquad (9.233)$$

where $\mathscr{P}^{(N)} = \mathscr{P}^{(N)}(\mathbf{r}^N)$ and only coordinate integrations, $\int d\mathbf{r}^N$, are included in the notation $\int_{\{\mathbf{r}\}}$. Equations 9.223 and 9.224 are changed in the analogous manner. At equilibrium we have

$$\mathscr{P}^{(N)}(\mathbf{r}^N) = \mathcal{P}^{(N)}(\mathbf{r}^N) = \frac{e^{-\beta[\breve{U}_N^{\text{pot}}(\mathbf{r}^N)-\mu N]}}{\Xi\,\Lambda^{3N}N!} \qquad \text{(equilibrium)}$$

and Equation 9.233 gives $\mathcal{A} = \Theta + \mu\bar{N} - \int d\mathbf{r}\, n(\mathbf{r})v(\mathbf{r})$.

The fundamental principle in DFT is that if we know the unique functional $\mathcal{A}^{ex} = \mathcal{A}^{ex}[n]$, we can calculate a large variety of properties for an equilibrium system. For example, by taking the functional derivative $\delta\mathcal{A}^{\text{ex}}/\delta n(\mathbf{r})$ we obtain $c^{(1)}(\mathbf{r})$ and from it we can calculate the number density distribution $n(\mathbf{r})$. We have also seen that from the second derivative $\delta^2\mathcal{A}^{\text{ex}}/\delta n(\mathbf{r})\delta n(\mathbf{r}')$ we obtain $c^{(2)}(\mathbf{r}',\mathbf{r})$ (Equation 9.127) and from it we can calculate $h^{(2)}(\mathbf{r}',\mathbf{r})$ via the Ornstein Zernike equation 9.40. Having obtained $n(\mathbf{r})$ and $h^{(2)}(\mathbf{r}',\mathbf{r})$ we can calculate several thermodynamical properties provided that the interparticle interactions are pairwise additive. If these interactions contain three-particle and higher terms, one can do the calculations using higher order distribution functions obtained from the higher functional derivatives described in Section 9.2.3. Alternatively (and often preferably), one can obtain several thermodynamical properties from Θ, which is given by Equation 9.227, for instance by taking ordinary derivatives. This was explored in Section 2.5.2 in Volume I.

In short, there exist many different ways of calculating equilibrium properties from $\mathcal{A}^{ex}[n]$ – those mentioned here are just examples.

This is great in principle, but in practice one stumbles on the fact that the functional $\mathcal{A}^{ex}[n]$ is unknown in general and one cannot write it down as a closed formula. There are a few simple systems for which the functional is known, for example the ideal gas, but for the vast majority of systems it is completely unknown. As we have seen, the functional *does* exist and *is* unique, but if we do not know it, we cannot do anything. The solution to this dilemma is the same as for other formulations of exact statistical mechanics: one makes approximations. By constructing an approximate functional $\mathcal{A}^{ex}[n]$ and using the machinery to extract various properties from it, one can assess its accuracy by comparing the results with, for example, simulations. The construction process can be described as a combination of educated guesses, experience and trial and error. Later, in Sections 10.2 and 10.3, we will give examples of some kinds of functionals.

APPENDICES

Appendix 9A
The variation of density profiles and average number of particles in a slit of various widths at constant chemical potential

In this appendix we treat the density distribution of a fluid between the two walls, I and II, in equilibrium with a bulk fluid with a given density (constant chemical potential). The main objective is to explore some consequences of Equation 9.49, namely how the density profile in the slit changes when the surface separation D^S is varied and how the average number of particles per unit area, $\bar{\mathscr{N}}$, in the slit changes with D^S at constant chemical potential.

The interaction potentials between the particles and walls are $v_{\text{I}}(\ell)$ and $v_{\text{II}}(\ell)$, where ℓ is the distance from each wall surface. We place the origin of the coordinate system at the surface of wall I with the z axis in the perpendicular direction. In the application of Equation 9.49 we have $v(\mathbf{r}) = v_{\text{I}}(z)$ and the potential $v_{\text{B}}(\mathbf{r}-\mathbf{r}_{\text{B}})$ is replaced by $v_{\text{II}}(D^S - z)$ since a point with coordinate z is located at the distance $D^S - z$ from surface II.[62] The total external potential of the fluid in Equation 9.48 is thus given by

$$v(\mathbf{r};\mathbf{r}_{\text{B}}) = v(z;D^S) \equiv v_{\text{I}}(z) + v_{\text{II}}(D^S - z)$$

and depends on the surface separation. Likewise, the density distribution in the slit between the walls is written as $n(z;D^S)$ and the pair correlation function as $h^{(2)}(\mathbf{r}_1,\mathbf{r}_2;D^S) = h^{(2)}(s_{12},z_1,z_2;D^S)$, which in the present geometry

[62] Alternatively we may argue as follows: We let the point $\mathbf{r}_{\text{B}}$ be located at the surface of wall II (instead of its center of mass) so the z component of $\mathbf{r}_{\text{B}}$ is equal to D^S. Since v_{B} depends solely on z, we have $v_{\text{B}}(\mathbf{r}-\mathbf{r}_{\text{B}}) = v_{\text{B}}(z - D^S) = v_{\text{II}}(D^S - z)$ since v_{II} is defined in the local coordinate system of wall II, which has its z axis in the negative direction.

depends on the three independent coordinates z_1, z_2 and $s_{12} = |\mathbf{s}_{12}|$, where $\mathbf{s}_{12} = (x_2 - x_1, y_2 - y_1)$. Equation 9.49 becomes[63]

$$\begin{aligned}\frac{\partial n(z_1; D^S)}{\partial D^S} &= -\beta n(z_1; D^S)\left[\frac{\partial v(z_1; D^S)}{\partial D^S}\right. \\ &+ \left.\int dz_2 d\mathbf{s}_{12}\, h^{(2)}(s_{12}, z_1, z_2; D^S) n(z_2; D^S)\frac{\partial v(z_2; D^S)}{\partial D^S}\right],\end{aligned} \tag{9.234}$$

where

$$\frac{\partial v(z; D^S)}{\partial D^S} = v'_{\mathrm{II}}(D^S - z)$$

and $v'_{\mathrm{II}}(\ell) = dv_{\mathrm{II}}(\ell)/d\ell$. Equation 9.234 can readily be generalized to cases where the surfaces are not smooth so the external potential depends on x and y as well as on z. Thereby, one simply applies Equation 9.49 to those cases.

The mean number of particles per unit area can now be evaluated from $\bar{\mathscr{N}}(D^S) = \int_0^{D^S} dz_1\, n(z_1; D^S)$, where we assume that the points $z = 0$ and $z = D^S$ at the surfaces of I and II, respectively, are selected so that the density n is zero for $z_1 \leq 0$ and $z_1 \geq D^S$ (it is zero since the potential $v(z_1; D^S)$ is infinite there). We obtain

$$\frac{d\bar{\mathscr{N}}(D^S)}{dD^S} = \int_0^{D^S} dz_1\, \frac{\partial n(z_1; D^S)}{\partial D^S}$$

because $n(D^S; D^S) = 0$, and by inserting Equation 9.234 we obtain the derivative of $\bar{\mathscr{N}}(D^S)$ *under the condition of constant chemical potential* for the fluid in the slit

$$\begin{aligned}\frac{d\bar{\mathscr{N}}(D^S)}{dD^S} &= -\beta\int_0^{D^S} dz_1\, n(z_1; D^S)\left[v'_{\mathrm{II}}(D^S - z_1)\right. \\ &+ \left.\int dz_2 d\mathbf{s}_{12}\, h^{(2)}(s_{12}, z_1, z_2; D^S) n(z_2; D^S) v'_{\mathrm{II}}(D^S - z_2)\right].\end{aligned} \tag{9.235}$$

If wall II is soft, that is, $v(z)$ goes continuously to infinity as function of z when the wall is approached, the use of this formula is straightforward, but when the wall has a hard core, there arises contributions that require special attention.

When wall II has a hard core, the potential $v_{\mathrm{II}}(\ell)$ is infinitely large for ℓ inside the core and is finite outside. More precisely, when the distance of closest approach to the wall is $d^{\mathrm{h}}_{\mathrm{II}}$ for the centers of the fluid particles, the potential $v_{\mathrm{II}}(D^S - z)$ has a jump to infinity at $z = D^S - d^{\mathrm{h}}_{\mathrm{II}}$, which implies that the density has a jump to zero there.

---

[63] Equation 9.234 was first derived by J. R. Henderson, Mol. Phys. **59** (1986) 89 (https://doi.org/10.1080/00268978600101931). The present derivation, where this equation is a special case of Equation 9.49, is taken from *loc. cit.* in footnote 20 on page 23.

For a *hard wall* the potential $v_{\rm II}(D^{\mathcal{S}}-z)$ is zero outside the core, i.e., for $z < D^{\mathcal{S}} - d^{\rm h}_{\rm II}$, and its derivative is, of course, also zero there. This implies that the integrands in Equation 9.235 are zero apart from the contributions from the infinite jump of $v_{\rm II}(D^{\mathcal{S}}-z)$ at $z = D^{\mathcal{S}} - d^{\rm h}_{\rm II}$, which gives rise to a Dirac delta function. We obtain[64]

$$\frac{d\bar{\mathcal{N}}(D^{\mathcal{S}})}{dD^{\mathcal{S}}} = \bar{\mathcal{N}}'_{\rm core,II}(D^{\mathcal{S}}) \qquad \text{(hard wall)}, \tag{9.236}$$

where the rhs is the contact contribution given by

$$\begin{aligned}\bar{\mathcal{N}}'_{\rm core,II}(D^{\mathcal{S}}) &= n^{\rm cont}_{\rm II}(D^{\mathcal{S}})\left\{1 + \int_0^{D^{\mathcal{S}}-d^{\rm h}_{\rm II}} dz_1\, n(z_1;D^{\mathcal{S}})\right.\\ &\quad \times \left.\int d\mathbf{s}_{12}\, h^{(2)}(s_{12}, z_1, D^{\mathcal{S}} - d^{\rm h}_{\rm II}; D^{\mathcal{S}})\right\}\end{aligned} \tag{9.237}$$

and $n^{\rm cont}_{\rm II}(D^{\mathcal{S}})$ is the contact density at wall II for surface separation $D^{\mathcal{S}}$, that is, the limit of $n(z_1;D^{\mathcal{S}})$ when $z_1 \to [D^{\mathcal{S}} - d^{\rm h}_{\rm II}]^-$, where superscript $-$ means that one approaches the hard wall from smaller z_1 values, $z_1 < D^{\mathcal{S}} - d^{\rm h}_{\rm II}$.

For the case of a *hard core wall*, the integrands in Equation 9.235 are in general nonzero for $0 < z < D^{\mathcal{S}} - d^{\rm h}_{\rm II}$, so the integrals are evaluated with these limits and we obtain

$$\begin{aligned}&\frac{d\bar{\mathcal{N}}(D^{\mathcal{S}})}{dD^{\mathcal{S}}}\\ &= \bar{\mathcal{N}}'_{\rm core,II}(D^{\mathcal{S}}) - \beta\int_0^{D^{\mathcal{S}}-d^{\rm h}_{\rm II}} dz_1\, n(z_1;D^{\mathcal{S}})\left[v'_{\rm II}(D^{\mathcal{S}}-z_1)\right.\\ &+ \left.\int_0^{D^{\mathcal{S}}-d^{\rm h}_{\rm II}} dz_2 \int d\mathbf{s}_{12}\, h^{(2)}(s_{12}, z_1, z_2; D^{\mathcal{S}}) n(z_2;D^{\mathcal{S}}) v'_{\rm II}(D^{\mathcal{S}}-z_2)\right].\end{aligned} \tag{9.238}$$

Analogous results can be obtained for wall I.

---

[64]The contact contribution $\bar{\mathcal{N}}'_{\rm core,II}(D^{\mathcal{S}})$ given by Equation 9.237 can be obtained as follows: Let us denote the contact point as $z^{\rm cont} \equiv D^{\mathcal{S}} - d^{\rm h}_{\rm II}$ and write $n(z;D^{\mathcal{S}}) = e^{-\beta v_{\rm II}(D^{\mathcal{S}}-z)}Y(z;D^{\mathcal{S}})$, where Y is continuous at $z = z^{\rm cont}$ (Y is an analogue to the cavity function in Equation 9.26). Then $n(z;D^{\mathcal{S}})v'_{\rm II}(D^{\mathcal{S}}-z) = \beta^{-1}[\partial e^{-\beta v_{\rm II}(D^{\mathcal{S}}-z)}/\partial z]\,Y(z;D^{\mathcal{S}})$. The function $e^{-\beta v_{\rm II}(D^{\mathcal{S}}-z)}$ has a step discontinuity at $z = z^{\rm cont}$ and is zero for $z > z^{\rm cont}$. The derivative of the step generates a Dirac delta function (with a minus sign since it is a downwards step), which implies that there are contributions to the z_1 and z_2 integrals in Equation 9.235 that contain the factor $-\beta^{-1}\delta(z^{\rm cont}-z)e^{-\beta v_{\rm II}(D^{\mathcal{S}}-z^{\rm cont})}Y(z;D^{\mathcal{S}}) = -\beta^{-1}\delta(z^{\rm cont}-z)n(z^{\rm cont};D^{\mathcal{S}})$ with $z = z_1$ and $z = z_2$, respectively, in the integrands. The integrations give rise to the result in Equation 9.237, where $n^{\rm cont}_{\rm II}(D^{\mathcal{S}}) = n(z^{\rm cont};D^{\mathcal{S}})$.

Appendix 9B
The Fourier transform in two and three dimensions

Three-dimensional case

We define the **Fourier transform** $\tilde{f}(\mathbf{k})$ of a function $f(\mathbf{r})$ in three-dimensional space, $\mathbf{r} = (x, y, z)$, as

$$\tilde{f}(\mathbf{k}) = \int d\mathbf{r}\, f(\mathbf{r}) e^{-\mathrm{i}\mathbf{k}\cdot\mathbf{r}}, \tag{9.239}$$

where i is the imaginary unit (distinguish this from the species index i), $\mathbf{k} = (k_x, k_y, k_z)$ is the wave vector and $k = |\mathbf{k}|$ the wave number. The unit vector $\hat{\mathbf{k}} = \mathbf{k}/k$ is the direction of propagation of a planar wave[65] with a spatial part equal to

$$e^{-\mathrm{i}\mathbf{k}\cdot\mathbf{r}} = \cos(\frac{2\pi\hat{\mathbf{k}}\cdot\mathbf{r}}{\lambda}) - \mathrm{i}\,\sin(\frac{2\pi\hat{\mathbf{k}}\cdot\mathbf{r}}{\lambda}),$$

where the wave length λ is given by $k = 2\pi/\lambda$ (note that if we set the direction of $\mathbf{r}$ equal to that of $\hat{\mathbf{k}}$, i.e., if $\hat{\mathbf{r}} = \hat{\mathbf{k}}$, we have $2\pi\hat{\mathbf{k}}\cdot\mathbf{r}/\lambda = 2\pi r/\lambda$).

For a spherically symmetric function $f(r)$ we can write the Fourier transform as[66]

$$\tilde{f}(k) = \int d\mathbf{r}\, f(r)\frac{\sin(kr)}{kr} = \frac{4\pi}{k}\int_0^\infty dr\, f(r) r \sin(kr). \tag{9.240}$$

The Fourier transform of a real, spherically symmetric function is real, while Fourier transform of a real function $f(\mathbf{r})$ in general is complex valued. (In this treatise we will only consider Fourier transforms of real functions.) Since $e^{-\mathrm{i}\mathbf{k}\cdot\mathbf{r}} = \cos(\mathbf{k}\cdot\mathbf{r}) - \mathrm{i}\sin(\mathbf{k}\cdot\mathbf{r})$, the real part of $\tilde{f}(\mathbf{k})$ is the cosine transform

$$\Re\left(\tilde{f}(\mathbf{k})\right) = \int d\mathbf{r}\, f(\mathbf{r}) \cos(\mathbf{k}\cdot\mathbf{r})$$

and the imaginary part is the negative sine transform

$$\Im\left(\tilde{f}(\mathbf{k})\right) = -\int d\mathbf{r}\, f(\mathbf{r}) \sin(\mathbf{k}\cdot\mathbf{r}),$$

where $\Re(\cdot)$ and $\Im(\cdot)$ extract the real and imaginary parts, respectively.

For radial functions $f(r)$, the variable r is real and positive by definition, but we can consider $\tilde{f}(k)$ for other values of k, for example

$$\tilde{f}(-k) = \tilde{f}(k), \tag{9.241}$$

---

[65] A sinusoidal planar wave with wave vector k, angular frequency ω and phase ϕ is described by $\sin(\mathbf{k}\cdot\mathbf{r} - \omega t + \phi) = \sin(k\hat{\mathbf{k}}\cdot\mathbf{r} - \omega t + \phi) = \sin(2\pi[\hat{\mathbf{k}}\cdot\mathbf{r}/\lambda - \nu t] + \phi)$, where its wave length is $\lambda = 2\pi/k$ and frequency is $\nu = \omega/2\pi$. The direction of propagation of the wave is given by $\hat{\mathbf{k}}$. The complex-valued counterpart of such a planar wave is $\exp(\mathrm{i}[\mathbf{k}\cdot\mathbf{r} - \omega t + \phi]) = \exp(\mathrm{i}\mathbf{k}\cdot\mathbf{r})\exp(-\mathrm{i}\omega t)\exp(\mathrm{i}\phi)$.

[66] This follows from $\int d\hat{\mathbf{r}}\, e^{-\mathrm{i}\mathbf{k}\cdot\mathbf{r}} = 4\pi\sin(kr)/(kr)$, where $\int d\hat{\mathbf{r}}$ denotes the angular integration for $\mathbf{r}$ in spherical polar coordinates.

which follows directly from Equation 9.240. If we take $k = \mathrm{i}\alpha$ and use the relationship $\sin(\mathrm{i}x)/(\mathrm{i}x) = \sinh(x)/x$, we likewise get

$$\tilde{f}(\mathrm{i}\alpha) = \int d\mathbf{r}\, f(r)\frac{\sinh(\alpha r)}{\alpha r}, \tag{9.242}$$

which converges if $f(r)$ decays sufficiently fast with increasing r.

The **inverse Fourier transform** is given by

$$f(\mathbf{r}) = \frac{1}{(2\pi)^3}\int d\mathbf{k}\, \tilde{f}(\mathbf{k})e^{\mathrm{i}\mathbf{k}\cdot\mathbf{r}}. \tag{9.243}$$

This shows that the Fourier transform can be regarded as the representation of the function $f(\mathbf{r})$ by a superposition of planar waves of different wave lengths and directions and with prefactor $\tilde{f}(\mathbf{k})/(2\pi)^3$. For a spherically symmetric function $\tilde{f}(k)$, we have

$$f(r) = \frac{1}{(2\pi)^3}\int d\mathbf{k}\, \tilde{f}(k)\frac{\sin(kr)}{kr} = \frac{1}{2\pi^2 r}\int_0^\infty dk\, \tilde{f}(k)k\sin(kr) \tag{9.244}$$

which, alternatively, can be interpreted as a representation of $f(r)$ by spherical waves $\sin(kr)/(kr)$.

It follows that the behavior of $\tilde{f}(\mathbf{k})$ for small k describes features of $f(\mathbf{r})$ that are described by a superposition of waves with long wave lengths. Such features vary slowly with position $\mathbf{r}$. On the other hand $\tilde{f}(\mathbf{k})$ for very large k correspond to a superposition of waves with very short wave lengths, so this part of $\tilde{f}(\mathbf{k})$ describes features that vary rapidly with $\mathbf{r}$.

Some useful properties:

The value of the Fourier transform at $\mathbf{k} = 0$ is the integral of $f(\mathbf{r})$ over the entire space

$$\tilde{f}(0) = \int d\mathbf{r}\, f(\mathbf{r}), \tag{9.245}$$

which follows directly from the definition (9.239). The Fourier transform of the Dirac function $\delta^{(3)}(\mathbf{r})$ is unity

$$\widetilde{\delta^{(3)}}(\mathbf{k}) = 1, \tag{9.246}$$

which follows from $\int d\mathbf{r}\,\delta^{(3)}(\mathbf{r})e^{-\mathrm{i}\mathbf{k}\cdot\mathbf{r}} = e^0 = 1$. The Fourier transform of $f_{(\mathbf{a})}(\mathbf{r}) \equiv f(\mathbf{r}+\mathbf{a})$, where $\mathbf{a}$ is constant vector, is

$$\tilde{f}_{(\mathbf{a})}(\mathbf{k}) = e^{\mathrm{i}\mathbf{k}\cdot\mathbf{a}}\tilde{f}(\mathbf{k}), \tag{9.247}$$

which follows from

$$\int d\mathbf{r}\, f(\mathbf{r}+\mathbf{a})e^{-\mathrm{i}\mathbf{k}\cdot\mathbf{r}} = \int d\mathbf{s}\, f(\mathbf{s})e^{-\mathrm{i}\mathbf{k}\cdot(\mathbf{s}-\mathbf{a})} = e^{\mathrm{i}\mathbf{k}\cdot\mathbf{a}}\int d\mathbf{s}\, f(\mathbf{s})e^{-\mathrm{i}\mathbf{k}\cdot\mathbf{s}}$$

where we have made the variable substitution $\mathbf{s} = \mathbf{r}+\mathbf{a}$.

To take the gradient of a function in $\mathbf{r}$ space, $\nabla f(\mathbf{r})$, corresponds to a multiplication of its Fourier transform by $i\mathbf{k}$,

$$\widetilde{\nabla f} = i\mathbf{k}\tilde{f}(\mathbf{k}), \tag{9.248}$$

which can be derived by taking the gradient of $f(\mathbf{r})$ in Equation 9.243. To take the Laplace operator of a function, $\nabla^2 f(\mathbf{r})$, corresponds to a multiplication of the transform by $-k^2$

$$\widetilde{\nabla^2 f} = -k^2\tilde{f}(\mathbf{k}), \tag{9.249}$$

which is a consequence of $\nabla^2 \equiv \nabla \cdot \nabla$ and $(i\mathbf{k}) \cdot (i\mathbf{k}) = -k^2$.

The Fourier transform of the **convolution integral** of two functions $f_1(\mathrm{r})$ and $f_2(\mathrm{r})$

$$f(\mathbf{r}_{12}) = \int d\mathbf{r}_3\, f_1(\mathbf{r}_{13}) f_2(\mathbf{r}_{32}), \tag{9.250}$$

which also can be written as

$$f(\mathbf{r}) = \int d\mathbf{r}'\, f_1(\mathbf{r}-\mathbf{r}') f_1(\mathbf{r}'),$$

is equal to the product of their Fourier transforms

$$\tilde{f}(\mathbf{k}) = \tilde{f}_1(\mathbf{k})\tilde{f}_2(\mathbf{k}). \tag{9.251}$$

An integral like

$$F(\mathbf{r}_{12}) = \int d\mathbf{r}_3\, f_1(\mathbf{r}_{13}) f_2(\mathbf{r}_{23})$$

is not a convolution since the argument of the function f_2 is $\mathbf{r}_{23}$ instead of $\mathbf{r}_{32}$. Since $\mathbf{r}_{32} = -\mathbf{r}_{23}$, the Fourier transform is

$$\tilde{F}(\mathbf{k}) = \tilde{f}_1(\mathbf{k})\tilde{f}_2(-\mathbf{k}). \tag{9.252}$$

We can alternatively write $\tilde{f}_2(-\mathbf{k}) = \underline{\tilde{f}_2(\mathbf{k})}$ where underline means complex conjugate.

Another useful equation is **Parseval's formula**

$$\int d\mathbf{r}\, f_1(\mathbf{r}) f_2(\mathbf{r}) = \frac{1}{(2\pi)^3}\int d\mathbf{k}\, \tilde{f}_1(\mathbf{k})\tilde{f}_2(-\mathbf{k}), \tag{9.253}$$

which also can be written as

$$\int d\mathbf{r}\, f_1(\mathbf{r}) f_2(\mathbf{r}) = \frac{1}{(2\pi)^3}\int d\mathbf{k}\, \tilde{f}_1(\mathbf{k})\underline{\tilde{f}_2(\mathbf{k})}.$$

Two-dimensional case

We define the **Fourier transform** $\underset{\sim}{f}(\mathbf{k})$ of a function $f(\mathbf{s})$ in two-dimensional space, $\mathbf{s} = (x, y)$, as

$$\underset{\sim}{f}(\mathbf{k}) = \int d\mathbf{s}\, f(\mathbf{s}) e^{-i\mathbf{k}\cdot\mathbf{s}}, \tag{9.254}$$

where $\mathbf{k} = (k_x, k_y)$ is the two-dimensional wave vector and $k = |\mathbf{k}|$ the wave number. The inverse Fourier transform is in this case given by

$$f(\mathbf{s}) = \frac{1}{(2\pi)^2} \int d\mathbf{k}\, \underset{\sim}{f}(\mathbf{k}) e^{i\mathbf{k}\cdot\mathbf{s}}. \tag{9.255}$$

For a radially symmetric function $f(s)$, where $s = |\mathbf{s}| = [x^2 + y^2]^{1/2}$, we can write the Fourier transform as[67]

$$\underset{\sim}{f}(k) = \int d\mathbf{s}\, f(s) J_0(ks) = 2\pi \int_0^\infty ds\, s f(s) J_0(ks), \tag{9.256}$$

where $J_0(x)$ is the zeroth order Bessel function. This kind of transform is also called a Hankel transform.

The two-dimensional Fourier transform has properties analogous to those of the three-dimensional transform, for example

$$\underset{\sim}{f}(0) = \int d\mathbf{s}\, f(\mathbf{s})$$

and

$$\underset{\sim}{f}_{(\mathbf{a})}(\mathbf{k}) = e^{i\mathbf{k}\cdot\mathbf{a}} \underset{\sim}{f}(\mathbf{k}),$$

where $\mathbf{a}$ is constant two-dimensional vector. Furthermore, the Fourier transform of a convolution integral

$$f(\mathbf{s}_{12}) = \int d\mathbf{s}_3\, f_1(\mathbf{s}_{13}) f_2(\mathbf{s}_{32})$$

is given by

$$\underset{\sim}{f}(\mathbf{k}) = \underset{\sim}{f}_1(\mathbf{k}) \underset{\sim}{f}_2(\mathbf{k}).$$

and the transform of the two-dimensional gradient $\nabla_\mathbf{s} = (\partial/\partial x, \partial/\partial y)$ of a function $f(\mathbf{s})$ is

$$\underset{\sim}{\nabla_\mathbf{s} f} = i\mathbf{k} \underset{\sim}{f}(\mathbf{k}) \tag{9.257}$$

as follows from the gradient of Equation 9.255.

[67] This follows from $\int_0^{2\pi} d\phi\, e^{-i\mathbf{k}\cdot\mathbf{s}} = \int_0^{2\pi} d\phi\, e^{-iks\cos\phi} = \int_0^{2\pi} d\phi\, \cos(ks\cos\phi) = 2\pi J_0(ks)$, where ϕ is the polar angle for $\mathbf{s}$ in two-dimensional polar coordinates.

Appendix 9C
Proofs of Equations 9.205 and 9.206.

In this Appendix we will first evaluate the integral

$$\int_0^1 d\xi \int d\mathbf{r}\, n(\mathbf{r}) h^{(2)}(\mathbf{r},\mathbf{r}';\xi) \frac{\partial \gamma^{(2)}(\mathbf{r},\mathbf{r}';\xi)}{\partial \xi} \equiv \mathfrak{I}_{h\gamma'}, \tag{9.258}$$

which we denote as $\mathfrak{I}_{h\gamma'}$, and obtain Equation 9.205. Then, starting on the facing page, we will prove Equation 9.206 for the function $c^{(1)\mathrm{R}}$.

For the first task we need to be able to treat a two-component fluid because when ξ lies in the interval $0 < \xi < 1$, the system contains two "species," namely the original particles of the fluid and the added particle that is partially coupled to the extent ξ. The OZ equation for a bulk mixture is (cf. Equation 9.70)

$$h_{ij}^{(2)}(r_{12}) = c_{ij}^{(2)}(r_{12}) + \sum_l n_l^{\mathrm{b}} \int d\mathbf{r}_3\, c_{il}^{(2)}(r_{13}) h_{lj}^{(2)}(r_{32}) \quad \text{(bulk mixture)}$$

the corresponding equation for an inhomogeneous mixture is

$$h_{ij}^{(2)}(\mathbf{r},\mathbf{r}') = c_{ij}^{(2)}(\mathbf{r},\mathbf{r}') + \sum_l \int d\mathbf{r}''\, c_{il}^{(2)}(\mathbf{r},\mathbf{r}'') n_l(\mathbf{r}'') h_{lj}^{(2)}(\mathbf{r}'',\mathbf{r}') \tag{9.259}$$

(these equations will be strictly derived for the general case in Section 11.5). In the current case, the sum over l runs over the two species: the original particles (species 1) and the partially coupled one (species 2). Since the number density $n_2(\mathbf{r}'')$ of the latter is zero, only one term in the sum is nonzero and we have

$$\begin{aligned} h_{11}^{(2)}(\mathbf{r},\mathbf{r}') &= c_{11}^{(2)}(\mathbf{r},\mathbf{r}') + \int d\mathbf{r}''\, c_{11}^{(2)}(\mathbf{r},\mathbf{r}'') n_1(\mathbf{r}'') h_{11}^{(2)}(\mathbf{r}'',\mathbf{r}') \\ h_{12}^{(2)}(\mathbf{r},\mathbf{r}') &= c_{12}^{(2)}(\mathbf{r},\mathbf{r}') + \int d\mathbf{r}''\, c_{11}^{(2)}(\mathbf{r},\mathbf{r}'') n_1(\mathbf{r}'') h_{12}^{(2)}(\mathbf{r}'',\mathbf{r}'). \end{aligned}$$

The first equation is equal to the ordinary OZ equation 9.40 for the pure fluid. Since we deal with a pure one-component fluid here, we skip the indices for h_{11}, c_{11} and n_1 in accordance with the notation used earlier. The second equation deals with the correlation functions between the original particles and the partially coupled one. We can identify $h_{12}^{(2)}$ as being the correlation function between the original particles and the partially coupled one, that is, $h_{12}^{(2)}(\mathbf{r},\mathbf{r}') = h^{(2)}(\mathbf{r},\mathbf{r}';\xi)$, and likewise for $c_{12}^{(2)}$, so the latter equation can be written as

$$\begin{aligned} \gamma^{(2)}(\mathbf{r},\mathbf{r}';\xi) &= h^{(2)}(\mathbf{r},\mathbf{r}';\xi) - c^{(2)}(\mathbf{r},\mathbf{r}';\xi) \\ &= \int d\mathbf{r}''\, c^{(2)}(\mathbf{r},\mathbf{r}'') n(\mathbf{r}'') h^{(2)}(\mathbf{r}'',\mathbf{r}';\xi), \end{aligned} \tag{9.260}$$

where the function $c^{(2)}(\mathbf{r}, \mathbf{r}'')$ in the rhs is the direct correlation function of the original pure fluid. Again we have skipped the indices.

We are now ready to evaluate $\mathfrak{I}_{h\gamma'}$ and by inserting $\gamma^{(2)}$ from Equation 9.260 into Equation 9.258 we obtain

$$\begin{aligned}\mathfrak{I}_{h\gamma'} &= \int_0^1 d\xi \int d\mathbf{r}\, d\mathbf{r}'' n(\mathbf{r})n(\mathbf{r}'')h^{(2)}(\mathbf{r}, \mathbf{r}'; \xi)\frac{\partial h^{(2)}(\mathbf{r}'', \mathbf{r}'; \xi)}{\partial \xi} c^{(2)}(\mathbf{r}, \mathbf{r}'') \\ &= \frac{1}{2}\int_0^1 d\xi \int d\mathbf{r}\, d\mathbf{r}'' n(\mathbf{r})n(\mathbf{r}'')\frac{\partial}{\partial \xi}\left(h^{(2)}(\mathbf{r}, \mathbf{r}'; \xi)h^{(2)}(\mathbf{r}'', \mathbf{r}'; \xi)\right) c^{(2)}(\mathbf{r}, \mathbf{r}''),\end{aligned}$$

where the second line is obtained by using the fact that when we perform the derivative in this line (using the product rule) and integrate with respect to $\mathbf{r}$ and $\mathbf{r}''$, we obtain two terms that have equal values (just the names of the integration variables are different). One of these terms has the same integrand as the second line and the other is obtained if we swap $\mathbf{r} \leftrightarrow \mathbf{r}''$ in this line so that

$$h^{(2)}(\mathbf{r}, \mathbf{r}'; \xi)\frac{\partial h^{(2)}(\mathbf{r}'', \mathbf{r}'; \xi)}{\partial \xi} \Rightarrow h^{(2)}(\mathbf{r}'', \mathbf{r}'; \xi)\frac{\partial h^{(2)}(\mathbf{r}, \mathbf{r}'; \xi)}{\partial \xi},$$

while everything else stays the same. Since there are two equal contributions we need to multiply the integral in the third line by 1/2 in order to obtain the correct result.

The ξ integration in the second line can easily be performed and we obtain

$$\begin{aligned}\mathfrak{I}_{h\gamma'} &= \frac{1}{2}\int d\mathbf{r}\, d\mathbf{r}''\, n(\mathbf{r})n(\mathbf{r}'')\left(h^{(2)}(\mathbf{r}, \mathbf{r}'; \xi)h^{(2)}(\mathbf{r}'', \mathbf{r}'; \xi)\right)\Big|_{\xi=1} c^{(2)}(\mathbf{r}, \mathbf{r}'') \\ &= \frac{1}{2}\int d\mathbf{r}\, n(\mathbf{r})h^{(2)}(\mathbf{r}, \mathbf{r}') \int d\mathbf{r}''\, h^{(2)}(\mathbf{r}'', \mathbf{r}')n(\mathbf{r}'')c^{(2)}(\mathbf{r}, \mathbf{r}'') \\ &= \frac{1}{2}\int d\mathbf{r}\, n(\mathbf{r})h^{(2)}(\mathbf{r}, \mathbf{r}')\left[h^{(2)}(\mathbf{r}, \mathbf{r}') - c^{(2)}(\mathbf{r}, \mathbf{r}')\right] \\ &= \frac{1}{2}\int d\mathbf{r}\, n(\mathbf{r})\left[h^{(2)}(\mathbf{r}, \mathbf{r}')\right]^2 - \frac{1}{2}\int d\mathbf{r}\, h^{(2)}(\mathbf{r}, \mathbf{r}')n(\mathbf{r})c^{(2)}(\mathbf{r}, \mathbf{r}'),\end{aligned}$$

where we have used the OZ equation to obtain the third line. The OZ equation can also be used for the last term in the fourth line, resulting in $-[h^{(2)}(\mathbf{r}', \mathbf{r}') - c^{(2)}(\mathbf{r}', \mathbf{r}')]/2$, which is equal to $-\gamma^{(2)}(\mathbf{r}', \mathbf{r}')/2$. Finally, by inserting this into $\mathfrak{I}_{h\gamma'}$ we obtain Equation 9.205.

Next, we will derive Equation 9.206. To do this we need to do a slight generalization of the functional expansion 9.160 in order to obtain the corresponding expansion for the function $b^{(2)}(\mathbf{r}, \mathbf{r}'; \xi)$. We obtained Equation 9.160 from Equation 9.159 by inserting $\Delta n(\mathbf{s}_\nu|\mathbf{r}') = n(\mathbf{s}_\nu)h^{(2)}(\mathbf{s}_\nu, \mathbf{r}')$ and the only difference in the present case is to insert $n(\mathbf{s}_\nu)h^{(2)}(\mathbf{s}_\nu, \mathbf{r}'; \xi)$ instead, because this is the density change when one inserts a partially coupled particle at $\mathbf{r}'$.

We obtain

$$
\begin{aligned}
b^{(2)}(\mathbf{r},\mathbf{r}';\xi) &= \sum_{m=3}^{\infty} \frac{1}{(m-1)!} \int d\mathbf{s}_1 \ldots d\mathbf{s}_{m-1}\, c^{(m)}(\mathbf{r},\mathbf{s}_1,\ \ldots\ ,\mathbf{s}_{m-1}) \\
&\times\ n(\mathbf{s}_1)h^{(2)}(\mathbf{s}_1,\mathbf{r}';\xi)\ldots n(\mathbf{s}_{m-1})h^{(2)}(\mathbf{s}_{m-1},\mathbf{r}';\xi), \qquad (9.261)
\end{aligned}
$$

where all $c^{(m)}$ for $m \geq 3$, as before, are evaluated for the system *before* the change in density Δn. By inserting this expression into the definition of the residual part of $c^{(1)}$, Equation 9.202, we obtain after rearrangement

$$
\begin{aligned}
c^{(1)\mathrm{R}}(\mathbf{r}') &= \int_0^1 d\xi \sum_{m=3}^{\infty} \frac{1}{(m-1)!} \int d\mathbf{s}_1 \ldots d\mathbf{s}_{m-1} d\mathbf{r}\, c^{(m)}(\mathbf{r},\mathbf{s}_1,\ldots,\mathbf{s}_{m-1}) \\
&\times\ n(\mathbf{s}_1)h^{(2)}(\mathbf{s}_1,\mathbf{r}';\xi)\ldots n(\mathbf{s}_{m-1})h^{(2)}(\mathbf{s}_{m-1},\mathbf{r}';\xi) \\
&\times\ n(\mathbf{r})\frac{\partial h^{(2)}(\mathbf{r},\mathbf{r}';\xi)}{\partial \xi}. \qquad (9.262)
\end{aligned}
$$

We can freely choose integration path for the coupling-parameter integration, as long as it satisfies the requirements in Equation 9.197. The simplest choice in the present case is to use the linear path where $n(\mathbf{r}|\mathbf{r}';\xi) = n(\mathbf{r})[1 + \xi h^{(2)}(\mathbf{r},\mathbf{r}')]$, that is, $h^{(2)}(\mathbf{r},\mathbf{r}';\xi) = \xi h^{(2)}(\mathbf{r},\mathbf{r}')$. When this is inserted for all factors $h^{(2)}$ in the expression for $c^{(1)\mathrm{R}}$, we obtain a factor ξ^{m-1} in the integrand, while all other parts of the integrand are independent of ξ. Since $\int_0^1 d\xi\, \xi^{m-1} = 1/m$ we finally obtain Equation 9.206, where we have written $\mathbf{s}_m$ instead of $\mathbf{r}$ and utilized the fact that $c^{(m)}$ is symmetric with respect to all its variables.

It is of interest to see that Equation 9.206 actually is obtained when any coupling-parameter integration path is used, not just the linear one. In fact, we have

$$
\begin{aligned}
c^{(1)\mathrm{R}}(\mathbf{r}') &= \int_0^1 d\xi \sum_{m=3}^{\infty} \frac{1}{(m-1)!} \cdot \frac{1}{m} \int d\mathbf{s}_1 \ldots d\mathbf{s}_{m-1}\, d\mathbf{r} \\
&\times\ c^{(m)}(\mathbf{r},\mathbf{s}_1,\ \ldots\ ,\mathbf{s}_{m-1})n(\mathbf{s}_1)\ldots n(\mathbf{s}_{m-1})n(\mathbf{r}) \\
&\times\ \frac{\partial}{\partial \xi}\left[h^{(2)}(\mathbf{s}_1,\mathbf{r}';\xi)\ldots h^{(2)}(\mathbf{s}_{m-1},\mathbf{r}';\xi)h^{(2)}(\mathbf{r},\mathbf{r}';\xi)\right] \qquad (9.263)
\end{aligned}
$$

because the differentiation of the square bracket with respect to ξ, using the product rule, yields m terms in the integrand, each containing one factor $\partial h^{(2)}/\partial \xi$ and $m-1$ factors $h^{(2)}$. The value of the integral with respect to the spatial variables of each of these terms is the same (only the names of the integration variables differ), so the value of one of the integrals [for instance the one that contains $\partial h^{(2)}(\mathbf{r},\mathbf{r}';\xi)/\partial\xi$, which is the one that we have above] is $1/m$ times the sum of all. To deduce this fact, one has to use the symmetry of $c^{(m)}$ with respect to all its variables. The ξ integration can be easily performed and gives Equation 9.206.

From these observations it follows that if the coupling-parameter path independence of $c^{(1)\text{R}}$ (and hence of $c^{(1)}$ and the chemical potential) is to be maintained in approximate theories, the approximation used for $c^{(m)}$ has to be symmetric with respect to all its variables. If this symmetry is violated, the values of $c^{(1)}$ and the chemical potential depend on the path used to calculate them, which leads to a thermodynamic inconsistency of the approximation, for example that the chemical potential is not a state function.

Appendix 9D
Uniqueness of the equilibrium density distribution in the grand canonical ensemble

In this Appendix we will show the following for given interparticle interaction potential $\breve{U}^{\text{intr}}$, temperature T and chemical potential μ in the grand canonical ensemble: *For each external potential $v(\mathbf{r})$ there exists a unique equilibrium density distribution $n(\mathbf{r}')$ that no other external potential gives rise to.* Inversely, for each (possible) $n(\mathbf{r})$ there exists a unique external potential $v(\mathbf{r})$ that gives rise to it, so there is a 1-1 relationship between $v(\mathbf{r}')$ and $n(\mathbf{r})$. This is an important aspect of one of the Hohenberg-Kohn-Mermin theorems that are dealt with in Section 9.3.

In the grand canonical ensemble the configurational probability density for N fluid particles is (Equation I:3.41)

$$\mathcal{P}^{(N)}(\mathbf{r}^N) = \frac{\zeta^N}{N!\,\Xi} e^{-\beta \breve{U}_N^{\text{pot}}(\mathbf{r}^N)} \tag{9.264}$$

and $\mathcal{P}^{(N)}$ is normalized, $\sum_{N=0}^{\infty} \int d\mathbf{r}^N \mathcal{P}^{(N)}(\mathbf{r}^N) = 1$. The activity $\zeta = e^{\beta\mu}/\Lambda^3$ is constant because μ is constant. The average of a quantity X with coordinate dependence $\breve{X}_N(\mathbf{r}^N)$ is given by (Equation I:3.42)

$$\langle X \rangle = \sum_{N=0}^{\infty} \int d\mathbf{r}^N \mathcal{P}^{(N)}(\mathbf{r}^N) \breve{X}_N(\mathbf{r}^N) \equiv \sum_N \int_{\{\mathbf{r}\}} \left(\mathcal{P}^{(N)} \breve{X}_N \right),$$

where we have introduced a simplified notation for the average expression in the rhs. For quantity f with coordinate dependence $\sum_{\nu=1}^{N} f(\mathbf{r}_\nu)$ this implies

$$\langle f \rangle = \sum_{N=1}^{\infty} \int d\mathbf{r}^N \mathcal{P}^{(N)}(\mathbf{r}^N) \sum_{\nu=1}^{N} f(\mathbf{r}_\nu) = \int d\mathbf{r}_1 f(\mathbf{r}_1) n(\mathbf{r}_1),$$

as shown in Section 5.10 of Volume I (Equation I:5.113).

Consider a fluid for given μ and T that is exposed to an external potential $v(\mathbf{r}) = v_{\text{A}}(\mathbf{r})$ labeled A. The system has $\Xi = \Xi_{\text{A}}$ and the potential energy for N fluid particles is $\breve{U}_N^{\text{pot}}(\mathbf{r}^N) = \breve{U}_N^{\text{intr}}(\mathbf{r}^N) + \sum_{\nu=1}^{N} v_{\text{A}}(\mathbf{r}_\nu)$. We denote $\mathcal{P}^{(N)}(\mathbf{r}^N)$ for this case as $\mathcal{P}_{\text{A}}^{(N)}(\mathbf{r}^N)$. Equation 9.264 yields

$$\ln \mathcal{P}_{\text{A}}^{(N)}(\mathbf{r}^N) = \ln \frac{\zeta^N}{N!} + \beta\Theta_{\text{A}} - \beta \breve{U}_N^{\text{intr}}(\mathbf{r}^N) - \beta \sum_{\nu=1}^{N} v_{\text{A}}(\mathbf{r}_\nu)$$

where the grand potential $\Theta_{\mathrm{A}} = -k_B T \ln \Xi_{\mathrm{A}}$. We also consider the same fluid (same T, μ and function $\breve{U}^{\mathrm{intr}}$ of the particle coordinates) exposed to a different external potential $v(\mathbf{r}) = v_{\mathrm{B}}(\mathbf{r})$ labeled B. For this system $\Xi = \Xi_{\mathrm{B}}$, $\Theta = \Theta_{\mathrm{B}}$ and $\mathcal{P}^{(N)}(\mathbf{r}^N) = \mathcal{P}_{\mathrm{B}}^{(N)}(\mathbf{r}^N)$. We have

$$\begin{aligned}
\ln \mathcal{P}_{\mathrm{A}}^{(N)}(\mathbf{r}^N) - \beta\Theta_{\mathrm{A}} &= \ln \frac{\zeta^N}{N!} - \beta \breve{U}_N^{\mathrm{intr}}(\mathbf{r}^N) - \beta \sum_{\nu=1}^{N} v_{\mathrm{A}}(\mathbf{r}_\nu) \\
\ln \mathcal{P}_{\mathrm{B}}^{(N)}(\mathbf{r}^N) - \beta\Theta_{\mathrm{B}} &= \ln \frac{\zeta^N}{N!} - \beta \breve{U}_N^{\mathrm{intr}}(\mathbf{r}^N) - \beta \sum_{\nu=1}^{N} v_{\mathrm{B}}(\mathbf{r}_\nu)
\end{aligned}$$

and by taking the difference between these equations we obtain

$$\ln \mathcal{P}_{\mathrm{A}}^{(N)}(\mathbf{r}^N) - \ln \mathcal{P}_{\mathrm{B}}^{(N)}(\mathbf{r}^N) + \beta[\Theta_{\mathrm{B}} - \Theta_{\mathrm{A}}] = \beta \sum_{\nu=1}^{N} [v_{\mathrm{B}}(\mathbf{r}_\nu) - v_{\mathrm{A}}(\mathbf{r}_\nu)]. \quad (9.265)$$

We now take the average of each side of this expression with respect to the probability density $\mathcal{P}_{\mathrm{B}}^{(N)}$ and obtain

$$\begin{aligned}
&\sum_N \int_{\{\mathbf{r}\}} \left(\mathcal{P}_{\mathrm{B}}^{(N)} \ln \mathcal{P}_{\mathrm{A}}^{(N)}\right) - \sum_N \int_{\{\mathbf{r}\}} \left(\mathcal{P}_{\mathrm{B}}^{(N)} \ln \mathcal{P}_{\mathrm{B}}^{(N)}\right) + \beta[\Theta_{\mathrm{B}} - \Theta_{\mathrm{A}}] \\
&= \beta \int d\mathbf{r}_1 [v_{\mathrm{B}}(\mathbf{r}_1) - v_{\mathrm{A}}(\mathbf{r}_1)] n_{\mathrm{B}}(\mathbf{r}_1),
\end{aligned} \quad (9.266)$$

where $n_{\mathrm{B}}(\mathbf{r}_1)$ is the density distribution when the external potential is v_{B} and where we have used the fact that Θ_{A} and Θ_{B} are constant.

Now, the first two terms in Equation 9.266 can be written

$$\begin{aligned}
\sum_N \int_{\{\mathbf{r}\}} \left(\mathcal{P}_{\mathrm{B}}^{(N)} \ln \mathcal{P}_{\mathrm{A}}^{(N)}\right) - \sum_N \int_{\{\mathbf{r}\}} \left(\mathcal{P}_{\mathrm{B}}^{(N)} \ln \mathcal{P}_{\mathrm{B}}^{(N)}\right) &= \sum_N \int_{\{\mathbf{r}\}} \left(\mathcal{P}_{\mathrm{B}}^{(N)} \ln[\mathcal{P}_{\mathrm{A}}^{(N)}/\mathcal{P}_{\mathrm{B}}^{(N)}]\right) \\
&\leq \sum_N \int_{\{\mathbf{r}\}} \left(\mathcal{P}_{\mathrm{B}}^{(N)} \{[\mathcal{P}_{\mathrm{A}}^{(N)}/\mathcal{P}_{\mathrm{B}}^{(N)}] - 1\}\right) \\
&= \sum_N \int_{\{\mathbf{r}\}} \left(\mathcal{P}_{\mathrm{A}}^{(N)}\right) - \sum_N \int_{\{\mathbf{r}\}} \left(\mathcal{P}_{\mathrm{B}}^{(N)}\right) = 1 - 1 = 0
\end{aligned}$$

where the inequality follows from the fact that $\ln x \leq x - 1$ with equality only for $x = 1$, as apparent from Figure 9.4. Thus we can conclude that

$$\sum_N \int_{\{\mathbf{r}\}} \left(\mathcal{P}_{\mathrm{B}}^{(N)} \ln \mathcal{P}_{\mathrm{A}}^{(N)}\right) - \sum_N \int_{\{\mathbf{r}\}} \left(\mathcal{P}_{\mathrm{B}}^{(N)} \ln \mathcal{P}_{\mathrm{B}}^{(N)}\right) \leq 0 \quad (9.267)$$

where equality occurs only when $\mathcal{P}_{\mathrm{A}}^{(N)}(\mathbf{r}^N) = \mathcal{P}_{\mathrm{B}}^{(N)}(\mathbf{r}^N)$ for all $\mathbf{r}^N$. This inequality is known as **Gibbs' inequality** and is true for any kind of (normalized) probability density.

Since $\mathcal{P}_{\mathrm{A}}$ and $\mathcal{P}_{\mathrm{B}}$ are different in our case we can conclude from Equation 9.266 that

$$\beta[\Theta_{\mathrm{B}} - \Theta_{\mathrm{A}}] < \beta \int d\mathbf{r}_1 [v_{\mathrm{B}}(\mathbf{r}_1) - v_{\mathrm{A}}(\mathbf{r}_1)] n_{\mathrm{B}}(\mathbf{r}_1).$$

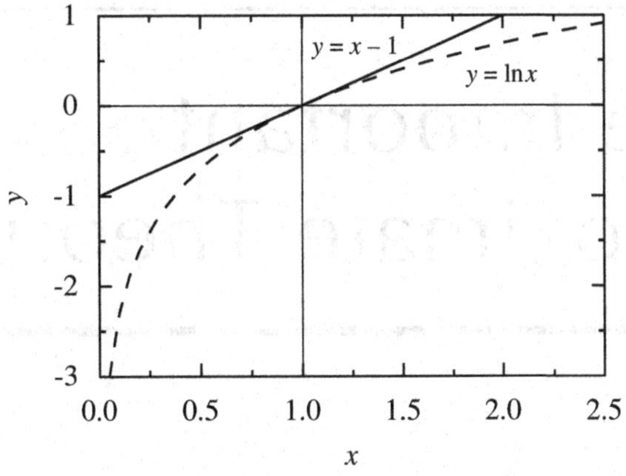

Figure 9.4 A plot that illustrates the inequality $\ln x \leq x - 1$.

By doing the same derivation with A and B interchanged [e.g. by taking the average of Equation 9.265 with respect to $\mathcal{P}_{\mathrm{A}}^{(N)}$] we obtain

$$\beta[\Theta_{\mathrm{A}} - \Theta_{\mathrm{B}}] < \beta \int d\mathbf{r}_1 [v_{\mathrm{A}}(\mathbf{r}_1) - v_{\mathrm{B}}(\mathbf{r}_1)] n_{\mathrm{A}}(\mathbf{r}_1).$$

The sum of these two equations yields

$$0 < \beta \int d\mathbf{r}_1 [v_{\mathrm{B}}(\mathbf{r}_1) - v_{\mathrm{A}}(\mathbf{r}_1)]\, [n_{\mathrm{B}}(\mathbf{r}_1) - n_{\mathrm{A}}(\mathbf{r}_1)],$$

which implies that $n_{\mathrm{B}}(\mathbf{r}_1)$ and $n_{\mathrm{A}}(\mathbf{r}_1)$ cannot be equal everywhere since, otherwise, the rhs would be identically equal to zero leading to $0 < 0$. Thus we have a 1-1 relationship between $n(\mathbf{r})$ and $v(\mathbf{r}')$ as we set out to prove. Since we, at least in principle, can determine $v(\mathbf{r}')$ uniquely once we know $n(\mathbf{r})$, we can formally regard v as being a functional of n, which we write $v = v(\mathbf{r}'; [n])$. This is a very important result in liquid state theory.

CHAPTER 10

Some Important Approximate Theories

In the previous chapter we laid the framework for liquid state theory and in the present one we will treat several important approximations that are used in this theory. First, we will deal with integral equation approximations in a more in-depth manner than we did in the previous chapter. Such approximations have a long history in the statistical mechanics of fluids and there exist a huge number of different ones suggested in the literature; here we will only consider some of the most prominent and useful. We will divide them into two groups depending on whether or not they use some consistency criteria; that is, some exact relationships that the theory should fulfill and is used to determine some parameter in the closure relationship.

Various ways to assess the accuracy of the closures for bulk fluids are explored. Since the value of the pressure for a bulk fluid has been extensively used for such assessments, we will also treat some other types of approximations that are commonly used to calculate the pressure of a hard sphere fluid. They are included in Section 10.1.1. A more demanding way to assess the performance of a closure is to calculate the excess chemical potential for the fluid and the value of the bridge function at the center of a particle, i.e., at $r = 0$. As we will see, there exist a series of consistency criteria, for instance so-called zero-separation theorems (for values of various functions at $r = 0$), that can be used to do such assessments and to improve closures in a systematic manner. This described in Section 10.1.2.

Thereafter, integral theories for inhomogeneous fluids are presented. Section 10.1.3 deal with such fluids at the singlet distribution level and Section 10.1.4 at the pair distribution level. In Section 10.1.5 some practical expressions for the chemical potential and the singlet direct-correlation function are derived. They are particularly useful for the treatment of inhomogeneous fluids. A range of alternative methods to calculate density profiles is thereby explored.

 DOI: 10.1201/9781003286882-10

DFT approximations constitute an important topic and they are introduced in Section 10.2. Many of them have been superseded by the Fundamental Measure Theory (FMT) for hard sphere mixtures and Section 10.3 is devoted to it. Various refinements of the original version of FMT are presented as well as a more consistent approach. The rather intricate formalism of FMT is presented, first for the bulk pressure, then for the pair correlations of the bulk fluid and finally for the inhomogeneous case. This allows for a gradual introduction with a stepwise increase in complexity of the FMT concepts and entities that are necessary for the complete theory.

10.1 INTEGRAL EQUATION APPROXIMATIONS, A CLOSER LOOK

In this section we will make a more thorough treatment of integral equation approximations, which were introduced in Sections 9.1.3.2 and 9.1.3.3. Each approximation constitutes a choice of closure relationship and, as we saw, it can usually be formulated as some expression for the bridge function. For pair correlation theories this amounts to a choice for $b^{(2)}$.

The main topic in the first part of the present section is the important group of integral equation theories that use either the Ansatz $b^{(2)} \approx f(h^{(2)} - c^{(2)}) = f(\gamma^{(2)})$ (Equation 9.88) or $b^{(2)} \approx f(\gamma^{(2)} - \varphi^{\mathrm{R}}) = f(\gamma^{*(2)})$ (Equation 9.92). The function $f(x)$ is selected differently depending on the closure and $\varphi^{\mathrm{R}}(r)$ can be selected to be equal to $\beta u^{\mathrm{L}}(r)$ or some other appropriate function, where u^{L} is the long-range part of the pair potential suitably defined. The quality of the resulting approximation depends to a large extent on the physical principles used in its formulation. We will start with some approximations that do not use any consistency criteria and in Section 10.1.2 we explore the use of such criteria, which substantially improve the quality of the theory.

10.1.1 Inconsistent integral equation theories for bulk fluids

The simplest option is the HNC approximation, where one simply sets $b^{(2)} = 0$. The physical interpretation of this approximation has been discussed earlier in connection to Equations 9.159 and 9.184. The HNC approximation is often very useful for fluids where the particles interact with long-range interactions, like charged particles, while it in general is lousy for the description of effects of short-range interactions, like the packing of hard spheres. This implies, for example, that the HNC approximation is quite accurate for dilute to medium-concentrated electrolyte solutions in the primitive model and for thin classical plasmas, but not so good for dense ionic fluids, where both electrostatic interactions and packing effects are important. In the present treatment we will, however, focus on fluids consisting of uncharged particles, while electrolytes will be treated in Chapters 12 and 13. Since we only deal with bulk fluids here, we drop the superscript b, indicating bulk, for the functions. We maintain, however, the notation n^{b} for the bulk density.

We will start with the *Percus-Yevick approximation* given by $b^{(2)} = \ln(1+\gamma^{(2)}) - \gamma^{(2)}$ (Equation 9.80) or, equivalently, in terms of the cavity function

$$y^{(2)} = 1 + h^{(2)} - c^{(2)} = 1 + \gamma^{(2)} \quad \text{(PY)} \tag{10.1}$$

(Equation 9.139), which is the same as Equation 9.78. A justification of this approximation was given in Example 9.B.

For a bulk fluid of hard spheres with diameter d^{h} we always have

$$g^{(2)}(r) = e^{-\beta u(r)} y^{(2)}(r) = \begin{cases} 0, & r < d^{\mathrm{h}} \\ y^{(2)}(r) & r > d^{\mathrm{h}} \end{cases}$$

and the PY approximation 10.1 implies that

$$c^{(2)}(r) = g^{(2)}(r) - y^{(2)}(r) = \begin{cases} -y^{(2)}(r), & r < d^{\mathrm{h}} \\ 0, & r > d^{\mathrm{h}} \end{cases} \quad \text{(PY, hard spheres).}$$

As mentioned in Section 9.1.3.2, the Mean Spherical Approximation and the PY approximation are identical for hard spheres, so we are actually treating both approximations here.

Recall that the function $y^{(2)}(r)$ is continuous. From the relationships above it follows that $g^{(2)}(d^{\mathrm{h}+}) = y^{(2)}(d^{\mathrm{h}})$ and $c^{(2)}(d^{\mathrm{h}-}) = -y^{(2)}(d^{\mathrm{h}})$, where $d^{\mathrm{h}+}$ and $d^{\mathrm{h}-}$ denotes the limit when $r \to d^{\mathrm{h}}$ from the right and the left, respectively. The value $g^{(2)}(d^{\mathrm{h}+})$ is called the contact value of $g^{(2)}(r)$ and $n^{\mathrm{b}} g^{(2)}(d^{\mathrm{h}+})$ the contact density at the hard sphere surface.

The OZ equation 9.72 can be written as

$$\gamma^{(2)}(r) = n^{\mathrm{b}} \int d\mathbf{r}'\, c^{(2)}(r')[g^{(2)}(|\mathbf{r} - \mathbf{r}'|) - 1].$$

By inserting the expressions above for $g^{(2)}$ and $c^{(2)}$ in terms of $y^{(2)}$ we obtain from Equation 10.1 the following integral equation for the cavity function of a hard sphere fluid

$$y^{(2)}(r) = 1 - n^{\mathrm{b}} \int\limits_{\substack{r' < d^{\mathrm{h}} \\ |\mathbf{r}-\mathbf{r}'| > d^{\mathrm{h}}}} d\mathbf{r}'\, y^{(2)}(r') y^{(2)}(|\mathbf{r} - \mathbf{r}'|) + n^{\mathrm{b}} \int\limits_{r' < d^{\mathrm{h}}} d\mathbf{r}'\, y^{(2)}(r'),$$

where $g^{(2)}$ and $-c^{(2)}$ in the OZ equation have been replaced by $y^{(2)}$ whenever the former functions are nonzero and where the limits on the integrals have been set accordingly. This equation has an exact solution in terms of elementary functions[1] and for $r \leq d^{\mathrm{h}}$ the function $y^{(2)}(r)$ is simply a third order

[1] E. Thiele, J. Chem. Phys., **39** (1963) 474 (https://doi.org/10.1063/1.1734272), M. S. Wertheim, Phys. Rev. Lett., **10** (1963) 321 (https://doi.org/10.1103/PhysRevLett.10.321) and M. S. Wertheim, J. Math. Phys., **5** (1964) 643 (https://doi.org/10.1063/1.1704158).

polynomial in the reduced radial variable $r_{\rm R} \equiv r/d^{\rm h}$, so we have

$$-c^{(2)}(r) = \begin{cases} a_0 + a_1 r_{\rm R} + a_2 r_{\rm R}^2 + a_3 r_{\rm R}^3, & r_{\rm R} < 1 \\ 0, & r_{\rm R} > 1 \end{cases} \quad \text{(PY, hard spheres)}, \tag{10.2}$$

where the coefficients have been determined as

$$\begin{aligned} a_0 &= \frac{(1+2\eta^{\rm hs})^2}{(1-\eta^{\rm hs})^4}, & a_1 &= -\frac{3\eta^{\rm hs}(2+\eta^{\rm hs})^2}{2(1-\eta^{\rm hs})^4}, \\ a_2 &= 0, & a_3 &= \frac{\eta^{\rm hs} a_0}{2} = \frac{\eta^{\rm hs}(1+2\eta^{\rm hs})^2}{2(1-\eta^{\rm hs})^4} \qquad \text{(PY)}, \end{aligned} \tag{10.3}$$

where $\eta^{\rm hs} = n^{\rm b}\mathrm{V}^{\rm h} = \bar{N}\mathrm{V}^{\rm h}/V$ is the **hard sphere packing fraction**, $\mathrm{V}^{\rm h}$ is the volume of a hard sphere with diameter $d^{\rm h}$ and $\bar{N}$ is the average number of sphere centers in the volume V. Thus, we have

$$\eta^{\rm hs} = \frac{\pi (d^{\rm h})^3 n^{\rm b}}{6}.$$

Distinguish between $\eta^{\rm hs}$ and η, which has a different meaning elsewhere in this treatise.

For $r > d^{\rm h}$ the solution for $y^{(2)}(r)$, which is equal to $g^{(2)}(r)$ there, can be written as a superposition of exponentially damped, oscillatory functions, $\sum_{\nu=1}^{\infty} C_\nu e^{-\mathfrak{a}_\nu r}\cos(\mathfrak{b}_\nu r + \mathfrak{c}_\nu)/r$ where C_ν, $\mathfrak{a}_\nu$, $\mathfrak{b}_\nu$ and $\mathfrak{c}_\nu$ are constants (we will call such functions "oscillatory Yukawa functions"). Alternatively, $g^{(2)}(r)$ can be written as some rather complicated analytic expressions for each interval $md^{\rm h} < r < (m+1)d^{\rm h}$ with integer m for $m \geq 1$.[2] These constants and expressions, which will not be given here, can be determined from the OZ equation with Equation 10.2 inserted. The resulting $g^{(2)}(r)$ is a smooth function of r for $r > d^{\rm h}$.

The PY solution 10.2 for $c(r)$ can be used to obtain the **equation of state**, which gives the pressure P as a function of density $n^{\rm b}$. The pressure of a fluid can be calculated by using the **virial equation** (Equation I:5.73)

$$\frac{\beta P}{n^{\rm b}} = 1 - \frac{\beta n^{\rm b}}{6}\int d\mathbf{r}\, r \frac{du(r)}{dr} g^{(2)}(r). \tag{10.4}$$

For a hard sphere fluid this reduces to the expression (Equation I:5.77)

$$\frac{\beta P}{n^{\rm b}} = 1 + \frac{2\pi (d^{\rm h})^3}{3} n^{\rm b} g^{(2)}(d^{\rm h+}) = 1 + 4\eta^{\rm hs} g^{(2)}(d^{\rm h+}) \quad \text{(hard spheres)}.$$

The quantity $\beta P/n^{\rm b}$ is called the **compressibility factor**, which has the value 1 for an ideal gas. Its nonideal part, the second term, is for a hard

[2] W. R. Smith and D. Henderson, Mol. Phys., **19** (1970) 411 (https://doi.org/10.1080/00268977000101421).

sphere fluid proportional to the contact value $g^{(2)}(d^{\mathrm{h}+})$. Since $g^{(2)}(d^{\mathrm{h}+}) = y^{(2)}(d^{\mathrm{h}}) = -c^{(2)}(d^{\mathrm{h}-})$ and $-c^{(2)}(d^{\mathrm{h}-}) = a_0 + a_1 + a_2 + a_3$ (from Equation 10.2) we obtain after simplification

$$\frac{\beta P^{\mathrm{v}}}{n^{\mathrm{b}}} = \frac{1 + 2\eta^{\mathrm{hs}} + 3(\eta^{\mathrm{hs}})^2}{(1 - \eta^{\mathrm{hs}})^2} \quad \text{(PY, hard spheres)}, \tag{10.5}$$

where superscript "v" indicates that the pressure is obtained from the virial equation (the "virial route").

The pressure of a fluid can alternatively be calculated from the isothermal compressibility χ_T (the "compressibility route"). From Equation 9.189 it follows that $(\partial P/\partial n^{\mathrm{b}})_T = 1/(n^{\mathrm{b}}\chi_T)$, so we have

$$P = \int_0^{n^{\mathrm{b}}} dn'^{\mathrm{b}} \left(\frac{\partial P}{\partial n'^{\mathrm{b}}}\right)_T = \int_0^{n^{\mathrm{b}}} \frac{dn'^{\mathrm{b}}}{n'^{\mathrm{b}}\chi_T(n'^{\mathrm{b}})}, \tag{10.6}$$

where χ_T is evaluated at density n'^{b}. We can calculate χ_T from $c^{(2)}(r)$ by using the compressibility equation 9.190. Taking $c^{(2)}(r)$ from Equation 10.2 we obtain after integration

$$\frac{\beta}{n^{\mathrm{b}}\chi_T^{\mathrm{c}}} = \frac{(1 + 2\eta^{\mathrm{hs}})^2}{(1 - \eta^{\mathrm{hs}})^4} \quad \text{(PY, hard spheres)}, \tag{10.7}$$

where superscript "c" indicates that χ_T is obtained from the compressibility equation. By inserting this into Equation 10.6 and integrating we get

$$\frac{\beta P^{\mathrm{c}}}{n^{\mathrm{b}}} = \frac{1 + \eta^{\mathrm{hs}} + (\eta^{\mathrm{hs}})^2}{(1 - \eta^{\mathrm{hs}})^3} \quad \text{(PY, hard spheres)}. \tag{10.8}$$

We see that the viral and compressibility routes yield different expressions 10.5 and 10.8 for the pressure. Thus, we obtain different values of the pressure depending on how it is calculated. In exact theory the pressure or any other quantity will have the same value irrespectively of the manner it is calculated, but, as has been mentioned earlier, in approximate theories this need not be the case and inconsistencies like this occur. As pointed out earlier, *it is unavoidable that approximate theories are inconsistent in some ways*, but the better the approximation the fewer inconsistencies there are, or, alternatively, the remaining inconsistencies cause less deviations from the correct values.

{**Exercise 10.1:** a) Perform the calculations mentioned above to obtain Equations 10.5, 10.7 and 10.8. b) Starting from Equation 10.5, derive the expression for χ_T in the PY approximation as obtained from the virial equation. Compare the result to Equation 10.7.}

Each of the Equations 10.5 and 10.8 for the pressure can be analyzed by writing it as a **virial expansion**. Such an expansion is (see Section 3.4 in Volume I)

$$\frac{\beta P}{n^{\mathrm{b}}} = 1 + \sum_{\nu\geq 2} B_{\nu}\,[n^{\mathrm{b}}]^{\nu-1} = 1 + \sum_{\nu\geq 2} B_{\nu}^{*}\,[\eta^{\mathrm{hs}}]^{\nu-1}, \tag{10.9}$$

where the coefficients $B_\nu = B_\nu(T)$ are the **virial coefficients** that depend on temperature only and $B_\nu^* = B_\nu/(V^{\mathrm{h}})^{\nu-1}$ are dimensionless virial coefficients. B_ν can be calculated from the pair interaction potential, at least in principle. For example, we have (from Equations I:3.54 and I:3.56)

$$\begin{aligned} B_2(T) &= -\frac{1}{2V}\int d\mathbf{r}_1 d\mathbf{r}_2 f_{\mathrm{M}}(r_{12}) \\ B_3(T) &= -\frac{1}{3V}\int d\mathbf{r}_1 d\mathbf{r}_2 d\mathbf{r}_2 f_{\mathrm{M}}(r_{12}) f_{\mathrm{M}}(r_{13}) f_{\mathrm{M}}(r_{23}), \end{aligned} \tag{10.10}$$

where $f_{\mathrm{M}}(r_{ij}) \equiv \exp[-\beta u(r_{ij})] - 1$ is called the **Mayer f-function.**[3] The higher virial coefficients are complicated sums of cluster integrals involving $f_{\mathrm{M}}(r_{ij})$.

By expanding P from Equations 10.5 and 10.8 in power series in η^{hs}, we obtain the PY versions of the virial expansion

$$\begin{aligned} \frac{\beta P^{\mathrm{v}}}{n^{\mathrm{b}}} &= 1 + 4\eta^{\mathrm{hs}} + 10(\eta^{\mathrm{hs}})^2 + 16(\eta^{\mathrm{hs}})^3 + 22(\eta^{\mathrm{hs}})^4 + 28(\eta^{\mathrm{hs}})^5 + \ldots \\ \frac{\beta P^{\mathrm{c}}}{n^{\mathrm{b}}} &= 1 + 4\eta^{\mathrm{hs}} + 10(\eta^{\mathrm{hs}})^2 + 19(\eta^{\mathrm{hs}})^3 + 31(\eta^{\mathrm{hs}})^4 + 46(\eta^{\mathrm{hs}})^5 + \ldots . \end{aligned}$$

Before continuing with integral equations, we will treat the virial expansion for hard sphere fluids in more detail and present some very useful formulas for P.

Accurate expressions for P, μ^{ex} and χ_T of hard sphere fluids

As a starting point, we consider the correct virial expansion for a hard sphere fluid, which is

$$\begin{aligned} \frac{\beta P}{n^{\mathrm{b}}} &= 1 + 4\eta^{\mathrm{hs}} + 10(\eta^{\mathrm{hs}})^2 + 18.36(\eta^{\mathrm{hs}})^3 + 28.22(\eta^{\mathrm{hs}})^4 \\ &+ 39.82(\eta^{\mathrm{hs}})^5 + \ldots \qquad \text{(hard spheres)}, \end{aligned} \tag{10.11}$$

[3]In Volume I after Equation I:3.56, there is a misprint in the definition of f_{M} that lacks a minus sign in the exponent. The same error occurs in the shaded text box after Equation I:3.56.

Table 10.1 The dimensionless virial coefficients B_ν^* for a hard sphere fluid calculated from the Carnahan-Starling formula 10.12 and the Liu formula 10.13 compared with the best estimates obtained by Schultz and Kofke.[4]

ν	C-S values	Liu values	Best estimates
2	4	4	4
3	10	10	10
4	18	18.385	18.3648
5	28	28.154	28.2244
6	40	39.808	39.8152
7	54	53.346	53.3421
8	70	68.769	68.529
9	88	86.077	85.826
10	108	105.27	105.7
11	130	126.35	126.5

(see Table 10.1 for more precise virial coefficients up to B_{11}^*),[4] where the first few coefficients B_2^* to B_4^* are known analytically[5] and the rest are calculated numerically from the explicit expressions for B_ν in terms of the pair potential. We see that the PY predictions of the virial expansion differ substantially from the correct one from the cubic term and onwards.

There exist several accurate approximations for P of a hard sphere fluid, the most well-known is the **Carnahan and Starling (C-S) formula** for P.[6] Carnahan and Starling approximated the coefficients B_2^* to B_6^* in the virial expansion by integers

$$\frac{\beta P}{n^{\mathrm{b}}} \approx 1 + 4\eta^{\mathrm{hs}} + 10(\eta^{\mathrm{hs}})^2 + 18(\eta^{\mathrm{hs}})^3 + 28(\eta^{\mathrm{hs}})^4 + 40(\eta^{\mathrm{hs}})^5 + \ldots$$

and noted that they can be written as $B_\nu^* = \nu^2 + \nu - 2$. They furthermore assumed that B_ν^* for $\nu > 5$ are given by the same expression and obtained the simple formula

$$\frac{\beta P}{n^{\mathrm{b}}} = \frac{1 + \eta^{\mathrm{hs}} + (\eta^{\mathrm{hs}})^2 - (\eta^{\mathrm{hs}})^3}{(1 - \eta^{\mathrm{hs}})^3} \quad \text{(C-S)} \tag{10.12}$$

[4]The values in the fourth column of Table 10.1 have been obtained by A. J. Schultz and D. A. Kofke, Phys. Rev. E, **90** (2014) 023301 (https://doi.org/10.1103/PhysRevE.90.023301).

[5]$B_2^* = 4$, $B_3^* = 10$ and $B_4^* = [2707 + (438\sqrt{2} - 4131\arccos(1/3))/\pi]/70 \approx 18.364768383$

[6]N. F. Carnahan and K. E. Starling, J. Chem. Phys. **51** (1969) 635 (https://doi.org/10.1063/1.1672048).

that has a power series expansion with coefficients $B_\nu^* = \nu^2 + \nu - 2$. Also the C-S values of B_ν^* for $\nu > 5$ are, in fact, in quite good agreement with the precise values as can be seen in Table 10.1. Equation 10.12 gives surprisingly good values of the pressure as function of $\eta^{\rm hs}$ in excellent agreement with simulations results for the hard sphere fluid. As we can see, the virial coefficients for C-S lie between those for $P^{\rm V}$ and $P^{\rm C}$ in the PY approximation and, in fact, $P^{\text{C-S}} = (P^{\rm v,PY} + 2P^{\rm c,PY})/3$.

An even better, simple approximation for P of a hard sphere fluid is the **Liu formula**[7]

$$\frac{\beta P}{n^{\rm b}} = \frac{1 + \eta^{\rm hs} + (\eta^{\rm hs})^2 - \frac{8}{13}(\eta^{\rm hs})^3 - (\eta^{\rm hs})^4 + \frac{1}{2}(\eta^{\rm hs})^5}{(1-\eta^{\rm hs})^3} \quad \text{(Liu)}. \tag{10.13}$$

Its power series expansion is

$$\begin{aligned} \frac{\beta P}{n^{\rm b.}} &= 1 + 4\eta^{\rm hs} + 10(\eta^{\rm hs})^2 + 18.38(\eta^{\rm hs})^3 + 28.15(\eta^{\rm hs})^4 \\ &+ 39.81(\eta^{\rm hs})^5 + \dots, \quad \text{(Liu)} \end{aligned}$$

and the higher order virial coefficients are listed in Table 10.1. As we can see, the virial coefficients are considerably more accurate than those in the C-S approach. However, the actual values of the pressure from the Liu and C-S formulas differ very little. For example, for $\eta^{\rm hs} = 0.491$, where the hard sphere fluid has a phase transition to the solid state, $\beta P/n^{\rm b}$ is 12.24 from the C-S formula and it is 12.25 from the Liu formula. For many practical purposes, the C-S formula can be considered to give virtually the exact value of $\beta P/n^{\rm b}$ although the Liu formula is more accurate. In general, one should use the entire formula rather than a truncated virial expansion thereof because many terms of the expansion must be included in order to obtain the highly accurate values.

Since the C-S expression for $\beta P/n^{\rm b}$ is very accurate for all densities, one can obtain precise values of the excess chemical potential from Equation 10.12 via an integration of the Gibbs-Duhem equation $n^{\rm b} d\mu = dP$ when T is constant and likewise one can obtain the isothermal compressibility by using $(\partial P/\partial n^{\rm b})_T = 1/(n^{\rm b}\chi_T)$ (from Equation 9.189), see Exercise 10.2. The results are

$$\beta\mu^{\rm ex} = \frac{8\eta^{\rm hs} - 9(\eta^{\rm hs})^2 + 3(\eta^{\rm hs})^3}{(1-\eta^{\rm hs})^3} \quad \text{(C-S)}. \tag{10.14}$$

$$\frac{\beta}{n^{\rm b}\chi_T} = 1 + \frac{8\eta^{\rm hs} - 2(\eta^{\rm hs})^2}{(1-\eta^{\rm hs})^4} \quad \text{(C-S)}. \tag{10.15}$$

[7] H. Liu, Mol. Phys., **119** (2021) e1886364 (https://doi.org/10.1080/00268976.2021.1886364).

{**Exercise 10.2:** Use the Gibbs-Duhem equation and $(\partial P/\partial n^{\mathrm{b}})_T = 1/(n^{\mathrm{b}}\chi_T)$ to derive Equations 10.14 and 10.15 from Equation 10.12 Derive also the expansion of the excess chemical potential for hard spheres

$$\beta\mu^{\mathrm{ex}} = \sum_{\nu\geq 2} \frac{\nu B_\nu^*}{\nu - 1} [\eta^{\mathrm{hs}}]^{\nu-1} = 8\eta^{\mathrm{hs}} + 15[\eta^{\mathrm{hs}}]^2 + \ldots \qquad (10.16)$$

using the Gibbs-Duhem equation.}

One can derive the corresponding expressions from the Liu formula 10.13, which gives even more correct values. However, the values obtained from the C-S and Liu expressions differ very little; for $\eta^{\mathrm{hs}} = 0.491$ the difference in $\beta\mu^{\mathrm{ex}}$ is less than 0.2 % and in $\beta/(n^{\mathrm{b}}\chi_T)$ less than 0.3 %. The Liu formulas for these quantities are therefore not shown here.

There are other expressions for thermodynamical quantities of hard sphere fluids that are very accurate, at parity with the Liu formula. The **Hansen-Goos (H-G) formula** for the isothermal compressibility is[8]

$$\frac{\beta}{n^{\mathrm{b}}\chi_T} = \frac{[1 + 2\eta^{\mathrm{hs}} - a_3(\eta^{\mathrm{hs}})^3 + a_4(\eta^{\mathrm{hs}})^4]^2}{(1 - \eta^{\mathrm{hs}})^4} \qquad \text{(H-G)},$$

where the constants a_3 and a_4 can be explicitly expressed analytically[9] and are approximately given by $a_3 \approx 1.270463$ and $a_4 \approx 0.694605$. From this expression one can obtain the corresponding formulas for P and μ^{ex}, which will not be given here.

In Figure 10.1 $\beta P/n^{\mathrm{b}}$ is plotted as function of the packing fraction and we see that the PY values at high η^{hs} deviate quite a lot from the accurate ones given by the C-S and Liu formulas. The substantial inconsistency of the PY results for P^{c} and P^{v} is apparent. The corresponding results from the HNC approximation (not shown) are even worse with P^{c} lying quite far below the accurate P and with P^{v} quite far above, i.e., on the opposite side compared to the corresponding PY predictions. The figure also show the results of the *Martynov-Sarkisov (M-S) approximation*[10] and we see that they are much better, in particular the values obtained via the compressibility route (the upper full curve). The M-S result from the virial route is also good, so there is a rather small inconsistency for P in this approximation. Results from the *Bomont-Bretonnet (B-B)*[11] and the *Zero-separation (ZSEP)* approximations[12]

---

[8]H. Hansen-Goos, J. Chem. Phys., **144** (2016) 164506 (https://doi.org/10.1063/1.4947534).

[9]The expressions for the constants are $a_3 = 2(19 - B_4^*)$, where B_4^* is given in footnote 5, and $a_4 = -3/7 + 5a_3/8 + 4[1 - 21a_3/32 + 105a_3^2/1024]^{1/2}/7$.

[10]G. A. Martynov and G. N. Sarkisov, Mol. Phys. **49** (1983) 1495 (https://doi.org/10.1080/00268978300102111).

[11]J.-M. Bomont and J.-L. Bretonnet, J. Chem. Phys., **121** (2004) 1548 (https://doi.org/10.1063/1.1764772).

[12]L. L. Lee, J. Chem. Phys., **110** (1999) 7589 (https://doi.org/10.1063/1.478661).

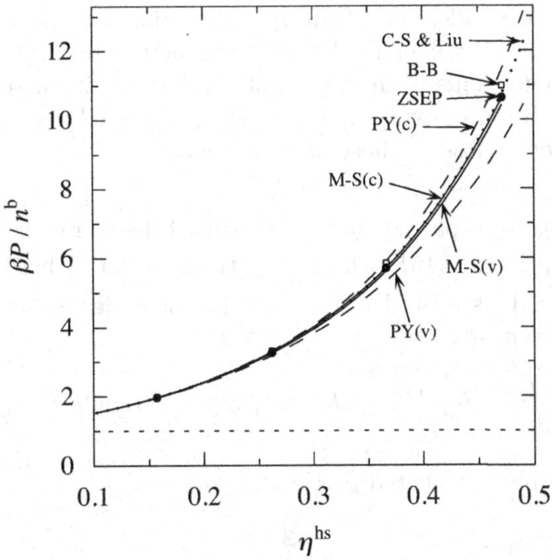

Figure 10.1 The compressibility factor $\beta P/n^{\text{b}}$ of a hard sphere fluid plotted as a function of the packing fraction η^{hs}. The PY predictions are shown as dashed curves, the upper and the lower ones are calculated via the compressibility (c) and virial (v) routes, respectively. The values from the very accurate Carnahan-Starling (C-S) and Liu formulas, Equations 10.12 and 10.13, are shown as dotted curves. They are indistinguishable from each other on the scale of the figure (every second dot shows the C-S value and every second the Liu value, apart from the last (upper) one which show the virtually overlapping dots from both formulas). The full curves show the predictions from the Martynov-Sarkisov (M-S) approximation, the upper and the lower ones are calculated via c and v routes, respectively (data taken from Martynov et al.[10]). The open squares show the results from the B-B approximation and full circles those from the ZSEP approximation (data taken from Bomont et al.[11] and Lee[12] respectively).

are also shown. We will first investigate the MS approximation in more detail and in the next section we will treat B-B and ZSEP.

The M-S approximation is given by the closure $b^{(2)} = [1+2\gamma^{(2)}]^{1/2} - \gamma^{(2)} - 1$ (Equation 9.89) or, equivalently, by

$$b^{(2)} = -\tfrac{1}{2}[\beta w^{(2)\text{intr}}]^2 \qquad \text{(M-S)} \tag{10.17}$$

(Equation 9.90, see also footnote 32 on page 35 in Section 9.1). Recall that $w^{(2)} = u + w^{(2)\text{intr}}$ (Equation 9.23) and that $\beta w^{(2)\text{intr}} = \gamma^{(2)} + b^{(2)}$ (Equation 9.69). This approximation is intended also for other fluids than hard spheres. Martynov and Sarkisov[13] were guided to the closure 10.17 both from

[13]See reference in footnote 10.

an analysis at low densities and from arguments valid for conditions near the critical point. The verification of an approximation for two different conditions often increases the likelihood that it will be a good approximation also for other conditions. This is an important observation in the construction of approximations. Here, we will pursue a low density analysis for hard spheres based on arguments close to those used by them.

We will utilize the expansion of $b^{(2)}$ in terms of the pair correlation function $h^{(2)}$ given in Equation 9.161, where only the first term, here called $b_{\mathrm{I}}^{(2)}$, is written explicitly. This term dominates for low densities, so $b^{(2)} \approx b_{\mathrm{I}}^{(2)}$. For a bulk fluid it is given by

$$b_{\mathrm{I}}^{(2)}(r_{12}) = \frac{1}{2}\int d\mathbf{r}_3 d\mathbf{r}_4 h^{(2)}(r_{13})h^{(2)}(r_{14})n^{\mathrm{b}}h^{(2)}(r_{34})n^{\mathrm{b}}h^{(2)}(r_{32})h^{(2)}(r_{42}). \tag{10.18}$$

The integral can be represented by the diagram

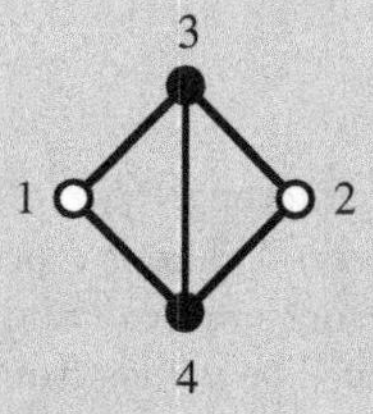

where each line represents $h^{(2)}$, the open circles represent the points $\mathbf{r}_1$ and $\mathbf{r}_2$ and the filled circles represent $\mathbf{r}_3$ and $\mathbf{r}_4$. Each of the latter has a factor n^{b} and is integrated over the whole space. The rhs of Equation 10.17 constitutes an approximation for $b_{\mathrm{I}}^{(2)}$ that can be justified in the following manner.

For a hard sphere fluid the function $h^{(2)}(r)$ is equal to -1 when $r < d^{\mathrm{h}}$ and it decays exponentially to zero (in an oscillatory manner) when r increases. Consider the factor $h^{(2)}(r_{34})$ in the integrand of Equation 10.18 (the vertical line in the diagram). A major contribution to $b_{\mathrm{I}}^{(2)}(r_{12})$ comes, in fact, from the integration region where $r_{34} < d^{\mathrm{h}}$ and $h^{(2)}(r_{34}) = -1$. When $r_{34} > d^{\mathrm{h}}$, the contributions from most of the integration region are relatively small, in particular when r_{12} is large, because the products $h^{(2)}(r_{13})h^{(2)}(r_{34})h^{(2)}(r_{32})$ and $h^{(2)}(r_{14})h^{(2)}(r_{34})h^{(2)}(r_{42})$ then contain up to three exponentially small factors there (each of these factors is represented by the lines that connect to the filled circle 3 or 4, respectively). It is therefore a reasonable approximation to set $h^{(2)}(r_{34}) = -1$ in the integral and we obtain

$$\begin{aligned} b_{\mathrm{I}}^{(2)}(r_{12}) &\approx -\frac{1}{2}\int d\mathbf{r}_3 d\mathbf{r}_4 h^{(2)}(r_{13})h^{(2)}(r_{14})(n^{\mathrm{b}})^2 h^{(2)}(r_{32})h^{(2)}(r_{42}) \\ &= -\frac{1}{2}\left[\int d\mathbf{r}_3 h^{(2)}(r_{13})n^{\mathrm{b}}h^{(2)}(r_{32})\right]^2. \end{aligned}$$

As we will see, $\beta w^{(2)\mathrm{intr}}(r_{12}) \approx \int d\mathbf{r}_3 h^{(2)}(r_{13}) n^{\mathrm{b}} h^{(2)}(r_{32})$ for low densities, so we find that $b^{(2)} \approx b_{\mathrm{I}}^{(2)} \approx -\frac{1}{2}[\beta w^{(2)\mathrm{intr}}]^2$, that is, Equation 10.17.

The approximation for $\beta w^{(2)\mathrm{intr}}$ that we used here can be obtained by writing the OZ equation 9.70 as

$$\begin{aligned} c^{(2)}(r_{12}) &= h^{(2)}(r_{12}) - n^{\mathrm{b}} \int d\mathbf{r}_3\, c^{(2)}(r_{13}) h^{(2)}(r_{32}) \\ &= h^{(2)}(r_{12}) - n^{\mathrm{b}} \int d\mathbf{r}_3\, h^{(2)}(r_{13}) h^{(2)}(r_{32}) \\ &+ (n^{\mathrm{b}})^2 \int d\mathbf{r}_3 \left[\int d\mathbf{r}_4\, c^{(2)}(r_{14}) h^{(2)}(r_{43}) \right] h^{(2)}(r_{32}) \end{aligned}$$

(cf. the manipulations that lead to Equation 9.41), so we have $\gamma^{(2)}(r_{12}) \approx n^{\mathrm{b}} \int d\mathbf{r}_3\, h^{(2)}(r_{13}) h^{(2)}(r_{32})$ to the lowest order in n^{b}. In $\beta w^{(2)\mathrm{intr}} = \gamma^{(2)} + b^{(2)}$ the term $b^{(2)}$ is a higher order contribution in n^{b} because its terms contain $(n^{\mathrm{b}})^{\nu}$ with $\nu \geq 2$, so we obtain $\beta w^{(2)\mathrm{intr}} \approx \int d\mathbf{r}_3 h^{(2)}(r_{13}) n^{\mathrm{b}} h^{(2)}(r_{32})$ to the lowest order.

10.1.2 The use of consistency criteria in integral equation theories for bulk fluids

As we have seen, the pressure P^{v} calculated via the *virial route* (from the virial equation 10.4) and P^{c} via the *compressibility route* (from Equation 10.6 with χ_T obtained from the compressibility equation 9.190) are often different in approximate theories, while they must be the same in exact statistical mechanics. In integral equation approximations, a common approach is to enforce the equality $P^{\mathrm{v}} = P^{\mathrm{c}}$. The application of this consistency criterium and others that will be introduced below can make integral equation theories very accurate. Such criteria therefore constitute important tools for the design of approximations in statistical mechanics of fluids.

One can enforce the condition $P^{\mathrm{v}} = P^{\mathrm{c}}$ by introducing a suitable parameter in the closure approximation and selecting its value such that P^{v} and P^{c} become equal. The closure and the choice of parameter must, of course, be such that it is possible to get $P^{\mathrm{v}} = P^{\mathrm{c}}$. A classical example is the Roger-Young closure 9.99 that makes an interpolation between the PY and HNC approximations. As noted earlier, the pressure of a hard sphere fluid from the PY and HNC approximations lie on opposite sides of the accurate pressure, so one can obtain a better result for the pressure by means of such an interpolation. The R-Y closure has a parameter α in the switching function $\mathfrak{f}(r) = 1 - e^{-\alpha r}$, which is used to implement the interpolation. The criterion in the selection of α is accordingly that $P^{\mathrm{v}} = P^{\mathrm{c}}$ and by selecting α in this manner one obtains a better closure than the PY and HNC closures. The R-Y closure gives, for example, improved $g^{(2)}(r)$ and P compared to the latter closures. For soft core fluids, the Z-H approximation (Equation 9.98), which interpolates between the SMSA and HNC approximations, does the same trick. There is, however, no guarantee that the values of other entities become better.

A better way to asses the quality of a closure is to investigate the accuracy of the chemical potential obtained from the theory because it is more demanding to obtain an accurate μ^{ex} than the pressure. The chemical potential is sensitive to the over-all quality of the functions $c^{(2)}(r)$, $y^{(2)}(r)$ and $b^{(2)}(r)$, in particular their values inside the core region. As we have seen from

$$\beta\mu^{\mathrm{ex,b}} = n^{\mathrm{b}} \int d\mathbf{r} \left[\tfrac{1}{2}[h^{(2)\mathrm{b}}(r)]^2 - c^{(2)\mathrm{b}}(r) + g^{(2)\mathrm{b}}(r) b^{(2)\mathrm{b}}(r) \right] - \frac{\gamma^{(2)\mathrm{b}}(0)}{2} - c^{(1)\mathrm{R,b}}$$

(Equation 9.203), μ^{ex} has contributions from the various pair functions for *all* distances $0 \le r \le \infty$. This is in contrast to the virial pressure P^{v} (Equation 10.4), which is entirely determined by $g^{(2)}(r)$ outside the core because $g^{(2)}$ is zero inside. Also the excess internal energy $\bar{U}^{\mathrm{ex}}$ (Equation I:5.70)

$$\bar{U}^{\mathrm{ex}} = \frac{\bar{N} n^{\mathrm{b}}}{2} \int d\mathbf{r}\, u(r) g^{(2)}(r) \tag{10.19}$$

is determined by $g^{(2)}(r)$ outside the core (except for hard spheres where $\bar{U}^{\mathrm{ex}}$ is zero).

The importance of the values of various pair functions inside the core region is further illustrated by the following exact formula for the chemical potential of hard sphere fluids

$$\beta\mu^{\mathrm{ex,b}} = \ln y^{(2)}(0) \equiv \gamma^{(2)}(0) + b^{(2)}(0) \quad \text{(hard spheres)}, \tag{10.20}$$

which gives $\mu^{\mathrm{ex,b}}$ entirely in terms of the value of these pair functions at $r = 0$. The first equality in this equation, $\ln y^{(2)}(0) = \beta\mu^{\mathrm{ex,b}}$, is called the ***first zero-separation theorem for*** $\mathbf{y^{(2)}}$, where "zero separation" refers to $r = 0$. It is derived in Appendix 10A (Equation 10.186) and is a further consistency criterium that can be used for integral equation theories. (For fluids other than hard spheres this theorem is given below in Equation 10.29.) The value of $\gamma^{(2)}(r)$ for $r = 0$ is given by the **first zero-separation theorem for** $\boldsymbol{\gamma^{(2)}}$

$$\gamma^{(2)}(0) = \frac{\beta}{n^{\mathrm{b}} \chi_T} - 1 + n^{\mathrm{b}} \int d\mathbf{r}'\, c^{(2)}(r') g^{(2)}(r'), \tag{10.21}$$

which is a consequence of the OZ equation as shown in Appendix 10A (Equation 10.189). The last term comprises only a small fraction of the value of $\gamma^{(2)}(0)$ for hards spheres (in the PY approximation the integral is equal to zero), so the other terms dominate. Incidentally we note that the expression for $b^{(2)}(0)$ that is obtained when one inserts Equation 10.21 into Equation 10.20 is called the *first zero-separation theorem for* $b^{(2)}$ for hard sphere fluids (see Equation 10.190).

As we saw earlier, all contributions to the chemical potential apart from the term $c^{(1)\mathrm{R,b}}$ can be expressed in terms of the pair functions of the system. The term $c^{(1)\mathrm{R,b}}$ contains complicated bridge diagram contributions (cf. Equation 9.206). However, for integral equation approximations that assume that $b^{(2)} = f(\gamma^{(2)})$ one can express this term explicitly in terms of the pair functions[14]

$$c^{(1)\mathrm{R,b}} = n^{\mathrm{b}} \int d\mathbf{r} \frac{h^{(2)}(r)}{\gamma^{(2)}(r)} \int_0^{\gamma^{(2)}(r)} dx\, f(x)$$

(this is derived in Section 10.1.5, Equation 10.57). The integral of $f(x)$ is straightforward to calculate. For example, in the PY approximation, where $f(\gamma^{(2)}) = \ln(1 + \gamma^{(2)}) - \gamma^{(2)}$, we have $\int_0^{\gamma^{(2)}} dx\, f(x) = (1 + \gamma^{(2)}) \ln(1 + \gamma^{(2)}) - [\gamma^{(2)}]^2/2 - \gamma^{(2)}$.

It follows that

$$\begin{aligned} \beta\mu^{\mathrm{ex,b}} &= n^{\mathrm{b}} \int d\mathbf{r} \left[\tfrac{1}{2}[h^{(2)b}(r)]^2 - c^{(2)}(r) + g^{(2)}(r) b^{(2)}(r) \right. \qquad (10.22) \\ &\left. - \frac{h^{(2)}(r)}{\gamma^{(2)}(r)} \int_0^{\gamma^{(2)}(r)} dx\, f(x) \right] - \frac{\gamma^{(2)}(0)}{2} \qquad \text{if } b^{(2)} = f(\gamma^{(2)}). \end{aligned}$$

This is very useful since *one can determine excess chemical potential* without doing any coupling-parameter integration like in

$$\mu = k_B T \ln(\Lambda^3 n^{\mathrm{b}}) + n^{\mathrm{b}} \int_0^1 d\xi \int d\mathbf{r}\, g^{(2)}(r;\xi) \frac{\partial u(r;\xi)}{\partial \xi} \qquad (10.23)$$

(cf. Equation 9.200), where the last term is $\mu^{\mathrm{ex,b}}$. Thus one does not need to calculate $g^{(2)}$ for all states with $0 \leq \xi \leq 1$. One neither need to calculate it for a range of densities as in Equation 9.187.

We will now assess the quality of integral equation approximations by comparing their predictions for $\beta\mu^{\mathrm{ex,b}}$ with accurate values. The predictions are obtained from Equation 10.20 and from Equation 10.22. For a hard sphere fluid of density $n^{\mathrm{b}} = 0.85(d^{\mathrm{h}})^{-3}$, the value of $\beta\mu^{\mathrm{ex,b}}$ from Equation 10.20 in various approximations is: HNC 30.50, PY 3.63, R-Y 3.56 and M-S 7.34, while the accurate value from the Carnahan-Starling approximation (Equation 10.14) is 11.95. Here, we see that the R-Y approximation does not succeed in interpolating between the HNC and the PY approximations for this quantity and that all four theories give very poor results. The best one is the M-S approximation, which nevertheless is not good. If one calculates $\beta\mu^{\mathrm{ex,b}}$ using Equation 10.22, one instead obtains the values HNC 14.86, PY 8.83, R-Y 9.53 and M-S 10.18, which are better but still not good, so one must improve the closure approximations so they give good pair functions inside the core region.

Before proceeding, we note that Equation 10.21 implies that approximations that give accurate results for χ_T and hence P^{c} also give good values

[14]R. Kjellander and S. Sarman, J. Chem. Phys., **90** (1989) 2768 (https://doi.org/10.1063/1.455924).

of $\gamma^{(2)}(0)$. Therefore, one way to distinguish the quality of various closures is to compare the values obtained for $b^{(2)}(0)$, which appears in Equation 10.20, instead of $\beta\mu^{\mathrm{ex,b}}$ itself.

Table 10.2 The values of $b^{(2)}(0)$ for a hard sphere bulk fluid at various densities n^{b} and packing fractions η^{hs} obtained from the PY, M-S, B-B and ZSEP approximations compared with the estimated exact value.
[a]Data from Lee.[15] [b]Results from Bomont and Bretonnet.[16] [c]Results from Lee.[17]

$n^{\mathrm{b}}/(d^{\mathrm{h}})^3$	η^{hs}	$b^{(2)}(0)$: PY[a]	M-S[a]	B-B[b]	ZSEP[c]	"Exact"[a]
0.3	0.1571	–1.19	–1.00	–0.693	–0.7257	–0.683
0.5	0.2618	–4.65	–3.90	–3.04	–3.044	–2.905
0.7	0.3665	–14.0	–12.0	–9.83	–9.8261	–9.549
0.9	0.4712	–40.1	–35.5	–29.00	–30.363	–29.42

For hard sphere fluids one can obtain an accurate estimate of the exact value of $b^{(2)}(0)$ from Equation 10.20 by using the Carnahan-Starling formulas for $\beta\mu^{\mathrm{ex,b}}$ and $\beta/(n^{\mathrm{b}}\chi_T)$ (Equations 10.14 and 10.15) and taking the value the last integral in Equation 10.21 from fits to simulations.[15] In Table 10.2 these values of $b^{(2)}(0)$ are denoted as "Exact" and they are compared to the results of some integral equation theories, including PY and M-S. We can see that the latter two theories give far too low values of $b^{(2)}(0)$. As we have seen in Figure 10.1, the M-S approximation nearly fulfills the requirement that $P^{\mathrm{v}} = P^{\mathrm{c}}$ and gives very good values for P^{c}. However, it does not give a good $\beta\mu^{\mathrm{ex,b}}$, which is mainly due to the incorrect values of $b^{(2)}(0)$ shown in Table 10.2.

The ***Bomont-Bretonnet (B-B) approximation*** (presented in Section 9.1.3.3) can for hard sphere fluids be regarded as an extension of the M-S approximation and as seen in the table, its results for $b^{(2)}(0)$ are very good. In the B-B closure, Equation 9.97, one uses $\gamma^{*(2)} = \gamma^{(2)} - \beta u^{\mathrm{L}}$ and since $u^{\mathrm{L}} = 0$ for hard spheres, one has $\gamma^{*(2)} = \gamma^{(2)}$, so the closure becomes[16]

$$b^{(2)} = \left[1 + 2\gamma^{(2)} + \tau(\gamma^{(2)})^2\right]^{1/2} - \gamma^{(2)} - 1 \qquad \text{(B-B, hard spheres)},$$

which differs from the M-S closure only by the term $\tau(\gamma^{(2)})^2$ in the square bracket. The parameter τ is determined from the consistency criterium that

$$\left(\frac{\partial P^{\mathrm{v}}}{\partial n^{\mathrm{b}}}\right)_T = \left(\frac{\partial P^{\mathrm{c}}}{\partial n^{\mathrm{b}}}\right)_T \equiv \frac{1}{n^{\mathrm{b}}\chi_T}. \tag{10.24}$$

[15]L. L. Lee, J. Chem. Phys. **103** (1995) 9388 (https://doi.org/10.1063/1.469998).
[16]J.-M. Bomont and J.-L. Bretonnet, J. Chem. Phys., **121** (2004) 1548 (https://doi.org/10.1063/1.1764772).

This is equivalent to $P^{\mathrm{v}} = P^{\mathrm{c}}$ provided that Equation 10.24 holds for all n^{b} because one obtains the pressures P^{v} and P^{c} by integrating the derivatives from $n^{\mathrm{b}} = 0$ (cf. Equation 10.6). The determination of τ is done in an iterative manner during the numerical calculations. The advantage with the condition 10.24 compared to $P^{\mathrm{v}} = P^{\mathrm{c}}$ is that the former is directly applicable for the actual system under investigation, so one does not need to calculate χ_T as a function of density from zero to n^{b} in order to calculate P^{c} when implementing $P^{\mathrm{v}} = P^{\mathrm{c}}$. As we can see in Figure 10.1, this approximation gives very good results for the pressure. Furthermore, it gives very good results for $\beta\mu^{\mathrm{ex,b}}$ (not shown), for instance when this quantity is calculated via Equation 10.22. This theory also predicts several other entities accurately.

Another approximation that also gives very good results for the pressure, $b^{(2)}(0)$, $\beta\mu^{\mathrm{ex,b}}$ and many other entities is the ***Zero-separation (ZSEP) approximation*** developed by Lloyd L. Lee. The results for P and $b^{(2)}(0)$ of hard sphere fluids[17] are shown in Figure 10.1 and Table 10.2, respectively. As mentioned in Section 9.1.3.2, the ZSEP closure is a refinement of the Verlet closure, Equation 9.91, and reads[18]

$$b^{(2)} = -\frac{\tau_1[\gamma^{*(2)}]^2}{2} \cdot \frac{1+\tau_2\gamma^{*(2)}}{1+\tau_3\gamma^{*(2)}} \qquad \text{(ZSEP)}, \tag{10.25}$$

where $\gamma^{*(2)} \equiv \gamma^{(2)} - \varphi^{\mathrm{R}}$ and the values of the parameters τ_1, τ_2 and τ_3 are selected such that the resulting theory satisfies some consistency criteria, see below. The function φ^{R} is not set equal to u^{L} in this approximation (the latter is zero for hard sphere fluids), so another function is used for φ^{R}.[19]

The philosophy of this theory is quite different from the M-S and B-B approximations: one makes a mathematically flexible ansatz for the closure (the Padé approximant in Equation 10.25 with adjustable parameters) and determines these parameters by using exact criteria. For hard sphere fluids these consistency criteria are

(i) Equation 10.24 for the P^{v} and P^{c} derivatives,

(ii) the first zero-separation theorem 10.20 for $y^{(2)}$ and

---

[17] L. L. Lee, J. Chem. Phys., **110** (1999) 7589 (https://doi.org/10.1063/1.478661).

[18] L. L. Lee uses a different set of parameters ζ, α and ϕ in the closure and our parameters are $\tau_1 = \zeta$, $\tau_2 = (1-\phi)\alpha$ and $\tau_3 = \alpha$.

[19] For hard sphere fluids, the function φ^{R} is in Lee's paper (cited in footnotes 15 and 17) selected as proportional to a Mayer f-function, $\varphi^{\mathrm{R}}(r) = n^{\mathrm{b}}[1-e^{-\beta^* u^*(r)}]/2$, where u^* in the first paper (footnote 15) is selected equal the hard sphere potential and in the second paper (footnote 17) is selected as a repulsive soft potential. The latter is equal to the short-range part $u^s(r)$ of the Lennard-Jones (LJ) potential defined in Equation 9.85 (the LJ potential is defined in Equation 9.86) and the parameter β^* is set such that $\beta^*\varepsilon^{\mathrm{LJ}} = 1/9$. Note that β^* should not be confused with β because there is no temperature dependence involved for the hard sphere fluid. The reason to choose a repulsive soft potential instead of the hard sphere potential is to avoid the discontinuity at $r = d^{\mathrm{h}}$. The choice of a Lennard-Jones potential is not essential and other repulsive soft potentials can alternatively be used. If u^* is set equal to the hard sphere potential the value of β^* does not matter.

(iii) the **second zero-separation theorem for b$^{(2)}$**

$$d^{\mathrm{h}} \left. \frac{db^{(2)}(r)}{dr} \right|_{r=0} = 6\eta^{\mathrm{hs}} y^{(2)}(d^{\mathrm{h}}) \left[y^{(2)}(d^{\mathrm{h}}) - 1 \right] \quad \text{(hard spheres)}, \qquad (10.26)$$

which is derived in Appendix 10A (Equation 10.198). The latter gives the derivative of $b^{(2)}$ at $r = 0$ in terms of the contact value of $y^{(2)}$ at $r = d^{\mathrm{h}}$. These three exact conditions are sufficient to determine τ_1, τ_2 and τ_3 and it is found that $\tau_3 = 1.0$ is a suitable value for all densities while τ_1 and τ_2 vary with density. The combination of the thermodynamic consistency (i) and the two pair function (structural) consistencies (ii) and (iii) is a characteristic feature of the ZSEP theory.

For completeness, we mention the **second zero-separation theorems for y$^{(2)}$** and **$\gamma^{(2)}$** that are derived in Appendix 10A (Equations 10.197 and 10.196)

$$d^{\mathrm{h}} \left. \frac{d \ln y^{(2)}(r)}{dr} \right|_{r=0} = -6\eta^{\mathrm{hs}} y^{(2)}(d^{\mathrm{h}}) \qquad \text{(hard spheres)} \qquad (10.27)$$

$$d^{\mathrm{h}} \left. \frac{d\gamma^{(2)}(r)}{dr} \right|_{r=0} = -6\eta^{\mathrm{hs}} \left[g^{(2)}(d^{\mathrm{h}+}) \right]^2 \qquad \text{(hard core fluids)} \qquad (10.28)$$

which relate the derivatives at $r = 0$ to the $y^{(2)}$ and $g^{(2)}$ values at $r = d^{\mathrm{h}}$. For systems where the pair potential and therefore the correlation functions are continuous functions of r, the second theorem for $\gamma^{(2)}$ says that $d\gamma^{(2)}(r)/dr|_{r=0} = 0$ (Equation 10.195).

Both the first and the second zero-separation theorems for $\gamma^{(2)}$ are consequences of the OZ equation, so they are always fulfilled for integral equation theories based on the OZ equation. Given Equation 10.27, we see that Equations 10.26 and 10.28 follow from each other because $\ln y^{(2)}(r) = \gamma^{(2)}(r) + b^{(2)}(r)$. This means that only one of them can be used as an independent consistency criterium and the corresponding fact applies to the first zero-separation theorems. Thus, the ZSEP approximation for hard spheres utilizes the maximum amount of constraints given by these theorems.

The fact that the B-B and ZSEP approximations perform about equally well despite that the former only contains one parameter, gives the B-B approach a clear advantage. However, the generality of a closure based on a Padé approximant like the ZSEP approximation together with the use of exact consistency criteria makes this latter approach very interesting from a fundamental point of view.

Let us now proceed with other fluids. The general **first zero-separation theorem for y$^{(2)}$** (derived in Appendix 10A, Equation 10.188) is given by

$$\ln y^{(2)}(0) = 2\beta\mu^{\mathrm{ex,b}} - \beta\mu_2^{\mathrm{ex,b}}, \qquad (10.29)$$

where $\mu_2^{\rm ex,b}$ is the excess chemical potential for the insertion of a single particle that interacts twice as strongly with the particles in the fluid as they do amongst themselves, that is, its interaction potential is $2u(r)$ (see Appendix 10A). This form of the theorem is useful for fluids other than hard spheres, but it is valid also for the latter because $\mu_2^{\rm ex,b} = \mu^{\rm ex,b}$ for hard spheres as explained in the Appendix.

The B-B and ZSEP approximations have been used with excellent results for Lennard-Jones (LJ) fluids, where $u(r) = u^{\rm LJ}(r)$ given in Equation 9.86. We again start with the ***Bomont-Bretonnet approximation***. As we have seen in Section 9.1.3.3, this approach is an improvement of the Vompe-Martynov (V-M) approximation and from its performance in the Duh-Haymet plots, Figures 9.3 and 9.2, one expects it to work well. The B-B closure is

$$b^{(2)} = \left[1 + 2\gamma^{*(2)} + \tau(\gamma^{*(2)})^2\right]^{1/2} - \gamma^{*(2)} - 1 \qquad \text{(B-B)}$$

(Equation 9.97), where $\gamma^{*(2)} = \gamma^{(2)} - \beta u^{\rm L}$. The parameter τ is selected such that the pressure condition 10.24 is fulfilled, like in the case of hard sphere fluids. In this theory, the long-range part of the pair potential for LJ fluids is selected so that $u^{\rm L}(r)$ agrees with $u^{\rm LJ}(r)$ for large r but differs for small r from the Weeks-Chandler-Anderson definition of $u^{\ell}(r)$ in Equation 9.84. The functions $u^{\rm L}(r)$ and $u^{\rm S}(r) = u(r) - u^{\rm L}(r)$ used in the B-B approach are plotted in Figure 9.1, where they are compared to $u^{\ell}(r)$ and $u^{s}(r)$. In the B-B choice of $u^{\rm L}(r)$, it is constant for small r and joins in a smooth manner to $u^{\rm LJ}(r)$ for increasing r as a fourth order polynomial

$$u^{\rm L}(r) = \begin{cases} -2\varepsilon^{\rm LJ}, & r \le r_1 \\ a_1 + a_2 r + a_3 r^2 + a_3 r^4, & r_1 < r \le r_2 \\ u^{\rm LJ}(r), & r > r_2, \end{cases} \qquad (10.30)$$

where $r_1 = 0.88d^{\rm h}$, $r_2 = 1.6d^{\rm h}$ and the coefficients a_1 to a_4 are chosen such that $u^{\rm L}(r)$ is continuous and has continuous derivative at $r = r_1$ and $r = r_2$.[20] An important difference from $u^{\ell}(r)$ in Equation 9.84, which used in the V-M closure, is that $u^{\ell}(0) = -\varepsilon^{\rm LJ}$ while here we have $u^{\rm L}(0) = -2\varepsilon^{\rm LJ}$. The factor of 2 has been motivated by the fact that $b^{(2)}(r) = \ln y^{(2)}(r) - \gamma^{(2)}(r)$ at $r = 0$ is connected via Equation 10.29 to the insertion of a single particle with twice the strength of the interaction, and, as we will see, the choice of $u^{\rm L}$ instead of u^{ℓ} in the definition of $\gamma^{*(2)}$ leads to significantly better results.

Like in the case of hard sphere fluids, a difficulty in integral equation theories for LJ fluids and other soft core fluids is to obtain accurate values for the bridge function for small r where the potential is strongly repulsive.

[20]For details see J.-M. Bomont and J.-L. Bretonnet, J. Chem. Phys. **114** (2001) 4141 (https://doi.org/10.1063/1.1344610), where explicit expressions for a_1 to a_4 are given in Equation 11 of this publication. There are two printing errors in these expressions: the fourth term inside the parenthesis in the numerator of a_2 should be $-6r_2 u(r_2)$ and the sixth term in the numerator of a_3 should be $-3r_2 u(r_2)$.

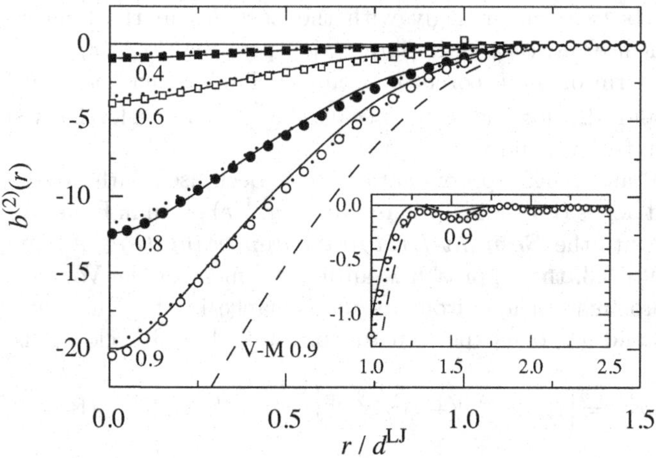

Figure 10.2 The pair bridge function $b^{(2)}(r)$ for a Lennard-Jones fluid at reduced temperature $T^* = k_BT/\varepsilon^{LJ} = 1.5$ and with reduced density $n^{b*} = n^b(d^{LJ})^3 = 0.9$, 0.8, 0.6 and 0.4 calculated in the ZSEP approximation (full curves), the B-B approximation (dotted curves) and MC simulation (symbols).[21] The results from the Vompe-Martynov approximation (dashed curve) for $n^{b*} = 0.9$ is also shown. In the inset, a magnified view of $b^{(2)}(r)$ for $n^{b*} = 0.9$ is shown for $r/d^{LJ} \geq 1$.

As discussed earlier, an accurate $b^{(2)}(r)$ for small r is crucial in order to yield correct thermodynamical quantities like the chemical potential.

The bridge function obtained from different approximations is plotted for various densities in Figure 10.2[21] together with results from MC simulations.[22] The insert shows a magnified view of $b^{(2)}(r)$ for $r/d^{LJ} \geq 1$ for $n^{b*} = 0.9$. We see that the B-B approximation results, which have been obtained by Bomont and Bretonnet,[23] agree very well with the MC data for all r. In Figure 10.3[24] we see that this approximation gives excellent $g^{(2)}(r)$ and it gives also very good results for thermodynamical quantities, including the chemical potential (not shown).

[21] Figure 10.2 is based on MC data taken from the work by Llano-Restrepo and Chapman (see footnote 22), B-B and V-M data from the work by Bomont and Bretonnet (see footnote 23) and ZSEP data from the work of Lee (see footnote 25). Jean-Marc Bomont and Lloyd L. Lee are acknowledged for kindly providing the original data.

[22] M. Llano-Restrepo and W. G. Chapman, J. Chem. Phys. **97** (1992) 2046 (https://doi.org/10.1063/1.463142).

[23] J.-M. Bomont and J.-L. Bretonnet, Mol. Phys. **101** (2003) 3249 (https://doi.org/10.1080/00268970310001619313).

[24] Figure 10.3 is based on data from the same publications as Figure 10.2, see footnote 21. Jean-Marc Bomont and Lloyd L. Lee are acknowledged for kindly providing the original data.

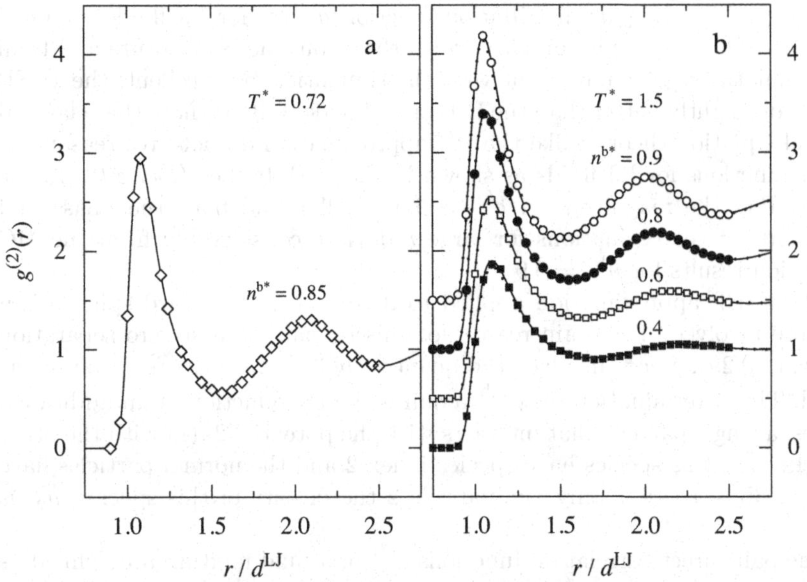

Figure 10.3 (a) The pair distribution function $g^{(2)}(r)$ for an LJ fluid with reduced density $n^{b*} = 0.85$ at reduced temperature $T^* = 0.72$ calculated in the B-B approximation (full curve) and MC simulation (symbols). (b) The same for $n^{b*} = 0.9$, 0.8, 0.6 and 0.4 at $T^* = 1.5$ calculated in the ZSEP approximation (full curves), B-B approximation (dotted curves) and MC simulations (symbols).[24]

For comparison, Figure 10.2 also shows $b^{(2)}(r)$ for $n^{b*} = 0.9$ obtained from the V-M approximation, where $\tau = 0$ (cf. Equations 9.95 and 9.96). We see that V-M gives inferior results for $b^{(2)}(r)$ for small r compared to B-B, while for $r/d^{LJ} \gtrsim 1.3$ these approximations give quite similar results (see the insert in the figure). There are two reasons for the better performance of the B-B approximation for small r, namely the selection of a value of $\tau \neq 0$ given by the consistency criterium 10.24 and the use of $u^{L}(r)$ given by Equation 10.30 rather than $u^{\ell}(r)$ from the WCA definition, Equation 9.84. As shown in the work by Bomont and Bretonnet (*loc. cit.*), if τ is determined from the same consistency criterium but $u^{\ell}(r)$ is used, the $b^{(2)}(r)$ curve for $n^{b*} = 0.9$ lies about halfway between the V-M and B-B approximation results for small r in Figure 10.2, which is not good enough.

In Figure 10.2 the bridge function $b^{(2)}(r)$ for various densities from the ***ZSEP approximation*** obtained by L. L. Lee[25] is also shown. These results

[25] L. L. Lee, J. Chem. Phys. **107** (1997) 7360 (https://doi.org/10.1063/1.474974).

are equally good or slightly better as those from the B-B approximation. In Figure 10.3b the pair distribution function $g^{(2)}(r)$ for LJ fluids obtained from the ZSEP closure is compared with that from the B-B closure and from MC simulations. The agreement with the simulations is excellent; the ZSEP results are slightly better than the B-B ones. Incidentally we note that classical integral equation theories like the PY approximation do not give reasonable bridge functions for LJ fluids as shown in the work by Lee (*loc. cit.*). In the case of PY, the magnitude of $b^{(2)}(r)$ for small r and both the phase and magnitude of the oscillations for large r deviate considerably from the MC simulation results for $n^{\mathrm{b}*} = 0.9$.

The ZSEP approximation applied to fluids other than hard spheres is a bit more involved. The main reason for this is that the first zero separation theorem 10.29 for $y^{(2)}$ involves the quantity $\mu_2^{\mathrm{ex,b}}$, which needs to be determined. The determination of $\mu_2^{\mathrm{ex,b}}$ requires the introduction of an additional species, a single particle that interacts with the potential $2u(r)$ with the other particles. Let this species have species index 2 and the normal particles have index 1. Since there is only one particle 2 the density of this species, n_2^{b}, is zero.

The pair direct-correlation functions $c_{ij}^{(2)}$ in a fluid mixture are defined as

$$c_{ij}^{(2)}(\mathbf{r}',\mathbf{r}) = \frac{\delta c_i^{(1)}(\mathbf{r}')}{\delta n_j(\mathbf{r})} = -\beta \frac{\delta^2 A^{\mathrm{ex}}}{\delta n_j(\mathbf{r})\delta n_i(\mathbf{r}')}, \tag{10.31}$$

which constitutes the generalization of Equation 9.127 to a system with several species. The OZ equation for such a system is (cf. Equation 9.70)

$$h_{ij}^{(2)}(r_{12}) = c_{ij}^{(2)}(r_{12}) + \sum_l n_l^{\mathrm{b}} \int d\mathbf{r}_3\, c_{il}^{(2)}(r_{13}) h_{lj}^{(2)}(r_{32}) \quad \text{(bulk mixture)} \tag{10.32}$$

(these equations will be derived for the general case in Section 11.5). Since $n_2^{\mathrm{b}} = 0$ there is only one term in the sum and the OZ equation yields

$$\begin{aligned} \gamma_{11}^{(2)}(r_{12}) &= n_1^{\mathrm{b}} \int d\mathbf{r}_3\, c_{11}^{(2)}(r_{13}) h_{11}^{(2)}(r_{32}) \\ \gamma_{12}^{(2)}(r_{12}) &= n_1^{\mathrm{b}} \int d\mathbf{r}_3\, c_{11}^{(2)}(r_{13}) h_{12}^{(2)}(r_{32}), \end{aligned} \tag{10.33}$$

where $\gamma_{ij}^{(2)} = h_{ij}^{(2)} - c_{ij}^{(2)}$. Note that since $n_2^{\mathrm{b}} = 0$ we have $c_{11}^{(2)} = c^{(2)}$ and $h_{11}^{(2)} = h^{(2)}$ in the original notation for the system without species 2.

The direct correlation functions $c_{11}^{(2)}$ and $c_{12}^{(2)}$ are involved an additional zero-separation theorem that is derived in Appendix 10A (Equation 10.194)

$$\frac{\partial \ln y_{11}^{(2)}(0)}{\partial n^{\mathrm{b}}} = -2 \int d\mathbf{r}\, c_{11}^{(2)}(r) + \int d\mathbf{r}\, c_{12}^{(2)}(r), \tag{10.34}$$

where $y_{11}^{(2)} = y^{(2)}$ for the system without species 2. This is the **zero-separation theorem for $\mathbf{dy^{(2)}/dn^b}$**. It is used in the ZSEP approximation.

The main principles for the ZSEP approximation remain the same as for the hard sphere fluid. The closures for the 11 and 12 correlation functions are

$$\left.\begin{aligned} b_{11}^{(2)} &= -\frac{\tau_1[\gamma_{11}^{*(2)}]^2}{2}\cdot\frac{1+\tau_2\gamma_{11}^{*(2)}}{1+\tau_3\gamma_{11}^{*(2)}} \\ b_{12}^{(2)} &= -\frac{\tau_4[\gamma_{12}^{*(2)}]^2}{2}\cdot\frac{1+\tau_5\gamma_{12}^{*(2)}}{1+\tau_6\gamma_{12}^{*(2)}} \end{aligned}\right\} \quad \text{(ZSEP)}$$

where the parameters τ_1 to τ_6 are determined by consistency criteria. For Lennard Jones fluids one must select a different function φ^{R} in $\gamma_{1j}^{*(2)} \equiv \gamma_{1j}^{(2)} - \varphi_{1j}^{\mathrm{R}}$ than for hard spheres. In the work of Lee,[26] a density dependent form of the long-range part of the LJ potential was used

$$\varphi_{1j}^{\mathrm{R}}(r) = -4\beta\varepsilon_{1j}^{\mathrm{LJ}}\left(\frac{d^{\mathrm{LJ}}}{r}\right)^6\mathfrak{F}(r),$$

where $\varepsilon_{11}^{\mathrm{LJ}} = \varepsilon^{\mathrm{LJ}}$, $\varepsilon_{12}^{\mathrm{LJ}} = 2\varepsilon^{\mathrm{LJ}}$ and $\mathfrak{F}(r)$ depends on density.[27] When $r \to \infty$ the function $\mathfrak{F}(r) \to 1$ so $\varphi_{1j}^{\mathrm{R}}(r) \sim u^{\mathrm{LJ}}(r)$ in this limit and $\mathfrak{F}(r)$ goes exponentially fast to zero when $r \to 0$ so that $\varphi_{1j}^{\mathrm{R}}(r) \to 0$ there. The function $\varphi_{1j}^{\mathrm{R}}(r)$ has a negative minimum near $r = d^{\mathrm{LJ}}$ that depends on density.

The consistency criteria used in the ZSEP closure are:

(i) Equation 10.24 for the P^{v} and P^{c} derivatives,
(ii) the first zero-separation theorem 10.29 for $y^{(2)}$,
(iii) Equation 9.188 for $\partial\mu/\partial n^{\mathrm{b}}$ of pure fluid without species 2,
(iv) Equation 10.34 for $\partial \ln y^{(2)}(0)/\partial n^{\mathrm{b}}$ and
(v) the thermodynamic Maxwell relationship[28]

$$\left(\frac{\partial[\bar{U}/\bar{N}]}{\partial n^{\mathrm{b}}}\right)_T = \frac{1}{[n^{\mathrm{b}}]^2}\left(\frac{\partial[\beta P]}{\partial\beta}\right)_V, \tag{10.35}$$

[26] See the reference in footnote 25.

[27] The function $\mathfrak{F}(r)$ is defined as

$$\mathfrak{F}(r) = \exp\left[-\frac{\delta^{\mathrm{LJ}}}{n^{\mathrm{b}}(d^{\mathrm{h}})^3}\left(\frac{d^{\mathrm{LJ}}}{r}\right)^{6n^{\mathrm{b}}(d^{\mathrm{h}})^3}\right],$$

where δ^{LJ} is a damping parameter. For the low reduced temperature $k_BT/\varepsilon^{\mathrm{LJ}} = 0.81$ Lee used $\delta^{\mathrm{LJ}} = 1.07$ and otherwise $\delta^{\mathrm{LJ}} = 1.0$. This form of the long-range part of the LJ potential (without the damping parameter δ^{LJ}) was suggested in D.-M. Duh and D. Henderson, J. Chem. Phys. **104** (1996) 6742 (https://doi.org/10.1063/1.471391).

[28] We have $(\partial[U/N]/\partial n^{\mathrm{b}})_{T,N} = N^{-2}(\partial U/\partial[1/V])_{T,N} = -[n^{\mathrm{b}}]^{-2}(\partial U/\partial V)_{T,N}$. One obtains Equation 10.35 from this equality by insertion of the Maxwell relationship $(\partial U/\partial V)_{T,N} = -(\partial[\beta P]/\partial\beta)_{V,N}$, which follows from the facts that $(\partial[\beta A]/\partial\beta)_{V,N} = U$ [from Equation I:2.49], $(\partial[\beta A]/\partial V)_{\beta,N} = -\beta P$ [from Equation I:2.51] and the equality of the cross-derivatives ($\partial/\partial V$ and $\partial/\partial\beta$, respectively), which are equal to $(\partial^2[\beta A]/\partial V\partial\beta)_N$.

which relates the density dependence of the internal energy to the temperature dependence of the pressure.

The internal energy per particle, $\bar{U}/\bar{N}$, is calculated with $\bar{U}^{\mathrm{ex}}/\bar{N}$ from Equation 10.19 and $\mu^{\mathrm{ex,b}}$ is calculated from a generalization[29] of Equation 10.22 with the 11-correlation functions inserted. The same equation with the 12-correlation functions inserted are used to calculate $\mu_2^{\mathrm{ex,b}}$.

The criteria (i)–(v) are not sufficient to determine the six parameters τ_1 to τ_6, but practical experience for LJ fluids has shown that τ_1 is virtually independent of temperature and this fact can be used to evaluate τ_4, which plays the same role for $b_{12}^{(2)}$ as τ_1 does for $b_{11}^{(2)}$. The pair potential occurs as $u(r)/k_BT$, so a doubling of the pair potential (which is relevant for the 1-2 interactions) is the same as a decrease in temperature with a factor of two for the 1-1 interactions. For each density n^{b}, one can therefore set $\tau_4 = \tau_1$ as a good approximation and then there are only five independent parameters to be determined. The criteria (i)–(v) are sufficient for this.

10.1.3 Singlet-level integral equation theories for inhomogeneous fluids

Let us next turn to integral equation approximations for inhomogeneous fluids. In this case one needs an equation for the density distribution $n(\mathbf{r})$. In the vast majority of studies where integral equation theories for inhomogeneous fluids are applied in the literature, one uses Equations 9.59 and 9.61, that is

$$\begin{aligned} n(\mathbf{r}_1) &= n^{\mathrm{b}} e^{-\beta v(\mathbf{r}_1)+\gamma^{(1)}(\mathbf{r}_1)+b^{(1)}(\mathbf{r}_1)}, \\ &= n^{\mathrm{b}} e^{-\beta v(\mathbf{r}_1)+\int d\mathbf{r}_3\, c^{(2)\mathrm{b}}(r_{13})\left[n(\mathbf{r}_3)-n^{\mathrm{b}}\right]+b^{(1)}(\mathbf{r}_1)}, \end{aligned} \tag{10.36}$$

with an approximate singlet bridge function $b^{(1)}(\mathbf{r})$. These are the **singlet-level integral equation theories**, which are approximations for the singlet distribution function $g^{(1)}(\mathbf{r}_1) = n(\mathbf{r}_1)/n_i^{\mathrm{b}}$ of the inhomogeneous fluid. A common choice is to set $b^{(1)}(\mathbf{r}_1) = 0$, which is the **singlet HNC approximation**, whereby one has

$$n(\mathbf{r}_1) = n^{\mathrm{b}} e^{-\beta v(\mathbf{r}_1)+\int d\mathbf{r}_3\, c^{(2)\mathrm{b}}(r_{13})\left[n(\mathbf{r}_3)-n^{\mathrm{b}}\right]} \quad \text{(singlet HNC)}, \tag{10.37}$$

where $c^{(2)\mathrm{b}}(r_{13}) = c^{(2)\mathrm{b}}(\mathbf{r}_1, \mathbf{r}_3)$ is the pair direct-correlation function for the bulk fluid of density n^{b} that is in equilibrium with the inhomogeneous fluid. One can obtain $c^{(2)\mathrm{b}}$ by using any of the integral equation closures in Sections 9.1.3.2 and 9.1.3.3 and once this is done, one can calculate the density profile $n(\mathbf{r})$ by solving Equation 10.37 numerically, for instance by iteration. If, for example, the PY approximation is used for the bulk pair-correlation

[29]The generalization can be obtained from Equation 10.22 by replacing $\int_0^{\gamma^{(2)}(r)} dx\, f(x)$ by $\int_{-\varphi^{\mathrm{R}}(r)}^{\gamma^{(2)}(r)-\varphi^{\mathrm{R}}(r)} dx\, f(x)$. One can derive it from Equation 9.203 by inserting $c^{(1)\mathrm{R,b}}$ from Equation 10.62 applied to the bulk phase with $\varphi_0^{\mathrm{R}} = \varphi^{\mathrm{R}}$ and $\varphi_1^{\mathrm{R}} = 0$.

function, the resulting approximation for the inhomogeneous fluid is called the **HNC/PY approximation.**

By selecting some other approximation for $b^{(1)}(\mathbf{r})$ in Equation 10.36, one obtains another closure for the singlet density distribution, which may be combined with any closure for the pair distributions in bulk used to calculate $c^{(2)\mathrm{b}}(r)$. Thereby, one tries to use a combination of approximations that gives as high accuracy as possible for $n(\mathbf{r})$; which ones are the best usually depends on the kind of fluid one has in the system.

A possibility for a singlet distribution closure is to use an approximation of the kind $b^{(1)} \approx F(\gamma^{(1)})$ for some function F. This is analogous to the use of the approximation $b^{(2)} \approx f(\gamma^{(2)})$ in Equation 9.88 for the pair distribution. An example is the **singlet Verlet approximation** for the bridge function $b^{(1)}(\mathbf{r})$, where one uses the closure

$$b^{(1)} = -\frac{\tau_1'[\gamma^{(1)}]^2}{2} \cdot \frac{1+\tau_2'\gamma^{(1)}}{1+\tau_3'\gamma^{(1)}} \qquad \text{(singlet Verlet)} \qquad (10.38)$$

with constants τ_1', τ_2' and τ_3' that are chosen such that some consistency criteria are satisfied for the system. This particular closure is used here solely for the purpose to illustrate the principles involved and for no other reason.

In order to understand the relationship between $b^{(1)}$ and $b^{(2)}$, it is very illustrative to see how the singlet distribution theory for inhomogeneous fluids is related to the pair distribution theory for bulk fluids. In particular, we will deal with the cases of spherical or planar geometry, i.e., when the external potential of the inhomogeneous fluid originates from a spherical body or a smooth planar wall in contact with the fluid, respectively. We start with the spherical case.

If we have a spherical body (a large spherical particle) immersed in a fluid that consists of much smaller spherical particles, the latter is inhomogeneous due to the interactions with the body. This system can be formed in the following manner. We start with a bulk fluid composed of a binary mixture of large and small spherical particles with number densities $n_{\mathrm{L}}^{\mathrm{b}}$ and $n_{\mathrm{S}}^{\mathrm{b}}$, respectively, where subscript L stands for large and S for small. We assume that the former are very large compared to the latter. We now decrease the number of large particles so $n_{\mathrm{L}}^{\mathrm{b}} \to 0$. In the end we have only one of them left whereby $n_{\mathrm{L}}^{\mathrm{b}} = 0$ and we have obtained a fluid consisting of small particles in contact with a single spherical body. Then we have an inhomogeneous fluid of small particles in spherical geometry and the large particle can be considered as being an external body in relation to the system consisting of the small particles. This is a trick that we will use to understand the relationship between $b^{(1)}$ and $b^{(2)}$.

For the binary bulk mixture with densities $n_{\mathrm{L}}^{\mathrm{b}}$ and $n_{\mathrm{S}}^{\mathrm{b}}$, the interaction between a particle of species i located at $\mathbf{r}_1$ and a particle of species j located at $\mathbf{r}_2$, where $i, j = \mathrm{L}$ or S, is given by the pair interaction potential $u_{ij}(r_{12})$. The density of species i at distance r_{12} from the center of a particles of species j is equal to $n_i^{\mathrm{b}} g_{ij}^{(2)}(r_{12})$, where $g_{ij}^{(2)}$ is the pair distribution function. The latter is given by (cf. Equations 9.65 and 9.67)

$$\begin{aligned} g_{ij}^{(2)}(r_{12}) &= e^{-\beta u_{ij}(r_{12})+\gamma_{ij}^{(2)}(r_{12})+b_{ij}^{(2)}(r_{12})} \quad (10.39) \\ &= e^{-\beta u_{ij}(r_{12})+h_{ij}^{(2)}(r_{12})-c_{ij}^{(2)}(r_{12})+b_{ij}^{(2)}(r_{12})} \quad \text{(bulk mixture)}, \end{aligned}$$

where $c_{ij}^{(2)}$ is the pair direct-correlation function for species i and j, $\gamma_{ij}^{(2)} = h_{ij}^{(2)} - c_{ij}^{(2)}$ and $b_{ij}^{(2)}$ is the bridge function in the mixture. The OZ equation for a bulk mixture is given by Equation 10.32. In the present case, the sum over l in the OZ equation runs over the species S and L and we have, for example,

$$h_{\mathrm{SS}}^{(2)}(r_{12}) = c_{\mathrm{SS}}^{(2)}(r_{12}) + n_{\mathrm{S}}^{\mathrm{b}} \int d\mathbf{r}_3\, c_{\mathrm{SS}}^{(2)}(r_{13})h_{\mathrm{SS}}^{(2)}(r_{32}) + n_{\mathrm{L}}^{\mathrm{b}} \int d\mathbf{r}_3\, c_{\mathrm{SL}}^{(2)}(r_{13})h_{\mathrm{LS}}^{(2)}(r_{32})$$

and

$$h_{\mathrm{SL}}^{(2)}(r_{12}) = c_{\mathrm{SL}}^{(2)}(r_{12}) + n_{\mathrm{S}}^{\mathrm{b}} \int d\mathbf{r}_3\, c_{\mathrm{SS}}^{(2)}(r_{13})h_{\mathrm{SL}}^{(2)}(r_{32}) + n_{\mathrm{L}}^{\mathrm{b}} \int d\mathbf{r}_3\, c_{\mathrm{SL}}^{(2)}(r_{13})h_{\mathrm{LL}}^{(2)}(r_{32}).$$

In the limit $n_{\mathrm{L}}^{\mathrm{b}} \to 0$ the first equation becomes

$$h_{\mathrm{SS}}^{(2)}(r_{12}) = c_{\mathrm{SS}}^{(2)}(r_{12}) + n_{\mathrm{S}}^{\mathrm{b}} \int d\mathbf{r}_3\, c_{\mathrm{SS}}^{(2)}(r_{13})h_{\mathrm{SS}}^{(2)}(r_{32}) \quad \text{(bulk fluid)} \quad (10.40)$$

and the second can be written

$$\begin{aligned} \gamma_{\mathrm{SL}}^{(2)}(r_{12}) &= h_{\mathrm{SL}}^{(2)}(r_{12}) - c_{\mathrm{SL}}^{(2)}(r_{12}) = n_{\mathrm{S}}^{\mathrm{b}} \int d\mathbf{r}_3\, c_{\mathrm{SS}}^{(2)}(r_{13})h_{\mathrm{SL}}^{(2)}(r_{32}) \\ &= \int d\mathbf{r}_3\, c_{\mathrm{SS}}^{(2)}(r_{13}) \left[n_{\mathrm{S}}(\mathbf{r}_3|\mathbf{r}_2;\mathrm{L}) - n_{\mathrm{S}}^{\mathrm{b}}\right] \quad (10.41) \end{aligned}$$

where $n_{\mathrm{S}}(\mathbf{r}_3|\mathbf{r}_2;\mathrm{L}) \equiv n_{\mathrm{S}}^{\mathrm{b}} g_{\mathrm{SL}}^{(2)}(r_{32})$ is the density of small particles at $\mathbf{r}_3$ in the surroundings of a large particle with its center located at $\mathbf{r}_2$.

We can recognize Equation 10.40 as the OZ equation for a pure bulk fluid of small particles. Furthermore, Equation 10.39 for ij = SS is the same as for the bulk fluid. Thus, the various pair functions for ij = SS in the limit $n_{\mathrm{L}}^{\mathrm{b}} \to 0$ are those of the pure bulk phase. In order to maintain the notation used earlier, we will attach the superscript b to these functions, for example $h_{\mathrm{SS}}^{(2)\mathrm{b}}(r_{12})$, $c_{\mathrm{SS}}^{(2)\mathrm{b}}(r_{12})$ and $b_{\mathrm{SS}}^{(2)\mathrm{b}}(r_{12})$.

By inserting Equation 10.41 into Equation 10.39 for i = S and j = L and multiplying by $n_{\mathrm{S}}^{\mathrm{b}}$ we obtain

$$\begin{aligned} n_{\mathrm{S}}(\mathbf{r}_1|\mathbf{r}_2;\mathrm{L}) &\equiv n_{\mathrm{S}}^{\mathrm{b}}\, g_{\mathrm{SL}}^{(2)}(r_{12}) \\ &= n_{\mathrm{S}}^{\mathrm{b}}\, e^{-\beta u_{\mathrm{SL}}(r_{12})+\int d\mathbf{r}_3\, c_{\mathrm{SS}}^{(2)\mathrm{b}}(r_{13})\left[n_{\mathrm{S}}(\mathbf{r}_3|\mathbf{r}_2;\mathrm{L})-n_{\mathrm{S}}^{\mathrm{b}}\right]+b_{\mathrm{SL}}^{(2)}(r_{12})}, \end{aligned}$$

where we have entered superscript b in $c_{\mathrm{SS}}^{(2)\mathrm{b}}(r_{13})$ for the reasons just given. If we now regard the single large particle as being external to the fluid of small particles and place the origin of the coordinate system at its center, i.e., we set $\mathbf{r}_2 = 0$ and hence $r_{12} = r_1$, the external potential for the small particles is equal to $v_\mathrm{S}(r_1) = u_\mathrm{SL}(r_1)$ and we can identify $n_\mathrm{S}(\mathbf{r}_1|0;\mathrm{L})$ as the inhomogeneous density $n_\mathrm{S}(\mathbf{r}_1)$ of the fluid in the presence of an external large particle. Thereby, the equation becomes

$$n_\mathrm{S}(r_1) = n_\mathrm{S}^\mathrm{b} e^{-\beta v_\mathrm{S}(r_1)+\int d\mathbf{r}_3\, c_\mathrm{SS}^{(2)\mathrm{b}}(r_{13})[n_\mathrm{S}(r_3)-n_\mathrm{S}^\mathrm{b}]+b_\mathrm{S}^{(1)}(r_1)}, \tag{10.42}$$

where we have written $b_\mathrm{S}^{(1)}(r_1)$ instead of $b_\mathrm{SL}^{(2)}(r_1)$ because we can identify it with the singlet bridge function in Equation 10.36, which has the same form as Equation 10.42. Thus, $b_\mathrm{SL}^{(2)}(r_1)$ in the limit $n_\mathrm{L}^\mathrm{b} \to 0$ is equal to the singlet bridge function $b_\mathrm{S}^{(1)}(r_1)$ for small particles in the presence of the single large particle.

If we, for example, treat the binary mixture in the Verlet approximation, Equation 9.91 applies for each pair of species i and j, so we have

$$b_{ij}^{(2)} = -\frac{\tau_{1,ij}[\gamma_{ij}^{(2)}]^2}{2} \cdot \frac{1+\tau_{2,ij}\gamma_{ij}^{(2)}}{1+\tau_{3,ij}\gamma_{ij}^{(2)}} \quad \text{(Verlet, bulk mixture)}, \tag{10.43}$$

where the constants $\tau_{\nu,ij}$ for $\nu = 1,2,3$ in general depends on the species index i and j. In the limit $n_\mathrm{L}^\mathrm{b} \to 0$ with only one large particle remaining, only the combinations $ij = \mathrm{SS}$ and $ij = \mathrm{SL}$ need to be considered. For $ij = \mathrm{SS}$ we thereby use the Verlet approximation 9.91 for $b_\mathrm{SS}^{(2)} = b^{(2)}$ in a pure bulk phase of small particles, so the constants $\tau_{\nu,\mathrm{SS}}$ are τ_ν of the bulk for $\nu = 1,2,3$. Furthermore, it follows that in the singlet Verlet closure 10.38 we have $\tau'_\nu = \tau_{\nu,\mathrm{SL}}$, where $\tau_{\nu,SL}$ is evaluated in the limit $n_\mathrm{L}^\mathrm{b} \to 0$.

Finally, we proceed to the case of planar geometry, which can be obtained from the formalism just presented in the limit where the radius $\mathcal{R}_\mathrm{L}$ of the single large spherical particle goes to infinity, whereby the large particle turns into a planar body – a planar wall. Before doing this, we need to move the origin of the coordinate system to the surface of the large particle whereby we use primed coordinates (x', y', z') in order to distinguish them from the previous ones. We now let $\mathcal{R}_\mathrm{L} \to \infty$ and obtain the planar wall, whereby we select the z' coordinate perpendicular to the wall surface and have the x' and y' axes in the lateral directions. The density n_S and the other singlet functions then depends only on z' and we have from Equation 10.42

$$n_\mathrm{S}(z'_1) = n_\mathrm{S}^\mathrm{b} e^{-\beta v_\mathrm{S}(z'_1)+\int d\mathbf{r}'_3\, c_\mathrm{SS}^{(2)\mathrm{b}}(r'_{13})[n_\mathrm{S}(z'_3)-n_\mathrm{S}^\mathrm{b}]+b_\mathrm{S}^{(1)}(z'_1)}, \tag{10.44}$$

which is Equation 10.36 in planar geometry for a fluid of the small particles.

10.1.4 Pair-level integral equation theories for inhomogeneous fluids

A more refined kind of theory for inhomogeneous fluids are **pair-level integral equation theories** where one uses a closure approximation on the pair level like the ones mentioned in Sections 9.1.3.2 and 9.1.3.3. The crucial difference from the singlet level theories in Section 10.1.3 is that one explicitly calculates the pair correlation functions for the inhomogeneous fluid instead of using pair functions like $c^{(2)\mathrm{b}}$ for the homogeneous bulk fluid. Therefore, the pair level theories are much more accurate than the singlet level ones when the same kind of closure approximation is used. One should keep in mind, though, that with an exact bridge function $b^{(1)}(\mathbf{r})$ in Equation 9.59 and exact $b^{(2)}(\mathbf{r}_1, \mathbf{r}_2)$ in Equation 9.67, both the singlet and pair level theories are exact. It is the quality of the closure approximation that ultimately determines the accuracy of the integral equation theory.

As mentioned earlier, pair functions like $h^{(2)}(\mathbf{r}_1, \mathbf{r}_2)$ are anisotropic in inhomogeneous fluids even when the pair interaction potential $u(\mathbf{r}_1, \mathbf{r}_2) = u(r_{12})$ is spherically symmetric. The unknown functions to be determined in this case are $h^{(2)}(\mathbf{r}_1, \mathbf{r}_2)$ and $c^{(2)}(\mathbf{r}_1, \mathbf{r}_2)$ and the density distribution $n(\mathbf{r})$. The set of three equations needed to obtain them can consist of the closure approximation for the pair distribution functions, the OZ Equation 9.40 and an equation for the singlet density distribution $n(\mathbf{r})$, like the BGY equation 9.10 or any of the TZLMBW equations 9.46 and 9.47. The latter two are equivalent to each other and as a rule, they are better to use than the BGY equation.

As we have seen earlier, in planar geometry like a fluid near a planar smooth surface or between two such surfaces, the singlet functions depends solely on the z coordinate perpendicular to the surfaces, like $n(\mathbf{r}) = n(z)$, and the pair distribution functions depend on three independent coordinates, like $g^{(2)}(\mathbf{r}_1, \mathbf{r}_2) = g^{(2)}(s_{12}, z_1, z_2)$, where $s_{12} = |\mathbf{s}_2 - \mathbf{s}_1|$ with $\mathbf{s}_\nu = (x_\nu, y_\nu)$, and we have $r_{12} = [s_{12}^2 + (z_2 - z_1)^2]^{1/2}$. A system of equations for pair-level theories of inhomogeneous fluids in planar geometry can accordingly be

$$h^{(2)}(s_{12}, z_1, z_2) = c^{(2)}(s_{12}, z_1, z_2) + \int dz_3\, n(z_3) \int d\mathbf{s}_3\, c^{(2)}(s_{13}, z_1, z_3) h^{(2)}(s_{32}, z_3, z_2) \quad (10.45)$$

$$g^{(2)}(s_{12}, z_1, z_2) = e^{-\beta u(r_{12}) + h^{(2)}(s_{12}, z_1, z_2) - c^{(2)}(s_{12}, z_1, z_2) + b^{(2)}(s_{12}, z_1, z_2)} \quad (10.46)$$

$$\frac{1}{n(z_1)} \cdot \frac{dn(z_1)}{dz_1} = -\beta \frac{dv(z_1)}{dz_1} + \int dz_2 \frac{dn(z_2)}{dz_2} \int d\mathbf{s}_{12}\, c^{(2)}(s_{12}, z_1, z_2), \quad (10.47)$$

which we can identity as the OZ Equation 9.40, the equation 9.67 and the TZLMBW equation 9.47. The latter can be replaced by another equation for the singlet density distribution as mentioned earlier and, as we will see, in some cases there exist other options as well. In Equation 10.46, an approximate closure approximation for $b^{(2)}$ is to be inserted, for instance any of those given in Sections 9.1.3.2 and 9.1.3.3. This is *the only approximation made*, everything else is exact. For closures that contain one or more parameters that are determined for a bulk fluid, the same value(s) can in many cases also be used for an inhomogeneous fluid in equilibrium with that bulk fluid. Like the other pair functions (apart from the pair potential $u(r_{12})$), the pair bridge function $b^{(2)}(\mathbf{r}_1, \mathbf{r}_2)$ is anisotropic.

The system of equations 10.45–10.47 can be solved numerically for $h^{(2)}$, $c^{(2)}$ and n in a self-consistent manner. A brief description of how this can be done is given in the shaded text box below, which can be skipped in the first reading.

★ A suitable iterative procedure is the following: Step A is to solve Equations 10.45 and 10.46 for $h^{(2)}$ and $c^{(2)}$ with a trial density profile $n(z)$ inserted. In step B the resulting $c^{(2)}$ is inserted into Equation 10.47, which then is solved for $n(z)$. Thereafter, this new $n(z)$ is inserted into Equation 10.45 and step A is repeated, whereby one obtains new $h^{(2)}$ and $c^{(2)}$. Step B is then done with the new $c^{(2)}$ inserted and the whole procedure is repeated as many times as necessary. Thereby one must use procedures like those described in footnote 30 in order to facilitate the convergence.

As in the case of a bulk fluid, it is suitable to do a Fourier transform of the OZ equation 10.45 in order to simplify the integral over $\mathbf{s}_3$, which is a convolution integral in two-dimensional (x, y) space (see Appendix 9B). The Fourier transform of Equation 10.45 is

$$\underset{\sim}{h}^{(2)}(k, z_1, z_2) = \underset{\sim}{c}^{(2)}(k, z_1, z_2) + \int dz_3\, n(z_3)\underset{\sim}{c}^{(2)}(k, z_1, z_3)\underset{\sim}{h}^{(2)}(k, z_3, z_2),$$

where $\underset{\sim}{h}^{(2)}(k, z_1, z_2)$ is the two-dimensional Fourier transform over $\mathbf{s} = (x, y)$ of $h^{(2)}(s, z_1, z_2)$ and analogously for $\underset{\sim}{c}^{(2)}$ (cf. Equation 9.76 in bulk, where a three-dimensional Fourier transformation was used). Like in the bulk case, one switches between Fourier space and ordinary space during the process of solving the system of equations.[30] We may note that in Equation 10.47 we have $\int d\mathbf{s}_{12}\, c^{(2)}(s_{12}, z_1, z_2) = \underset{\sim}{c}^{(2)}(0, z_1, z_2)$, so the value of the Fourier transform of $c^{(2)}$ at $k = 0$ can be used there.

When one solves the set of equations 10.45–10.47 one needs boundary conditions. For a bulk fluid in contact with a planar wall that causes the inhomogeneity in its vicinity, one can use as boundary condition for Equation 10.47 the fact that $n(z) \to n^{\mathrm{b}}$ when the distance from the wall increases to infinity. Furthermore, one uses the fact that the pair functions $h^{(2)}$ and $c^{(2)}$ approach the respective bulk functions $h^{(2)\mathrm{b}}$ and $c^{(2)\mathrm{b}}$ in the same limit.

It is in general more involved to treat the case of a fluid in a slit between two planar surfaces in equilibrium with a bulk fluid of given density n^{b} and thereby with a given chemical potential μ^{b}. One must then ensure that the chemical potential μ of the inhomogeneous fluid in the slit has the same value, $\mu = \mu^{\mathrm{b}}$. In the general case this is not easy and it would be much simpler if

[30] Unfortunately, the fast Fourier transform techniques are not available in two dimensions, so one has to resort less rapid procedures.

one could calculate the density distribution $n(\mathbf{r})$ directly from

$$n(\mathbf{r}) = n^{\mathrm{b}} e^{-\beta v(\mathbf{r}) + c^{(1)}(\mathbf{r}) - c^{(1)\mathrm{b}}} \tag{10.48}$$

(Equation 9.15) for a given $c^{(1)\mathrm{b}}$ (which is equal to $-\beta\mu^{\mathrm{ex,b}}$), but this requires that $c^{(1)}(\mathbf{r})$ can be directly calculated, which is not feasible in general. However, as will be shown in the next section (10.1.5), in some common approximations it is possible to calculate $c^{(1)}(\mathbf{r})$ directly for a given chemical potential. Then one does not need to solve Equation 10.47, which instead is replaced by a much simpler equation. In these cases it is, as we will see, straightforward to treat both a fluid in contact with a single planar wall and a fluid in a slit between two walls. In order to obtain the pair functions one still needs to solve Equation 10.45 and 10.46, whereafter $c^{(1)}(\mathbf{r})$ can be calculated directly from these functions.

In the general case one must, however, make sure that the chemical potential μ of the inhomogeneous fluid in the slit is equal to that of the bulk, $\mu = \mu^{\mathrm{b}}$. In the shaded text box below it is described how this can be done.

Equation 10.47 can be solved for a certain surface separation D^S if one knows, for example, the value of the density $n(z)$ at some point, say, $z = z_0$. It can also be solved if the average number of particles per unit area, $\bar{\mathscr{N}} = \int dz\, n(z)$, for the given separation is known. The problem is that one does not know in general which value of $n(z_0)$ or $\bar{\mathscr{N}}$ corresponds to the given chemical potential.

When D^S is large, the density in the middle part of the slit is very close to the bulk value n^{b}, so one can solve Equation 10.47 for $n(z_1)$ by selecting z_0 in the middle of the slit and set $n(z_0) = n^{\mathrm{b}}$. Thereby it is guaranteed that $\mu = \mu^{\mathrm{b}}$. Once this is done, one can proceed to smaller D^S by calculating $\bar{\mathscr{N}}$ as a function of D^S under the condition of constant chemical potential. The derivative $d\bar{\mathscr{N}}(D^S)/dD^S$ is available from Equation 9.52 so $\bar{\mathscr{N}}(D^S)$ can be obtained by integration starting from the large separation where $n(z_1)$ has been calculated and decreasing D^S stepwise. Then the system of equations 10.45–10.47 must be solved for each intermediate step. In the end, one has obtained $n(z)$, $h^{(2)}$ and $c^{(2)}$ for all D^S in the interval that is covered and $\mu = \mu^{\mathrm{b}}$ throughout.[31]

Let us now turn to some examples of closure approximations that have been applied for inhomogeneous fluids on the pair level. Like in the case of

[31] Alternatively it is possible to start the procedure from very small D^S, where $n(z)$ can be obtained for $\mu = \mu^{\mathrm{b}}$, and do the integration of $d\bar{\mathscr{N}}(D^S)/dD^S$ for *increasing* D^S values, see *loc. cit.* given in footnote 21. For cases where there is a phase separation for some D^S value, it is important to be able to start from both a large surface separation and a very small one. In either case, a metastable branch will be followed for slit widths after the D^S value where the transition occurs. One branch can thereby be obtained from the procedure with decreasing D^S and the other one from the procedure with increasing D^S.

bulk fluids, the simplest choice of pair bridge function for inhomogeneous fluids is the HNC approximation $b^{(2)} = 0$, Equation 9.73, to be used in the system of equations 10.45–10.47. This is called the **Anisotropic Hypernetted Chain (AHNC) approximation,**[32] which is quite good for fluids of particles that interact with long-ranged potentials. In this approximation, the density distribution $n(\mathbf{r})$ can, in fact, be calculated directly from Equation 10.48 and there is no need to use the procedure in the shaded text box above. This is the case because in the AHNC approximation one can express $c^{(1)}(\mathbf{r})$ explicitly in terms of $h^{(2)}(\mathbf{r}_1, \mathbf{r}_2)$, $c^{(2)}(\mathbf{r}_1, \mathbf{r}_2)$ and $n(\mathbf{r})$ without any further approximation (see Equation 10.51). This result will be shown later in Section 10.1.5, where we will obtain the equation for the density distribution, Equation 10.52, that replaces Equation 10.47. The value of $c^{(1)\mathrm{b}}$ in Equation 10.48, which is equal to $-\beta\mu^{\mathrm{ex,b}}$, is determined by the chemical potential of the bulk phase that is in equilibrium with the inhomogeneous fluid. This is the way that the chemical potential is set for the fluid in the slit in the calculations.

Any of the closure approximations for the pair functions in Sections 9.1.3.2 and 9.1.3.3 can be used in Equation 10.46. An empirical observation is that a closure that works well for bulk fluid usually also works well for an inhomogeneous fluid consisting of the same particles. Here, we will only mention a few of the approximations where the set of equations 10.45–10.47 are used.

By applying the PY closure 9.79 in Equation 10.46, one obtains the **Anisotropic PY (APY) approximation.** Alternative names of this approximation are the **Inhomogeneous PY approximation** and the **PY2 approximation.** This approximation works well for the anisotropic pair functions and density profiles of inhomogeneous hard sphere fluids; for the highest densities the accuracy is only semiquantitative like in the bulk fluid. The APY approximation was used to obtain the density profile $n(z)$ and the pair functions $n(\mathbf{r}_2|\mathbf{r}_1) = n(z_2)g^{(2)}(\mathbf{r}_2, \mathbf{r}_1)$ for an inhomogeneous hard sphere fluid between two hard walls shown in Figures 7.8, 7.11 and 7.13 in Volume I of this treatise. For Lennard-Jones fluids, the **Zerah-Hansen closure** (Equation 9.98) has likewise been applied to the inhomogeneous case. It was used in the calculations for Figure 5.8 in Volume I, which shows the anisotropic pair density distribution function for a LJ fluid enclosed between two planar walls.

In the **Anisotropic RHNC (ARHNC) approximation** one inserts an approximate $b^{(2)} \approx b^{(2)}_{\mathrm{Ref}}$ from a reference system into Equation 10.46. For particles with a hard core, like primitive model electrolytes that we dealt with in Sections 7.2.2 and 8.5 in Volume I, a suitable reference system is an inhomogeneous hard sphere fluid with the same density profile as the actual system (the electrolyte). The anisotropic bridge function $b^{(2)}_{\mathrm{Ref}}$ can then be extracted from a solution of Equations 10.45 and 10.46 in the APY approximation for the hard sphere reference fluid. Since the profile of the reference system is

---

[32]This approximation for pair correlations may also be called the "Inhomogeneous HNC approximation," but this is not preferred because this designation can be confused with the very common singlet HNC approximation that uses the HNC closure on the singlet level rather than on the pair level.

set equal to that of the actual system, one does not use Equation 10.47 for the reference. It is only used for the actual system.[33] Results from this approximation can be found in Figures 7.14 and 7.15 in Volume I, which show charge density distributions around ions in an inhomogeneous electrolyte near a charged surface.

The numerical solution of the system of equations 10.45–10.47 for inhomogeneous fluids may appear complicated, but at least in planar geometry this is feasible and with modern computers, $n(z_1)$ and the entire $g^{(2)}(s_{12}, z_1, z_2)$ for all s_{12}, z_1 and z_2 can be quite rapidly calculated. These functions can also be calculated by computer simulations, see Section 5.13.3.2 in Volume I where some strategies are described. However, to map out the *entire* $g^{(2)}(s_{12}, z_1, z_2)$ for all s_{12}, z_1 and z_2 by MC or MD simulations is extremely time consuming. It is, nevertheless, straightforward to calculate $g^{(2)}$ point-wise by simulation for selected values of the coordinates by means of test particle insertions described in the cited section, which works well in practice provided the fluid is not too dense.

In Volume I there are several examples of results from pair level integral equation theories showing plots of density profiles $n(z)$ and pair density distributions $n^{(1)}(\mathbf{r}_2|\mathbf{r}_1) \equiv g^{(2)}(\mathbf{r}_1, \mathbf{r}_2)n(\mathbf{r}_2)$ near one planar surface or between two such surfaces for hard sphere fluids, Lennard-Jones fluids and simple electrolytes (as mentioned earlier, examples of pair density distributions are shown in Figures 5.8, 7.11, 7.13, 7.14 and 7.15; see also Figure 5.7, which contains an explanation of how the pair density results are plotted). Experimental and theoretical results for the anisotropic structure factor $\mathrm{S}(\mathbf{k})$, where $\mathbf{k}$ is the wave vector, for a hard sphere fluid confined between two planar walls are also shown (see Figure 7.9). The structure factor $\mathrm{S}(\mathbf{k})$ for an inhomogeneous system is given by a Fourier transform of $n(\mathbf{r}_1)n(\mathbf{r}_2)h^{(2)}(\mathbf{r}_1, \mathbf{r}_2)$, as explained in Section 5.6 (see Equation I:5.58). In planar geometry, $\mathrm{S}(\mathbf{k})$ depends on the $\mathbf{k}$ components in the perpendicular and lateral directions, $k_\perp$ and $k_\parallel$, respectively.

10.1.5 The evaluation of $c^{(1)}(\mathbf{r})$ and μ in common integral equation theories

In the theory of inhomogeneous fluids, the singlet direct-correlation function $c^{(1)}(\mathbf{r})$ has a particularly important role since it determines the density distribution $n(\mathbf{r})$ via the relationship $\mu = k_BT \ln(\Lambda^3 n(\mathbf{r})) + v(\mathbf{r}) - k_BT\, c^{(1)}(\mathbf{r})$, which

---

[33] The use of a reference system implies that there is an extra step in the iterative procedure described in the shaded box above on page 127. In step A one needs $b^{(2)}_{\mathrm{Ref}}$ and this is obtained for the hard sphere reference fluid by using the trial density profile $n(z)$ of the actual system. Thus Equations 10.45 and 10.46 are solved twice with the same profile $n(z)$ at step A: once for the reference system and once for the actual system. This is repeated for each round of the iterations and eventually $h^{(2)}$ and $c^{(2)}$ become the final pair correlation functions and $n(z)$ the final profile for the actual system, while $b^{(2)}_{\mathrm{Ref}}$ becomes the bridge function for the reference with this profile.

expresses that the chemical potential is equal everywhere at equilibrium. This relationship can be expressed as

$$n(\mathbf{r}) = \zeta e^{-\beta v(\mathbf{r})+c^{(1)}(\mathbf{r})} = n^{\mathrm{b}} e^{-\beta v(\mathbf{r})+c^{(1)}(\mathbf{r})-c^{(1)\mathrm{b}}} \tag{10.49}$$

(Equation 9.207).

For both inhomogeneous and homogeneous fluids, the determination of an accurate μ is an important and quite demanding task for integral equation theories. This entity is crucial in many applications. It can also be used to assess the reliability of the closure as we did in Section 10.1.1.

The evaluations of $c^{(1)}(\mathbf{r})$ and μ are, however, not simple in general. One can determine them by means of a coupling-parameter integration of the pair distribution function as in Equations 9.199 and 9.200, at least in principle. In practice, such an integration can always be used for the bulk phase, Equation 10.23. In the bulk, an alternative is to use Equation 9.187, where μ is calculated by from $c^{(2)\mathrm{b}}$ and an integration is done over the density. This is the same as an integration involving the isothermal compressibility χ_T obtained from $c^{(2)\mathrm{b}}$ via Equation 9.190, whereby one uses the fact (from Equation 9.189) that $(\partial\mu/\partial n^{\mathrm{b}})_T = 1/([n^{\mathrm{b}}]^2\chi_T)$.

In any of these cases one needs to determine the pair functions for a series of systems; either $g^{(2)\mathrm{b}}(r;\xi)$ for a fluid with an inserted coupled particle of various coupling strengths $0 \leq \xi \leq 1$ or the direct correlation function $c^{(2)\mathrm{b}}(r;\{n'^{\mathrm{b}}\})$ for a fluid of various densities $0 < n'^{\mathrm{b}} \leq n^{\mathrm{b}}$. It would be advantageous to be able to calculate μ only from the pair functions for the actual system itself, that is, for the system with fully coupled particles at density n^{b}.

For inhomogeneous fluids, if we can express $c^{(1)}(\mathbf{r})$ in terms of the pair functions for the actual system itself, Equation 10.49 provides the expression that gives the density distribution $n(\mathbf{r})$ in terms of pair functions that are available in practice in many cases. This simplifies the theory considerably, as mentioned earlier.

As we saw in Section 9.2.4.3, most of the contributions from the coupling-parameter integration can, in fact, be expressed in terms of the pair and singlet functions for the actual system, see Equations 9.201 and 9.203, but there is a residual contribution, $c^{(1)\mathrm{R}}$, that originates from the bridge function according to

$$c^{(1)\mathrm{R}}(\mathbf{r}') = \int_0^1 d\xi \int d\mathbf{r}\, n(\mathbf{r}) \frac{\partial h^{(2)}(\mathbf{r},\mathbf{r}';\xi)}{\partial \xi} b^{(2)}(\mathbf{r},\mathbf{r}';\xi) \tag{10.50}$$

(Equation 9.202). This contribution is in general complicated to evaluate. (The same apply to Equations 9.212 and 9.217 in the Martynov formalism, where $B_{\mathrm{M}}^{(2)}$ is difficult to obtain. The evaluation of $B_{\mathrm{M}}^{(2)}$ is treated in Appendix 10B, where some approximations for it are presented.)

In the HNC approximation, however, this complication does not arise because $b^{(2)} = 0$ and hence $c^{(1)\mathrm{R}} = 0$. We then see from Equation 9.201 that

$$c^{(1)}(\mathbf{r}') = -\int d\mathbf{r}\, n(\mathbf{r}) \left(\frac{1}{2} \left[h^{(2)}(\mathbf{r},\mathbf{r}') \right]^2 - c^{(2)}(\mathbf{r},\mathbf{r}') \right) + \frac{\gamma^{(2)}(\mathbf{r}',\mathbf{r}')}{2} \quad \text{(HNC)}, \tag{10.51}$$

which means that $c^{(1)}(\mathbf{r}')$ and μ can be expressed explicitly in the HNC approximation in terms of $h^{(2)}(\mathbf{r},\mathbf{r}')$, $c^{(2)}(\mathbf{r},\mathbf{r}')$ and $n(\mathbf{r})$ for the actual fluid with fully coupled particles. For a bulk fluid we have

$$\begin{aligned} \mu &= k_B T \ln(\Lambda^3 n^{\mathrm{b}}) + k_B T n^{\mathrm{b}} \int d\mathbf{r} \left(\frac{1}{2} \left[h^{(2)b}(r) \right]^2 - c^{(2)\mathrm{b}}(r) \right) \\ &- \frac{k_B T \gamma^{(2)\mathrm{b}}(0)}{2} \qquad \text{(HNC)} \end{aligned}$$

(cf. Equation 9.203).

Equation 10.51 implies that in the AHNC approximation for inhomogeneous fluids, which uses the HNC approximation for the pair correlations (see Section 10.1.4), one can calculate the singlet density $n(\mathbf{r})$ much more easily than using equations like the BGY equation or any of the TZLMBW equations, the use of which are discussed Section 10.1.4. By inserting Equation 10.51 for the HNC approximation into Equation 9.207, one immediately obtains $n(\mathbf{r}')$ for a given activity ζ

$$n(\mathbf{r}') = \zeta e^{-\beta v(\mathbf{r}') - \int d\mathbf{r}\, n(\mathbf{r}) \left(\left[h^{(2)}(\mathbf{r},\mathbf{r}') \right]^2 /2 - c^{(2)}(\mathbf{r},\mathbf{r}') \right) + \gamma^{(2)}(\mathbf{r}',\mathbf{r}')/2} \quad \text{(AHNC)} \tag{10.52}$$

when the pair correlations are calculated. Note that one completely avoids the need to satisfy the boundary conditions for the BGY or the TZLMBW equations as discussed in Section 10.1.4. This constitutes a considerable simplification, in particular for fluids in confined geometry like a fluid in a slit between two planar surfaces in equilibrium with a bulk fluid with a given chemical potential μ. Another advantage with Equation 10.52 is that it is valid also in cases when the interaction between the particles in the fluid depend on their position relative to the wall or walls, like in cases where charged particles induce polarization charges in the latter, see Section 7.1.4 in Volume I (cf. footnote 18 on page 21).

In some other common approximations it is, in fact, also possible to obtain $c^{(1)}(\mathbf{r})$ and $\mu^{\mathrm{ex,b}}$ as explicit expressions in terms of the pair functions for the actual system,[34] like for the group of integral equation closures where one has made the approximation $b^{(2)} \approx f(h^{(2)} - c^{(2)}) = f(\gamma^{(2)})$ and f is some suitable function (Equation 9.88). The task is thereby to obtain an expression for $c^{(1)\mathrm{R}}$ that is valid in these approximations. This is done in the following shaded text box and the result is Equation 10.57.

[34] Reference cited in footnote 14.

One can expand the function $f(\gamma^{(2)})$ in a Taylor expansion in $\gamma^{(2)}$ as

$$b^{(2)} \approx f(\gamma^{(2)}) = \sum_{\nu=2}^{\infty} \frac{a_\nu}{\nu!}[\gamma^{(2)}]^\nu, \tag{10.53}$$

where the constants $a_i = [d^\ell f(x)/dx^\ell]_{x=0}$. This expansion starts with the square term because the bridge function expresses the nonlinear effects of correlations and it converges at least if $\gamma^{(2)}$ is sufficiently small. The OZ equation 9.40 written in the form

$$\gamma^{(2)}(\mathbf{r},\mathbf{r}') = \int d\mathbf{s}_i\, c^{(2)}(\mathbf{r},\mathbf{s}_i)n(\mathbf{s}_i)h^{(2)}(\mathbf{s}_i,\mathbf{r}') \tag{10.54}$$

implies, for example that

$$\begin{aligned}[\gamma^{(2)}(\mathbf{r},\mathbf{r}')]^2 &= \int d\mathbf{s}_1\, c^{(2)}(\mathbf{r},\mathbf{s}_1)n(\mathbf{s}_1)h^{(2)}(\mathbf{s}_1,\mathbf{r}') \\ &\times \int d\mathbf{s}_2\, c^{(2)}(\mathbf{r},\mathbf{s}_2)n(\mathbf{s}_2)h^{(2)}(\mathbf{s}_2,\mathbf{r}').\end{aligned}$$

Therefore, Equation 10.53 with $\nu = m-1$ can be expressed as

$$\begin{aligned}b^{(2)}(\mathbf{r},\mathbf{r}') &\approx \sum_{m=3}^{\infty} \frac{a_{m-1}}{(m-1)!}\int d\mathbf{s}_1 d\mathbf{s}_2 \ldots d\mathbf{s}_{m-1} \\ &\times c^{(2)}(\mathbf{r},\mathbf{s}_1)c^{(2)}(\mathbf{r},\mathbf{s}_2)\ldots c^{(2)}(\mathbf{r},\mathbf{s}_{m-1}) \\ &\times n(\mathbf{s}_1)h^{(2)}(\mathbf{s}_1,\mathbf{r}')n(\mathbf{s}_2)h^{(2)}(\mathbf{s}_2,\mathbf{r}')\ldots n(\mathbf{s}_{m-1})h^{(2)}(\mathbf{s}_{m-1},\mathbf{r}')\end{aligned}$$

and by comparison with Equation 9.160 we see that one obtain the approximation $b^{(2)} \approx f(\gamma^{(2)})$ by assuming that

$$c^{(m)}(\mathbf{r},\mathbf{s}_1,\mathbf{s}_2,\ldots,\mathbf{s}_{m-1}) \approx a_{m-1}c^{(2)}(\mathbf{r},\mathbf{s}_1)c^{(2)}(\mathbf{r},\mathbf{s}_2)\ldots c^{(2)}(\mathbf{r},\mathbf{s}_{m-1}). \tag{10.55}$$

While $c^{(m)}$ in the lhs is symmetric with respect to interchange of all variables, the rhs is not symmetric with respect to any interchange $\mathbf{r} \leftrightarrow \mathbf{s}_i$ in all factors. As we will see, this is a deficiency that has negative consequences for the thermodynamic consistency of the integral equation approximations of this kind.

By entering Equation 10.55 in the form

$$c^{(m)}(\mathbf{s}_1,\mathbf{s}_2,\ldots,\mathbf{s}_{m-1},\mathbf{s}_m) \approx a_{m-1}c^{(2)}(\mathbf{s}_1,\mathbf{s}_2)c^{(2)}(\mathbf{s}_1,\mathbf{s}_3)\ldots c^{(2)}(\mathbf{s}_1,\mathbf{s}_m)$$

into the functional expansion of $c^{(1)\mathrm{R}}$ in Equation 9.206 we obtain the approximation

$$\begin{aligned} c^{(1)\mathrm{R}}(\mathbf{r}') &\approx \sum_{m=3}^{\infty} \frac{a_{m-1}}{m!} \int d\mathbf{s}_1 \dots d\mathbf{s}_m \, c^{(2)}(\mathbf{s}_1, \mathbf{s}_2) c^{(2)}(\mathbf{s}_1, \mathbf{s}_3) \dots c^{(2)}(\mathbf{s}_1, \mathbf{s}_m) \\ &\times \; n(\mathbf{s}_1) h^{(2)}(\mathbf{s}_1, \mathbf{r}') n(\mathbf{s}_2) h^{(2)}(\mathbf{s}_2, \mathbf{r}') \dots n(\mathbf{s}_m) h^{(2)}(\mathbf{s}_m, \mathbf{r}') \\ &= \sum_{m=3}^{\infty} \frac{a_{m-1}}{m!} \int d\mathbf{s}_1 \, n(\mathbf{s}_1) h^{(2)}(\mathbf{s}_1, \mathbf{r}') \left[\gamma^{(2)}(\mathbf{s}_1, \mathbf{r}')\right]^{m-1}, \end{aligned} \tag{10.56}$$

where we have used the fact that each of the $m-1$ integrals of the form $\int d\mathbf{s}_i \, c^{(2)} n h^{(2)}$ for $i = 2$ to m gives the same result, namely $\gamma^{(2)}(\mathbf{s}_1, \mathbf{r}')$, according to Equation 10.54.

Now, from Equation 10.53 we see that

$$\int_0^{\gamma^{(2)}} dx \, f(x) = \sum_{\nu=2}^{\infty} \frac{a_\nu}{(\nu+1)!} [\gamma^{(2)}]^{\nu+1} = \gamma^{(2)} \sum_{m=3}^{\infty} \frac{a_{m-1}}{m!} [\gamma^{(2)}]^{m-1}.$$

The coefficient $a_{m-1}/m!$ in the rhs is the same as in Equation 10.56, which can be written as

$$c^{(1)\mathrm{R}}(\mathbf{r}') \approx \int d\mathbf{s}_1 \, n(\mathbf{s}_1) h^{(2)}(\mathbf{s}_1, \mathbf{r}') \left[\sum_{m=3}^{\infty} \frac{a_{m-1}}{m!} \left[\gamma^{(2)}(\mathbf{s}_1, \mathbf{r}')\right]^{m-1} \right],$$

and Equation 10.57 follows.

The function $c^{(1)\mathrm{R}}$ can be calculated from

$$c^{(1)\mathrm{R}}(\mathbf{r}') = \int d\mathbf{s}_1 \, n(\mathbf{s}_1) \frac{h^{(2)}(\mathbf{s}_1, \mathbf{r}')}{\gamma^{(2)}(\mathbf{s}_1, \mathbf{r}')} \int_0^{\gamma^{(2)}(\mathbf{s}_1, \mathbf{r}')} dx \, f(x) \qquad \text{if } b^{(2)} = f(\gamma^{(2)}). \tag{10.57}$$

Thus, for approximations that assume that $b^{(2)} = f(\gamma^{(2)})$, we have expressed $c^{(1)\mathrm{R}}$ explicitly in terms of the pair functions and the density distribution for the actual system. For example, in the PY and M-S approximations, where $f(\gamma^{(2)}) = \ln(1+\gamma^{(2)}) - \gamma^{(2)}$ and $f(\gamma^{(2)}) = [1+2\gamma^{(2)}]^{1/2} - \gamma^{(2)} - 1$, respectively, we have

$$\int_0^{\gamma^{(2)}} dx \, f(x) = \begin{cases} (1+\gamma^{(2)}) \ln(1+\gamma^{(2)}) - \dfrac{[\gamma^{(2)}]^2}{2} - \gamma^{(2)} & \text{(PY)} \\[2ex] \dfrac{[1+2\gamma^{(2)}]^{3/2}}{3} - \dfrac{[\gamma^{(2)}]^2}{2} - \gamma^{(2)} - \dfrac{1}{3} & \text{(M-S).} \end{cases}$$

For a bulk fluid we obtain the explicit expression for μ in Equation 10.22. Note that these results for $c^{(1)}(\mathbf{r})$ and μ were obtained from the exact expression 9.206 for $c^{(1)\mathrm{R}}$ by simply inserting the approximation 10.55 that corresponds to $b^{(2)} \approx f(\gamma^{(2)})$. Thus, Equations 10.57 and 10.22 are valid in these approximations *without any further approximation than the integral equation closure itself.*

Thus, we can either calculate the values of $c^{(1)}(\mathbf{r})$ and μ by using Equation 10.57 or by a coupling-parameter integration like in Equations 9.199 and 9.200. There is, however, no guarantee that the values obtained in these two different manners are the same even when the distribution functions are calculated in the same integral equation theory. This is an example of the inconsistencies for thermodynamic quantities and other entities that are prevalent in approximate theories. However, by constructing the integral equation closure in a judicious manner, one can strive for a minimization of the inconsistencies. This is a fruitful strategy for such approximation as discussed in Section 10.1. It is unfortunately not possible to remove *all* inconsistencies in approximate theories. In exact theory, the values of the various entities are independent of how they are calculated; they should, for example, be independent of the path taken when one performs a coupling-parameter integration.[35]

In the derivation of Equation 10.57 it was assumed that $\gamma^{(2)}$ is sufficiently small so that the expansion in Equation 10.53 converges. This condition is, however, not needed because we can obtain the same result directly from the definition 10.50 of $c^{(1)\mathrm{R}}$ by selecting $h^{(2)}(\mathbf{r}, \mathbf{r}'; \xi) = \xi h^{(2)}(\mathbf{r}, \mathbf{r}')$, that is, a linear integration path for the coupling parameter. This is shown in the following shaded text box.

In order to show Equation 10.57, we note that for the linear path we have $\partial h^{(2)}(\mathbf{r}, \mathbf{r}'; \xi)/\partial \xi = h^{(2)}(\mathbf{r}, \mathbf{r}')$ and from the OZ equation in the form $\gamma^{(2)}(\mathbf{r}, \mathbf{r}'; \xi) = \int d\mathbf{r}''\, c^{(2)}(\mathbf{r}, \mathbf{r}'')n(\mathbf{r}'')h^{(2)}(\mathbf{r}'', \mathbf{r}'; \xi)$ (Equation 9.260) it follows that $\gamma^{(2)}(\mathbf{r}, \mathbf{r}'; \xi) = \xi\gamma^{(2)}(\mathbf{r}, \mathbf{r}')$, so we have

$$b^{(2)}(\mathbf{r}, \mathbf{r}'; \xi) = f\left(\gamma^{(2)}(\mathbf{r}, \mathbf{r}'; \xi)\right) = f\left(\xi\gamma^{(2)}(\mathbf{r}, \mathbf{r}')\right).$$

[35] In order to understand why the inconsistency for $c^{(1)}(\mathbf{r})$ and μ occurs in these kinds of approximations, we can investigate the derivation of Equation 9.206 for $c^{(1)\mathrm{R}}(\mathbf{r})$ in Appendix 9C. In exact theory, the path independence of the coupling-parameter integration that leads to this equation is explicitly shown in the appendix when Equation 9.263 is obtained from Equation 9.262. Thereby, the symmetry of $c^{(m)}(\mathbf{s}_1, \mathbf{s}_2, \ldots, \mathbf{s}_m)$ with respect to all its variables is used. This symmetry is a crucial requirement in the derivation. For the closures that assume that $b^{(2)} \approx f(\gamma^{(2)})$, we have, however, seen in Equation 10.55 that the approximation for $c^{(m)}$ in these closures is *not* symmetric, so the function $c^{(1)\mathrm{R}}(\mathbf{r})$ obtained from Equation 9.262 with the approximate $c^{(m)}$ depends on the path taken. Therefore, it *does* matter whether the closure approximation is used *before* or *after* the coupling-parameter integration (in the derivation of Equation 10.57 it was used after). For a more general discussion about path dependence of $c^{(1)}$ and μ in integral equation theories see the reference in footnote 14 on page 113.

By inserting this into the definition of $c^{(1)\mathrm{R}}(\mathbf{r}')$ (Equation 10.50) we obtain[36]

$$
\begin{aligned}
c^{(1)\mathrm{R}}(\mathbf{r}') &= \int_0^1 d\xi \int d\mathbf{r}\, n(\mathbf{r}) h^{(2)}(\mathbf{r},\mathbf{r}') f\left(\xi\gamma^{(2)}(\mathbf{r},\mathbf{r}')\right) \\
&= \int d\mathbf{r}\, n(\mathbf{r}) h^{(2)}(\mathbf{r},\mathbf{r}') \left[\int_0^1 d\xi\, f(\xi\tau)\right]_{\tau=\gamma^{(2)}(\mathbf{r},\mathbf{r}')} \\
&= \int d\mathbf{r}\, n(\mathbf{r}) h^{(2)}(\mathbf{r},\mathbf{r}') \left[\frac{1}{\tau}\int_0^{\tau} dx\, f(x)\right]_{\tau=\gamma^{(2)}(\mathbf{r},\mathbf{r}')} \\
&= \int d\mathbf{r}\, n(\mathbf{r}) \frac{h^{(2)}(\mathbf{r},\mathbf{r}')}{\gamma^{(2)}(\mathbf{r},\mathbf{r}')} \int_0^{\gamma^{(2)}(\mathbf{r},\mathbf{r}')} dx\, f(x) \quad \text{if } b^{(2)} = f(\gamma^{(2)}), \quad (10.58)
\end{aligned}
$$

where we have made the variable substitution $x = \xi\tau$ in the third line. This result is the same as Equation 10.57. Note that the selection of the linear integration path is not an additional approximation (i.e., in addition to the the closure) because if this path is selected in exact statistical mechanics one obtains an exact result.

The same result is obtained if we assume that $h^{(2)}(\mathbf{r},\mathbf{r}';\xi) = \alpha(\xi)h^{(2)}(\mathbf{r},\mathbf{r}')$, where the function $\alpha(\xi)$ is differentiable and satisfies $\alpha(0) = 0$ and $\alpha(1) = 1$, for example $\alpha(\xi) = \xi^l$ for any $l > 0$ (see Exercise 10.3). However, if we select some other integration path for the coupling parameter, the result can be different because the ξ integration for this class of closures is path dependent.[37] How much $c^{(1)\mathrm{R}}(\mathbf{r})$ varies due to the selection of integration paths depends on the quality of the closure and as mentioned earlier, one strives for closures that minimize the inconsistencies. In exact theory there is no path dependence.

{**Exercise 10.3:** Show that if one selects $h^{(2)}(\mathbf{r},\mathbf{r}';\xi) = \alpha(\xi)h^{(2)}(\mathbf{r},\mathbf{r}')$ with $\alpha(0) = 0$ and $\alpha(1) = 1$ in Equation 10.50 with $b^{(2)}(\mathbf{r},\mathbf{r}';\xi) = f(\gamma^{(2)}(\mathbf{r},\mathbf{r}';\xi))$, one obtains the result in Equation 10.57 after the ξ integration.}

Next, we consider the integral equation closures of the type introduced in Equation 9.92, that is,

$$b^{(2)} = f(\gamma^{(2)} - \varphi^{\mathrm{R}}) = f(\gamma^{*(2)}),$$

where φ^{R} can be selected to be equal to βu^{L} or some other suitable function; u^{L} is the long range part of the pair potential. Various choices of φ^{R} have

[36] Reference cited in footnote 14 on page 113.

[37] A different integration path for the coupling parameter is, for example, to set $u(\mathbf{r},\mathbf{r}';\xi) = \xi u(\mathbf{r},\mathbf{r}')$. For this path the ξ-dependence of $h^{(2)}(\mathbf{r},\mathbf{r}';\xi)$ and $\gamma^{(2)}(\mathbf{r},\mathbf{r}';\xi)$ is complicated, so the calculations we did in the derivation of Equation 10.58 do not apply. (For particles with a hard core this $u(\mathbf{r},\mathbf{r}';\xi)$ can be used for $r > d^{\mathrm{h}}$ and for $r < d^{\mathrm{h}}$ we can, for instance, take $u(\mathbf{r},\mathbf{r}';\xi) = \xi/(1-\xi)$).

been discussed in Section 10.1.2. For this kind of closure we can obtain expressions for $c^{(1)}(\mathbf{r})$ analogous to that in Equation 10.57, but this requires an approximation that is additional to the integral equation closure.

If one uses the linear integration path for the coupling-parameter integration with $h^{(2)}(\mathbf{r},\mathbf{r}';\xi) = \xi h^{(2)}(\mathbf{r},\mathbf{r}')$, one has $\gamma^{(2)}(\mathbf{r},\mathbf{r}';\xi) = \xi\gamma^{(2)}(\mathbf{r},\mathbf{r}')$ as before. As explained in Appendix 10B, one can set $\varphi^{\mathrm{R}}(\mathbf{r},\mathbf{r}';\xi) \approx \xi\varphi^{\mathrm{R}}(\mathbf{r},\mathbf{r}')$ as a reasonable approximation. Then, $\gamma^{*(2)}(\mathbf{r},\mathbf{r}';\xi) = \xi\gamma^{*(2)}(\mathbf{r},\mathbf{r}')$ and in the same manner as in Equation 10.58 one obtains[38]

$$c^{(1)\mathrm{R}}(\mathbf{r}') \approx \int d\mathbf{r}\, n(\mathbf{r}) \frac{h^{(2)}(\mathbf{r},\mathbf{r}')}{\gamma^{*(2)}(\mathbf{r},\mathbf{r}')} \int\limits_0^{\gamma^{*(2)}(\mathbf{r},\mathbf{r}')} dx\, f(x). \tag{10.59}$$

(Alternatively, one can take $h^{(2)}$ to be proportional to $\alpha(\xi)$, but this gives the same result.) For a bulk fluid we have

$$\begin{aligned}\beta\mu^{\mathrm{ex,b}} \;=\; -c^{(1)\mathrm{b}} \;&\approx\; n^{\mathrm{b}} \int d\mathbf{r} \left(\frac{1}{2}\left[h^{(2)b}(r)\right]^2 - c^{(2)\mathrm{b}}(r) + g^{(2)\mathrm{b}}(r) b^{(2)\mathrm{b}}(r) \right.\\ &\quad \left. - \; \frac{h^{(2)\mathrm{b}}(r)}{\gamma^{*(2)\mathrm{b}}(r)} \int_0^{\gamma^{*(2)\mathrm{b}}(r)} dx\, f(x) \right) - \frac{\gamma^{(2)\mathrm{b}}(0)}{2},\end{aligned} \tag{10.60}$$

which corresponds to Equation 10.22.

Let us consider the case where one selects $\varphi^{\mathrm{R}} = \beta u^{\mathrm{L}}$ and has $b^{(2)} = f(\gamma^{*(2)})$ with $\gamma^{*(2)} = \gamma^{(2)} - \beta u^{\mathrm{L}}$. Another approximation that has been used[39] is to neglect the ξ dependence of u^{L} and to set $\gamma^{*(2)}(\mathbf{r},\mathbf{r}';\xi) = \xi\gamma^{(2)}(\mathbf{r},\mathbf{r}') - \beta u^{\mathrm{L}}(\mathbf{r},\mathbf{r}')$. Then one obtains instead

$$c^{(1)\mathrm{R}}(\mathbf{r}') \approx \int d\mathbf{r}\, n(\mathbf{r}) \frac{h^{(2)}(\mathbf{r},\mathbf{r}')}{\gamma^{(2)}(\mathbf{r},\mathbf{r}')} \int\limits_{-\beta u^{\mathrm{L}}(\mathbf{r},\mathbf{r}')}^{\gamma^{(2)}(\mathbf{r},\mathbf{r}') - \beta u^{\mathrm{L}}(\mathbf{r},\mathbf{r}')} dx\, f(x)\,. \tag{10.61}$$

This approximation violates the condition that $b^{(2)}(\mathbf{r},\mathbf{r}';\xi)$ is zero for $\xi = 0$, but one has to keep in mind that the approximation for u^{L} is used only in the evaluation of the coupling-parameter integral.

When one uses the more general closure $b^{(2)} = f(\gamma^{*(2)}) = f(\gamma^{(2)} - \varphi^{\mathrm{R}})$, one may as an approximation assume that $\varphi^{\mathrm{R}}(\mathbf{r},\mathbf{r}';\xi) \approx \xi\varphi_1^{\mathrm{R}}(\mathbf{r},\mathbf{r}') + \varphi_0^{\mathrm{R}}(\mathbf{r},\mathbf{r}')$, where φ_1^{R} and φ_0^{R} are suitable functions. Then one obtains

$$c^{(1)\mathrm{R}}(\mathbf{r}') \approx \int d\mathbf{r}\, n(\mathbf{r}) \frac{h^{(2)}(\mathbf{r},\mathbf{r}')}{\gamma^{(2)}(\mathbf{r},\mathbf{r}') - \varphi_1^{\mathrm{R}}(\mathbf{r},\mathbf{r}')} \int\limits_{-\varphi_0^{\mathrm{R}}(\mathbf{r},\mathbf{r}')}^{\gamma^{(2)}(\mathbf{r},\mathbf{r}') - \varphi^{\mathrm{R}}(\mathbf{r},\mathbf{r}')} dx\, f(x)\,. \tag{10.62}$$

---

[38]E. Lomba and L. L. Lee, Int. J. Thermophys. **17** (1996) 663 (https://doi.org/10.1007/BF01441512).

[39]L. L. Lee, D. Ghonasgi, and E. Lomba J. Chem. Phys., **104** (1996) 8058 (https://doi.org/10.1063/1.471522);
L. L. Lee, J. Chem. Phys., **107** (1997) 7360 (https://doi.org/10.1063/1.474974).

The corresponding result is valid for $\beta\mu^{\mathrm{ex,b}} = -c^{(1)\mathrm{b}}$ in the bulk phase. The condition that $b^{(2)}(\mathbf{r},\mathbf{r}';\xi)$ is zero for $\xi = 0$ is fulfilled only when $\varphi_0^{\mathrm{R}} = 0$ and one then obtains the same result as in Equation 10.59 and 10.60.

10.2 INTRODUCTION TO DFT APPROXIMATIONS

10.2.1 Mean field theory; Density expansion approximations

In Section 9.3 we saw that the grand potential Θ and Helmholtz free energy A in the grand canonical ensemble are unique functionals of the density distribution $n(\mathbf{r})$. The grand potential equals

$$\Theta = \Theta[n] = A[n] - \mu\bar{N} = \mathcal{A}[n] + \int d\mathbf{r}\, n(\mathbf{r})v(\mathbf{r}) - \mu\int d\mathbf{r}\, n(\mathbf{r})$$

(Equation 9.227), where $\mathcal{A} = \mathcal{A}^{\mathrm{ideal}} + \mathcal{A}^{\mathrm{ex}}$ is the intrinsic Helmholtz free energy given in Equation 9.230, so we have

$$\Theta[n] = \int d\mathbf{r}\, n(\mathbf{r})\left(k_BT\left[\ln(\Lambda^3 n(\mathbf{r})) - 1\right] + v(\mathbf{r}) - \mu\right) + \mathcal{A}^{\mathrm{ex}}[n].$$

We also saw that the DFT variational principle 9.228

$$\frac{\delta\Theta}{\delta n(\mathbf{r})} = 0 \tag{10.63}$$

yields the Euler-Lagrange equation in Equation 9.231, which is the same as the equilibrium condition for the inhomogeneous fluid

$$\mu = \mu^{\mathrm{ideal}}(\mathbf{r}) + \mu^{\mathrm{ex}}(\mathbf{r}) = k_BT\ln\left(\Lambda^3 n(\mathbf{r})\right) + \mu^{\mathrm{ex}}(\mathbf{r}) \tag{10.64}$$

(cf. Equation 9.5), where

$$\mu^{\mathrm{ex}}(\mathbf{r}) = v(\mathbf{r}) + \frac{\delta\mathcal{A}^{\mathrm{ex}}}{\delta n(\mathbf{r})} = v(\mathbf{r}) - k_BT\, c^{(1)}(\mathbf{r}).$$

Alternatively the Euler-Lagrange equation can be written as

$$n(\mathbf{r}) = \zeta e^{-\beta v(\mathbf{r}) + c^{(1)}(\mathbf{r})} = n^{\mathrm{b}} e^{-\beta v(\mathbf{r}) + c^{(1)}(\mathbf{r}) - c^{(1)\mathrm{b}}} \tag{10.65}$$

(cf. Equations 9.13 and 9.15), where $c^{(1)\mathrm{b}}$ is the value of $c^{(1)}$ in a bulk fluid of density n^{b} that is in equilibrium with the inhomogeneous fluid with density distribution $n(\mathbf{r})$.

Since the functional $\mathcal{A}^{\mathrm{ex}}[n]$ is unknown in general, one has to make some approximation for it. In Section 9.2.4.1, Equation 9.180, we saw that

$$\mathcal{A}^{ex} = \frac{1}{2}\int d\mathbf{r}_1 d\mathbf{r}_2\, u(r_{12})n(\mathbf{r}_1)n(\mathbf{r}_2)\int_0^1 d\xi\, g^{(2)}(\mathbf{r}_1,\mathbf{r}_2;\xi).$$

It is not known how one can write the integral as a functional of $n(\mathbf{r})$. A very simple approximation is to neglect the pair correlations between the particles and set $g^{(2)} = 1$. We obtain

$$\mathcal{A}^{\text{ex}} \approx \frac{1}{2}\int d\mathbf{r}_1 d\mathbf{r}_2\, u(r_{12}) n(\mathbf{r}_1) n(\mathbf{r}_2), \tag{10.66}$$

which yields

$$\frac{\delta \mathcal{A}^{\text{ex}}}{\delta n(\mathbf{r}_3)} \approx \frac{1}{2}\int d\mathbf{r}_1 d\mathbf{r}_2\, u(r_{12}) \left[\delta^{(3)}(\mathbf{r}_3 - \mathbf{r}_1) n(\mathbf{r}_2) + n(\mathbf{r}_1)\delta^{(3)}(\mathbf{r}_3 - \mathbf{r}_2)\right]$$

where we have used Equation 9.105. Since the contribution from the last term in the square bracket is the same as that from the first one (only the names of the integration variables differ), we have

$$\mu^{\text{ex}}(\mathbf{r}_3) = v(\mathbf{r}_3) + \frac{\delta \mathcal{A}^{\text{ex}}}{\delta n(\mathbf{r}_3)} \approx v(\mathbf{r}_3) + \int d\mathbf{r}_2\, u(r_{32}) n(\mathbf{r}_2)$$

and we see that a particle placed at $\mathbf{r}_3$ interacts with the other particles in the system as if the latter were not correlated with the first one. The integral in $\mu^{\text{ex}}(\mathbf{r}_3)$ gives the *mean potential* at $\mathbf{r}_3$ from the surrounding particles. The approximation 10.66 is called the **Mean field approximation (MFA)**.

This approximation is obviously not applicable as it stands when the pair potential contains a hard core repulsion or a soft repulsion that is not integrable at the origin. One can then split the pair potential in two parts $u = u^s + u^\ell$ such that u^ℓ is a long-range part (usually attractive) and u^s is the short-range repulsive part, like we did in Equations 9.84 and 9.85. For a hard core fluid one takes $u^s(r) = u^{\text{core}}(r)$, which is ∞ for $r < d^{\text{h}}$ and zero otherwise. Equation 10.66 can then be applied for the free energy contributions from u^ℓ, while those from hard core correlations, $\mathcal{A}^{\text{ex,core}}$, are approximated in some other manner, for example as the free energy for a hard sphere fluid with density $n(\mathbf{r})$. One thereby takes

$$\mathcal{A}^{\text{ex}} \approx \frac{1}{2}\int d\mathbf{r}_1 d\mathbf{r}_2\, u^\ell(r_{12}) n(\mathbf{r}_1) n(\mathbf{r}_2) + \mathcal{A}^{\text{ex,core}} \tag{10.67}$$

and obtains

$$\mu^{\text{ex}}(\mathbf{r}_3) \approx v(\mathbf{r}_3) + \int d\mathbf{r}_2\, u^\ell(r_{32}) n(\mathbf{r}_2) + \mu^{\text{ex,core}}(\mathbf{r}_3),$$

where $\mu^{\text{ex,core}}(\mathbf{r}_3) = \delta \mathcal{A}^{\text{ex,core}}/\delta n(\mathbf{r}_3)$, and hence

$$n(\mathbf{r}_3) = \zeta e^{-\beta[v(\mathbf{r}_3) + \int d\mathbf{r}_2\, u^\ell(r_{32}) n(\mathbf{r}_2) + \mu^{\text{ex,core}}(\mathbf{r}_3)]}. \tag{10.68}$$

It is, in fact, very common in DFT for fluids to use Equation 10.67 together with $\mathcal{A}^{\text{ex,core}}$ taken from an accurate DFT theory for hard sphere fluids (the latter will be a major topic later). Thereby the attractions between the fluid particles are treated in the MFA, while the hard core correlations are handled more accurately. This is a quite good approximation in many cases.

{**Exercise 10.4:** Obtain an expression for $g(\mathbf{r}_3, \mathbf{r}_1)$ in an inhomogeneous hard-core fluid by applying Equation 10.68 to the density distribution around a particle placed at $\mathbf{r}_1$, whereby $v(\mathbf{r}_3)$ is replaced by $v(\mathbf{r}_3|\mathbf{r}_1) = v(\mathbf{r}_3)+u(|\mathbf{r}_3-\mathbf{r}_1|)$ and $n(\mathbf{r}_3)$ by $n(\mathbf{r}_3|\mathbf{r}_1) = n(\mathbf{r}_3)g(\mathbf{r}_3, \mathbf{r}_1)$. Apply this expression to a bulk fluid of density n^{b} and obtain the following approximation for $g(r_{13})$

$$g(r_{13}) \approx e^{-\beta[u(r_{13})+n^{\mathrm{b}} \int d\mathbf{r}_2\, u^{\ell}(r_{32})h(r_{21})+w^{(2)\mathrm{core}}(r_{13})]},$$

where $w^{(2)\mathrm{core}}(r_{13}) = \mu^{\mathrm{ex,core}}(\mathbf{r}_3|\mathbf{r}_1) - \mu^{\mathrm{ex,core,b}}$ and $\mu^{\mathrm{ex,core,b}}$ is the value of $\mu^{\mathrm{ex,core}}$ in the bulk fluid. Note that this is the *mean field approximation* for bulk fluids given in Equations I:5.102 and I:5:125, where u^{ℓ} is denoted u^{rest}. In this case the correlation between the particle at $\mathbf{r}_1$ (the central particle) and each particle in the neighborhood *is* included in the theory, but the correlations *between* the surrounding particles are neglected in the contribution from the interaction potential u^{ℓ} given by the integral in the exponent.}

To obtain another simple example of an approximate functional $\mathcal{A}^{\mathrm{ex}}[n]$, let us consider an inhomogeneous fluid with density distribution $n(\mathbf{r})$ that deviates only a little from the bulk density n^{b}, so the difference $\Delta n(\mathbf{r}) = n(\mathbf{r}) - n^{\mathrm{b}}$ is rather small everywhere. In this case, we can make a functional Taylor expansion $\mathcal{A}^{ex}[n]$ in terms of $\Delta n(\mathbf{r})$ and use only the first few terms of the expansion as an approximation. This is like an ordinary Taylor expansion of a function $f(x)$ in terms of $\Delta x = x - x_0$ that we truncate after, say, the square term and obtain a good approximation for $f(x)$ when x deviates only a little from x_0. Recall that the coefficients in this Taylor expansion are the derivatives $f'(s)$ and $f''(s)/2$ evaluated at $s = x_0$.

In our case we can use the first two terms of the functional expansion of $\mathcal{A}^{ex}[n]$ according to Equation 9.148, where we set $\Delta\mathcal{F} = \Delta\mathcal{A}^{\mathrm{ex}} \equiv \mathcal{A}^{\mathrm{ex}} - \mathcal{A}^{\mathrm{ex,b}}$ and $\Delta f(\mathbf{r}_\nu) = \Delta n(\mathbf{r}_\nu) \equiv n(\mathbf{r}_\nu) - n^{\mathrm{b}}$. We thereby obtain

$$\Delta\mathcal{A}^{\mathrm{ex}} \approx \int d\mathbf{r}_1 \frac{\delta\mathcal{A}^{\mathrm{ex}}}{\delta n(r_1)}\Delta n(\mathbf{r}_1) + \frac{1}{2}\int d\mathbf{r}_1 d\mathbf{r}_2 \frac{\delta^2\mathcal{A}^{\mathrm{ex}}}{\delta n(r_1)\delta n(r_2)}\Delta n(\mathbf{r}_1)\Delta n(\mathbf{r}_2),$$

where the derivatives of $\mathcal{A}^{\mathrm{ex}}$ are evaluated for the bulk phase of density n^{b}. These derivatives are $-k_B T c^{(1)\mathrm{b}}$ (a constant) and $-k_B T c^{(2)\mathrm{b}}(r_{12})$, respectively (Equations 9.126 and 9.127 evaluated in the bulk). Hence we obtain

$$\begin{aligned}\beta\left[\mathcal{A}^{\mathrm{ex}}[n] - \mathcal{A}^{\mathrm{ex,b}}\right] \approx\ & -c^{(1)\mathrm{b}} \int d\mathbf{r}_1 \left(n(\mathbf{r}_1) - n^{\mathrm{b}}\right) \\ & - \frac{1}{2}\int d\mathbf{r}_1 d\mathbf{r}_2\, c^{(2)\mathrm{b}}(r_{12}) \left(n(\mathbf{r}_1) - n^{\mathrm{b}}\right)\left(n(\mathbf{r}_2) - n^{\mathrm{b}}\right).\end{aligned} \tag{10.69}$$

By taking the functional derivative of this expression with respect to $n(\mathbf{r}_3)$ we obtain

$$\begin{aligned}-c^{(1)}(\mathbf{r}_3) \equiv \beta\frac{\delta\mathcal{A}^{\mathrm{ex}}}{\delta n(\mathbf{r}_3)} \quad &\approx \quad -c^{(1)\mathrm{b}}\int d\mathbf{r}_1\delta^{(3)}(\mathbf{r}_3-\mathbf{r}_1)\\ &- \quad \frac{1}{2}\int d\mathbf{r}_1 d\mathbf{r}_2\, c^{(2)\mathrm{b}}(r_{12})\delta^{(3)}(\mathbf{r}_3-\mathbf{r}_1)\left(n(\mathbf{r}_2)-n^{\mathrm{b}}\right)\\ &- \quad \frac{1}{2}\int d\mathbf{r}_1 d\mathbf{r}_2\, c^{(2)\mathrm{b}}(r_{12})\left(n(\mathbf{r}_1)-n^{\mathrm{b}}\right)\delta^{(3)}(\mathbf{r}_3-\mathbf{r}_2),\end{aligned}$$

which gives

$$c^{(1)}(\mathbf{r}_3) \approx c^{(1)\mathrm{b}} + \int d\mathbf{r}_2\, c^{(2)\mathrm{b}}(r_{32})\left(n(\mathbf{r}_2)-n^{\mathrm{b}}\right). \tag{10.70}$$

The Euler-Lagrange equation in the form of Equation 10.65 now yields

$$n(\mathbf{r}_3) \approx n^{\mathrm{b}}e^{-\beta v(\mathbf{r}_3)+\int d\mathbf{r}_3\, c^{(2)\mathrm{b}}(r_{32})\left(n(\mathbf{r}_3)-n^{\mathrm{b}}\right)},$$

which is the same as the *singlet HNC approximation* for the inhomogeneous fluid (Equation 10.37). Hence the HNC approximation is obtained by using the DFT variational principle and minimizing the grand potential with the approximate $\mathcal{A}^{\mathrm{ex}}[n]$ used here.

Incidentally we may note that the higher order terms in the expansion 10.69 involve the direct correlation functions $c^{(m)}$ for $m \geq 3$, which are defined in Equation 9.156. The sum of the corresponding contributions to Equation 10.70 constitutes the singlet bridge function $b^{(1)}(\mathbf{r}_3)$.

10.2.2 Free energy density; The local density and weighted density approximations

For a bulk fluid, the **excess free energy density** is given by $A^{\mathrm{ex,b}}/V$. Note that $A^{\mathrm{ex,b}} = \mathcal{A}^{\mathrm{ex,b}}$ since external potential $v(\mathbf{r})$ is zero in the bulk. For convenience we introduce the notation $\Phi^{\mathrm{ex,b}} = \beta\mathcal{A}^{\mathrm{ex,b}}/V$, which has the dimension 1/volume. $\Phi^{\mathrm{ex,b}}$ is a well-defined entity and for simplicity it will also be called the "free energy density" despite the factor β. For inhomogeneous fluids, we will consider a free energy density Φ^{ex} that is a function of the location $\mathbf{r}$. From Equation 9.180 we see that it is possible to write

$$\Phi^{\mathrm{ex}}(\mathbf{r}) = \frac{\beta}{2}n^{(1)}(\mathbf{r})\int d\mathbf{r}'\, u(\mathbf{r},\mathbf{r}')n^{(1)}(\mathbf{r}')\int_0^1 d\xi\, g^{(2)}(\mathbf{r},\mathbf{r}';\xi)$$

because $\int d\mathbf{r}\, \Phi^{\mathrm{ex}}(\mathbf{r}) = \beta\mathcal{A}^{\mathrm{ex}}$ with $\mathcal{A}^{\mathrm{ex}}$ given by Equation 9.180. However, there are many exact expressions for $\beta\mathcal{A}^{\mathrm{ex}}$ that can be written as $\beta\mathcal{A}^{\mathrm{ex}} = \int d\mathbf{r}\, F(\mathbf{r})$ with some function $F(\mathbf{r})$ and we can then set $\Phi^{\mathrm{ex}}(\mathbf{r}) = F(\mathbf{r})$ despite that the value of Φ^{ex} at point $\mathbf{r}$ may be different from that of the first choice.

For instance, if one adds any function $f(\mathbf{r})$ with $\int d\mathbf{r}\, f(\mathbf{r}) = 0$ to the original $\Phi^{\rm ex}(\mathbf{r})$ and set $\Phi^{\rm ex}_{\rm other}(\mathbf{r}) = \Phi^{\rm ex}(\mathbf{r}) + f(\mathbf{r})$ one still obtains the same $\beta\mathcal{A}^{\rm ex}$ by integration of $\Phi^{\rm ex}_{\rm other}$. In fact, $\Phi^{\rm ex}(\mathbf{r})$ *is not a uniquely defined entity* and one cannot assign a unique number $\Phi^{\rm ex}(\mathbf{r})$ at each point r. Its integral over the whole system $\int d\mathbf{r}\, \Phi^{\rm ex}(\mathbf{r}) = \beta\mathcal{A}^{\rm ex}$, is, however, well defined and unique. Nevertheless, $\Phi^{\rm ex}(\mathbf{r})$ is an important entity in various approximations and if two approximate theories with different $\Phi^{\rm ex}(\mathbf{r})$ give the same value $\beta\mathcal{A}^{\rm ex}$ of its integral, the theories are equivalent.

In the theories that we will discuss next, it is assumed that one knows entities like the free energy density for the bulk fluids, at least approximately, and one uses this knowledge to build a theory for inhomogeneous fluids. In some cases other entities for the bulk, such as the pair direct-correlation function $c^{(2)\rm b}(r)$, is also be assumed to be known.

In the **Local Density Approximation (LDA)** one assumes that $\Phi^{\rm ex}$ at point $\mathbf{r}$ has the same value as $\Phi^{\rm ex,b}(n^{\rm b})$ for a bulk fluid of a density $n^{\rm b}$ equal to the local density $n(\mathbf{r})$, that is,

$$\Phi^{\rm ex}(\mathbf{r};[n]) = \beta\left.\frac{\mathcal{A}^{\rm ex,b}}{V}\right|_{n^{\rm b}=n(\mathbf{r})} = \Phi^{\rm ex,b}(n(\mathbf{r})) \quad \text{(LDA)}. \tag{10.71}$$

Hence the functional $\mathcal{A}^{\rm ex}[n]$ for the inhomogeneous fluid is given by

$$\beta\mathcal{A}^{\rm ex}[n] = \int d\mathbf{r}\, \Phi^{\rm ex,b}(n(\mathbf{r})) \quad \text{(LDA)}. \tag{10.72}$$

We have

$$\beta\frac{\delta\mathcal{A}^{\rm ex}}{\delta n(\mathbf{r}')} = \int d\mathbf{r}\, \frac{\partial\Phi^{\rm ex,b}(n(\mathbf{r}))}{\partial n(\mathbf{r})}\frac{\delta n(\mathbf{r})}{\delta n(\mathbf{r}')} = \int d\mathbf{r}\, \frac{\partial\Phi^{\rm ex,b}(n(\mathbf{r}))}{\partial n(\mathbf{r})}\delta^{(3)}(|\mathbf{r}-\mathbf{r}'|)$$

and since

$$\frac{\partial\Phi^{\rm ex,b}(n^{\rm b})}{\partial n^{\rm b}} = \frac{\beta}{V}\cdot\frac{\partial\mathcal{A}^{\rm ex,b}(n^{\rm b})}{\partial n^{\rm b}} = \beta\frac{\partial\mathcal{A}^{\rm ex,b}(n^{\rm b})}{\partial\bar{N}^{\rm b}} = \beta\mu^{\rm ex,b}(n^{\rm b}),$$

where $\mu^{\rm ex,b}(n^{\rm b})$ is the excess chemical potential for a bulk phase of density $n^{\rm b}$, we obtain

$$\mu^{\rm ex}(\mathbf{r}') = v(\mathbf{r}') + \mu^{\rm ex,b}(n(\mathbf{r}')). \quad \text{(LDA)} \tag{10.73}$$

As we have seen in Equation 9.200, where the sum of the last two terms equals $\mu^{\rm ex}(\mathbf{r}')$, the correct value of $\mu^{\rm ex}(\mathbf{r}')$ depends on the density $n(\mathbf{r})$ and the pair distribution function $g^{(2)}(\mathbf{r},\mathbf{r}';\xi)$ for all points $\mathbf{r}$ in the entire neighborhood of the point $\mathbf{r}'$. In order for Equation 10.73 to be a good approximation, this neighborhood must be similar to a bulk fluid with a constant density given by $n^{\rm b} = n(\mathbf{r}')$. Otherwise, the reversible work of adding a particle at $\mathbf{r}'$ (i.e. $\mu^{\rm ex}$) will be substantially different when a particle is immersed in the inhomogeneous fluid or in the homogeneous fluid. The LDA is therefore useful only for systems where the density $n(\mathbf{r})$ varies very smoothly with changing **r**. In addition, near a surface where the density distribution often has a large gradient or is strongly oscillatory, the local density $n(\mathbf{r}')$ can be larger than the maximal density possible for the bulk fluid[40] and $\Phi^{\mathrm{ex,b}}(n(\mathbf{r}'))$ is not defined. For these reasons, it is much better to use an average density over the neighborhood of the point $\mathbf{r}'$ in the argument of $\Phi^{\mathrm{ex,b}}$ in Equations 10.71 and 10.72.

Therefore, one way to proceed is to introduce the *excess free energy per particle* of the bulk phase

$$\varphi^{\mathrm{ex,b}}(n^{\mathrm{b}}) = \frac{\Phi^{\mathrm{ex,b}}(n^{\mathrm{b}})}{n^{\mathrm{b}}}$$

and to assume that

$$\beta\mathcal{A}^{\mathrm{ex}}[n] \approx \int d\mathbf{r}\, n(\mathbf{r})\varphi^{\mathrm{ex,b}}(\bar{n}(\mathbf{r})), \quad \text{(WDA)} \tag{10.74}$$

where $\bar{n}(\mathbf{r})$ is an average density over the neighborhood of the point **r** in the inhomogeneous fluid, defined as a *weighted density*

$$\bar{n}(\mathbf{r}) = \int d\mathbf{r}'\, \omega(|\mathbf{r} - \mathbf{r}'|)n(\mathbf{r}'), \tag{10.75}$$

where $\omega(r)$ is a weight function that is normalized, that is,

$$\int d\mathbf{r}\, \omega(r) = 1$$

(distinguish ω from bold $\boldsymbol{\omega}$, which has a different meaning elsewhere in this treatise). This approximation is called the **Weighted Density Approximation (WDA)**. A crucial element in this approximation is the selection of the weight function $\omega(r)$ and one can do this in several different manners; here we will only treat a few. Incidentally we may note that if we select $\omega(r) = \delta^{(3)}(r)$, Equations 10.74 and 10.75 yield the LDA, Equation 10.72.

A simple choice of $\omega(r)$ is to take $\bar{n}(\mathbf{r})$ as the average density inside a sphere with radius d centered at **r**, which means that

$$\omega(r) = \frac{3}{4\pi d^3}\mathcal{H}(d - r), \tag{10.76}$$

where $\mathcal{H}(x)$ is the Heaviside step function that is equal to 0 for $x < 0$ and 1 for $x \geq 0$, so $\mathcal{H}(d - r) = 1$ for $r \leq d$ and zero otherwise. For hard spheres, a reasonable choice is to take $d = d^{\mathrm{h}}$, the hard sphere diameter, which is the radius of the excluded volume around each sphere.

[40] For a fluid outside a planar wall, say that s particles per unit area have their centers within a thin slice of width h parallel to the surface. The average three-dimensional density s/h can then be very large even if s is rather low provided that h is sufficiently small. Under such circumstances s/h can exceed the *three-dimensional* packing limit for a bulk fluid while the value of s is lower than the *two-dimensional* packing limit (laterally in the slice) so the packing is physically possible.

As shown in Exercise 10.5, Equation 10.74 implies that in the bulk phase we have the pair direct-correlation function

$$\begin{aligned} c^{(2)}(|\mathbf{r}-\mathbf{r}'|) &\approx -2\varphi'^{\,\mathrm{ex,b}}(n^{\mathrm{b}})\omega(|\mathbf{r}-\mathbf{r}'|) \\ &- n^{\mathrm{b}}\varphi''^{\,\mathrm{ex,b}}(n^{\mathrm{b}})\int d\mathbf{r}''\,\omega(|\mathbf{r}-\mathbf{r}''|)\omega(|\mathbf{r}'-\mathbf{r}''|), \end{aligned} \quad (10.77)$$

where $\varphi'^{\,\mathrm{ex,b}}(n)$ and $\varphi''^{\,\mathrm{ex,b}}(n)$ are the first and second derivatives of $\varphi^{\mathrm{ex,b}}(n)$ with respect to n.

{**Exercise 10.5:** Use the definition of $c^{(2)}$ in Equation 9.127 to show that Equation 10.74 with $\bar{n}(\mathbf{r})$ given by Equation 10.75 implies Equation 10.77 for a bulk fluid of density n^{b}. Note that

$$\delta\bar{n}(\mathbf{r})/\delta n(\mathbf{r}') = \omega(|\mathbf{r}-\mathbf{r}'|), \quad (10.78)$$

which follows from Equation 10.75.}

For a hard sphere fluid we may select $\omega(r)$ as in Equation 10.76 with $d = d^{\mathrm{h}}$. Then, the first term in the rhs of Equation 10.77 is zero outside the hard core region. In the Percus-Yevick approximation, which works quite well for hard spheres as we saw in Section 10.1, $c^{(2)}(r)$ is indeed zero for $r > d^{\mathrm{h}}$. This justifies the selection of $d = d^{\mathrm{h}}$ for hard spheres in the present approximation. The second term in Equation 10.77 is proportional to $\int d\mathbf{r}''\,\mathcal{H}(d^{\mathrm{h}}-|\mathbf{r}-\mathbf{r}''|)\mathcal{H}(d^{\mathrm{h}}-|\mathbf{r}'-\mathbf{r}''|)$, which is equal to the intersection volume of two spheres of radii d^{h} centered at $\mathbf{r}$ and $\mathbf{r}'$ (the integrand is one inside the intersection and zero outside). This is the only contribution to $c^{(2)}(r)$ that varies with r for $r < d^{\mathrm{h}}$ in the current approximation; the contribution from the first term is constant. The second term is zero when the two spheres do not intersect, which occurs for center-to-center distances larger than $2d^{\mathrm{h}}$. Thus, $c^{(2)}(r)$ is nonzero for $d^{\mathrm{h}} < r < 2d^{\mathrm{h}}$, which signals that this approximation is not very accurate, which is confirmed in practical use, in particular for large densities.

We may note that it would be better instead to have a contribution proportional $\int d\mathbf{r}''\,\mathcal{H}(\mathrm{R}^{\mathrm{h}}-|\mathbf{r}-\mathbf{r}''|)\mathcal{H}(\mathrm{R}^{\mathrm{h}}-|\mathbf{r}'-\mathbf{r}''|)$, where R^{h} is the hard sphere *radius*, because it will only give contributions to $c^{(2)}(r)$ inside the hard core region. In the following section 10.3 we will see that this kind of contribution occurs both in the PY approximation (cf. Equations 10.130 and 10.136) and the *Fundamental Measure Theory* (FMT) for hard sphere fluids. FMT is in general a very accurate approximation that has superseded the approximations presented in the current section. Therefore, other DFT theories have only briefly been introduced here and they mainly constitute background material for the FMT.

Before finishing the current section, we will mention a much better way to select a weight function in the WDA that allows one to have an accurate pair direct-correlation function for the bulk phase. One thereby assumes that ω is density dependent, $\omega(r,n)$, and instead of Equation 10.75 one takes[41]

$$\bar{n}(\mathbf{r}) = \int d\mathbf{r}'\, \omega\left(|\mathbf{r}-\mathbf{r}'|, \bar{n}(\mathbf{r})\right) n(\mathbf{r}'), \tag{10.79}$$

which is an equation for $\bar{n}(\mathbf{r})$ because the latter occurs on both sides. The pair direct-correlation function can be obtained from Equation 10.74 as before, but it will contain additional contributions due to the density dependence of $\omega\left(|\mathbf{r}-\mathbf{r}'|, \bar{n}(\mathbf{r})\right)$. By requiring that $c^{(2)\mathrm{b}}(r)$ for the bulk phase agrees with that obtained from an accurate approximation for the same phase, one obtains a much better weight function that can be used in WDA for inhomogeneous fluids.

10.3 FUNDAMENTAL MEASURE THEORY (FMT)

The Fundamental Measure Theory is an approximation that in many respects have superseded other DFT approximations for hard sphere systems. It is also useful for fluids that have an attractive part of the pair interaction potential in addition to the hard core potential, but for such systems one usually includes the effects of the attractive contributions as a perturbation on the mean field level for a hard sphere fluid treated with FMT. Fundamental Measure Theory was originally suggested by Yaakov Rosenfeld[42] in 1989 and has subsequently been improved in many ways. We will treat various versions of FMT and also an extension of it that is based on a more consistent free energy expression. We will start with bulk mixtures of hard spheres and later extend the treatment to inhomogeneous mixtures. The original FMT was set up to treat the latter, but by starting with bulk fluids we will be able to gradually introduce the various key elements of FMT that can appear to be rather complicated.

10.3.1 Hard sphere bulk mixtures and an introduction to FMT

10.3.1.1 The bulk pressure and free energy density

For a bulk fluid we have $\bar{U} + PV - TS - \sum_i \mu_i \bar{N}_i = 0$ (cf. Equation I:2.123 generalized to many components) and since $A = \bar{U} - TS$ this implies that

$$P = -\frac{A}{V} + \beta \sum_i \mu_i n_i^{\mathrm{b}} = k_B T \sum_i n_i^{\mathrm{b}} - \frac{A^{\mathrm{ex}}}{V} + \beta \sum_i \mu_i^{\mathrm{ex}} n_i^{\mathrm{b}},$$

[41] P. Tarazona, Phys. Rev. A **31** (1985) 2672 (https://doi.org/10.1103/PhysRevA.31.2672); W. A. Curtin and N. W. Ashcroft, Phys. Rev. A **32** (1985) 2909 (https://doi.org/10.1103/PhysRevA.32.2909); P. Tarazona, J. A. Cuesta, and Y. Martínez-Ratón, Lect. Notes Phys. **753** (2008) 247 (https://doi.org/10.1007/978-3-540-78767-9_7).

[42] Y. Rosenfeld, Phys. Rev. Lett. **63** (1989) 980 (https://doi.org/10.1103/PhysRevLett.63.980).

where we have used

$$\begin{aligned} A &= A^{\text{ideal}} + A^{\text{ex}} = k_B T \sum_i \bar{N}_i \left[\ln(\Lambda_i^3 n_i^{\text{b}}) - 1\right] + A^{\text{ex}} \\ \mu_i &= \mu_i^{\text{ideal}} + \mu_i^{\text{ex}} = k_B T \ln(\Lambda_i^3 n_i^{\text{b}}) + \mu_i^{\text{ex}}, \end{aligned}$$

which are Equations 9.230 and 10.64 generalized to many-component systems and applied to a bulk phase (in the bulk $A = \mathcal{A}$ since external potentials $v_i(\mathbf{r})$ are zero).

The excess free energy density $A^{\text{ex,b}}/V = \mathcal{A}^{\text{ex,b}}/V$ plays an important role in the theory and as before we use the notation $\Phi^{\text{ex,b}} = \beta\mathcal{A}^{\text{ex,b}}/V$, which has the dimension 1/volume. Thus we have

$$\beta P = \sum_i n_i^{\text{b}} - \Phi^{\text{ex,b}} + \beta \sum_i \mu_i^{\text{ex}} n_i^{\text{b}}, \tag{10.80}$$

where the first term on the rhs is the ideal contribution to βP. We will continue to use superscript "b" to indicate that $A^{\text{ex,b}}$ and $\Phi^{\text{ex,b}}$ are for a bulk phase, but all other entities for the bulk apart from n_i^{b} will not be marked in this manner.

Let us consider a bulk mixture of hard spheres of radii R_i^{h}. The distance of closest approach for the centers of two spheres of species i and j is $d_{ij}^{\text{h}} = \mathsf{R}_i^{\text{h}} + \mathsf{R}_j^{\text{h}}$. The Percus-Yevick approximation for mixtures is (cf. Equation 9.78)

$$g_{ij}^{(2)}(r) = e^{-\beta u_{ij}(r_{12})}[1 + h_{ij}^{(2)}(r) - c_{ij}^{(2)}(r)]. \qquad \text{(PY)} \tag{10.81}$$

For hard spheres this implies that $c_{ij}^{(2)}(r) = 0$ for $r > d_{ij}^{\text{h}}$ and, as always, $g_{ij}^{(2)}(r) = 0$ for $r < d_{ij}^{\text{h}}$. The solution to the PY integral equation, which is obtained when Equation 10.81 is inserted in the OZ equation, has been obtained by Lebowitz.[43] He showed that $c_{ij}^{(2)}(r)$ for $r < d_{ij}^{\text{h}}$ is a rational function: for $r < |\mathsf{R}_i^{\text{h}} - \mathsf{R}_j^{\text{h}}|$ it is constant and for $|\mathsf{R}_i^{\text{h}} - \mathsf{R}_j^{\text{h}}| < r < d_{ij}^{\text{h}}$ it is a fourth order polynomial divided by r. We will present this solution in more detail in Section 10.3.1.3.

As we have seen in Section 10.1, the pressure for the one-component case, expressed as $\beta P/n^{\text{b}}$, can in the PY approximation be written in terms of the hard sphere packing fraction η^{hs}. Different results are obtained from the compressibility and the virial routes, P^{c} and P^{v} (Equations 10.5 and 10.8). The corresponding expressions for the pressure of a hard sphere mixture are a bit more complicated and involve the following weighted-density parameters[44]

$$\xi_0 = \sum_i n_i^{\text{b}}, \quad \xi_{\text{R}} = \sum_i \mathsf{R}_i^{\text{h}} n_i^{\text{b}}, \quad \xi_{\text{A}} = \sum_i \mathsf{A}_i^{\text{h}} n_i^{\text{b}}, \quad \xi_{\text{V}} = \sum_i \mathsf{V}_i^{\text{h}} n_i^{\text{b}}, \tag{10.82}$$

[43] J. L. Lebowitz, Phys. Rev. **133** (1964) A895 (https://doi.org/10.1103/PhysRev.133.A895).

[44] A common alternative notation is ξ_0, $\xi_1 \equiv \xi_{\text{R}}$, $\xi_2 \equiv \xi_{\text{A}}$ and $\xi_3 \equiv \xi_{\text{V}}$. They are also called SPT variables (from Scaled Particle Theory).

where $\mathsf{A}_i^{\mathrm{h}} = 4\pi(\mathsf{R}_i^{\mathrm{h}})^2$ and $\mathsf{V}_i^{\mathrm{h}} = 4\pi(\mathsf{R}_i^{\mathrm{h}})^3/3$ are the area and the volume of a hard sphere of radius $\mathsf{R}_i^{\mathrm{h}}$. The first and the last quantities are familiar: ξ_0 is the total number density $n_{\mathrm{tot}}^{\mathrm{b}}$ and ξ_{V} is the total hard sphere packing fraction $\eta_{\mathrm{tot}}^{\mathrm{hs}} = \sum_i \eta_i^{\mathrm{hs}}$, where $\eta_i^{\mathrm{hs}} = \mathsf{V}_i^{\mathrm{h}} n_i^{\mathrm{b}}$. The remaining quantities, ξ_{R} and ξ_{A}, contain the density weighted by other geometrical entities, the sphere radii and area, respectively. The dimensions of the four quantities in Equation 10.82 are $1/(\text{length})^3$, $1/(\text{length})^2$, 1/length and dimensionless, respectively.

The PY pressure for the mixture obtained via the compressibility and the virial routes, respectively, is given by[45]

$$\beta P^{\mathrm{c}} = \frac{\xi_0}{1-\xi_{\mathrm{V}}} + \frac{\xi_{\mathrm{R}}\,\xi_{\mathrm{A}}}{(1-\xi_{\mathrm{V}})^2} + \frac{\xi_{\mathrm{A}}^3}{12\pi(1-\xi_{\mathrm{V}})^3} \quad \text{(PY, hard spheres)} \qquad (10.83)$$

and

$$\begin{aligned} \beta P^{\mathrm{v}} &= \frac{\xi_0}{1-\xi_{\mathrm{V}}} + \frac{\xi_{\mathrm{R}}\,\xi_{\mathrm{A}}}{(1-\xi_{\mathrm{V}})^2} + \frac{\xi_{\mathrm{A}}^3}{12\pi(1-\xi_{\mathrm{V}})^2} \quad \text{(PY, hard spheres)} \\ &= \beta P^{\mathrm{c}} - \frac{\xi_{\mathrm{A}}^3\,\xi_{\mathrm{V}}}{12\pi(1-\xi_{\mathrm{V}})^3}. \end{aligned} \qquad (10.84)$$

Note the difference in power in the denominator of the last term for βP^{c} and the corresponding term for βP^{v}. Considering that the excluded volume for i-particles around each j-particle, or vice versa, is a spherical void of radius $d_{ij}^{\mathrm{h}} = \mathsf{R}_i^{\mathrm{h}} + \mathsf{R}_j^{\mathrm{h}}$, it is remarkable that the pressure is entirely expressed in terms of geometrical entities of the individual species, $\mathsf{R}_i^{\mathrm{h}}$, $\mathsf{A}_i^{\mathrm{h}}$ and $\mathsf{V}_i^{\mathrm{h}}$, in the PY approximation. The PY pressure is quite accurate except for high densities, where there are significant deviations. For a one-component system of density n^{b}, we have

$$\begin{aligned} \xi_0 &= n^{\mathrm{b}}, \quad \xi_{\mathrm{V}} = \eta^{\mathrm{hs}} \\ \xi_{\mathrm{R}}\xi_{\mathrm{A}} &= \mathsf{R}^{\mathrm{h}} n^{\mathrm{b}} \times 4\pi(\mathsf{R}^{\mathrm{h}})^2 n^{\mathrm{b}} = 3n^{\mathrm{b}}\eta^{\mathrm{hs}} \quad \text{(one-component)} \\ \xi_{\mathrm{A}}^3 &= [4\pi(\mathsf{R}^{\mathrm{h}})^2 n^{\mathrm{b}}]^3 = 36\pi n^{\mathrm{b}}(\eta^{\mathrm{hs}})^2, \end{aligned} \qquad (10.85)$$

and Equations 10.83 and 10.84 become Equations 10.5 and 10.8 after rearrangements. We see that the appearance of η^{hs} in the rhs of Equation 10.85 for $\xi_{\mathrm{R}}\xi_{\mathrm{A}}$ and ξ_{A}^3 is a special feature of the one-component case since Equation 10.82 for a single component lacks the sums over species.

For the one-component hard sphere fluid we saw in Section 10.1 that the much more accurate Carnahan-Starling formula for the pressure (Equation 10.86)

$$\frac{\beta P}{n^{\mathrm{b}}} = \frac{1+\eta^{\mathrm{hs}} + (\eta^{\mathrm{hs}})^2 - (\eta^{\mathrm{hs}})^3}{(1-\eta^{\mathrm{hs}})^3} \quad \text{(C-S, bulk)} \qquad (10.86)$$

[45]These results were obtained by Lebowitz, see footnote 43, who used a different notation.

can be obtained by taking $P^{\text{C-S}} = (P^{\text{v,PY}} + 2P^{\text{c,PY}})/3$. For a mixture, one can likewise obtain a more accurate pressure by taking 2/3 of Equation 10.83 and 1/3 of Equation 10.84, which yields

$$\beta P = \frac{\xi_0}{1-\xi_V} + \frac{\xi_R\,\xi_A}{(1-\xi_V)^2} + \frac{\xi_A^3}{12\pi(1-\xi_V)^3}\left[1-\xi_V/3\right] \quad \text{(BMCSL)} \qquad (10.87)$$

This is the Boublík-Mansoori-Carnahan-Starling-Leland (BMCSL)[46] expression for the pressure of a hard sphere mixture.

Hansen-Goos and Roth (HG-R)[47] have derived an even more accurate expression for the pressure of hard sphere mixtures

$$\begin{aligned}\beta P &= \frac{\xi_0}{1-\xi_V} + \frac{\xi_R\,\xi_A}{(1-\xi_V)^2}\left[1+\xi_V^2/3\right] + \\ &+ \frac{\xi_A^3}{12\pi(1-\xi_V)^3}\left[1-2\xi_V/3+\xi_V^2/3\right] \quad \text{(HG-R).} \qquad (10.88)\end{aligned}$$

Like Equation 10.87, this equation reduces to the C-S equation of state 10.86 for a one-component system. In Section 10.3.1.2 we will encounter yet another expression for the pressure of hard sphere mixtures (Equation 10.129) that for high densities has a similar accuracy as Equation 10.88. It too involves only the entities ξ_0, ξ_R, ξ_A and ξ_V and is based on the C-S expression for the one-component case, but it can easily be generalized to incorporate even more accurate equations of state. For the time being, we will, however, limit ourselves to the PY, BMCSL and HG-R expressions.

Equation 10.88 can be derived by using the formalism of the **Fundamental Measure Theory** and its derivation will serve as an introduction to that theory. The most important use of FMT is for inhomogeneous fluids, but we will start by exploring it for the bulk. A basic assumption in FMT for the bulk phase is that an approximate $\Phi^{\text{ex,b}}$ for a hard-sphere mixture can be expressed solely in terms of ξ_0, ξ_R, ξ_A and ξ_V, that is,

$$\Phi^{\text{ex,b}} = \Phi^{\text{ex,b}}(\xi_0, \xi_R, \xi_A, \xi_V) \quad \text{(FMT).} \qquad (10.89)$$

Note that ξ_0, $\xi_R\xi_A$ and ξ_A^3 are the only combinations with dimension $1/(\text{length})^3$ that can be formed from products of the three first entities. Rosenfeld made the Ansatz that $\Phi^{\text{ex,b}}$, which has the same dimension, can be written as the following power series expansion in ξ_0, ξ_R and ξ_A up to third order

$$\Phi^{\text{ex,b}} = \mathsf{f}_0(\xi_V)\xi_0 + \mathsf{f}_1(\xi_V)\xi_R\xi_A + \mathsf{f}_2(\xi_V)\xi_A^3 \quad \text{(Rosenfeld),} \qquad (10.90)$$

[46] T. Boublík, J. Chem. Phys. **53** (1970) 471 (https://doi.org/10.1063/1.1673824) and G. A. Mansoori, N. F. Carnahan, K. E. Starling, and T. W. Leland, J. Chem. Phys. **54** (1971) 1523 (https://doi.org/10.1063/1.1675048).

[47] H. Hansen-Goos and R. Roth, J. Chem. Phys. **124** (2006) 154506 (https://doi.org/10.1063/1.2187491).

where the coefficients f_ℓ are, as yet, unknown functions of the dimensionless ξ_V. This means that $\Phi^{\mathrm{ex,b}}$ has a functional form that is similar to that of the pressure expressions above. The appearance of ξ_V, ξ_A, ξ_R and ξ_0 in $\Phi^{\mathrm{ex,b}}$ means that the free energy density is expressed in terms of fundamental measures, *volume*, *surface area* and *radius*, of the individual spheres that constitute the fluid. This is the reason for the name Fundamental Measure Theory.

As we will see, there are several FMT versions that use Rosenfeld's Ansatz 10.90. They differ in the functions $\mathsf{f}_\ell(\xi_\mathrm{V})$. The original choice of these functions by Rosenfeld will be designated "original FMT" (oFMT) below (Equation 10.99). To distinguish the Ansatz itself from the choice of functions we will designate the former as "Rosenfeld," which hence includes several variants – one of which is oFMT.

It follows from $\Phi^{\mathrm{ex,b}} = \Phi^{\mathrm{ex,b}}(\xi_0, \xi_\mathrm{R}, \xi_\mathrm{A}, \xi_\mathrm{V})$ that

$$
\begin{aligned}
\beta\mu_i^{\mathrm{ex}} &= \frac{\partial\Phi^{\mathrm{ex,b}}}{\partial n_i^{\mathrm{b}}} = \sum_{\nu\in\{0,\mathrm{R,A,V}\}} \frac{\partial\Phi^{\mathrm{ex,b}}}{\partial\xi_\nu}\frac{\partial\xi_\nu}{\partial n_i^{\mathrm{b}}} \\
&= \frac{\partial\Phi^{\mathrm{ex,b}}}{\partial\xi_0} + \frac{\partial\Phi^{\mathrm{ex,b}}}{\partial\xi_\mathrm{R}}\mathrm{R}_i^{\mathrm{h}} + \frac{\partial\Phi^{\mathrm{ex,b}}}{\partial\xi_\mathrm{A}}\mathrm{A}_i^{\mathrm{h}} + \frac{\partial\Phi^{\mathrm{ex,b}}}{\partial\xi_\mathrm{V}}\mathrm{V}_i^{\mathrm{h}}, \qquad (10.91)
\end{aligned}
$$

which implies that

$$
\beta\sum_i \mu_i^{\mathrm{ex}} n_i^{\mathrm{b}} = \sum_{\nu\in\{0,\mathrm{R,A,V}\}} \frac{\partial\Phi^{\mathrm{ex,b}}}{\partial\xi_\nu}\xi_\nu,
$$

where we have used Equation 10.82. From Equation 10.80 it hence follows that

$$
\beta P = \xi_0 - \Phi^{\mathrm{ex,b}} + \sum_{\nu\in\{0,\mathrm{R,A,V}\}} \frac{\partial\Phi^{\mathrm{ex,b}}}{\partial\xi_\nu}\xi_\nu. \qquad (10.92)
$$

Using Equation 10.90 one therefore obtains after simplification

$$
\begin{aligned}
\beta P &= \left[1+\xi_\mathrm{V}\mathsf{f}_0'(\xi_\mathrm{V})\right]\xi_0 + \left[\mathsf{f}_1(\xi_\mathrm{V})+\xi_\mathrm{V}\mathsf{f}_1'(\xi_\mathrm{V})\right]\xi_\mathrm{R}\xi_\mathrm{A} \\
&+ \left[2\mathsf{f}_2(\xi_\mathrm{V})+\xi_\mathrm{V}\mathsf{f}_2'(\xi_\mathrm{V})\right]\xi_\mathrm{A}^3, \qquad (10.93)
\end{aligned}
$$

where $\mathsf{f}_\ell'(\xi) = d\mathsf{f}_\ell(\xi)/d\xi$.

Another expression for βP can be obtained from the fact that μ_i^{ex} is the reversible work when a hard sphere is inserted and then held at some position in the fluid. For a very large sphere this work is dominated by the work against the pressure to create a void large enough to fit the sphere, so we have the exact relationship

$$
\lim_{R_i\to\infty}\frac{\mu_i^{\mathrm{ex}}}{V_i} = P.
$$

In FMT, Equation 10.91 implies that this means that

$$
\beta P = \frac{\partial\Phi^{\mathrm{ex,b}}}{\partial\xi_\mathrm{V}} \qquad (10.94)
$$

and when Equation 10.90 is inserted we obtain

$$\beta P \;=\; \mathrm{f}_0'(\xi_\mathrm{V})\xi_0 + \mathrm{f}_1'(\xi_\mathrm{V})\xi_\mathrm{R}\xi_\mathrm{A} + \mathrm{f}_2'(\xi_\mathrm{V})\xi_\mathrm{A}^3. \tag{10.95}$$

We see from Equations 10.93 and 10.95 that βP and $\Phi^{\mathrm{ex,b}}$ indeed have similar functional forms in FMT as inferred earlier. By multiplying both sides of Equation 10.95 by ξ_V and subtracting it from Equation 10.93 we obtain after rearrangement

$$\beta P = \frac{\xi_0 + \mathrm{f}_1(\xi_\mathrm{V})\xi_\mathrm{R}\xi_\mathrm{A} + 2\mathrm{f}_2(\xi_\mathrm{V})\xi_\mathrm{A}^3}{1-\xi_\mathrm{V}}, \tag{10.96}$$

which is considerably simpler than Equation 10.93.

In fact, we can determine the functions f_0, f_1 and f_2 uniquely by requiring thermodynamic consistency in the sense that for a given free energy density, Equations 10.92 and 10.94 should give the same pressure. When $\Phi^{\mathrm{ex,b}}$ is given by Ansatz 10.90, this is equivalent to require that Equations 10.95 and 10.96 give the same pressure. By setting the coefficients in front of ξ_0, $\xi_\mathrm{R}\xi_\mathrm{A}$ and ξ_A^3 equal in the two expressions, we obtain the differential equations

$$\mathrm{f}_0'(\xi_\mathrm{V}) = \frac{1}{1-\xi_\mathrm{V}}, \quad \mathrm{f}_1'(\xi_\mathrm{V}) = \frac{\mathrm{f}_1(\xi_\mathrm{V})}{1-\xi_\mathrm{V}}, \quad \mathrm{f}_2'(\xi_\mathrm{V}) = \frac{2\mathrm{f}_2(\xi_\mathrm{V})}{1-\xi_\mathrm{V}}.$$

These equations are readily solved and by inserting the solutions into Equation 10.90 we get

$$\Phi^{\mathrm{ex,b}} = -\ln(1-\xi_\mathrm{V})\xi_0 + K_0\xi_0 + \frac{K_1\xi_\mathrm{R}\xi_\mathrm{A}}{1-\xi_\mathrm{V}} + \frac{K_2\xi_\mathrm{A}^3}{(1-\xi_\mathrm{V})^2}, \tag{10.97}$$

where the constants K_0, K_1 and K_2 originate from the integration constants. We obtain $K_0 = 0$ from the fact that μ_i^{ex} must be zero when the density is zero. The remaining two constants can be determined by applying the result to a one-component system and requiring that the pressure obtained from $\Phi^{\mathrm{ex,b}}$ gives the correct second and third virial coefficients, which are given Table 10.1. This gives $K_1 = 1$ and $K_2 = 1/(24\pi)$ as shown in the following exercise.

{**Exercise 10.6:** Determine βP from $\Phi^{\mathrm{ex,b}}$ in Equation 10.97. Apply it to a one-component system and make a virial expansion in the packing fraction η^{hs}. Use this expansion to determine K_1 and K_2 so that the correct second and third virial coefficients are obtained.}

The final result is

$$\Phi^{\mathrm{ex,b}} = -\ln(1-\xi_\mathrm{V})\xi_0 + \frac{\xi_\mathrm{R}\xi_\mathrm{A}}{1-\xi_\mathrm{V}} + \frac{\xi_\mathrm{A}^3}{24\pi(1-\xi_\mathrm{V})^2} \quad \text{(oFMT)}, \tag{10.98}$$

which is the free energy density obtained in Rosenfeld's original version of FMT (oFMT) applied to bulk. We accordingly have

$$\mathfrak{f}_0(x) = -\ln(1-x); \quad \mathfrak{f}_1(x) = \frac{1}{1-x}, \quad \mathfrak{f}_2(x) = \frac{1}{24\pi(1-x)^2} \quad \text{(oFMT)} \tag{10.99}$$

with $x = \xi_V$.

From both Equations 10.92 and 10.94 we obtain using Equation 10.98

$$\beta P = \frac{\xi_0}{1-\xi_V} + \frac{\xi_R \xi_A}{(1-\xi_V)^2} + \frac{\xi_A^3}{12\pi(1-\xi_V)^3} \quad \text{(oFMT).} \tag{10.100}$$

This is the same as the PY pressure from the compressibility route given by Equation 10.83. Thus, when the Ansatz 10.90 is combined with the thermodynamic consistency for the pressure (i.e., the equality of βP from Equations 10.92 and 10.94), one can never get any better results for a mixture than the PY pressure P^c, which is not very good for high densities.

It is therefore essential to obtain a better $\Phi^{ex,b}$ than Equation 10.98. The expression for βP in Equation 10.93, which was obtained from the Ansatz 10.90 and Equation 10.92, is a sum of terms containing ξ_0, $\xi_R\xi_A$ or ξ_A^3, each term having a coefficient that is a function of ξ_V. The BMCSL expression 10.87 for βP, which is more accurate than the PY expression, has the same form. One can therefore identify the coefficients in front of ξ_0, $\xi_R\xi_A$ and ξ_A^3 as being equal in equations 10.87 and 10.93 and determine the functions $\mathfrak{f}_0$, $\mathfrak{f}_1$ and $\mathfrak{f}_2$ by integration. The terms with ξ_0 and $\xi_R\xi_A$ in Equation 10.87 are the same as in oFMT, Equation 10.100, and $\mathfrak{f}_0(x)$ and $\mathfrak{f}_1(x)$ are also the same as in oFMT. The determination of $\mathfrak{f}_2$ is left as an exercise (Exercise 10.7) and the final result is

$$\mathfrak{f}_2(x) = \frac{1}{36\pi}\left[\frac{1}{x(1-x)^2} + \frac{\ln(1-x)}{x^2}\right] \quad \text{(wbFMT),} \tag{10.101}$$

which is used in the so-called *White Bear version of FMT* (wbFMT).[48] This FMT version is further discussed in Section 10.3.2.3.

{**Exercise 10.7:** Determine $\mathfrak{f}_2(\xi_V)$ from an identification of the coefficients in front of ξ_A^3 in equations 10.87 and 10.93 and integration of the resulting equation. HINT: One may use the identity $2f(x) + xf'(x) = [x^2 f(x)]'/x$.}

When these $\mathfrak{f}_0$, $\mathfrak{f}_1$ and $\mathfrak{f}_2$ are inserted in Equation 10.90 one obtains the following expression for $\Phi^{ex,b}$

$$\begin{aligned}\Phi^{ex,b} &= -\ln(1-\xi_V)\xi_0 + \frac{\xi_R\xi_A}{1-\xi_V} \\ &+ \left[\frac{1}{\xi_V(1-\xi_V)^2} + \frac{\ln(1-\xi_V)}{(\xi_V)^2}\right]\frac{\xi_A^3}{36\pi} \quad \text{(wbFMT).}\end{aligned} \tag{10.102}$$

[48] R. Roth, R. Evans, A. Lang, and G. Kahl, J. Phys.: Condens. Matter **14** (2002) 12063 (https://doi.org/10.1088/0953-8984/14/46/313).

Inserting this $\Phi^{\text{ex,b}}$ into Equation 10.94, one gets a different pressure than the BMCSL pressure that one started from, so this theory is inconsistent. This is true not only for a bulk mixture, but also for the one-component case.

Of course, the same thermodynamic inconsistency occurs for $\Phi^{\text{ex,b}}$ associated with the HG-R pressure expression 10.88 for mixtures; it is only the PY pressure P^c and the corresponding $\Phi^{\text{ex,b}}$ that do not have such an inconsistency when $\Phi^{\text{ex,b}}$ is given by the Ansatz 10.90. However, for the one-component hard sphere fluid, the HG-R theory has, in fact, thermodynamic consistency; it is constructed so that its $\Phi^{\text{ex,b}}$ gives the same pressure from Equations 10.92 and 10.94 for the one-component case. As mentioned earlier, the pressure obtained in the HG-R theory for a single component is equal to the C-S pressure, which is very accurate. Thus, this theory has a partial consistency (for one component systems, but not for mixtures) and this is a major reason why HG-R pressure expression is more accurate for mixtures that the other ones. As discussed earlier, all approximate theories have inconsistencies of one kind or another and the most one can do is to make them as small as possible if they cannot be eliminated.

Derivation of the HG-R pressure expression

To derive Equation 10.88 and the associated $\Phi^{\text{ex,b}}$, we will determine the functions f_0, f_1 and f_2 in Equation 10.90 that give the C-S pressure for a bulk phase, Equation 10.86, when they are used in both Equation 10.95 and Equation 10.96. By inserting Equation 10.85 into the latter two equations we obtain

$$\frac{\beta P}{n^{\text{b}}} = \mathsf{f}_0'(\eta^{\text{hs}}) + 3\eta^{\text{hs}}\mathsf{f}_1'(\eta^{\text{hs}}) + 36\pi(\eta^{\text{hs}})^2\mathsf{f}_2'(\eta^{\text{hs}})$$

and

$$\frac{\beta P}{n^{\text{b}}} = \frac{1 + 3\eta^{\text{hs}}\mathsf{f}_1(\eta^{\text{hs}}) + 72\pi(\eta^{\text{hs}})^2\mathsf{f}_2(\eta^{\text{hs}})}{1-\eta^{\text{hs}}},$$

respectively. For simplicity in notation in the following we introduce $\mathsf{F}_2(\eta^{\text{hs}}) = 36\pi\mathsf{f}_2(\eta^{\text{hs}})$.

We will keep f_0 the same as before, so $\mathsf{f}_0'(\eta^{\text{hs}}) = 1/(1-\eta^{\text{hs}})$. By setting $\beta P/n^{\text{b}}$ equal to the C-S pressure in both equations we obtain after simplification, writing x instead of η^{hs},

$$\begin{aligned} 3\mathsf{f}_1'(x) + x\mathsf{F}_2'(x) &= \frac{3-x^2}{(1-x)^3} \\ 3\mathsf{f}_1(x) + 2x\mathsf{F}_2(x) &= \frac{3-x^2}{(1-x)^2}. \end{aligned}$$

By taking the derivative of the second equation and subtracting the first one, we obtain after simplification the following differential equation

$$2\mathsf{F}_2(x) + x\mathsf{F}_2'(x) \equiv \frac{(x^2\mathsf{F}_2(x))'}{x} = \frac{3-2x+x^2}{(1-x)^3}.$$

The last equality can easily be integrated and we obtain

$$x^2\mathsf{F}_2(x) = -\frac{x(1-3x+x^2)}{(1-x)^2} - \ln(1-x).$$

We can now obtain $\mathsf{f}_1(x)$ from $\mathsf{F}_2(x)$ by using any of the relationships between these functions above. The final result is

$$\mathsf{f}_1(x) = \frac{1}{3}\left[\frac{5-x}{1-x} + \frac{2\ln(1-x)}{x}\right] \quad \text{(HG-R)}$$

$$\mathsf{f}_2(x) = -\frac{1}{36\pi}\left[\frac{1-3x+x^2}{x(1-x)^2} + \frac{\ln(1-x)}{x^2}\right]. \quad (10.103)$$

We can now determine $\Phi^{\mathrm{ex,b}}$ by inserting this result into Equation 10.90 and we obtain, using the original notation for mixtures,

$$\begin{aligned}\Phi^{\mathrm{ex,b}} &= -\ln(1-\xi_{\mathrm{V}})\xi_0 + \left[\frac{5-\xi_{\mathrm{V}}}{1-\xi_{\mathrm{V}}} + \frac{2\ln(1-\xi_{\mathrm{V}})}{\xi_{\mathrm{V}}}\right]\frac{\xi_{\mathrm{R}}\xi_{\mathrm{A}}}{3} \\ &- \left[\frac{1-3\xi_{\mathrm{V}}+\xi_{\mathrm{V}}^2}{\xi_{\mathrm{V}}(1-\xi_{\mathrm{V}})^2} + \frac{\ln(1-\xi_{\mathrm{V}})}{\xi_{\mathrm{V}}^2}\right]\frac{\xi_{\mathrm{A}}^3}{36\pi} \quad \text{(HG-R).} \quad (10.104)\end{aligned}$$

From Equation 10.96 we can finally obtain the HG-R expression for the pressure, Equation 10.88. This free energy density is used in the so-called *White Bear Mark II FMT*, see Section 10.3.2.3.

{**Exercise 10.8:** Use Equation 10.92 to determine βP from $\Phi^{\mathrm{ex,b}}$ in Equation 10.104 and verify Equation 10.88. Determine βP also from Equation 10.94 with the same $\Phi^{\mathrm{ex,b}}$ and verify that the pressure is different.}

10.3.1.2 A more consistent free energy approach

The inconsistency of P for hard sphere mixtures obtained so far from Equations 10.92 and 10.94 in FMT (apart from the oFMT case) is a consequence of Rosenfeld's Ansatz that $\Phi^{\mathrm{ex,b}}$ is given by the power series expansion shown in Equation 10.90. By making a more general FMT Ansatz for $\Phi^{\mathrm{ex,b}}$ in terms of ξ_0, ξ_{R}, ξ_{A} and ξ_{V}, Andrés Santos[49] obtained an approximate $\Phi^{\mathrm{ex,b}}$ that gives a fully consistent P for hard sphere mixtures from the two pressure expressions

[49] A. Santos, J. Chem. Phys. **136** (2012) 136102 (https://doi.org/10.1063/1.3702439) and Phys. Rev. E **86** (2012) 040102(R) (https://doi.org/10.1103/PhysRevE.86.040102).

10.92 and 10.94. Such a consistency requires that

$$\frac{\partial \Phi^{\text{ex,b}}}{\partial \xi_{\text{V}}} = \xi_0 - \Phi^{\text{ex,b}} + \sum_{\nu \in \{0,\text{R},\text{A},\text{V}\}} \frac{\partial \Phi^{\text{ex,b}}}{\partial \xi_\nu} \xi_\nu, \tag{10.105}$$

which hence has to be fulfilled by the function $\Phi^{\text{ex,b}} = \Phi^{\text{ex,b}}(\xi_0, \xi_{\text{R}}, \xi_{\text{A}}, \xi_{\text{V}})$.

In **Santos' Approach to FMT**, the key quantities are[50]

$$\mathring{\xi}_{\text{R}} = \sum_i \mathsf{R}_i^{\text{h}} \mathsf{x}_i = \frac{\xi_{\text{R}}}{\xi_0}, \quad \mathring{\xi}_{\text{A}} = \sum_i \mathsf{A}_i^{\text{h}} \mathsf{x}_i = \frac{\xi_{\text{A}}}{\xi_0}, \quad \mathring{\xi}_{\text{V}} = \sum_i \mathsf{V}_i^{\text{h}} \mathsf{x}_i = \frac{\xi_{\text{V}}}{\xi_0}$$

(cf. Equation 10.82), where $\mathsf{x}_i = n_i^{\text{b}}/n_{\text{tot}}^{\text{b}}$ is the mole fraction of species i in the bulk mixture (recall that $\xi_0 = \sum_i n_i^{\text{b}} = n_{\text{tot}}^{\text{b}}$). The dimensions of these three quantities are length, area and volume, respectively. In this approach one makes an Ansatz for the free energy per particle, $\varphi^{\text{ex,b}}$, in terms of the dimensionless combinations

$$\xi_{\text{A}}^* \equiv \frac{\mathring{\xi}_{\text{A}}}{(\mathring{\xi}_{\text{R}})^2}, \quad \xi_{\text{V}}^* \equiv \frac{\mathring{\xi}_{\text{V}}}{(\mathring{\xi}_{\text{R}})^3}$$

and the dimensionless $\xi_{\text{V}} \equiv \eta_{\text{tot}}^{\text{hs}}$. The Ansatz is that $\varphi^{\text{ex,b}}$ is a function solely of these three entities

$$\varphi^{\text{ex,b}} = \varphi^{\text{ex,b}}(\xi_{\text{V}}, \xi_{\text{A}}^*, \xi_{\text{V}}^*) \quad \text{(Santos)} \tag{10.106}$$

and since $\varphi^{\text{ex,b}} = \Phi^{\text{ex,b}}/n_{\text{tot}}^{\text{b}} = \Phi^{\text{ex,b}}/\xi_0$, we have

$$\Phi^{\text{ex,b}} = \xi_0 \varphi^{\text{ex,b}}(\xi_{\text{V}}, \xi_{\text{A}}^*, \xi_{\text{V}}^*) \quad \text{(Santos)}. \tag{10.107}$$

Furthermore, one requires that the theory applies irrespectively of the sizes of the spheres, including the case where $\mathsf{R}_i^{\text{h}} \to 0$ for one species ($i = 0$), that is, when there are point particles present (spheres with zero radius) in addition to the spheres with finite radii.

In the shaded text box below it is shown that these requirements lead to

$$\Phi^{\text{ex,b}} = -\xi_0 \ln(1 - \xi_{\text{V}}) + \frac{\xi_{\text{R}} \xi_{\text{A}}}{1 - \xi_{\text{V}}} \mathsf{F}(\tau) \quad \text{(Santos)}, \tag{10.108}$$

where $\mathsf{F}(\tau)$ is an as yet undetermined function and

$$\tau = \frac{\xi_{\text{A}}^2/\xi_{\text{R}}}{12\pi(1 - \xi_{\text{V}})}. \tag{10.109}$$

[50] In the original formulation of the theory, Andrés Santos instead used the moments M_ℓ of the sphere size distribution defined as $\mathsf{M}_\ell = \sum_i (d_i^{\text{h}})^\ell \mathsf{x}_i$, which are proportional to $\mathring{\xi}_\nu$. We have $\mathring{\xi}_{\text{R}} = \mathsf{M}_1/2$, $\mathring{\xi}_{\text{A}} = \pi \mathsf{M}_2$, and $\mathring{\xi}_{\text{V}} = \pi \mathsf{M}_3/6$. His approach was originally formulated for polydispersive hard sphere mixtures with continuous size distributions with a *"truncatable" structure.* A model free energy for polydisperse systems is said to have such a structure when it depends on the size distribution only through a *finite* number of moments M_ν; in the present case of hard spheres M_1, M_2, and M_3.

It is straightforward to verify that $\Phi^{\mathrm{ex,b}}$ given by Equation 10.108 fulfills the differential equation 10.105 for *any* differentiable function $\mathsf{F}(\tau)$, but in the text box below it is shown that there does not exist any other possibility. Note that Rosenfeld's original version of FMT has the form 10.108 with $\mathsf{F}(\tau) = 1 + \tau/2$, so oFMT is a special case of the Santos approach. However, the wbFMT and HG-R approximations for $\Phi^{\mathrm{ex,b}}$ do not have this form, which is expected since they do not give a consistent pressure for hard sphere mixtures.

We will now show that Equation 10.108 constitutes the necessary form of the free energy expression consistent with requirements above. First, we note that when one adds point particles to a hard sphere fluid, the values of ξ_{R}, ξ_{A} and ξ_{V} are not changed, but $\xi_0 = \sum_{i>0} n_i^{\mathrm{b}}$ is changed to $\xi_0^{\mathsf{P}} \equiv \xi_0 + n_0^{\mathrm{b}}$, where n_0^{b} is the number density of point particles and superscript "P" indicates that such particles are present. It follows from the definitions of ξ_ν^* and ξ_ν° that

$$\xi_{\mathrm{A}}^* = \frac{\xi_{\mathrm{A}}}{(\xi_{\mathrm{R}})^2}\xi_0 \quad \text{and} \quad \xi_{\mathrm{V}}^* = \frac{\xi_{\mathrm{V}}}{(\xi_{\mathrm{R}})^3}\xi_0^2$$

and hence

$$\frac{\xi_{\mathrm{A}}^{*\mathsf{P}}}{\xi_{\mathrm{A}}^*} = \frac{\xi_0^{\mathsf{P}}}{\xi_0} \quad \text{and} \quad \frac{\xi_{\mathrm{V}}^{*\mathsf{P}}}{\xi_{\mathrm{V}}^*} = \left[\frac{\xi_0^{\mathsf{P}}}{\xi_0}\right]^2, \tag{10.110}$$

so ξ_{A}^* changes by a factor ξ_0^{P}/ξ_0 and ξ_{V}^* by a factor $(\xi_0^{\mathsf{P}}/\xi_0)^2$ when point particles are added.

Consider a bulk system of volume V that contains N hard spheres of various finite sizes and N_0 point particles. The point particles cannot penetrate the former, but since they do not interact with each other they can move freely in the voids between the hard spheres, that is, in a space of volume $V^{\mathrm{free}} = V - V_{\mathrm{tot}}^{\mathrm{hs}} = V(1 - \xi_{\mathrm{V}})$, where $V_{\mathrm{tot}}^{\mathrm{hs}}$ is the total volume of the N hard spheres and we have used the fact that $V_{\mathrm{tot}}^{\mathrm{hs}}/V = \eta_{\mathrm{tot}}^{\mathrm{hs}} = \xi_{\mathrm{V}}$. The excess Helmholtz free energy A_N^{ex} of the system in the *absence* of the point particles is given by (cf. Equation I:3.23)

$$\beta A_N^{\mathrm{ex}} = -\ln\frac{Z_N}{V^N} = -\ln\left[\frac{1}{V^N}\int d\mathbf{r}^N\, e^{-\beta\breve{U}_N^{\mathrm{pot}}(\mathbf{r}^N)}\right]$$

and in their *presence*

$$\begin{aligned}\beta A_{N+N_0}^{\mathrm{ex}} &= -\ln\left[\frac{1}{V^{N+N_0}}\int d\mathbf{r}^N \int d\mathbf{r}'^{N_0}\, e^{-\beta\breve{U}_{N+N_0}^{\mathrm{pot}}(\mathbf{r}^N,\mathbf{r}'^{N_0})}\right]\\ &= -\ln\left[\frac{1}{V^N}\int d\mathbf{r}^N e^{-\beta\breve{U}_N^{\mathrm{pot}}(\mathbf{r}^N)}\left(\frac{1}{V^{N_0}}\int d\mathbf{r}'^{N_0}\, e^{-\beta\Delta\breve{U}_{N_0}(\mathbf{r}'^{N_0}|\mathbf{r}^N)}\right)\right],\end{aligned}$$

where $\mathbf{r}'^{N_0}$ is the set of coordinates of the point particles and we have used the fact that $\breve{U}_{N+N_0}^{\mathrm{pot}}(\mathbf{r}^N,\mathbf{r}'^{N_0}) = \breve{U}_N^{\mathrm{pot}}(\mathbf{r}^N) + \Delta\breve{U}_{N_0}(\mathbf{r}'^{N_0}|\mathbf{r}^N)$, where the last term is the interaction potential between the point particles and the hard spheres when the coordinates of the latter are given by $\mathbf{r}^N$ (recall that the point particles do not interact with each other). $\Delta\breve{U}_{N_0}$ is zero when *all* point particles are located in the voids between the hard spheres and it is ∞ otherwise (i.e., when any point particle overlaps with a hard sphere). We therefore have

$$\int d\mathbf{r}'^{N_0}\, e^{-\beta\Delta\breve{U}_{N_0}(\mathbf{r}'^{N_0}|\mathbf{r}^N)} = (V^{\text{free}})^{N_0} = V^{N_0}(1-\xi_{\text{V}})^{N_0},$$

which is independent of $\mathbf{r}^N$. It follows that $\beta A^{\text{ex}}_{N+N_0} = \beta A^{\text{ex}}_N - N_0\ln(1-\xi_{\text{V}})$.

The excess free energy density in presence of the point particles, $\Phi^{\text{ex,b;P}}$, is hence equal to $\Phi^{\text{ex,b;P}} = \Phi^{\text{ex,b}} - n_0^{\text{b}}\ln(1-\xi_{\text{V}})$, which we can write as

$$\xi_0^{\text{P}}\left[\varphi^{\text{ex,b;P}} + \ln(1-\xi_{\text{V}})\right] = \xi_0\left[\varphi^{\text{ex,b}} + \ln(1-\xi_{\text{V}})\right] \tag{10.111}$$

since $\Phi^{\text{ex,b;P}} = \xi_0^{\text{P}}\varphi^{\text{ex,b;P}}$ and $\xi_0^{\text{P}} = \xi_0 + n_0^{\text{b}}$. This is an exact result. Using the approximate Ansatz 10.106, which implies that $\varphi^{\text{ex,b;P}} = \varphi^{\text{ex,b}}\left(\xi_{\text{V}}, \xi_{\text{A}}^{*\text{P}}, \xi_{\text{V}}^{*\text{P}}\right)$, and inserting $\xi_\nu^{*\text{P}}$ from Equation 10.110 we obtain

$$a\left[\varphi^{\text{ex,b}}\left(\xi_{\text{V}}, a\xi_{\text{A}}^*, a^2\xi_{\text{V}}^*\right) + \ln(1-\xi_{\text{V}})\right] = \varphi^{\text{ex,b}}\left(\xi_{\text{V}}, \xi_{\text{A}}^*, \xi_{\text{V}}^*\right) + \ln(1-\xi_{\text{V}}), \tag{10.112}$$

where $a = \xi_0^{\text{P}}/\xi_0$. This is a scaling relationship that the function $\varphi^{\text{ex,b}}(\cdot)$ must fulfill for all $a \geq 1$ because n_0^{b} can have any value ≥ 0. By selecting $a = C/\xi_{\text{A}}^*$, where C is any constant that makes $a \geq 1$, Equation 10.112 becomes

$$\frac{C}{\xi_{\text{A}}^*}\left[\varphi^{\text{ex,b}}\left(\xi_{\text{V}}, C, C^2\frac{\xi_{\text{V}}^*}{(\xi_{\text{A}}^*)^2}\right) + \ln(1-\xi_{\text{V}})\right] = \varphi^{\text{ex,b}}\left(\xi_{\text{V}}, \xi_{\text{A}}^*, \xi_{\text{V}}^*\right) + \ln(1-\xi_{\text{V}}). \tag{10.113}$$

Since

$$\frac{\xi_{\text{V}}^*}{(\xi_{\text{A}}^*)^2} = \xi_{\text{V}} \times \frac{\xi_{\text{R}}}{\xi_{\text{A}}^2} \quad \text{and} \quad \frac{1}{\xi_{\text{A}}^*} = \frac{\xi_{\text{R}}\xi_{\text{A}}}{\xi_0} \times \frac{\xi_{\text{R}}}{\xi_{\text{A}}^2},$$

the lhs of Equation 10.113 is equal to $\xi_{\text{R}}\xi_{\text{A}}/\xi_0$ times a function of the variables ξ_{V} and $\xi_{\text{R}}/\xi_{\text{A}}^2$. It is, in fact, more suitable to use the variable $\xi_{\text{A}}^2/\xi_{\text{R}}$ instead its inverse $\xi_{\text{R}}/\xi_{\text{A}}^2$, whereby Equation 10.113 can be written as

$$\frac{\xi_{\text{R}}\xi_{\text{A}}}{\xi_0}\mathcal{F}_1\left(\xi_{\text{V}}, \frac{\xi_{\text{A}}^2}{\xi_{\text{R}}}\right) = \varphi^{\text{ex,b}}\left(\xi_{\text{V}}, \xi_{\text{A}}^*, \xi_{\text{V}}^*\right) + \ln(1-\xi_{\text{V}}),$$

where the dimensionless function $\mathcal{F}_1(\cdot)$ is, as yet, undetermined. We can conclude using Equation 10.107 that

$$\Phi^{\text{ex,b}} = \xi_0\varphi^{\text{ex,b}} = -\xi_0\ln(1-\xi_{\text{V}}) + \xi_{\text{R}}\xi_{\text{A}}\mathcal{F}_1\left(\xi_{\text{V}}, \frac{\xi_{\text{A}}^2}{\xi_{\text{R}}}\right). \tag{10.114}$$

The result that $\Phi^{\text{ex,b}}$ depends in this fashion on the variables ξ_0, ξ_{R}, ξ_{A} and ξ_{V} is a consequence of the exact relationship 10.111 and the Santos Ansatz

10.106. Note that the Rosenfeld's FMT Ansatz 10.90 with $\mathsf{f}_0(x) = -\ln(1-x)$ would be obtained if

$$\mathcal{F}_1\left(\xi_V, \frac{\xi_A^2}{\xi_R}\right) = \mathsf{f}_1(\xi_V) + \mathsf{f}_2(\xi_V)\frac{\xi_A^2}{\xi_R} \quad \text{(Rosenfeld).} \tag{10.115}$$

Thus *Rosenfeld's Ansatz can be regarded as a special case* of Equation 10.114. It follows that the oFMT, wbFMT and HG-R approximations for $\Phi^{\text{ex,b}}$, which are all based on Rosenfeld's Ansatz, are given by Equation 10.114 with particular functions $\mathcal{F}_1(\cdot)$ that can be obtained by inserting their respective $\mathsf{f}_1(\xi_V)$ and $\mathsf{f}_2(\xi_V)$ into Equation 10.115.

We will now determine the necessary form of the function $\mathcal{F}_1(\cdot)$ so that $\Phi^{\text{ex,b}}$ gives a consistent pressure for hard sphere mixtures, which means that Equation 10.105 must be fulfilled. By inserting $\Phi^{\text{ex,b}}$ from Equation 10.114 into this equation we obtain after simplification the partial differential equation

$$\mathcal{F}_1(\xi_V, t) = (1-\xi_V)\frac{\partial \mathcal{F}_1(\xi_V, t)}{\partial \xi_V} - t\frac{\partial \mathcal{F}_1(\xi_V, t)}{\partial t}, \tag{10.116}$$

where $t = \xi_A^2/\xi_R$. In order to solve this equation, let us make the variable substitution $s = 1 - \xi_V$ and introduce the function $\mathcal{F}_2(s,t) \equiv \mathcal{F}_1(1-s,t) = \mathcal{F}_1(\xi_V, t)$. Then the differential equation is transformed into

$$\mathcal{F}_2(s,t) = -s\frac{\partial \mathcal{F}_2(s,t)}{\partial s} - t\frac{\partial \mathcal{F}_2(s,t)}{\partial t}. \tag{10.117}$$

We now write the unknown function $\mathcal{F}_2(s,t)$ as a two-dimensional Laurent series, which is a generalized Taylor expansion that involves both positive and negative powers of the variables s and t,

$$\mathcal{F}_2(s,t) = \sum_{\ell=-\infty}^{\infty}\sum_{\ell'=-\infty}^{\infty} a_{\ell\ell'} s^\ell t^{\ell'},$$

where the coefficients $a_{\ell\ell'}$ are, as yet, unknown. By inserting this into Equation 10.117 we obtain

$$\sum_{\ell=-\infty}^{\infty}\sum_{\ell'=-\infty}^{\infty} a_{\ell\ell'} s^\ell t^{\ell'} = -\sum_{\ell=-\infty}^{\infty}\sum_{\ell'=-\infty}^{\infty} (\ell+\ell') a_{\ell\ell'} s^\ell t^{\ell'},$$

which is satisfied by any Laurent series with coefficients that fulfill the condition $a_{\ell\ell'} = -(\ell+\ell')a_{\ell\ell'}$, so $a_{\ell\ell'} = 0$ unless $\ell + \ell' = -1$. For the nonzero coefficients we have $\ell = -(\ell'+1)$, whereby we can define $b_{\ell'} \equiv a_{-(\ell'+1),\ell'}$ and write[51]

$$\mathcal{F}_2(s,t) = \sum_{\ell'=-\infty}^{\infty} \frac{b_{\ell'}}{s}\left[\frac{t}{s}\right]^{\ell'} = \frac{1}{s}\mathsf{f}\left(\frac{t}{s}\right),$$

where $\mathfrak{f}(y) = \sum_{\ell'=-\infty}^{\infty} b_{\ell'} y^{\ell'}$ is an undetermined one-dimensional function. Thus, we conclude that $\mathcal{F}_1$ has the form

$$\mathcal{F}_1\left(\xi_V, \frac{\xi_A^2}{\xi_R}\right) = \frac{1}{1-\xi_V}\,\mathfrak{f}\left(\frac{\xi_A^2/\xi_R}{1-\xi_V}\right) \tag{10.118}$$

with some function $\mathfrak{f}(\cdot)$. Note that oFMT would be obtained if we set $\mathfrak{f}(y) = 1 + y/24\pi$ because we then have $\mathfrak{f}_1(\xi_V) = 1/(1-\xi_V)$ and $\mathfrak{f}_2(\xi_V) = 1/[24\pi(1-\xi_V)]$ in Equation 10.115 (cf. Equation 10.99). However, the wbFMT and HG-R approximations *do not* have $\mathcal{F}_1$ of the form 10.118 (cf. Equations 10.101 and 10.103, respectively).

Let us now consider a pure one-component fluid with density n^b and hard sphere packing fraction η^{hs}. It follows from Equation 10.85 that the argument of the function $\mathfrak{f}$ in Equation 10.118 in this case is

$$\frac{\xi_A^2/\xi_R}{1-\xi_V} = 12\pi\frac{\eta^{hs}}{1-\eta^{hs}} \quad \text{(one-component)}.$$

Therefore, it is more convenient to write Equation 10.118 in terms of a function $\mathsf{F}(\tau) = \mathfrak{f}(12\pi\tau)$, where

$$\tau = \frac{\xi_A^2/\xi_R}{12\pi(1-\xi_V)}, \tag{10.119}$$

so that the function argument is $\tau = \eta^{hs}/(1-\eta^{hs})$ in the one-component case. Thereby, instead of Equation 10.118 we have in the general multicomponent case

$$\mathcal{F}_1\left(\xi_V, \frac{\xi_A^2}{\xi_R}\right) = \frac{1}{1-\xi_V}\mathsf{F}(\tau) \quad \text{(Santos)}.$$

Of course, $\mathsf{F}(\cdot)$ is an undetermined function because $\mathfrak{f}(\cdot)$ is, as yet, undetermined. From these results and Equation 10.114 we obtain Equation 10.108.

From the results in the text box we can conclude that if $\Phi^{ex,b}$ is a function solely of the quantities ξ_0, ξ_R, ξ_A and ξ_V in accordance with the Ansatz 10.106–10.107, *a consistent excess free energy density for a bulk hard sphere mixture must have the form*

$$\Phi^{ex,b} = -\xi_0 \ln(1-\xi_V) + \frac{\xi_R\xi_A}{1-\xi_V}\mathsf{F}(\tau) \quad \text{(Santos)}, \tag{10.120}$$

(Equation 10.108) with an as yet undermined function $\mathsf{F}(\tau)$ and with τ from Equation 10.109, which is the same as Equation 10.119.

[51] Equivalently, we can have $\mathcal{F}_2(s,t) = f(s/t)/t$, where the undetermined function is $f(x) = \mathfrak{f}(1/x)/x$.

The function $\mathsf{F}(\tau)$ is the same irrespectively of the composition of the hard sphere fluid. If we apply Equation 10.120 to the pure one-component case and use Equation 10.85, we obtain

$$\varphi_{\mathrm{o}}^{\mathrm{ex,b}}\left(\eta^{\mathrm{hs}}\right)=\frac{\Phi_{\mathrm{o}}^{\mathrm{ex,b}}}{n^{\mathrm{b}}}=-\ln\left(1-\eta^{\mathrm{hs}}\right)+\frac{3\eta^{\mathrm{hs}}}{1-\eta^{\mathrm{hs}}}\,\mathsf{F}\left(\eta^{\mathrm{hs}}/(1-\eta^{\mathrm{hs}})\right),$$

where subscript o (= one) indicates that $\varphi_{\mathrm{o}}^{\mathrm{ex,b}}$ and $\Phi_{\mathrm{o}}^{\mathrm{ex,b}}$ are excess free energy quantities for the one-component case. Since $\tau=\eta^{\mathrm{hs}}/(1-\eta^{\mathrm{hs}})$ implies that $\eta^{\mathrm{hs}}=\tau/(1+\tau)$ and $1-\eta^{\mathrm{hs}}=1/(1+\tau)$, we obtain from this relationship

$$\mathsf{F}(\tau)=\frac{1}{3\tau}\left[\varphi_{\mathrm{o}}^{\mathrm{ex,b}}\left(\tau/(1+\tau)\right)-\ln\left(1+\tau\right)\right]\qquad\text{(Santos).}\qquad(10.121)$$

A very important consequence of this relationship is that if $\varphi_{\mathrm{o}}^{\mathrm{ex,b}}\left(\eta^{\mathrm{hs}}\right)$ for the pure one-component fluid is given, the function $\mathsf{F}(\tau)$ *for any hard sphere mixture is known* when the Ansatz 10.106 is used as an approximation. This is a great advantage of the Santos approach.

Since $\varphi_{\mathrm{o}}^{\mathrm{ex,b}}=\beta\mu^{\mathrm{ex}}+1-\beta P/n^{\mathrm{b}}$ for the one-component fluid, we can use the expansions 10.9 and 10.16 for $\beta P/n^{\mathrm{b}}$ and $\beta\mu^{\mathrm{ex}}$ to obtain

$$\varphi_{\mathrm{o}}^{\mathrm{ex,b}}\left(\eta^{\mathrm{hs}}\right)=4\eta^{\mathrm{hs}}+5[\eta^{\mathrm{hs}}]^2+\mathcal{O}\left([\eta^{\mathrm{hs}}]^3\right),$$

where the exact values of the second and third virial coefficients have been used. By inserting this with $\eta^{\mathrm{hs}}=\tau/(1+\tau)$ in Equation 10.121 we get $\mathsf{F}(\tau)=1+\tau/2+\mathcal{O}\left(\tau^2\right)$ for small τ. Therefore, we can always write

$$\mathsf{F}(\tau)=1+\frac{\tau}{2}\left[1+\mathsf{f}(\tau)\right],\qquad(10.122)$$

where $\mathsf{f}(\tau)\sim\mathsf{a}\tau$ when $\tau\to 0$ for some constant a. If we set $\mathsf{f}(\tau)=0$, which means all nonlinear contributions are neglected in $\mathsf{F}(\tau)$, we get Rosenfeld's original version of FMT, where $\mathsf{F}(\tau)=1+\tau/2$ as we have seen. From Equation 10.120 we obtain by inserting Equation 10.122 and the definition of τ

$$\begin{aligned}\Phi^{\mathrm{ex,b}} &= -\xi_0\ln\left(1-\xi_{\mathrm{V}}\right)+\frac{\xi_{\mathrm{R}}\xi_{\mathrm{A}}}{1-\xi_{\mathrm{V}}}\\ &+ \frac{\xi_{\mathrm{A}}^3}{24\pi(1-\xi_{\mathrm{V}})^2}\left[1+\mathsf{f}\left(\frac{\xi_{\mathrm{A}}^2/\xi_{\mathrm{R}}}{12\pi(1-\xi_{\mathrm{V}})}\right)\right]\qquad\text{(Santos),}\qquad(10.123)\end{aligned}$$

which is the most general form of $\Phi^{\mathrm{ex,b}}$ for the multicomponent fluid. The last term has a different structure than the FMT Ansatz 10.90 due to the presence of $\mathsf{f}(\tau)$, which depends on $\xi_{\mathrm{A}}^2/\xi_{\mathrm{R}}$ and ξ_{V}.

If we, for example, use the Carnahan-Starling equation of state 10.12 and the corresponding equation 10.14 for $\beta\mu^{\mathrm{ex}}$ to obtain $\varphi_{\mathrm{o}}^{\mathrm{ex,b}}=\beta\mu^{\mathrm{ex}}+1-\beta P/n^{\mathrm{b}}$, we get

$$\varphi_{\mathrm{o}}^{\mathrm{ex,b}}\left(\eta^{\mathrm{hs}}\right)=\frac{4\eta^{\mathrm{hs}}-3(\eta^{\mathrm{hs}})^2}{(1-\eta^{\mathrm{hs}})^2}\qquad\text{(C-S),}\qquad(10.124)$$

whereby Equation 10.121 becomes

$$\mathsf{F}(\tau) = \frac{1}{3}\left[4 + \tau - \frac{1}{\tau}\ln(1+\tau)\right] \qquad \text{(C-S).} \qquad (10.125)$$

From Equation 10.120 we obtain for the general multicomponent case

$$\begin{aligned} \Phi^{\mathrm{ex,b}} &= -\ln(1-\xi_{\mathrm{V}})\,\xi_0 + \frac{4\xi_{\mathrm{R}}\xi_{\mathrm{A}}}{3(1-\xi_{\mathrm{V}})} + \frac{\xi_{\mathrm{A}}^3}{36\pi(1-\xi_{\mathrm{V}})^2} \\ &- \frac{4\pi\,\xi_{\mathrm{R}}^2}{\xi_{\mathrm{A}}}\ln\left(1 + \frac{\xi_{\mathrm{A}}^2/\xi_{\mathrm{R}}}{12\pi(1-\xi_{\mathrm{V}})}\right) \qquad \text{(C-S + Santos)} \qquad (10.126) \end{aligned}$$

where the last term contains the contributions that are nonlinear in $\xi_{\mathrm{A}}^2/\xi_{\mathrm{R}}$.

The pressure can be obtained from either Equations 10.92 or 10.94 and we get in the general case

$$\beta P = \frac{\xi_0}{1-\xi_{\mathrm{V}}} + \frac{\xi_{\mathrm{R}}\xi_{\mathrm{A}}}{(1-\xi_{\mathrm{V}})^2}\left[\mathsf{F}(\tau) + \tau\mathsf{F}'(\tau)\right] \qquad \text{(Santos),} \qquad (10.127)$$

where $\mathsf{F}'(\tau) = d\mathsf{F}(\tau)/d\tau$. If we insert Equation 10.125, we obtain

$$\beta P = \frac{\xi_0}{1-\xi_{\mathrm{V}}} + \frac{\xi_{\mathrm{R}}\xi_{\mathrm{A}}}{(1-\xi_{\mathrm{V}})^2}\cdot\frac{3+6\tau+2\tau^2}{3(1+\tau)} \qquad \text{(C-S + Santos),} \qquad (10.128)$$

which yields

$$\begin{aligned} \beta P &= \frac{\xi_0}{1-\xi_{\mathrm{V}}} + \frac{4\xi_{\mathrm{R}}\xi_{\mathrm{A}}}{3(1-\xi_{\mathrm{V}})^2} + \frac{\xi_{\mathrm{A}}^3}{18\pi(1-\xi_{\mathrm{V}})^3} \\ &- \frac{4\pi\,\xi_{\mathrm{R}}^2\xi_{\mathrm{A}}}{(1-\xi_{\mathrm{V}})[\xi_{\mathrm{A}}^2 + 12\pi\xi_{\mathrm{R}}(1-\xi_{\mathrm{V}})]} \qquad \text{(C-S + Santos).} \qquad (10.129) \end{aligned}$$

This expression is considerably more accurate than the BMCSL equation of state 10.87. For cases that have been tested,[52] Equation 10.129 has a similar accuracy for high densities as the HG-R approximation, Equation 10.88, which is also based on the C-S approximation, but for low densities the HG-R predictions are better. However, in Equation 10.121 one can use more accurate free energy expressions for $\varphi_{\mathrm{o}}^{\mathrm{ex,b}}(\eta^{\mathrm{hs}})$ in the one-component case than that from the C-S approximation to build an even better theory.

In the general case, the Santos expression 10.120 for $\Phi^{\mathrm{ex,b}}$ gives the excess chemical potential μ_i^{ex} as

$$\begin{aligned} \beta\mu_i^{\mathrm{ex}} &= \frac{\partial\Phi^{\mathrm{ex,b}}}{\partial n_i^{\mathrm{b}}} = -\ln(1-\xi_{\mathrm{V}}) + \frac{\xi_{\mathrm{A}}}{1-\xi_{\mathrm{V}}}\left[\mathsf{F}(\tau) - \tau\mathsf{F}'(\tau)\right]\mathsf{R}_i^{\mathrm{h}} \\ &+ \frac{\xi_{\mathrm{R}}}{1-\xi_{\mathrm{V}}}\left[\mathsf{F}(\tau) + 2\tau\mathsf{F}'(\tau)\right]\mathsf{A}_i^{\mathrm{h}} + \beta P\mathsf{V}_i^{\mathrm{h}} \qquad \text{(Santos),} \end{aligned}$$

[52]See 2nd reference in footnote 49 and reference in footnote 47.

where we have used Equations 10.91 and 10.94. When this is applied to the C-S case with $\mathsf{F}(\tau)$ from Equation 10.125, the resulting $\beta\mu_i^{\text{ex}}$ is better than or practically equal to that obtained in the HG-R approximation (with $\Phi^{\text{ex,b}}$ from Equation 10.104) for the cases investigated.[53]

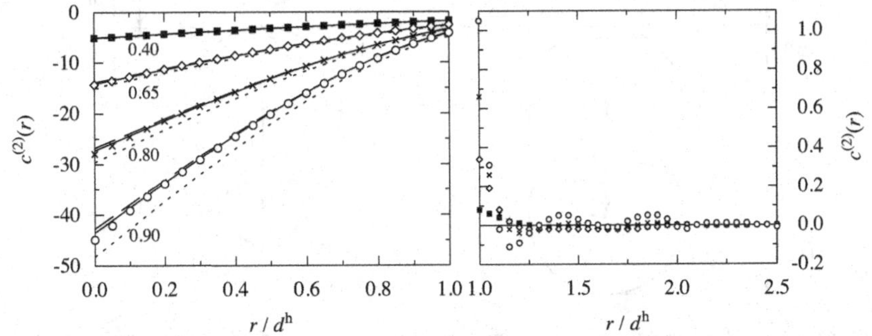

Figure 10.4 The pair direct-correlation function $c^{(2)}(r)$ for a one-component hard sphere fluid in bulk at various reduced densities $n^{\text{b}}(d^{\text{h}})^3$ plotted as functions of r/d^{h}: from top to bottom $n^{\text{b}}(d^{\text{h}})^3 = 0.40$, 0.65, 0.80 and 0.90, which correspond to hard sphere packing fractions $\eta^{\text{hs}} = 0.209$, 0.340, 0.419 and 0.471, respectively. The symbols show the results from MC simulations[54] for $0 \le r/d^{\text{h}} \le 1$ in the left frame and for $r/d^{\text{h}} \ge 1$ in the right frame. Note the large difference in ordinate scale for the two frames. At $r/d^{\text{h}} = 1$ the function has a jump that is equal to the contact value $g^{(2)}(d^{\text{h}+})$ since $h^{(2)}(r) - c^{(2)}(r)$ is continuous for $r = d^{\text{h}}$. In the left frame the results from the PY approximation are shown as short dashes, from the wbFMT approximation as long dashes and from the approximation "Santos-v + CS" as full curves (Equation 10.182).

10.3.1.3 *Pair direct-correlation functions in bulk*

Let us now turn to correlation functions for hard sphere fluids. For the one-component case, $c^{(2)}(r)$ is in the PY approximation explicitly given by Equation 10.2 and it is plotted as short dashes in Figure 10.4, where it is compared with results from MC simulations[54] and some more accurate approximations (the latter results will be discussed later). We see that $c^{(2)}(r)$ is a negative, monotonically increasing function for $r < d^{\text{h}}$. For $r > d^{\text{h}}$ the MC results show that $c^{(2)}(r)$ is oscillatory, but much smaller in magnitude than for $r < d^{\text{h}}$. In the approximations considered here, $c^{(2)}(r)$ is zero for $r > d^{\text{h}}$, but this is not unreasonable because the correct $c^{(2)}(r)$ is small there.

We see that the PY approximation is not very accurate for large densities, which is in agreement with what we have found earlier. In the figure,

---

[53] See 2nd reference in footnote 49.

[54] R. D. Groot, J. P. van der Eerden and N. M. Faber, J. Chem. Phys. **87** (1987) 2263 (https://doi.org/10.1063/1.453155).

the maximal density is $n^{\mathrm{b}} = 0.9\,(d^{\mathrm{h}})^{-3}$, which corresponds to a hard sphere packing fraction η^{hs} =0.471. For a slightly higher density where $\eta^{\mathrm{hs}} = 0.491$, the fluid has a phase transition to the solid state.

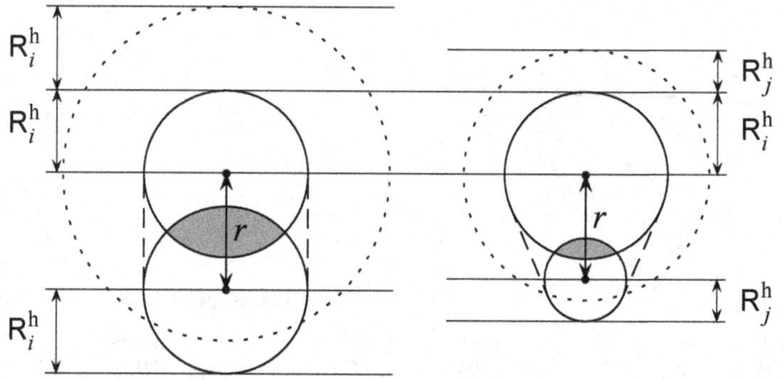

Figure 10.5 Two overlapping spheres of equal (left) and different (right) radii with separation r between their centers. The large dashed circle shows the exclusion sphere of radius $d_{ij}^{\mathrm{h}} = \mathsf{R}_i^{\mathrm{h}} + \mathsf{R}_j^{\mathrm{h}}$ around the upper sphere. The gray area shows the overlap region of the spheres, which has volume $\Delta\mathsf{V}_{ij}(r)$ (the overlap volume) and area $\Delta\mathsf{A}_{ij}(r)$ (the overlap area), i.e., the total area of its top and bottom curved surfaces. The convex envelope of the two spheres, which is used in the calculation of $\overline{\mathsf{R}_{ij}}(r)$ and $\Delta\mathsf{R}_{ij}(r)$, consists of the surface formed when the sphere surfaces are linked with the surface indicated by the long-dashed lines in the figure (the lines are common tangents to the circles). It is the surface of a spherocylinder when the sphere radii are equal and for different radii it is the surface of a conical frustum connected with spherical caps at the top and bottom (sometimes called a spherocone).

For the one-component hard-sphere fluid, $c^{(2)}(r)$ is a polynomial for $r \le d^{\mathrm{h}}$. The coefficients of the polynomial can be written in terms of the hard sphere packing fraction η^{hs} alone, Equation 10.3. For mixtures, $c_{ij}^{(2)}(r)$ is, as mentioned earlier, instead a rational function for $r < d_{ij}^{\mathrm{h}}$ and zero for $r > d_{ij}^{\mathrm{h}}$. The coefficients of this function are a bit more complicated and can, as we will see, be expressed in terms of ξ_0, ξ_{R}, ξ_{A} and ξ_{V}. A very useful result[55] is that the expression for $c_{ij}^{(2)}(r)$ in the Percus-Yevick approximation can be written as a sum of geometrical features of two overlapping spheres of radii $\mathsf{R}_i^{\mathrm{h}}$ and $\mathsf{R}_j^{\mathrm{h}}$ as functions of the separation r between their centers

$$-c_{ij}^{(2)}(r) = a_0\mathcal{H}(d_{ij}^{\mathrm{h}} - r) + a_{\mathrm{R}}\Delta\mathsf{R}_{ij}(r) + a_{\mathrm{A}}\Delta\mathsf{A}_{ij}(r) + a_{\mathrm{V}}\Delta\mathsf{V}_{ij}(r) \quad \text{(PY)}, \tag{10.130}$$

where a_ν, $\nu \in \{0, \mathrm{R}, \mathrm{A}, \mathrm{V}\}$, are density-dependent coefficients, $\Delta\mathsf{R}_{ij}(r)$ is the so-called "overlap mean radius" of the two spheres (see below), $\Delta\mathsf{A}_{ij}(r)$ is their overlap area and $\Delta\mathsf{V}_{ij}(r)$ is their overlap volume (see Figure 10.5). We

[55]Y. Rosenfeld, J. Chem. Phys. 89 (1988) 4272 (https://doi.org/10.1063/1.454810).

have

$$
\begin{aligned}
\mathcal{H}(d_{ij}^{\mathrm{h}}-r) &= \begin{cases} 1, & r < d_{ij}^{\mathrm{h}} \\ 0 & r > d_{ij}^{\mathrm{h}} \end{cases} \\
\Delta \mathsf{R}_{ij}(r) &= \left[\mathsf{R}_i^{\mathrm{h}} + \mathsf{R}_i^{\mathrm{h}} - \overline{\mathsf{R}_{ij}}(r)\right] \mathcal{H}(d_{ij}^{\mathrm{h}}-r) \\
\Delta \mathsf{A}_{ij}(r) &= \mathsf{A}_i^{\mathrm{h}} + \mathsf{A}_i^{\mathrm{h}} - \mathsf{A}_{(i)\cup(j)}^{\mathrm{h}}(r) \\
\Delta \mathsf{V}_{ij}(r) &= \mathsf{V}_i^{\mathrm{h}} + \mathsf{V}_i^{\mathrm{h}} - \mathsf{V}_{(i)\cup(j)}^{\mathrm{h}}(r),
\end{aligned}
\qquad (10.131)
$$

where $\overline{\mathsf{R}_{ij}}(r)$ is the mean radius of the convex envelope of the two spheres,[56] $\mathsf{A}_{(i)\cup(j)}^{\mathrm{h}}(r)$ is the area and $\mathsf{V}_{(i)\cup(j)}^{\mathrm{h}}(r)$ is the volume of the union of the two spheres. $\mathcal{H}(d_{ij}^{\mathrm{h}}-r)$ is equal to one when the two spheres overlap and zero otherwise. Note that all functions in Equation 10.131 are zero when $r > d_{ij}^{\mathrm{h}}$, that is, outside the exclusion sphere for i-particles around a j-particle and vice versa.

As we will see later (after Equation 10.163), the coefficients in Equation 10.130 can be obtained from

$$
a_\nu = \frac{\partial^2 \Phi^{\mathrm{ex,b}}}{\partial \xi_\nu \partial \xi_{\mathrm{V}}} \quad \text{for } \nu \in \{0, \mathrm{R}, \mathrm{A}, \mathrm{V}\} \quad (\mathrm{PY}) \qquad (10.132)
$$

with $\Phi^{\mathrm{ex,b}}$ from Equation 10.98, which yields

$$
\begin{aligned}
a_0 &= \frac{1}{1-\xi_{\mathrm{V}}}, \qquad a_{\mathrm{R}} = \frac{\xi_{\mathrm{A}}}{(1-\xi_{\mathrm{V}})^2} \\
a_{\mathrm{A}} &= \frac{\xi_{\mathrm{R}}}{(1-\xi_{\mathrm{V}})^2} + \frac{\xi_{\mathrm{A}}^2}{4\pi(1-\xi_{\mathrm{V}})^3} \qquad (\mathrm{PY}) \\
a_{\mathrm{V}} &= \frac{\xi_0}{(1-\xi_{\mathrm{V}})^2} + \frac{2\xi_{\mathrm{R}}\,\xi_{\mathrm{A}}}{(1-\xi_{\mathrm{V}})^3} + \frac{\xi_{\mathrm{A}}^3}{4\pi(1-\xi_{\mathrm{V}})^4}.
\end{aligned}
\qquad (10.133)
$$

For general i and j, the purely geometrical entities $\Delta\mathsf{R}_{ij}$, $\Delta\mathsf{A}_{ij}$ and $\Delta\mathsf{V}_{ij}$ are

---

[56] The mean radius $\overline{\mathsf{R}_{ij}}(r)$ is a property of the surface formed by the convex envelope of the two overlapping spheres of species i and j, see Figure 10.5, whereby r is the separation between the sphere centers. $\overline{\mathsf{R}_{ij}}$ is defined as

$$
\overline{\mathsf{R}_{ij}} = \frac{1}{4\pi} \int_{\substack{\text{envelope}\\ \text{surface}}} d\mathsf{A}_{ij}\, \frac{1}{2}\left[\frac{1}{\mathcal{R}_{\max}} + \frac{1}{\mathcal{R}_{\min}}\right],
$$

where the integral is the surface integral of the mean curvature $[1/\mathcal{R}_{\max}+1/\mathcal{R}_{\min}]/2$, where $\mathcal{R}_{\max}$ and $\mathcal{R}_{\min}$ are the principal radii of curvature at each point of the surface.

For example, for two spheres of equal radii ($i = j$) illustrated in the left part of Figure 10.5, one has $\overline{\mathsf{R}_{ii}}(r) = \mathsf{R}_i^{\mathrm{h}} + r/4$, where the first term, $\mathsf{R}_i^{\mathrm{h}}$, originates from the integral over the two spherical caps (half spheres in this case), where $\mathcal{R}_{\max} = \mathcal{R}_{\min} = \mathsf{R}_i^{\mathrm{h}}$, and the second term, $r/4$, originates from the integral over the cylindrical portion of length r, where $\mathcal{R}_{\max} = \mathsf{R}_i^{\mathrm{h}}$ and $\mathcal{R}_{\min} = 0$. Thus one has $\Delta\mathsf{R}_{ii}(r) = \mathsf{R}_i^{\mathrm{h}} - r/4$ in this case.

simple rational functions of r and we give the formulas for them without any proof. For $|\mathsf{R}_i^{\mathrm{h}} - \mathsf{R}_j^{\mathrm{h}}| \le r \le d_{ij}^{\mathrm{h}}$ we have

$$\begin{aligned} \Delta\mathsf{R}_{ij}(r) &= \frac{2d_{ij}^{\mathrm{h}}r - r^2 - \left(\mathsf{R}_i^{\mathrm{h}} - \mathsf{R}_j^{\mathrm{h}}\right)^2}{4r} \\ \Delta\mathsf{A}_{ij}(r) &= \frac{\pi\left(d_{ij}^{\mathrm{h}} - r\right)\left(d_{ij}^{\mathrm{h}}r - \left(\mathsf{R}_i^{\mathrm{h}} - \mathsf{R}_j^{\mathrm{h}}\right)^2\right)}{r} \\ \Delta\mathsf{V}_{ij}(r) &= \frac{\pi\left(d_{ij}^{\mathrm{h}} - r\right)^2\left(2d_{ij}^{\mathrm{h}}r + r^2 - 3\left(\mathsf{R}_i^{\mathrm{h}} - \mathsf{R}_j^{\mathrm{h}}\right)^2\right)}{12r} \end{aligned} \qquad (10.134)$$

and for $0 < r < |\mathsf{R}_i^{\mathrm{h}} - \mathsf{R}_j^{\mathrm{h}}|$ the values of these functions are the smallest of $\{\mathsf{R}_i^{\mathrm{h}}, \mathsf{R}_j^{\mathrm{h}}\}$, $\{\mathsf{A}_i^{\mathrm{h}}, \mathsf{A}_j^{\mathrm{h}}\}$ and $\{\mathsf{V}_i^{\mathrm{h}}, \mathsf{V}_j^{\mathrm{h}}\}$, respectively. Furthermore, we have the relationship

$$\Delta\mathsf{R}_{ij}(r) = \frac{\Delta\mathsf{A}_{ij}(r)}{4\pi[\mathsf{R}_i^{\mathrm{h}} + \mathsf{R}_j^{\mathrm{h}}]} + \frac{\mathsf{R}_i^{\mathrm{h}}\mathsf{R}_j^{\mathrm{h}}}{\mathsf{R}_i^{\mathrm{h}} + \mathsf{R}_j^{\mathrm{h}}}\mathcal{H}(d_{ij}^{\mathrm{h}} - r). \qquad (10.135)$$

The verification of it is left as an exercise.

{**Exercise 10.9:** Use Equation 10.134 to verify Equation 10.135 for $|\mathsf{R}_i^{\mathrm{h}} - \mathsf{R}_j^{\mathrm{h}}| \le r \le d_{ij}^{\mathrm{h}}$ and verify it also for $0 < r < |\mathsf{R}_i^{\mathrm{h}} - \mathsf{R}_j^{\mathrm{h}}|$.}

We saw earlier that the PY pressure for a hard sphere mixture, Equations 10.83 and 10.84, can be entirely expressed in terms of geometrical properties of the individual spheres, $\mathsf{R}_i^{\mathrm{h}}$, $\mathsf{A}_i^{\mathrm{h}}$ and $\mathsf{V}_i^{\mathrm{h}}$, via the entities ξ_0, ξ_{R}, ξ_{A} and ξ_{V} defined in Equation 10.82. As regards $c_{ij}^{(2)}(r)$, Equation 10.130 is written in terms of functions associated with pairs of spheres rather than those for individual spheres. However, one can easily write the overlap entities $\Delta\mathsf{A}_{ij}(r)$ and $\Delta\mathsf{V}_{ij}(r)$ in terms of functions that solely involve individual spheres. The same is true for $\mathcal{H}(d_{ij}^{\mathrm{h}} - r)$ and $\Delta\mathsf{R}_{ij}(r)$, but this requires a bit more effort, so we first deal with the former two.

We start with the overlap volume of two spheres of radii $\mathsf{R}_i^{\mathrm{h}}$ and $\mathsf{R}_j^{\mathrm{h}}$ placed with their centers at $\mathbf{r}_1$ and $\mathbf{r}_2$, respectively, and $\mathbf{r}_3$ is an arbitrary point. Let us consider the product $\mathcal{H}(\mathsf{R}_i^{\mathrm{h}} - r_{13})\mathcal{H}(\mathsf{R}_j^{\mathrm{h}} - r_{23})$ when we vary $\mathbf{r}_3$. The first function is equal to one when $\mathbf{r}_3$ is inside the i-sphere and zero outside and the corresponding is true for the second function. Therefore, the product is equal to one inside the region where the spheres overlap and zero otherwise, so we have

$$\Delta\mathsf{V}_{ij}(r_{12}) = \int d\mathbf{r}_3\mathcal{H}(\mathsf{R}_i^{\mathrm{h}} - r_{13})\mathcal{H}(\mathsf{R}_j^{\mathrm{h}} - r_{23}). \qquad (10.136)$$

Next, we treat the surface area of the overlap region and we will show that

$$\Delta\mathsf{A}_{ij}(r_{12}) = \int d\mathbf{r}_3\delta(\mathsf{R}_i^{\mathrm{h}} - r_{13})\mathcal{H}(\mathsf{R}_j^{\mathrm{h}} - r_{23}) + \int d\mathbf{r}_3\mathcal{H}(\mathsf{R}_i^{\mathrm{h}} - r_{13})\delta(\mathsf{R}_j^{\mathrm{h}} - r_{23}). \qquad (10.137)$$

In order to understand this formula we start with the part the i-sphere's surface that is inside the j-sphere. For this purpose we consider the delta function $\delta(\mathsf{R}_i^{\mathrm{h}} - r_{13})$, which is zero outside the surface of the i-sphere and has a "delta-function peak" at the surface, $r_{13} = \mathsf{R}_i^{\mathrm{h}}$. This peak has the radial integral equal to one. The product $\delta(\mathsf{R}_i^{\mathrm{h}} - r_{13})\mathcal{H}(\mathsf{R}_j^{\mathrm{h}} - r_{23})$ is nonzero only when $\mathbf{r}_3$ lies at the part of the i-sphere's surface that is inside the j-sphere. The integral of this product over $\mathbf{r}_3$ gives the area of this part inside the j-sphere. The corresponding is true when we switch $i \leftrightarrow j$, so the total area of the two curved surfaces of the overlap region is given by Equation 10.137.

For both $\Delta\mathsf{V}_{ij}$ and $\Delta\mathsf{A}_{ij}$, each factor of the integrands deals only with a single sphere of species i or j. In order to also write $\mathcal{H}(d_{ij}^{\mathrm{h}} - r_{12})$ in terms of functions that solely involve individual spheres, we will start with case $i = j$, so we are dealing with two spheres with radius $\mathsf{R}_i^{\mathrm{h}}$, and then generalize the result to $i \neq j$. First, we will show that

$$\begin{aligned}\mathcal{H}(d_{ii}^{\mathrm{h}} - r_{12}) &= \frac{1}{2\pi(\mathsf{R}_i^{\mathrm{h}})^2}\int d\mathbf{r}_3\delta(\mathsf{R}_i^{\mathrm{h}} - r_{13})\mathcal{H}(\mathsf{R}_i^{\mathrm{h}} - r_{23}) \\ &+ \frac{1}{2\pi\mathsf{R}_i^{\mathrm{h}}}\int d\mathbf{r}_3\delta(\mathsf{R}_i^{\mathrm{h}} - r_{13})\delta(\mathsf{R}_i^{\mathrm{h}} - r_{23})\left[1 - \frac{\mathbf{r}_{13}\cdot\mathbf{r}_{23}}{r_{13}r_{23}}\right].\end{aligned} \quad (10.138)$$

This is done quite easily as shown in the following shaded text box.

The rhs of Equation 10.138 is equal to zero when $r_{12} > d_{ii}^{\mathrm{h}}$ because the integrands are zero there and we need to show that it is equal to one when $0 < r_{12} \leq d_{ii}^{\mathrm{h}}$. The first term has an integral of same form as those in Equation 10.137 and using Equation 10.134 we can easily see that its value for $0 < r_{12} \leq d_{ii}^{\mathrm{h}}$ when $j = i$ is equal to $\pi\left(d_{ii}^{\mathrm{h}} - r_{12}\right)d_{ii}^{\mathrm{h}}/2$ (the factor 1/2 arises because there are two equal terms in $\Delta\mathsf{A}_{ii}$). Since $d_{ii}^{\mathrm{h}} = 2\mathsf{R}_i^{\mathrm{h}}$, the entire first term is $1 - r_{12}/[2\mathsf{R}_i^{\mathrm{h}}]$ and it remains to show that the second term has the value $r_{12}/[2\mathsf{R}_i^{\mathrm{h}}]$ when $0 < r_{12} \leq d_{ii}^{\mathrm{h}}$.

The square bracket in the second term equals $1 - \hat{\mathbf{r}}_{13}\cdot\hat{\mathbf{r}}_{23} = 1 - \cos(\alpha)$, where α is the angle between the vectors $\mathbf{r}_{13}$ and $\mathbf{r}_{23}$. Due to the two delta functions in the integrand, the only contribution from the integral arises when $r_{13} = r_{23} = \mathsf{R}_i^{\mathrm{h}}$. From the law of cosine in geometry it follows that $r_{12}^2 = r_{13}^2 + r_{23}^2 - 2r_{13}r_{23}\cos\alpha$, see Figure 10.6, and since $r_{13} = r_{23} = \mathsf{R}_i^{\mathrm{h}}$ we conclude that $r_{12}^2/[2(\mathsf{R}_i^{\mathrm{h}})^2] = 1 - \cos(\alpha)$. Thus the square bracket can be replaced by $r_{12}^2/[2(\mathsf{R}_i^{\mathrm{h}})^2]$, which is independent of $\mathbf{r}_3$ and can be moved outside the integral. It remains to evaluate the integral of the two delta functions. In order to do this, we note that for any two functions $F_1(r)$ and $F_2(r)$, we have[57]

$$\int d\mathbf{r}_3 F_1(r_{13})F_2(r_{23}) = \frac{2\pi}{r_{12}}\int_0^{\infty} dr_{13}r_{13}F_1(r_{13})\int_{|r_{12}-r_{13}|}^{r_{12}+r_{13}} dr_{23}r_{23}F_2(r_{23}). \quad (10.139)$$

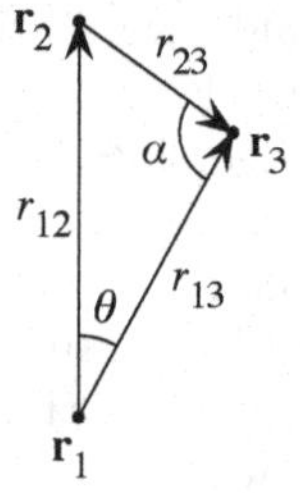

Figure 10.6 Illustration of two coordinate points $\mathbf{r}_1$ and $\mathbf{r}_2$ separated by distance r_{12} and a third point $\mathbf{r}_3$ that serves as integration variable in, for example, Equation 10.138. The law of cosine in geometry says that $r_{12}^2 = r_{13}^2 + r_{23}^2 - 2r_{13}r_{23}\cos\alpha$ and $r_{23}^2 = r_{12}^2 + r_{13}^2 - 2r_{12}r_{13}\cos\theta$.

We apply this formula to $F_1(r_{13}) = \delta(\mathsf{R}_i^{\mathrm{h}} - r_{13})$ and $F_2(r_{23}) = \delta(\mathsf{R}_i^{\mathrm{h}} - r_{23})$ and since both integrals on the rhs then equal $\mathsf{R}_i^{\mathrm{h}}$ we obtain

$$\int d\mathbf{r}_3 \delta(\mathsf{R}_i^{\mathrm{h}} - r_{13})\delta(\mathsf{R}_i^{\mathrm{h}} - r_{23}) = \frac{2\pi}{r_{12}}\mathsf{R}_i^{\mathrm{h}}\mathsf{R}_i^{\mathrm{h}}$$

when $0 < r_{12} \leq d_{ii}^{\mathrm{h}}$. Gathering these results we can conclude that the second term in Equation 10.138 is indeed equal to $r_{12}/[2\mathsf{R}_i^{\mathrm{h}}]$ for the same r_{12} values, which finalizes the proof of Equation 10.138.

Before we generalize Equation 10.138 to $i \neq j$, we will introduce another notation for $\mathcal{H}(\mathsf{R}_i^{\mathrm{h}} - r)$, $\delta(\mathsf{R}_i^{\mathrm{h}} - r)$ and the variants thereof that appear in the rhs of that equation. As we will see, this notation makes the generalization to $i \neq j$ easier as it makes clear that the rhs of Equation 10.138 only contains functions that solely involve individual spheres. We define[58]

$$\omega_i^{\mathrm{V}}(r) = \mathcal{H}(\mathsf{R}_i^{\mathrm{h}} - r), \quad \omega_i^{\mathrm{A}}(r) = \delta(\mathsf{R}_i^{\mathrm{h}} - r),$$
$$\omega_i^{\mathrm{R}}(r) = \frac{\delta(\mathsf{R}_i^{\mathrm{h}} - r)}{4\pi\mathsf{R}_i^{\mathrm{h}}}, \quad \omega_i^{0}(r) = \frac{\delta(\mathsf{R}_i^{\mathrm{h}} - r)}{4\pi(\mathsf{R}_i^{\mathrm{h}})^2}. \tag{10.140}$$

[57] If we place the origin at r_1, the integral in the lhs of Equation 10.139 can be written in spherical polar coordinates as

$$\int d\mathbf{r}_{13} F_1(r_{13})F_2(r_{23}) = 2\pi \int_0^\infty dr_{13}\, r_{13}^2 F_1(r_{13}) \int_0^\pi d\theta \sin\theta\, F_2(r_{23}),$$

where θ is the angle between $\mathbf{r}_{12}$ and $\mathbf{r}_{13}$. By making the variable substitution from θ to r_{23} at constant r_{13} via the relationship $r_{23}^2 = r_{12}^2 + r_{13}^2 - 2r_{12}r_{13}\cos\theta$, which follows from the law of cosine (see Figure 10.6) and yields $2r_{23}dr_{23} = 2r_{12}r_{13}\sin\theta\, d\theta$, we obtain Equation 10.139.

[58] A common alternative notation is $\omega_i^{(0)}(r) \equiv \omega_i^0(r)$, $\omega_i^{(1)}(r) \equiv \omega_i^{\mathrm{R}}(r)$, $\omega_i^{(2)}(r) \equiv \omega_i^{\mathrm{A}}(r)$ and $\omega_i^{(3)}(r) \equiv \omega_i^{\mathrm{V}}(r)$. This is not used here because in this treatise superscript (1), (2) etc. have the meaning of "singlet," "pair" etc.

Note that

$$\omega_i^{\mathrm{A}}(r) = \mathrm{A}_i^{\mathrm{h}}\omega_i^0(r), \quad \omega_i^{\mathrm{R}}(r) = \mathrm{R}_i^{\mathrm{h}}\omega_i^0(r), \tag{10.141}$$

which motivates the superscripts A and R. These function will later be used as weight functions in FMT for inhomogeneous systems. The function $\omega_i^{\mathrm{V}}(r)$ is dimensionless while $\omega_i^{\mathrm{A}}(r)$, $\omega_i^{\mathrm{R}}(r)$ and $\omega_i^0(r)$ have dimensions 1/length, $1/(\text{length})^2$ and $1/(\text{length})^3$, respectively. A rule of thumb for the notation is that if one multiplies the respective function by volume $= (\text{length})^3$, the resulting dimension is volume for ω_i^{V}, area for ω_i^{A}, length for ω_i^{R} and dimensionless for ω_i^0.

We also introduce[59]

$$\begin{aligned}\vec{\omega}_i^{\mathrm{A}^{\hat{\mathbf{r}}}}(\mathbf{r}) &= \hat{\mathbf{r}}\,\delta(\mathrm{R}_i^{\mathrm{h}} - r) = \hat{\mathbf{r}}\,\mathrm{A}_i^{\mathrm{h}}\omega_i^0(r) \\ \vec{\omega}_i^{\mathrm{R}^{\hat{\mathbf{r}}}}(\mathbf{r}) &= \hat{\mathbf{r}}\,\frac{\delta(\mathrm{R}_i^{\mathrm{h}} - r)}{4\pi\mathrm{R}_i^{\mathrm{h}}} = \hat{\mathbf{r}}\,\mathrm{R}_i^{\mathrm{h}}\omega_i^0(r),\end{aligned} \tag{10.142}$$

where $\vec{\omega}$ is a vector (distinguish this from $\boldsymbol{\omega}$, which has a different meaning elsewhere in this treatise) and superscript $\hat{\mathbf{r}}$ on $\mathrm{A}^{\hat{\mathbf{r}}}$ and $\mathrm{R}^{\hat{\mathbf{r}}}$ indicates the presence of the factor $\hat{\mathbf{r}} = \mathbf{r}/r$. Note that

$$\vec{\omega}_i^{\mathrm{A}^{\hat{\mathbf{r}}}}(r) = -\nabla\mathcal{H}(\mathrm{R}_i^{\mathrm{h}} - r) = -\nabla\omega_i^{\mathrm{V}}(r), \tag{10.143}$$

because $d\mathcal{H}(r)/dr = \delta(r)$ and $\nabla r = \mathbf{r}/r$.

In this notation we can write Equation 10.136 as

$$\Delta\mathrm{V}_{ij}(r_{12}) = \int d\mathbf{r}_3\,\omega_i^{\mathrm{V}}(\mathrm{R}_i^{\mathrm{h}} - r_{13})\omega_j^{\mathrm{V}}(\mathrm{R}_j^{\mathrm{h}} - r_{23}) = \omega_i^{\mathrm{V}} \otimes \omega_j^{\mathrm{V}}(r_{12}), \tag{10.144}$$

where $\otimes$ denotes the **convolution operator**, which is defined as (cf. Equation 9.250)

$$f_1 \otimes f_2(\mathbf{r}_{12}) \equiv \int d\mathbf{r}_3 f_1(\mathbf{r}_{13}) f_2(\mathbf{r}_{32}).$$

Remember that $\mathbf{r}_{12} = \mathbf{r}_2 - \mathbf{r}_1$ and note the order of the indices in $\mathbf{r}_{13}$ and $\mathbf{r}_{32}$ in the arguments of the functions in the integrand. For radial functions $\mathrm{f}(r)$ that depends only on $r = |\mathbf{r}|$ like in Equation 10.144 the order does not matter since $r_{32} = r_{23}$.

Equation 10.137 is in this notation

$$\Delta\mathrm{A}_{ij}(r_{12}) = \omega_i^{\mathrm{A}} \otimes \omega_j^{\mathrm{V}}(r_{12}) + \omega_i^{\mathrm{V}} \otimes \omega_j^{\mathrm{A}}(r_{12}) \tag{10.145}$$

and Equation 10.138 is

$$\begin{aligned}&\mathcal{H}(d_{ii}^{\mathrm{h}} - r_{12}) \\ &\quad = 2\left[\omega_i^0 \otimes \omega_i^{\mathrm{V}}(r_{12}) + \omega_i^{\mathrm{A}} \otimes \omega_i^{\mathrm{R}}(r_{12}) + \vec{\omega}_i^{\mathrm{A}^{\hat{\mathbf{r}}}} \otimes \vec{\omega}_i^{\mathrm{R}^{\hat{\mathbf{r}}}}(r_{12})\right],\end{aligned} \tag{10.146}$$

---

59 It is common to use the notation $\vec{\omega}_i^{(1)}(\mathbf{r}) \equiv \vec{\omega}_i^{\mathrm{R}^{\hat{\mathbf{r}}}}(\mathbf{r})$ and $\vec{\omega}_i^{(2)}(\mathbf{r}) \equiv \vec{\omega}_i^{\mathrm{A}^{\hat{\mathbf{r}}}}(\mathbf{r})$. Sometimes $\omega_i^{(\nu)}$ is used instead of $\vec{\omega}_i^{(\nu)}$ for $\nu = 1, 2$.

where $\otimes$ in the last term also includes the vector product $\vec{\omega}_i^{\mathrm{A}^{\mathrm{r}}} \cdot \vec{\omega}_i^{\mathrm{R}^{\mathrm{r}}}$. The plus sign in front of this term arises because the product $\mathbf{r}_{13} \cdot \mathbf{r}_{23}$ in Equation 10.138 must be changed to $-\mathbf{r}_{13} \cdot \mathbf{r}_{32}$ in order for the integral to be a convolution.

The generalization of Equation 10.146 to $i \neq j$ is[60]

$$\begin{aligned}\mathcal{H}(d_{ij}^{\mathrm{h}} - r_{12}) \\ &= \omega_i^0 \otimes \omega_j^{\mathrm{V}}(r_{12}) + \omega_i^{\mathrm{V}} \otimes \omega_j^0(r_{12}) + \omega_i^{\mathrm{A}} \otimes \omega_j^{\mathrm{R}}(r_{12}) + \omega_i^{\mathrm{R}} \otimes \omega_j^{\mathrm{A}}(r_{12}) \\ &+ \vec{\omega}_i^{\mathrm{A}^{\mathrm{r}}} \otimes \vec{\omega}_j^{\mathrm{R}^{\mathrm{r}}}(r_{12}) + \vec{\omega}_i^{\mathrm{R}^{\mathrm{r}}} \otimes \vec{\omega}_j^{\mathrm{A}^{\mathrm{r}}}(r_{12})\end{aligned} \tag{10.147}$$

that contains pairs of terms where one term becomes the other when $i \leftrightarrow j$. Each pair becomes the corresponding term in Equation 10.146 when $i = j$. Equation 10.147 can be proven by using geometrical arguments or, more simply, it can be verified in Fourier space.[61]

It remains to find the corresponding expression for $\Delta\mathsf{R}_{ij}(r_{12})$. It can be derived from the relationship 10.135; this task is left as an exercise (Exercise 10.10). One obtains

$$\begin{aligned}\Delta\mathsf{R}_{ij}(r_{12}) &= \omega_i^{\mathrm{R}} \otimes \omega_j^{\mathrm{V}}(r_{12}) + \omega_i^{\mathrm{V}} \otimes \omega_j^{\mathrm{R}}(r_{12}) \\ &+ \frac{1}{4\pi}\left[\omega_i^{\mathrm{A}} \otimes \omega_j^{\mathrm{A}}(r_{12}) + \vec{\omega}_i^{\mathrm{A}^{\mathrm{r}}} \otimes \vec{\omega}_j^{\mathrm{A}^{\mathrm{r}}}(r_{12})\right].\end{aligned} \tag{10.148}$$

{**Exercise 10.10:** Derive Equation 10.148 from Equations 10.135, 10.145 and 10.147. The definitions 10.140 and 10.142 need also to be used.}

We have now expressed all parts of $c_{ij}^{(2)}$ in Equation 10.130 in terms of functions that solely involve individual spheres of species i and j and we can write

$$-c_{ij}^{(2)}(r) = \sum_{\nu,\nu' \in \{0,\mathrm{R},\mathrm{A},\mathrm{V}\}} \mathsf{a}_{\nu\nu'}\, \omega_i^{\nu} \otimes \omega_j^{\nu'}(r) + \sum_{\nu,\nu' \in \{\mathrm{R}^{\mathrm{r}},\mathrm{A}^{\mathrm{r}}\}} \mathsf{a}_{\nu\nu'}\, \vec{\omega}_i^{\nu} \otimes \vec{\omega}_j^{\nu'}(r) \tag{10.149}$$

where the coefficients $\mathsf{a}_{\nu\nu'}$ can be expressed in terms of a_ℓ in Equation 10.130 and vice versa. Inserting Equations 10.144, 10.145, 10.147 and 10.148 into

---

[60] In the FMT literature it quite common to define a "convolution integral" as $\int d\mathbf{r}_3 f_1(\mathbf{r}_{13}) f_2(\mathbf{r}_{23})$ with $\mathbf{r}_{23}$ instead of $\mathbf{r}_{32}$. Then the last two terms in Equation 10.147 have minus signs. However, this integral is not a convolution in the mathematical sense and its Fourier transform is $\tilde{f}_1(\mathbf{k})\underline{\tilde{f}_2(\mathbf{k})}$, where underline denotes complex conjugate, instead of $\tilde{f}_1(\mathbf{k})\tilde{f}_2(\mathbf{k})$ (cf. Equation 9.251). For radial functions the order does not matter since $r_{32} = r_{23}$. The Fourier transforms of such functions are real.

[61] The Fourier transform of $\mathcal{H}(\mathsf{R}-r)$ is $4\pi[\sin(kR) - kR\cos(kr)]/k^3$ and that of $\delta(\mathsf{R}-r)$ is $4\pi R\sin(kR)/k$. The transform of the vectorial functions in Equation 10.142 can be obtained via Equation 10.143 by using Equation 9.248 (together, the two gradients give a factor of $-k^2$). By utilizing Equation 9.251, collecting terms with factors $1/k^2$, $1/k^3$ and $1/k^4$ and using basic trigonometric identities, one can readily verify Equation 10.147.

Equation 10.130 and comparing with Equation 10.149, we find that $\mathsf{a}_{\nu\nu'} = \mathsf{a}_{\nu'\nu}$ and

$$\begin{aligned} a_0 &= \mathsf{a}_{0\mathrm{V}} = \mathsf{a}_{\mathrm{AR}} = \mathsf{a}_{\mathrm{A}^{\hat{\mathbf{r}}}\mathrm{R}^{\hat{\mathbf{r}}}}, \\ a_{\mathrm{R}} &= \mathsf{a}_{\mathrm{RV}} = 4\pi\mathsf{a}_{\mathrm{AA}} = 4\pi\mathsf{a}_{\mathrm{A}^{\hat{\mathbf{r}}}\mathrm{A}^{\hat{\mathbf{r}}}}, \quad \text{(PY)} \\ a_{\mathrm{A}} &= \mathsf{a}_{\mathrm{AV}}, \quad a_{\mathrm{V}} = \mathsf{a}_{\mathrm{VV}}. \end{aligned} \tag{10.150}$$

The rest of the $\mathsf{a}_{\nu\nu'}$ coefficients are zero because they do not correspond to any combination of ν and ν' that occur in the convolutions above. The coefficients a_ℓ have the values shown in Equation 10.133 in the PY approximation. When $c_{ij}^{(2)}(r)$ is known the pair distribution functions $g_{ij}^{(2)}(r)$ can be calculated from the OZ equation.

HINT: In order to see which two functions can occur in each convolution in Equation 10.144-10.148, the following rule of thumb may be helpful. One must match dimensions on both sides of the equality, whereby the $\otimes$ operator has the dimension volume because of the volume element $d\mathbf{r}_3$ in the convolution integral. Recall the previous rule of thumb about multiplication by volume, so if both sides of the equation are multiplied by an extra factor of volume one gets the following matching in Equation 10.144: $\Delta\mathrm{V}\times$volume $\Leftrightarrow$ superscripts V & V in the rhs. Likewise, in Equation 10.145 one has $\Delta\mathrm{A}\times$volume $\Leftrightarrow$ superscripts A & V, Equation 10.147: $\mathcal{H}\times$volume $\Leftrightarrow$ 0 & V and A & R and in Equation 10.148: $\Delta\mathrm{R}\times$volume $\Leftrightarrow$ R & V and A & A.

Equation 10.149 can be written in a simplified notation

$$-c_{ij}^{(2)}(r) = \sum_{\nu,\nu'} \mathsf{a}_{\nu\nu'}\, \omega_i^{\nu} \otimes \omega_j^{\nu'}(r), \tag{10.151}$$

where the indices ν and ν' runs over $0, \mathrm{R}, \mathrm{A}, \mathrm{V}, \mathrm{R}^{\hat{\mathbf{r}}}, \mathrm{A}^{\hat{\mathbf{r}}}$ and the vectors $\vec{\omega}_i^{\mathrm{R}^{\hat{\mathbf{r}}}}$ and $\vec{\omega}_i^{\mathrm{A}^{\hat{\mathbf{r}}}}$ are included as factors in the sum. *This convention will be used from now on.*

10.3.2 FMT for inhomogeneous hard sphere mixtures

10.3.2.1 The weighted density distributions in FMT

As we saw in Section 10.3.1.1, FMT is based on an approximate free energy density. For a bulk phase we have $\Phi^{\mathrm{ex,b}} = \beta\mathcal{A}^{\mathrm{ex,b}}/V$ and for inhomogeneous fluids the free energy density depends on location, so one has $\Phi^{\mathrm{ex}} = \Phi^{\mathrm{ex}}(\mathbf{r})$. As mentioned in Section 10.2.2, $\Phi^{\mathrm{ex}}(\mathbf{r})$ is not uniquely defined and one cannot assign a unique number to $\Phi^{\mathrm{ex}}(\mathbf{r})$ at each point r, but $\int d\mathbf{r}\, \Phi^{\mathrm{ex}}(\mathbf{r}) = \beta\mathcal{A}^{\mathrm{ex}}$, is well defined and unique. Theories with different $\Phi^{\mathrm{ex}}(\mathbf{r})$ are equivalent if they give the same value $\beta\mathcal{A}^{\mathrm{ex}}$ of its integral.

Analogously to the case of a one-component system, $\mathcal{A}^{\mathrm{ex}}$ for a system with M components is a unique, but unknown, functional of the density distributions $\mathcal{A}^{\mathrm{ex}} = \mathcal{A}^{\mathrm{ex}}[\{n_i(\mathbf{r})\}_{i=1}^M]$, where $\{n_i(\mathbf{r})\}_{i=1}^M = n_1(\mathbf{r}), n_2(\mathbf{r}), \ldots, n_M(\mathbf{r})$. A basic assumption in FMT for hard sphere mixtures is the following approximation. It is assumed that the free energy functional can be written in terms of a density $\Phi^{\text{ex}}(\mathbf{r})$ that is a *function* (not a functional) of a set of **weighted density distributions** $\mathrm{n}_\nu(\mathbf{r})$, that is,

$$\beta \mathcal{A}^{\text{ex}}[\{n_i\}_i] = \int d\mathbf{r}\, \Phi^{\text{ex}}(\{\mathrm{n}_\nu(\mathbf{r})\}_\nu), \quad \text{(FMT)} \tag{10.152}$$

where i runs over all species and we have for all values of ν

$$\mathrm{n}_\nu(\mathbf{r}) = \int d\mathbf{r}' \sum_i \omega_i^\nu(\mathbf{r}-\mathbf{r}')n_i(\mathbf{r}') \tag{10.153}$$

with various weight functions $\omega_i^\nu(r)$, cf. Equation 10.75. In Rosenfeld's FMT Ansatz for inhomogeneous hard sphere mixtures these weight function are exactly those we encountered in Equations 10.140 and 10.142: the scalar functions $\omega_i^0(r)$, $\omega_i^{\mathrm{R}}(r)$, $\omega_i^{\mathrm{A}}(r)$ and $\omega_i^{\mathrm{V}}(r)$ and the vector functions $\vec{\omega}_i^{\mathrm{R}^{\hat{\mathbf{r}}}}(\mathbf{r})$ and $\vec{\omega}_i^{\mathrm{A}^{\hat{\mathbf{r}}}}(\mathbf{r})$. Thus, we have the following six weighted densities: $\mathrm{n}_0(\mathbf{r})$, $\mathrm{n}_{\mathrm{R}}(\mathbf{r})$, $\mathrm{n}_{\mathrm{A}}(\mathbf{r})$, $\mathrm{n}_{\mathrm{V}}(\mathbf{r})$, $\vec{\mathrm{n}}_{\mathrm{R}^{\hat{\mathbf{r}}}}(\mathbf{r})$ and $\vec{\mathrm{n}}_{\mathrm{A}^{\hat{\mathbf{r}}}}(\mathbf{r})$ that are defined according to Equation 10.153[62]

$$\begin{aligned}
\mathrm{n}_0(\mathbf{r}) &= \int d\mathbf{r}' \sum_i \omega_i^0(|\mathbf{r}-\mathbf{r}'|)n_i(\mathbf{r}') \\
&\vdots \\
\vec{\mathrm{n}}_{\mathrm{R}^{\hat{\mathbf{r}}}}(\mathbf{r}) &= \int d\mathbf{r}' \sum_i \vec{\omega}_i^{\mathrm{R}^{\hat{\mathbf{r}}}}(\mathbf{r}-\mathbf{r}')n_i(\mathbf{r}') \\
\vec{\mathrm{n}}_{\mathrm{A}^{\hat{\mathbf{r}}}}(\mathbf{r}) &= \int d\mathbf{r}' \sum_i \vec{\omega}_i^{\mathrm{A}^{\hat{\mathbf{r}}}}(\mathbf{r}-\mathbf{r}')n_i(\mathbf{r}')
\end{aligned} \tag{10.154}$$

and include the two "vectorial densities" $\vec{\mathrm{n}}_{\mathrm{R}^{\hat{\mathbf{r}}}}(\mathbf{r})$ and $\vec{\mathrm{n}}_{\mathrm{A}^{\hat{\mathbf{r}}}}(\mathbf{r})$. Equation 10.153 implies that (cf. Equation 10.78)

$$\frac{\delta \mathrm{n}_\nu(\mathbf{r})}{\delta n_i(\mathbf{r}')} = \omega_i^\nu(\mathbf{r}-\mathbf{r}'), \quad \frac{\delta \vec{\mathrm{n}}_\nu(\mathbf{r})}{\delta n_i(\mathbf{r}')} = \vec{\omega}_i^\nu(\mathbf{r}-\mathbf{r}'), \tag{10.155}$$

where the latter applies for the vectorial entities. Note that the dimensions of the weighted densities are different from each other; they have the same dimension as the corresponding weight function.

Let us interpret what the weighted densities means. For any function $f(\mathbf{r}')$ the integral

$$\int d\mathbf{r}' \omega_i^0(|\mathbf{r}-\mathbf{r}'|)f(\mathbf{r}') = \int d\mathbf{r}' \frac{\delta(\mathsf{R}_i^{\mathrm{h}} - |\mathbf{r}-\mathbf{r}'|)}{\mathsf{A}_i^{\mathrm{h}}} f(\mathbf{r}') = \langle f \rangle_{\mathsf{A}_i}(\mathbf{r})$$

[62] A common alternative notation is $n_0 \equiv \mathrm{n}_0$, $n_1 \equiv \mathrm{n}_{\mathrm{R}}$, $n_2 \equiv \mathrm{n}_{\mathrm{A}}$, $n_3 \equiv \mathrm{n}_{\mathrm{V}}$, $\vec{n}_1 \equiv \vec{\mathrm{n}}_{\mathrm{R}^{\hat{\mathbf{r}}}}$ and $\vec{n}_2 \equiv \vec{\mathrm{n}}_{\mathrm{A}^{\hat{\mathbf{r}}}}$. The last two are often denoted $\mathbf{n}_1$ and $\mathbf{n}_2$ or $\mathbf{n}_{V1}$ and $\mathbf{n}_{V2}$, respectively.

gives the average of $f(\mathbf{r}')$ over the surface of a sphere of radius $\mathsf{R}_i^{\mathrm{h}}$ located at $\mathbf{r}$. Likewise,

$$\int d\mathbf{r}'\, \omega_i^{\mathrm{V}}(|\mathbf{r}-\mathbf{r}'|)f(\mathbf{r}') = \int d\mathbf{r}'\, \mathsf{V}_i^{\mathrm{h}} \frac{\mathcal{H}(\mathsf{R}_i^{\mathrm{h}} - |\mathbf{r}-\mathbf{r}'|)}{\mathsf{V}_i^{\mathrm{h}}} f(\mathbf{r}') = \mathsf{V}_i^{\mathrm{h}} \langle f \rangle_{\mathsf{V}_i}(\mathbf{r})$$

gives $\mathsf{V}_i^{\mathrm{h}}$ times the average of $f(\mathbf{r}')$ over the volume of such a sphere. Note that this sphere is only a spherical part of space, not a particle.

We see from the definition of $\mathrm{n}_0(\mathbf{r})$ in Equation 10.154 that $\mathrm{n}_0(\mathbf{r}) = \sum_i \langle n_i \rangle_{\mathsf{A}_i}(\mathbf{r})$. From the presence of the delta function in the definition of the average, it follows that the hard spheres that contribute for each i have their centers at the sphere surface of radius $\mathsf{R}_i^{\mathrm{h}}$, which is centered at $\mathbf{r}$. For each of them the hard sphere surface touches the point $\mathbf{r}$, so $\mathrm{n}_0(\mathbf{r})$ is the *average density of hard spheres with surfaces that touch* $\mathbf{r}$. The same is true for $\mathrm{n}_{\mathrm{R}}(\mathbf{r})$ and $\mathrm{n}_{\mathrm{A}}(\mathbf{r})$ except that the density is weighted by the radius $\mathsf{R}_i^{\mathrm{h}}$ and area $\mathsf{A}_i^{\mathrm{h}}$, respectively, of each hard sphere that contributes. We have, for example

$$\mathrm{n}_{\mathrm{R}}(\mathbf{r}) = \sum_i \int d\mathbf{r}' \mathsf{R}_i^{\mathrm{h}}\, \omega_i^0(|\mathbf{r}-\mathbf{r}'|)n_i(\mathbf{r}')$$

(cf. Equation 10.141).

The function $\mathrm{n}_{\mathrm{V}}(\mathbf{r}) = \sum_i \mathsf{V}_i^{\mathrm{h}} \langle n_i \rangle_{\mathsf{V}_i}(\mathbf{r})$ is equal to the local packing fraction at the point $\mathbf{r}$ because only spheres that overlap with the point $\mathbf{r}$ contribute to the average (interpreted as an ensemble average) and each of them has volume $\mathsf{V}_i^{\mathrm{h}}$. This function is also equal to the probability that the point $\mathbf{r}$ happens to be inside a hard sphere for the ensemble of particle configurations of the system at equilibrium.

The vectorial densities $\vec{\mathrm{n}}_{\mathrm{R}^{\hat{\mathrm{r}}}}(\mathbf{r})$ and $\vec{\mathrm{n}}_{\mathrm{A}^{\hat{\mathrm{r}}}}(\mathbf{r})$ are more tricky to interpret. The weight functions $\vec{\omega}_i^{\mathrm{A}^{\hat{\mathrm{r}}}}$ and $\vec{\omega}_i^{\mathrm{R}^{\hat{\mathrm{r}}}}$ arise for geometrical reasons as seen in Equation 10.138 and its derivation. From the definitions 10.142 and 10.154 it follows, for example,

$$\vec{\mathrm{n}}_{\mathrm{R}^{\hat{\mathrm{r}}}}(\mathbf{r}) = \sum_i \int d\mathbf{r}' \frac{\mathbf{r}-\mathbf{r}'}{|\mathbf{r}-\mathbf{r}'|} \mathsf{R}_i^{\mathrm{h}}\, \omega_i^0(|\mathbf{r}-\mathbf{r}'|)n_i(\mathbf{r}'),$$

which differ from the expression for $\vec{\mathrm{n}}_{\mathrm{R}}(\mathbf{r})$ by the unit vector $(\mathbf{r}-\mathbf{r}')/|\mathbf{r}-\mathbf{r}'|$. Like for $\mathrm{n}_{\mathrm{R}}(\mathbf{r})$ and $\mathrm{n}_{\mathrm{A}}(\mathbf{r})$, each hard sphere whose surface touches the point $\mathbf{r}$ contributes to the average, but with the peculiarity that its contribution is a vector in the radial direction toward the point $\mathbf{r}$. The length of this vector is the same as the corresponding scalar contribution to $\mathrm{n}_{\mathrm{R}}(\mathbf{r})$ and $\mathrm{n}_{\mathrm{A}}(\mathbf{r})$. As a consequence, the direction and length of the total vector $\vec{\mathrm{n}}_{\mathrm{R}^{\hat{\mathrm{r}}}}(\mathbf{r})$ and $\vec{\mathrm{n}}_{\mathrm{A}^{\hat{\mathrm{r}}}}(\mathbf{r})$ reflect the anisotropy of the distribution of hard spheres that touch $\mathbf{r}$, whereby the importance of each hard sphere in the average depends on its size via the factors $\mathsf{R}_i^{\mathrm{h}}$ and $\mathsf{A}_i^{\mathrm{h}}$, respectively.

In the bulk phase, the vectorial densities are zero (because the fluid is isotropic), while the other four weighted densities are

$$
\begin{aligned}
\mathrm{n_V} &= \int d\mathbf{r}' \sum_i \omega_i^{\mathrm{V}}(|\mathbf{r}-\mathbf{r}'|)) n_i^{\mathrm{b}} = \sum_i \mathrm{V}_i^{\mathrm{h}} n_i^{\mathrm{b}} = \xi_{\mathrm{V}} \quad \text{(bulk)} \\
\mathrm{n_A} &= \int d\mathbf{r}' \sum_i \mathrm{A}_i^{\mathrm{h}} \omega_i^0(|\mathbf{r}-\mathbf{r}'|)) n_i^{\mathrm{b}} = \sum_i \mathrm{A}_i^{\mathrm{h}} n_i^{\mathrm{b}} = \xi_{\mathrm{A}} \quad \text{(bulk)} \\
\mathrm{n_R} &= \int d\mathbf{r}' \sum_i \mathrm{R}_i^{\mathrm{h}} \omega_i^0(|\mathbf{r}-\mathbf{r}'|)) n_i^{\mathrm{b}} = \sum_i \mathrm{R}_i^{\mathrm{h}} n_i^{\mathrm{b}} = \xi_{\mathrm{R}} \quad \text{(bulk)} \\
\mathrm{n_0} &= \int d\mathbf{r}' \sum_i \omega_i^0(|\mathbf{r}-\mathbf{r}'|)) n_i^{\mathrm{b}} = \sum_i n_i^{\mathrm{b}} = \xi_0 \quad \text{(bulk)} \qquad (10.156)
\end{aligned}
$$

where we have used Equation 10.141 and the definitions 10.82. Thus these weighted densities in bulk are equal to the entities ξ_ν that occur in the expressions for the bulk pressure and free energy density that we investigated in Section 10.3.1.1 and the coefficients for $c_{ij}^{(2)}(r)$ in Equation 10.133.

10.3.2.2 Direct correlation functions

The singlet and pair direct-correlation functions for a mixture are given by

$$
c_i^{(1)}(\mathbf{r}_1) = -\beta \frac{\delta \mathcal{A}^{\mathrm{ex}}}{\delta n_i(\mathbf{r}_1)}, \quad c_{ij}^{(2)}(\mathbf{r}_1, \mathbf{r}_2) = -\beta \frac{\delta^2 \mathcal{A}^{\mathrm{ex}}}{\delta n_i(\mathbf{r}_1) \delta n_j(\mathbf{r}_2)} \qquad (10.157)
$$

(cf. Equations 9.126 and 9.127). Inserting the FMT Ansatz 10.152 we obtain

$$
c_i^{(1)}(\mathbf{r}_1) = -\frac{\delta \int d\mathbf{r}\, \Phi^{\mathrm{ex}}(\{\mathrm{n}_\nu(\mathbf{r})\}_\nu)}{\delta n_i(\mathbf{r}_1)} \quad \text{(FMT)}. \qquad (10.158)
$$

Since $\Phi^{\mathrm{ex}}(\{\mathrm{n}_\nu(\mathbf{r})\}_\nu)$ is a function of all $\mathrm{n}_\nu(\mathbf{r})$ given in Equation 10.154, we have

$$
\frac{\delta \Phi^{\mathrm{ex}}(\{\mathrm{n}_\nu(\mathbf{r})\}_\nu)}{\delta n_i(\mathbf{r}_1)} = \sum_\nu \frac{\partial \Phi^{\mathrm{ex}}(\mathbf{r})}{\partial \mathrm{n}_\nu(\mathbf{r})} \frac{\delta \mathrm{n}_\nu(\mathbf{r})}{\delta n_i(\mathbf{r}_1)} = \sum_\nu \frac{\partial \Phi^{\mathrm{ex}}(\mathbf{r})}{\partial \mathrm{n}_\nu(\mathbf{r})} \omega_i^\nu(\mathbf{r}-\mathbf{r}_1), \qquad (10.159)
$$

where we have utilized Equation 10.155. In Rosenfeld's FMT Ansatz the index ν runs over $0, \mathrm{R}, \mathrm{A}, \mathrm{V}, \mathrm{R}^{\hat{\mathbf{r}}}, \mathrm{A}^{\hat{\mathbf{r}}}$, so the vectors $\vec{\omega}_i^{\mathrm{R}}$ and $\vec{\omega}_i^{\mathrm{A}}$ are included in the sum. The vectorial terms in the last sum should be interpreted as

$$
\frac{\partial \Phi^{\mathrm{ex}}(\mathbf{r})}{\partial \vec{\mathrm{n}}_\nu(\mathbf{r})} \cdot \vec{\omega}_i^\nu(\mathbf{r}-\mathbf{r}_1) \equiv \sum_{\alpha \in \{x,y,x\}} \frac{\partial \Phi^{\mathrm{ex}}(\mathbf{r})}{\partial \mathrm{n}_{\alpha,\nu}(\mathbf{r})} \omega_{\alpha,i}^\nu(\mathbf{r}-\mathbf{r}_1),
$$

where $\mathrm{n}_{\alpha,\nu}$ for $\alpha \in \{x,y,x\}$ are the x, y, z components of $\vec{\mathrm{n}}_\nu$ and analogously for $\omega_{\alpha,i}^\nu$.

From Equations 10.158 and 10.159 we obtain

$$-c_i^{(1)}(\mathbf{r}_1) = \int d\mathbf{r} \sum_{\nu} \left. \frac{\partial \Phi^{\text{ex}}}{\partial \mathrm{n}_\nu} \right|_{\mathrm{n}_\nu = \mathrm{n}_\nu(\mathbf{r})} \omega_i^\nu(\mathbf{r} - \mathbf{r}_1) \tag{10.160}$$

and by taking the second functional derivative with respect to $\delta n_j(\mathbf{r}_2)$ we get

$$-c_{ij}^{(2)}(\mathbf{r}_1, \mathbf{r}_2) = \int d\mathbf{r} \sum_{\nu,\nu'} \left. \frac{\partial^2 \Phi^{\text{ex}}}{\partial \mathrm{n}_\nu \partial \mathrm{n}_{\nu'}} \right|_{\substack{\mathrm{n}_\nu = \mathrm{n}_\nu(\mathbf{r}) \\ \mathrm{n}_{\nu'} = \mathrm{n}_{\nu'}(\mathbf{r})}} \omega_i^\nu(\mathbf{r} - \mathbf{r}_1) \omega_j^{\nu'}(\mathbf{r} - \mathbf{r}_2) \quad \text{(FMT).} \tag{10.161}$$

Higher order direct correlations functions can be obtained in the same fashion by taking further functional derivatives.

The second derivative of Φ^{ex} in the vectorial terms in Equation 10.161, $\partial^2 \Phi^{\text{ex}} / \partial \vec{\mathrm{n}}_\nu(\mathbf{r}) \partial \vec{\mathrm{n}}_{\nu'}(\mathbf{r})$, is a tensor and has nine spatial components $\partial^2 \Phi^{\text{ex}} / \partial \mathrm{n}_{\alpha,\nu} \partial \mathrm{n}_{\alpha',\nu'}$ for $\alpha, \alpha' \in \{x, y, x\}$. The product with the vectors $\vec{\omega}_i^\nu$ and $\vec{\omega}_j^\nu$ can be written as

$$\vec{\omega}_i^\nu \cdot \frac{\partial^2 \Phi^{\text{ex}}}{\partial \vec{\mathrm{n}}_\nu \partial \vec{\mathrm{n}}_{\nu'}} \cdot \vec{\omega}_j^{\nu'} \equiv \sum_{\alpha,\alpha' \in \{x,y,x\}} \omega_{\alpha,i}^\nu \frac{\partial^2 \Phi^{\text{ex}}}{\partial \mathrm{n}_{\alpha,\nu} \partial \mathrm{n}_{\alpha',\nu'}} \omega_{\alpha',j}^{\nu'}$$

where the functions depend on the coordinates as shown in Equation 10.161.

In the application of Equations 10.160 and 10.161 one needs to consider the following. As we will see, the vectorial contributions to Φ^{ex} in Rosenfeld's Ansatz always occur as vector products $\vec{\mathrm{n}}_\nu \cdot \vec{\mathrm{n}}_{\nu'}$,[63] which implies that $\partial \Phi^{\text{ex}} / \partial \vec{\mathrm{n}}_\nu$ will contain the factor $\partial(\vec{\mathrm{n}}_\nu \cdot \vec{\mathrm{n}}_{\nu'}) / \partial \vec{\mathrm{n}}_\nu$ (from the inner-derivative factor when the chain rule is applied). Since $\partial \vec{\mathrm{n}}_\nu / \partial \vec{\mathrm{n}}_\nu = \mathbb{I}$, where $\mathbb{I}$ is the unit tensor that has the components $\mathbb{I}_{\alpha\alpha'} = \delta_{\alpha\alpha'}$ (the Kronecker delta), we have $\partial(\vec{\mathrm{n}}_\nu \cdot \vec{\mathrm{n}}_{\nu'}) / \partial \vec{\mathrm{n}}_\nu = \mathbb{I} \cdot \vec{\mathrm{n}}_{\nu'} = \vec{\mathrm{n}}_{\nu'}$. Therefore, the contribution from $\partial \Phi^{\text{ex}} / \partial \vec{\mathrm{n}}_\nu$ to $c_i^{(1)}(\mathbf{r}_1)$ contains the factor $\vec{\mathrm{n}}_{\nu'}(\mathbf{r}) \cdot \vec{\omega}_i^\nu(\mathbf{r} - \mathbf{r}_1)$ in the integrand of Equation 10.160. All terms in the integrand of $c_{ij}^{(2)}$ in Equation 10.161 that originate from the mixed vector-nonvector derivatives $\partial^2 \Phi^{\text{ex}} / \partial \vec{\mathrm{n}}_\nu \partial \mathrm{n}_{\nu'}$ (for $\nu' \in \{0, \mathrm{R}, \mathrm{A}, \mathrm{V}\}$) contain this factor $\vec{\mathrm{n}}_{\nu'} \cdot \vec{\omega}_i^\nu$. They vanish in the bulk phase because $\vec{\mathrm{n}}_{\nu'} = 0$ there. The $\partial / \partial \vec{\mathrm{n}}_{\nu'}$ derivative of $\vec{\mathrm{n}}_{\nu'} \cdot \vec{\omega}_i^\nu$ is equal to $\vec{\omega}_i^\nu$, so it gives rise to a contribution to $\partial^2 \Phi^{\text{ex}} / \partial \vec{\mathrm{n}}_\nu \partial \vec{\mathrm{n}}_{\nu'}$ in $c_{ij}^{(2)}$ that is proportional to the factor $\vec{\omega}_i^\nu \cdot \vec{\omega}_j^{\nu'}$. All other contributions from $\partial^2 \Phi^{\text{ex}} / \partial \vec{\mathrm{n}}_\nu \partial \vec{\mathrm{n}}_{\nu'}$ (if any) vanish in the bulk since they contain the factor $(\vec{\mathrm{n}}_{\nu'} \cdot \vec{\omega}_i^\nu)(\vec{\mathrm{n}}_\nu \cdot \vec{\omega}_j^{\nu'})$.

Let us apply these results to a bulk phase where $\Phi^{\text{ex}} = \Phi^{\text{ex,b}}$ is independent of position. Then $\mathrm{n}_\nu = \xi_\nu$ for $\nu = 0, \mathrm{R}, \mathrm{A}$ and V are also constants independent

[63] As we will see in Section 10.3.2.3, this also is the case for the Santos-v approach but not for the tensorial extensions of Rosenfeld's version of FMT.

of position and so is $\partial\Phi^{\mathrm{ex}}/\partial\mathrm{n}_\nu\partial\mathrm{n}_{\nu'}$. In this case, Equation 10.161 becomes for all FMT versions based on Rosenfeld's Ansatz

$$\begin{aligned}-c_{ij}^{(2)}(r_{12}) &= \sum_{\nu,\nu'} \left.\frac{\partial^2\Phi^{\mathrm{ex}}}{\partial\mathrm{n}_\nu\partial\mathrm{n}_{\nu'}}\right|_{\mathrm{bulk}} \int d\mathbf{r}\, \omega_i^\nu(|\mathbf{r}-\mathbf{r}_1|)\omega_j^{\nu'}(|\mathbf{r}-\mathbf{r}_2|) \\ &= \sum_{\nu,\nu'\in\{0,\mathrm{R},\mathrm{A},\mathrm{V}\}} \frac{\partial^2\Phi^{\mathrm{ex,b}}}{\partial\xi_\nu\partial\xi_{\nu'}}\,\omega_i^\nu\otimes\omega_j^{\nu'}(r_{12}) \qquad (10.162) \\ &+ \sum_{\nu,\nu'\in\{\mathrm{R}^{\mathfrak{f}},\mathrm{A}^{\mathfrak{f}}\}} \left.\frac{\partial^2\Phi^{\mathrm{ex}}}{\partial\vec{\mathrm{n}}_\nu\partial\vec{\mathrm{n}}_{\nu'}}\right|_{\mathrm{bulk}} \vec{\omega}_i^\nu\otimes\vec{\omega}_j^{\nu'}(r_{12}). \quad \text{(bulk, Rosenfeld)}\end{aligned}$$

Thus, for the bulk fluid FMT always gives $c_{ij}^{(2)}(r_{12})$ of the same form as in Equation 10.149 and we can identify

$$\mathrm{a}_{\nu\nu'} = \frac{\partial^2\Phi^{\mathrm{ex,b}}}{\partial\xi_\nu\partial\xi_{\nu'}} \qquad \text{(bulk, Rosenfeld)} \qquad (10.163)$$

for $\nu,\nu'\in\{0,\mathrm{R},\mathrm{A},\mathrm{V}\}$. The convolutions $\omega_i^\nu\otimes\omega_j^{\nu'}(r_{12})$ are nonzero only for distances r_{12} where the hard spheres located at $\mathbf{r}_1$ and $\mathbf{r}_2$ overlap or touch. Thus, the approximate $c_{ij}^{(2)}(r_{12})$ obtained in FMT is always zero outside the region of overlap between the various species of spheres, like it is the PY approximation. It follows from our findings in Section 10.1.1 that the values of $c^{(2)}(r_{12})$ inside the core region are very important. The actual $c^{(2)}(r_{12})$ is, as we saw from the MC results shown in Figure 10.4, nonzero outside the core, although considerably smaller than inside. This latter feature cannot be obtained in FMT based of the weight functions in Equations 10.140 and 10.142. However, since $c^{(2)}(r_{12})$ is small outside the core, it can be a reasonable approximation to set it equal to zero there.

The set of equations 10.150 show how the different coefficients $\mathrm{a}_{\nu\nu'}$ obtained in the PY approximation are related to the coefficients a_ℓ that appear in the expression for $c_{ij}^{(2)}(r)$ in Equation 10.130 for $\ell = 0,\mathrm{R},\mathrm{A}$ and V. From the first equalities in this set of equations we see that these relationships can be summarized as $a_\nu = \mathrm{a}_{\nu\mathrm{V}}$ and using Equation 10.163 we see that $a_\nu = \partial\Phi^{\mathrm{ex,b}}/\partial\xi_\nu\partial\xi_{\mathrm{V}}$. We have thereby shown Equation 10.132.

The first two equations in 10.150 also show how some of the different coefficients $\mathrm{a}_{\nu\nu'}$ obtained in the PY approximation are related to each other, namely $\mathrm{a}_{0\mathrm{V}} = \mathrm{a}_{\mathrm{AR}} = \mathrm{a}_{\mathrm{A}^{\mathfrak{f}}\mathrm{R}^{\mathfrak{f}}}$ and $\mathrm{a}_{\mathrm{RV}} = 4\pi\mathrm{a}_{\mathrm{AA}} = 4\pi\mathrm{a}_{\mathrm{A}^{\mathfrak{f}}\mathrm{A}^{\mathfrak{f}}}$. The relationships $\mathrm{a}_{0\mathrm{V}} = \mathrm{a}_{\mathrm{AR}}$ and $\mathrm{a}_{\mathrm{RV}} = 4\pi\mathrm{a}_{\mathrm{AA}}$ can alternatively be shown from Equation 10.163 and $\Phi^{\mathrm{ex,b}}$ given in Equation 10.98, which is valid in both Rosenfeld's original version of FMT for the bulk and the PY approximation. We have

$$\begin{aligned}\mathrm{a}_{0\mathrm{V}} &= \frac{\partial^2\Phi^{\mathrm{ex,b}}}{\partial\xi_0\partial\xi_{\mathrm{V}}} = \frac{1}{1-\xi_{\mathrm{V}}} = \frac{\partial^2\Phi^{\mathrm{ex,b}}}{\partial\xi_{\mathrm{A}}\partial\xi_{\mathrm{R}}} = \mathrm{a}_{\mathrm{AR}} \\ \mathrm{a}_{\mathrm{RV}} &= \frac{\partial^2\Phi^{\mathrm{ex,b}}}{\partial\xi_{\mathrm{R}}\partial\xi_{\mathrm{V}}} = \frac{\xi_{\mathrm{A}}}{(1-\xi_{\mathrm{V}})^2} = 4\pi\frac{\partial^2\Phi^{\mathrm{ex,b}}}{\partial\xi_{\mathrm{A}}^2} = 4\pi\mathrm{a}_{\mathrm{AA}}.\end{aligned} \qquad \text{(bulk, oFMT)}$$

From the relationships we also obtain

$$\mathsf{a}_{\mathrm{A^{\hat{r}}R^{\hat{r}}}} = \frac{1}{1-\xi_{\mathrm{V}}}, \quad \mathsf{a}_{\mathrm{A^{\hat{r}}A^{\hat{r}}}} = \frac{\xi_{\mathrm{A}}}{4\pi(1-\xi_{\mathrm{V}})^2} \quad \text{(bulk, oFMT)}, \tag{10.164}$$

which will be useful later.

10.3.2.3 The free energy density in FMT

Equations 10.152–10.154 imply that for inhomogeneous systems, we have

$$\Phi^{\mathrm{ex}}(\mathbf{r}) = \Phi^{\mathrm{ex}}(\mathrm{n_0}(\mathbf{r}), \mathrm{n_R}(\mathbf{r}), \mathrm{n_A}(\mathbf{r}), \mathrm{n_V}(\mathbf{r}), \vec{\mathrm{n}}_{\mathrm{R^{\hat{r}}}}(\mathbf{r}), \vec{\mathrm{n}}_{\mathrm{A^{\hat{r}}}}(\mathbf{r})) \quad \text{(Rosenfeld)}, \tag{10.165}$$

which is Rosenfeld's Ansatz for Φ^{ex} in such systems. As a further approximation in this Ansatz one assumes that $\Phi^{\mathrm{ex}}(\mathbf{r})$ can be written as a power series expansion in the weighted density distributions up to third order. Analogously to the case of bulk phases, where we have the corresponding expansion shown in Equation 10.90, one forms products of the weighted densities with the same dimension as Φ^{ex}, that is, $1/(\mathrm{length})^3$. These products are $\mathrm{n_0}$, $\mathrm{n_R n_A}$, $\mathrm{n_A^3}$, $\vec{\mathrm{n}}_{\mathrm{R}} \cdot \vec{\mathrm{n}}_{\mathrm{A}}$ and $\mathrm{n_A}(\vec{\mathrm{n}}_{\mathrm{A}} \cdot \vec{\mathrm{n}}_{\mathrm{A}})$, so we have

$$\begin{aligned} \Phi^{\mathrm{ex}}(\{\mathrm{n}_\nu\}_\nu) &= \mathsf{f}_0(\mathrm{n_V})\mathrm{n_0} + \mathsf{f}_1(\mathrm{n_V})\mathrm{n_R n_A} + \mathsf{f}_2(\mathrm{n_V})\mathrm{n_A^3} \\ &+ \mathsf{f}_3(\mathrm{n_V})\vec{\mathrm{n}}_{\mathrm{R^{\hat{r}}}} \cdot \vec{\mathrm{n}}_{\mathrm{A^{\hat{r}}}} + \mathsf{f}_4(\mathrm{n_V})\mathrm{n_A}(\vec{\mathrm{n}}_{\mathrm{A^{\hat{r}}}} \cdot \vec{\mathrm{n}}_{\mathrm{A^{\hat{r}}}}) \quad \text{(Rosenfeld)}, \end{aligned} \tag{10.166}$$

where the coefficients f_ℓ are, as yet, unknown functions of the dimensionless $\mathrm{n_V}$. Remember that all n_ν are functions of $\mathbf{r}$ and that Φ^{ex} depends on $\mathbf{r}$ via the n_ν functions. In the bulk, this expression (with Equation 10.156 inserted) becomes equal to Equation 10.90 because the vectorial densities are equal to zero there.

In Rosenfeld's original FMT (oFMT)[64] for inhomogeneous hard-sphere mixtures one identifies the functions $\mathsf{f}_0(x)$, $\mathsf{f}_1(x)$ and $\mathsf{f}_2(x)$ with those in bulk, Equation 10.99, but with $x = \mathrm{n_V}(\mathbf{r})$ inserted. The two functions $\mathsf{f}_3(x)$ and $\mathsf{f}_4(x)$ remain to be determined. To this end we take the derivatives of $\Phi^{\mathrm{ex}}(\{\mathrm{n}_\nu\}_\nu)$ with respect to $\vec{\mathrm{n}}_{\mathrm{R}}$ and $\vec{\mathrm{n}}_{\mathrm{A}}$

$$\begin{aligned} \frac{\partial \Phi^{\mathrm{ex}}}{\partial \vec{\mathrm{n}}_{\mathrm{A^{\hat{r}}}}} &= \mathsf{f}_3(\mathrm{n_V})\vec{\mathrm{n}}_{\mathrm{R^{\hat{r}}}} + 2\mathsf{f}_4(\mathrm{n_V})\mathrm{n_A}\vec{\mathrm{n}}_{\mathrm{A^{\hat{r}}}} \\ \frac{\partial \Phi^{\mathrm{ex}}}{\partial \vec{\mathrm{n}}_{\mathrm{R^{\hat{r}}}}} &= \mathsf{f}_3(\mathrm{n_V})\vec{\mathrm{n}}_{\mathrm{A^{\hat{r}}}} \end{aligned}$$

and we have the second derivatives

$$\frac{\partial^2 \Phi^{\mathrm{ex}}}{\partial \vec{\mathrm{n}}_{\mathrm{R^{\hat{r}}}} \partial \vec{\mathrm{n}}_{\mathrm{A^{\hat{r}}}}} = \mathsf{f}_3(\mathrm{n_V})\mathbb{I}, \quad \frac{\partial^2 \Phi^{\mathrm{ex}}}{\partial^2 \vec{\mathrm{n}}_{\mathrm{A^{\hat{r}}}}} = 2\mathsf{f}_4(\mathrm{n_V})\mathrm{n_A}\mathbb{I},$$

[64]Reference in Footnote 42.

where we have used $\partial\vec{\mathrm{n}}_\nu/\partial\vec{\mathrm{n}}_\nu = \mathbb{I}$. The corresponding contributions to $-c_{ij}^{(2)}(\mathbf{r}_1,\mathbf{r}_2)$ in Equation 10.161 are hence

$$\int d\mathbf{r}\, \mathsf{f}_3(\mathrm{n_V}(\mathbf{r}))\, \omega_i^{\mathrm{R}^{\hat{\mathbf{r}}}}(\mathbf{r}-\mathbf{r}_1)\cdot\omega_j^{\mathrm{A}^{\hat{\mathbf{r}}}}(\mathbf{r}-\mathbf{r}_2)$$

and

$$\int d\mathbf{r}\, 2\mathsf{f}_4(\mathrm{n_V}(\mathbf{r}))\,\mathrm{n_A}(\mathbf{r})\, \omega_i^{\mathrm{A}^{\hat{\mathbf{r}}}}(\mathbf{r}-\mathbf{r}_1)\cdot\omega_j^{\mathrm{A}^{\hat{\mathbf{r}}}}(\mathbf{r}-\mathbf{r}_2).$$

For a bulk fluid, where $\mathrm{n_V} = \xi_\mathrm{V}$ and $\mathrm{n_A} = \xi_\mathrm{V}$ are independent of position, these contributions become as shown in the lhs of the following equations

$$\mathsf{f}_3(\xi_\mathrm{V}) \int d\mathbf{r}\, \omega_i^{\mathrm{R}^{\hat{\mathbf{r}}}}(\mathbf{r}-\mathbf{r}_1)\cdot\omega_j^{\mathrm{A}^{\hat{\mathbf{r}}}}(\mathbf{r}-\mathbf{r}_2) \equiv \mathsf{a}_{\mathrm{A}^{\hat{\mathbf{r}}}\mathrm{R}^{\hat{\mathbf{r}}}}\, \omega_i^{\mathrm{R}^{\hat{\mathbf{r}}}} \otimes \omega_j^{\mathrm{A}^{\hat{\mathbf{r}}}}(r_{12})$$

and

$$2\mathsf{f}_4(\xi_\mathrm{V})\xi_\mathrm{A} \int d\mathbf{r}\, \omega_i^{\mathrm{A}^{\hat{\mathbf{r}}}}(\mathbf{r}-\mathbf{r}_1)\cdot\omega_j^{\mathrm{A}^{\hat{\mathbf{r}}}}(\mathbf{r}-\mathbf{r}_2) \equiv \mathsf{a}_{\mathrm{A}^{\hat{\mathbf{r}}}\mathrm{A}^{\hat{\mathbf{r}}}}\, \omega_i^{\mathrm{A}^{\hat{\mathbf{r}}}} \otimes \omega_j^{\mathrm{A}^{\hat{\mathbf{r}}}}(r_{12}).$$

In the rhs we have made the identification with the corresponding terms in Equation 10.149 for the bulk phase. The integrals on the lhs are equal to the convolution integrals on the rhs with reversed signs (to turn the integrals into convolutions $\mathbf{r}-\mathbf{r}_1$ has to be changed to $\mathbf{r}_1-\mathbf{r}$). Thus it follows that

$$-\mathsf{f}_3(\xi_\mathrm{V}) = \mathsf{a}_{\mathrm{A}^{\hat{\mathbf{r}}}\mathrm{R}^{\hat{\mathbf{r}}}}, \quad -2\mathsf{f}_4(\xi_\mathrm{V})\xi_\mathrm{A} = \mathsf{a}_{\mathrm{A}^{\hat{\mathbf{r}}}\mathrm{A}^{\hat{\mathbf{r}}}}$$

and from Equation 10.164 we see that

$$\mathsf{f}_3(x) = -\frac{1}{1-x}, \quad \mathsf{f}_4(x) = -\frac{1}{8\pi(1-x)^2} \quad \text{(oFMT)},$$

so we have

$$\mathsf{f}_3(x) = -\mathsf{f}_1(x) \quad \text{and} \quad \mathsf{f}_4(x) = -3\mathsf{f}_2(x). \tag{10.167}$$

Hence, Equation 10.166 becomes[65]

$$\begin{aligned}\Phi^{\mathrm{ex}}(\{\mathrm{n}_\nu(\mathbf{r})\}_\nu) &= -\ln(1-\mathrm{n_V})\mathrm{n}_0 + \frac{\mathrm{n_R n_A} - \vec{\mathrm{n}}_{\mathrm{R}^{\hat{\mathbf{r}}}}\cdot\vec{\mathrm{n}}_{\mathrm{A}^{\hat{\mathbf{r}}}}}{1-\mathrm{n_V}} \\ &+ \frac{\left(\mathrm{n_A^2} - 3\vec{\mathrm{n}}_{\mathrm{A}^{\hat{\mathbf{r}}}}\cdot\vec{\mathrm{n}}_{\mathrm{A}^{\hat{\mathbf{r}}}}\right)\mathrm{n_A}}{24\pi(1-\mathrm{n_V})^2} \quad \text{(oFMT)},\end{aligned} \tag{10.168}$$

which reduces to Equation 10.98 in the bulk. It follows that we can obtain Equation 10.168 from the corresponding expression 10.98 for the bulk fluid by substituting

$$\begin{aligned}\xi_\mathrm{V} &\to \mathrm{n_V} \\ \xi_0 &\to \mathrm{n}_0 \\ \xi_\mathrm{R}\xi_\mathrm{A} &\to \mathrm{n_R n_A} - \vec{\mathrm{n}}_{\mathrm{R}^{\hat{\mathbf{r}}}}\cdot\vec{\mathrm{n}}_{\mathrm{A}^{\hat{\mathbf{r}}}} \\ \xi_\mathrm{A}^3 &\to \left(\mathrm{n_A^2} - 3\vec{\mathrm{n}}_{\mathrm{A}^{\hat{\mathbf{r}}}}\cdot\vec{\mathrm{n}}_{\mathrm{A}^{\hat{\mathbf{r}}}}\right)\mathrm{n_A}.\end{aligned} \tag{10.169}$$

[65]In some early papers on FMT in the literature, the vectorial contributions to Φ^{ex} had the wrong sign.

We can write Equation 10.168 as

$$\Phi^{\text{ex}}(\{\mathrm{n}_\nu(\mathbf{r})\}_\nu) = \Phi_1^{\text{ex}} + \Phi_2^{\text{ex}} + \Phi_3^{\text{ex}}, \tag{10.170}$$

where

$$\begin{aligned} \Phi_1^{\text{ex}} &= -\mathrm{n}_0 \ln(1 - \mathrm{n}_\mathrm{V}) \\ \Phi_2^{\text{ex}} &= \frac{\mathrm{n}_\mathrm{R}\mathrm{n}_\mathrm{A} - \vec{\mathrm{n}}_{\mathrm{R}\hat{\mathbf{r}}} \cdot \vec{\mathrm{n}}_{\mathrm{A}\hat{\mathbf{r}}}}{1 - \mathrm{n}_\mathrm{V}} \\ \Phi_3^{\text{ex}} &= \frac{\left(\mathrm{n}_\mathrm{A}^2 - 3\vec{\mathrm{n}}_{\mathrm{A}\hat{\mathbf{r}}} \cdot \vec{\mathrm{n}}_{\mathrm{A}\hat{\mathbf{r}}}\right)\mathrm{n}_\mathrm{A}}{24\pi(1 - \mathrm{n}_\mathrm{V})^2} \qquad \text{(oFMT)}. \end{aligned} \tag{10.171}$$

These equations constitute the free energy density expressions in Rosenfeld's original version of FMT.

Φ^{ex} can be used to calculate properties of inhomogeneous hard sphere fluids, for example, to calculate density profiles and surface forces for such fluids confined between two planar walls. Such properties for the one-component case were treated in other approximations and by simulation in Chapters 7 and 8 in Volume I of this treatise. As we saw in Volume I, for a hard sphere fluid in contact with a hard wall the contact density at the wall surface, n^{cont}, is determined by the bulk pressure P^{b} via the relationship[66]

$$k_B T n^{\text{cont}} = P^{\text{b}}. \tag{10.172}$$

The disjoining pressure $P^{\mathcal{S}}$ (the surface force per unit area) between two such surfaces at separation $D^{\mathcal{S}}$ is given by (Equation I:8.45)

$$P^{\mathcal{S}} = k_B T \left[n^{\text{cont}}(D^{\mathcal{S}}) - n^{\text{cont}}(\infty) \right],$$

that is, the difference in contact value of the density at the wall surface at separation $D^{\mathcal{S}}$ and at infinite separation ($k_B T n^{\text{cont}}(\infty) = P^{\text{b}}$).

Thus, in order to obtain an accurate density profile near a surface and accurate surface forces, it is important to have a theory that gives a very good bulk pressure. Since the PY approximation and hence oFMT give inaccurate pressures at high densities, the density profile near the surface and the disjoining pressure will also be inaccurate in oFMT. Therefore, it is important to use FMT versions that give more accurate pressures. As we have seen in Section 10.3.1.1, one such version is the **White Bear FMT**,[67] which is based on the BMCSL pressure for bulk, Equation 10.87. It uses f_2 according to Equation 10.101 but the same f_0 and f_1 as oFMT. In wbFMT Equation 10.167 is also

---

[66]The contact density is the value of the density at the distance of closest approach of the hard sphere centers to the surface. It is illustrated in Figure I:5.5, which shows the density profile of a hard sphere fluid between two hard walls. Equation 10.172 follows from Equations I:8.1 and I:8.45, as shown in the text below the latter equation (n^{cont} for a single wall is equal to $n^{\text{cont}}(\infty)$ for two walls at infinite separation).

[67]R. Roth et al. (see reference in footnote 48) and Y.-X. Yu and J. Wu, J. Chem. Phys. **117** (2002) 10156 (https://doi.org/10.1063/1.1520530).

used, so the free energy density is given by Equation 10.170 where $\Phi_1^{\rm ex}$ and $\Phi_2^{\rm ex}$ are the same as in oFMT (Equation 10.171), but

$$\Phi_3^{\rm ex} = \frac{\left(\mathrm{n}_{\rm A}^2 - 3\vec{\mathrm{n}}_{\rm A\hat{r}} \cdot \vec{\mathrm{n}}_{\rm A\hat{r}}\right)\mathrm{n}_{\rm A}}{36\pi}\left[\frac{1}{\mathrm{n}_{\rm V}(1-\mathrm{n}_{\rm V})^2} + \frac{\ln(1-\mathrm{n}_{\rm V})}{\mathrm{n}_{\rm V}^2}\right] \quad \text{(wbFMT)} \tag{10.173}$$

It performs better than $\Phi^{\rm ex}$ from oFMT. We see that we can obtain the total $\Phi^{\rm ex}$ from the bulk expression for $\Phi^{\rm ex,b}$ in Equation 10.102 by using the substitution 10.169.

An even better version of FMT is the so-called **White Bear Mark II FMT** (wb2FMT),[68] which is based on the more accurate Hansen-Goos and Roth pressure expression 10.88. In this approach f_1 and f_2 are, as we have seen, given by Equation 10.103 while f_0 is the same as in oFMT. Like wbFMT, Equation 10.167 and hence Equation 10.169 are also used. The free energy density is accordingly given by Equation 10.170 with the same same $\Phi_1^{\rm ex}$ as in oFMT (Equation 10.171), but with

$$\begin{aligned}\Phi_2^{\rm ex} &= \frac{\mathrm{n}_{\rm R}\mathrm{n}_{\rm A} - \vec{\mathrm{n}}_{\rm R\hat{r}} \cdot \vec{\mathrm{n}}_{\rm A\hat{r}}}{3}\left[\frac{5-\mathrm{n}_{\rm V}}{1-\mathrm{n}_{\rm V}} + \frac{2\ln(1-\mathrm{n}_{\rm V})}{\mathrm{n}_{\rm V}}\right] \quad \text{(wb2FMT)}\\ \Phi_3^{\rm ex} &= -\frac{\left(\mathrm{n}_{\rm A}^2 - 3\vec{\mathrm{n}}_{\rm A\hat{r}} \cdot \vec{\mathrm{n}}_{\rm A\hat{r}}\right)\mathrm{n}_{\rm A}}{36\pi}\left[\frac{1-3\mathrm{n}_{\rm V}+\mathrm{n}_{\rm V}^2}{\mathrm{n}_{\rm V}(1-\mathrm{n}_{\rm V})^2} + \frac{\ln(1-\mathrm{n}_{\rm V})}{\mathrm{n}_{\rm V}^2}\right].\end{aligned} \tag{10.174}$$

Its predictions for density distributions differ very little from those of wbFMT. A major advantage of wb2FMT compared to wbFMT lies in its predictions for the solvation free energy for convex bodies immersed in the fluid, in particular the values of the thermodynamic coefficients: pressure, planar wall surface tension and bending rigidities, which describe the free energy changes for bending deformations of a surface.

A versatile free energy functional can be used to treat various types of inhomogeneous fluids, like near the surface of an immersed particle, near a wall in contact with the fluid or in confinement. It is therefore of interest to see how FMT can handle cases of different degrees of confinement. A one-component fluid confined between two planar walls in the limit of small surface separation, where only one layer of fluid particles can be present between the surfaces, acts like a two-dimensional fluid, that is, a fluid of hard disks that are located in a plane at, say, $z = 0$, see the left part of Figure 10.7 (we have selected the z axis is perpendicular to the surfaces). This limit is an example of a "**dimensional crossover**," where a three-dimensional fluid with density $n^{[3D]}(\mathbf{r}) = n^{[3D]}(x, y, z)$ is turned into a two-dimensional one with density $n^{[2D]}(x, y)$ and the sphere centers can move solely in a single plane. In the latter case, the three-dimensional density distribution is given by $n^{[3D]}(x, y, z) = n^{[2D]}(x, y)\delta(z)$. The FMT versions treated so far in the

[68] H. Hansen-Goos and R. Roth, J. Phys.: Condens. Matter **18** (2006) 8413 (https://doi.org/10.1088/0953-8984/18/37/002).

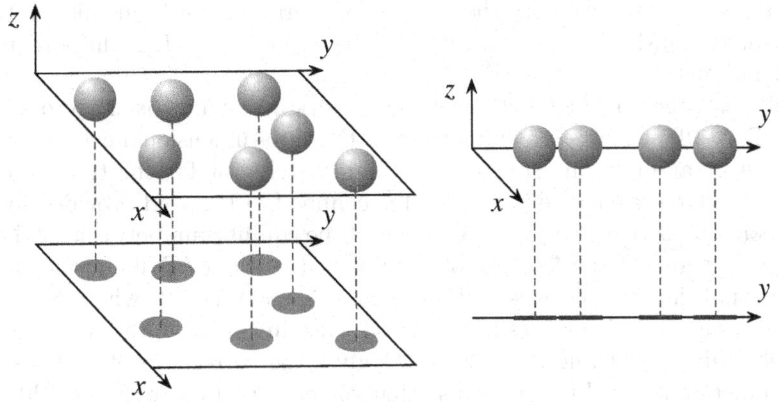

Figure 10.7 Illustration of "dimensional crossover" from 3D to 2D to the left and from 3D to 1D to the right. The left part shows hard spheres located with their centers on the (x, y)-plane and with density $n^{[3D]}(x, y, z) = n^{[2D]}(x, y)\delta(z)$, which behave as two-dimensional hard disks with density $n^{[2D]}(x, y)$ shown at the bottom. The right part shows shows hard spheres located with their centers on the y axis and with density $n^{[3D]}(x, y, z) = n^{[1D]}(y)\delta(x)\delta(z)$, which behave as one-dimensional hard rods with density $n^{[1D]}(y)$.

present section, oFMT, wbFMT and wb2FMT, are useful in this limit for both one-component and multicomponent fluids.

An even more confined case is a hard sphere fluid in a cylindrical pore. In the limit of small pore diameter, where the sphere centers are restricted to solely move along the cylinder's central line (which we select as the y-axis), the spheres behave like one-dimensional hard rods with the one-dimensional density $n^{[1D]}(y)$, see the right part of Figure 10.7. Then, the three-dimensional density distribution of the spheres is given by $n^{[3D]}(x, y, z) = n^{[1D]}(y)\delta(x)\delta(z)$. The exact statistical mechanics of one-dimensional fluids of hard rods is given by an analytic theory derived by Percus and coworkers.[69] Any accurate and versatile density functional for the three-dimensional fluid must in the dimensional cross-over from 3D $\to$ 1D yield the free energy given by this exact theory. However, this is not the case with the oFMT, wbFMT and wb2FMT functionals above, which behave quite badly in this limit. The culprit is the contribution from Φ_3^{ex} to the free energy, $k_B T \int d\mathbf{r}\, \Phi^{\mathrm{ex}}(\mathbf{r})$, which diverges due to the factor $\left(\mathrm{n}_{\mathrm{A}}^2 - 3\overrightarrow{\mathrm{n}}_{\mathrm{A}^{\hat{r}}} \cdot \overrightarrow{\mathrm{n}}_{\mathrm{A}^{\hat{r}}}\right)\mathrm{n}_{\mathrm{A}}$.[70] Nevertheless, the contributions from

[69] For the single-component fluid of one-dimensional hard rods see J. K. Percus, J. Stat. Phys. **15** (1976) 505 (https://doi.org/10.1007/BF01020803) and for the multicomponent case see T. K. Vanderlick, H. T. Davis and J. K. Percus, J. Chem. Phys. **91** (1989) 7136 (https://doi.org/10.1063/1.457329).

[70] For details see Y. Rosenfeld, M. Schmidt, H. Löwen, and P. Tarazona, Phys. Rev. E **55** (1997) 4245 (https://doi.org/10.1103/PhysRevE.55.4245).

the sum of Φ_1^{ex} and Φ_2^{ex} give the entire free energy correctly in this limit, so Φ_3^{ex} should vanish. The FMT functionals need to be fixed so these requirements are met.

Likewise, these FMT functionals do not work in the dimensional cross-over from 3D $\to$ 0D, where the confinement in the zero-dimensional case is a void that can fit at most one sphere. This limit is relevant for the treatment of solids, where each particle in the solid is confined in the void provided by its neighbors, in particular for the treatment of the equilibrium between solid and fluid phases and of the freezing of a fluid. In the strict 0D limit, the three-dimensional density becomes $n^{[3D]}(x,y,z) = \bar{\mathsf{N}}\delta(x)\delta(y)\delta(z)$, where $\bar{\mathsf{N}} \leq 1$ is the mean number of particles in the cavity. Also in this case, the contribution from Φ_3^{ex} diverges, while the rest of Φ^{ex} gives the correct result.[71] Thus the FMT functionals need to be fixed so that the contribution to Φ^{ex} in addition to Φ_1^{ex} and Φ_2^{ex} vanishes in the 0D limit, but retains the appropriate behavior for the 3D case.

For the one-component hard sphere fluid, Pedro Tarazona[72] rectified these problems by complementing $\left(\mathrm{n}_{\mathrm{A}}^2 - 3\vec{\mathrm{n}}_{\mathrm{A}\hat{\mathbf{r}}} \cdot \vec{\mathrm{n}}_{\mathrm{A}\hat{\mathbf{r}}}\right)\mathrm{n}_{\mathrm{A}}$ with contributions from a tensorial weighted density distribution $\overleftrightarrow{\mathrm{n}}_{\mathrm{A}\hat{\mathbf{r}}\hat{\mathbf{r}}}(\mathbf{r}) = \int d\mathbf{r}' \overleftrightarrow{\omega}^{\mathrm{A}\hat{\mathbf{r}}\hat{\mathbf{r}}}(\mathbf{r}-\mathbf{r}')n(\mathbf{r}')$, where the tensor weight function is $\overleftrightarrow{\omega}^{\mathrm{A}\hat{\mathbf{r}}\hat{\mathbf{r}}}(\mathbf{r}) = (\hat{\mathbf{r}}\hat{\mathbf{r}}^{\mathsf{T}} - \mathbb{I}/3)\,\delta(\mathsf{R}^{\mathrm{h}} - r)$.[73] The solution he found was to add $9(\vec{\mathrm{n}}_{\mathrm{A}\hat{\mathbf{r}}} \cdot \overleftrightarrow{\mathrm{n}}_{\mathrm{A}\hat{\mathbf{r}}\hat{\mathbf{r}}} \cdot \vec{\mathrm{n}}_{\mathrm{A}\hat{\mathbf{r}}} - \mathrm{Tr}[(\overleftrightarrow{\mathrm{n}}_{\mathrm{A}\hat{\mathbf{r}}\hat{\mathbf{r}}})^3])/2$, where $\mathrm{Tr}[\,\cdot\,]$ denotes the trace, that is, the sum of the diagonal elements of the tensor. This means that in the tensorial version of oFMT (oFMT-t) one uses

$$\Phi_3^{\mathrm{ex,T}} = \frac{\left(\mathrm{n}_{\mathrm{A}}^2 - 3\vec{\mathrm{n}}_{\mathrm{A}\hat{\mathbf{r}}} \cdot \vec{\mathrm{n}}_{\mathrm{A}\hat{\mathbf{r}}}\right)\mathrm{n}_{\mathrm{A}} + 9\left(\vec{\mathrm{n}}_{\mathrm{A}\hat{\mathbf{r}}} \cdot \overleftrightarrow{\mathrm{n}}_{\mathrm{A}\hat{\mathbf{r}}\hat{\mathbf{r}}} \cdot \vec{\mathrm{n}}_{\mathrm{A}\hat{\mathbf{r}}} - \mathrm{Tr}[(\overleftrightarrow{\mathrm{n}}_{\mathrm{A}\hat{\mathbf{r}}\hat{\mathbf{r}}})^3]\right)/2}{24\pi(1-\mathrm{n}_{\mathrm{V}})^2} \quad \text{(oFMT-t)} \tag{10.175}$$

instead of Φ_3^{ex} in Equation 10.171. $\Phi_3^{\mathrm{ex,T}}$ is designed so that it vanishes for any 1D density distribution of hard spheres and also in the strict 0D limit. Furthermore, like the vectorial weighted density distributions, $\overleftrightarrow{\mathrm{n}}_{\mathrm{A}\hat{\mathbf{r}}\hat{\mathbf{r}}}$ vanishes

---

[71] See reference in footnote 70. The individual terms with factors $\mathrm{n}_{\mathrm{R}}\mathrm{n}_{\mathrm{A}}$ and $\vec{\mathrm{n}}_{\mathrm{R}\hat{\mathbf{r}}} \cdot \vec{\mathrm{n}}_{\mathrm{A}\hat{\mathbf{r}}}$ in Φ_2^{ex} give divergent contributions in the zero-dimensional limit, but these divergences cancel each other in $\mathrm{n}_{\mathrm{R}}\mathrm{n}_{\mathrm{A}} - \vec{\mathrm{n}}_{\mathrm{R}\hat{\mathbf{r}}} \cdot \vec{\mathrm{n}}_{\mathrm{A}\hat{\mathbf{r}}}$, which become zero in this limit. Such a cancellation does not occur in $\mathrm{n}_{\mathrm{A}}\left(\mathrm{n}_{\mathrm{A}}^2 - 3\vec{\mathrm{n}}_{\mathrm{A}\hat{\mathbf{r}}} \cdot \vec{\mathrm{n}}_{\mathrm{A}\hat{\mathbf{r}}}\right)$. The term Φ_1^{ex} goes to the exact result for the 0D case.

[72] P. Tarazona, Phys. Rev. Lett. **84** (2000) 694 (https://doi.org/10.1103/PhysRevLett.84.694) and Physica A **306** (2002) 243 (https://doi.org/10.1016/S0378-4371(02)00501-0).

[73] We have $\hat{\mathbf{r}}\hat{\mathbf{r}}^{\mathsf{T}} = \frac{1}{r^2}\begin{bmatrix} xx & xy & xz \\ xy & yy & yz \\ xz & yz & zz \end{bmatrix}$ and the term $\mathbb{I}/3$ makes the tensors $\overleftrightarrow{\omega}^{\mathrm{A}\hat{\mathbf{r}}\hat{\mathbf{r}}}$ and $\overleftrightarrow{\mathrm{n}}_{\mathrm{A}\hat{\mathbf{r}}\hat{\mathbf{r}}}$ traceless, that is, the sum of the diagonal elements is equal to zero. In the references in footnote 72 Tarazona used instead the weight function $\hat{\mathbf{r}}\hat{\mathbf{r}}\,\delta(\mathsf{R}^{\mathrm{h}} - r)$, which leads to a tensorial weighted density distribution with non-zero trace.

in the bulk, so $\Phi_3^{\mathrm{ex,T}}$ becomes equal to ξ_{A}^3 in the bulk phase. For the bulk phase the theory gives the PY $c^{(2)}(r)$ like the original oFMT.

In wbFMT and wb2FMT one makes the same addition of tensorial contributions and changes Φ_3^{ex} in Equations 10.173 and 10.174 in the same fashion, for example

$$\begin{aligned}
&\Phi_3^{\mathrm{ex,T}} \\
&\quad = -\frac{\left(\mathbf{n}_{\mathrm{A}}^2 - 3\vec{\mathbf{n}}_{\mathrm{A}\hat{\mathbf{r}}} \cdot \vec{\mathbf{n}}_{\mathrm{A}\hat{\mathbf{r}}}\right)\mathbf{n}_{\mathrm{A}} + 9\left(\vec{\mathbf{n}}_{\mathrm{A}\hat{\mathbf{r}}} \cdot \overleftrightarrow{\mathbf{n}}_{\mathrm{A}\hat{\mathbf{r}}\hat{\mathbf{r}}} \cdot \vec{\mathbf{n}}_{\mathrm{A}\hat{\mathbf{r}}} - \mathrm{Tr}[(\overleftrightarrow{\mathbf{n}}_{\mathrm{A}\hat{\mathbf{r}}\hat{\mathbf{r}}})^3]\right)/2}{36\pi} \\
&\quad \times \left[\frac{1 - 3\mathbf{n}_{\mathrm{V}} + \mathbf{n}_{\mathrm{V}}^2}{\mathbf{n}_{\mathrm{V}}(1 - \mathbf{n}_{\mathrm{V}})^2} + \frac{\ln(1 - \mathbf{n}_{\mathrm{V}})}{\mathbf{n}_{\mathrm{V}}^2}\right] \qquad \text{(wb2FMT-t)}
\end{aligned} \tag{10.176}$$

for wb2FMT. For hard sphere mixtures, it is not sufficient to use these second order tensorial contributions and one normally includes contributions from a third order tensor.[74]

Santos' version of FMT is more flexible and, in fact, one does not need to use tensorial contributions in order to avoid the singularities in the 1D and strict 0D limits. It is sufficient with vectorial contributions. However, in contrast to the previous FMT versions, in Santos' expression for $\Phi^{\mathrm{ex,b}}$

$$\begin{aligned}
\Phi^{\mathrm{ex,b}} &= -\xi_0 \ln(1 - \xi_{\mathrm{V}}) + \frac{\xi_{\mathrm{R}}\xi_{\mathrm{A}}}{1 - \xi_{\mathrm{V}}} \\
&+ \frac{\xi_{\mathrm{A}}^3}{24\pi(1 - \xi_{\mathrm{V}})^2}\left[1 + \mathrm{f}\left(\frac{\xi_{\mathrm{A}}^2/\xi_{\mathrm{R}}}{12\pi(1 - \xi_{\mathrm{V}})}\right)\right] \qquad \text{(Santos)},
\end{aligned} \tag{10.177}$$

(Equation 10.123) it is *never* suitable to make the replacements for the various combinations of ξ_ν as in Equation 10.169, because the contribution from the function f with this replacement inserted leads to an unphysical singularity for $c^{(2)}(r)$ at $r = 0$ in the bulk phase.[75] Instead, following Hansen-Goos et al.,[76] one can complement the usual weighted density distributions $\mathbf{n}_0(\mathbf{r})$, $\mathbf{n}_{\mathrm{R}}(\mathbf{r})$, $\mathbf{n}_{\mathrm{A}}(\mathbf{r})$ and $\mathbf{n}_{\mathrm{V}}(\mathbf{r})$ by the following two new weighted densities

$$\mathbf{n}_{\mathrm{R^v}} \equiv \mathbf{n}_{\mathrm{R}} - \frac{\vec{\mathbf{n}}_{\mathrm{R}\hat{\mathbf{r}}} \cdot \vec{\mathbf{n}}_{\mathrm{A}\hat{\mathbf{r}}}}{\mathbf{n}_{\mathrm{A}}} \quad \text{and} \quad \mathbf{n}_{\mathrm{A^v}} \equiv \mathbf{n}_{\mathrm{A}} - \frac{\vec{\mathbf{n}}_{\mathrm{A}\hat{\mathbf{r}}} \cdot \vec{\mathbf{n}}_{\mathrm{A}\hat{\mathbf{r}}}}{\mathbf{n}_{\mathrm{A}}}, \tag{10.178}$$

which involve the vectorial densities $\vec{\mathbf{n}}_{\mathrm{R}\hat{\mathbf{r}}}(\mathbf{r})$ and $\vec{\mathbf{n}}_{\mathrm{A}\hat{\mathbf{r}}}(\mathbf{r})$. The superscript "v" on $\mathrm{R^v}$ and $\mathrm{A^v}$ indicates that the vectorial contributions are present and we will call this theory the "Santos-v" approach. Furthermore, one requires

[74] J. A. Cuesta, Y. Martínez-Ratón and P. Tarazona, J. Phys.: Condens. Matter **14** (2002) 11965 (https://doi.org/10.1088/0953-8984/14/46/307).
[75] J. F. Lutsko, Phys. Rev. E **87** (2013) 014103 (https://doi.org/10.1103/PhysRevE.87.014103).
[76] H. Hansen-Goos, M. Mortazavifar, M. Oettel, and R. Roth, Phys. Rev. E **91** (2015) 052121 (https://doi.org/10.1103/PhysRevE.91.052121).

that $\Phi^{ex}(\mathbf{r})$ satisfies the following generalization of the pressure consistency criterium 10.105

$$\frac{\partial \Phi^{ex}}{\partial n_V} = n_0 - \Phi^{ex,b} + \sum_{\nu \in \{0,R,A,V,R^v,A^v\}} \frac{\partial \Phi^{ex}}{\partial n_\nu} n_\nu, \tag{10.179}$$

where the sum now includes $\nu = R^v$ and A^v. This condition can be used for an inhomogeneous system that deviates very little from bulk; in the limit of vanishing inhomogeneity (i.e., vanishing external potential, the "bulk limit") one has $n_{R^v} \to n_R \to \xi_R$ and $n_{A^v} \to n_A \to \xi_R$ and Equation 10.156 applies. It is straightforward to verify (by differentiation of Φ^{ex} and insertion into the equation) that Equation 10.179 has the solution

$$\begin{aligned} \Phi^{ex} &= -n_0 \ln(1-n_V) + \frac{n_R^v n_A}{1-n_V} \qquad \text{(Santos-v)} \\ &+ \frac{(n_{A^v})^2 n_A}{24\pi(1-n_V)^2}\left[1+\mathcal{F}\left(\frac{(n_{A^v})^2/n_{R^v}}{12\pi(1-n_V)}, \frac{n_{R^v}}{n_R}, \frac{n_{A^v}}{n_A}\right)\right], \end{aligned} \tag{10.180}$$

where $\mathcal{F}(X, Y, Z)$ is *any* differentiable function of X, Y, Z. Note that this expression for Φ^{ex} can be obtained, apart from the last two arguments of $\mathcal{F}$, by making the following replacements in Equation 10.177

$$\begin{aligned} \xi_V &\to n_V \\ \xi_0 &\to n_0 \\ \xi_R \xi_A &\to n_{R^v} n_A \equiv n_R n_A - \vec{n}_{R^f} \cdot \vec{n}_{A^f} \\ \xi_A^3 &\to (n_{A^v})^2 n_A \\ \frac{\xi_A^2}{\xi_R} &\to \frac{(n_{A^v})^2}{n_{R^v}} \qquad \text{(Santos-v)}, \end{aligned}$$

where only the last two lines differ from Equation 10.169. Note that the first two terms in Equation 10.180 are the same as Φ_1^{ex} and Φ_2^{ex} in oFMT, Equation 10.171, while the last term is different

$$\Phi_3^{ex} = \frac{(n_{A^v})^2 n_A}{24\pi(1-n_V)^2}\left[1+\mathcal{F}\left(\frac{(n_{A^v})^2/n_{R^v}}{12\pi(1-n_V)}, \frac{n_{R^v}}{n_R}, \frac{n_{A^v}}{n_A}\right)\right] \qquad \text{(Santos-v)} \tag{10.181}$$

In the limit of vanishing density, the first argument of the function $\mathcal{F}$ goes to zero, i.e., one has $X \to 0$. For physical reasons one requires that $\mathcal{F}(X, Y, Z)$ for $X \to 0$ remains finite, but otherwise $\mathcal{F}$ is, as yet, arbitrary. The requirements on $\mathcal{F}$ ensures that Φ^{ex} in Equation 10.180 gives the exact free energy in the strict 0D limit as well as in the low density limit.[77]

For a bulk phase one has $Y = n_{R^v}/n_R = 1$ and $Z = n_{A^v}/n_A = 1$. According to Equation 10.177, $\mathcal{F}(X, Y, Z)$ in Equation 10.180 applied to a bulk phase becomes $\mathcal{F}(\tau, 1, 1) = f(\tau)$, where $\tau = \xi_A^2/[\xi_R 12\pi(1-\xi_V)]$ (Equation 10.109).

[77] This is shown in the reference in footnote 76.

The function $\mathsf{f}(\tau)$ is uniquely determined from $\varphi_{\mathrm{o}}^{\mathrm{ex,b}}$ for the one-component bulk fluid via Equations 10.121 and 10.122. For example, one can, as we have seen, use the Carnahan-Starling expression for $\varphi_{\mathrm{o}}^{\mathrm{ex,b}}$, Equation 10.124. Thus we have the additional requirement $\mathcal{F}(X,Y,Z) \to \mathsf{f}(X)$ when $Y \to 1$ and $Z \to 1$. One can use the remaining arbitrariness of $\mathcal{F}(X,Y,Z)$ in order to fulfill other physical requirements, like avoiding the unphysical singularity for $c^{(2)}(r)$ mentioned earlier and making the theory yield the exact result for the homogeneous 1D dimensional cross-over limit. There are many possibilities due to the flexibility of the theory and some of them have successfully been explored by Hansen-Goos et al.[78] For example, using $\mathsf{f}(\tau)$ from the CS expression 10.125 for $\mathsf{F}(\tau) = 1 + \tau\,[1 + \mathsf{f}(\tau)]\,/2$, they obtained the following explicit expression for $c^{(2)}(r)$ for the one-component hard-sphere bulk fluid

$$-c^{(2)}(r) = \begin{cases} a_0 + a_1 r_{\mathrm{R}} + a_2 r_{\mathrm{R}}^2 + a_3 r_{\mathrm{R}}^3, & r_{\mathrm{R}} < 1 \\ 0, & r_{\mathrm{R}} > 1 \end{cases}$$

where $r_{\mathrm{R}} \equiv r/d^{\mathrm{h}}$ and

$$\begin{aligned} a_0 &= \frac{1 + 4\eta^{\mathrm{hs}} + 4(\eta^{\mathrm{hs}})^2 - 4(\eta^{\mathrm{hs}})^3 + (\eta^{\mathrm{hs}})^4}{(1-\eta^{\mathrm{hs}})^4}, \\ a_1 &= -\frac{\eta^{\mathrm{hs}}[18 + 20\eta^{\mathrm{hs}} - 12(\eta^{\mathrm{hs}})^2 + (\eta^{\mathrm{hs}})^4]}{3(1-\eta^{\mathrm{hs}})^4}, \\ a_2 &= 0, \quad a_3 = \frac{\eta^{\mathrm{hs}} a_0}{2} \qquad \text{(Santos-v + CS)}, \end{aligned} \tag{10.182}$$

cf. the corresponding expressions 10.3 in the PY approximation. In Figure 10.4, $c^{(2)}(r)$ from this expression is compared to the results from the PY and wbFMT approximations and MC simulations. We see that the Santos-v + CS results are very good, somewhat better than those from wbFMT.[79] The Santos-v approach is a promising kind of theory that is based on the very sound principles summarized here and we will not take space to describe any further details of the work done on it.

10.3.2.4 *The singlet and pair distributions for inhomogeneous fluids in FMT*

In Sections 10.1.3 and 10.1.4 we saw how singlet and pair density distributions for inhomogeneous fluids can be calculated by integral equation theories. Here, we are going to see how such distributions can be calculated by DFT, in particular in FMT (similar considerations apply for the weighted DFT in Section 10.2.2). As a starting point, any of the approximate expressions for

[78] See reference in footnote 76.

[79] The wb2FMT values of $c^{(2)}(r)$ (not shown) are about equally good as the Santos-v + CS results. They lie within 1 % of the latter for all densities.

$\Phi^{\mathrm{ex}}(\{\mathbf{n}_\nu(\mathbf{r})\}_\nu)$ as function of a set of weighted density distributions

$$\mathbf{n}_\nu(\mathbf{r}) = \int d\mathbf{r}' \sum_i \omega_i^\nu(\mathbf{r}-\mathbf{r}')n_i(\mathbf{r}')$$

(Equation 10.153) presented in the previous Section 10.3.2.3 can be used. Here, we use the convention that vectorial and tensorial weighted density distributions $\overrightarrow{\mathbf{n}}_\nu(\mathbf{r})$, $\overleftrightarrow{\mathbf{n}}_\nu$ and weight functions $\overrightarrow{\omega}_i^\nu$ and $\overleftrightarrow{\omega}^\nu$ are also included when we write $\mathbf{n}_\nu$ and ω_i^ν.

The singlet density distribution $n_i(\mathbf{r})$ for species i is given by

$$n_i(\mathbf{r}) = \zeta_i e^{-\beta v_i(\mathbf{r})+c_i^{(1)}(\mathbf{r})} = n_i^{\mathrm{b}} e^{-\beta v_i(\mathbf{r})+c_i^{(1)}(\mathbf{r})-c_i^{(1)\mathrm{b}}} \tag{10.183}$$

(cf. Equation 10.65), where

$$c_i^{(1)}(\mathbf{r}_1) = -\frac{\delta \int d\mathbf{r}\, \Phi^{\mathrm{ex}}(\{\mathbf{n}_\nu(\mathbf{r})\}_\nu)}{\delta n_i(\mathbf{r}_1)} \qquad \text{(FMT)}$$

(Equation 10.158), which leads to

$$-c_i^{(1)}(\mathbf{r}_1) = \int d\mathbf{r} \sum_\nu \left.\frac{\partial \Phi^{\mathrm{ex}}}{\partial \mathbf{n}_\nu}\right|_{\mathbf{n}_\nu=\mathbf{n}_\nu(\mathbf{r})} \omega_i^\nu(\mathbf{r}-\mathbf{r}_1) \qquad \text{(FMT)} \tag{10.184}$$

(Equation 10.160). For the vectorial and tensorial contributions, the derivative $\partial\Phi^{\mathrm{ex}}/\partial\mathbf{n}_\nu$ has a vectorial or tensorial character and the appropriate product is performed (cf. the discussion after Equation 10.159).

In Equation 10.184 one inserts the derivatives of the explicit expression for $\Phi^{\mathrm{ex}} = \Phi_1^{\mathrm{ex}} + \Phi_2^{\mathrm{ex}} + \Phi_3^{\mathrm{ex}}$ of the various approximations, like Rosenfeld's original FMT (oFMT, Equation 10.171), the White Bear FMT (wbFMT, Equation 10.173), the White Bear Mark II FMT (wb2FMT, 10.174), the tensorial versions of FMT (like Equations 10.175 and 10.176) or the Santos-v approach (Equation 10.181). This gives $c_i^{(1)}$ in terms of the singlet densities $n_j(\mathbf{r})$ for all j, which can be inserted in Equation 10.183. Thereby one obtains an equation for the singlet density distributions $n_j(\mathbf{r})$ that can be solved numerically in various geometries, for instance for fluids near one wall or between two planar walls, around a spherical body or in cylindrical pore. An advantage with the FMT approach is that it can be applied also for quite complicated geometries like wedge-shaped spaces.

Figure 10.8 show examples of ***density profiles*** calculated in the oFMT, wbFMT and wb2FMT approximations and Molecular Dynamics (MD) simulations by Davidchack et al.[80] for hard sphere fluids in contact with a planar wall at bulk densities $n^{\mathrm{b}} = 0.701$, 0.857 and $0.938d^{-3}$, where d is the sphere

[80]Data taken from R. L. Davidchack, B. B. Laird and R. Roth, Cond. Matter Phys. **19** (2016) 23001 (https://doi.org/10.5488/CMP.19.23001). Prof. Davidchack is acknowledged for kindly providing the original data.

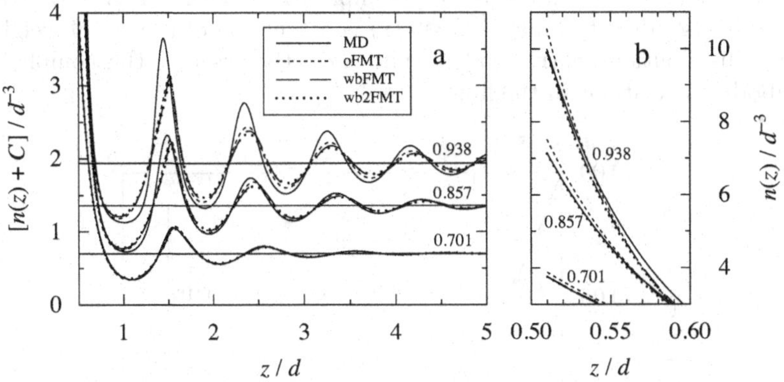

Figure 10.8 a) Density profiles $n(z)$ for hard sphere fluids in contact with a planar hard wall calculated by various FMT approximations (oFMT, wbFMT and wb2FMT) and by Molecular Dynamics.[80] The spheres have diameter d, the wall surface is located at $z = 0$ and for large z the density approaches the bulk value $n^{\mathrm{b}} = 0.701$, 0.857 and $0.938d^{-3}$, respectively, shown by the horizontal line in each set of curves. The second and third set of curves counted from the bottom have for clarity been shifted upwards and are plotted as $n(z) + C$ with $C = 0.5$ and 1.0, respectively. b) A magnified view of the profiles for small z. In this plot the curves are not shifted ($C = 0$).

diameter. This corresponds to hard sphere packing fractions $\eta^{\mathrm{hs}} = 0.367$, 0.449 and 0.492, respectively. In the latter case the fluid is close to bulk freezing.

For bulk densities up to $n^{\mathrm{b}} \approx 0.7d^{-3}$ the agreement between the results of the FMT approximations and those of MD simulations is excellent, except for the original FMT by Rosenfeld (oFMT) very close to the wall where there are increasing deviations for $n(z)$ when n^{b} is increased. This is expected because the contact density $n^{\mathrm{cont}} = n(0.5d)$ is proportional to the bulk pressure P^{b} (see Equation 10.172) and P^{b} from oFMT is, as we have seen, not very good except for low densities where it works well.

For $n^{\mathrm{b}} > 0.7d^{-3}$ the results from the wbFMT and wb2FMT approximations very close to the surface remain very good as we can see in frame (b), but the oscillatory structure of $n(z)$ for $z \gtrsim 0.7d$ shown in frame (a) is not very well described in these approximations. In particular, the peaks of the oscillations are considerably underestimated by all three FMT approximations at very high densities. In fact, the original approximation, oFMT, is slightly better than the other two for z away from the surface, but the differences between the FMT results are small compared to the deviation from the MD results. The wbFMT and wb2FMT approximations were developed to give good results for the bulk pressure and therefore they give excellent contact densities, but this does obviously not help away from the surface when the density is high. All three FMT approximations suffer at high densities from the restrictions posed by the basic assumptions in the FMT Ansatz, where the free-energy density functional $\Phi^{\rm ex}(\mathbf{r})$ is a function of the set of weighted density distributions $\mathbf{n}_\nu(\mathbf{r})$, and do not correctly describe the complicated multibody correlations in the fluid.

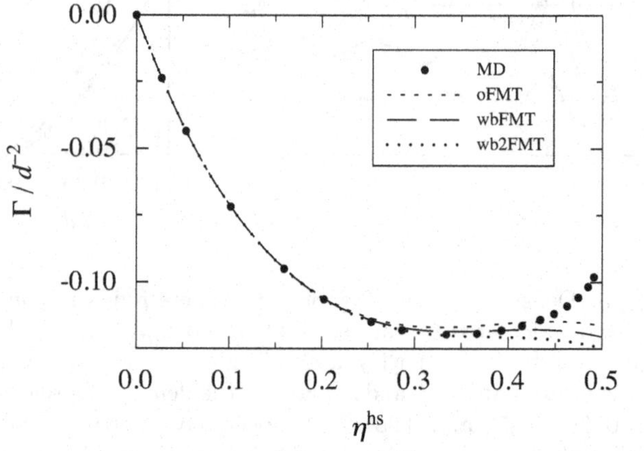

Figure 10.9 The excess adsorption Γ of the hard sphere fluid outside a hard wall plotted as a function of the hard sphere packing fraction $\eta^{\rm hs}$ of the bulk phase. Γ is calculated from the density profiles in Figure 10.8 for the oFMT, wbFMT and wb2FMT approximations and Molecular Dynamics simulations (the data are taken from the same source).

Consequences of the systematic deviations from the MD results for the FMT profiles at high density can be clearly seen in excess adsorption $\Gamma = \int_0^\infty dz\, [n(z) - n^{\rm b}]$ as shown in Figure 10.9.[81] Γ from the wb2FMT approximation agrees very well with the MD results up to $\eta^{\rm hs} \approx 0.35$ that corresponds to $n^{\rm b} \approx 0.67 d^{-3}$, but already when $\eta^{\rm hs} \gtrsim 0.25$ the other FMT results start to deviate from the MD data and from each other.

Let us now turn to the ***pair distribution functions*** for inhomogeneous hard sphere mixtures. Once the density distributions $n_i(\mathbf{r})$ for the various species have been determined, one can obtain the pair direct-correlation function $c_{ij}^{(2)}(\mathbf{r}, \mathbf{r}')$ from Equation 10.161, which involves $\partial^2 \Phi^{\rm ex} / \partial \mathbf{n}_\nu \partial \mathbf{n}_{\nu'}$ calculated from the respective explicit expression for $\Phi^{\rm ex}$, whereby one inserts the obtained weighted density distributions $\mathbf{n}_\nu(\mathbf{r})$. Having $c_{ij}^{(2)}(\mathbf{r}, \mathbf{r}')$ and $n_i(\mathbf{r})$, one can determine the pair correlation function $h_{ij}^{(2)}(\mathbf{r}, \mathbf{r}')$ from the Ornstein-Zernike equation 9.40 (in planar geometry the OZ equation has the form given in Equation 10.45). Despite that oFMT gives the Percus-Yevick pair functions in the bulk phase, it does *not* yield the corresponding PY pair functions for the inhomogeneous fluid, that is, from the Anisotropic PY

[81] The dividing surface between the fluid and the wall is set at $z = 0$.

approximation described in Section 10.1.4. While the latter enforces the core condition $g_{ij}^{(2)} = 0$ in the core region, FMT does not do this so $g_{ij}^{(2)} \neq 0$ there, although it may approximately be zero.

A considerable numerical advantage of FMT is that $c_{ij}^{(2)}$ can be directly obtained in Fourier space when the external field has planar or spherical symmetry,[82] whereby it is straightforward to solve the OZ equation in that space in order to obtain $h_{ij}^{(2)}$ (recall that convolutions become products in Fourier space, so the convolution integrals are avoided there). This is possible because the scalar and vectorial weight functions for planar and spherical geometries are explicitly known in Fourier space and the appropriate products with $\partial^2\Phi^{\text{ex}}/\partial \mathbf{n}_\nu \partial \mathbf{n}_{\nu'}$ can also be worked out there.

For integral equation theories like the APY approximation the situation is different because the closure is expressed in real space, so in order to solve the OZ equation in Fourier space, one needs to repeatedly pass between the real and Fourier spaces during the numerical iterations. It is thereby time consuming to do the numerical Fourier transforms and back-transforms. However, the accuracy of the results from FMT is always limited by the Ansatz that the free energy functional $\mathcal{A}^{\text{ex}} = \mathcal{A}^{\text{ex}}[\{n_i(\mathbf{r})\}_{i=1}^{M}]$ can be written in terms of a free energy density $\Phi^{\text{ex}}(\mathbf{r})$ that is a *function* (not a functional) of a set of weighted density distributions $\mathbf{n}_\nu(\mathbf{r})$. Distribution function theories based on closure approximations are limited in a different manner because they rely on the quality of the approximation for the bridge function, whereby the limits are set for practical reasons rather than principal ones. In Volume I there are several plots of anisotropic pair-distribution functions for inhomogeneous systems obtained by integral equation theory (see the list of plots at the end of Section 10.1.4).

As examples of anisotropic pair-distribution functions calculated by FMT, we will show results obtained by Härtel et al.[83] for binary mixtures of large and small hard spheres in contact with a planar wall. The spheres have diameters d_1 and $d_2 = 1.4d_1$ and the number-density of both species in bulk is the same (a 50:50 mixture). Density profiles for this system are shown in Figure 10.10 for total bulk packing fractions $\eta_{\text{tot}}^{\text{hs}} \equiv \eta_1^{\text{hs}} + \eta_2^{\text{hs}} = 0.3$ and 0.54. They are calculated using the wb2FMT-t approximation that includes a tensorial correction, see Equation 10.176. The spheres shown in the figure illustrate the origin of the major peaks of the curves for $\eta_{\text{tot}}^{\text{hs}} = 0.54$. The leftmost circles depict a large sphere (red) and a small sphere (blue) in contact with the wall

[82] A. Härtel, M. Kohl, and M. Schmiedeberg, Phys. Rev. E **92** (2015) 042310 (https://doi.org/10.1103/PhysRevE.92.042310); S. M. Tschopp and J. M. Brader, Phys. Rev. E **103** (2021) 042103 (https://doi.org/10.1103/PhysRevE.103.042103); A. Härtel, 2013, *Density Functional Theory of Hard Colloidal Particles: From Bulk to Interfaces,* Aachen: Shaker Verlag.

[83] A. Härtel, M. Kohl, and M. Schmiedeberg, Phys. Rev. E **92** (2015) 042310 (https://doi.org/10.1103/PhysRevE.92.042310). Dr. Andreas Härtel is acknowledged for kindly providing the original data for Figure 10.10 and the contour plots in Figures 10.11 and 10.12.

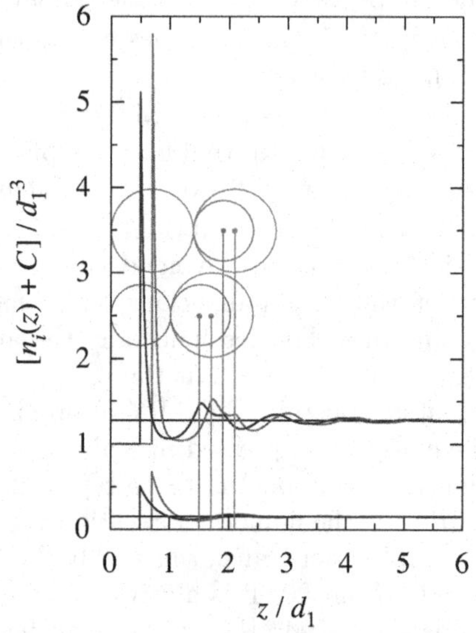

Figure 10.10 **(See color insert).** Density profiles $n_i(z)$ for a binary 50:50 fluid mixture of small and large hard spheres ($i = 1$ and 2, respectively) outside a planar hard wall in contact with a bulk fluid, which lies to the right for large z. The wall surface is located at $z = 0$. The results were obtained from the wb2FMT-t approximation by A. Härtel et al.[83] The small spheres have diameters d_1 (blue curves) and the large ones $d_2 = 1.4d_1$ (red curves). For the lower set of curves the total packing fraction of the bulk fluid is $\eta_{\text{tot}}^{\text{hs}} = 0.3$ and for the upper set of curves $\eta_{\text{tot}}^{\text{hs}} = 0.54$. The latter are shifted upwards for clarity and are plotted as $n(z) + C$ where $C = 1$, while $C = 0$ for the lower curves. The circles and vertical lines illustrate the origin of the major peaks in the curves (see text).

surface and their centers coincide with the main peak of the respective curve. Each of these spheres can have a large or a small sphere to the right that touches it, here represented by the two circles drawn in the figure. There are vertical lines drawn downwards from the centers of these circles. As seen in the figure, the lines are close to the two secondary peaks of each curve in the upper set of curves (the red lines correspond to the red curve and the blue lines correspond to the blue one). Thus, the peaks correspond approximately to these local particle configurations.

Anisotropic pair correlation functions $h_{ij}^{(2)}(\mathbf{r}, \mathbf{r}') = h_{ij}^{(2)}(s, z, z')$, where $s = ((x - x')^2 + (y - y')^2)^{1/2}$ (cf. Equation 10.45), were calculated in the same approximation and also by Brownian Dynamics (BD)[84] simulations. Examples of these functions are shown for $\eta_{\text{tot}}^{\text{hs}} = 0.5$ in Figure 10.11 and for $\eta_{\text{tot}}^{\text{hs}} = 0.3$

[84]See the original reference in footnote 83 for a description of the BD simulations, which were done with soft spheres with strongly repulsive pair potentials rather than hard spheres.

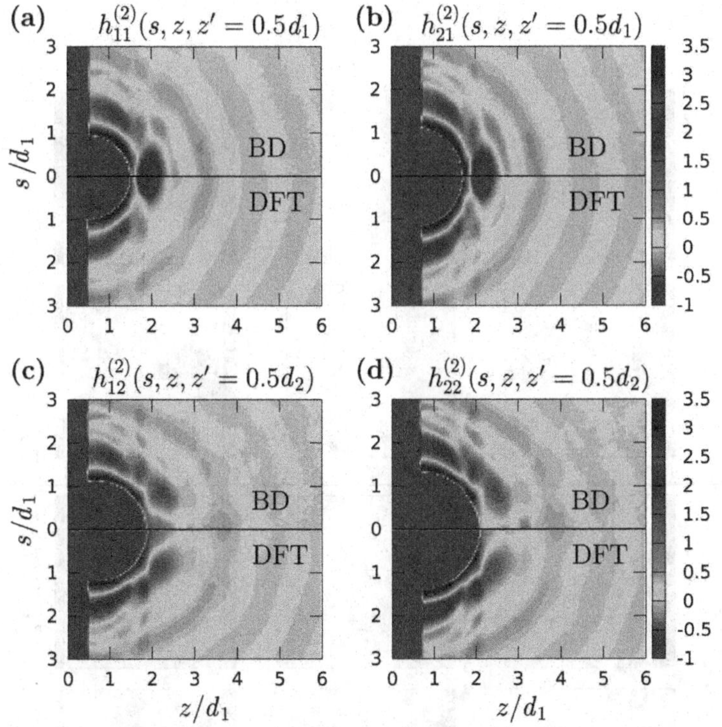

Figure 10.11 **(See color insert).** Anisotropic pair correlation functions $h_{ij}^{(2)}(s, z, z')$ for a binary 50:50 mixture of small and large hard spheres outside a planar hard wall and plotted in a cross-section plane perpendicular to the wall surface, which is located at $z = 0$. The hard spheres have diameters d_1 and $d_2 = 1.4d_1$. The four frames show the correlation function for various species i when a fixed particle of species j is placed in contact with the wall surface: (a) i = small and j = small, (b) i = large and j = small, (c) i = small and j = large and (d) i = large and j = large. The fixed particle thereby has its center at $z' = 0.5d_j$ and the pair correlations are plotted as functions of the cylindrical coordinates r, z (the system has cylindrical symmetry around the horizontal black line in each frame). The fluid for large z to the right is in its bulk state with a total packing fraction $\eta_{\mathrm{tot}}^{\mathrm{hs}} = 0.5$. Each frame is split up into results from Brownian dynamics (BD) simulations at top and Density Functional Theory (DFT) results at bottom from the wb2FMT-t approximation (for DFT the correlation functions inside the wall and inside the fixed particle have been set to -1, see text). [Reprinted figure with permission from A. Härtel, M. Kohl, and M. Schmiedeberg, Phys. Rev. E **92** (2015) 042310 (http://dx.doi.org/10.1103/PhysRevE.92.042310). Copyright 2015 by the American Physical Society.]

and 0.54 in Figure 10.12.[85] The density profiles in Figure 10.10 are those of the latter two systems. As mentioned above, $h_{ij}^{(2)}$ calculated via the OZ equation

[85]An explanation of how to interpret this kind of contour plots of pair functions is given in Figures 5.7 and 5.8 in Volume I.

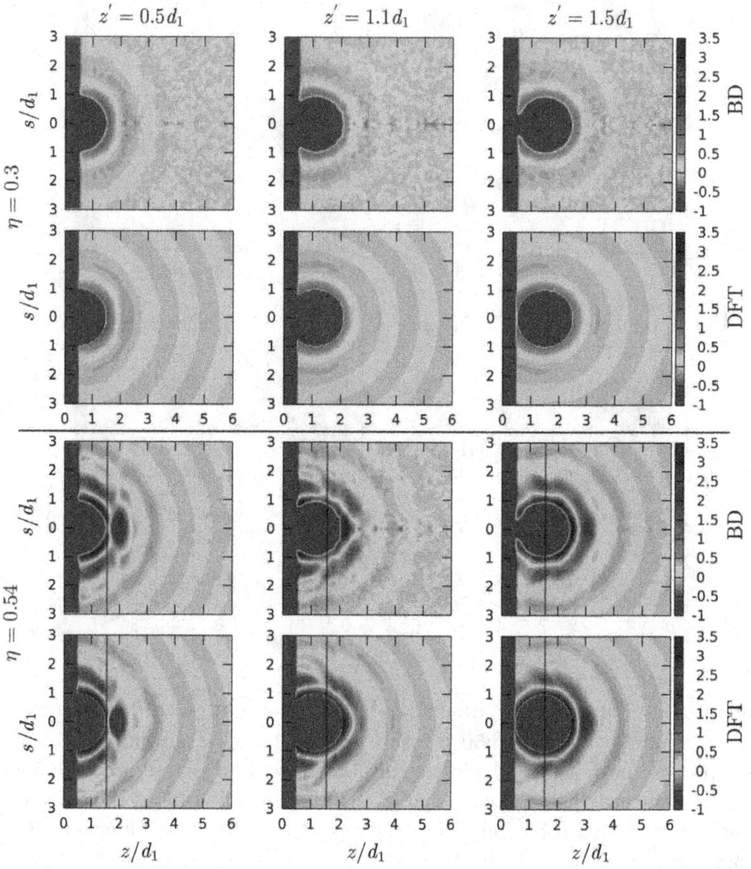

Figure 10.12 **(See color insert).** Anisotropic pair correlation functions for the same binary hard-sphere mixtures as in Figure 10.10. The plot shows the function $h_{11}^{(2)}(s, z, z')$ for small spheres of diameter d_1 when a small sphere is placed with its center fixed at various distances z' from the wall surface. The total packing fraction $\eta_{\rm tot}^{\rm hs}$ of the bulk (denoted η in the figure) is 0.3 for the top two rows of frames and 0.54 for the bottom two rows. Each column of frames shows a different position z' for the center of the fixed particle ($z' = 0.5$, 1.1, or $1.5d_1$). BD simulation results are shown in the first and third rows and DFT results (from the wb2FMT-t approximation) in the second and forth row ($h_{11}^{(2)}$ inside the wall and inside the fixed particle has been set to -1 for the DFT results, see text). The speckled pattern for the BD results is due to poor statistics where $h_{11}^{(2)}$ is small. [Reprinted figure with permission from A. Härtel, M. Kohl, and M. Schmiedeberg, Phys. Rev. E **92** (2015) 042310 (http://dx.doi.org/10.1103/PhysRevE.92.042310). Copyright 2015 by the American Physical Society.]

from $c_{ij}^{(2)}$ given by the FMT formula 10.161 does not satisfy the condition that $h_{ij}^{(2)} = -1$ inside the core. This value of $h_{ij}^{(2)}$ has instead been enforced in the figures.

In Figure 10.11, the upper half of each frame shows the BD results and the lower half shows the DFT results in the wb2FMT-t approximation for $h_{ij}^{(2)}(\mathbf{r}, \mathbf{r}')$ when a fixed sphere of species j (placed at $\mathbf{r}'$) is in contact with the wall surface (its center has coordinates $z' = 0.5d_j$ and $s = 0$). The rather complicated structure of the correlation function is the result of a compromise between the planar arrangement induced by the wall and spherical arrangement induced by the fixed sphere (the same applies to the anisotropic pair distribution plots in Volume I). The four frames show $h_{11}^{(2)}$, $h_{21}^{(2)}$, $h_{12}^{(2)}$ and $h_{22}^{(2)}$, respectively, and the BD and DFT results are in good over-all agreement with each other. However, in frames (a) and (b) the BD results show somewhat more distinct peaks (orange-red color) around $s \approx 1.0d_1, z \approx 2.1d_1$, that is, for the spheres in the second layer outside the fixed sphere. These peaks have been attributed to the formation of a local fcc or bcc structure (face-centered or body-centered cubic structure).

Figure 10.12 shows the correlation function $h_{11}^{(2)}$ between small spheres, where one such sphere is placed with its center fixed at various distances z' from the wall surface. The upper four frames are for the system with $\eta_{\text{tot}}^{\text{hs}} = 0.3$ and the lower four for $\eta_{\text{tot}}^{\text{hs}} = 0.54$, whereby alternating rows of frames show either BD or DFT results. There is a good quantitative agreement between the BD and DFT results, although for $\eta_{\text{tot}}^{\text{hs}} = 0.54$ there are the same kind of small deviations in the second layer around each fixed sphere as in the previous figure.[86] For this high packing fraction glassy dynamics has set in.

Thus there are techniques available to explicitly calculate anisotropic pair distribution functions of inhomogeneous fluids by either FMT or integral equation theory. As we have seen in Volume I, this allows a rather detailed analysis of the structural mechanisms in action in the fluids and also of the interparticle interactions and surface forces.

APPENDICES

Appendix 10A
Zero-separation theorems for pair correlation functions

In this Appendix we will present and derive some exact relationships for various pair correlation functions evaluated at zero separation, $r = 0$, like $y^{(2)}(0)$ and $b^{(2)}(0)$, and their derivatives at $r = 0$, like $dy^{(2)}(0)/dr$ and $db^{(2)}(0)/dr$. These relationships constitute consistency criteria that the correlation functions must fulfill in exact statistical mechanics of bulk fluids and they can, for

[86] In Figure 10 of the original paper (reference in footnote 83) there are more detailed comparisons between the BD and DFT results for for $\eta_{\text{tot}}^{\text{hs}} = 0.54$ along the vertical black lines at $z = 1.55d_1$ drawn in Figure 10.12.

example, be used to construct good approximate theories for such functions by enforcing these criteria.

The pair distribution function is given by (Equations 9.26 and 9.65)

$$g^{(2)}(r) = e^{-\beta u(r)} y^{(2)}(r) = e^{-\beta u(r) + \gamma^{(2)}(r) + b^{(2)}(r)}, \tag{10.185}$$

where $y^{(2)}(r) = e^{\gamma^{(2)}(r)+b^{(2)}(r)}$ and $\gamma^{(2)}(r) = h^{(2)}(r) - c^{(2)}(r)$. For hard core fluids, where $g^{(2)}(r) = 0$ when $r < d^{\mathrm{h}}$ because $u(r) = \infty$ there, the values of $y^{(2)}(r)$ and $b^{(2)}(r)$ inside the hard core are important for many properties of the fluid. For instance, various thermodynamic quantities like the compressibility calculated via the compressibility equation 9.190 and the chemical potential depend in general quite strongly on the values of these functions inside the core region. Also for other fluids, the values of these function near $r = 0$ are important.

The **first zero-separation theorem for $\mathbf{y}^{(2)}$ for hard spheres** says that

$$\ln y^{(2)}(0) = \beta\mu^{\mathrm{ex,b}} \quad \text{(hard spheres)}. \tag{10.186}$$

This relationship follows from the general result obtained from Equations 9.32 and 9.25

$$\ln y^{(2)}(\mathbf{r}, \mathbf{r}') = -\beta w^{(2)\mathrm{intr}}(\mathbf{r}, \mathbf{r}') = -\beta[\mu^{(2)\mathrm{ex}\text{-}u}(\mathbf{r}, \mathbf{r}') - \mu^{\mathrm{ex}}(\mathbf{r}) - \mu^{\mathrm{ex}}(\mathbf{r}')], \tag{10.187}$$

where $\mu^{(2)\mathrm{ex}\text{-}u}(\mathbf{r}, \mathbf{r}')$ is the reversible work against interactions when one simultaneously inserts two (imagined) particles at $\mathbf{r}$ and $\mathbf{r}'$, respectively, that do not interact with each other but interact fully with all other particles in the fluid. In the limit $\mathbf{r}' \to \mathbf{r}$, the two particles will coincide with each other and then their added interaction with the other particles is the same as for a single particle located at $\mathbf{r}$ with twice the strength of the interaction potential, that is, with the potential $2u$. For a hard sphere fluid, however, the interaction potential with two coinciding hard spheres and a single hard sphere is exactly the same, namely, infinity inside the core and zero outside ($2 \times \infty = 1 \times \infty$ and $2 \times 0 = 1 \times 0$). In this case, the surrounding hard spheres in the fluid cannot distinguish a two-particle insertion from a single-particle insertion at the same place, so $\mu^{(2)\mathrm{ex}\text{-}u}(\mathbf{r}, \mathbf{r}) = \mu^{\mathrm{ex}}(\mathbf{r})$ and we have

$$\ln y^{(2)}(\mathbf{r}, \mathbf{r}) = \beta\mu^{\mathrm{ex}}(\mathbf{r}) \quad \text{(hard spheres)},$$

which for a bulk fluid gives Equation 10.186 since $y^{(2)}(|\mathbf{r} - \mathbf{r}|) = y^{(2)}(0)$ and $\mu^{\mathrm{ex}}(\mathbf{r}) = \mu^{\mathrm{ex,b}}$ for all $\mathbf{r}$ in the bulk phase.

For fluids other than hard spheres we have instead $\lim_{\mathbf{r}' \to \mathbf{r}} \mu^{(2)\mathrm{ex}\text{-}u}(\mathbf{r}, \mathbf{r}') = \mu_2^{\mathrm{ex}}(\mathbf{r})$, where μ_2^{ex} is the excess chemical potential for the insertion of a single particle that interacts twice as strongly with the other particles as they do among themselves. Then, Equation 10.187 applied to the bulk phase gives

$$\ln y^{(2)}(0) = 2\beta\mu^{\mathrm{ex,b}} - \beta\mu_2^{\mathrm{ex,b}}, \tag{10.188}$$

which is the **first zero-separation theorem for $\mathbf{y}^{(2)}$** for the general case.

By setting $r = 0$ in the OZ equation 9.72 we get

$$\begin{aligned}\gamma^{(2)}(0) &= h^{(2)}(0) - c^{(2)}(0) = n^{\mathrm{b}} \int d\mathbf{r}'\, c^{(2)}(r')h^{(2)}(r') \\ &= n^{\mathrm{b}} \int d\mathbf{r}'\, c^{(2)}(r')g^{(2)}(r') - n^{\mathrm{b}} \int d\mathbf{r}'\, c^{(2)}(r').\end{aligned}$$

Using he compressibility equation 9.190 we obtain the **first zero-separation theorem for $\gamma^{(2)}$**

$$\gamma^{(2)}(0) = n^{\mathrm{b}} \int d\mathbf{r}'\, c^{(2)}(r')g^{(2)}(r') + \frac{\beta}{n^{\mathrm{b}}\chi_T} - 1, \tag{10.189}$$

which is valid in general.

We can now derive an expression for $b^{(2)}(0)$. Since $\ln y^{(2)}(r) = \gamma^{(2)}(r) + b^{(2)}(r)$ we have $b^{(2)}(0) = \ln y^{(2)}(0) - \gamma^{(2)}(0)$. By combining Equations 10.186 and 10.189 we obtain the **first zero-separation theorem for $\mathbf{b}^{(2)}$ for hard spheres**

$$\begin{aligned}b^{(2)}(0) &= \beta\mu^{\mathrm{ex,b}} - n^{\mathrm{b}} \int\limits_{r'>d^{\mathrm{h}}} d\mathbf{r}'\, c^{(2)}(r')g^{(2)}(r') \\ &+ 1 - \frac{\beta}{n^{\mathrm{b}}\chi_T} \quad \text{(hard spheres)},\end{aligned} \tag{10.190}$$

where we have used the fact that $g^{(2)}(r') = 0$ for $r' < d^{\mathrm{h}}$. The **first zero-separation theorem for $\mathbf{b}^{(2)}$** in the general case is given by

$$\begin{aligned}b^{(2)}(0) &= 2\beta\mu^{\mathrm{ex,b}} - \beta\mu_2^{\mathrm{ex,b}} \\ &- n^{\mathrm{b}} \int d\mathbf{r}'\, c^{(2)}(r')g^{(2)}(r') + 1 - \frac{\beta}{n^{\mathrm{b}}\chi_T},\end{aligned} \tag{10.191}$$

which is obtained from Equations 10.188 and 10.189.

There is a also zero-separation theorem for the density derivative of $y^{(2)}(0)$, which involves the chemical potential $\mu_2^{\mathrm{ex,b}}$ of the added single particle that has the interaction potential $2u$ with the other particles. In this case we consider a two-species system where the particles of the original fluid is species 1 and the added particle is species 2 that has density $n_2^{\mathrm{b}} = 0$ (cf. the OZ equation 10.33). We can derive the zero-separation theorem by using the definition 10.31 of $c_{ij}^{(2)}$, which implies that

$$\delta c_i^{(1)}(\mathbf{r}') = \int d\mathbf{r}\, c_{ij}^{(2)}(\mathbf{r}', \mathbf{r})\delta n_j(\mathbf{r}).$$

Applying this to a bulk fluid where the density of species $j = 1$ is changed by dn_1^{b} we obtain

$$dc_i^{(1)\mathrm{b}} = \left[\int d\mathbf{r}\, c_{i1}^{(2)\mathrm{b}}(r)\right] dn_1^{\mathrm{b}},$$

so, using the fact that $\beta\mu_i^{\mathrm{ex,b}} = -c_i^{(1)\mathrm{b}}$, we have

$$\beta\left(\frac{\partial\mu_1^{\mathrm{ex,b}}}{\partial n_1^{\mathrm{b}}}\right)_{T,n_2^{\mathrm{b}}} = -\left(\frac{\partial c_1^{(1)\mathrm{b}}}{\partial n_1^{\mathrm{b}}}\right)_{T,n_2^{\mathrm{b}}} = -\int d\mathbf{r}\, c_{11}^{(2)\mathrm{b}}(r) \quad (10.192)$$

$$\beta\left(\frac{\partial\mu_2^{\mathrm{ex,b}}}{\partial n_1^{\mathrm{b}}}\right)_{T,n_2^{\mathrm{b}}} = -\left(\frac{\partial c_2^{(1)\mathrm{b}}}{\partial n_1^{\mathrm{b}}}\right)_{T,n_2^{\mathrm{b}}} = -\int d\mathbf{r}\, c_{21}^{(2)\mathrm{b}}(r), \quad (10.193)$$

where $\mu_1^{\mathrm{ex,b}}$ is the same as $\mu^{\mathrm{ex,b}}$ in Equation 10.188 and Equation 10.192 is the same as the excess part of Equation 9.188. The derivative of Equation 10.188 with respect to n_1^{b} is

$$\frac{\partial \ln y_{11}^{(2)}(0)}{\partial n_1^{\mathrm{b}}} = 2\beta\frac{\partial\mu_1^{\mathrm{ex,b}}}{\partial n_1^{\mathrm{b}}} - \beta\frac{\partial\mu_2^{\mathrm{ex,b}}}{\partial n_1^{\mathrm{b}}},$$

where $y_{11}^{(2)} = y^{(2)}$ in the original notation for the system without species 2. By inserting Equations 10.192 and 10.193 we obtain (skipping superscript b on $c^{(2)}$)

$$\frac{\partial \ln y_{11}^{(2)}(0)}{\partial n_1^{\mathrm{b}}} = -2\int d\mathbf{r}\, c_{11}^{(2)}(r) + \int d\mathbf{r}\, c_{12}^{(2)}(r), \quad (10.194)$$

which is the **zero-separation theorem for the density derivative of $\mathbf{y}^{(2)}$**.

Next, we turn to theorems for the spacial derivatives of the various functions evaluated at $r = 0$. We will first state them and then present the proofs. The **second zero-separation theorem for $\gamma^{(2)}$** for fluids with a continuous pair potential, for example LJ fluids, is

$$\left.\frac{d\gamma^{(2)}(r)}{dr}\right|_{r=0} = 0 \qquad \text{(when } u(r) \text{ is continuous).} \quad (10.195)$$

For fluids with a hard core potential we have instead

$$d^{\mathrm{h}}\left.\frac{d\gamma^{(2)}(r)}{dr}\right|_{r=0} = -6\eta^{\mathrm{hs}}\left[g^{(2)}(d^{\mathrm{h}+})\right]^2 \quad \text{(hard core fluids),} \quad (10.196)$$

where $\eta^{\mathrm{hs}} = \pi(d^{\mathrm{h}})^3 n^{\mathrm{b}}/6$ is the hard sphere packing fraction. Apart from the discontinuity of the pair potential at $r = d^{\mathrm{h}+}$ in this case, $u(r)$ is assumed to be continuous for $r > d^{\mathrm{h}+}$. For a fluid of hard spheres we can replace $g^{(2)}(d^{\mathrm{h}+})$ by $y^{(2)}(d^{\mathrm{h}})$ in this equation because $g^{(2)}(r) = y^{(2)}(r)$ for $r > d^{\mathrm{h}}$.

The **second zero-separation theorems for $\mathbf{y}^{(2)}$ and $\mathbf{b}^{(2)}$ for hard spheres** are

$$d^{\mathrm{h}}\left.\frac{d\ln y^{(2)}(r)}{dr}\right|_{r=0} = -6\eta^{\mathrm{hs}}y^{(2)}(d^{\mathrm{h}}) \qquad \text{(hard spheres)} \quad (10.197)$$

$$d^{\mathrm{h}}\left.\frac{db^{(2)}(r)}{dr}\right|_{r=0} = 6\eta^{\mathrm{hs}}y^{(2)}(d^{\mathrm{h}})\left[y^{(2)}(d^{\mathrm{h}}) - 1\right] \quad \text{(hard spheres).} \quad (10.198)$$

The last equation follows from Equation 10.196 and 10.197 because $db^{(2)}(r)/dr = d\ln y^{(2)}(r)/dr - d\gamma^{(2)}(r)/dr$.

First, we will prove Equations 10.195 and 10.196. The proof of Equation 10.197 starts on page 198. The second zero-separation theorem for $\gamma^{(2)}$ is a consequence of the OZ equation 9.70

$$\gamma^{(2)}(r_{12}) = h^{(2)}(r_{12}) - c^{(2)}(r_{12}) = n^{\mathrm{b}} \int d\mathbf{r}_3\, c^{(2)}(r_{13})h^{(2)}(r_{32}). \qquad (10.199)$$

Our task is to evaluate the derivative of $\gamma^{(2)}(r_{12})$ when $r_{12} \to 0$. To start, we investigate some general properties of the convolution of two functions $f_a(r)$ and $f_b(r)$

$$F(r_{12}) = \int d\mathbf{r}_3\, f_a(r_{13})f_b(r_{32}) \qquad (10.200)$$

in the limit $r_{12} \to 0$. We will show that the derivative $F'(0)$ is zero unless both $f_a(r)$ and $f_b(r)$ are discontinuous *for the same* r value. First, we will show that $F'(0) = 0$ if both $f_a(r)$ and $f_b(r)$ are continuous functions and thereafter that the same is true if only one of the functions is discontinuous or if discontinuities occur at different r values for the two functions. When we have shown this, we will evaluate the derivative when *both* functions are discontinuous for $r = \mathcal{R}$, where $\mathcal{R} > 0$ is arbitrary.

First, we treat the case when $f_a(r)$ and $f_b(r)$ are continuous functions of r and have continuous derivatives. We start by taking the gradient of $F(r_{12}) = F(|\mathbf{r}_1 - \mathbf{r}_2|)$. Using the fact that $\nabla_i f(|\mathbf{r}_i - \mathbf{r}_j|) = \hat{\mathbf{r}}_{ji} f'(r_{ji})$, where $\hat{\mathbf{r}}_{ji} = \mathbf{r}_{ji}/r_{ji}$ and $\mathbf{r}_{ji} = \mathbf{r}_i - \mathbf{r}_j$, we obtain

$$\nabla_1 F(r_{12}) = \int d\mathbf{r}_3\, \hat{\mathbf{r}}_{31} f_a'(r_{13}) f_b(r_{32}).$$

Next, we transform $\nabla_1 F$ into dF/dr_{12}, which is easily done by a multiplication with $\hat{\mathbf{r}}_{21}$ as follows from the general relationship

$$\hat{\mathbf{r}}_{21} \cdot \nabla_i f(|\mathbf{r}_1 - \mathbf{r}_2|) = [\hat{\mathbf{r}}_{21}]^2 f'(r_{12}) = f'(r_{12}). \qquad (10.201)$$

We obtain

$$\frac{dF(r_{12})}{dr_{12}} = \hat{\mathbf{r}}_{21} \cdot \nabla_1 F(r_{12}) = \int d\mathbf{r}_3\, \hat{\mathbf{r}}_{21} \cdot \hat{\mathbf{r}}_{31}\, f_a'(r_{13}) f_b(r_{32}). \qquad (10.202)$$

In order to evaluate the integral in the limit $r_{12} \to 0$, we select $\mathbf{r}_1$ as the origin of the coordinate system and place $\mathbf{r}_2$ nearby; we will later let $\mathbf{r}_2 \to \mathbf{r}_1$. Since $\mathbf{r}_1$ is the origin, the integration variable is $\mathbf{r}_3 = \mathbf{r}_{13}$. Furthermore, we select the z axis to have the same direction as $\mathbf{r}_{21} = \mathbf{r}_1 - \mathbf{r}_2$, see Figure 10.13a. The angle between the vector $\mathbf{r}_{13}$ and the z axis is denoted θ, which is also the angle between the vectors $\mathbf{r}_{13}$ and $\mathbf{r}_{21}$. Equation 10.202 can now be written as

$$\frac{dF(r_{12})}{dr_{12}} = -\int d\mathbf{r}_{13}\, \cos\theta\, f_a'(r_{13}) f_b(r_{32})$$

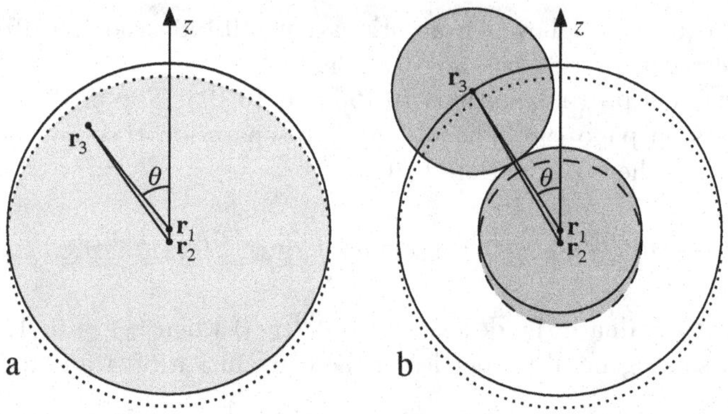

Figure 10.13 (a) Illustration used in the proof of the second zero-separation theorem for $\gamma^{(2)}$. The point $\mathbf{r}_3$ is the integration variable in Equation 10.202 and the points $\mathbf{r}_1$ and $\mathbf{r}_2$ occur in the argument of $F(r_{12}) = F(|\mathbf{r}_1 - \mathbf{r}_2|)$. The plane of the figure is selected to coincide with the common plane of these three points. The z axis is selected to have the same direction as $\mathbf{r}_{21} = \mathbf{r}_1 - \mathbf{r}_2$ and the angle between the vector $\mathbf{r}_{13}$ and the z axis is θ . The circles indicate two overlapping spheres with the same radius $\mathcal{R}$ centered at $\mathbf{r}_1$ and $\mathbf{r}_2$, respectively. The gray part shows the intersection of the two spheres. (b) A cross-section through the centers of three spheres (gray) with diameter d^{h}. Spheres 1 and 2 are centered at $\mathbf{r}_1$ and $\mathbf{r}_2$, respectively, and are overlapping, while sphere 3 is centered at $\mathbf{r}_3$. The two large circles show the cross-sections of two spheres with radii d^{h} that are centered at $\mathbf{r}_1$ and $\mathbf{r}_2$, respectively. The z axis and the angle θ are selected analogously to subfigure (a).

because $\hat{\mathbf{r}}_{21} \cdot \hat{\mathbf{r}}_{31} = -\hat{\mathbf{r}}_{21} \cdot \hat{\mathbf{r}}_{13} = -\cos\theta$. Let us use spherical polar coordinates for the vector $\mathbf{r}_{13}$, which has the polar angle θ. We take the limit $\mathbf{r}_2 \to \mathbf{r}_1$, whereby $r_{32} \to r_{31} = r_{13}$, and obtain

$$\left.\frac{dF(r_{12})}{dr_{12}}\right|_{r_{12}=0} = -2\pi \left[\int_0^{\pi} d\theta \, \sin\theta \cos\theta\right] \int_0^{\infty} dr_{13} \, r_{13}^2 \, f_a'(r_{13}) f_b(r_{13}) = 0$$

because the square bracket is zero. Thus, the derivative $F'(0)$ is equal to zero in this case.

If $f_a(r)$ has a discontinuity and/or a discontinuous derivative at, say, $r = \mathcal{R}_a$, we can split the radial integration in two parts $0 < r_{13} < \mathcal{R}_a$ and $\mathcal{R}_a < r_{13} < \infty$ and conclude that each part is zero for the same reason, so we have $F'(0) = 0$ in this case too. In case $f_b(r)$ has a discontinuity at, say, $r = \mathcal{R}_b \neq \mathcal{R}_a$, we can split the integration at $\mathcal{R}_b$ and likewise draw the same conclusion for each part. Since $f_b'(r)$ does not appear in the integrand, this derivative can be discontinuous at, for example, $r = \mathcal{R}_a$ or $\mathcal{R}_b$ without affecting the conclusion. (Note that since Equation 10.200 is symmetrical in f_a and f_b, the same considerations must apply to both functions.) To conclude, we have shown the assertion that $F'(0) = 0$ when the two functions do *not* have discontinuities for the same r.

Next, we treat the case where both functions have a discontinuity for $r = \mathcal{R}$ but are otherwise continuous. Say that the discontinuity step for $f_a(r)$ is $\Delta f_a(\mathcal{R}) \equiv f_a(\mathcal{R}^+) - f_a(\mathcal{R}^-)$ and that for $f_b(r)$ is $\Delta f_b(\mathcal{R})$. We will now use the fact that each discontinuity can be eliminated by adding a function that has an equal discontinuity in the opposite direction. Let us therefore consider the spherical step-down function defined by $s_{\mathcal{R}}(r) = 1$ for $r \leq \mathcal{R}$ and $s_{\mathcal{R}}(r) = 0$ for $r < \mathcal{R}$ and we multiply this function by $\Delta f_\alpha(\mathcal{R})$ for $\alpha = a, b$. The function $f_\alpha^{\mathrm{c}}(r) \equiv f_\alpha(r) + \Delta f_\alpha(\mathcal{R}) s_{\mathcal{R}}(r)$ is continuous for all r, so for the functions $f_a^{\mathrm{c}}(r)$ and $f_b^{\mathrm{c}}(r)$ the previous result $F'(0) = 0$ applies.

Let us now consider

$$
\begin{aligned}
& f_a(r_{13}) f_b(r_{32}) \\
= \; & \left(f_a^{\mathrm{c}}(r_{13}) - \Delta f_a(\mathcal{R}) s_{\mathcal{R}}(r_{13})\right)\left(f_b^{\mathrm{c}}(r_{32}) - \Delta f_b(\mathcal{R}) s_{\mathcal{R}}(r_{32})\right) \\
= \; & f_a^{\mathrm{c}}(r_{13}) f_b^{\mathrm{c}}(r_{32}) - \left[f_a^{\mathrm{c}}(r_{13}) s_{\mathcal{R}}(r_{32})\right] \Delta f_b(\mathcal{R}) \\
- \; & \left[s_{\mathcal{R}}(r_{13}) f_b^{\mathrm{c}}(r_{32})\right] \Delta f_a(\mathcal{R}) + \left[s_{\mathcal{R}}(r_{13}) s_{\mathcal{R}}(r_{32})\right] \Delta f_a(\mathcal{R}) \, \Delta f_b(\mathcal{R}).
\end{aligned}
$$

Since all terms on the rhs apart from the last one contain either two continuous functions or only one discontinuous function, they give the contribution zero to $F'(0)$. We can therefore conclude that

$$
\begin{aligned}
\left.\frac{dF(r_{12})}{dr_{12}}\right|_{r_{12}=0} &= \left[\frac{d}{dr_{12}} \int d\mathbf{r}_3 \, f_a(r_{13}) f_b(r_{32})\right]_{r_{12}=0} \\
&= \left[\frac{d}{dr_{12}} \int d\mathbf{r}_3 \, s_{\mathcal{R}}(r_{13}) s_{\mathcal{R}}(r_{32})\right]_{r_{12}=0} \Delta f_a(\mathcal{R}) \, \Delta f_b(\mathcal{R}).
\end{aligned}
$$

It remains to determine the value of the last square bracket. Since $s_{\mathcal{R}}(r_{i3})$ for $i = 1, 2$ is equal to 1 for all points $\mathbf{r}_3$ inside a sphere of radius $\mathcal{R}$ centered at $\mathbf{r}_i$ and is zero outside, the product $s_{\mathcal{R}}(r_{13}) s_{\mathcal{R}}(r_{32})$ is equal to one inside the intersection of two such spheres centered at $\mathbf{r}_1$ and $\mathbf{r}_2$, respectively and is zero outside, see Figure 10.13a where the intersection is shown in light gray. Thus the integral over $\mathbf{r}_3$ is equal to the volume of the intersection of two spheres of radius $\mathcal{R}$ with centers separated by distance r_{12}. The intersection volume of the two equal-sized spheres is equal to $\pi(4\mathcal{R} + r_{12})(2\mathcal{R} - r_{12})^2/12$, so we have

$$
\begin{aligned}
\frac{d}{dr_{12}} \int d\mathbf{r}_3 \, s_{\mathcal{R}}(r_{13}) s_{\mathcal{R}}(r_{32}) &= \frac{d}{dr_{12}} \left[\frac{\pi}{12}(4\mathcal{R} + r_{12})(2\mathcal{R} - r_{12})^2\right] \\
&= -\frac{\pi}{4}\left[4\mathcal{R}^2 - r_{12}^2\right].
\end{aligned}
$$

Therefore, we can finally conclude that

$$
\left.\frac{dF(r_{12})}{dr_{12}}\right|_{r_{12}=0} = -\pi\mathcal{R}^2 \Delta f_a(\mathcal{R}) \, \Delta f_b(\mathcal{R}). \tag{10.203}
$$

This equation includes the previous cases too since the rhs is zero unless both $f_a(r)$ and $f_b(r)$ have a discontinuity for the same r value, here denoted $\mathcal{R}$.

Now, we will apply this general result for convolutions to the OZ equation 10.199, where $F = \gamma^{(2)}$, $f_a = c^{(2)}$ and $f_b = h^{(2)}$. For systems where the pair potential $u(r)$ is continuous, both $c^{(2)}(r)$ and $h^{(2)}(r)$ are continuous so we obtain $dF(r)/dr|_{r=0} = d\gamma^{(2)}(r)/dr|_{r=0} = 0$, which is Equation 10.195. If $u(r)$ has a discontinuity for $r = \mathcal{R}$, $g^{(2)}(r)$ has a discontinuity there due to the factor $e^{-\beta u(r)}$ in Equation 10.185, while $y^{(2)}(r)$, $\gamma^{(2)}(r)$ and $b^{(2)}(r)$ are continuous. If the discontinuity step for $g^{(2)}(r)$ and therefore for $h^{(2)}(r)$ is equal to $\Delta g^{(2)}(\mathcal{R})$, the function $c^{(2)}(r)$ must also have the same discontinuity step for $r = \mathcal{R}$ because $c^{(2)}(r) = h^{(2)}(r) - \gamma^{(2)}(r)$. Applying Equation 10.203 we can conclude that

$$\left.\frac{d\gamma^{(2)}(r)}{dr}\right|_{r=0} = -\pi\mathcal{R}^2[\Delta g^{(2)}(\mathcal{R})]^2. \tag{10.204}$$

For a hard core fluid this result applies with $\mathcal{R} = d^{\mathrm{h}}$ and then we have

$$\left.\frac{d\gamma^{(2)}(r)}{dr}\right|_{r=0} = -\pi(d^{\mathrm{h}})^2[g^{(2)}(d^{\mathrm{h}+})]^2 \quad \text{(hard core fluid)} \tag{10.205}$$

because the step at $r = d^{\mathrm{h}}$ has the height $g^{(2)}(d^{\mathrm{h}+})$, i.e., the contact value for $g^{(2)}$. This completes the proof of Equation 10.196.

Finally we turn to the derivation of the second zero-separation theorem for $y^{(2)}$ for a hard sphere fluid, Equation 10.197. As a starting point we use the definition of the pair density distribution function $n^{(2)}(\mathbf{r}_1, \mathbf{r}_2)$ in the grand canonical ensemble (Equation I:5.105)

$$n^{(2)}(\mathbf{r}_1, \mathbf{r}_2) = \frac{1}{\Xi}\sum_{N=2}^{\infty}\frac{\zeta^N}{(N-2)!}\int d\mathbf{r}_3 \dots d\mathbf{r}_N e^{-\beta\breve{U}_N^{\mathrm{pot}}(\mathbf{r}_1,\mathbf{r}_2,\dots,\mathbf{r}_N)}, \tag{10.206}$$

where $\breve{U}_N^{\mathrm{pot}}(\mathbf{r}_1, \mathbf{r}_2, \dots, \mathbf{r}_N) = \sum_{\nu=1}^{N}\sum_{\nu'=\nu+1}^{N} u(r_{\nu\nu'})$ in the bulk fluid (cf. Equation 9.1) and Ξ is the grand canonical partition function (Equation I:3.37)

$$\Xi = \sum_{N=0}^{\infty}\frac{\zeta^N}{N!}\int d\mathbf{r}_1 \dots d\mathbf{r}_N e^{-\beta\breve{U}_N^{\mathrm{pot}}(\mathbf{r}_1,\mathbf{r}_2,\dots,\mathbf{r}_N)}. \tag{10.207}$$

In both equations we use the convention that there is no integral sign in the first term; the $N = 2$ term in Equation 10.206 is therefore $\zeta^2 e^{-\beta\breve{U}_2^{\mathrm{pot}}}/\Xi = \zeta^2 e^{-\beta u(r_{12})}/\Xi$ and the $N = 0$ term in Equation 10.207 is equal to 1.

Since $n^{(2)}(r_{12}) = (n^{\mathrm{b}})^2 g^{(2)}(r_{12}) = (n^{\mathrm{b}})^2 y^{(2)}(r_{12})e^{-\beta u(r_{12})}$, Equation 10.206 implies that

$$y^{(2)}(|\mathbf{r}_1 - \mathbf{r}_2|) = \frac{1}{(n^{\mathrm{b}})^2\,\Xi}\left[\zeta^2 + \sum_{N=3}^{\infty}\frac{\zeta^N}{(N-2)!}\int d\mathbf{r}_3 \dots d\mathbf{r}_N e^{-\beta\breve{U}_N^{\mathrm{rest}}(\mathbf{r}_1,\mathbf{r}_2,\dots,\mathbf{r}_N)}\right],$$

where $\breve{U}_N^{\text{rest}}(\mathbf{r}_1, \mathbf{r}_2, \ldots, \mathbf{r}_N) = \breve{U}_N^{\text{pot}}(\mathbf{r}_1, \mathbf{r}_2, \ldots, \mathbf{r}_N) - u(r_{12})$, that is,

$$\begin{aligned}\breve{U}_N^{\text{rest}}(\mathbf{r}_1, \mathbf{r}_2, \ldots, \mathbf{r}_N) &= \sum_{\nu'=3}^{N} u(r_{1\nu'}) + \sum_{\nu=2}^{N} \sum_{\nu'=\nu+1}^{N} u(r_{\nu\nu'}) \\ &= \sum_{\nu'=3}^{N} u(r_{1\nu'}) + \breve{U}_{N-1}^{\text{pot}}(\mathbf{r}_2, \mathbf{r}_3, \ldots, \mathbf{r}_N).\end{aligned}$$

Here, the last term is the total interaction energy between the $N-1$ particles numbered from 2 to N.

We start by calculating $\nabla_1 y^{(2)}(|\mathbf{r}_1 - \mathbf{r}_2|)$, whereby we will use the fact that $\nabla_i f(|\mathbf{r}_i - \mathbf{r}_j|) = \hat{\mathbf{r}}_{ji} f'(r_{ji})$, where $\hat{\mathbf{r}}_{ji} = \mathbf{r}_{ji}/r_{ji}$ and $\mathbf{r}_{ji} = \mathbf{r}_i - \mathbf{r}_j$. The integrand in $y^{(2)}$ can be written as

$$e^{-\beta \breve{U}_N^{\text{rest}}(\mathbf{r}_1, \ldots, \mathbf{r}_N)} = \left[\prod_{\nu'=3}^{N} e^{-\beta u(r_{1\nu'})}\right] e^{-\beta \breve{U}_{N-1}^{\text{pot}}(\mathbf{r}_2, \mathbf{r}_3, \ldots, \mathbf{r}_N)}.$$

Since $u(r) = \infty$ for $r < d^{\text{h}}$ and zero for $r > d^{\text{h}}$, we have $e^{-\beta u(r)} = \mathcal{H}(r - d^{\text{h}})$, where $\mathcal{H}(x)$ is the Heaviside step function that is equal to 0 for $x < 0$ and 1 for $x > 0$. The radial derivative of each factor $e^{-\beta u(r)}$ is therefore $de^{-\beta u(r)}/dr = \delta(r - d^{\text{h}})$, i.e., the Dirac delta function (the functions $\mathcal{H}(x)$ and $\delta(r) = d\mathcal{H}(x)/dx$ are treated Appendix 5A in Volume I). By taking the gradient of the integrand in $y^{(2)}$ and using the product rule we obtain

$$\begin{aligned}\nabla_1 e^{-\beta \breve{U}_N^{\text{rest}}(\mathbf{r}_1, \mathbf{r}_2, \ldots, \mathbf{r}_N)} &= \sum_{\nu=3}^{N} \hat{\mathbf{r}}_{\nu 1} \delta(r_{1\nu} - d^{\text{h}}) \left[\prod_{\substack{\nu'=3 \\ \nu' \neq \nu}}^{N} e^{-\beta u(r_{1\nu'})}\right] \\ &\times \; e^{-\beta \breve{U}_{N-1}^{\text{pot}}(\mathbf{r}_2, \mathbf{r}_3, \ldots, \mathbf{r}_N)}.\end{aligned}$$

Since $\breve{U}_{N-1}^{\text{pot}}$ is symmetric with respect to permutations of its variables, the integral over $\mathbf{r}_3 \ldots \mathbf{r}_N$ of each of these $N-2$ terms have exactly the same value (only the names of the integration variables differ), so the sum of the integrated terms is equal to $N-2$ times the value of the first one (with $\nu = 3$). Thus we conclude that

$$\begin{aligned}\nabla_1 y^{(2)}(|\mathbf{r}_1 - \mathbf{r}_2|) &= \int d\mathbf{r}_3 \, \hat{\mathbf{r}}_{31} \delta(r_{13} - d^{\text{h}}) \left[\frac{1}{(n^{\text{b}})^2 \, \Xi} \sum_{N=3}^{\infty} \frac{\zeta^N (N-2)}{(N-2)!}\right. \\ &\times \left. \int d\mathbf{r}_4 \ldots d\mathbf{r}_N e^{-\beta\left\{\sum_{\nu'=4}^{N} u(r_{1\nu'}) + \breve{U}_{N-1}^{\text{pot}}(\mathbf{r}_2, \mathbf{r}_3, \ldots, \mathbf{r}_N)\right\}}\right].\end{aligned}$$

The exponent in the integrand contains the sum $\sum_{\nu'=4}^{N} [u(r_{1\nu'}) + u(r_{2\nu'})]$, where $u(r_{2\nu'})$ appears in $\breve{U}_{N-1}^{\text{pot}}$. When $\mathbf{r}_1 \to \mathbf{r}_2$ this sum goes to $\sum_{\nu'=4}^{N} u(r_{2\nu'})$ due to the fact that hard spheres 1 and 2 then coincide and the presence of sphere 1 then cannot be experienced by the surrounding particles (cf. the earlier discussion). Hence, the exponent goes to $-\beta\breve{U}^{\text{pot}}_{N-1}(\mathbf{r}_2, \mathbf{r}_3, \ldots, \mathbf{r}_N)$ in this limit. This implies that the entire square bracket goes to $\zeta g^{(2)}(r_{23}) = \zeta y^{(2)}(r_{23})e^{-\beta u(r_{23})}$ when $\mathbf{r}_1 \to \mathbf{r}_2$, as can be realized from Equation 10.206 by means of a substitution where N is replaced by $N' - 1$. The square bracket contains the factor $e^{-\beta u(r_{23})}$, where $u(r_{23})$ appears in $\breve{U}^{\text{pot}}_{N-1}$. This factor, which is equal to $\mathcal{H}(r_{23} - d^{\text{h}})$, is present also before the limit is taken, so the entire expression for $\nabla_1 y^{(2)}$ when $r_{12} \neq 0$ has the structure

$$\nabla_1 y^{(2)}(|\mathbf{r}_1 - \mathbf{r}_2|) = \int d\mathbf{r}_3\, \hat{\mathbf{r}}_{31}\delta(r_{13} - d^{\text{h}})\mathcal{H}(r_{23} - d^{\text{h}})F(\mathbf{r}_1, \mathbf{r}_2, \mathbf{r}_3),$$

where $F(\mathbf{r}_1, \mathbf{r}_2, \mathbf{r}_3) \to \zeta y^{(2)}(r_{23})$ when $\mathbf{r}_1 \to \mathbf{r}_2$. By multiplying by $\hat{\mathbf{r}}_{21}$ and using Equation 10.201 we obtain

$$\frac{dy^{(2)}(r_{12})}{dr_{12}} = \int d\mathbf{r}_3\, \hat{\mathbf{r}}_{21} \cdot \hat{\mathbf{r}}_{31}\, \delta(r_{13} - d^{\text{h}})\mathcal{H}(r_{23} - d^{\text{h}})F(\mathbf{r}_1, \mathbf{r}_2, \mathbf{r}_3). \quad (10.208)$$

For the same reasons as in the evaluation of Equation 10.202, we have $\hat{\mathbf{r}}_{21} \cdot \hat{\mathbf{r}}_{31} = -\cos\theta$, whereby we have selected the z-axis to be directed along the unit vector $\hat{\mathbf{r}}_{21}$ and the angle between $\hat{\mathbf{r}}_{21}$ and $\hat{\mathbf{r}}_{13}$ is denoted θ, cf. Figure 10.13a. The relevant geometry for the present case is illustrated in Figure 10.13b, which shows three gray spheres: two overlapping spheres centered at $\mathbf{r}_1$ and $\mathbf{r}_2$, respectively, and a third sphere centered at $\mathbf{r}_3$. The points $\mathbf{r}_1$ and $\mathbf{r}_2$ are close to each other because we will later take the limit $\mathbf{r}_2 \to \mathbf{r}_1$. Sphere 2 is shown as the small dashed circle in the figure.

We are now ready to evaluate the integral over $\mathbf{r}_3$ in Equation 10.208. The integrand is zero except when $r_{13} = d^{\text{h}}$ and $r_{23} \geq d^{\text{h}}$ due to the factors $\delta(r_{13} - d^{\text{h}})$ and $\mathcal{H}(r_{23} - d^{\text{h}})$. In order to see what this implies, we turn to Figure 10.13b where the sphere 3 touches sphere 1, that is, we have $r_{13} = d^{\text{h}}$ as implied by the delta function. When $\mathbf{r}_3$ is changed such that sphere 3 moves along the periphery of sphere 1, the integrand is nonzero provided that $r_{23} \geq d^{\text{h}}$ due to the factor $\mathcal{H}(r_{23} - d^{\text{h}})$. This occurs whenever $\mathbf{r}_3$ lies outside the dotted circle centered at $\mathbf{r}_2$ in the figure. Hence, $r_{23} \geq d^{\text{h}}$ and $r_{13} = d^{\text{h}}$ are fulfilled for the part of the full large circle that lies above the dotted one. When $\mathbf{r}_3$ lies on the full large circle and the angle θ is changed, we have $r_{23} \geq d^{\text{h}}$ until sphere 3 starts to overlap with sphere 2, which is not permitted. This occurs when $\mathbf{r}_3$ reaches a crossing point between the full and the dotted large circles. Thus, the values of θ are restricted in this fashion. We now let $\mathbf{r}_2$ approach $\mathbf{r}_1$ and when their difference is infinitesimally small, the permitted range for θ becomes $|\theta| \leq \pi/2$, that is, the upper half circle.

Let us select point $\mathbf{r}_1$ as the origin of the coordinate system and take the limit $\mathbf{r}_2 \to \mathbf{r}_1$. If we express $\mathbf{r}_3$ in spherical polar coordinates, we accordingly have $0 \leq \phi \leq 2\pi$ and $0 \leq \theta \leq \pi/2$, where ϕ is the azimuthal angle of $\mathbf{r}_3$. It follows that in the limit $r_{12} \to 0$, we have from Equation 10.208

$$\left.\frac{dy^{(2)}(r_{12})}{dr_{12}}\right|_{r_{12}=0} = -2\pi \int_0^{\pi/2} d\theta \, \sin\theta \cos\theta \int_0^{\infty} dr_{13}\, r_{13}^2 \, \delta(r_{13} - d^{\rm h})\zeta y^{(2)}(r_{13})$$
$$= -\pi (d^{\rm h})^2 \zeta y^{(2)}(d^{\rm h})$$

where we have used the facts that $\mathbf{r}_3 = \mathbf{r}_{13}$ and $r_{23} = r_{13}$ (since $\mathbf{r}_1 = \mathbf{r}_2$ is the selected origin) and $\int_0^{\pi/2} d\theta \, \sin\theta \cos\theta = 1/2$. Equation 10.197 follows since $d \ln y^{(2)}(r)/dr|_{r=0} = [dy^{(2)}(r)/dr]_{r=0}/y^{(2)}(0)$ and, from Equation 10.186, $y^{(2)}(0) = e^{\beta\mu^{\rm ex,b}} = \zeta/n^{\rm b}$. This completes the proof of Equation 10.197.

Appendix 10B
Coupling-parameter dependence of bridge functions and the evaluation of $c^{(1)}(\mathbf{r})$

In this Appendix we treat some issues regarding the coupling-parameter integration for the evaluation of the function $c^{(1)\rm R}(\mathbf{r})$ in Section 10.1.5. This function is the residual part of the direct correlation function $c^{(1)}(\mathbf{r})$ that is not expressed explicitly in terms of pair correlation functions in Equation 9.201. We will also treat the corresponding integration in the evaluation of the bridge function $B_{\rm M}^{(2)}(\mathbf{r},\mathbf{r}')$ used in the alternative formalism by Martynov[87] and we will consider various approximations for this function (this is done in the text that is marked with ★ below).

The evaluation of $c^{(1)\rm R}(\mathbf{r}')$, which is defined in Equation 9.202, involves a coupling-parameter integration for a single particle that is partially coupled to the rest of the particles. For the integral equation closures of the type $b^{(2)} = f(\gamma^{*(2)})$, where $\gamma^{*(2)} = \gamma^{(2)} - \varphi^{\rm R}$, we can use the linear integration path and take $h^{(2)}(\mathbf{r},\mathbf{r}';\xi) = \xi h^{(2)}(\mathbf{r},\mathbf{r}')$, as explained in Section 10.1.5, so we have $\gamma^{(2)}(\mathbf{r},\mathbf{r}';\xi) = \xi\gamma^{(2)}(\mathbf{r},\mathbf{r}')$ from Equation 9.260 (alternatively we may take $h^{(2)}$ to be proportional to $\alpha(\xi)$, but this gives the same results). We then have $b^{(2)}(\mathbf{r},\mathbf{r}';\xi) = f(\xi\gamma^{(2)}(\mathbf{r},\mathbf{r}') - \varphi^{\rm R}(\mathbf{r},\mathbf{r}';\xi))$, where the ξ dependence of $\varphi^{\rm R}$ is unknown in the general case. From the definition of $c^{(1)\rm R}(\mathbf{r}')$ in Equation 9.202 we obtain

$$c^{(1)\rm R}(\mathbf{r}') = \int d\mathbf{r}\, n(\mathbf{r}) h^{(2)}(\mathbf{r},\mathbf{r}') \int_0^1 d\xi \, f\left(\xi\gamma^{(2)}(\mathbf{r},\mathbf{r}') - \varphi^{\rm R}(\mathbf{r},\mathbf{r}';\xi)\right) \quad (10.209)$$

and we cannot come much further unless we make some additional approximation. We will show that it reasonable to take $\varphi^{\rm R}(\mathbf{r},\mathbf{r}';\xi) \approx \xi\varphi^{\rm R}(\mathbf{r},\mathbf{r}')$, which implies that $\gamma^{*(2)}(\mathbf{r},\mathbf{r}';\xi) = \xi\gamma^{*(2)}(\mathbf{r},\mathbf{r}')$.

Let us first consider the case where we select $\varphi^{\rm R} = \beta u^{\rm L}$, so we have $b^{(2)} = f(\gamma^{(2)} - \beta u^{\rm L})$ and we let $h^{(2)}$ and $\gamma^{(2)}$ vary linearly with ξ. The pair

[87] This formalism is described in the shaded text box on page 77.

potential $u(\mathbf{r},\mathbf{r}';\xi)$ between the partially coupled particle and the other particles *does not* vary linearly with ξ, because it is the pair potential that makes the density around the partially coupled particle to be precisely equal to $n(\mathbf{r})[\xi h^{(2)}(\mathbf{r},\mathbf{r}')+1]$.[88] This is explained for the general case in connection to Equation 9.181, whereby in the current case (with a partially coupled particle at $\mathbf{r}'$) the external potential is $v(\mathbf{r}|\mathbf{r}';\xi)=v(\mathbf{r})+u(\mathbf{r},\mathbf{r}';\xi)$ and the density is $n(\mathbf{r}|\mathbf{r}';\xi)=n(\mathbf{r})g^{(2)}(\mathbf{r},\mathbf{r}';\xi)$, which are subject to the conditions 9.197 at $\xi=0$ and $\xi=1$.

However, when ξ is small $u(\mathbf{r},\mathbf{r}';\xi)$ varies linearly with ξ as a good approximation. In order to see this and find $u(\mathbf{r},\mathbf{r}';\xi)$ for small ξ we ask: What potential $\delta v(\mathbf{r})$ gives rise to the deviation $\delta n(\mathbf{r}'')$ from the density $n(\mathbf{r}'')$, where $\delta n(\mathbf{r}'')$ in the present case is given by $\delta n(\mathbf{r}''|\mathbf{r}')=\xi n(\mathbf{r}'')h^{(2)}(\mathbf{r}'',\mathbf{r}')$? The second Yvon equation 9.43 gives the answer because it tells what variation in potential gives rise to a prescribed small change in density. When we insert $\delta n(\mathbf{r}''|\mathbf{r}')$ into this equation we obtain the variation $\delta v(\mathbf{r}|\mathbf{r}')$, which equals $u(\mathbf{r},\mathbf{r}';\xi)$ for small ξ, so we have

$$\begin{aligned}\delta v(\mathbf{r}|\mathbf{r}') &= -\frac{1}{\beta}\left[\frac{\xi n(\mathbf{r})h^{(2)}(\mathbf{r},\mathbf{r}')}{n(\mathbf{r})} - \int d\mathbf{r}''\, c^{(2)}(\mathbf{r}'',\mathbf{r})\left\{\xi n(\mathbf{r}'')h^{(2)}(\mathbf{r}'',\mathbf{r}')\right\}\right]\\ &= -\frac{\xi c^{(2)}(\mathbf{r},\mathbf{r}')}{\beta},\end{aligned}$$

where we have used the OZ equation 9.40. Thus we have

$$u(\mathbf{r},\mathbf{r}';\xi)\approx -\xi k_B T c^{(2)}(\mathbf{r},\mathbf{r}') \quad \text{for small } \xi.$$

Now, since u^{L} is the long-range part of u and since we have $-k_BTc^{(2)}(\mathbf{r},\mathbf{r}')\sim u(\mathbf{r},\mathbf{r}')$ for large $|\mathbf{r}-\mathbf{r}'|$ (Equation 9.42), it follows that $u^{\mathrm{L}}(\mathbf{r},\mathbf{r}';\xi)\approx\xi u(\mathbf{r},\mathbf{r}')$ when $|\mathbf{r}-\mathbf{r}'|$ is sufficiently large. Since the tails of u, $c^{(2)}$ and $h^{(2)}$ for large $|\mathbf{r}-\mathbf{r}'|$ are small, we have $u^{\mathrm{L}}(\mathbf{r},\mathbf{r}';\xi)\approx\xi u(\mathbf{r},\mathbf{r}')$ in the tail not only for small ξ, but for all values $0\le\xi\le1$ because the deviation in density from the bulk value is small. In general, the ξ dependence of $u^{\mathrm{L}}(\mathbf{r},\mathbf{r}';\xi)$ is unknown away from the tail region. However, since the splitting of u into its short and long range parts, $u=u^{\mathrm{S}}+u^{\mathrm{L}}$, contains some freedom, we may as an approximation *define* $u^{\mathrm{L}}(\mathbf{r},\mathbf{r}';\xi)=\xi u^{\mathrm{L}}(\mathbf{r},\mathbf{r}')$ for all $|\mathbf{r}-\mathbf{r}'|$ and for $0\le\xi\le1$. We thereby let the short-range part be equal to $u^{\mathrm{S}}(\mathbf{r},\mathbf{r}';\xi)=u(\mathbf{r},\mathbf{r}';\xi)-\xi u^{\mathrm{L}}(\mathbf{r},\mathbf{r}')$. Thus it is reasonable to take $u^{\mathrm{L}}(\mathbf{r},\mathbf{r}';\xi)\approx\xi u^{\mathrm{L}}(\mathbf{r},\mathbf{r}')$. These results are used in Section 10.1.5, Equation 10.59, for the case $\varphi^{\mathrm{R}}=\beta u^{\mathrm{L}}$.

Second, we consider the more general closure $b^{(2)}=f(\gamma^{*(2)})=f(\gamma^{(2)}-\varphi^{\mathrm{R}})$ with a different choice of φ^{R} than βu^{L}. Also in this case one may as an approximation assume that $\varphi^{\mathrm{R}}(\mathbf{r},\mathbf{r}';\xi)\approx\xi\varphi^{\mathrm{R}}(\mathbf{r},\mathbf{r}')$, which implies that $\gamma^{*(2)}(\mathbf{r},\mathbf{r}';\xi)=\xi\gamma^{*(2)}(\mathbf{r},\mathbf{r}')$. This is a sensible approximation for the following reasons.

[88] Likewise, if one uses a linear ξ dependence for $u(\mathbf{r},\mathbf{r}';\xi)$, the pair correlation function $h^{(2)}(\mathbf{r},\mathbf{r}';\xi)$ does not vary linearly with ξ.

For the linear coupling-parameter integration path where $h^{(2)}(\mathbf{r},\mathbf{r}';\xi) = \xi h^{(2)}(\mathbf{r},\mathbf{r}')$, we see from the expansion of $b^{(2)}(\mathbf{r},\mathbf{r}';\xi)$ in Equation 9.261 of Appendix 9C that the exact $b^{(2)}$ can be written

$$b^{(2)}(\mathbf{r},\mathbf{r}';\xi) = \sum_{m=3}^{\infty} \frac{A_m(\mathbf{r},\mathbf{r}')\xi^{m-1}}{(m-1)!} = \frac{A_3\xi^2}{2} + \frac{A_4\xi^3}{6} + \frac{A_5\xi^4}{24} + \ldots, \quad (10.210)$$

where the functions A_m are given by

$$A_m = A_m(\mathbf{r},\mathbf{r}') = \int d\mathbf{s}_1 \ldots d\mathbf{s}_{m-1}\, c^{(m)}(\mathbf{r},\mathbf{s}_1,\ldots,\mathbf{s}_{m-1}) \prod_{i=1}^{m-1} n(\mathbf{s}_i)h^{(2)}(\mathbf{s}_i,\mathbf{r}').$$

For the approximation $b^{(2)} \approx f(\gamma^{*(2)})$ with $\varphi^{\mathrm{R}}(\mathbf{r},\mathbf{r}';\xi) = \xi\varphi^{\mathrm{R}}(\mathbf{r},\mathbf{r}')$ one has $\gamma^{*(2)}(\mathbf{r},\mathbf{r}';\xi) = \xi\gamma^{*(2)}(\mathbf{r},\mathbf{r}')$ for the linear path. The function $f(\gamma^{*(2)})$ can be expanded in a power series and one obtains, cf. Equation 10.53,

$$b^{(2)} \approx f(\gamma^{*(2)}) = \sum_{\nu=2}^{\infty} \frac{a_\nu}{\nu!}[\gamma^{*(2)}]^\nu$$

that always starts with a square term. By inserting $\gamma^{*(2)}(\mathbf{r},\mathbf{r}';\xi) = \xi\gamma^{*(2)}(\mathbf{r},\mathbf{r}')$ and comparing with the exact expansion 10.210, it follows for this type of closures that A_m is approximated by

$$A_m(\mathbf{r},\mathbf{r}') \approx a_{m-1}[\gamma^{*(2)}(\mathbf{r},\mathbf{r}')]^{m-1} \quad (10.211)$$

for $m \geq 3$. Both the exact and the approximate $b^{(2)}(\mathbf{r},\mathbf{r}';\xi)$ can hence be written as a power series in ξ that starts with a square term when one uses the linear path $h^{(2)}(\mathbf{r},\mathbf{r}';\xi) = \xi h^{(2)}(\mathbf{r},\mathbf{r}')$. Thus it is sensible to use a $\varphi^{\mathrm{R}}(\mathbf{r},\mathbf{r}';\xi)$ that is proportional to ξ in these closures.

★ In the rest of this Appendix we will treat the function $B_{\mathrm{M}}^{(2)}(\mathbf{r},\mathbf{r}')$ used in the alternative formalism by Martynov, which was introduced in the shaded text box on page 77. This function can also be evaluated by means of a coupling parameter integration.

One can determine $c^{(1)}(\mathbf{r})$ from Equation 9.212, which is equivalent to Equations 9.201 (likewise μ^{ex} can be determined Equation 9.217 that is equivalent to Equation 9.203). As we have seen, the difficulty in the calculation of $c^{(1)}(\mathbf{r})$ and μ^{ex} lies in the evaluation of $c^{(1)\mathrm{R}}(\mathbf{r})$ and we have

$$c^{(1)\mathrm{R}}(\mathbf{r}') = \frac{1}{2}\int d\mathbf{r}\, n(\mathbf{r})h^{(2)}(\mathbf{r},\mathbf{r}')\left[b^{(2)}(\mathbf{r},\mathbf{r}') - B_{\mathrm{M}}^{(2)}(\mathbf{r},\mathbf{r}')\right] \quad (10.212)$$

(Equation 9.215). The function $B_{\mathrm{M}}^{(2)}$ is explicitly given by the expansion shown in Equation 9.214, which differs from the corresponding expansion for $b^{(2)}$ in Equation 9.160 only by the coefficients in front of the integrals.

We can obtain the functional expansion for the exact $B_{\mathrm{M}}^{(2)}$ that corresponds to the expansion of $b^{(2)}$ in Equation 10.210 by generalizing Equation 9.214 to any ξ value, whereby we get

$$B_{\mathrm{M}}^{(2)}(\mathbf{r},\mathbf{r}';\xi) = \sum_{m=3}^{\infty} \frac{A_m(\mathbf{r},\mathbf{r}')(m-2)\xi^{m-1}}{m!} = \frac{A_3\xi^2}{6} + \frac{A_4\xi^3}{12} + \frac{A_5\xi^4}{40} + \dots ,$$

where A_m is the same as for $b^{(2)}$ in Equation 10.210. The only difference between the expansions for $b^{(2)}$ and $B_{\mathrm{M}}^{(2)}$ is that each term of $b^{(2)}$ equals A_m times the factor $\xi^{m-1}/(m-1)!$, while each term of $B_{\mathrm{M}}^{(2)}$ equals A_m times the factor $(m-2)\xi^{m-1}/m!$. It is easy to manipulate a power series with the former factors into one with the latter be means of a series of integrations and differentiations. This can be done in the following manner

$$\xi^2 \frac{d}{d\xi}\left[\frac{1}{\xi^2}\int_0^{\xi} d\xi' \frac{\xi'^{\,m-1}}{(m-1)!}\right] = \frac{(m-2)\xi^{m-1}}{m!}.$$

Therefore, it follows that

$$\xi^2 \frac{d}{d\xi}\left[\frac{1}{\xi^2}\int_0^{\xi} d\xi'\, b^{(2)}(\mathbf{r},\mathbf{r}';\xi')\right] = B_{\mathrm{M}}^{(2)}(\mathbf{r},\mathbf{r}';\xi),$$

which for $\xi = 1$ can be written as

$$B_{\mathrm{M}}^{(2)}(\mathbf{r},\mathbf{r}') = b^{(2)}(\mathbf{r},\mathbf{r}') - 2\int_0^1 d\xi'\, b^{(2)}(\mathbf{r},\mathbf{r}';\xi') \quad \text{(linear path)}. \tag{10.213}$$

This expression is exact, but is valid only for the linear coupling-parameter integration path.

If the ξ expansion for $b^{(2)}$ in Equation 10.210 is truncated after the first term so that $b^{(2)}(\mathbf{r},\mathbf{r}';\xi) \approx A_3(\mathbf{r},\mathbf{r}')\xi^2/2$, we get from Equation 10.213 $B_{\mathrm{M}}^{(2)}(\mathbf{r},\mathbf{r}') \approx A_3(\mathbf{r},\mathbf{r}')/6 \approx b^{(2)}(\mathbf{r},\mathbf{r}')/3$. Thereby we have obtained the approximation

$$B_{\mathrm{M}}^{(2)} \approx b^{(2)}/3, \tag{10.214}$$

which was first suggested by Kiselyov and Martynov.[89] This implies that if we use the approximation $b^{(2)} \approx f(\gamma^{*(2)})$, we have $B_{\mathrm{M}}^{(2)} \approx f(\gamma^{*(2)})/3$. By inserting this approximation into Equation 10.212 we obtain $c^{(1)\mathrm{R}}$ in terms of the pair functions of the actual system. This approximation is useful in many cases, but in general it is not highly accurate.

[89] O. E. Kiselyov and G. A. Martynov J. Chem. Phys., **93** (1990) 1942 (https://doi.org/10.1063/1.459071). They based the approximation on the coefficient for the first bridge diagram in $b^{(2)}$, which is defined in Equation 9.161.

It is better to use Equation 10.213 as it stands in order to obtain an explicit expression for $B_{\mathrm{M}}^{(2)}$ in terms of the pair function $\gamma^{(2)}$. When we use the approximation $b^{(2)}(\mathbf{r},\mathbf{r}';\xi) = f(\gamma^{*(2)}(\mathbf{r},\mathbf{r}';\xi))$ with $\gamma^{*(2)}(\mathbf{r},\mathbf{r}';\xi) = \xi\gamma^{*(2)}(\mathbf{r},\mathbf{r}')$, we get

$$B_{\mathrm{M}}^{(2)}(\mathbf{r},\mathbf{r}') = f(\gamma^{*(2)}(\mathbf{r},\mathbf{r}')) - \frac{2}{\gamma^{*(2)}(\mathbf{r},\mathbf{r}')} \int\limits_{0}^{\gamma^{*(2)}(\mathbf{r},\mathbf{r}')} dx\, f(x). \qquad (10.215)$$

This result can be directly used in practical calculations. The same result is obtained if we assume that $\gamma^{*(2)}(\mathbf{r},\mathbf{r}';\xi) = \alpha(\xi)\gamma^{*(2)}(\mathbf{r},\mathbf{r}')$, where the function $\alpha(\xi)$ is differentiable and satisfies $\alpha(0) = 0$ and $\alpha(1) = 1$. Note that one obtains Equation 10.57 for $c^{(1)\mathrm{R}}$ when Equation 10.215 is inserted into Equation 10.212. A corresponding expression for the singlet function $B_{\mathrm{M}}^{(1)}(\mathbf{r}')$ can be obtained in these approximations by inserting Equation 10.215 into Equation 9.213.

There are other approximate ways to obtain $B_{\mathrm{M}}^{(2)}$ and hence $c^{(1)}$ and μ^{ex} without doing any explicit coupling parameter integration. For bulk fluids a generalization of Equation 10.214 has been suggested by Bomont el al.[90] namely

$$B_{\mathrm{M}}^{(2)\mathrm{b}}(r) \approx \tau'\, b^{(2)\mathrm{b}}(r), \qquad (10.216)$$

where τ' is a parameter that depends on T and n^{b}. This approximation has successfully been used for Lennard-Jones and hard sphere fluids, whereby τ' has been determined from the thermodynamic consistency criterium that the Gibbs-Duhems relationship $n^{\mathrm{b}}d\mu = dP$ is fulfilled, which via Equations 9.189 and 9.190 can be expressed as the condition that $\beta\,\partial\mu^{\mathrm{ex}}/dn^{\mathrm{b}} = -\int d\mathbf{r}\, c^{(2)\mathrm{b}}(r)$.

Let us see why this approximation can work. From the mean value theorem[91] of integral calculus it follows from Equation 10.213 (applied to the bulk phase) that

$$B_{\mathrm{M}}^{(2)\mathrm{b}}(r) = b^{(2)\mathrm{b}}(r) - 2b^{(2)\mathrm{b}}\left(r;\bar{\xi}(r)\right) \quad \text{(linear path)},$$

where $\bar{\xi}$ lies in the interval $0 < \bar{\xi} < 1$. This expression is exact, but $\bar{\xi}(r)$ is in general unknown. We accordingly have

$$\frac{B_{\mathrm{M}}^{(2)\mathrm{b}}(r)}{b^{(2)\mathrm{b}}(r)} = 1 - 2\,\frac{b^{(2)\mathrm{b}}\left(r;\bar{\xi}(r)\right)}{b^{(2)\mathrm{b}}(r)} \quad \text{(linear path)}, \qquad (10.217)$$

where the last ratio involves $b^{(2)\mathrm{b}}$ for two different coupling strengths for each r. In Equation 10.216 it is assumed that the rhs of Equation 10.217 is

[90] J.-M. Bomont, J. Chem. Phys., **119** (2003) 11484 (https://doi.org/10.1063/1.1623184); J.-M. Bomont and J.-L. Bretonnet J. Chem. Phys., **121** (2004) 1548 (https://doi.org/10.1063/1.1764772).

[91] The mean value theorem states that for a continuous function $f(x)$ we have $\int_a^b dx\, f(x) = (b-a)f(\bar{x})$, where $a < \bar{x} < b$.

approximately equal to a constant τ' independent of r, which is unlikely the case. However, in this approximation, τ' is determined from the consistency criterium involving μ^{ex}, which is the quantity that is evaluated when one uses the assumption 10.216. Thus an interpretation of this approximation is that the value of τ' corresponds to the approximation that

$$\tau' \approx 1 - 2 \left\langle \frac{b^{(2)\mathrm{b}}\left(r; \bar{\xi}(r)\right)}{b^{(2)\mathrm{b}}(r; 1)} \right\rangle_{\text{most important } r},$$

where the average is taken over the r values that matters the most for the contributions from $B_{\mathrm{M}}^{(2)\mathrm{b}}(r)$ in the evaluation of μ^{ex}. These r values are, of course, not known a priori and neither is $\bar{\xi}(r)$, but this does not matter since τ' is determined by other means.

CHAPTER 11

Distribution Function Theory for Nonspherical Particles

So far we have treated distribution and correlation functions for fluids consisting of spherical particles. In this chapter we will generalize the distribution function theory to systems with nonspherical rigid particles. For such particles the interparticle interactions depend on the distances between the particles and their orientations relative to each other. The interactions with an external field depend on the positions and orientations of the particles relative to the sources of this field. Initially we will set up the formal basis for various probability distributions and correlation functions for one-component systems. Distribution function theory in the canonical and the grand canonical ensembles is thereby explored in considerable detail, including probability densities, particle density distributions and different types of correlation functions of various orders. Functional derivative relationships between the various correlation functions, density distributions and free energy entities are established. Many-body chemical potentials are introduced and utilized in the formulation of the theory. Finally, we will generalize the formalism to many components and introduce some theories for effective solute-solute interactions in solutions, where the solvent is treated implicitly.

11.1 CLASSICAL STATISTICAL MECHANICS FOR NONSPHERICAL PARTICLES

We will consider systems containing rigid particles of any size and shape (the particles we treat here may be rigid molecules or any other type of particles). The particles interact with each other and may also interact with an external field. At a certain instance of time the center of mass of particle ν is located at

DOI: 10.1201/9781003286882-11

coordinates $\mathbf{r}_\nu = (x_\nu, y_\nu, z_\nu)$ and the particle has a certain orientation relative to the coordinate frame (the "laboratory frame"). The orientation is specified by a variable $\boldsymbol{\omega}_\nu$ that describes the orientational angles of a body-fixed "particle frame" relative to the laboratory frame of reference, see Appendix 11A. The particle frame is attached to the particle and follows its translational and rotational motions. We will use the particle frame with principal axes, in which the moment of inertia tensor of the particle takes the form of a diagonal matrix. The orientational variable is normalized so that $\int d\boldsymbol{\omega}_\nu = 1$.

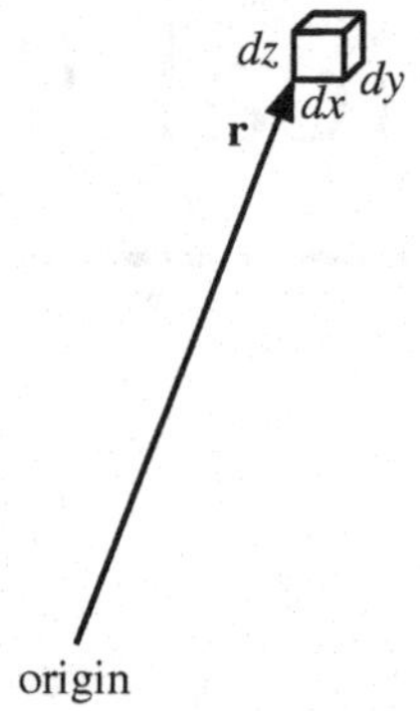

Figure 11.1 A volume element $d\mathbf{r} = dx\,dy\,dz$ located at position $\mathbf{r}$.

The number density of particles at point $\mathbf{r}$, irrespectively of their orientations, is equal to $n^{(1)}(\mathbf{r}) \equiv n(\mathbf{r})$, so the average number of particles with centers in the volume element $d\mathbf{r}$ at position $\mathbf{r}$ is equal to $n(\mathbf{r})d\mathbf{r}$ (cf. figure 11.1). In order to include information regarding the particle orientations, we introduce the number density $\mathfrak{n}^{(1)}(\mathbf{r}, \boldsymbol{\omega})$, which is defined so that $\mathfrak{n}^{(1)}(\mathbf{r}, \boldsymbol{\omega})d\mathbf{r}\,d\boldsymbol{\omega}$ is equal to the average number of particles with centers within $d\mathbf{r}$ from $\mathbf{r}$ and with orientations within $d\boldsymbol{\omega}$ from $\boldsymbol{\omega}$. We have the singlet density distribution

$$n^{(1)}(\mathbf{r}) = \int d\boldsymbol{\omega}\, \mathfrak{n}^{(1)}(\mathbf{r}, \boldsymbol{\omega}).$$

Note the difference in the letter "n" in $\mathfrak{n}^{(1)}$ and $n^{(1)}$. Throughout this treatise $\mathfrak{n}$ includes the orientational variable and n does not. The same is true for the higher order density distributions $\mathfrak{n}^{(l)}$ and $n^{(l)}$ that we will introduce later.

The *instantaneous potential energy* of the system depends on the positions and orientations of the particles, $\breve{U}^{\text{pot}} = \breve{U}^{\text{pot}}(\mathbf{r}_1, \boldsymbol{\omega}_1, \mathbf{r}_2, \boldsymbol{\omega}_2, \mathbf{r}_3, \boldsymbol{\omega}_3, \ldots)$. The total energy is given by $\breve{U}^{\text{tot}} = \breve{U}^{\text{pot}} + \breve{U}^{\text{kin}}$, where $\breve{U}^{\text{kin}}$ is the *total kinetic energy* that is the sum of the translational and rotational kinetic energies of each particle, $\mathfrak{u}_\nu^{\text{tr}}$ and $\mathfrak{u}_\nu^{\text{rot}}$, respectively. The translational energy is given by

$$\breve{U}^{\text{tr}} = \sum_\nu \mathfrak{u}_\nu^{\text{tr}} = \sum_\nu \frac{p_\nu^2}{2m_\nu},$$

where ν runs over all particles in the system, $p_\nu = |\mathbf{p}_\nu|$ is the momentum and m_ν is the mass of particle ν. The rotational energy is

$$\breve{U}^{\rm rot} = \sum_\nu \mathrm{u}_\nu^{\rm rot} = \sum_\nu \left[\frac{j_{x,\nu}^2}{2I_{x,\nu}} + \frac{j_{y,\nu}^2}{2I_{y,\nu}} + \frac{j_{z,\nu}^2}{2I_{z,\nu}} \right],$$

where $\mathbf{j}_\nu = (j_{x,\nu}, j_{y,\nu}, j_{z,\nu})$ is the angular momentum and $I_{\alpha,\nu}$ for $\alpha = x, y, z$ are the $\alpha\alpha$ components of the moment of inertia tensor (in the principal particle frame) of particle ν. The interaction between the particles can be a pair interaction potential $u(\mathbf{r}, \boldsymbol{\omega}, \mathbf{r}', \boldsymbol{\omega}')$, whereby we have

$$\breve{U}^{\rm pot}(\mathbf{r}_1, \boldsymbol{\omega}_1, \mathbf{r}_2, \boldsymbol{\omega}_2, \ldots) = \sum_{\nu,\nu'>\nu} u(\mathbf{r}_\nu, \boldsymbol{\omega}_\nu, \mathbf{r}_{\nu'}, \boldsymbol{\omega}_{\nu'}) + \sum_\nu v(\mathbf{r}_\nu, \boldsymbol{\omega}_\nu),$$

where $v(\mathbf{r}, \boldsymbol{\omega})$ is the external potential, but this will *not* be assumed in general.

We can obtain the mean values of the properties of the system by following the motion of the particles in time, i.e. $(\mathbf{r}_\nu(t), \boldsymbol{\omega}_\nu(t))$ for each particle ν, and form the time average. As we did for spherical particles in Volume I of this treatise, we can alternatively calculate the ensemble average of the same property, which according to the ergodic hypothesis gives the same result. In the latter case, we need the probability to observe each possible outcome of values for the set of coordinates $(\mathbf{r}_1, \boldsymbol{\omega}_1, \mathbf{r}_2, \boldsymbol{\omega}_2, \mathbf{r}_3, \boldsymbol{\omega}_3, \ldots)$ and momenta $(\mathbf{p}_1, \mathbf{j}_1, \mathbf{p}_2, \mathbf{j}_2, \mathbf{p}_3, \mathbf{j}_3, \ldots)$.

To start with we assume that there are N particles in the system. The spatial positions for these particles is denoted $\mathbf{r}^N \equiv \mathbf{r}_1, \mathbf{r}_2, \ldots, \mathbf{r}_N$ and the orientational coordinates $\boldsymbol{\omega}^N \equiv \boldsymbol{\omega}_1, \boldsymbol{\omega}_2, \ldots, \boldsymbol{\omega}_N$. Likewise we have $\mathbf{p}^N \equiv \mathbf{p}_1, \mathbf{p}_2, \ldots, \mathbf{p}_N$ and $\mathbf{j}^N = \mathbf{j}_1, \mathbf{j}_2, \ldots, \mathbf{j}_N$. The **probability density** $\mathcal{P}_N^{(N)\rm tot}$ to observe the N particles at specific positions and with specific orientations and momenta is in the canonical ensemble given by the Boltzmann distribution[1]

$$\mathcal{P}_N^{(N)\rm tot}(\mathbf{r}^N, \boldsymbol{\omega}^N, \mathbf{p}^N, \mathbf{j}^N) = C_{\rm tot} e^{-\beta\left[\breve{U}_N^{\rm pot}(\mathbf{r}^N, \boldsymbol{\omega}^N) + \breve{U}_N^{\rm kin}(\mathbf{p}^N, \mathbf{j}^N)\right]},$$

where $C_{\rm tot}$ is the normalization constant given by

$$C_{\rm tot}^{-1} = \int d\mathbf{r}^N d\boldsymbol{\omega}^N d\mathbf{p}^N d\mathbf{j}^N e^{-\beta\left[\breve{U}_N^{\rm pot} + \breve{U}_N^{\rm kin}\right]} \qquad (11.1)$$

with differential $d\boldsymbol{\omega}^N \equiv d\boldsymbol{\omega}_1 d\boldsymbol{\omega}_2 \ldots d\boldsymbol{\omega}_N$ and likewise for $d\mathbf{j}^N$. As before, subscript N for the various entities, for example $\mathcal{P}_N^{(N)\rm tot}$ and $\breve{U}_N^{\rm pot}$, indicates that there are N particles present in the system.

The interpretation of $\mathcal{P}_N^{(N)\rm tot}$ is that the product $\mathcal{P}_N^{(N)\rm tot} d\mathbf{r}^N d\boldsymbol{\omega}^N d\mathbf{p}^N d\mathbf{j}^N$ is equal to the probability to *simultaneously* observe

[1] The corresponding expression for spherical particles is $\mathcal{P}_N^{(N)\rm tot}(\mathbf{r}^N, \mathbf{p}^N) = C_{\rm tot} e^{-\beta\left[\breve{U}_N^{\rm pot}(\mathbf{r}^N) + \breve{U}_N^{\rm kin}(\mathbf{p}^N)\right]}$, Equation I:3.2.

particle 1 with its center in volume element $d\mathbf{r}_1$ at position $\mathbf{r}_1$,

particle 2 with its center in volume element $d\mathbf{r}_2$ at position $\mathbf{r}_2$,

... ,

and particle N with its center in volume element $d\mathbf{r}_N$ at position $\mathbf{r}_N$ ***when***

particle ν, for $\nu = 1, 2, \ldots, N$, has an orientation within $d\boldsymbol{\omega}_\nu$ from $\boldsymbol{\omega}_\nu$, ***and***

momenta within $d\mathbf{p}_\nu$ and $d\mathbf{j}_\nu$ from $\mathbf{p}_\nu$ and $\mathbf{j}_\nu$, respectively.

The degrees of freedom of the particles include both spatial and rotational coordinates. The latter are two per particle for linear molecules and three otherwise (a rotation around the axis of symmetry in the linear case does not count).

The canonical partition function is

$$Q_N = \frac{\omega_\circ^N}{N!h^{N\varpi}} \int d\mathbf{r}^N d\boldsymbol{\omega}^N d\mathbf{p}^N d\mathbf{j}^N e^{-\beta[\check{U}_N^{\mathrm{pot}} + \check{U}_N^{\mathrm{kin}}]},$$

where $\omega_\circ = 4\pi$ for the linear case and $\omega_\circ = 8\pi^2$ otherwise (see Appendix 11A) and ϖ is the number of (classical) degrees of freedom for each particle; $\varpi = 5$ for linear molecules and $\varpi = 6$ otherwise. Q_N differ from the inverse normalization factor C_{tot}^{-1} in Equation 11.1 by the prefactor, where $\omega_\circ^N$ originates from the normalization of the orientation variables $\boldsymbol{\omega}_\nu$ and $N!h^{N\varpi}$ arises from quantum effects, whereby $N!$ accounts for the indistinguishability of the particles (cf. Section 3.2 in Volume I).

Like for the spherical case, $\mathcal{P}_N^{(N)\mathrm{tot}}$ factorizes into a potential and a kinetic part $\mathcal{P}_N^{(N)\mathrm{tot}} = \mathcal{P}_N^{(N)\mathrm{pot}} \mathcal{P}_N^{(N)\mathrm{kin}}$ since $\check{U}_N^{\mathrm{pot}}$ and $\check{U}_N^{\mathrm{kin}}$ are additive. $\mathcal{P}_N^{(N)\mathrm{kin}}$ can also be factorized[2] and we have

$$\mathcal{P}_N^{(N)\mathrm{kin}} = \prod_{\nu=1}^{N} \mathcal{P}_\nu^{(1)\mathrm{tr}}(\mathbf{p}_\nu)\mathcal{P}_\nu^{(1)\mathrm{rot}}(\mathbf{j}_\nu),$$

where (Equation I:3.7)

$$\mathcal{P}_\nu^{(1)\mathrm{tr}}(\mathbf{p}_\nu) = \frac{e^{-\beta(p_{\nu,x}^2 + p_{\nu,y}^2 + p_{\nu,z}^2)/(2m_\nu)}}{(2\pi m_\nu k_B T)^{3/2}}$$

and

$$\mathcal{P}_\nu^{(1)\mathrm{rot}}(\mathbf{j}_\nu) = \frac{e^{-\beta j_{x,\nu}^2/(2I_{x,\nu})}}{(2\pi I_{x,\nu} k_B T)^{1/2}} \cdot \frac{e^{-\beta j_{y,\nu}^2/(2I_{y,\nu})}}{(2\pi I_{y,\nu} k_B T)^{1/2}} \cdot \frac{e^{-\beta j_{z,\nu}^2/(2I_{z,\nu})}}{(2\pi I_{z,\nu} k_B T)^{1/2}}.$$

[2]See Chapter 3 in: C. G. Gray and K. E. Gubbins, 1984, *Theory of molecular fluids, Volume 1: Fundamentals*, Oxford: Clarendon Press.

For linear molecules the last factor in $\mathcal{P}_\nu^{(1)\mathrm{rot}}$ is missing (the z axis of the particle frame has been placed along the symmetry axis).

The probability density $\mathcal{P}_N^{(N)\mathrm{pot}}(\mathbf{r}^N, \boldsymbol{\omega}^N)$ for positions and orientations of the particles is the entity that we will be concerned with in this treatise. Like before, we can include the kinetic probability density whenever we need it since it is explicitly known. The set of coordinates $(\mathbf{r}^N, \boldsymbol{\omega}^N)$ constitutes the **configuration** of the particles. For simplicity in notation we will introduce the composite variable

$$\mathbf{R}_\nu = (\mathbf{r}_\nu, \boldsymbol{\omega}_\nu)$$

for the position and orientation of particle ν and likewise $\mathbf{R}^N \equiv \mathbf{R}_1, \mathbf{R}_2, \ldots, \mathbf{R}_N$ and $d\mathbf{R}^N \equiv d\mathbf{R}_1 d\mathbf{R}_2 \ldots d\mathbf{R}_N$. This means that $\mathbf{R}^N$ corresponds to $(\mathbf{r}^N, \boldsymbol{\omega}^N)$ and $d\mathbf{R}^N$ corresponds to $d\mathbf{r}^N d\boldsymbol{\omega}^N$ in the previous formulas. We will, however, explicitly write $(\mathbf{r}_\nu, \boldsymbol{\omega}_\nu)$ instead of $\mathbf{R}_\nu$ when this is needed for clarity.

Like in the spherical case we will drop "pot" in the superscript for the configurational probability density, $\mathcal{P}_N^{(N)} \equiv \mathcal{P}_N^{(N)\mathrm{pot}}$, and take

$$\mathcal{P}_N^{(N)}(\mathbf{R}^N) \equiv \frac{1}{Z_N} e^{-\beta \check{U}_N^{\mathrm{pot}}(\mathbf{R}^N)}, \tag{11.2}$$

where the configurational partition function is

$$Z_N = C_{\mathrm{pot}}^{-1} = \int d\mathbf{R}^N e^{-\beta \check{U}_N^{\mathrm{pot}}(\mathbf{R}^N)}. \tag{11.3}$$

Thus Equation 9.166 for $\mathcal{P}_N^{(N)}(\mathbf{r}^N)$ in the spherical case and the associated expression for Z_N become generalized to the nonspherical one by simply replacing $\mathbf{r}$ by $\mathbf{R}$. The same is true for many other results. For instance, the average of a quantity X that depends on particle positions and orientation and has the instantaneous value $\check{X} = \check{X}(\mathbf{r}^N, \boldsymbol{\omega}^N) \equiv \check{X}(\mathbf{R}^N)$ can be calculated from

$$\bar{X}_N = \int d\mathbf{R}^N \check{X}(\mathbf{R}^N) \mathcal{P}_N^{(N)}(\mathbf{R}^N) \tag{11.4}$$

(cf. Equation 9.165).

Let us now turn to the canonical partition function. Both the $\mathbf{p}$ and $\mathbf{j}$ integrals in Q_N can be performed analytically and Q_N can expressed as

$$Q_N = \frac{\mathfrak{q}_\mathrm{r}^N}{N!\Lambda^{3N}} \int d\mathbf{R}^N e^{-\beta \check{U}_N^{\mathrm{pot}}(\mathbf{R}^N)} = \frac{Z_N \mathfrak{q}_\mathrm{r}^N}{N!\Lambda^{3N}} = \frac{Z_N \eta^N}{N!}, \tag{11.5}$$

where

$$\eta = \frac{\mathfrak{q}_\mathrm{r}}{\Lambda^3}, \qquad \mathfrak{q}_\mathrm{r} = \begin{cases} 8\pi^2 I k_B T/(\varsigma h^2), & \text{linear molecules} \\ [(8\pi^2 k_B T)^3 \pi I_x I_y I_z]^{1/2}/(\varsigma h^3), & \text{otherwise} \end{cases} \tag{11.6}$$

and ς is a so-called *symmetry number* that is usually equal to one, but differs from one for molecules that have some symmetry. This number it is introduced in order to account for the fact that a symmetrical molecule can be rotated so that identical atoms change places with each other, whereby the new orientation cannot be distinguished from the original one.[3] Do not confuse η with the entity $\eta^{\rm hs}$, which is the hard sphere packing fraction.

In the canonical ensemble, the Helmholtz free energy is

$$A_N = -k_B T \ln Q_N = k_B T \ln \frac{N!}{Z_N \eta^N} \tag{11.7}$$

and the chemical potential is

$$\begin{aligned} \mu &= k_B T \left[\ln \frac{N!}{Z_N \eta^N} - \ln \frac{(N-1)!}{Z_{N-1} \eta^{N-1}} \right] \\ &= k_B T \ln \frac{N Z_{N-1}}{Z_N \eta}. \end{aligned} \tag{11.8}$$

These equations are the same as for spherical particles, Equations I:3.19 and I:3.24, respectively. The chemical potential of a uniform ideal gas is

$$\mu = k_B T \ln \left(\frac{N}{V\eta} \right) = k_B T \ln \left(\frac{n^{\rm b}}{\eta} \right) \quad \text{(uniform ideal gas)} \tag{11.9}$$

(cf. Equation I:3.25), since for such a gas (when $\breve{U}_N^{\rm pot} = 0$), we have

$$Z_N = \int d\mathbf{R}^N = \int d\mathbf{r}^N d\omega^N = \left[\int d\mathbf{r} \right]^N \left[\int d\omega \right]^N = V^N,$$

where we have used $\int d\omega = 1$. The activity ζ is defined as

$$\zeta = \eta e^{\beta\mu}, \tag{11.10}$$

which can be expressed as

$$\zeta = \frac{N Z_{N-1}}{Z_N} \tag{11.11}$$

in the canonical ensemble.

In the **grand canonical ensemble** we have the partition function

$$\Xi = \sum_{N=0}^{\infty} Q_N e^{\beta\mu N} = \sum_{N=0}^{\infty} \frac{\zeta^N Z_N}{N!} = \sum_{N=0}^{\infty} \frac{\zeta^N}{N!} \int d\mathbf{R}^N e^{-\beta \breve{U}_N^{\rm pot}(\mathbf{R}^N)} \tag{11.12}$$

and we define

$$\mathcal{P}^{(N)}(\mathbf{R}^N) \equiv \mathscr{P}(N) \mathcal{P}_N^{(N)}(\mathbf{R}^N) = \frac{\zeta^N}{N!\,\Xi} e^{-\beta \breve{U}_N^{\rm pot}(\mathbf{R}^N)} \tag{11.13}$$

[3]The symmetry number is equal to the number of different ways one can rotate a molecule in such a manner. For example, ς is equal to 1 for CO and other unsymmetrical cases, ς is 2 for O_2, C_2H_2 and H_2O, 3 for NH_3, 4 for C_2H_4 and 12 for CH_4.

(cf. Equation I:3.41), where

$$\mathscr{P}(N) = \frac{Q_N e^{\beta\mu N}}{\Xi} = \frac{\varsigma^N Z_N}{N!\,\Xi} \tag{11.14}$$

is the probability that there are N particles in the system (Equation 9.218). The average of the quantity X is obtained from

$$\begin{aligned} \bar{X} \equiv \langle X \rangle &= \sum_{N=0}^{\infty} \int d\mathbf{R}^N \breve{X}_N(\mathbf{R}^N) \mathcal{P}^{(N)}(\mathbf{R}^N) \\ &= \sum_{N=0}^{\infty} \mathscr{P}(N) \int d\mathbf{R}^N \breve{X}_N(\mathbf{R}^N) \mathcal{P}_N^{(N)}(\mathbf{R}^N) = \sum_{N=0}^{\infty} \mathscr{P}(N) \bar{X}_N, \end{aligned} \tag{11.15}$$

which corresponds to Equations I:3.40 and I:3.42 for spherical particles.

To proceed to the expressions for the distribution functions, we will start with systems containing a constant number of particles, N, in the canonical ensemble and then continue with open systems in the grand canonical ensemble, where N can vary while the chemical potential μ is constant.

11.2 DENSITY DISTRIBUTIONS IN THE CANONICAL ENSEMBLE

In the canonical ensemble the probability density $\mathcal{P}_N^{(N)}(\mathbf{R}^N) \equiv \mathcal{P}_N^{(N)}(\mathbf{r}^N, \boldsymbol{\omega}^N)$ to observe a certain particle configuration with positions $\mathbf{r}^N \equiv \mathbf{r}_1, \mathbf{r}_2, \ldots, \mathbf{r}_N$ and orientations $\boldsymbol{\omega}^N \equiv \boldsymbol{\omega}_1, \boldsymbol{\omega}_2, \ldots, \boldsymbol{\omega}_N$ is given by Equation 11.2. The potential energy $\breve{U}_N^{\text{pot}}(\mathbf{R}^N)$ is (cf. Equation 9.167)

$$\breve{U}_N^{\text{pot}}(\mathbf{R}^N) = \breve{U}_N^{\text{intr}}(\mathbf{R}^N) + \sum_{\nu=1}^{N} v(\mathbf{R}_\nu), \tag{11.16}$$

where $\breve{U}_N^{\text{intr}}(\mathbf{R}^N)$ is the interaction energy between the particles in the fluid (the intrinsic potential energy) and $v(\mathbf{R}_\nu)$ gives the interaction with the external field.

Like in the case of spherical particles (Section 5.5.3 in Volume I), we have to take into account of that the particles are treated as if they were distinguishable in $\mathcal{P}_N^{(N)}$, while they are indistinguishable in reality. Therefore, we multiply $\mathcal{P}_N^{(N)}$ by $N!$ and define the physically relevant function

$$\mathfrak{n}_N^{(N)}(\mathbf{r}^N, \boldsymbol{\omega}^N) \equiv \mathfrak{n}_N^{(N)}(\mathbf{R}^N) = N!\,\mathcal{P}_N^{(N)}(\mathbf{R}^N) = \frac{N!}{Z_N} e^{-\beta \breve{U}_N^{\text{pot}}(\mathbf{R}^N)}, \tag{11.17}$$

the ***N*-point density distribution function**. If we only care about the particles positions and not their orientations, we can integrate over all possible orientations of the particles and obtain

$$n_N^{(N)}(\mathbf{r}^N) = \int d\boldsymbol{\omega}^N \mathfrak{n}_N^{(N)}(\mathbf{r}^N, \boldsymbol{\omega}^N). \tag{11.18}$$

Both $\mathfrak{n}_N^{(N)}$ and $n_N^{(N)}$ have the dimension (number density)^N, that is, (length)$^{-3N}$. For spherically symmetric particles the orientational variables do not matter and there is no difference between $\mathfrak{n}_N^{(N)}$ and $n_N^{(N)}$ since $\int d\omega^N = 1^N = 1$.

The **l-point density distribution function** is defined from

$$\mathfrak{n}_N^{(l)}(\mathbf{r}^l, \omega^l) \equiv \mathfrak{n}_N^{(l)}(\mathbf{R}^l) = \frac{N!}{(N-l)!} \mathcal{P}_N^{(l)}(\mathbf{R}^l) \tag{11.19}$$

(cf. Equation I:5.51), where

$$\begin{aligned} \mathcal{P}_N^{(l)}(\mathbf{R}^l) &= \int d\mathbf{R}_{l+1} \ldots d\mathbf{R}_N \mathcal{P}_N^{(N)}(\mathbf{R}^l) \\ &= \frac{1}{Z_N} \int d\mathbf{R}_{l+1} \ldots d\mathbf{R}_N e^{-\beta \breve{U}_N^{\mathrm{pot}}(\mathbf{R}^l)} \end{aligned} \tag{11.20}$$

is the probability density to observe l particles in a certain configuration $\mathbf{R}^l$ irrespectively of the configurations of the remaining $N-l$ particles. Furthermore

$$n_N^{(l)}(\mathbf{r}^l) = \int d\omega^l \mathfrak{n}_N^{(l)}(\mathbf{r}^l, \omega^l) \tag{11.21}$$

gives the **density** distribution regardless of the orientations of the l particles. Both $\mathfrak{n}_N^{(l)}$ and $n_N^{(l)}$ have the dimension (number density)^l. Note that $\mathfrak{n}_N^{(l)}$ and $n_N^{(l)}$ are not usual probability densities since they are not normalized to 1. Their integral over the whole space is given by

$$\int d\mathbf{r}^l d\omega^l \mathfrak{n}_N^{(l)}(\mathbf{r}^l, \omega^l) = \int d\mathbf{r}^l n_N^{(l)}(\mathbf{r}^l) = \frac{N!}{(N-l)!}, \tag{11.22}$$

which follows from the normalization of $\mathcal{P}_N^{(l)}$.

11.2.1 The singlet density distribution $\mathfrak{n}_N^{(1)}$

The cases with the lowest l values are of most interest in practice. For $l = 1$ we have

$$\mathfrak{n}_N^{(1)}(\mathbf{R}_1) = N\mathcal{P}_N^{(1)}(\mathbf{R}_1) = \frac{N}{Z_N} \int d\mathbf{R}_2 \ldots d\mathbf{R}_N e^{-\beta \breve{U}_N^{\mathrm{pot}}(\mathbf{R}^N)}, \tag{11.23}$$

where $\mathcal{P}_N^{(1)}(\mathbf{R}_1) \equiv \mathcal{P}_N^{(1)}(\mathbf{r}_1, \omega_1)$ is the probability density to observe particle 1 with orientation ω_1 at position $\mathbf{r}_1$ irrespectively of the configurations of the remaining particles. The product $\mathcal{P}_N^{(1)}(\mathbf{r}_1, \omega_1) d\mathbf{r}_1 d\omega_1$ is accordingly the *probability* to observe particle 1 with orientation within $d\omega_1$ from ω_1 and located inside the volume element $d\mathbf{r}_1$ at position $\mathbf{r}_1$ (cf. figure 11.1). Since there are N particles with equal probability to be in the volume element $d\mathbf{r}_1$, the product $N\mathcal{P}_N^{(1)}(\mathbf{r}_1, \omega_1) d\mathbf{r}_1 d\omega_1$ gives the *mean number of particles* with the prescribed orientation in the volume element, so if we divide by $d\mathbf{r}_1$ we obtain the *mean number density*. It follows that $\mathfrak{n}_N^{(1)}(\mathbf{R}_1)d\boldsymbol{\omega}_1$ is the number density of particles with orientation within $d\boldsymbol{\omega}_1$ from $\boldsymbol{\omega}_1$ at position $\mathbf{r}_1$. As before we have $n_N^{(1)}(\mathbf{r}_1) = \int d\boldsymbol{\omega}_1 \mathfrak{n}_N^{(1)}(\mathbf{r}_1, \boldsymbol{\omega}_1)$.

From Equation 11.22 with $l = 1$, we have $\int d\mathbf{r}_1 d\boldsymbol{\omega}_1 \mathfrak{n}_N^{(1)}(\mathbf{r}_1, \boldsymbol{\omega}_1) = \int d\mathbf{r}_1 n_N^{(1)}(\mathbf{r}_1) = N$. For a macroscopic bulk fluid $n_N^{(1)\mathrm{b}} = N/V = n^{\mathrm{b}}$ where $V = \int d\mathbf{r}_1$ is the volume of the whole system. All orientations are equally likely in the isotropic bulk phase, so we have $\mathfrak{n}_N^{(1)\mathrm{b}} = n^{\mathrm{b}}$.

Like in the case of spherical particles (see Section 5.8 in Volume I), the density distribution $\mathfrak{n}_N^{(1)}(\mathbf{R})$ can be obtained as the average of the microscopic density distribution $\breve{\mathfrak{n}}_{\{\mathbf{R}^N\}}^{(1)}(\mathbf{R})$, which defined for each instantaneous configuration $\mathbf{R}^N$ of the particles (with coordinates $\mathbf{R}_\nu = (\mathbf{r}_\nu, \boldsymbol{\omega}_\nu)$, $1 \leq \nu \leq N$) as

$$\breve{\mathfrak{n}}_{\{\mathbf{R}^N\}}^{(1)}(\mathbf{R}) \equiv \sum_{\nu=1}^{N} \delta^{(3)}(\mathbf{R} - \mathbf{R}_\nu), \tag{11.24}$$

where $\delta^{(3)}(\mathbf{R} - \mathbf{R}_\nu) \equiv \delta^{(3)}(\mathbf{r} - \mathbf{r}_\nu)\delta^{\omega}(\boldsymbol{\omega} - \boldsymbol{\omega}_\nu)$ is the Dirac delta function for the variable $\mathbf{R} = (\mathbf{r}, \boldsymbol{\omega})$ (cf. Equation 9.164). Here, we have introduced the delta function $\delta^{\omega}(\boldsymbol{\omega})$ in orientational space, which has the property that for a function $f(\boldsymbol{\omega})$ of the orientational variable $\boldsymbol{\omega}$ one has $\int d\boldsymbol{\omega}\, \delta^{\omega}(\boldsymbol{\omega} - \boldsymbol{\omega}')f(\boldsymbol{\omega}) = f(\boldsymbol{\omega}')$. The function $\breve{\mathfrak{n}}_{\{\mathbf{R}^N\}}^{(1)}(\mathbf{R})$ is zero everywhere except at each point $\mathbf{r}$ where a particle center is located; the function has a peak at each such point. The function is also zero unless the orientation $\boldsymbol{\omega}$ is equal to that of the particle located the point in question. We have, using the canonical average defined in Equation 11.4

$$\begin{aligned} \left\langle \breve{\mathfrak{n}}_{\{\mathbf{R}^N\}}^{(1)}(\mathbf{R}) \right\rangle &= \int d\mathbf{R}^N \breve{\mathfrak{n}}_{\{\mathbf{R}^N\}}^{(1)}(\mathbf{R}) \mathcal{P}_N^{(N)}(\mathbf{R}^N) \\ &= \sum_{\nu=1}^{N} \int d\mathbf{R}^N \delta^{(3)}(\mathbf{R} - \mathbf{R}_\nu) \mathcal{P}_N^{(N)}(\mathbf{R}^N) \\ &= \sum_{\nu=1}^{N} \int d\mathbf{R}^N \Big|_{\text{not } d\mathbf{R}_\nu} \mathcal{P}_N^{(N)}(\mathbf{R}^N)\Big|_{\mathbf{R}_\nu = \mathbf{R}} = N\mathcal{P}_N^{(1)}(\mathbf{R}) \end{aligned} \tag{11.25}$$

where, for example, $d\mathbf{R}^N\big|_{\text{not } d\mathbf{R}_1} = d\mathbf{R}_2 \ldots d\mathbf{R}_N$ (i.e., all differentials except $d\mathbf{R}_1$) and where the last equality follows from the fact that all N terms in the sum are identical (only the names of the integration variables differ). We have also used Equation 11.20 for $l = 1$ to obtain $\mathcal{P}_N^{(1)}$. From Equations 11.23 and 11.25 it follows that

$$\mathfrak{n}_N^{(1)}(\mathbf{R}) = \left\langle \breve{\mathfrak{n}}_{\{\mathbf{R}^N\}}^{(1)}(\mathbf{R}) \right\rangle. \tag{11.26}$$

This way to obtain a density distribution is elegant mathematically, physically appealing and often convenient to use.

11.2.2 The pair density distribution $\mathfrak{n}_N^{(2)}$

The case $l = 2$ for Equation 11.19 is particularly important since it concerns the pair distribution function. We have

$$\begin{aligned}\mathfrak{n}_N^{(2)}(\mathbf{R}_1, \mathbf{R}_2) &= N(N-1)\mathcal{P}_N^{(2)}(\mathbf{R}_1, \mathbf{R}_2) \\ &= \frac{N(N-1)}{Z_N} \int d\mathbf{R}_3 \ldots d\mathbf{R}_N e^{-\beta \breve{U}_N^{\mathrm{pot}}(\mathbf{R}^N)} \end{aligned} \tag{11.27}$$

(cf. Equation I:5.45). Let us consider the situation when one of the N particles in the system is fixed and has a fixed orientation. We select particle 2 to be fixed while the other particles are free to move. The distribution around the fixed particle with orientation $\boldsymbol{\omega}_2$ and located at $\mathbf{r}_2$ is denoted as $\mathfrak{n}_N^{(1)}(\mathbf{R}_1|\mathbf{R}_2) \equiv \mathfrak{n}_N^{(1)}(\mathbf{r}_1, \boldsymbol{\omega}_1|\mathbf{r}_2, \boldsymbol{\omega}_2)$. In analogy to the discussion for $\mathfrak{n}_N^{(1)}$ above, the product $\mathfrak{n}_N^{(1)}(\mathbf{R}_1|\mathbf{R}_2)d\boldsymbol{\omega}_1$ is the number density of particles with orientation within $d\boldsymbol{\omega}_1$ from $\boldsymbol{\omega}_1$ at position $\mathbf{r}_1$ in presence of the fixed particle. We have

$$\mathfrak{n}_N^{(1)}(\mathbf{R}_1|\mathbf{R}_2) = \frac{\mathfrak{n}_N^{(2)}(\mathbf{R}_1, \mathbf{R}_2)}{\mathfrak{n}_N^{(1)}(\mathbf{R}_2)}, \tag{11.28}$$

which can be realized by using the same argument with conditional probabilities as when we obtained Equation I:5.47 in Volume I for spherical particles. It is, however, a good exercise to show Equation 11.28 from the definitions of the distribution functions, as done in the Example below.

Example 11.A: Prove Equation 11.28 explicitly from the definitions of $\mathfrak{n}_N^{(1)}$ and $\mathfrak{n}_N^{(2)}$.

We will eventually consider an N particle system where one of the particles is fixed and the remaining $N-1$ particles are mobile, but if the fixed particle is not present, we have an ordinary system with $N-1$ mobile particles. Its density distribution would be (from Equation 11.23 with $N-1$ instead of N)

$$\mathfrak{n}_{N-1}^{(1)}(\mathbf{R}_1) = \frac{N-1}{Z_{N-1}} \int d\mathbf{R}_2 \ldots d\mathbf{R}_{N-1} e^{-\beta \breve{U}_{N-1}^{\mathrm{pot}}(\mathbf{R}^{N-1})}, \tag{11.29}$$

where $\breve{U}_{N-1}^{\mathrm{pot}}$ is the instantaneous interaction energy for the $N-1$ particles, $\mathbf{R}^{N-1} \equiv \mathbf{R}_1, \mathbf{R}_2, \ldots, \mathbf{R}_{N-1}$, and

$$Z_{N-1} = \int d\mathbf{R}_1 \ldots d\mathbf{R}_{N-1} e^{-\beta \breve{U}_{N-1}^{\mathrm{pot}}(\mathbf{R}^{N-1})} \tag{11.30}$$

(cf. Equation 11.3).

Let us now add a fixed particle with orientation ω_0 at position $\mathbf{r}_0$. For convenience, we call this particle number "0" rather than number "2" since particle 2 is one of the mobile particles in the equations above (the number does not matter since all particles are equal). The original $N-1$ particles in the system are free to move and their distribution in presence of the fixed particle is given by $\mathfrak{n}_N^{(1)}(\mathbf{R}_1|\mathbf{R}_0)$. To obtain the new potential energy we have to add the interaction energy between the new particle and the $N-1$ particles and also its interaction energy with the external field. The instantaneous configuration of all N particles is given by $(\mathbf{R}_0, \mathbf{R}^{N-1})$ and the total interaction energy is $\breve{U}_N^{\mathrm{pot}}(\mathbf{R}_0, \mathbf{R}^{N-1})$ which of course is the same as $\breve{U}_N^{\mathrm{pot}}(\mathbf{R}^N)$ except the names of the variables. Thus, $\mathfrak{n}_N^{(1)}(\mathbf{R}_1|\mathbf{R}_0)$ is obtained from the expression for $\mathfrak{n}_{N-1}^{(1)}(\mathbf{R}_1)$ in Equation 11.29 by simply inserting the new energy $\breve{U}_N^{\mathrm{pot}}$ in Equations 11.29 and 11.30. We obtain

$$\mathfrak{n}_N^{(1)}(\mathbf{R}_1|\mathbf{R}_0) = \frac{N-1}{\mathsf{Z}_N^{(1)}(\mathbf{R}_0)} \int d\mathbf{R}_2 \dots d\mathbf{R}_{N-1} e^{-\beta \breve{U}_N^{\mathrm{pot}}(\mathbf{R}_0, \mathbf{R}^{N-1})}, \tag{11.31}$$

where

$$\mathsf{Z}_N^{(1)}(\mathbf{R}_0) = \int d\mathbf{R}_1 \dots d\mathbf{R}_{N-1} e^{-\beta \breve{U}_N^{\mathrm{pot}}(\mathbf{R}_0, \mathbf{R}^{N-1})} \tag{11.32}$$

and where we have written subscript N instead of $N-1$ for the functions $\mathfrak{n}_N^{(1)}$ and $\mathsf{Z}_N^{(1)}$ since there are now N particles in the system. Distinguish the function $\mathsf{Z}_N^{(1)}$ from Z_N. For $\mathsf{Z}_N^{(1)}(\mathbf{R}_0)$ one of the particles is fixed, while all N particles are mobile for Z_N.

A comparison between Equation 11.32 and the rhs of Equation 11.23 shows that

$$\frac{N}{Z_N} \mathsf{Z}_N^{(1)}(\mathbf{R}_0) = \mathfrak{n}_N^{(1)}(\mathbf{R}_0) \tag{11.33}$$

(only the names of the variables are different). By inserting this result in Equation 11.31 and using Equation 11.27 we obtain

$$\mathfrak{n}_N^{(1)}(\mathbf{R}_1|\mathbf{R}_0) = \frac{\mathfrak{n}_N^{(2)}(\mathbf{R}_1, \mathbf{R}_0)}{\mathfrak{n}_N^{(1)}(\mathbf{R}_0)},$$

which is the same as Equation 11.28.

As usual, we define the **pair distribution function** $g_N^{(2)}$ from

$$\mathfrak{n}_N^{(1)}(\mathbf{R}_1|\mathbf{R}_2) = \mathfrak{n}_N^{(1)}(\mathbf{R}_1)\, g_N^{(2)}(\mathbf{R}_1, \mathbf{R}_2) \tag{11.34}$$

and it follows from Equation 11.28 that

$$\mathfrak{n}_N^{(2)}(\mathbf{R}_1,\mathbf{R}_2) = \mathfrak{n}_N^{(1)}(\mathbf{R}_1)\,\mathfrak{n}_N^{(1)}(\mathbf{R}_2)\,g_N^{(2)}(\mathbf{R}_1,\mathbf{R}_2). \tag{11.35}$$

As before, the **pair correlation function** is $h_N^{(2)}(\mathbf{R}_1,\mathbf{R}_2) = g_N^{(2)}(\mathbf{R}_1,\mathbf{R}_2)-1$. When the separation r_{12} between $\mathbf{r}_1$ and $\mathbf{r}_2$ is large, the particle at $\mathbf{r}_2$ has very little influence the density ar $\mathbf{r}_1$, $\mathfrak{n}_N^{(2)}(\mathbf{R}_1|\mathbf{R}_2) \approx \mathfrak{n}_N^{(1)}(\mathbf{R}_1)$, so we have $\mathfrak{n}_N^{(2)}(\mathbf{R}_1,\mathbf{R}_2) \approx \mathfrak{n}_N^{(1)}(\mathbf{R}_1)\,\mathfrak{n}_N^{(1)}(\mathbf{R}_2)$ and $g_N^{(2)}(\mathbf{R}_1,\mathbf{R}_2) \approx 1$, which means that $h_N^{(2)}(\mathbf{R}_1,\mathbf{R}_2) \approx 0$. Even if the points $\mathbf{r}_1$ and $\mathbf{r}_2$ are very far apart and the interactions between the particles at $\mathbf{r}_1$ and $\mathbf{r}_2$ are completely negligible $g_N^{(2)}(\mathbf{R}_1,\mathbf{R}_2)$ is not exactly one and $h_N^{(2)}(\mathbf{R}_1,\mathbf{R}_2)$ not exactly zero. The reason is that we are dealing with a finite system with N particles and the density distribution $\mathfrak{n}_N^{(1)}(\mathbf{R}_1|\mathbf{R}_2)$ in Equation 11.34 is formed by $N-1$ particles distributed in space; one particle is fixed at $\mathbf{r}_2$ (the integral $\int d\mathbf{R}_1\mathfrak{n}_N^{(1)}(\mathbf{R}_1|\mathbf{R}_2) = N-1$ as can be seen from Equation 11.31). For example, for a uniform ideal gas

$$g_N^{(2)}(\mathbf{R}_1,\mathbf{R}_2) = \frac{\mathfrak{n}_N^{(1)}(\mathbf{R}_1|\mathbf{R}_2)}{\mathfrak{n}_N^{(1)\mathrm{b}}} = \frac{\frac{N-1}{V}}{\frac{N}{V}} = 1-\frac{1}{N} \neq 1 \qquad \text{(uniform ideal gas).}$$

In the general case, for a homogeneous bulk system when the separation r_{12} is very large but $\mathbf{r}_1$ and $\mathbf{r}_2$ are still in the interior of the fluid, we have[4]

$$g_N^{(2)}(\mathbf{R}_1,\mathbf{R}_2) \sim 1-\frac{n^{\mathrm{b}}k_BT\chi_T}{N} \qquad \text{(bulk fluid)} \tag{11.36}$$

in the limit of large N, where χ_T is the isothermal compressibility of the system defined in Equation 9.189. This means that the deviation of the correlation function from zero for large r_{12} is very small when N is large.[5]

We can also obtain the pair density distribution $\mathfrak{n}_N^{(2)}(\mathbf{R},\mathbf{R}')$ from the average of the microscopic pair density distribution $\breve{\mathfrak{n}}_{\{\mathbf{R}^N\}}^{(2)}(\mathbf{R},\mathbf{R}')$ defined by (cf. Equation 11.24)

$$\breve{\mathfrak{n}}_{\{\mathbf{R}^N\}}^{(2)}(\mathbf{R},\mathbf{R}') \equiv \sum_{\nu=1}^{N}\sum_{\substack{\nu'=1\\(\nu'\neq\nu)}}^{N}\delta^{(3)}(\mathbf{R}-\mathbf{R}_\nu)\delta^{(3)}(\mathbf{R}'-\mathbf{R}_{\nu'}),$$

which is zero unless $\mathbf{r}$ is located at the center of one particle and $\mathbf{r}'$ is located at the center of another one and the two particles have orientations $\boldsymbol{\omega}$ and $\boldsymbol{\omega}'$ respectively. We have

$$\mathfrak{n}_N^{(2)}(\mathbf{R},\mathbf{R}') = \left\langle \breve{\mathfrak{n}}_{\{\mathbf{R}^N\}}^{(2)}(\mathbf{R},\mathbf{R}')\right\rangle, \tag{11.37}$$

[4] J. L. Lebowitz and J. K. Percus, Phys. Rev. **122** (1961) 1675 (https://doi.org/10.1103/PhysRev.122.1675).

[5] See also footnote 20 on page 245.

which follows from

$$\begin{aligned}\left\langle \breve{\mathfrak{n}}^{(2)}_{\{\mathbf{R}^N\}}(\mathbf{R},\mathbf{R}')\right\rangle &\\ &= \sum_{v=1}^{N}\sum_{\substack{\nu'=1\\(\nu'\neq\nu)}}^{N}\int d\mathbf{R}^N \delta^{(3)}(\mathbf{R}-\mathbf{R}_\nu)\delta^{(3)}(\mathbf{R}'-\mathbf{R}_{\nu'})\mathcal{P}_N^{(N)}(\mathbf{R}^N) \qquad (11.38)\\ &= N(N-1)\mathcal{P}_N^{(2)}(\mathbf{R},\mathbf{R}'),\end{aligned}$$

where we have used the fact the all $N(N-1)$ terms in the sum are equal numerically.

{Exercise 11.1: Fill in the details in the proof of Equation 11.38.**}**

Note that the product $\breve{\mathfrak{n}}^{(1)}_{\{\mathbf{R}^N\}}(\mathbf{R})\breve{\mathfrak{n}}^{(1)}_{\{\mathbf{R}^N\}}(\mathbf{R}')$ is nearly the same as $\breve{\mathfrak{n}}^{(2)}_{\{\mathbf{R}^N\}}(\mathbf{R},\mathbf{R}')$; the difference is that the former contains contributions where $\nu=\nu'$. Thus

$$\breve{\mathfrak{n}}^{(1)}_{\{\mathbf{R}^N\}}(\mathbf{R})\breve{\mathfrak{n}}^{(1)}_{\{\mathbf{R}^N\}}(\mathbf{R}') = \breve{\mathfrak{n}}^{(2)}_{\{\mathbf{R}^N\}}(\mathbf{R},\mathbf{R}') + \sum_{\nu=1}^{N}\delta^{(3)}(\mathbf{R}-\mathbf{R}_\nu)\delta^{(3)}(\mathbf{R}'-\mathbf{R}_\nu)$$

and we have

$$\left\langle \breve{\mathfrak{n}}^{(1)}_{\{\mathbf{R}^N\}}(\mathbf{R})\breve{\mathfrak{n}}^{(1)}_{\{\mathbf{R}^N\}}(\mathbf{R}')\right\rangle = \mathfrak{n}_N^{(2)}(\mathbf{R},\mathbf{R}') + \mathfrak{n}_N^{(1)}(\mathbf{R})\delta^{(3)}(\mathbf{R}'-\mathbf{R}), \qquad (11.39)$$

where we have obtained the last term by using the fact that[6] $\delta^{(3)}(\mathbf{r}-\mathbf{r}_\nu)\delta^{(3)}(\mathbf{r}'-\mathbf{r}_\nu) = \delta^{(3)}(\mathbf{r}-\mathbf{r}_\nu)\delta^{(3)}(\mathbf{r}'-\mathbf{r})$, the corresponding equality for the delta functions in $\boldsymbol{\omega}$ and a reasoning analogous to that in Equation 11.25. We can write the result as

$$\begin{aligned}\left\langle \breve{\mathfrak{n}}^{(1)}_{\{\mathbf{R}^N\}}(\mathbf{R})\breve{\mathfrak{n}}^{(1)}_{\{\mathbf{R}^N\}}(\mathbf{R}')\right\rangle &= \mathfrak{n}_N^{(1)}(\mathbf{R}')\left(\mathfrak{n}_N^{(1)}(\mathbf{R}|\mathbf{R}') + \delta^{(3)}(\mathbf{R}'-\mathbf{R})\right)\\ &= \mathfrak{n}_N^{(1)}(\mathbf{R})\left(\mathfrak{n}_N^{(1)}(\mathbf{R}'|\mathbf{R}) + \delta^{(3)}(\mathbf{R}-\mathbf{R}')\right).\end{aligned}$$

While $\mathfrak{n}_N^{(1)}(\mathbf{R}|\mathbf{R}')$ is, as before, the density distribution in the presence of a fixed particle with coordinates $\mathbf{R}'$, the term $\delta^{(3)}(\mathbf{R}'-\mathbf{R})$ in the first line can be interpreted as the density distribution for the fixed particle itself and is a "self-term" in the density distribution. The corresponding interpretation can be made of the second line in the equation.

11.2.3 The triplet density distribution $\mathfrak{n}_N^{(3)}$

We now turn to the case $l=3$ of Equation 11.19, which yields

$$\mathfrak{n}_N^{(3)}(\mathbf{R}_1,\mathbf{R}_2,\mathbf{R}_3) = N(N-1)(N-2)\mathcal{P}_N^{(3)}(\mathbf{R}_1,\mathbf{R}_2,\mathbf{R}_3).$$

[6]This fact follows from $\int d\mathbf{r}_\nu f(\mathbf{r}_\nu)\delta^{(3)}(\mathbf{r}-\mathbf{r}_\nu)\delta^{(3)}(\mathbf{r}'-\mathbf{r}_\nu) = [f(\mathbf{r}_\nu)\delta^{(3)}(\mathbf{r}'-\mathbf{r}_\nu)]_{\mathbf{r}_\nu=\mathbf{r}} = f(\mathbf{r})\delta^{(3)}(\mathbf{r}'-\mathbf{r}) = \int d\mathbf{r}_\nu f(\mathbf{r}_\nu)\delta^{(3)}(\mathbf{r}-\mathbf{r}_\nu)\delta^{(3)}(\mathbf{r}'-\mathbf{r})$.

The triplet distribution function $g_N^{(3)}$ is defined from

$$\mathfrak{n}_N^{(3)}(\mathbf{R}_1,\mathbf{R}_2,\mathbf{R}_3) = \mathfrak{n}_N^{(1)}(\mathbf{R}_1)\,\mathfrak{n}_N^{(1)}(\mathbf{R}_2)\,\mathfrak{n}_N^{(1)}(\mathbf{R}_3)\,g_N^{(3)}(\mathbf{R}_1,\mathbf{R}_2,\mathbf{R}_3). \quad (11.40)$$

From its definition we see that $\mathfrak{n}_N^{(3)}$ is symmetric with respect to permutations of 1, 2 and 3. It follows that $g_N^{(3)}$ is also symmetric in this manner.

It is useful to consider the pair density distribution function for a system where one particle, say particle 3, is fixed. We denote this as $\mathfrak{n}_N^{(2)}(\mathbf{R}_1,\mathbf{R}_2|\mathbf{R}_3)$. One can show that (cf. Equation 11.28)

$$\mathfrak{n}_N^{(2)}(\mathbf{R}_1,\mathbf{R}_2|\mathbf{R}_3) = \frac{\mathfrak{n}_N^{(3)}(\mathbf{R}_1,\mathbf{R}_2,\mathbf{R}_3)}{\mathfrak{n}_N^{(1)}(\mathbf{R}_3)}, \quad (11.41)$$

in analogy to the demonstration in Example 11:A.

{**Exercise 11.2:** Do the derivation of Equation 11.41 in analogy to the proof in Example 11:A.}

From Equations 11.40 and 11.41 it follows that

$$\mathfrak{n}_N^{(2)}(\mathbf{R}_1,\mathbf{R}_2|\mathbf{R}_3) = \mathfrak{n}_N^{(1)}(\mathbf{R}_1)\,\mathfrak{n}_N^{(1)}(\mathbf{R}_2)\,g_N^{(3)}(\mathbf{R}_1,\mathbf{R}_2,\mathbf{R}_3). \quad (11.42)$$

If the point $\mathbf{r}_3$ is very far from $\mathbf{r}_1$ and $\mathbf{r}_2$, we have $\mathfrak{n}_N^{(2)}(\mathbf{R}_1,\mathbf{R}_2|\mathbf{R}_3) \approx \mathfrak{n}_N^{(2)}(\mathbf{R}_1,\mathbf{R}_2)$ since the influence of a particle at $\mathbf{r}_3$ on the density distribution at $\mathbf{r}_1$ and $\mathbf{r}_2$ then is very small. Therefore, $\mathfrak{n}_N^{(3)}(\mathbf{R}_1,\mathbf{R}_2,\mathbf{R}_3) \approx \mathfrak{n}_N^{(2)}(\mathbf{R}_1,\mathbf{R}_2)\,\mathfrak{n}_N^{(1)}(\mathbf{R}_3)$ and $g_N^{(3)}(\mathbf{R}_1,\mathbf{R}_2,\mathbf{R}_3) \approx g_N^{(2)}(\mathbf{R}_1,\mathbf{R}_2)$ under these conditions. If all three points $\mathbf{r}_1$, $\mathbf{r}_2$ and $\mathbf{r}_3$ are very far from each other it follows that $\mathfrak{n}_N^{(3)}(\mathbf{R}_1,\mathbf{R}_2,\mathbf{R}_3) \approx \mathfrak{n}_N^{(1)}(\mathbf{R}_1)\,\mathfrak{n}_N^{(1)}(\mathbf{R}_2)\,\mathfrak{n}_N^{(1)}(\mathbf{R}_3)$ and $g_N^{(3)}(\mathbf{R}_1,\mathbf{R}_2,\mathbf{R}_3) \approx 1$. For large N, the approximation sign here is virtually an equality since the deviation is of the order $1/N$.

Another useful formula is

$$\mathfrak{n}_N^{(2)}(\mathbf{R}_1,\mathbf{R}_2|\mathbf{R}_3) = \mathfrak{n}_N^{(1)}(\mathbf{R}_1|\mathbf{R}_2,\mathbf{R}_3)\,\mathfrak{n}_N^{(1)}(\mathbf{R}_2|\mathbf{R}_3), \quad (11.43)$$

where $\mathfrak{n}_N^{(1)}(\mathbf{R}_1|\mathbf{R}_2,\mathbf{R}_3)$ is the singlet density distribution when two particles, both 2 and 3, are fixed. This relationship can be obtained as follows. Equation 11.28 for a system with $N-1$ particles (in absence of one of the N particles, say, particle 3) can be written $\mathfrak{n}_{N-1}^{(2)}(\mathbf{R}_1,\mathbf{R}_2) = \mathfrak{n}_{N-1}^{(1)}(\mathbf{R}_1|\mathbf{R}_2)\mathfrak{n}_{N-1}^{(1)}(\mathbf{R}_2)$. If we add particle 3 to the system and keep it fixed, the density distributions in presence of this fixed particle are by definition those of Equation 11.43 (as before we may think that the fixed particle contributes to the external potential).

From Equations 11.41 and 11.43 it follows that

$$\mathfrak{n}_N^{(3)}(\mathbf{R}_1,\mathbf{R}_2,\mathbf{R}_3) = \mathfrak{n}_N^{(1)}(\mathbf{R}_1|\mathbf{R}_2,\mathbf{R}_3) \times \mathfrak{n}_N^{(1)}(\mathbf{R}_2|\mathbf{R}_3) \times \mathfrak{n}_N^{(1)}(\mathbf{R}_3) \quad (11.44)$$

We can write this in terms of conditional probabilities as

$$\mathfrak{n}_N^{(3)}(\mathbf{R}_1, \mathbf{R}_2, \mathbf{R}_3) = (N-2)\mathcal{P}_N^{(1)}(\mathbf{R}_1|\mathbf{R}_2, \mathbf{R}_3) \times (N-1)\mathcal{P}_N^{(1)}(\mathbf{R}_2|\mathbf{R}_3) \times N\mathcal{P}_N^{(1)}(\mathbf{R}_3).$$

The corresponding relationships with any permutation of 1, 2 and 3 are also valid.

In analogy to Equation 11.37, we have

$$\mathfrak{n}_N^{(3)}(\mathbf{R}, \mathbf{R}', \mathbf{R}'') = \left\langle \breve{\mathfrak{n}}_{\{\mathbf{R}^N\}}^{(3)}(\mathbf{R}, \mathbf{R}', \mathbf{R}'') \right\rangle, \tag{11.45}$$

where

$$\breve{\mathfrak{n}}_{\{\mathbf{R}^N\}}^{(3)}(\mathbf{R}, \mathbf{R}', \mathbf{R}'') = \sum_{\nu=1}^{N} \sum_{\substack{\nu'=1 \\ (\nu'\neq\nu)}}^{N} \sum_{\substack{\nu''=1 \\ (\nu''\neq\nu,\nu')}}^{N} \delta^{(3)}(\mathbf{R} - \mathbf{R}_\nu)\delta^{(3)}(\mathbf{R}' - \mathbf{R}_{\nu'})\delta^{(3)}(\mathbf{R}'' - \mathbf{R}_{\nu''}).$$

{Exercise 11.3: Show Equation 11.45.**}**

Furthermore, one can show that (cf. Equation 11.39)[7]

$$\begin{aligned} \left\langle \breve{\mathfrak{n}}_N^{(1)}(\mathbf{R})\breve{\mathfrak{n}}_N^{(1)}(\mathbf{R}')\breve{\mathfrak{n}}_N^{(1)}(\mathbf{R}'') \right\rangle &= \mathfrak{n}_N^{(3)}(\mathbf{R}, \mathbf{R}', \mathbf{R}'') \\ &+ \left[\mathfrak{n}_N^{(2)}(\mathbf{R}, \mathbf{R}')\delta^{(3)}(\mathbf{R}' - \mathbf{R}'') + \{\mathbf{R}, \mathbf{R}', \mathbf{R}'' \text{ permutations}\}\right] \\ &+ \mathfrak{n}_N^{(1)}(\mathbf{R})\delta^{(3)}(\mathbf{R} - \mathbf{R}')\delta^{(3)}(\mathbf{R} - \mathbf{R}''), \end{aligned} \tag{11.46}$$

which differs from Equation 11.45 only when at least two of $\mathbf{R}, \mathbf{R}', \mathbf{R}''$ coincide (the delta function contributions originate from the terms that were left out from $\breve{\mathfrak{n}}_N^{(3)}(\mathbf{R}, \mathbf{R}', \mathbf{R}'')$ due to the restrictions $\nu' \neq \nu$ and $\nu'' \neq \nu, \nu'$).

The relationship between the **triplet correlation function** $h_N^{(3)}(\mathbf{R}_1, \mathbf{R}_2, \mathbf{R}_3)$ and the triplet distribution function $g_N^{(3)}(\mathbf{R}_1, \mathbf{R}_2, \mathbf{R}_3)$ is somewhat more complicated than for the pair functions $h_N^{(2)}(\mathbf{R}_1, \mathbf{R}_2) = g_N^{(2)}(\mathbf{R}_1, \mathbf{R}_2) - 1$. We define

$$\begin{aligned} h_N^{(3)}(\mathbf{R}_1, \mathbf{R}_2, \mathbf{R}_3) &= g_N^{(3)}(\mathbf{R}_1, \mathbf{R}_2, \mathbf{R}_3) \\ &- g_N^{(2)}(\mathbf{R}_1, \mathbf{R}_2) - g_N^{(2)}(\mathbf{R}_1, \mathbf{R}_3) - g_N^{(2)}(\mathbf{R}_2, \mathbf{R}_3) + 2. \end{aligned} \tag{11.47}$$

[7]The terms in square brackets are $\mathfrak{n}_N^{(2)}(\mathbf{R}, \mathbf{R}')\delta^{(3)}(\mathbf{R}' - \mathbf{R}'') + \mathfrak{n}_N^{(2)}(\mathbf{R}', \mathbf{R}'')\delta^{(3)}(\mathbf{R}'' - \mathbf{R}) + \mathfrak{n}_N^{(2)}(\mathbf{R}'', \mathbf{R})\delta^{(3)}(\mathbf{R} - \mathbf{R}')$.

This function has the property that $h_N^{(3)}(\mathbf{R}_1, \mathbf{R}_2, \mathbf{R}_3) \approx 0$ when *at least one* of the points $\mathbf{r}_1$, $\mathbf{r}_2$ and $\mathbf{r}_3$ is far away from the other two (see Exercise 11.4). The deviation from 0 is of the order of $1/N$ so $h_N^{(3)}(\mathbf{R}_1, \mathbf{R}_2, \mathbf{R}_3)$ is virtually zero under these conditions provided that N is large.

{**Exercise 11.4:** Show that $h_N^{(3)}(\mathbf{r}_1, \omega_1, \mathbf{r}_2, \omega_2, \mathbf{r}_3, \omega_3) \approx 0$ when at least one of the points $\mathbf{r}_1$, $\mathbf{r}_2$ and $\mathbf{r}_3$ is far away from the other two.}

11.3 FREE ENERGY RELATIONSHIPS IN THE CANONICAL ENSEMBLE

As we have utilized many time earlier, there are intimate relationships between the density distribution functions and variations in free energy of the system. Here, we will generalize these relationships to nonspherical particles and obtain some new ones. For the special case of spherical particles, we are now in a position to obtain some of them more rigorously than before.

11.3.1 Density distributions, forces and free energy changes

Let us consider Helmholtz free energy, $A_N = -k_BT \ln Q_N$, with Q_N from Equation 11.5,

$$\begin{aligned} A_N &= k_BT \ln(N!/\eta^N) \\ &- k_BT \ln \int d\mathbf{R}^N \exp\left(-\beta\left[\breve{U}_N^{\text{intr}}(\mathbf{R}^N) + \sum_{\nu=1}^{N} v(\mathbf{R}_\nu)\right]\right), \end{aligned}$$

where the integral is equal to Z_N and we have written $\breve{U}_N^{\text{pot}}$ as in Equation 11.16. Note that we do not make any assumption about pair-wise additivity in $\breve{U}_N^{\text{intr}}$. The only place where v enters is in the exponential function, so if we take the functional derivative of A_N with respect to $v(\mathbf{R})$ we obtain

$$\begin{aligned} \frac{\delta A_N}{\delta v(\mathbf{R})} &= \frac{1}{Z_N}\int d\mathbf{R}^N \sum_{\nu=1}^{N} \frac{\delta v(\mathbf{R}_\nu)}{\delta v(\mathbf{R})} e^{-\beta \breve{U}_N^{\text{pot}}(\mathbf{R}^N)} \\ &= \sum_{\nu=1}^{N} \int d\mathbf{R}^N \delta^{(3)}(\mathbf{R} - \mathbf{R}_\nu) \mathcal{P}_N^{(N)}(\mathbf{R}^N), \end{aligned}$$

where we have used[8]

$$\frac{\delta v(\mathbf{R}_\nu)}{\delta v(\mathbf{R})} = \delta^{(3)}(\mathbf{R} - \mathbf{R}_\nu), \tag{11.48}$$

which is a the generalization of Equation 9.105 to both positional and orientational variables. Now, following the same reasoning as in Equations 11.25

[8]Since $\int d\mathbf{r} d\omega\, \delta^{(3)}(\mathbf{r} - \mathbf{r}')\delta^{\omega}(\omega - \omega')\delta f(\mathbf{r}, \omega) = \delta f(\mathbf{r}', \omega')$ it follows that $\delta f(\mathbf{r}', \omega')$ / $\delta f(\mathbf{r}, \omega) = \delta^{(3)}(\mathbf{r} - \mathbf{r}')\delta^{\omega}(\omega - \omega')$.

and 11.26 we obtain

$$\frac{\delta A_N}{\delta v(\mathbf{R})} = \mathfrak{n}_N^{(1)}(\mathbf{R}), \tag{11.49}$$

which can write as $dA_N = \int d\mathbf{R}\, \mathfrak{n}_N^{(1)}(\mathbf{R})\delta v(\mathbf{R})$. This free energy change is equal to the reversible work done when changing the external potential, so the latter equation corresponds to Equation 9.34; in the canonical ensemble we have dA_N instead of $d\Theta$ for the reversible work, see the discussion in Footnote 11 on page 15 in Section 9.1.2. These results also verify Equation 9.118 for the case where $dN = 0$ (where the last term in the equation is zero).

Next, we turn to an investigation of the excess chemical potential $\mu_N^{\text{ex}}(\mathbf{R}) = \mu_N^{\text{ex}}(\mathbf{r}, \boldsymbol{\omega})$ that is defined at each point in space and for each particle orientation. Let us consider a system with $N-1$ mobile particles. We insert one more particle and place it fixed at position $\mathbf{r}$ and with fixed orientation $\boldsymbol{\omega}$. The change in free energy of the system is given by

$$\Delta A = k_B T \ln \frac{(N-1)!}{\mathsf{Z}_N^{(1)}(\mathbf{R})\eta^{N-1}} - k_B T \ln \frac{(N-1)!}{Z_{N-1}\eta^{N-1}} = k_B T \ln \frac{Z_{N-1}}{\mathsf{Z}_N^{(1)}(\mathbf{R})}, \tag{11.50}$$

where $\mathsf{Z}_N^{(1)}(\mathbf{R})$ is defined in Equation 11.32. By inserting Equation 11.33 we obtain

$$\Delta A = k_B T \ln \frac{N Z_{N-1}}{Z_N \mathfrak{n}_N^{(1)}(\mathbf{R})} = \mu - k_B T \ln(\mathfrak{n}_N^{(1)}(\mathbf{R})/\eta), \tag{11.51}$$

where we have used Equation 11.8 to obtain the last equality.

ΔA is the reversible work required to insert the new particle at position $\mathbf{r}$ and with orientation $\boldsymbol{\omega}$; it is the work against interactions, that is, the excess chemical potential $\mu_N^{\text{ex}}(\mathbf{R})$, which depends on both the position and the orientation. Equation 11.50 can therefore be written as

$$\mu_N^{\text{ex}}(\mathbf{R}) = \Delta A = k_B T \ln \frac{Z_{N-1}}{\mathsf{Z}_N^{(1)}(\mathbf{R})}. \tag{11.52}$$

and Equation 11.51 as

$$\mu = k_B T \ln(\mathfrak{n}_N^{(1)}(\mathbf{R})/\eta) + \mu_N^{\text{ex}}(\mathbf{R}). \tag{11.53}$$

The latter equation expresses the equilibrium condition that the chemical potential must be *equal everywhere and for all orientations.* This is a generalization of Equation 9.5 (where $\Lambda^3 = 1/\eta$) to nonspherical particles.

The first term on the right hand side in Equation 11.53 is the *local value of the ideal chemical potential*; it is the difference between μ and $\mu_N^{\text{ex}}(\mathbf{R})$. The excess part $\mu_N^{\text{ex}}(\mathbf{R})$ is the change in free energy when we insert the particle into the system at the fixed position $\mathbf{r}$ and orientation $\boldsymbol{\omega}$, while μ is the change when we insert the new particle and let it be freely mobile. Thus, the ideal chemical potential at $\mathbf{R}$, $k_B T \ln(\mathfrak{n}_N^{(1)}(\mathbf{R})/\eta)$, is equal to the change in free energy when the new particle is released from its fixed position and orientation and is allowed to move around freely.

Using $\zeta = \eta e^{\beta\mu}$ we can write Equation 11.53 as

$$\mathfrak{n}_N^{(1)}(\mathbf{R}) = \zeta e^{-\beta\mu_N^{\mathrm{ex}}(\mathbf{R})}. \tag{11.54}$$

For a bulk fluid with density n^{b} Equation 11.53 becomes $\mu = k_B T \ln(n^{\mathrm{b}}/\eta) + \mu^{\mathrm{ex,b}}$ and we have

$$\zeta = n^{\mathrm{b}} e^{\beta\mu^{\mathrm{ex,b}}} = n^{\mathrm{b}}\gamma^{\mathrm{b}} \tag{11.55}$$

where $\gamma^{\mathrm{b}} = \exp(\beta\mu^{\mathrm{ex,b}})$ is the activity coefficient for the bulk phase. We can write Equation 11.54 as

$$\mathfrak{n}_N^{(1)}(\mathbf{R}) = n^{\mathrm{b}}\gamma^{\mathrm{b}} e^{-\beta\mu_N^{\mathrm{ex}}(\mathbf{R})} = n^{\mathrm{b}} e^{-\beta[\mu_N^{\mathrm{ex}}(\mathbf{R}) - \mu^{\mathrm{ex,b}}]}. \tag{11.56}$$

As before, the *singlet potential of mean force* $w_N^{(1)}$ can be written as (cf. $w(\mathbf{r})$ in Table 9.1)

$$w_N^{(1)}(\mathbf{R}_1) = \mu_N^{\mathrm{ex}}(\mathbf{R}_1) - \mu^{\mathrm{ex,b}} \tag{11.57}$$

so we have[9]

$$\mathfrak{n}_N^{(1)}(\mathbf{R}_1) = n^{\mathrm{b}} e^{-\beta w_N^{(1)}(\mathbf{R}_1)}, \tag{11.58}$$

with the usual convention of zero value of $w^{(1)}$ in the bulk phase, $w^{(1)\mathrm{b}} = 0$.

By inserting the definition of $\mathfrak{n}_N^{(1)}$, Equation 11.23, we obtain

$$e^{-\beta w_N^{(1)}(\mathbf{R}_1)} = \frac{N}{n^{\mathrm{b}} Z_N} \int d\mathbf{R}_2 \dots d\mathbf{R}_N e^{-\beta \breve{U}_N^{\mathrm{pot}}(\mathbf{R}^N)} \tag{11.59}$$

and hence we have

$$w_N^{(1)}(\mathbf{R}_1) = -k_B T \ln\left[\int d\mathbf{R}_2 \dots d\mathbf{R}_N e^{-\beta \breve{U}_N^{\mathrm{pot}}(\mathbf{R}^N)}\right] - k_B T \ln\left[\frac{N}{n^{\mathrm{b}} Z_N}\right],$$

where the last term does not depend on $\mathbf{R}_1$. Let us take the gradient with respect to $\mathbf{r}_1$ and obtain

$$\begin{aligned}
\nabla_1 w_N^{(1)}(\mathbf{R}_1) &= \frac{\int d\mathbf{R}_2 \dots d\mathbf{R}_N e^{-\beta \breve{U}_N^{\mathrm{pot}}(\mathbf{R}^N)} \nabla_1 \breve{U}_N^{\mathrm{pot}}(\mathbf{R}^N)}{\int d\mathbf{R}_2 \dots d\mathbf{R}_N e^{-\beta \breve{U}_N^{\mathrm{pot}}(\mathbf{R}^N)}} \\
&= \int d\mathbf{R}_2 \dots d\mathbf{R}_N \mathcal{P}_N^{(N-1)}(\mathbf{R}_2, \dots, \mathbf{R}_N | \mathbf{R}_1) \nabla_1 \breve{U}_N^{\mathrm{pot}}(\mathbf{R}^N) \\
&= \left\langle \nabla_1 \breve{U}_N^{\mathrm{pot}} \right\rangle\Big|_{\mathbf{R}_1 \mathrm{fix}},
\end{aligned} \tag{11.60}$$

[9]If no bulk is present in the system n^{b} is the density of a bulk phase with the same chemical potential as the fluid in the system. Any other choice of zero for $w_N^{(1)}$ just changes the latter by an additive constant. If we, for example, would select $w_N^{(1)} = \mu_N^{\mathrm{ex}}$, the prefactor in Equation 11.58 would be ζ as in Equation 11.54.

where

$$\mathcal{P}_N^{(N-1)}(\mathbf{R}_2, \dots, \mathbf{R}_N|\mathbf{R}_1) = \frac{e^{-\beta \breve{U}_N^{\text{pot}}(\mathbf{R}^N)}}{\int d\mathbf{R}'_2 \dots d\mathbf{R}'_N e^{-\beta \breve{U}_N^{\text{pot}}(\mathbf{R}_1, \mathbf{R}'_2, \dots, \mathbf{R}'_N)}} = \frac{e^{-\beta \breve{U}_N^{\text{pot}}(\mathbf{R}^N)}}{\mathrm{Z}_N^{(1)}(\mathbf{R}_1)}$$

is the probability density to find particles 2, ... , N in a certain configuration when particle 1 is located at $\mathbf{r}_1$ and has orientation $\boldsymbol{\omega}_1$.

Now, $-\nabla_1 \breve{U}_N^{\text{pot}}$ is the instantaneous force acting on particle 1. The average $\langle \cdot \rangle_{\mathbf{R}_1\,\text{fix}}$ in the rhs of Equation 11.60 is taken over all possible configurations for the mobile particles 2, ... , N when particle 1 is located at $\mathbf{r}_1$ with orientation $\boldsymbol{\omega}_1$. Thus,

$$\mathbf{F}_N(\mathbf{R}_1) = -\left\langle \nabla_1 \breve{U}_N^{\text{pot}} \right\rangle\Big|_{\mathbf{R}_1\text{fix}}$$

is the mean force acting in particle 1 from the interactions with the rest of the system, so Equation 11.60 yields

$$-\nabla_1 w_N^{(1)}(\mathbf{R}_1) = \mathbf{F}_N(\mathbf{R}_1). \tag{11.61}$$

The particle also experiences a torque that can be calculated from the gradient of $w_N^{(1)}$ with respect to the orientational variable $\boldsymbol{\omega}_1$.

Since $\breve{U}_N^{\text{pot}}(\mathbf{R}^N) = \breve{U}_N^{\text{intr}}(\mathbf{R}^N) + \sum_{\nu=1}^{N} v(\mathbf{R}_\nu)$ it follows that

$$\mathbf{F}_N(\mathbf{R}_1) = -\left\langle \nabla_1 \breve{U}_N^{\text{intr}} \right\rangle\Big|_{\mathbf{R}_1\text{fix}} - \nabla_1 v(\mathbf{R}_1) = \mathbf{F}_N^{\text{intr}}(\mathbf{R}_1) - \nabla_1 v(\mathbf{R}_1)$$

and we have

$$w_N^{(1)}(\mathbf{R}_1) = w_N^{(1)\text{intr}}(\mathbf{R}_1) + v(\mathbf{R}_1).$$

When the interactions between the particles is given by a pairwise additive potential, we have

$$\breve{U}^{\text{intr}}(\mathbf{R}^N) = \sum_{\nu,\nu'>\nu} u(\mathbf{R}_\nu, \mathbf{R}_{\nu'}) \tag{11.62}$$

and hence

$$\mathbf{F}_N^{\text{intr}}(\mathbf{R}_1) = -\nabla_1 w_N^{(1)\text{intr}}(\mathbf{R}_1) = \left\langle -\sum_{\nu'=2}^{N} \nabla_1 u(\mathbf{R}_1, \mathbf{R}_{\nu'}) \right\rangle\Bigg|_{\mathbf{R}_1\text{fix}},$$

which is the average force on particle 1 from the pair interactions with the other particles.

11.3.2 The l-particle chemical potential and potential of mean force

We can define an ***l*-particle (*l*-point) potential of mean force** $w_N^{(l)}$ from[10]

$$\mathfrak{n}_N^{(l)}(\mathbf{R}^l) = (n^{\mathrm{b}})^l e^{-\beta w_N^{(l)}(\mathbf{R}^l)}, \tag{11.63}$$

or, equivalently,

$$e^{-\beta w_N^{(l)}(\mathbf{R}^l)} = \frac{N!}{(n^{\mathrm{b}})^l (N-l)!\, Z_N} \int d\mathbf{R}_{l+1} \dots d\mathbf{R}_N \, e^{-\beta \breve{U}_N^{\mathrm{pot}}(\mathbf{R}^N)}.$$

For example, the **pair potential of mean force** $w_N^{(2)}(\mathbf{R}_1, \mathbf{R}_2)$ has the property that $-\nabla_1 w_N^{(2)}(\mathbf{R}_1, \mathbf{R}_2)$ is the mean force on particle 1 located at $\mathbf{r}_1$ and with orientation $\boldsymbol{\omega}_1$ when particle 2 is fixed at $\mathbf{r}_2$ and has orientation $\boldsymbol{\omega}_2$ and all other particles are free to move. When the interparticle interaction potential is pairwise additive according to Equation 11.62, we have

$$\begin{aligned} -\nabla_1 w_N^{(2)}(\mathbf{R}_1, \mathbf{R}_2) &= \left. \left\langle - \sum_{\nu'=3}^{N} \nabla_1 u(\mathbf{R}_1, \mathbf{R}_{\nu'}) \right\rangle \right|_{\mathbf{R}_1, \mathbf{R}_2 \text{ fix}} \\ &- \nabla_1 u(\mathbf{R}_1, \mathbf{R}_2) - \nabla_1 v(\mathbf{R}_1), \end{aligned} \tag{11.64}$$

as shown Exercise 11.5. Likewise, $-\nabla_2 w_N^{(2)}$ is the force on particle 2 when particle 1 is held fixed.

{**Exercise 11.5:** Show Equation 11.64 from the definition of $w_N^{(2)}$, that is, Equation 11.63 for $l = 2$.}

For a macroscopic bulk fluid where $\mathfrak{n}_N^{(1)\mathrm{b}} = n^{\mathrm{b}}$ we see from Equations 11.35 and 11.63 for $l = 2$ that[11]

$$g_N^{(2)}(\mathbf{R}_1, \mathbf{R}_2) = e^{-\beta w_N^{(2)}(\mathbf{R}_1, \mathbf{R}_2)} \qquad \text{(bulk fluid)},$$

so $w_N^{(2)}(\mathbf{R}_1, \mathbf{R}_2) = -k_B T \ln g_N^{(2)}(\mathbf{R}_1, \mathbf{R}_2)$ and hence the force on particle 1 in presence of a fixed particle 2 is $-\nabla_1 w_N^{(2)}(\mathbf{R}_1, \mathbf{R}_2) = \nabla_1 [k_B T \ln g_N^{(2)}(\mathbf{R}_1, \mathbf{R}_2)]$. This is however not the case for an inhomogeneous fluid as shown in the following exercise.

{**Exercise 11.6:**
Show that for an inhomogeneous fluid, $\nabla_1 [k_B T \ln g_N^{(2)}(\mathbf{R}_1, \mathbf{R}_2)]$ is equal to the difference between the force on particle 1 in the presence of a fixed particle

---

[10] The prefactor $(n^{\mathrm{b}})^l$ in this definition can be changed to some other reference density to the power l, for example $(N/V)^l$ or $(\zeta)^l$. A change in prefactor changes $w_N^{(l)}$ by an additive constant. See also footnotes 9 and 11.

[11] This relationship justifies the selection of $(n^{\mathrm{b}})^l$ as the prefactor in Equation 11.63.

2 and the force on particle 1 when *all* other particles including particle 2 are freely mobile.}

In the presence of particles $2 \ldots l$ located at positions $\mathbf{r}_\nu$ and with orientations $\boldsymbol{\omega}_\nu$, $2 \le \nu \le l$, the mean force acting on a particle at $\mathbf{r}_1$ with orientation $\boldsymbol{\omega}_1$ is given by $-\nabla_1 w_N^{(l)}(\mathbf{r}^l, \boldsymbol{\omega}^l)$, which we can write

$$\mathbf{F}_N(\mathbf{R}_1|\mathbf{R}_2, \ldots, \mathbf{R}_l) = -\nabla_1 w_N^{(l)}(\mathbf{R}^l). \tag{11.65}$$

For pairwise additive interactions, the force on each particle can be written as the sum of pair forces from the $l-1$ fixed particles, the averages over the pair forces from the mobile particles and the force from the external field.

We have seen that the change in free energy ΔA due to the insertion of one particle with coordinates $\mathbf{R}_1$, i.e., $\Delta A = \mu_N^{\text{ex}}(\mathbf{R}_1)$, can be used to calculate $\mathfrak{n}_N^{(1)}(\mathbf{R}_1)$ (cf. Equations 11.52 and 11.54). Likewise, $w_N^{(2)}$ and $\mathfrak{n}_N^{(2)}$ can be calculated from the free energy change for a two-particle insertion (this has been treated for spherical particles in Sections 5.13.3.2 and 6.1.4 of Volume I). Let us therefore consider a system with $N-2$ mobile particles. We insert two more particles and place them fixed at positions $\mathbf{r}_1$ and $\mathbf{r}_2$ and with fixed orientations $\boldsymbol{\omega}_1$ and $\boldsymbol{\omega}_2$, respectively. The change in free energy of the system is a **two-particle excess chemical potential** $\mu_N^{(2)\text{ex}}$ that is given by

$$\mu_N^{(2)\text{ex}}(\mathbf{R}_1, \mathbf{R}_2) = k_B T \ln \frac{Z_{N-2}}{\mathsf{Z}_N^{(2)}(\mathbf{R}_1, \mathbf{R}_2)} \tag{11.66}$$

(cf. Equation 11.52), where

$$\mathsf{Z}_N^{(2)}(\mathbf{R}_1, \mathbf{R}_2) \equiv \int d\mathbf{R}_3 \ldots d\mathbf{R}_N \, e^{-\beta \breve{U}_N^{\text{pot}}(\mathbf{R}^N)}$$

(cf. Equation 11.32). This integral occurs in the definition of $\mathfrak{n}_N^{(2)}(\mathbf{R}_1, \mathbf{R}_2)$ (Equation 11.27), which can be written as

$$\mathfrak{n}_N^{(2)}(\mathbf{R}_1, \mathbf{R}_2) = \frac{N(N-1)}{Z_N} \mathsf{Z}_N^{(2)}(\mathbf{R}_1, \mathbf{R}_2) = \mathcal{C} e^{-\beta \mu^{(2)\text{ex}}(\mathbf{R}_1, \mathbf{R}_2)},$$

where $\mathcal{C} = N(N-1)Z_{N-2}/Z_N$ and we have used Equation 11.66 to obtain the rhs.

The constant $\mathcal{C}$ can be expressed in terms of activities as $\mathcal{C} = \zeta_N \zeta_{N-1}$, where $\zeta_N = N Z_{N-1}/Z_N$ is the activity for a system consisting of N particles and $\zeta_{N-1} = (N-1)Z_{N-2}/Z_{N-1}$ is the same for $N-1$ particles (see Equation 11.11, where $\zeta = \zeta_N$). Hence,

$$\mathfrak{n}_N^{(2)}(\mathbf{R}_1, \mathbf{R}_2) = \zeta_N \zeta_{N-1} e^{-\beta \mu^{(2)\text{ex}}(\mathbf{R}_1, \mathbf{R}_2)}. \tag{11.67}$$

When N is large the difference between ζ_{N-1} and ζ_N is very small so we may set $\zeta_N\zeta_{N-1} \approx \zeta_N^2$ to a very good approximation. If a part of the system is a bulk phase with density $n^{\rm b}$ we can write

$$\mathfrak{n}_N^{(2)}(\mathbf{R}_1, \mathbf{R}_2) = (n^{\rm b})^2 e^{-\beta[\mu^{(2)\rm ex}(\mathbf{R}_1,\mathbf{R}_2) - 2\mu^{\rm ex,b}]}$$

as a very good approximation, where we have used $\zeta = n^{\rm b} e^{\beta\mu^{\rm ex,b}}$ (Equation 11.55). Since $\mathfrak{n}_N^{(2)}(\mathbf{R}_1, \mathbf{R}_2) = (n^{\rm b})^2 e^{-\beta w_N^{(2)}(\mathbf{R}_1,\mathbf{R}_2)}$ (Equation 11.63 with $l = 2$) we have $w_N^{(2)}(\mathbf{R}_1, \mathbf{R}_2) = \mu_N^{(2)\rm ex}(\mathbf{R}_1, \mathbf{R}_2) - 2\mu^{\rm ex,b}$ under the same conditions.

In an analogous manner, one can define an ***l*-particle chemical potential** and use it to calculate $w_N^{(l)}$ and $\mathfrak{n}_N^{(l)}$.

11.3.3 The potential distribution theorem

The excess chemical potential $\mu_N^{\rm ex}$ and its two-particle counterpart $\mu_N^{(2)\rm ex}$ and hence $\mathfrak{n}_N^{(1)}$ and $\mathfrak{n}_N^{(2)}$ can be calculated from some rather simple formulas that are generalizations to nonspherical particles of the corresponding results in Section 5.13 of Volume I.[12] When we insert a particle at $\mathbf{r}$ with orientation $\boldsymbol{\omega}$ into a fluid consisting of $N-1$ mobile particles and keep it fixed there, the change in free energy is

$$\mu_N^{\rm ex}(\mathbf{R}) = -k_BT \ln \frac{\mathrm{Z}_N^{(1)}(\mathbf{R})}{Z_{N-1}} = -k_BT \ln \frac{\int d\mathbf{R}^{N-1} e^{-\beta \breve{U}_N^{\rm pot}(\mathbf{R},\mathbf{R}^{N-1})}}{\int d\mathbf{R}'^{N-1} e^{-\beta \breve{U}_{N-1}^{\rm pot}(\mathbf{R}'^{N-1})}} \tag{11.68}$$

(from Equations 11.32, 11.50 and 11.52), where

$$\breve{U}_N^{\rm pot}(\mathbf{R}, \mathbf{R}^{N-1}) = \breve{U}_N^{\rm pot}(\mathbf{R}, \mathbf{R}_1, \dots, \mathbf{R}_{N-1})$$

is the potential energy for all N particles and $\breve{U}_{N-1}^{\rm pot}(\mathbf{R}^{N-1})$ is that for the mobile particles $1 \dots N-1$ only. The contribution to the potential energy from the fixed particle will be denoted $\Delta\breve{u}$ and is equal to

$$\Delta\breve{u}_{\{\mathbf{R}^{N-1}\}}(\mathbf{R}) = \breve{U}_N^{\rm pot}(\mathbf{R}, \mathbf{R}^{N-1}) - \breve{U}_{N-1}^{\rm pot}(\mathbf{R}^{N-1}), \tag{11.69}$$

where subscript $\{\mathbf{R}^{N-1}\}$ indicates that $\Delta\breve{u}$ depends on the coordinates $\mathbf{R}_1, \dots, \mathbf{R}_{N-1}$ in addition to $\mathbf{R}$. If the interparticle interaction is pairwise additive according to Equation 11.62, we have

$$\Delta\breve{u}_{\{\mathbf{R}^{N-1}\}}(\mathbf{R}) = \sum_{\nu'=1}^{N-1} u(\mathbf{R}, \mathbf{R}_{\nu'}) + v(\mathbf{R}). \tag{11.70}$$

[12]See Equations I:5.143 and I:5.163, where $\mu^{\rm ex}(\mathbf{r}_0)$ in the first equation is equal to $\mu_{N+1}^{\rm ex}(\mathbf{r}_0)$ in the present notation and $\mu^{(2)\rm ex}(\mathbf{r}_0, \mathbf{r}_{0'})$ in the second equation is $\mu_{N+2}^{(2)\rm ex}(\mathbf{r}_0, \mathbf{r}_{0'})$. The starting point for the free energy difference obtained in Equations I:5.143 and I:5.163 was a system with N particles rather than a system with $N-1$ and $N-2$ particles, respectively, as used here. When N is large, the differences are very small.

In the general case we can write

$$\mu_N^{\text{ex}}(\mathbf{R}) = -k_B T \ln \frac{\int d\mathbf{R}^{N-1} e^{-\beta \Delta \breve{u}_{\{\mathbf{R}^{N-1}\}}(\mathbf{R})} e^{-\beta \breve{U}_{N-1}^{\text{pot}}(\mathbf{R}^{N-1})}}{\int d\mathbf{R}'^{N-1} e^{-\beta \breve{U}_{N-1}^{\text{pot}}(\mathbf{R}'^{N-1})}}$$

$$= -k_B T \ln \int d\mathbf{R}^{N-1} e^{-\beta \Delta \breve{u}_{\{\mathbf{R}^{N-1}\}}(\mathbf{R})} \mathcal{P}_{N-1}^{(N-1)}(\mathbf{R}^{N-1})$$

where $\mathcal{P}_{N-1}^{(N-1)}$ is the probability density to find the configuration $(\mathbf{R}_1, \ldots, \mathbf{R}_{N-1})$ for the $N-1$ mobile particles in *absence* of the fixed particle. Thus, the excess chemical potential can be expressed as

$$\mu_N^{\text{ex}}(\mathbf{R}) = -k_B T \ln \left\langle e^{-\beta \Delta \breve{u}_{\{\mathbf{R}^{N-1}\}}(\mathbf{R})} \right\rangle^{\circ} \Bigg|_{\mathbf{R}\text{ fix}} \tag{11.71}$$

where superscript "o" on the average indicates that the probability density for the $N-1$ mobile particles is not affected by particle 1. Thus, the average is taken for a situation when the fixed particle is treated as a "ghost" (often called a "test particle") that the mobile particles do not interact with. This result for $\mu^{\text{ex}}(\mathbf{R})$, which we can write as

$$e^{-\beta \mu_N^{\text{ex}}(\mathbf{R})} = \left\langle e^{-\beta \Delta \breve{u}_{\{\mathbf{R}^{N-1}\}}(\mathbf{R})} \right\rangle^{\circ} \Bigg|_{\mathbf{R}\text{ fix}},$$

is known as the **potential distribution theorem** or **Widom's formula.**[13]

Let us split $\Delta\breve{u}$ into the energy $\Delta\breve{u}^{\text{intr}}$ from the interactions with the mobile particles and the contribution from the external potential

$$\Delta \breve{u}_{\{\mathbf{R}^{N-1}\}}(\mathbf{R}) = \Delta \breve{u}_{\{\mathbf{R}^{N-1}\}}^{\text{intr}}(\mathbf{R}) + v(\mathbf{R}) \tag{11.72}$$

and write Equation 11.71 as

$$\mu_N^{\text{ex}}(\mathbf{R}) = -k_B T \ln \left\langle e^{-\beta \Delta \breve{u}_{\{\mathbf{R}^{N-1}\}}^{\text{intr}}(\mathbf{R})} \right\rangle^{\circ} \Bigg|_{\mathbf{R}\text{ fix}} + v(\mathbf{R}). \tag{11.73}$$

We can identify the first term on the rhs as $-k_B T c_N^{(1)}(\mathbf{R})$, where $c_N^{(1)}$ is the direct correlation function for a system with N particles (cf. Equation 9.8), so we have

$$c_N^{(1)}(\mathbf{R}) = \ln \left\langle e^{-\beta \Delta \breve{u}_{\{\mathbf{R}^{N-1}\}}^{\text{intr}}(\mathbf{R})} \right\rangle^{\circ} \Bigg|_{\mathbf{R}\text{ fix}}.$$

The hierarchy of direct correlation functions belongs, however, more naturally to the grand canonical ensemble since the density there is freely variable, which it is not when the number of particles are finite.

By using Widom's formula one can measure the excess chemical potential at a point $\mathbf{r}$ without disturbing the system and use the value to calculate the density at that point from Equation 11.56. We thereby measure $\mu_N^{\text{ex}}(\mathbf{R})$ for

[13] B. Widom, J. Chem. Phys., **39** (1963) 2808 (https://doi.org/10.1063/1.1734110).

a particle with a certain orientation ω and obtain the corresponding density $\mathfrak{n}_N^{(1)}(\mathbf{R})$. The argument $\exp(-\beta\Delta\breve{\mathfrak{u}}^{\mathrm{intr}})$ of the average can fluctuate a lot as a function of the coordinates of the mobile particles. For example, when $\Delta\breve{\mathfrak{u}}^{\mathrm{intr}}$ is negative and large, the values of the exponential function can be much higher than otherwise. Furthermore, for dense systems where there is little space for a new particle, $\Delta\breve{\mathfrak{u}}^{\mathrm{intr}}$ is very large or infinite for most values of $\mathbf{R}$ and the exponential function is close to or equal to zero for these $\mathbf{R}$. For these reasons, it can be difficult to obtain an accurate value of the average. In practice, Widom's formula is most useful, for example in simulations, if the fluid is not too dense since it is easier to obtain an accurate average in that case.

Example 11.B: Interpretation of Widom's formula for hard particles

Consider a fluid of hard particles ("hp") where the pair interaction is zero when the particles do not overlap and infinity otherwise. For such a system the average in Equation 11.73 has a simple interpretation. Consider a certain configuration for the $N-1$ mobile particles and let us try to insert a new particle with orientation ω at position $\mathbf{r}$. If there is space for the new particle there, which means that the configuration has a void where the new particle fits, we have $\Delta\breve{\mathfrak{u}}^{\mathrm{intr}}_{\{\mathbf{R}^{N-1}\}}(\mathbf{R}) = 0$ and $\exp(-\beta\Delta\breve{\mathfrak{u}}^{\mathrm{intr}}_{\{\mathbf{R}^{N-1}\}}(\mathbf{R})) = 1$. If the new particle does not fit, i.e. it would overlap with some other particle(s) when we try to insert it, we have $\Delta\breve{\mathfrak{u}}^{\mathrm{intr}}_{\{\mathbf{R}^{N-1}\}}(\mathbf{R}) = \infty$ since $u(\mathbf{R},\mathbf{R}_\nu) = \infty$ for at least one ν value in Equation 11.70 and hence $\exp(-\beta\Delta\breve{\mathfrak{u}}^{\mathrm{intr}}_{\{\mathbf{R}^{N-1}\}}(\mathbf{R})) = 0$. Since the contribution to the average is 1 when the new particle fits and 0 otherwise, the average is equal to the probability $\mathfrak{p}^{\mathrm{hp}}(\mathbf{R})$ to successfully insert a new hard particle with its center at the point $\mathbf{r}$ with the given orientation in the system with $N-1$ particles.

Thus $\mu_N^{\mathrm{ex}}(\mathbf{R}) = -k_BT\ln\mathfrak{p}^{\mathrm{hp}}(\mathbf{R}) + v(\mathbf{R})$ and the density distribution function of the fluid is given by

$$\mathfrak{n}_N^{(1)}(\mathbf{R}) = \zeta\mathfrak{p}^{\mathrm{hp}}(\mathbf{R})e^{-\beta v(\mathbf{R})}.$$

For a bulk fluid we have $n^{\mathrm{b}} = \zeta\mathfrak{p}^{\mathrm{hp,b}}$ and the activity coefficient $\gamma^{\mathrm{b}} = 1/\mathfrak{p}^{\mathrm{hp,b}}$ (for a bulk fluid $\mathfrak{p}$ is of course independent of orientation).

For a two-particle insertion at positions $\mathbf{r}$ and $\mathbf{r}'$ and with fixed orientations ω and ω', respectively, the two-particle excess chemical potential $\mu_N^{(2)\mathrm{ex}}$ from Equation 11.66 can likewise be expressed as

$$\mu_N^{(2)\mathrm{ex}}(\mathbf{R},\mathbf{R}') = -k_B T \ln \left\langle e^{-\beta\Delta\breve{\mathfrak{u}}^{(2)}_{\{\mathbf{R}^{N-2}\}}(\mathbf{R},\mathbf{R}')} \right\rangle^{\circ}\Bigg|_{\mathbf{R},\mathbf{R}'\text{ fix}}, \tag{11.74}$$

where

$$\Delta\breve{\mathfrak{u}}^{(2)}_{\{\mathbf{R}^{N-2}\}}(\mathbf{R},\mathbf{R}') = \breve{U}_N^{\mathrm{pot}}(\mathbf{R},\mathbf{R}',\mathbf{R}^{N-2}) - \breve{U}_{N-2}^{\mathrm{pot}}(\mathbf{R}^{N-2})$$

with subscript $\{\mathbf{R}^{N-2}\} \equiv \{\mathbf{R}_1, \ldots, \mathbf{R}_{N-2}\}$ and where $\breve{U}_{N-2}^{\mathrm{pot}}$ is the potential energy for the mobile particles $1 \ldots N-2$ only. If the interactions are pairwise additive, Equation 11.74 can be written as

$$\begin{aligned}\mu_N^{(2)\mathrm{ex}}(\mathbf{R},\mathbf{R}') &= v(\mathbf{R}) + v(\mathbf{R}') + u(\mathbf{R},\mathbf{R}') \\ &- k_B T \ln \left\langle e^{-\beta\Delta\breve{\mathfrak{u}}^{(2)\mathrm{intr}}_{\{\mathbf{R}^{N-2}\}}(\mathbf{R},\mathbf{R}')} \right\rangle^{\circ}\Bigg|_{\mathbf{R},\mathbf{R}'\text{ fix}},\end{aligned}$$

where

$$\breve{\mathfrak{u}}^{(2)\mathrm{intr}}_{\{\mathbf{R}^{N-2}\}}(\mathbf{R},\mathbf{R}') = \sum_{\nu'=1}^{N-2} \left[u(\mathbf{R},\mathbf{R}_{\nu'}) + u(\mathbf{R}',\mathbf{R}_{\nu'})\right].$$

By using these results one can measure $\mu^{(2)\mathrm{ex}}$ for a pair of points without disturbing the system and use the value to calculate the pair distribution function for these points from Equation 11.67.

{Exercise 11.7: Show Equation 11.74 starting from Equation 11.66.**}**

11.4 DENSITY DISTRIBUTIONS IN THE GRAND CANONICAL ENSEMBLE

For an open system in the grand canonical ensemble, we have constant T, V and μ, while the number of particles can vary. The probability $\mathscr{P}(N)$ that the system with volume V has N particles at a given μ and T is $\mathscr{P}(N) = \zeta^N Z_N/[N!\,\Xi]$ (Equation 11.14). The number density distribution $\mathfrak{n}^{(1)}(\mathbf{R})$ is given by

$$\mathfrak{n}^{(1)}(\mathbf{R}) = \sum_{N=1}^{\infty} \mathscr{P}(N)\mathfrak{n}_N^{(1)}(\mathbf{R}), \tag{11.75}$$

which corresponds to Equation 9.162 for spherical particles. We will now investigate the higher order distribution functions.

11.4.1 Distribution functions and mean forces of various orders

The ***l*-point density distribution function** $\mathfrak{n}^{(l)}$ for an open system is given by

$$\begin{aligned} \mathfrak{n}^{(l)}(\mathbf{R}^l) &= \sum_{N=l}^{\infty} \mathscr{P}(N)\mathfrak{n}_N^{(l)}(\mathbf{R}^l) \\ &= \frac{1}{\Xi}\sum_{N=l}^{\infty} \frac{\zeta^N}{(N-l)!} \int d\mathbf{R}_{l+1} \dots d\mathbf{R}_N e^{-\beta \breve{U}_N^{\mathrm{pot}}(\mathbf{R}^N)}, \end{aligned} \tag{11.76}$$

where the grand canonical partition function Ξ is

$$\Xi = \sum_{N=0}^{\infty} \frac{\zeta^N Z_N}{N!} = \sum_{N=0}^{\infty} \frac{\zeta^N}{N!} \int d\mathbf{R}^N e^{-\beta \breve{U}_N^{\mathrm{pot}}(\mathbf{R}^N)} \tag{11.77}$$

for nonspherical particles (cf. Equation 9.169). The contribution from the term with $N = l$ in Equation 11.76 does not contain any integral sign, which is the convention used throughout for the first term in similar expressions. The l-point density distribution irrespectively of orientations is

$$n^{(l)}(\mathbf{r}^l) = \int d\boldsymbol{\omega}^l \mathfrak{n}^{(l)}(\mathbf{r}^l, \boldsymbol{\omega}^l).$$

The spatial integrals of $n^{(l)}$ and $\mathfrak{n}^{(l)}$ are given by

$$\begin{aligned} &\int_{V^l} d\mathbf{r}^l n^{(l)}(\mathbf{r}^l) = \int_{V^l} d\mathbf{r}^l \int d\boldsymbol{\omega}^l \mathfrak{n}^{(l)}(\mathbf{r}^l, \boldsymbol{\omega}^l) \\ &= \int d\mathbf{R}^l \mathfrak{n}^{(l)}(\mathbf{R}^l) = \sum_{N=l}^{\infty} \mathscr{P}(N) \frac{N!}{(N-l)!} = \left\langle \frac{N!}{(N-l)!} \right\rangle, \end{aligned} \tag{11.78}$$

(from Equation 11.22), where the integration over V^l implies that the spatial integrals are taken over the volume V for each $\mathbf{r}_\nu$, $1 \le \nu \le l$. For example, for $l = 1$ we have

$$\int_V d\mathbf{r}_1\, n^{(1)}(\mathbf{r}_1) = \int_V d\mathbf{r}_1 \int d\omega_1\, \mathfrak{n}^{(1)}(\mathbf{r}_1, \omega_1) = \langle N \rangle,$$

that is, the average number of particles in the volume V, as expected for the usual density distribution.

The density distribution in the presence of a fixed particle with orientation ω_2 at position $\mathbf{r}_2$ is $\mathfrak{n}^{(1)}(\mathbf{R}_1|\mathbf{R}_2)$ and like in the canonical ensemble (cf. Equation 11.28) it satisfies

$$\mathfrak{n}^{(1)}(\mathbf{R}_1|\mathbf{R}_2) = \frac{\mathfrak{n}^{(2)}(\mathbf{R}_1, \mathbf{R}_2)}{\mathfrak{n}^{(1)}(\mathbf{R}_2)}, \tag{11.79}$$

It is a suitable exercise to show Equation 11.79 in a similar manner as in the corresponding result in Example 11:A; this is left to the reader to do.

{**Exercise 11.8:** The density distribution $\mathfrak{n}^{(1)}$ in the presence of a fixed particle with orientation $\boldsymbol{\omega}_0$ at position $\mathbf{r}_0$ can be obtained from Equation 11.76 for $l = 1$ analogously to Equation 11.31 in Example 11:A. First, do the substitution $N \to N' - 1$ in Equations 11.76 and 11.77. Then, insert the particle with coordinates $\mathbf{R}_0$ in order to obtain

$$\mathfrak{n}^{(1)}(\mathbf{R}_1|\mathbf{R}_0) = \frac{1}{\Xi^{(1)}(\mathbf{R}_0)} \sum_{N'=2}^{\infty} \frac{\zeta^{N'-1}}{(N'-2)!} \int d\mathbf{R}_2 \ldots d\mathbf{R}_{N'-1} e^{-\beta \check{U}_{N'}^{\text{pot}}(\mathbf{R}_0, \mathbf{R}^{N'-1})}, \tag{11.80}$$

where $\check{U}_{N'}^{\text{pot}}(\mathbf{R}_0, \mathbf{R}^{N'-1}) \equiv \check{U}_{N'}^{\text{pot}}(\mathbf{R}_0, \mathbf{R}_1, \ldots, \mathbf{R}_{N'-1})$ and $\Xi^{(1)}(\mathbf{R}_0)$ is the grand canonical partition function in the presence of the fixed particle

$$\Xi^{(1)}(\mathbf{R}_0) = \sum_{N'=1}^{\infty} \frac{\zeta^{N'-1}}{(N'-1)!} \int d\mathbf{R}_1 \ldots d\mathbf{R}_{N'-1} e^{-\beta \check{U}_{N'}^{\text{pot}}(\mathbf{R}_0, \mathbf{R}^{N'-1})}. \tag{11.81}$$

Finally, use Equation 11.80 to show Equation 11.79, but with $\mathbf{R}_0$ instead $\mathbf{R}_2$. You need to show that

$$\mathfrak{n}^{(1)}(\mathbf{R}_0) = \frac{\zeta\, \Xi^{(1)}(\mathbf{R}_0)}{\Xi}, \tag{11.82}$$

(cf. Equation 11.33).}

Like before, the **pair distribution function** $g^{(2)}$ is defined from

$$\mathfrak{n}^{(2)}(\mathbf{R}_1, \mathbf{R}_2) = \mathfrak{n}^{(1)}(\mathbf{R}_1)\, \mathfrak{n}^{(1)}(\mathbf{R}_2)\, g^{(2)}(\mathbf{R}_1, \mathbf{R}_2), \tag{11.83}$$

and it follows from Equation 11.79 that

$$\mathfrak{n}^{(1)}(\mathbf{R}_1|\mathbf{R}_2) = \mathfrak{n}^{(1)}(\mathbf{R}_1)\, g^{(2)}(\mathbf{R}_1, \mathbf{R}_2).$$

When $\mathbf{r}_1$ is moved far away from $\mathbf{r}_2$ we have $\mathfrak{n}^{(1)}(\mathbf{R}_1|\mathbf{R}_2) \to \mathfrak{n}^{(1)}(\mathbf{R}_1)$, so

$$g^{(2)}(\mathbf{R}_1, \mathbf{R}_2) \to 1 \qquad \text{when } r_{12} \to \infty.$$

Note that this holds strictly true in an open system with a large volume, while for a closed system with N particles we have $g_N^{(2)}(\mathbf{R}_1, \mathbf{R}_2) \approx 1$ for large r_{12} (see Equation 11.36).

For a general l value greater than 1 we define the ***l*-point distribution function** $g^{(l)}$ as

$$g^{(l)}(\mathbf{R}^l) = \frac{\mathfrak{n}^{(l)}(\mathbf{R}^l)}{\prod_{\nu=1}^{l} \mathfrak{n}^{(1)}(\mathbf{R}_\nu)} \qquad \text{for } l > 1. \tag{11.84}$$

The functions $\mathfrak{n}^{(l)}(\mathbf{R}^l)$ and $g^{(l)}(\mathbf{R}^l)$ are symmetric with respect to permutations of the different $\mathbf{R}_\nu$ in $\mathbf{R}^l \equiv (\mathbf{R}_1, \dots, \mathbf{R}_l)$. For $l = 1$ it is common to define $g^{(1)}(\mathbf{R}_1) = \mathfrak{n}^{(1)}(\mathbf{R}_1)/n^{\mathrm{b}}$ (cf. Equation 9.3).

When a group of spatial coordinates, say $\mathbf{r}_1, \dots \mathbf{r}_m$, is moved far away from another group, say $\mathbf{r}_{m+1}, \dots \mathbf{r}_l$, we have

$$g^{(l)}(\mathbf{R}_1, \dots, \mathbf{R}_l) \to g^{(m)}(\mathbf{R}_1, \dots, \mathbf{R}_m)\, g^{(l-m)}(\mathbf{R}_{m+1}, \dots, \mathbf{R}_l).$$

Hence, when *all* points $\mathbf{r}_1, \dots, \mathbf{r}_l$ are moved away for each other $g^{(l)}(\mathbf{R}^l) \to 1$. The correlation functions $h^{(l)}$ for $l = 2, 3$ are defined as before: $h^{(2)} = g^{(2)} - 1$ and the triplet correlation function is

$$\begin{aligned} h^{(3)}(\mathbf{R}_1, \mathbf{R}_2, \mathbf{R}_3) &= g^{(3)}(\mathbf{R}_1, \mathbf{R}_2, \mathbf{R}_3) \\ &- g^{(2)}(\mathbf{R}_1, \mathbf{R}_2) - g^{(2)}(\mathbf{R}_1, \mathbf{R}_3) - g^{(2)}(\mathbf{R}_2, \mathbf{R}_3) + 2 \end{aligned} \tag{11.85}$$

(cf. Equation 11.47). For an open system, $h^{(2)}(\mathbf{R}_1, \mathbf{R}_2) \to 0$ when $r_{12} \to \infty$ and $h^{(3)}(\mathbf{R}_1, \mathbf{R}_2, \mathbf{R}_3) \to 0$ when *at least one* of the points $\mathbf{r}_1$, $\mathbf{r}_2$ and $\mathbf{r}_3$ is moved sufficiently far away from the other two. The function $h^{(3)}$ only considers the correlations between three particles (at $\mathbf{r}_1$, $\mathbf{r}_2$ and $\mathbf{r}_3$, respectively) that are not already included in the pair correlations, so when $h^{(3)} = 0$ it is possible that two of the particles are correlated to each other if they are sufficiently close to each other.

The microscopic particle density $\breve{\mathfrak{n}}^{(1)}$ in an open system is the analogue to $\breve{\mathfrak{n}}_N^{(1)}$ defined in Equation 11.24 for an N-particle system,[14] and the grand canonical average of $\breve{\mathfrak{n}}^{(1)}$ is equal to the singlet density distribution function $\mathfrak{n}^{(1)}$

$$\mathfrak{n}^{(1)}(\mathbf{R}) = \left\langle \breve{\mathfrak{n}}^{(1)}(\mathbf{R}) \right\rangle \tag{11.86}$$

(i.e., the analogue of Equation 11.26). Likewise, the density distribution functions $\mathfrak{n}^{(2)}$ and $\mathfrak{n}^{(3)}$ can be written as the average of $\breve{\mathfrak{n}}^{(2)}$ and $\breve{\mathfrak{n}}^{(3)}$

$$\mathfrak{n}^{(2)}(\mathbf{R}, \mathbf{R}') = \left\langle \breve{\mathfrak{n}}^{(2)}(\mathbf{R}, \mathbf{R}') \right\rangle \tag{11.87}$$

$$\mathfrak{n}^{(3)}(\mathbf{R}, \mathbf{R}', \mathbf{R}'') = \left\langle \breve{\mathfrak{n}}^{(3)}(\mathbf{R}, \mathbf{R}', \mathbf{R}'') \right\rangle \tag{11.88}$$

(in analogy to Equations 11.37 and 11.45). Furthermore, we have

$$\left\langle \breve{\mathfrak{n}}^{(1)}(\mathbf{R}) \breve{\mathfrak{n}}^{(1)}(\mathbf{R}') \right\rangle = \mathfrak{n}^{(2)}(\mathbf{R}, \mathbf{R}') + \mathfrak{n}^{(1)}(\mathbf{R}) \delta^{(3)}(\mathbf{R} - \mathbf{R}') \tag{11.89}$$

---

[14]When the system contains N particles $\breve{\mathfrak{n}}^{(1)}(\mathbf{R})$ is equal to $\sum_{\nu=1}^{N} \delta^{(3)}(\mathbf{R} - \mathbf{R}_\nu)$ (the dependence of $\breve{\mathfrak{n}}^{(1)}$ on the particle coordinates $\mathbf{R}_\nu$ is not explicitly shown here). The grand canonical average of $\breve{\mathfrak{n}}^{(1)}(\mathbf{R})$ according to Equation 11.15, where we replace $\breve{X}_N(\mathbf{R}^N)$ by $\sum_{\nu=1}^{N} \delta^{(3)}(\mathbf{R} - \mathbf{R}_\nu) \equiv \breve{\mathfrak{n}}_{\{\mathbf{R}^N\}}^{(1)}(\mathbf{R})$, becomes

$$\left\langle \breve{\mathfrak{n}}^{(1)}(\mathbf{R}) \right\rangle = \sum_{N=0}^{\infty} \mathscr{P}(N) \int d\mathbf{R}^N \breve{\mathfrak{n}}_{\{\mathbf{R}^N\}}^{(1)}(\mathbf{R}) \mathcal{P}_N^{(N)}(\mathbf{R}^N) = \sum_{N=0}^{\infty} \mathscr{P}(N) \mathfrak{n}_N^{(1)}(\mathbf{R}) = \mathfrak{n}^{(1)}(\mathbf{R}),$$

where we have used Equations 11.24 and 11.75 and the canonical averages given in Equations 11.25 and 11.26.

(cf. Equation 11.39) and $\langle \breve{\mathfrak{n}}^{(1)}(\mathbf{R})\breve{\mathfrak{n}}^{(1)}(\mathbf{R}')\breve{\mathfrak{n}}^{(1)}(\mathbf{R}'')\rangle$ can likewise be written as in Equation 11.46 (without subscript N). For future reference we write this latter relationship as

$$\begin{aligned}\left\langle \breve{\mathfrak{n}}^{(1)}(\mathbf{R})\breve{\mathfrak{n}}^{(1)}(\mathbf{R}')\breve{\mathfrak{n}}^{(1)}(\mathbf{R}'')\right\rangle &= \mathfrak{n}^{(3)}(\mathbf{R})\mathfrak{n}^{(3)}(\mathbf{R}')\mathfrak{n}^{(3)}(\mathbf{R}'')g^{(3)}(\mathbf{R},\mathbf{R}',\mathbf{R}'') \\ &+ \left[\mathfrak{n}^{(3)}(\mathbf{R})\mathfrak{n}^{(3)}(\mathbf{R}')g^{(2)}(\mathbf{R},\mathbf{R}')\delta^{(3)}(\mathbf{R}'-\mathbf{R}'') + \{\mathbf{R},\mathbf{R}',\mathbf{R}''\ {}_{\text{permutations}}\}\right] \\ &+ \mathfrak{n}^{(1)}(\mathbf{R})\delta^{(3)}(\mathbf{R}-\mathbf{R}')\delta^{(3)}(\mathbf{R}-\mathbf{R}''),\end{aligned} \tag{11.90}$$

where we have inserted $\mathfrak{n}^{(3)}$ and $\mathfrak{n}^{(2)}$ in terms of $g^{(3)}$ and $g^{(2)}$, respectively.

We define the **l-point potential of mean force** $w^{(l)}$ as before

$$\mathfrak{n}^{(l)}(\mathbf{R}^l) = (n^{\mathrm{b}})^l e^{-\beta w^{(l)}(\mathbf{R}^l)}, \tag{11.91}$$

where n^{b} is the density of a bulk phase with the same chemical potential μ as our system. The choice of $(n^{\mathrm{b}})^l$ as prefactor implies that for a bulk fluid, we have

$$g^{(l)}(\mathbf{R}^l) = e^{-\beta w^{(l)}(\mathbf{R}^l)} \qquad \text{(bulk fluid).}$$

From Equation 11.91 and the definition of $\mathfrak{n}^{(l)}(\mathbf{R}^l)$ (Equation 11.76) we obtain[15]

$$\begin{aligned} e^{-\beta w^{(l)}(\mathbf{R}^l)} &= \frac{1}{(n^{\mathrm{b}})^l}\sum_{N=l}^{\infty}\mathscr{P}(N)\mathfrak{n}_N^{(l)}(\mathbf{R}^l) \\ &= \frac{(\gamma^{\mathrm{b}})^l}{\Xi}\sum_{N=l}^{\infty}\frac{\zeta^{N-l}}{(N-l)!}\int d\mathbf{R}_{l+1}\dots d\mathbf{R}_N\, e^{-\beta \breve{U}_N^{\mathrm{pot}}(\mathbf{R}^N)},\end{aligned} \tag{11.92}$$

where we have used $\zeta = n^{\mathrm{b}}\gamma^{\mathrm{b}}$ (Equation 11.55). The term with $N = l$ in the rhs of Equation 11.92 does not contain any integral sign and is equal to $(\gamma^{\mathrm{b}})^l\, e^{-\beta\breve{U}_l^{\mathrm{pot}}(\mathbf{R}^l)}/\Xi$. In the limit of zero activity, that is, in the zero density limit, we have $\gamma^{\mathrm{b}} \to 1$ and $\Xi \to 1$. Equation 11.92 implies that $w^{(l)}(\mathbf{R}^l) \to \breve{U}_l^{\mathrm{pot}}(\mathbf{R}^l)$ in this limit, which is expected since all particles apart from the l particles involved in $\breve{U}_l^{\mathrm{pot}}$ then do not contribute.

In an open system in the presence of fixed particles $2 \dots l$ with coordinates $\mathbf{R}_\nu$, $2 \le \nu \le l$, the **mean force** acting on a particle at $\mathbf{r}_1$ with orientation ω_1 is given by

$$\mathbf{F}(\mathbf{R}_1|\mathbf{R}_2,\dots,\mathbf{R}_l) = -\nabla_1 w^{(l)}(\mathbf{R}_1,\dots,\mathbf{R}_l) \tag{11.93}$$

as shown in the following shaded text box. Likewise the **mean torque** on the particle is obtained from the gradient of $w^{(l)}$ with respect to the orientational variable.

---

[15]The alternative selection of $(\zeta)^l$ as prefactor in Equation 11.91 makes the factor $(\gamma^{\mathrm{b}})^l$ to disappear in the rhs of Equation 11.92 and changes $w_N^{(l)}$ by an additive constant. Such a choice can be motivated for example at phase coexistence where there is no unique n^{b} value or when an inhomogeneous fluid is in equilibrium with a solid phase rather than a bulk fluid.

We know from Section 11.3.2 that in a system with N particles, the mean force on particle 1 when particles 2, ... , l are fixed at $\mathbf{R}_2, \ldots, \mathbf{R}_l$ is equal to $\mathbf{F}_N(\mathbf{R}_1|\mathbf{R}_2, \ldots, \mathbf{R}_l) = -\nabla_1 w_N^{(2)}(\mathbf{R}^l) = \nabla_1 \ln \mathfrak{n}_N^{(l)}(\mathbf{R}^l)/\beta$ (Equation 11.65). Particle 1 can be considered as being fixed at $\mathbf{R}_1$. We are going to show that

$$-\nabla_1 w^{(l)}(\mathbf{R}^l) = \sum_{N=l}^{\infty} \mathscr{P}^{(l)}(N;\mathbf{R}^l)\mathbf{F}_N(\mathbf{R}_1|\mathbf{R}_2, \ldots, \mathbf{R}_l) = \mathbf{F}(\mathbf{R}_1|\mathbf{R}_2, \ldots, \mathbf{R}_l), \tag{11.94}$$

where $\mathscr{P}^{(l)}(N;\mathbf{R}^l)$ is the probability that an open system with fixed particles 1, ... , l at coordinates $\mathbf{R}^l$ contains, in total, N particles. This probability can be obtained from a generalization of

$$\mathscr{P}(N) = \frac{\zeta^N}{N!} \cdot \frac{Z_N}{\Xi}$$

as applied to the situation with l fixed particles (the remaining $N - l$ of the N particles are mobile), namely,

$$\begin{aligned}\mathscr{P}^{(l)}(N;\mathbf{R}^l) &= \frac{\zeta^{N-l}}{(N-l)!} \cdot \frac{\mathsf{Z}_N^{(l)}(\mathbf{R}^l)}{\Xi^{(l)}(\mathbf{R}^l)} \\ &= \frac{\zeta^N}{N!} \cdot \frac{Z_N \mathfrak{n}_N^{(l)}(\mathbf{R}^l)}{\Xi\, \mathfrak{n}^{(l)}(\mathbf{R}^l)} = \mathscr{P}(N)\frac{\mathfrak{n}_N^{(l)}(\mathbf{R}^l)}{\mathfrak{n}^{(l)}(\mathbf{R}^l)},\end{aligned} \tag{11.95}$$

where[16]

$$\mathsf{Z}_N^{(l)}(\mathbf{R}^l) = \int d\mathbf{R}_{l+1} \ldots d\mathbf{R}_N \, e^{-\beta \check{U}_N^{\mathrm{pot}}(\mathbf{R}^N)}$$

and we have used

$$\mathfrak{n}_N^{(l)}(\mathbf{R}^l) = \frac{N!}{(N-l)!\, Z_N} \mathsf{Z}_N^{(l)}(\mathbf{R}^l),$$

(from the definition 11.19 of $\mathfrak{n}_N^{(l)}$) and

$$\mathfrak{n}^{(l)}(\mathbf{R}^l) = \zeta^l \, \frac{\Xi^{(l)}(\mathbf{R}^l)}{\Xi},$$

which for $l = 1$ is the same as Equation 11.82 and the general case can be shown in an analogous manner as the latter.

[16]We have subscript N on $\mathsf{Z}_N^{(1)}(\mathbf{R}^l)$ since there are in total N particles (l fixed and $N - l$ mobile).

Furthermore, from the first line in Equation 11.92 it follows that

$$\begin{aligned}
-\beta\nabla_1 w^{(l)}(\mathbf{R}^l) &= \nabla_1\left[\ln\sum_{N=l}^{\infty}\mathscr{P}(N)\mathfrak{n}_N^{(l)}(\mathbf{R}^l)+\text{const}\right]\\
&= \frac{\sum_{N=l}^{\infty}\mathscr{P}(N)\nabla_1\mathfrak{n}_N^{(l)}(\mathbf{R}^l)}{\sum_{N=l}^{\infty}\mathscr{P}(N)\mathfrak{n}_N^{(l)}(\mathbf{R}^l)}\\
&= \frac{\sum_{N=l}^{\infty}\mathscr{P}(N)\mathfrak{n}_N^{(l)}(\mathbf{R}^l)\nabla_1\ln\mathfrak{n}_N^{(l)}(\mathbf{R}^l)}{\mathfrak{n}^{(l)}(\mathbf{R}^l)}\\
&= \sum_{N=l}^{\infty}\mathscr{P}^{(l)}(N;\mathbf{R}^l)\nabla_1\ln\mathfrak{n}_N^{(l)}(\mathbf{R}^l),
\end{aligned}$$

where we have used Equation 11.76 to simplify the denominator of the second line and Equation 11.95 to obtain the last line. From this we obtain the desired result 11.94.

Instead of keeping some particles fixed when evaluating the mean force like above, we may instead consider the case where *all* particles are mobile and evaluate the average for those configurations where particles $1,\ldots,l$ *happen to be located at* $\mathbf{r}^l$ *and have the orientations* $\boldsymbol{\omega}^l$. The probability $\mathscr{P}^{(l)}(N;\mathbf{R}^l)$ in Equation 11.94 can, in fact, be interpreted in this fashion. It was defined as the probability for a system to have N particles when l particles are kept fixed with coordinates $\mathbf{R}^l$, but we will now show that it contains the probability that the system contains N mobile particles combined with the probability that the system has l of its mobile particles at $\mathbf{R}^l$.

Let us first consider the case $l=1$. Then

$$\mathscr{P}^{(1)}(N;\mathbf{R}_1)=\frac{\mathscr{P}(N)\mathfrak{n}_N^{(1)}(\mathbf{R}_1)}{\mathfrak{n}^{(1)}(\mathbf{R}_1)}=\frac{[\mathscr{P}(N)\mathcal{P}_N^{(1)}(\mathbf{R}_1)d\mathbf{R}_1]N}{\mathfrak{n}^{(1)}(\mathbf{R}_1)d\mathbf{R}_1},$$

where we have used Equation 11.23 in the right hand side and multiplied the numerator and denominator with the differential $d\mathbf{R}_1\equiv d\mathbf{r}_1 d\boldsymbol{\omega}_1$. The square bracket in the numerator is the probability that the system contains N particles and also, at the same time, has particle 1 inside $d\mathbf{r}_1$ at $\mathbf{r}_1$ and with an orientation within $d\boldsymbol{\omega}_1$ from $\boldsymbol{\omega}_1$. The factor N accounts for that any of the N particles could be located there and the denominator makes $\sum_{N=1}^{\infty}\mathscr{P}^{(1)}(N;\mathbf{R}_1)=1$.

In the general case

$$\mathscr{P}^{(l)}(N;\mathbf{R}^l)=\frac{\mathscr{P}(N)\mathfrak{n}_N^{(l)}(\mathbf{R}^l)}{\mathfrak{n}^{(l)}(\mathbf{R}^l)}=\frac{[\mathscr{P}(N)\mathcal{P}_N^{(l)}(\mathbf{R}^l)d\mathbf{R}^l]N!/(N-l)!}{\mathfrak{n}^{(l)}(\mathbf{R}^l)d\mathbf{R}^l}$$

with the analogous interpretation. This shows that *it does not matter* if one records the average force when the l particles happen to be located at the coordinates $\mathbf{R}^l$ or if they are fixed there. In the latter case we can imagine that at each moment in time there is an opposing force acting on the particle that keeps it in place.

11.4.2 Various chemical potentials

The relationship between the potential of mean force and the excess chemical potential is the usual one[17]

$$w^{(1)}(\mathbf{R}) = \mu^{\mathrm{ex}}(\mathbf{R}) - \mu^{\mathrm{ex,b}} \tag{11.96}$$

(cf. Table 9.1) and we have

$$\mathfrak{n}^{(1)}(\mathbf{R}) = n^{\mathrm{b}} e^{-\beta[\mu^{\mathrm{ex}}(\mathbf{R}) - \mu^{\mathrm{ex,b}}]} = \zeta e^{-\beta\mu^{\mathrm{ex}}(\mathbf{R})} \tag{11.97}$$

(cf. Equations 11.54 and 11.56). This can be written as the equilibrium condition of equal chemical potential for all $\mathbf{R}$ (for all positions and orientations)

$$k_B T \ln(\mathfrak{n}^{(1)}(\mathbf{R})/\eta) + \mu^{\mathrm{ex}}(\mathbf{R}) = k_B T \ln(\zeta/\eta) = \mu \tag{11.98}$$

(cf. canonical ensemble Equations 11.53 and 11.54). For a bulk phase this reduces to the usual expression $\mu = k_B T \ln(n^{\mathrm{b}}/\eta) + \mu^{\mathrm{ex,b}}$. Equation 11.98 can be obtained from $\mathfrak{n}^{(1)}(\mathbf{R}) = \zeta\, \Xi^{(1)}(\mathbf{R})/\Xi$ (Equation 11.82), which can be written as

$$k_B T \ln(\mathfrak{n}^{(1)}(\mathbf{R})/\eta) + k_B T \ln \frac{\Xi}{\Xi^{(1)}(\mathbf{R})} = k_B T \ln(\zeta/\eta),$$

where $\Xi^{(1)}(\mathbf{R})$ is the grand canonical partition function for a system with a fixed particle at $\mathbf{R}$. The last term on the lhs can be identified as the excess chemical potential because we have

$$\mu^{\mathrm{ex}}(\mathbf{R}) = -k_B T\, [\ln \Xi^{(1)}(\mathbf{R}) - \ln \Xi] = \Delta\Theta, \tag{11.99}$$

which is the change in grand potential when we insert a particle with orientation $\boldsymbol{\omega}$ at position $\mathbf{r}$ and keep it there.

The **two-particle excess chemical potential** is the change in grand potential when two particles are inserted and kept fixed

$$\mu^{(2)\mathrm{ex}}(\mathbf{R}_1, \mathbf{R}_2) = -k_B T\, [\ln \Xi^{(2)}(\mathbf{R}_1, \mathbf{R}_2) - \ln \Xi],$$

---

[17]Sometimes a more suitable convention for $w^{(1)}$ is to select $w^{(1)}(\mathbf{R}) = \mu^{\mathrm{ex}}(\mathbf{R})$, which corresponds to the selection of $(\zeta)^l$ as prefactor in Equation 11.91 as mentioned in footnote 15. For example at bulk phase coexistence this selections is more suitable and would imply that the value of $w^{(1)}$ in either bulk phase is $\mu^{\mathrm{ex,b}}$ for that phase.

where $\Xi^{(2)}(\mathbf{R}_1, \mathbf{R}_2)$ is the grand canonical partition function in the presence of two fixed particles with coordinates $\mathbf{R}_1$ and $\mathbf{R}_2$. One can show that

$$\mathfrak{n}^{(2)}(\mathbf{R}_1, \mathbf{R}_2) = \zeta^2 e^{-\beta\mu^{(2)\mathrm{ex}}(\mathbf{R}_1, \mathbf{R}_2)}. \tag{11.100}$$

The proof is left as an exercise for the reader (see Exercise 11.9). Note that the corresponding result in the canonical ensemble, Equation 11.67, goes over to this expression in the limit $N \to \infty$ when μ is kept constant. By comparison with $\mathfrak{n}^{(2)}(\mathbf{R}_1, \mathbf{R}_2) = (n^{\mathrm{b}})^2 e^{-\beta w^{(2)}(\mathbf{R}_1, \mathbf{R}_2)}$ (Equation 11.91 for $l = 2$), we see that $w^{(2)}(\mathbf{R}_1, \mathbf{R}_2) = \mu^{(2)\mathrm{ex}}(\mathbf{R}_1, \mathbf{R}_2) - 2\mu^{\mathrm{ex,b}}$.

{**Exercise 11.9:** Show that Ξ for a system where we have immersed two fixed particles, 1 and 2, can be written as

$$\Xi^{(2)}(\mathbf{R}_1, \mathbf{R}_2) = \sum_{N'=2}^{\infty} \frac{\zeta^{N'-2}}{(N'-2)!} \int d\mathbf{R}_3 \ldots d\mathbf{R}_{N'} e^{-\beta \breve{U}_{N'}^{\mathrm{pot}}(\mathbf{R}^{N'})}.$$

and use this to prove Equation 11.100. **Hint:** The potential energy for two fixed particles $(1,2)$ and N mobile particles $(3, \ldots, N+2)$ is $\breve{U}_{N+2}^{\mathrm{pot}}(\mathbf{R}_1, \ldots, \mathbf{R}_{N+2})$. Adapt in Equation 11.77 to the case of two fixed particles and make a substitution $N \to N' - 2$.}

The **potential distribution theorem** (**Widom's formula**) for the excess chemical potential is valid also in the grand canonical ensemble. In the shaded text box below it is shown that (cf. Equation 11.73)

$$\mu^{\mathrm{ex}}(\mathbf{R}) = -k_B T \ln \left\langle e^{-\beta \Delta \breve{u}^{\mathrm{intr}}(\mathbf{R})} \right\rangle^{\circ} \Big|_{\mathbf{R}\,\mathrm{fix}} + v(\mathbf{R}), \tag{11.101}$$

where $\Delta\breve{u}^{\mathrm{intr}}(\mathbf{R})$ is the interaction energy of particle 1 with all other particles present in the system. The coordinates of the latter are not explicitly shown in $\Delta\breve{u}^{\mathrm{intr}}$. Superscript "o" on the average indicates, as before, that the mobile particles are not affected by particle 1, which accordingly is a "ghost particle." The first term in the rhs of Equation 11.101 is the intrinsic part of the excess chemical potential, so the **singlet direct-correlation function** is given by

$$c^{(1)}(\mathbf{R}) = \ln \left\langle e^{-\beta \Delta \breve{u}^{\mathrm{intr}}(\mathbf{R})} \right\rangle^{\circ} \Big|_{\mathbf{R}\,\mathrm{fix}} \tag{11.102}$$

(cf. footnote 8 on page 10 in Chapter 9). In a similar manner one can generalize the formula for the two-particle excess chemical potential, Equation 11.74, to an open system.

{**Exercise 11.10:** Generalize Equation 11.74 to an open system.}

In order to show Equation 11.101, let us write the definition 11.81 of $\Xi^{(1)}$ as

$$\Xi^{(1)}(\mathbf{R}) = \sum_{N=1}^{\infty} \frac{\zeta^{N-1}}{(N-1)!} \int d\mathbf{R}^{N-1} e^{-\beta \breve{U}_N^{\text{pot}}(\mathbf{R},\mathbf{R}^{N-1})}.$$

We have $\Delta\breve{\mathfrak{u}}^{\text{intr}}_{\{\mathbf{R}^{N-1}\}}(\mathbf{R}) = \breve{U}_N^{\text{pot}}(\mathbf{R},\mathbf{R}^{N-1}) - \breve{U}_{N-1}^{\text{pot}}(\mathbf{R}^{N-1}) - v(\mathbf{R})$ (cf. Equations 11.69 and 11.72) and by using this we obtain

$$\begin{aligned}\Xi^{(1)}(\mathbf{R}) &= e^{-\beta v(\mathbf{R})} \sum_{N=1}^{\infty} \frac{\zeta^{N-1}}{(N-1)!} \int d\mathbf{R}^{N-1} e^{-\beta \Delta\breve{\mathfrak{u}}^{\text{intr}}_{\{\mathbf{R}^{N-1}\}}(\mathbf{R})} e^{-\beta \breve{U}_{N-1}^{\text{pot}}(\mathbf{R}^{N-1})} \\ &= \Xi\, e^{-\beta v(\mathbf{R})} \left[\sum_{N=1}^{\infty} \mathscr{P}(N-1) \right. \\ &\quad \times \left. \int d\mathbf{R}^{N-1} e^{-\beta \Delta\breve{\mathfrak{u}}^{\text{intr}}_{\{\mathbf{R}^{N-1}\}}(\mathbf{R})} \mathcal{P}_{N-1}^{(N-1)}(\mathbf{R}^{N-1}) \right], \end{aligned} \tag{11.103}$$

where $\mathcal{P}_{N-1}^{(N-1)}$ and $\mathscr{P}(N-1)$ are the probabilities for $N-1$ mobile particles in the absence of the fixed particle (given by the definitions 11.2 and 11.14 of $\mathcal{P}_N^{(N)}$ and $\mathscr{P}(N)$ applied to $N-1$ particles).

Now, $\Delta\breve{\mathfrak{u}}^{\text{intr}}(\mathbf{R})$ that we introduced above in Equation 11.101 is the interaction energy of particle at $\mathbf{R}$ with all other particles present in the system at each moment, so it corresponds to $\Delta\breve{\mathfrak{u}}^{\text{intr}}_{\{\mathbf{R}^{N-1}\}}(\mathbf{R})$ for each N. The square bracket in the rhs of Equation 11.103 is equal to the grand canonical average of $\exp(-\beta\Delta\breve{\mathfrak{u}}^{\text{intr}})$ over all possible numbers of mobile particles and over all configurations for them.[18] Since $\mu^{\text{ex}}(\mathbf{R}) = -k_BT\,[\ln \Xi^{(1)}(\mathbf{R}) - \ln \Xi]$ (see Equation 11.99), Equation 11.101 follows from these results.

11.4.3 Functional derivative relationships and $H^{(l)}$ correlation functions

The generalization of Equation 9.102 to nonspherical particles

$$\frac{\delta\Theta}{\delta v(\mathbf{R}_1)} = \mathfrak{n}^{(1)}(\mathbf{R}_1) \tag{11.104}$$

can be proved strictly in a similar manner to Equation 11.49. This is left as an exercise for the reader.

[18]Compare with the arguments in footnote 14.

{Exercise 11.11: Derive Equation 11.104 by taking the functional derivative with respect to $v(\mathbf{R})$ of $\Theta = -k_B T \ln \Xi$ with Ξ given by Equation 11.77 and by using Equation 11.76 for $l = 1$.**}**

We also have

$$\frac{\delta^2 \Theta}{\delta v(\mathbf{R}_2)\delta v(\mathbf{R}_1)} = \frac{\delta \mathfrak{n}^{(1)}(\mathbf{R}_1)}{\delta v(\mathbf{R}_2)} = -\beta H^{(2)}(\mathbf{R}_1, \mathbf{R}_2) \tag{11.105}$$

(cf. Equation 9.117) where

$$H^{(2)}(\mathbf{R}_1, \mathbf{R}_2) \equiv \mathfrak{n}^{(1)}(\mathbf{R}_1)\mathfrak{n}^{(1)}(\mathbf{R}_2)h^{(2)}(\mathbf{R}_1, \mathbf{R}_2) + \mathfrak{n}^{(1)}(\mathbf{R}_1)\delta^{(3)}(\mathbf{R}_1 - \mathbf{R}_2) \tag{11.106}$$

is the generalization of the corresponding function for spherical particles defined in Equation 9.111. The proof of Equation 11.105 is given in Example 11:C.

Example 11.C: Derivation of Equation 11.105

The derivation of Equation 11.105 from the definition of $\mathfrak{n}^{(1)}(\mathbf{R})$ is a good exercise in functional differentiation. We have

$$\mathfrak{n}^{(1)}(\mathbf{R}_1) = \frac{1}{\Xi}\sum_{N=1}^{\infty}\frac{\zeta^N}{(N-1)!}\int d\mathbf{R}_2 \ldots d\mathbf{R}_N e^{-\beta\left[\breve{U}_N^{\mathrm{intr}}(\mathbf{R}^N)+\sum_{\nu=1}^{N} v(\mathbf{R}_\nu)\right]} \tag{11.107}$$

(Equation 11.76 with $l = 1$). It is, in fact, easier to differentiate $\ln \mathfrak{n}^{(1)}$ and we have, using $\ln(1/\Xi) = \beta\Theta$,

$$\begin{aligned}\ln \mathfrak{n}^{(1)}(\mathbf{R}_1) &= \beta\Theta - \beta v(\mathbf{R}_1)\\ &+ \ln\left[\zeta + \sum_{N=2}^{\infty}\frac{\zeta^N}{(N-1)!}\int d\mathbf{R}_2 \ldots d\mathbf{R}_N e^{-\beta\left[\breve{U}_N^{\mathrm{intr}}(\mathbf{R}^N)+\sum_{\nu=2}^{N} v(\mathbf{R}_\nu)\right]}\right],\end{aligned} \tag{11.108}$$

where $-\beta v(\mathbf{R}_1)$ originates from a factor $\exp(-\beta v(\mathbf{R}_1))$ that is not involved in the integration in Equation 11.107 and where we have singled out the $N = 1$ term from the sum in the exponent. Functional differentiation with respect to $v(\mathbf{R})$ gives

$$\begin{aligned}\frac{\delta \ln \mathfrak{n}^{(1)}(\mathbf{R}_1)}{\delta v(\mathbf{R})} &= \beta\frac{\delta\Theta}{\delta v(\mathbf{R})} - \beta\frac{\delta v(\mathbf{R}_1)}{\delta v(\mathbf{R})}\\ &+ \frac{1}{\mathrm{SB}}\sum_{N=2}^{\infty}\frac{\zeta^N}{(N-1)!}\int d\mathbf{R}_2 \ldots d\mathbf{R}_N\\ &\times \left[-\beta\sum_{\nu=2}^{N}\frac{\delta v(\mathbf{R}_\nu)}{\delta v(\mathbf{R})}\right] e^{-\beta\left[\breve{U}_N^{\mathrm{intr}}(\mathbf{R}^N)+\sum_{\nu=2}^{N} v(\mathbf{R}_\nu)\right]},\end{aligned}$$

where SB is equal to the large square bracket in Equation 11.108. Since this bracket is equal to $\Xi\,\mathfrak{n}^{(1)}(\mathbf{R}_1)/\exp(-\beta v(\mathbf{R}_1))$ we can write the equation as

$$\begin{aligned}\frac{\delta \ln \mathfrak{n}^{(1)}(\mathbf{R}_1)}{\delta v(\mathbf{R})} &= \beta\mathfrak{n}^{(1)}(\mathbf{R}) - \beta\delta^{(3)}(\mathbf{R}-\mathbf{R}_1)\\ &- \frac{\beta}{\Xi\,\mathfrak{n}^{(1)}(\mathbf{R}_1)}\sum_{N=2}^{\infty}\frac{\zeta^N}{(N-1)!}\\ &\times \sum_{\nu=2}^{N}\int d\mathbf{R}_2\ldots d\mathbf{R}_N\delta^{(3)}(\mathbf{R}-\mathbf{R}_\nu)e^{-\beta\breve{U}_N^{\mathrm{pot}}(\mathbf{R}^N)}.\end{aligned}$$

The last term equals $-\beta$ times

$$\begin{aligned}&\frac{1}{\Xi\,\mathfrak{n}^{(1)}(\mathbf{R}_1)}\sum_{N=2}^{\infty}\frac{\zeta^N}{(N-2)!}\int d\mathbf{R}_3\ldots d\mathbf{R}_N e^{-\beta\breve{U}_N^{\mathrm{pot}}(\mathbf{R}^N)}\Big|_{\mathbf{R}_2=\mathbf{R}}\\ &= \frac{\mathfrak{n}^{(2)}(\mathbf{R}_1,\mathbf{R})}{\mathfrak{n}^{(1)}(\mathbf{R}_1)} = \mathfrak{n}^{(1)}(\mathbf{R})g^{(2)}(\mathbf{R}_1,\mathbf{R}),\end{aligned}$$

where we in the lhs have used the fact that all $(N-1)$ terms in the sum over ν are equal, so this sum is $(N-1)$ times the first term, and where the second line follows from Equation 11.76 for $l=2$. We thus obtain

$$\begin{aligned}\frac{\delta \ln \mathfrak{n}^{(1)}(\mathbf{R}_1)}{\delta v(\mathbf{R})} &= \beta\left[\mathfrak{n}^{(1)}(\mathbf{R}) - \delta^{(3)}(\mathbf{R}-\mathbf{R}_1) - \mathfrak{n}^{(1)}(\mathbf{R})g^{(2)}(\mathbf{R}_1,\mathbf{R})\right]\\ &= -\beta\left[\delta^{(3)}(\mathbf{R}-\mathbf{R}_1) + \mathfrak{n}^{(1)}(\mathbf{R})h^{(2)}(\mathbf{R}_1,\mathbf{R})\right],\end{aligned}$$

Since the lhs equals $\left[\delta\mathfrak{n}^{(1)}(\mathbf{R}_1)/\delta v(\mathbf{R})\right]/\mathfrak{n}^{(1)}(\mathbf{R}_1)$, Equation 11.105 now follows.

The $H^{(l+1)}$ correlation function is defined from $H^{(l)}$ as follows

$$H^{(l+1)}(\mathbf{R}_1,\ldots,\mathbf{R}_l) = \frac{\delta H^{(l)}(\mathbf{R}_1,\ldots,\mathbf{R}_{l-1})}{\delta[-\beta v(\mathbf{R}_l)]}. \tag{11.109}$$

This together with Equation 11.104 imply that $H^{(l)}$ for any l can be obtained as the functional derivative

$$\frac{\delta^l[-\beta\Theta]}{\delta v(\mathbf{R}_1)\ldots\delta v(\mathbf{R}_l)} = (-\beta)^l H^{(l)}(\mathbf{R}^l). \tag{11.110}$$

For instance, the function $H^{(3)}(\mathbf{R}_1,\mathbf{R}_2,\mathbf{R}_3)$ equals

$$H^{(3)}(\mathbf{R}_1,\mathbf{R}_2,\mathbf{R}_3) = \beta^{-2}\frac{\delta^3\Theta}{\delta v(\mathbf{R}_1)\delta v(\mathbf{R}_2)\delta v(\mathbf{R}_3)} = \beta^{-2}\frac{\delta\mathfrak{n}^{(1)}(\mathbf{R}_1)}{\delta v(\mathbf{R}_2)\delta v(\mathbf{R}_3)}.$$

One can show that[19]

$$H^{(3)}(\mathbf{R}_1,\mathbf{R}_2,\mathbf{R}_3) = \mathfrak{n}^{(1)}(\mathbf{R}_1)\mathfrak{n}^{(1)}(\mathbf{R}_2)\mathfrak{n}^{(1)}(\mathbf{R}_3)h^{(3)}(\mathbf{R}_1,\mathbf{R}_2,\mathbf{R}_3)$$
$$+ \left[\mathfrak{n}^{(1)}(\mathbf{R}_1)\mathfrak{n}^{(1)}(\mathbf{R}_2)h^{(2)}(\mathbf{R}_1,\mathbf{R}_2)\delta^{(3)}(\mathbf{R}_2-\mathbf{R}_3) + \{\mathbf{R}_1,\mathbf{R}_2,\mathbf{R}_3 \text{ permutations}\}\right]$$
$$+ \mathfrak{n}^{(1)}(\mathbf{R}_1)\delta^{(3)}(\mathbf{R}_1-\mathbf{R}_2)\delta^{(3)}(\mathbf{R}_1-\mathbf{R}_3). \tag{11.111}$$

Note that when $\mathbf{r}_1$, $\mathbf{r}_2$ and $\mathbf{r}_3$ are *all* different, $H^{(3)}(\mathbf{R}_1,\mathbf{R}_2,\mathbf{R}_3)$ is equal to $\mathfrak{n}^{(1)}(\mathbf{R}_1)\mathfrak{n}^{(1)}(\mathbf{R}_2)\mathfrak{n}^{(1)}(\mathbf{R}_3)h^{(3)}(\mathbf{R}_1,\mathbf{R}_2,\mathbf{R}_3)$.

If we compare $H^{(3)}$ in Equation 11.111 with $\langle\breve{\mathfrak{n}}^{(1)}(\mathbf{R})\breve{\mathfrak{n}}^{(1)}(\mathbf{R}')\breve{\mathfrak{n}}^{(1)}(\mathbf{R}'')\rangle$ given by Equation 11.90 we see a close similarity: $h^{(3)}$ and $h^{(2)}$ in the former equation are simply replaced by $g^{(3)}$ and $g^{(2)}$, respectively, in the latter.

11.4.3.1 $H^{(2)}$ and its functional inverse $C^{(2)}$; the OZ and Yvon equations for nonspherical particles

The functional inverse $[H^{(2)}]^{-1} \equiv C^{(2)}$ of the function $H^{(2)}(\mathbf{R}_1,\mathbf{R}_2)$ satisfies (cf. Equation 9.116)

$$\int d\mathbf{R}_3 H^{(2)}(\mathbf{R}_1,\mathbf{R}_3)C^{(2)}(\mathbf{R}_3,\mathbf{R}_2) = \delta^{(3)}(\mathbf{R}_1-\mathbf{R}_2) \tag{11.112}$$

and defines the **pair direct-correlation function** $c^{(2)}$ for nonspherical particles from (cf. Equation 9.113)

$$C^{(2)}(\mathbf{R}_1,\mathbf{R}_2) = \frac{1}{\mathfrak{n}^{(1)}(\mathbf{R}_1)}\delta^{(3)}(\mathbf{R}_1-\mathbf{R}_2) - c^{(2)}(\mathbf{R}_1,\mathbf{R}_2). \tag{11.113}$$

Equation 11.112 is the **Ornstein-Zernike equation** generalized to nonspherical particles and it can be written

$$\int d\mathbf{R}_3\, \mathfrak{n}^{(1)}(\mathbf{R}_1)\left[h^{(2)}(\mathbf{R}_1,\mathbf{R}_3)\mathfrak{n}^{(1)}(\mathbf{R}_3) + \delta^{(3)}(\mathbf{R}_1-\mathbf{R}_3)\right]$$
$$\times\left[\frac{1}{\mathfrak{n}^{(1)}(\mathbf{R}_3)}\delta^{(3)}(\mathbf{R}_3-\mathbf{R}_2) - c^{(2)}(\mathbf{R}_3,\mathbf{R}_2)\right] = \delta^{(3)}(\mathbf{R}_1-\mathbf{R}_2), \tag{11.114}$$

which implies

$$h^{(2)}(\mathbf{R}_1,\mathbf{R}_2) = c^{(2)}(\mathbf{R}_1,\mathbf{R}_2) + \int d\mathbf{R}_3 h^{(2)}(\mathbf{R}_1,\mathbf{R}_3)\mathfrak{n}^{(1)}(\mathbf{R}_3)c^{(2)}(\mathbf{R}_3,\mathbf{R}_2), \tag{11.115}$$

that is, the OZ equation in the most common form.

---

[19]The terms in square brackets are $\mathfrak{n}^{(1)}(\mathbf{R}_1)\mathfrak{n}^{(1)}(\mathbf{R}_2)h^{(2)}(\mathbf{R}_1,\mathbf{R}_2)\delta^{(3)}(\mathbf{R}_2-\mathbf{R}_3) + \mathfrak{n}^{(1)}(\mathbf{R}_2)\mathfrak{n}^{(1)}(\mathbf{R}_3)h^{(2)}(\mathbf{R}_2,\mathbf{R}_3)\delta^{(3)}(\mathbf{R}_3-\mathbf{R}_1) + \mathfrak{n}^{(1)}(\mathbf{R}_3)\mathfrak{n}^{(1)}(\mathbf{R}_1)h^{(2)}(\mathbf{R}_3,\mathbf{R}_1)\delta^{(3)}(\mathbf{R}_1-\mathbf{R}_2)$.

{Exercise 11.12: Show that Equation 11.114 implies Equation 11.115.**}**

The second equality of Equation 11.105 is the **first Yvon equation** for nonspherical particles

$$\frac{\delta \mathfrak{n}^{(1)}(\mathbf{R}_1)}{\delta v(\mathbf{R}_2)} = -\beta \mathfrak{n}^{(1)}(\mathbf{R}_1)\left[\delta^{(3)}(\mathbf{R}_1-\mathbf{R}_2)+h^{(2)}(\mathbf{R}_1,\mathbf{R}_2)\mathfrak{n}^{(1)}(\mathbf{R}_2)\right]$$
$$= -\beta H^{(2)}(\mathbf{R}_1,\mathbf{R}_2). \qquad (11.116)$$

Its inverse, the **second Yvon equation**

$$\frac{\delta v(\mathbf{R}_1)}{\delta \mathfrak{n}^{(1)}(\mathbf{R}_2)} = -\frac{1}{\beta}\left[\frac{1}{\mathfrak{n}^{(1)}(\mathbf{R}_1)}\delta^{(3)}(\mathbf{R}_1-\mathbf{R}_2)-c^{(2)}(\mathbf{R}_1,\mathbf{R}_2)\right]$$
$$= -\frac{1}{\beta}C^{(2)}(\mathbf{R}_1,\mathbf{R}_2) \qquad (11.117)$$

is a consequence of Equation 11.113.

In the same manner as in Section 9.1.3 we can derive (cf. Equation 9.67)

$$g^{(2)}(\mathbf{R}_1,\mathbf{R}_2) = e^{-\beta u(\mathbf{R}_1,\mathbf{R}_2)+h^{(2)}(\mathbf{R}_1,\mathbf{R}_2)-c^{(2)}(\mathbf{R}_1,\mathbf{R}_2)+b^{(2)}(\mathbf{R}_1,\mathbf{R}_2)}, \qquad (11.118)$$

where $b^{(2)}$ is the bridge function that constitutes the nonlinear part of the change in $c^{(1)}$ expressed in terms of $\Delta\mathfrak{n}$ when one immerses a new particle in the fluid, cf. Equation 9.57. As before the bridge function describes the nonlinear response due to the interactions with the added particle. The bridge function can, for example, be expressed in terms of higher order direct correlation functions $c^{(m)}$ by an expression that is analogous to Equation 9.160.

11.4.3.2 The $H^{(l)}$ correlation functions and fluctuations

Next, we shall investigate some aspects of fluctuations in number of particles and densities in the grand canonical ensemble. From Equation 11.78 for $l = 2$ it follows that

$$\int d\mathbf{R}_1 d\mathbf{R}_2\, \mathfrak{n}^{(2)}(\mathbf{R}_1,\mathbf{R}_2) = \langle N^2\rangle - \langle N\rangle.$$

Since $\int d\mathbf{R}_1 d\mathbf{R}_2\, \mathfrak{n}^{(1)}(\mathbf{R}_1)\mathfrak{n}^{(1)}(\mathbf{R}_2) = \langle N\rangle^2$ we have

$$\int d\mathbf{R}_1 d\mathbf{R}_2 \left[\mathfrak{n}^{(2)}(\mathbf{R}_1,\mathbf{R}_2) - \mathfrak{n}^{(1)}(\mathbf{R}_1)\mathfrak{n}^{(1)}(\mathbf{R}_2)\right] = \langle N^2\rangle - \langle N\rangle - \langle N\rangle^2.$$

This implies that

$$\int d\mathbf{R}_1 d\mathbf{R}_2 \left[\mathfrak{n}^{(1)}(\mathbf{R}_1)\mathfrak{n}^{(1)}(\mathbf{R}_2)h^{(2)}(\mathbf{R}_1,\mathbf{R}_2) + \mathfrak{n}^{(1)}(\mathbf{R}_1)\delta^{(3)}(\mathbf{R}_1-\mathbf{R}_2)\right]$$
$$= \langle N^2\rangle - \langle N\rangle^2 = \left\langle (N-\langle N\rangle)^2\right\rangle = \sigma_N^2,$$

where σ_N is the root mean square deviation for the number of particles in the volume V, which was introduced in Section 2.6.2 of Volume I. As we saw there, this quantity is a measure of the magnitude of the fluctuations in N in the open system. Since the function inside the square bracket is $H^{(2)}$ we see that

$$\int d\mathbf{R}_1 d\mathbf{R}_2 \, H^{(2)}(\mathbf{R}_1, \mathbf{R}_2) = \sigma_N^2. \tag{11.119}$$

For a bulk fluid we have $\sigma_N^2/\bar{N} = n^{\mathrm{b}} k_B T \chi_T$ (Equation I:2.132), where χ_T is the isothermal compressibility. Furthermore, $h^{(2)}(\mathbf{R}_1, \mathbf{R}_2) \equiv h^{(2)}(\mathbf{r}_1, \mathbf{r}_2, \omega_1, \omega_2) = h^{(2)}(\mathbf{r}_{12}, \omega_1, \omega_2)$ since only the relative separation vector $\mathbf{r}_{12}$ between two particles matters for an isotropic bulk fluid. From Equation 11.119 it follows that

$$n^{\mathrm{b}} \int d\mathbf{r}_{12} d\omega_1 d\omega_2 h^{(2)}(\mathbf{r}_{12}, \omega_1, \omega_2) + 1 = n^{\mathrm{b}} k_B T \chi_T \qquad \text{(bulk fluid)}. \tag{11.120}$$

This the **compressibility equation**, which corresponds to Equation 9.192 for spherical particles. One of the ω integrals in Equation 11.120 can optionally be performed and gives a factor of 1 due to the isotropy.

Incidentally, we note the significant difference from the pair correlation function in the canonical ensemble that satisfies

$$n^{\mathrm{b}} \int d\mathbf{r}_{12} d\omega_1 d\omega_2 h_N^{(2)}(\mathbf{r}_{12}, \omega_1, \omega_2) + 1 = 0,$$

which follows from

$$n^{\mathrm{b}} \int d\mathbf{r}_{12} d\omega_1 d\omega_2 g_N^{(2)}(\mathbf{r}_{12}, \omega_1, \omega_2) = \int d\mathbf{r}_1 d\omega_1 \mathrm{n}_N^{(1)}(\mathbf{r}_1, \omega_1 | \mathbf{r}_2, \omega_2) = N - 1,$$

where the last equality follows from the fact that $N-1$ particles surround the particle located at $\mathbf{r}_2$. We have also used $n^{\mathrm{b}} \int d\mathbf{r}_{12} d\omega_1 d\omega_2 = n^{\mathrm{b}} V = N$. The difference between the ensembles is due to the fact that the number of particles is fixed in the canonical ensemble, while it fluctuates in the grand canonical ensemble.[20]

[20] In the canonical ensemble we have seen that $h_N^{(2)}$ does not approach 0 for large distances since the number of particles N is constant. Instead, for a system with many particles we have according to Equation 11.36 $h_N^{(2)}(\mathbf{r}_{12}, \omega_1, \omega_2) \sim -n^{\mathrm{b}} k_B T \chi_T / N$ when r_{12} is large. In order to we get a function that approaches zero when r_{12} is large one can take $h^{(2)} \approx h_N^{(2)} + n^{\mathrm{b}} k_B T \chi_T / N$. This $h^{(2)}$ satisfies Equation 11.120 approximately. These facts illustrate the relationship between Equations 11.36 and 11.120.

However, in the presence of infinitely repulsive potentials for small r_{12} (e.g. a hard core potential) $h^{(2)}$ also has to satisfy the condition that $h^{(2)} = -1$ and hence $g^{(2)} = 0$ there. Therefore, one should instead take $h^{(2)} \approx h_N^{(2)} + \tau(r_{12}) n^{\mathrm{b}} k_B T \chi_T / N$ where τ satisfies the conditions that $\tau(r_{12}) \sim 1$ when r_{12} is large, $\tau(r_{12}) = 0$ for small r_{12} and $\int d\mathbf{r}_{12}[\tau(r_{12}) - 1] = 0$. The $h^{(2)}$ thus obtained from the canonical $h_N^{(2)}$ is a reasonable approximation to the grand canonical $h^{(2)}$ provided that N is sufficiently large.

Let us investigate the function $H^{(2)}$ a bit further. In terms of the *microscopic* particle density $\breve{\mathfrak{n}}^{(1)}$ it can be written as

$$\begin{aligned} H^{(2)}(\mathbf{R},\mathbf{R}') &= \left\langle \left[\breve{\mathfrak{n}}^{(1)}(\mathbf{R}) - \mathfrak{n}^{(1)}(\mathbf{R})\right] \left[\breve{\mathfrak{n}}^{(1)}(\mathbf{R}') - \mathfrak{n}^{(1)}(\mathbf{R}')\right] \right\rangle \\ &= \left\langle \breve{\mathfrak{n}}^{(1)}(\mathbf{R})\breve{\mathfrak{n}}^{(1)}(\mathbf{R}') \right\rangle - \mathfrak{n}^{(1)}(\mathbf{R})\mathfrak{n}^{(1)}(\mathbf{R}'). \end{aligned} \tag{11.121}$$

We can verify this relationship by inserting Equation 11.89 in the rhs to obtain $H^{(2)}$ in Equation 11.106. In the first line, the difference $\breve{\mathfrak{n}}^{(1)}(\mathbf{R}) - \mathfrak{n}^{(1)}(\mathbf{R})$ is the instantaneous deviation of the density at $\mathbf{r}$ from its average value (for particles with a certain orientation ω),[21] i.e., the instantaneous density fluctuation. This means that $H^{(2)}$ describes how the fluctuation at $\mathbf{r}$ correlates with the corresponding fluctuation at $\mathbf{r}'$ (for particles with orientation ω'). The self-term $\mathfrak{n}^{(1)}(\mathbf{R})\delta^{(3)}(\mathbf{R}-\mathbf{R}')$ of $H^{(2)}$ in Equation 11.106 describes the perfect correlation of a particle with itself. In the absence of interparticle interactions $h^{(2)}(\mathbf{R},\mathbf{R}') = 0$ and then only the self-term $\mathfrak{n}^{(1)}(\mathbf{R})\delta^{(3)}(\mathbf{R}-\mathbf{R}')$ remains.

Likewise, one can show that

$$\begin{aligned} &\left\langle \left[\breve{\mathfrak{n}}^{(1)}(\mathbf{R}) - \mathfrak{n}^{(1)}(\mathbf{R})\right] \left[\breve{\mathfrak{n}}^{(1)}(\mathbf{R}') - \mathfrak{n}^{(1)}(\mathbf{R}')\right] \left[\breve{\mathfrak{n}}^{(1)}(\mathbf{R}'') - \mathfrak{n}^{(1)}(\mathbf{R}'')\right] \right\rangle \\ &= H^{(3)}(\mathbf{R},\mathbf{R}',\mathbf{R}''), \end{aligned} \tag{11.122}$$

which describes correlations in fluctuations (deviations from the mean density) at three points simultaneously.

11.4.4 Distribution and partition functions at different activities

In this section we will derive a relationship between the values of the grand canonical partition function evaluated at two different activities. We will also derive the such relationships for the distribution function $\mathfrak{n}^{(l)}$ for any l. They are particularly useful when they are generalized to many-component systems as we will see in Section 11.6.1.

The expression 11.77 for the grand canonical partition function

$$\Xi = \Xi(\zeta) = \sum_{m=0}^{\infty} \frac{\zeta^m}{m!} \int d\mathbf{R}^m e^{-\beta \breve{U}_m^{\mathrm{pot}}(\mathbf{R}^m)} \tag{11.123}$$

(where we have written m instead of N) can formally be regarded as a Taylor series in ζ where the integrals divided by $m!$ are the coefficients (they do not depend on ζ). This is a Taylor expansion around $\zeta = 0$ so these coefficients are the derivatives of the left hand side evaluated at zero activity, that is, at zero density.

[21] More precisely $[\breve{\mathfrak{n}}^{(1)}(\mathbf{r},\omega) - \mathfrak{n}^{(1)}(\mathbf{r},\omega)]d\omega$ is the deviation in density of particles with an orientation within $d\omega$ from ω.

We will show that $\Xi(\zeta)$ alternatively can be obtained from the properties of the liquid at finite density, where it has an activity ζ' different from ζ, by using the expression

$$\Xi(\zeta) = \sum_{m=0}^{\infty} \frac{(\zeta - \zeta')^m}{m!} \left[\frac{\Xi(\zeta')}{(\gamma'^{\mathrm{b}})^m}\right] \int d\mathbf{R}^m \, e^{-\beta w^{(m)}(\mathbf{R}^m;\zeta')}, \tag{11.124}$$

where γ'^{b} is the activity coefficient for a bulk phase of the same liquid at activity ζ' and $w^{(m)}(\mathbf{R}^m;\zeta')$ is the m-point potential of mean force for the system at the same activity.[22] This expression has a similar structure as Equation 11.123 and is a Taylor expansion around the activity ζ', where the Taylor coefficient that multiplies $(\zeta - \zeta')^m$ for each m is evaluated at ζ'.

When $\zeta' \to 0$ we have $\Xi(\zeta') \to 1$ and $\gamma'^{\mathrm{b}} \to 1$, and from Equation 11.92 it follows that $w^{(m)}(\mathbf{R}^m;\zeta') \to \breve{U}_m^{\mathrm{pot}}(\mathbf{R}^m)$ in this limit. Then Equation 11.124 goes over into 11.123 as it must.

The proof of Equation 11.124 can be found in the shaded text box on the following page, but before we prove it let us consider what it says. Apart from the factor in square bracket, the main difference between Equations 11.123 and 11.124 is that the m-point potential of mean force $w^{(m)}$ in the latter equation takes the role that the potential energy $\breve{U}_m^{\mathrm{pot}}$ for m particles has in the former. While $\breve{U}_m^{\mathrm{pot}}$ for each m only considers interactions between m particles in the absence of other particles (at zero density), $w^{(m)}$ contains contributions from interactions with *all* particles that surround the m particles in a liquid at activity ζ', including the effects of correlations between all particles in the system. It is remarkable that these complicated effects can be expressed simply in terms of a Boltzmann factor containing the potential of mean force.

The physical meaning of the Taylor coefficients in Equation 11.124 can be further illustrated as follows. The function $e^{-\beta w^{(m)}(\mathbf{R}^m)}$ is given in Equation 11.92 and by using this expression we can write the coefficients as

$$\begin{aligned}
&\left[\frac{\Xi(\zeta')}{(\gamma^{\mathrm{b}\prime})^m}\right] \int d\mathbf{R}^m \, e^{-\beta w^{(m)}(\mathbf{R}^m;\zeta')} \\
&\quad = \int d\mathbf{R}^m \sum_{N=m}^{\infty} \frac{(\zeta')^{N-m}}{(N-m)!} \int d\mathbf{R}_{m+1} \ldots d\mathbf{R}_N \, e^{-\beta \breve{U}_N^{\mathrm{pot}}(\mathbf{R}^N)} \\
&\quad = \sum_{N=m}^{\infty} \frac{(\zeta')^{N-m}}{(N-m)!} \int d\mathbf{R}^N \, e^{-\beta \breve{U}_N^{\mathrm{pot}}(\mathbf{R}^N)},
\end{aligned}$$

whereby the Taylor coefficents are written in a manner that includes the interactions between N particles for all $N \geq m$.

An important fact is that potentials of mean force $w^{(m)}$ with $m \geq 3$ in Equation 11.124 are *not pairwise additive* even if the interaction potentials

[22] The alternative selection of $(\zeta)^l$ as prefactor in Equation 11.91 makes the factor $(\gamma'^{\mathrm{b}})^m$ to disappear in the rhs of Equation 11.124, see also Footnote 15.

between the particles in $\breve{U}_m^{\text{pot}}$ of Equation 11.123 are pairwise additive. The presence of, for example, a third particle in the vicinity of a pair of particles affect the mean force between the latter two in a complicated manner because of the interparticle correlations with all particles in the neighborhood of the three.

A similar result as that in Equation 11.124 applies for density distribution functions of the system. Let us consider the l-point distribution function $\mathfrak{n}^{(l)}(\mathbf{R}^l)$ at activity ζ, which according to Equation 11.76 can be written as

$$\mathfrak{n}^{(l)}(\mathbf{R}^l;\zeta)\frac{\Xi(\zeta)}{\zeta^l} = \sum_{m=0}^{\infty}\frac{\zeta^m}{m!}\int d\mathbf{R}_{l+1}\dots d\mathbf{R}_{l+m}e^{-\beta\breve{U}_{l+m}^{\text{pot}}(\mathbf{R}^{l+m})}, \tag{11.125}$$

where we have made a substitution $N = m + l$ and explicitly shown that $\mathfrak{n}^{(l)}$ and Ξ depend on ζ. Like in Equation 11.123, the rhs of this expression is a Taylor expansion in ζ around $\zeta = 0$. It reduces to Equation 11.123 when we set $l = 0$ and define $\mathfrak{n}^{(0)} = 1$. In the shaded text box below we will show that

$$\begin{aligned}\mathfrak{n}^{(l)}(\mathbf{R}^l;\zeta)\frac{\Xi(\zeta)}{\zeta^l} &= \sum_{m=0}^{\infty}\frac{(\zeta-\zeta')^m}{m!}\left[\frac{\Xi(\zeta')}{(\gamma'^{\text{b}})^{l+m}}\right]\\ &\times \int d\mathbf{R}_{l+1}\dots d\mathbf{R}_{l+m}\,e^{-\beta w^{(l+m)}(\mathbf{R}^{l+m};\zeta')}, \end{aligned}\tag{11.126}$$

which has a similar structure as Equation 11.124. The $(l+m)$-point potential of mean force $w^{(l+m)}$ in Equation 11.126 has the role that the potential energy $\breve{U}_{l+m}^{\text{pot}}$ for $l + m$ particles has in Equation 11.125.

A major application of this kind of results, generalized to many-component systems, is the *McMillan-Mayer theory* presented in Section 11.6.1, where the analogy between the potential of mean force and the potential energy is explored for solutions. For instance, Equation 11.124, which can be written as

$$\frac{\Xi(\zeta)}{\Xi(\zeta')} = \sum_{m=0}^{\infty}\frac{1}{m!}\left[\frac{\zeta-\zeta'}{\gamma'^{\text{b}}}\right]^m\int d\mathbf{R}^m\,e^{-\beta w^{(m)}(\mathbf{R}^m;\zeta')}, \tag{11.127}$$

will be generalized to many-component systems in Section 11.6.1 (Equation 11.153).

Let us now proceed with the proof of Equations 11.124 and 11.126, where the former equation is the special case of the latter for $l = 0$, so we will only show the latter. The rhs of Equation 11.125 is a function of ζ that we will denote as $f(\zeta)$, that is,

$$f(\zeta) = \sum_{m=0}^{\infty}\frac{\zeta^m}{m!}\int d\mathbf{R}_{l+1}\dots d\mathbf{R}_{l+m}e^{-\beta\breve{U}_{l+m}^{\text{pot}}(\mathbf{R}^{l+m})} \tag{11.128}$$

($f(\zeta)$ is, of course, also equal to the lhs of Equation 11.125, but we focus on the rhs here). We will make a Taylor expansion of $f(\zeta)$ around another activity value ζ'

$$f(\zeta) = \sum_{\nu=0}^{\infty} \frac{(\zeta-\zeta')^\nu}{\nu!} \left.\frac{\partial^\nu f(\zeta)}{\partial \zeta^\nu}\right|_{\zeta=\zeta'}. \tag{11.129}$$

From Equation 11.128 we obtain

$$\begin{aligned} \frac{\partial^\nu f(\zeta)}{\partial \zeta^\nu} &= \sum_{m=\nu}^{\infty} \frac{\zeta^{m-\nu}}{(m-\nu)!} \int d\mathbf{R}_{l+1} \ldots d\mathbf{R}_{l+m} e^{-\beta \breve{U}^{\mathrm{pot}}_{l+m}(\mathbf{R}^{l+m})} \\ &= \sum_{t=0}^{\infty} \frac{\zeta^t}{t!} \int d\mathbf{R}_{l+1} \ldots d\mathbf{R}_{l+\nu+t}\, e^{-\beta \breve{U}^{\mathrm{pot}}_{l+\nu+t}(\mathbf{R}^{l+\nu+t})}, \end{aligned} \tag{11.130}$$

where we have made the substitution $t = m - \nu$, so we have $l + m = l + \nu + t$.

Now, Equation 11.125 can be written in such a manner that its right hand side becomes similar to the right hand side of Equation 11.130. By writing t instead of m and making the replacement $l \to l + \nu$ in Equation 11.125, we obtain

$$\mathfrak{n}^{(l+\nu)}(\mathbf{R}^{l+\nu};\zeta)\frac{\Xi(\zeta)}{\zeta^{l+\nu}} = \sum_{t=0}^{\infty} \frac{\zeta^t}{t!} \int d\mathbf{R}_{l+\nu+1} \ldots d\mathbf{R}_{l+\nu+t}\, e^{-\beta \breve{U}^{\mathrm{pot}}_{l+\nu+t}(\mathbf{R}^{l+\nu+t})}, \tag{11.131}$$

which differs from Equation 11.130 only in the number of integrations performed. By doing the integrations $\int d\mathbf{R}_{l+1} \ldots d\mathbf{R}_{l+\nu}$ of both sides of Equation 11.131 it becomes

$$\begin{aligned} &\int d\mathbf{R}_{l+1} \ldots d\mathbf{R}_{l+\nu}\, \mathfrak{n}^{(l+\nu)}(\mathbf{R}^{l+\nu};\zeta)\frac{\Xi(\zeta)}{\zeta^{l+\nu}} \\ &= \sum_{t=0}^{\infty} \frac{\zeta^t}{t!} \int d\mathbf{R}_{l+1} \ldots d\mathbf{R}_{l+\nu+t}\, e^{-\beta \breve{U}^{\mathrm{pot}}_{l+\nu+t}(\mathbf{R}^{l+\nu+t})}, \end{aligned}$$

so from this result and Equation 11.130 we can conclude that

$$\frac{\partial^\nu f(\zeta)}{\partial \zeta^\nu} = \int d\mathbf{R}_{l+1} \ldots d\mathbf{R}_{l+\nu}\, \mathfrak{n}^{(l+\nu)}(\mathbf{R}^{l+\nu};\zeta)\frac{\Xi(\zeta)}{\zeta^{l+\nu}}.$$

Thus, since $f(\zeta) = \mathfrak{n}^{(l)}(\mathbf{R}^l;\zeta)\,\Xi(\zeta)/\zeta^l$ (that is, the lhs of Equation 11.125) we have from Equation 11.129, where the derivatives are evaluated at $\zeta = \zeta'$,

$$\begin{aligned} \mathfrak{n}^{(l)}(\mathbf{R}^l;\zeta)\frac{\Xi(\zeta)}{\zeta^l} &= f(\zeta) = \sum_{\nu=0}^{\infty} \frac{(\zeta-\zeta')^\nu}{\nu!} \\ &\times \int d\mathbf{R}_{l+1} \ldots d\mathbf{R}_{l+\nu}\, \mathfrak{n}^{(l+\nu)}(\mathbf{R}^{l+\nu};\zeta')\frac{\Xi(\zeta')}{(\zeta')^{l+\nu}}. \end{aligned} \tag{11.132}$$

This result implies that $\mathfrak{n}^{(l)}$ *at one activity* ζ *can be calculated from all* $\mathfrak{n}^{(m)}$ *with* $m \geq l$ *at another activity* ζ'.

From $\mathfrak{n}^{(l)}(\mathbf{R}^l) = (n^{\mathrm{b}})^l e^{-\beta w^{(l)}(\mathbf{R}^l)}$ (Equation 11.91) it follows that

$$\mathfrak{n}^{(l+\nu)}(\mathbf{R}^{l+\nu};\zeta')\frac{\Xi(\zeta')}{(\zeta')^{l+\nu}} = e^{-\beta w^{(l+\nu)}(\mathbf{R}^{l+\nu};\zeta')}\frac{\Xi(\zeta')}{(\gamma'^{\mathrm{b}})^{l+\nu}}$$

because $n'^{\mathrm{b}}/\zeta' = 1/\gamma'^{\mathrm{b}}$ for the primed state (this corresponds to a different bulk density than that associated with ζ). By inserting this in Equation 11.132 and writing m instead of ν we obtain Equation 11.126. Thereby, we have also proven Equation 11.124, which constitutes the case $l = 0$.

11.5 DISTRIBUTION AND CORRELATION FUNCTIONS FOR MANY-COMPONENT SYSTEMS; BASIC RELATIONSHIPS

So far in Chapter 11 we have restricted ourselves to systems with only one component. We now turn to distribution function theory of general many-component systems, i.e., mixtures of several components. A **solution** is a mixture where we regard one or several components as the **solvent**, usually the component(s) that have the highest concentration. For simplicity in notation we shall only have one component as solvent, but the basic theory presented here is general. There is in principle no restriction in size and shape of the particles, so the general theory is applicable to both solutions of small particles/molecules and dispersions of, for example, colloid particles. As we have seen several times already, the theory of solutions and mixtures is very similar to that of pure fluids apart from the fact that there are species index that denote which set of coordinates belongs to which component.

The quantity $\mathfrak{n}_i^{(1)}(\mathbf{R}_1)g_{ij}^{(2)}(\mathbf{R}_1,\mathbf{R}_2) \equiv \mathfrak{n}_i^{(1)}(\mathbf{r}_1,\boldsymbol{\omega}_1)g_{ij}^{(2)}(\mathbf{r}_1,\boldsymbol{\omega}_1,\mathbf{r}_2,\boldsymbol{\omega}_2)$ is the density at $\mathbf{r}_1$ of particles of species i with orientation $\boldsymbol{\omega}_1$ when a particle of species j is located with its center at $\mathbf{r}_2$ and has orientation $\boldsymbol{\omega}_2$. We will use the following notation

$$\mathfrak{n}_{i;j}^{(1)}(\mathbf{R}_1|\mathbf{R}_2) \equiv \mathfrak{n}_i^{(1)}(\mathbf{R}_1)g_{ij}^{(2)}(\mathbf{R}_1,\mathbf{R}_2), \tag{11.133}$$

where $\mathfrak{n}_{i;j}^{(1)}$ signifies the density distribution of i-particles when a j-particle is fixed. An alternative notation is

$$\mathfrak{n}_i^{(1)}(\mathbf{R}_1|\mathbf{R}_2;j) \equiv \mathfrak{n}_{i;j}^{(1)}(\mathbf{R}_1|\mathbf{R}_2), \tag{11.134}$$

where both the coordinates and the species index of the fixed particle appear after the vertical bar in the lhs, but we will normally use the more compact notation $\mathfrak{n}_{i;j}^{(1)}$. The pair density distribution is

$$\mathfrak{n}_{ij}^{(2)}(\mathbf{R}_1,\mathbf{R}_2) = \mathfrak{n}_i^{(1)}(\mathbf{R}_1)\mathfrak{n}_j^{(1)}(\mathbf{R}_2)g_{ij}^{(2)}(\mathbf{R}_1,\mathbf{R}_2),$$

where the $\mathbf{R}_1$ coordinates belong to species i and $\mathbf{R}_2$ to species j. The functions $\mathfrak{n}_{ij}^{(2)}$ and $g_{ij}^{(2)}$ are symmetric, for instance $g_{ij}^{(2)}(\mathbf{R}_1, \mathbf{R}_2) = g_{ji}^{(2)}(\mathbf{R}_2, \mathbf{R}_1)$, where both the species indices and the coordinates are swapped.

Likewise, in $\mathfrak{n}_{ijl}^{(3)}(\mathbf{R}_1, \mathbf{R}_2, \mathbf{R}_3)$ the $\mathbf{R}_1$ coordinates is for species i, $\mathbf{R}_2$ for j and $\mathbf{R}_3$ for species l and

$$\mathfrak{n}_{ijl}^{(3)}(\mathbf{R}_1, \mathbf{R}_2, \mathbf{R}_3) = \mathfrak{n}_i^{(1)}(\mathbf{R}_1)\mathfrak{n}_j^{(1)}(\mathbf{R}_2)\mathfrak{n}_l^{(1)}(\mathbf{R}_3) g_{ijl}^{(3)}(\mathbf{R}_1, \mathbf{R}_2, \mathbf{R}_3).$$

The functions $\mathfrak{n}_{ijl}^{(3)}$ and $g_{ijl}^{(3)}$ are symmetric with respect to permutations of the species indices provided the coordinates are likewise permuted at the same time, for example $\mathfrak{n}_{ijl}^{(3)}(\mathbf{R}_1, \mathbf{R}_2, \mathbf{R}_3) = \mathfrak{n}_{lji}^{(3)}(\mathbf{R}_3, \mathbf{R}_2, \mathbf{R}_1)$. The formal definitions of density distribution functions can be found in Appendix 11B in a somewhat condensed notation.

A functional derivative of a functional F that depends on the function $s_j(\mathbf{R})$ for some value(s) of the species index j is expressed as

$$\frac{\delta F}{\delta s_j(\mathbf{R})} = f_j(\mathbf{R}),$$

which means that

$$\delta F = \sum_j \int d\mathbf{R}\, f_j(\mathbf{R}) \delta s_j(\mathbf{R})$$

for infinitesimally small variations δs_j and the sum is taken over all species j in the system. This sum appears here since dF usually has a contribution from variations in s_j for every j. Likewise,

$$\frac{\delta f_i(\mathbf{R}')}{\delta s_j(\mathbf{R})} = t_{ij}(\mathbf{R}', \mathbf{R})$$

means that

$$\delta f_i(\mathbf{R}') = \sum_j \int d\mathbf{R}\, t_{ij}(\mathbf{R}', \mathbf{R}) \delta s_j(\mathbf{R}).$$

The functional differentiation rules applies to the case of many components in the following manner. Since the external potentials $v_j(\mathbf{R})$ for various j are independent of each other we have, for example,[23]

$$\frac{\delta v_i(\mathbf{R}')}{\delta v_j(\mathbf{R})} = \delta_{ij}^{(3)}(\mathbf{R} - \mathbf{R}'), \tag{11.135}$$

where

$$\delta_{ij}^{(3)}(\mathbf{R} - \mathbf{R}') \equiv \delta_{ij}\delta^{(3)}(\mathbf{r} - \mathbf{r}')\delta^{\omega}(\boldsymbol{\omega} - \boldsymbol{\omega}')$$

and δ_{ij} is the Kronecker symbol.[24]

---

[23]This follows from $\delta v_i(\mathbf{r}', \boldsymbol{\omega}') = \sum_j \int d\mathbf{r}\, d\boldsymbol{\omega}\, \delta_{ij}\delta^{(3)}(\mathbf{r} - \mathbf{r}')\delta^{\omega}(\boldsymbol{\omega} - \boldsymbol{\omega}')\delta v_j(\mathbf{r}, \boldsymbol{\omega})$.

[24]The Kronecker symbol is defined as $\delta_{ij} = 1$ when $i = j$ and $\delta_{ij} = 0$ otherwise.

The chain rule for functional differentiation is

$$\frac{\delta f_i(\mathbf{R}')}{\delta s_j(\mathbf{R})} = \sum_l \int d\mathbf{R}'' \frac{\delta f_i(\mathbf{R}')}{\delta t_l(\mathbf{R}'')} \frac{\delta t_l(\mathbf{R}'')}{\delta s_j(\mathbf{R})} \tag{11.136}$$

for three functions f_i, s_j and t_l that are functionally dependent of each other for various values of the species indices i, j and l. The sum over l occurs since a variation in s_j affects t_l for the various species l which in turn affect f_i. The definition of functional inverses is

$$\sum_l \int d\mathbf{R}'' \frac{\delta f_i(\mathbf{R}')}{\delta t_l(\mathbf{R}'')} \frac{\delta t_l(\mathbf{R}'')}{\delta f_j(\mathbf{R})} = \delta_{ij}^{(3)}(\mathbf{R} - \mathbf{R}'), \tag{11.137}$$

which expresses that $\delta f_i(\mathbf{R}')/\delta t_l(\mathbf{R}'')$ and $\delta t_l(\mathbf{R}'')/\delta f_j(\mathbf{R})$ are the functional inverses of each other.

In the same manner as done for the one-component case, functional derivative relationships like the following can be proven:

$$\frac{\delta \Theta}{\delta v_i(\mathbf{R}_1)} = \mathfrak{n}_i^{(1)}(\mathbf{R}_1) \tag{11.138}$$

$$\frac{\delta^2 \Theta}{\delta v_j(\mathbf{R}_2)\delta v_i(\mathbf{R}_1)} = \frac{\delta \mathfrak{n}_i^{(1)}(\mathbf{R}_1)}{\delta v_j(\mathbf{R}_2)} = -\beta H_{ij}^{(2)}(\mathbf{R}_1, \mathbf{R}_2) \tag{11.139}$$

where

$$H_{ij}^{(2)}(\mathbf{R}_1, \mathbf{R}_2) \equiv \mathfrak{n}_i^{(1)}(\mathbf{R}_1)\delta_{ij}^{(3)}(\mathbf{R}_1 - \mathbf{R}_2) + \mathfrak{n}_i^{(1)}(\mathbf{R}_1)\mathfrak{n}_j^{(1)}(\mathbf{R}_2)h_{ij}^{(2)}(\mathbf{R}_1, \mathbf{R}_2). \tag{11.140}$$

The pair direct-correlation functions $c_{ij}^{(2)}(\mathbf{R}_1, \mathbf{R}_2)$ is defined from the functional inverse of $H_{ij}^{(2)}(\mathbf{R}_1, \mathbf{R}_2)$

$$\begin{aligned} [\{H^{(2)}\}^{-1}]_{ij}(\mathbf{R}_1, \mathbf{R}_2) &= C_{ij}^{(2)}(\mathbf{R}_1, \mathbf{R}_2) \qquad (11.141) \\ &= \frac{1}{\mathfrak{n}_i^{(1)}(\mathbf{R}_2)} \delta_{ij}^{(3)}(\mathbf{R}_1 - \mathbf{R}_2) - c_{ij}^{(2)}(\mathbf{R}_1, \mathbf{R}_2). \end{aligned}$$

The relationship (cf. Equation 11.112)[25]

$$\sum_l \int d\mathbf{R}_3\, C_{il}^{(2)}(\mathbf{R}_1, \mathbf{R}_3) H_{lj}^{(2)}(\mathbf{R}_3, \mathbf{R}_2) = \delta_{ij}^{(3)}(\mathbf{R}_1 - \mathbf{R}_2) \tag{11.142}$$

[25] If one wants a relationship between entities like $H_{ij}^{(2)}$ and $C_{ij}^{(2)}$ that is valid also when some component of the mixture has zero density, one can instead define $\mathring{H}_{ij}^{(2)} \equiv \delta_{ij}^{(3)} + [\mathfrak{n}_i^{(1)}\mathfrak{n}_i^{(1)}]^{1/2} h_{ij}^{(2)} = [\mathfrak{n}_i^{(1)}\mathfrak{n}_i^{(1)}]^{-1/2} H_{ij}^{(2)}$ and $\mathring{C}_{ij}^{(2)} \equiv \delta_{ij} - [\mathfrak{n}_i^{(1)}\mathfrak{n}_i^{(1)}]^{1/2} c_{ij}^{(2)} = [\mathfrak{n}_i^{(1)}\mathfrak{n}_i^{(1)}]^{1/2} C_{ij}^{(2)}$, which satisfy $\sum_l \int d\mathbf{R}_3\, \mathring{C}_{il}^{(2)}(\mathbf{R}_1, \mathbf{R}_3)\mathring{H}_{lj}^{(2)}(\mathbf{R}_3, \mathbf{R}_2) = \delta_{ij}(\mathbf{R}_1 - \mathbf{R}_2)$; the same as Equation 11.142.

is equivalent to the Ornstein-Zernike equations for a mixture (cf. Equations 9.259 and 11.115)

$$h_{ij}^{(2)}(\mathbf{R}_1, \mathbf{R}_2) = c_{ij}^{(2)}(\mathbf{R}_1, \mathbf{R}_2) + \sum_l \int d\mathbf{R}_3\, c_{il}^{(2)}(\mathbf{R}_1, \mathbf{R}_3)\mathfrak{n}_l^{(1)}(\mathbf{R}_3) h_{lj}^{(2)}(\mathbf{R}_3, \mathbf{R}_2), \tag{11.143}$$

which is a system of equations for the various combinations of i and j values. Furthermore, we have the first Yvon equation

$$\begin{aligned}\frac{\delta \mathfrak{n}_i^{(1)}(\mathbf{R}_1)}{\delta v_j(\mathbf{R}_2)} &= -\beta \mathfrak{n}_i^{(1)}(\mathbf{R}_1)\left[\delta_{ij}^{(3)}(\mathbf{R}_1 - \mathbf{R}_2) + h_{ij}^{(2)}(\mathbf{R}_1, \mathbf{R}_2)\mathfrak{n}_j^{(1)}(\mathbf{R}_2)\right] \\ &= -\beta H_{ij}^{(2)}(\mathbf{R}_1, \mathbf{R}_2) \end{aligned} \tag{11.144}$$

which is the same as Equation 11.139, and its inverse, the second Yvon equation,

$$\begin{aligned}\frac{\delta v_j(\mathbf{R}_1)}{\delta \mathfrak{n}_i^{(1)}(\mathbf{R}_2)} &= -\frac{1}{\beta}\left[\frac{1}{\mathfrak{n}_i^{(1)}(\mathbf{R}_1)}\delta_{ij}^{(3)}(\mathbf{R}_1 - \mathbf{R}_2) - c_{ij}^{(2)}(\mathbf{R}_1, \mathbf{R}_2)\right] \\ &= -\frac{1}{\beta} C_{ij}^{(2)}(\mathbf{R}_1, \mathbf{R}_2), \end{aligned} \tag{11.145}$$

which follows from Equation 11.141.

The intrinsic Helmholtz free energy is defined as

$$\mathcal{A} \equiv A - \sum_i \int d\mathbf{R}_1\, \mathfrak{n}_i(\mathbf{R}_1) v_i(\mathbf{R}_1)$$

and the intrinsic chemical potential $\mu_i^{\text{intr}}(\mathbf{R}_1)$ for component i is its functional derivative

$$\frac{\delta \mathcal{A}}{\delta \mathfrak{n}_i(\mathbf{R}_1)} = \mu_i^{\text{intr}}(\mathbf{R}_1) \equiv \mu_i - v_i(\mathbf{R}_1).$$

$\mathcal{A}$ consists of the ideal and excess parts $\mathcal{A} = \mathcal{A}^{\text{ideal}} + \mathcal{A}^{ex}$, where

$$\mathcal{A}^{\text{ideal}} = k_B T \sum_i \int d\mathbf{R}_1\, \mathfrak{n}_i(\mathbf{R}_1)\left[\ln(\mathfrak{n}_i(\mathbf{R}_1)/\eta_i) - 1\right] \tag{11.146}$$

and $\eta_i = \mathfrak{q}_{\text{r},i}/\Lambda_i^3$ (Equation 11.6 applied to species i). We have

$$\frac{\delta \mathcal{A}^{\text{ideal}}}{\delta \mathfrak{n}_i(\mathbf{R}_1)} = \mu_i^{\text{ideal}}(\mathbf{R}_1) = k_B T \ln(\mathfrak{n}_i(\mathbf{R}_1)/\eta_i).$$

The singlet direct-correlation function $c_i^{(1)}$ is obtained from

$$\frac{\delta \mathcal{A}^{\text{ex}}}{\delta \mathfrak{n}_i(\mathbf{R}_1)} = -k_B T\, c_i^{(1)}(\mathbf{R}_1),$$

and the total chemical potential for component i is

$$\mu_i = k_B T \ln(\mathfrak{n}_i(\mathbf{R}_1)/\eta_i) - k_B T\, c_i^{(1)}(\mathbf{R}_1) + v_i(\mathbf{R}_1),$$

where $-k_B T\, c_i^{(1)}(\mathbf{R}_1) + v_i(\mathbf{R}_1)$ is the excess chemical potential at $\mathbf{R}_1$. The value of μ_i is the same for all positions and orientations of the i-particles at equilibrium.

The pair direct-correlation function is given by

$$c_{ij}^{(2)}(\mathbf{R}_1, \mathbf{R}_2) = -\beta \frac{\delta^2 \mathcal{A}^{\mathrm{ex}}}{\delta \mathfrak{n}_i(\mathbf{R}_1)\delta \mathfrak{n}_j(\mathbf{R}_2)} = \frac{\delta c_i^{(1)}(\mathbf{R}_1)}{\delta n_j(\mathbf{R}_2)} = \frac{\delta c_j^{(1)}(\mathbf{R}_2)}{\delta n_i(\mathbf{R}_1)}$$

(cf. Equation 9.127 and 10.31) and the higher order direct correlation functions are defined from

$$c_{i_1\ldots i_m}^{(m)}(\mathbf{R}_1, \ldots, \mathbf{R}_m) = -\beta \frac{\delta^m \mathcal{A}^{\mathrm{ex}}}{\delta \mathfrak{n}_{i_1}(\mathbf{R}_1)\ldots\delta \mathfrak{n}_{i_m}(\mathbf{R}_m)} \tag{11.147}$$

(cf. Equation 9.156).

The density distribution function for species i is given by

$$\mathfrak{n}_i(\mathbf{R}_1) = \zeta_i e^{-\beta v_i(\mathbf{R}_1) + c_i^{(1)}(\mathbf{R}_1)},$$

where the activity ζ_i is defined as

$$\zeta_i = \eta_i e^{\beta \mu_i}. \tag{11.148}$$

From this we obtain

$$\mathfrak{n}_i(\mathbf{R}_1) = \mathfrak{n}_i(\mathbf{R}_2) e^{-\beta[v_i(\mathbf{R}_1) - v_i(\mathbf{R}_2)] + c_i^{(1)}(\mathbf{R}_1) - c_i^{(1)}(\mathbf{R}_2)}.$$

In the case that the point $\mathbf{r}_2$ is located in a bulk phase that is in equilibrium with the system (or is a part of the system), we have $v_i(\mathbf{R}_2) = 0$ since v_i is zero in bulk and we get

$$\mathfrak{n}_i(\mathbf{R}_1) = n_i^{\mathrm{b}} e^{-\beta v_i(\mathbf{R}_1) + c_i^{(1)}(\mathbf{R}_1) - c_i^{(1)\mathrm{b}}} = n_i^{\mathrm{b}} e^{-\beta w_i(\mathbf{R}_1)},$$

where $w_i(\mathbf{R}_1)$ is the singlet potential of mean force for species i. As before we have $w_i(\mathbf{R}_1) = v_i(\mathbf{R}_1) + w_i^{\mathrm{intr}}(\mathbf{R}_1)$, where w_i^{intr} is the intrinsic part of w_i given by $w_i^{\mathrm{intr}}(\mathbf{R}_1) = -k_B T \Delta c_i^{(1)}(\mathbf{R}_1)$ and $\Delta c_i^{(1)}(\mathbf{R}_1) \equiv c_i^{(1)}(\mathbf{R}_1) - c_i^{(1)\mathrm{b}}$.[26] We also have

$$g_i^{(1)}(\mathbf{R}_1) \equiv \frac{\mathfrak{n}_i(\mathbf{R}_1)}{n_i^{\mathrm{b}}} = e^{-\beta[v_i(\mathbf{R}_1) + w_i^{\mathrm{intr}}(\mathbf{R}_1)]} = e^{-\beta v_i(\mathbf{R}_1) + \Delta c_i^{(1)}(\mathbf{R}_1)},$$

which defines $g_i^{(1)}(\mathbf{R}_1)$.

[26]The function $-\beta w_i^{\mathrm{intr}} = \Delta c_i^{(1)}$ is sometimes called the *thermal potential* as defined on page 78.

The pair distribution function is

$$g_{ij}^{(2)}(\mathbf{R}_1, \mathbf{R}_2) = e^{-\beta w_{ij}^{(2)}(\mathbf{R}_1,\mathbf{R}_2)} = e^{-\beta[u_{ij}(\mathbf{R}_1,\mathbf{R}_2)+w_{ij}^{(2)\text{intr}}(\mathbf{R}_1,\mathbf{R}_2)]} \tag{11.149}$$

where $w_{ij}^{(2)\text{intr}}$ is the intrinsic part of $w_{ij}^{(2)}$.[27] Recall from Section 9.1.1 that the designation "intrinsic" here originates from a point of view where one of the particles, for instance that at $\mathbf{r}_2$, is considered to be external to the system. Then, we can obtain in the same fashion as for a one-component system

$$w_{ij}^{(2)\text{intr}}(\mathbf{R}_1, \mathbf{R}_2) = -k_B T \Delta c_i^{(1)}(\mathbf{R}_1|\mathbf{R}_2; j) = -k_B T[c_i^{(1)}(\mathbf{R}_1|\mathbf{R}_2; j) - c_i^{(1)}(\mathbf{R}_1)]$$

(cf. Equation 9.24), where $c_i^{(1)}(\mathbf{R}_1|\mathbf{R}_2; j)$ is the direct correlation function in the presence of a fixed particle of species j at $\mathbf{r}_2$ with orientation ω_2. The function $w^{(2)\text{intr}}(\mathbf{R}_1, \mathbf{R}_2)$ is symmetric because $\Delta c_i^{(1)}(\mathbf{R}_1|\mathbf{R}_2; j) = \Delta c_j^{(1)}(\mathbf{R}_2|\mathbf{R}_1; i)$, which can be shown in the same manner as in the one-component case (see the arguments that lead to Equations 9.29 and 9.30). We also have $\beta w_{ij}^{(2)\text{intr}} = h_{ij}^{(2)} - c_{ij}^{(2)} + b_{ij}^{(2)}$ and hence

$$g_{ij}^{(2)}(\mathbf{R}_1, \mathbf{R}_2) = e^{-\beta u_{ij}(\mathbf{R}_1,\mathbf{R}_2)+h_{ij}^{(2)}(\mathbf{R}_1,\mathbf{R}_2)-c_{ij}^{(2)}(\mathbf{R}_1,\mathbf{R}_2)+b_{ij}^{(2)}(\mathbf{R}_1,\mathbf{R}_2)}, \tag{11.150}$$

where the bridge function $b_{ij}^{(2)}$ is defined analogously to that in the one-component case, see Section 9.1.3.1.

The l-point potential of mean force is defined from the l-point density distribution as

$$\mathfrak{n}_{i_1\ldots i_l}^{(l)}(\mathbf{R}_1, \ldots, \mathbf{R}_l) = e^{-\beta w_{i_1\ldots i_l}^{(l)}(\mathbf{R}_1,\ldots,\mathbf{R}_l)} \prod_{m=1}^{l} n_{i_m}^{\mathrm{b}} \tag{11.151}$$

(cf. Equation 11.91), which for the special case of a bulk fluid implies that

$$g_{i_1\ldots i_l}^{(l)}(\mathbf{R}_1, \ldots, \mathbf{R}_l) = e^{-\beta w_{i_1\ldots i_l}^{(l)}(\mathbf{R}_1,\ldots,\mathbf{R}_l)} \qquad \text{(bulk fluid)}.$$

As we can see from the relationships above, the formal theory for solutions has essentially the same mathematical form as that for a one-component system. To do calculations one need approximations and they are also basically the same in most cases. For example, the HNC approximation is to set $b_{ij} = 0$ in Equation 11.150 and this is combined with the Ornstein-Zernike equations 11.143. For a bulk solution the result is a set of equations for h_{ij} and c_{ij} for all pairs of components. There is an equal number of equations as unknowns (h_{ij} and c_{ij}) and the set of equations can be solved numerically. In the same manner one can use the integral equation approximations in Sections 9.1.3.2

[27] The function $-\beta w_{ij}^{(2)\text{intr}}$ is sometimes called the *thermal pair potential* as defined on page 78.

and 9.1.3.3, for example a closure of the type $b_{ij}^{(2)} \approx f(h_{ij}^{(2)} - c_{ij}^{(2)}) = f(\gamma_{ij}^{(2)})$ (cf. Equation 9.88), where f is some function.

For an inhomogeneous system one needs also a set of equations for the density distributions. For spherically symmetric particles, one may, for instance, use one of Triezenberg-Zwanzig-Lovett-Mou-Buff-Wertheim equations 9.46 or 9.47 adapted to many-component systems. For example, the latter then takes the form

$$\nabla_1 \ln n_i(\mathbf{r}_1) = -\beta \left[\nabla_1 v_i(\mathbf{r}_1) + \sum_j \int d\mathbf{r}_3 \, c_{ij}^{(2)}(\mathbf{r}_1, \mathbf{r}_3) \nabla_3 n_j(\mathbf{r}_3) \right]. \tag{11.152}$$

This equation together with the OZ equation and the closure relationship constitute a compete set of equations for h_{ij}, c_{ij} and n_i that one can solve numerically. For some closures one can obtain $c_i^{(1)}(\mathbf{r})$ and thereby $n_i^{(1)}(\mathbf{r})$ directly from the pair correlation functions as we saw in Section 10.1.5. Then one does not need to solve Equation 11.152 or one of the other differential equations for $n_i^{(1)}(\mathbf{r})$. This constitutes a major simplification.

11.6 IMPLICIT SOLVENT THEORIES

An important question regarding the properties of solutions is how the presence of the solvent affects the interactions between the solute particles/molecules. Therefore, we will focus on the distribution functions of the solute particles and the interaction between them. The goal is to express properties of the solution in terms of effective interactions between solvated particles without *explicit* consideration of the solvent. Two theories will be presented, the McMillan-Mayer and the Adelman theories.

11.6.1 The McMillan-Mayer theory

In order to acquire some important tools for the investigation of effective interactions between solvated particles, we will first investigate how the distribution functions for many components depend on the activities of the various species in the solution. This was done for a one-component system in Section 11.4.4 and here we will generalize the theory to several components, which is the basis of the **McMillan-Mayer theory of solutions**.

A major final result of this theory is that *various properties of the solution at any concentration,* including distribution functions for the solute particles, can be obtained from effective solute-solute interaction potentials that are evaluated for solute particles immersed in pure solvent, that is, *at infinite dilution.* These effective potentials are equal to the potential of mean force between finite numbers of solute particles immersed in the pure solvent. The McMillan-Mayer result is conceptually important since it brings out the interactions between sets of solvated particles as a basis for understanding the properties of the solution. This conceptual understanding is in many ways more important than the practical use of the theory.

Equation 11.127 expresses the grand canonical partition function $\Xi(\zeta)$ for a one-component liquid at activity ζ in terms of properties of the same liquid at a different activity ζ'. The ratio $\Xi(\zeta)/\Xi(\zeta')$ is thereby given as a Taylor expansion in $[\zeta-\zeta']^m$ with $m \geq 0$, where the coefficients contain Boltzmann factors with the potential of mean force $w^{(m)}$ at activity ζ'. For a many-component mixture, the derivation of the equations corresponding to those in Section 11.4.4 is just a generalization of the one-dimensional Taylor expansions to a multidimensional ones in the set of activities of the various components. For example, for a three-component system, where one component (s) is denoted as the solvent, we have three activities: ζ_s, ζ_1 and ζ_2, one for each component. The equation that corresponds to Equation 11.127 is derived in Appendix 11C and is given by

$$\begin{aligned}
&\frac{\Xi(\zeta_s,\zeta_1,\zeta_2)}{\Xi(\zeta_s',\zeta_1',\zeta_2')} \\
&= \sum_{m_s,m_1,m_2\geq 0} \frac{1}{m_s!\,m_1!\,m_2!}\left[\frac{\zeta_s-\zeta_s'}{\gamma_s'^{\mathrm{b}}}\right]^{m_s}\left[\frac{\zeta_1-\zeta_1'}{\gamma_1'^{\mathrm{b}}}\right]^{m_1}\left[\frac{\zeta_2-\zeta_2'}{\gamma_2'^{\mathrm{b}}}\right]^{m_2} \\
&\times \int d\mathbf{R}^{\mathbf{m}}\, e^{-\beta w^{(m)}_{\mathtt{I}(\mathbf{m})}(\mathbf{R}^{\mathbf{m}};\zeta_s',\zeta_1',\zeta_2')},
\end{aligned} \tag{11.153}$$

where $\mathbf{R}^{\mathbf{m}} \equiv (\mathbf{R}^{m_s},\mathbf{R}^{m_1},\mathbf{R}^{m_2})$ are the coordinates (positions and orientations) of m_s solvent molecules, m_1 particles of species 1, and m_2 of species 2 and where $\mathbf{m} = (m_s,m_1,m_2)$ and $m = m_s+m_1+m_2$. In $w^{(m)}_{\mathtt{I}(\mathbf{m})}$ the symbol $\mathtt{I}(\mathbf{m}) = \mathtt{I}(m_s,m_1,m_2)$ denotes a list of species indices for the m particles, for example for $w^{(2)}_{ss}$, $w^{(2)}_{s1}$, $w^{(2)}_{12}$ and $w^{(2)}_{22}$ we have $\mathtt{I} = ss, s1, 12$ and 22, respectively, and for $w^{(3)}_{s11}$, $w^{(3)}_{s12}$ and $w^{(3)}_{111}$ we have $\mathtt{I} = s11, s12$ and 111, respectively. In the general case $\mathtt{I}(\mathbf{m})$ has the form $ss\ldots s11\ldots 122\ldots 2$, whereby the entry s occurs m_s times, 1 occurs m_1 times and 2 occurs m_2 times in $\mathtt{I}(m_s,m_1,m_2)$. The first species has coordinates $\mathbf{R}_1$ and the second $\mathbf{R}_2$ etc. Equation 11.153 is derived in Appendix 11C[28] for any number of components (Equation 11.180); for simplicity in notation we here present the theory for three components only.

We now apply Equation 11.153 to the special case where $\zeta_s' = \zeta_s$, $\zeta_1' = 0$ and $\zeta_2' = 0$. This correspond to the situation where the primed state is pure solvent (s) at the same chemical potential as in the solution (the unprimed state). Then all terms that contain species s disappear from the right hand side of Equation 11.153 (since $\zeta_s - \zeta_s' = 0$) and we obtain

$$\frac{\Xi(\zeta_s,\zeta_1,\zeta_2)}{\Xi(\zeta_s,0,0)} = \sum_{m_1,m_2\geq 0}\frac{1}{m_1!\,m_2!}\left[\frac{\zeta_1}{\gamma_1^0}\right]^{m_1}\left[\frac{\zeta_2}{\gamma_2^0}\right]^{m_2}\int d\mathbf{R}^{\mathbf{m}}\, e^{-\beta w^{(m)}_{\mathtt{I}(\mathbf{m})}(\mathbf{R}^{\mathbf{m}};\zeta_s,0,0)}, \tag{11.154}$$

[28]Appendix 11C uses a different notation that is explained in Appendix 11B.

where $\mathbf{m} = (m_1, m_2)$ and $\gamma_i^0 = \gamma_i^{\text{bulk}}(\zeta_s, 0, 0)$ is the activity coefficient of solute i at infinite dilution in the bulk (i.e., for a single solute particle in contact with pure solvent). The potential of mean force $w_{\mathrm{I}(\mathbf{m})}^{(m)}$ in this case constitutes the potential for m_1 particles of species 1 and m_2 particles of species 2 immersed in pure solvent, whereby they are located at positions $\mathbf{r}^{\mathbf{m}}$ and with orientations $\boldsymbol{\omega}^{\mathbf{m}}$. Note that the solvent degrees of freedom are not explicitly given in the right hand side, but they are implicit in $w_{\mathrm{I}(\mathbf{m})}^{(m)}$, which describes the *interaction potential between solvated solute particles* at infinite dilution.

Equation 11.154 can be written in terms of the grand potential of the solution $\Theta^{\text{solution}} = \Theta(\zeta_s, \zeta_1, \zeta_2) = -k_B T \ln \Xi(\zeta_s, \zeta_1, \zeta_2)$. For this purpose it is useful to introduce a modified activity $\mathfrak{z}_i \equiv \zeta_i/\gamma_i^0$. This activity has the property that the corresponding activity coefficient in bulk, defined by $\mathfrak{f}_i^{\text{b}} \equiv \mathfrak{z}_i/n_i^{\text{b}}$, goes to 1 at infinite dilution, which follows from the facts that $\mathfrak{f}_i^{\text{b}} = [\zeta_i/\gamma_i^0]/n_i^{\text{b}} = [\zeta_i/n_i^{\text{b}}]/\gamma_i^0 = \gamma_i^{\text{b}}/\gamma_i^0$ and $\gamma_i^{\text{b}} \to \gamma_i^0$ in the limit of infinite dilution. Equation 11.154 yields

$$\begin{aligned}\Theta^{\text{solution}} &= \Theta^{\text{pure solvent}} \\ &- k_B T \ln \left[\sum_{m_1, m_2 \geq 0} \frac{\mathfrak{z}_1^{m_1} \mathfrak{z}_2^{m_2}}{m_1!\, m_2!} \int d\mathbf{R}^{\mathbf{m}}\, e^{-\beta w_{\mathrm{I}(\mathbf{m})}^{(m)}(\mathbf{R}^{\mathbf{m}}; \zeta_s, 0, 0)} \right],\end{aligned} \tag{11.155}$$

where $\Theta^{\text{pure solvent}} = -k_B T \ln \Xi(\zeta_s, 0, 0)$ is the grand potential of pure solvent at the same solvent activity as in the solution.

If we remove the solvent entirely, we have a gas consisting solely of solute particles. The grand potential for such a gas is given by

$$\begin{aligned}\Theta^{\text{gas}} &= \Theta^{\text{vacuum}} \\ &- k_B T \ln \left[\sum_{m_1, m_2 \geq 0} \frac{\zeta_1^{m_1} \zeta_2^{m_2}}{m_1!\, m_2!} \int d\mathbf{R}^{\mathbf{m}} e^{-\beta \breve{U}_{\mathbf{m}}^{\text{pot}}(\mathbf{R}^{\mathbf{m}})} \right],\end{aligned} \tag{11.156}$$

which follows from Equation 11.174 in Appendix 11B (here we have written m instead of N). In order to have the same structure as Equation 11.155, we have introduced Θ^{vacuum}, but it is equal to zero. By comparing these two expressions we see that $\Delta\Theta = \Theta^{\text{solution}} - \Theta^{\text{pure solvent}}$ for the solution has exactly the same mathematical form as $\Theta^{\text{gas}} - \Theta^{\text{vacuum}}$, but with different interaction potentials in the exponent. Furthermore, the activities $\mathfrak{z}_i$ in the solution correspond to ζ_i for the gas.

The remarkable conclusion is that *the potential of mean force* $w_{\mathrm{I}(\mathbf{m})}^{(m)}$ *between the solute particles plays the same role in a solution as the potential energy does in a gas*, whereby $w_{\mathrm{I}(\mathbf{m})}^{(m)}$ is evaluated for m solute particles *immersed in the pure solvent.* This is a major result in the McMillan-Mayer theory of solutions. The significance of this result is apparent when one considers that various equilibrium properties of a system can, in principle, be obtained from the grand potential by taking various derivatives (including functional derivatives). To calculate Θ of a solution, it is hence not necessary to *explicitly* consider the solvent species since Equation 11.155 implies that $\Theta(\zeta_s, \zeta_1, \zeta_2)$ for a solution of a given composition (as specified by giving the activities) can be calculated from the interaction potential $w^{(m)}$ between the solvated particles at infinite dilution and the grand potential for pure solvent $\Theta(\zeta_s, 0, 0)$.

Of course, when one calculates the potential of mean force for sets of solute particles immersed in pure solvent, one needs to include the solvent species. However, the strength of this result is that it is fundamentally correct to have the solvent degrees for freedom implicit in the interaction between the solute particles. One can therefore make approximations for the potential of mean force between the solvated solutes and be sure that there is nothing fundamentally wrong with such an approach. Furthermore, this means that if one is able to improve the interaction potential between the solvated particles (i.e., the approximate potential of mean force), one will obtain more accurate results. Further improvements will, at least in principle, eventually lead to the correct results.

To use the McMillan-Mayer theory as its stands without simplifications is virtually undoable in practical calculations since one needs in principle to calculate the potential of mean force between all possible clusters of m solute particles immersed in pure solvent to obtain the correct $\Theta^{\mathrm{solution}}$. This potential involves complicated many-body correlations between the solute and solvent particles. In practice one is limited to approximations, for example to assume that $w^{(m)}_{\mathrm{I(m)}}$ can be written as sums of interaction between small sets of solvated particles; the simplest approximation is thereby to assume that it can be expressed as a sum of pair interactions. This can, of course, be quite inaccurate because the behavior of the solvent molecules around two solute particles can change substantially when a third particle is present and the solute interactions in solutions cannot in general be expressed as a superposition of effective potentials from the various pairs. In some cases this may, however, be a reasonable approximation.

The classical application of Equation 11.155 is for the osmotic pressure of a bulk solution. If the solution is in equilibrium with a pure solvent over a semipermeable membrane that does not allow the solute particles to pass, then we have exactly the conditions $\zeta_s' = \zeta_s$, $\zeta_1' = 0$ and $\zeta_2' = 0$ fulfilled. For a bulk system $\Theta \equiv -PV$, where P is the pressure. Thus, for a given V we have $\Delta\Theta \equiv \Theta^{\mathrm{solution}} - \Theta^{\mathrm{pure\ solvent}} = -(P^{\mathrm{solution}} - P^{\mathrm{pure\ solvent}})V = -\Pi V$, where Π is the **osmotic pressure**. The pressure of a gas of the same solute particles in the *absence* of the solvent can be obtained from $\Theta^{\mathrm{gas}} \equiv -P^{\mathrm{gas}}V$. It follows that the osmotic pressure is calculated in exactly the same manner from the potential of mean force between the solvated particles (obtained from an infinite dilute system) as the pressure of a gas is calculated from the interaction potential between the particles!

In order to see how the density distribution functions of the solution can be handled in a similar manner as Θ, we will give as an example the equation for the pair density distribution in the solution for the same special case with $\zeta'_s = \zeta_s$, $\zeta'_1 = 0$ and $\zeta'_2 = 0$. It is, as obtained from Equation 11.179 in Appendix 11C,[29]

$$\mathfrak{n}_{ij}^{(2)}(\mathbf{R}_1,\mathbf{R}_2;\zeta_s,\zeta_1,\zeta_2)\frac{\Xi(\zeta_s,\zeta_1,\zeta_2)}{\mathfrak{z}_i\mathfrak{z}_j}$$
$$= \Xi(\zeta_s,0,0)\sum_{m_1,m_2\geq 0}\frac{\mathfrak{z}_1^{m_1}\mathfrak{z}_2^{m_2}}{m_1!\,m_2!}\int d\mathbf{R}'^{\mathbf{m}}\,e^{-\beta w_{ij,\mathrm{I}(\mathbf{m})}^{(2+m)}(\mathbf{R}_1,\mathbf{R}_2,\mathbf{R}'^{\mathbf{m}};\zeta_s,0,0)}, \tag{11.157}$$

where $\mathbf{R}'^{\mathbf{m}}$ are the positions and orientations of m_1 particles of species 1 and m_2 particles of species 2. This equation applies to any species i and j in the solution (also the solvent species) and $w_{ij,\mathrm{I}(\mathbf{m})}^{(2+m)}$ is the potential of mean force for two particles of these species together with $m = m_1 + m_2$ additional solute particles; all of these $2 + m$ particles are thereby immersed in pure solvent. Like in $w_{\mathrm{I}(\mathbf{m})}^{(m)}$ of Equation 11.155, in the current $w_{ij,\mathrm{I}(\mathbf{m})}^{(2+m)}$ all degrees of freedom of the solvent are implicit.

The McMillan-Mayer theory of solutions is valid for both homogeneous and inhomogeneous systems. An example of the latter is that one may want to calculate the interaction force between the surfaces of two bodies, for instance two planar walls or two macroparticles immersed in a solution, as a function of their separation. The interactions of the fluid particles with the immersed bodies make the solution inhomogeneous. The force between the two bodies is given by the negative derivative of the grand potential as a function of the separation between them. Say that we use a *model for the interaction potentials for solvated particles in the solution*, i.e., a model for the potential of mean force where we do not include the solvent species explicitly. How to obtain the correct force in a calculation that is based on such a model?

Equation 11.155 implies that the correct grand potential Θ^{solution} is the sum of the grand potential $\Delta\Theta$ calculated from $w_{\mathrm{I}(\mathbf{m})}^{(m)}$ for the solvated solute particles (i.e., the last term in Equation 11.155) and the grand potential $\Theta^{\text{pure solvent}}$ calculated for the pure solvent. Both $\Delta\Theta$ and $\Theta^{\text{pure solvent}}$ are thereby obtained in the presence of the two bodies, which make both the solution and the pure solvent inhomogeneous. Thus the correct force between the two bodies is the sum of the force calculated in the model for the solution (with implicit solvent) and the force between the same bodies *in pure solvent.* One must not forget the last contribution!

11.6.2 The Adelman effective potential and direct correlation function

The McMillan-Mayer theory presented in the previous section gives a recipe of how to calculate the properties of solutions by using interaction potentials between the solute particles/molecules with implicit solvent. It is an exact

[29]The notation in Appendix 11C is explained in Appendix 11B.

theory for the solution, but a restriction on its practical usefulness is, as we have seen, that one needs to have access to effective many-particle interaction potentials between solvated solute particles in order to obtain the exactly correct results. These effective potentials are the potential of mean force for all possible clusters of solute particles immersed in pure solvent. In practical applications one needs to introduce approximations that expresses these many-particle potentials in a simpler manner, for example in terms of approximate pair, triplet or other few-particle potentials between the solvated particles. In the current section we will present a theory due to Adelman[30] that only uses effective pair potentials between solvated particles. It is an *exact* theory for the solute-solute pair distribution functions in bulk solutions, but it is approximate for all properties that cannot be expressed solely in terms of these pair distributions.

11.6.2.1 Binary bulk system

We will start with the case of a binary system of spherically symmetric particles in the bulk phase and then proceed with more general cases. One of the two components, labelled s, is considered to be the solvent and the other, labelled 1, is the solute. The densities are n_s^{b} and n_1^{b}, respectively. The pair distribution functions of the solution are $g_{ij}^{(2)}(r)$ for the various i and j values and we will initially focus on the function $g_{11}^{(2)}(r)$ for the solute particles. The aim is to treat the solvated particles without having to include the solvent explicitly. Thereby we should be able to treat the solute as a one-component system where the actual pair interaction potential $u_{11}(r)$ between the solute particles is replaced by an "effective" potential $u_{11}^{\mathrm{eff}}(r)$ that incorporates the effect of the solvent. The latter potential should then be selected such that the pair distribution function $g_{11}^{(2)}(r)$ in the "effective" one-component system (only solute) is identical to $g_{11}^{(2)}(r)$ for the solute in the solution phase (in the presence of explicit solvent). Hence, the one-component system with the pair interaction potential u_{11}^{eff} should *exactly reproduce* $g_{11}^{(2)}$ in the solution phase, where the actual pair interaction potential u_{11} as well as solute-solvent and solvent-solvent interactions are in action. Is this feasible? In fact, the pair density distribution $n_1^{\mathrm{b}} g_{11}^{(2)}(r)$ in a one-component bulk phase is a unique functional of the pair interaction potential between the particles[31] (this is similar to the fact shown in Appendix 9D that there is a one-to-one relationship between the density distribution an inhomogeneous system and the external potential). Thus there is only one pair potential $u_{11}^{\mathrm{eff}}(r)$ that can give rise to

[30] S. A. Adelman, Chem. Phys. Lett. **38** (1976) 567 (https://doi.org/10.1016/0009-2614(76)80041-3); J. Chem. Phys. **64** (1976) 724 (https://doi.org/10.1063/1.432218).

[31] This has been shown by R. L. Henderson, Phys. Lett. **49A** (1974) 197 (https://doi.org/10.1016/0375-9601(74)90847-0). The pair potential contains an arbitrary constant, which can be fixed by the condition that the potential goes to zero when $r \to \infty$.

the desired $n_1^{\mathrm{b}} g_{11}^{(2)}(r)$ in the effective one-component system and hence the pair distribution that is sought for.

In order to determine $u_{11}^{\mathrm{eff}}(r)$, let us consider the OZ equation 10.32 for the bulk mixture

$$h_{ij}^{(2)}(r_{12}) = c_{ij}^{(2)}(r_{12}) + \sum_{l} \int d\mathbf{r}_3\, c_{il}^{(2)}(r_{13}) n_l^{\mathrm{b}} h_{lj}^{(2)}(r_{32}), \tag{11.158}$$

where l in the sum assumes the values 1 and s. It turns out to be more convenient to work with these equations in Fourier space rather than the ordinary space, so we will take the Fourier transform of them.[32] The integral in Equation 11.158 is a convolution, which simply becomes a product in Fourier space, and we obtain

$$\tilde{h}_{ij}^{(2)}(k) = \tilde{c}_{ij}^{(2)}(k) + \sum_{l} \tilde{c}_{il}^{(2)}(k) n_l^{\mathrm{b}} \tilde{h}_{lj}^{(2)}(k), \tag{11.159}$$

where $\tilde{f}(k)$ is the Fourier transform of $f(r)$ for $f = h^{(2)}$ and $c^{(2)}$ and where k is the wave number. For $i = j = 1$ we have explicitly

$$\tilde{h}_{11}^{(2)}(k) = \tilde{c}_{11}^{(2)}(k) + \tilde{c}_{11}^{(2)}(k) n_1^{\mathrm{b}} \tilde{h}_{11}^{(2)}(k) + \tilde{c}_{1s}^{(2)}(k) n_s^{\mathrm{b}} \tilde{h}_{s1}^{(2)}(k). \tag{11.160}$$

and for $i = s$ and $j = 1$ we have

$$\tilde{h}_{s1}^{(2)}(k) = \tilde{c}_{s1}^{(2)}(k) + \tilde{c}_{s1}^{(2)}(k) n_1^{\mathrm{b}} \tilde{h}_{11}^{(2)}(k) + \tilde{c}_{ss}^{(2)}(k) n_s^{\mathrm{b}} \tilde{h}_{s1}^{(2)}(k). \tag{11.161}$$

We shall show that these two equations imply that for the effective one-component system, we have the OZ equation

$$\tilde{h}_{11}^{(2)}(k) = \tilde{c}_{11}^{(2)\mathrm{eff}}(k) + \tilde{c}_{11}^{(2)\mathrm{eff}}(k) n_1^{\mathrm{b}} \tilde{h}_{11}^{(2)}(k), \tag{11.162}$$

where $\tilde{h}_{11}^{(2)}(k)$ is the same as in the solution phase and where $\tilde{c}_{11}^{(2)\mathrm{eff}}(k)$ is the Fourier transform an effective direct correlation function that can be expressed explicitly in terms of the direct correlation functions $c_{ij}^{(2)}$ of the solution phase.

Equation 11.160 can be written

$$\tilde{h}_{11}^{(2)}(k) = \tilde{c}_{11}^{(2)}(k) \left[1 + n_1^{\mathrm{b}} \tilde{h}_{11}^{(2)}(k)\right] + \tilde{c}_{1s}^{(2)}(k) n_s^{\mathrm{b}} \tilde{h}_{s1}^{(2)}(k),$$

and by multiplying with $[1 + n_1^{\mathrm{b}} \tilde{h}_{11}^{(2)}(k)]^{-1}$ we obtain

$$\tilde{h}_{11}^{(2)}(k) \left[1 + n_1^{\mathrm{b}} \tilde{h}_{11}^{(2)}(k)\right]^{-1} = \tilde{c}_{11}^{(2)}(k) + \tilde{c}_{1s}^{(2)}(k) n_s^{\mathrm{b}} \tilde{h}_{s1}^{(2)}(k) \left[1 + n_1^{\mathrm{b}} \tilde{h}_{11}^{(2)}(k)\right]^{-1}. \tag{11.163}$$

Likewise, Equation 11.161 can be written as

$$[1 - \tilde{c}_{ss}^{(2)}(k) n_s^{\mathrm{b}}] \tilde{h}_{s1}^{(2)}(k) = \tilde{c}_{s1}^{(2)}(k) [1 + n_1^{\mathrm{b}} \tilde{h}_{11}^{(2)}(k)]$$

[32] The definition of Fourier transforms and a summary of some of their properties can be found in Appendix 9B.

which yields

$$\tilde{h}_{s1}^{(2)}(k)[1+n_1^{\mathrm{b}}\tilde{h}_{11}^{(2)}(k)]^{-1}=[1-\tilde{c}_{ss}^{(2)}(k)n_s^{\mathrm{b}}]^{-1}\tilde{c}_{s1}^{(2)}(k).$$

By inserting this in the rhs of Equation 11.163 we obtain

$$\tilde{h}_{11}^{(2)}(k)\,[1+n_1^{\mathrm{b}}\tilde{h}_{11}^{(2)}(k)]^{-1}=\tilde{c}_{11}^{(2)}(k)+\tilde{c}_{1s}^{(2)}(k)n_s^{\mathrm{b}}[1-\tilde{c}_{ss}^{(2)}(k)n_s^{\mathrm{b}}]^{-1}\tilde{c}_{s1}^{(2)}(k),$$

which can be written as

$$\tilde{h}_{11}^{(2)}(k)\,[1+n_1^{\mathrm{b}}\tilde{h}_{11}^{(2)}(k)]^{-1}=\tilde{c}_{11}^{(2)\mathrm{eff}}(k), \tag{11.164}$$

where we have defined

$$\tilde{c}_{11}^{(2)\mathrm{eff}}(k)\;\equiv\;\tilde{c}_{11}^{(2)}(k)+\tilde{c}_{1s}^{(2)}(k)n_s^{\mathrm{b}}[1-\tilde{c}_{ss}^{(2)}(k)n_s^{\mathrm{b}}]^{-1}\tilde{c}_{s1}^{(2)}(k). \tag{11.165}$$

Equation 11.164 can be written as Equation 11.162, which thereby has been demonstrated. We see that $\tilde{c}_{11}^{(2)\mathrm{eff}}$ contains $\tilde{c}_{11}^{(2)}$ of the solution phase together with the solvent-solute and solvent-solvent direct correlations.

Incidentally, we note that by using the power series expansion $[1-\tilde{c}_{ss}^{(2)}(k)n_s^{\mathrm{b}}]^{-1}=1+\tilde{c}_{ss}^{(2)}(k)n_s^{\mathrm{b}}+[\tilde{c}_{ss}^{(2)}(k)n_s^{\mathrm{b}}]^2+\ldots$ we can write Equation 11.165 as

$$\begin{aligned}\tilde{c}_{11}^{(2)\mathrm{eff}}(k)\;&=\;\tilde{c}_{11}^{(2)}(k)+\tilde{c}_{1s}^{(2)}(k)n_s^{\mathrm{b}}\tilde{c}_{s1}^{(2)}(k)+\tilde{c}_{1s}^{(2)}(k)n_s^{\mathrm{b}}\tilde{c}_{ss}^{(2)}(k)n_s^{\mathrm{b}}\tilde{c}_{s1}^{(2)}(k)\\&+\;\tilde{c}_{1s}^{(2)}(k)n_s^{\mathrm{b}}\tilde{c}_{ss}^{(2)}(k)n_s^{\mathrm{b}}\tilde{c}_{ss}^{(2)}(k)n_s^{\mathrm{b}}\tilde{c}_{s1}^{(2)}(k)+\ldots\end{aligned}$$

which becomes in real space

$$\begin{aligned}c_{11}^{(2)\mathrm{eff}}(r_{12})\;&=\;c_{11}^{(2)}(r_{12})+\int d\mathbf{r}_3\,c_{1s}^{(2)}(r_{13})n_s^{\mathrm{b}}c_{s1}^{(2)}(r_{32})\\&+\;\int d\mathbf{r}_3 d\mathbf{r}_4\,c_{1s}^{(2)}(r_{13})n_s^{\mathrm{b}}c_{ss}^{(2)}(r_{34})n_s^{\mathrm{b}}c_{s1}^{(2)}(r_{42})\\&+\;\int d\mathbf{r}_3 d\mathbf{r}_4 d\mathbf{r}_5\,c_{1s}^{(2)}(r_{13})n_s^{\mathrm{b}}c_{ss}^{(2)}(r_{34})n_s^{\mathrm{b}}c_{ss}^{(2)}(r_{45})n_s^{\mathrm{b}}c_{s1}^{(2)}(r_{52})+\ldots,\end{aligned} \tag{11.166}$$

where each integrand starts with $c_{1s}^{(2)}$, ends with $c_{s1}^{(2)}$ and – for the second integrand and forward – have chains of $c_{ss}^{(2)}$ in between (cf. Equation 9.41 and the accompanying illustration). This means that $c_{11}^{(2)\mathrm{eff}}(r_{12})$ carries the direct correlations between the solvent and the two solute particle at $\mathbf{r}_1$ and $\mathbf{r}_2$ (via $c_{1s}^{(2)}$ and $c_{s1}^{(2)}$) together with all chains of direct correlations between the solvent molecules (the $c_{ss}^{(2)}$ chains).

Having defined $c_{11}^{(2)\mathrm{eff}}$ for the effective one-component system, we can now relate the effective pair interaction potential u_{11}^{eff} in this system to the pair correlation functions. This can be done by using exactly the same relationship between such functions and the pair potential as for a normal one-component system, Equation 9.67, which in the effective system becomes

$$g_{11}^{(2)}(r_{12})=e^{-\beta u_{11}^{\mathrm{eff}}(r_{12})+h_{11}^{(2)}(r_{12})-c_{11}^{(2)\mathrm{eff}}(r_{12})+b_{11}^{(2)}(r_{12};[u_{11}^{\mathrm{eff}}])}, \tag{11.167}$$

where $b_{11}^{(2)}(r_{12};[u_{11}^{\mathrm{eff}}])$ is the bridge function for a one-component system with pair potential u_{11}^{eff}. This is a key feature of the Adelman theory. Thus it is *fundamentally correct* to use u_{11}^{eff} for the calculation of the pair correlations of the solute particles in the effective one-component system and $g_{11}^{(2)}$ that is obtained is exactly the same as in the actual solution phase. The problem is, of course, that we in practice do not know u_{11}^{eff} to start with and in order to find it we need to know $g_{11}^{(2)}$ in the actual solution phase. So what is the point? Let us explain this before we proceed with the practical determination of u_{11}^{eff}.

One important point of the Adelman theory is that it says in what manner the solvent influences the potential of mean force $w_{11}^{(2)}(r_{12}) = -k_B T \ln g_{11}^{(2)}(r_{12})$ between the solutes via the effective pair interaction $u_{11}^{\mathrm{eff}}(r_{12})$, i.e., how the presence of solvent modifies $u_{11}(r_{12})$ between the solutes in solution. This is conceptually important and can lead to an increased understanding of the mechanisms in action in the solution phase.

In practical applications one can use u_{11}^{eff} determined for one system as an approximation for u_{11}^{eff} in another system. The latter must be reasonably similar to the original system for this to make sense. One can, for example use u_{11}^{eff}, which is obtained for a homogeneous bulk solution, as an approximation for the solute-solute interaction in an inhomogeneous solution provided it does not depart far too much from the homogenous one. In this case, one also needs an approximation for the interaction of solvated particles with external body that causes the inhomogeneity. Once one has extracted u_{11}^{eff} from a calculation with explicit solvent in one system, one can use the same u_{11}^{eff} in order to obtain approximate results for a range of systems without having to do complicated calculations with explicit solvent for them. The solvent effects are included in the effective pair potential in an approximate manner and one only needs to do calculations solely for the solute (with implicit solvent), which is much simpler than to include the solvent explicitly.

The fact that u_{11}^{eff} in the Adelman theory depends on the solution composition constitute a major difference from the interaction potentials for solvated solute particles used in the McMillan-Mayer theory. The interaction potentials in the latter theory are calculated at infinite dilution and once they are known they can be used for any composition of the solution and the results are exact, but the draw-back is that one needs effective many-particle interaction potentials to obtain the correct results.

Let us now turn to the question of how to determine $u_{11}^{\mathrm{eff}}(r)$ once $g_{11}^{(2)}(r)$ for the solute in the solution phase has been determined in the presence of explicit solvent. We will give a few examples of how to proceed. One way is to calculate the pair distribution function by computer simulation of a one-component system with a given pair potential $u_{11}^{\mathrm{trial}}(r)$ and vary the latter until the obtained pair distribution, $g_{11}^{(2)\mathrm{trial}}(r)$, agrees within sufficient accuracy with $g_{11}^{(2)}(r)$ obtained in the solution phase.[33] The optimal $u_{11}^{\rm trial}(r)$ thus obtained equals $u_{11}^{\rm eff}(r)$. As mentioned in the beginning of the current section, there does not exist more than one $u_{11}^{\rm trial}(r)$ that yields a prescribed $g_{11}^{(2)}(r)$. In practical calculations one thereby varies $u_{11}^{\rm trial}(r)$ in a systematic manner, for instance by expressing it as a sum of basis functions with variable coefficients and vary the latter until the difference between $g_{11}^{(2)\rm trial}(r)$ and $g_{11}^{(2)}(r)$ is minimized.

Alternatively one can calculate $g_{11}^{(2)\rm trial}(r)$ in some accurate theory, for example a suitable integral equation approximation, for both the actual solution and the effective one-component system. One can then use the same procedure with a varying $u_{11}^{\rm trial}(r)$. This is, however, unnecessarily complicated when one uses an integral equation theory like those described in Sections 9.1.3.2 and 9.1.3.3, where the approximate bridge function $b_{11}^{(2)}$ is expressed analytically in terms of the pair correlation functions, for example $b_{11}^{(2)} = f(\gamma_{11}^{(2)}) = f(h_{11}^{(2)} - c_{11}^{(2)})$ with some function f. Then Equation 11.167 yields

$$\beta u_{11}^{\rm eff} = -\ln(h_{11}^{(2)} + 1) + h_{11}^{(2)} - c_{11}^{(2)\rm eff} + f(h_{11}^{(2)} - c_{11}^{(2)\rm eff}),$$

where $c_{11}^{(2)\rm eff}$ can be obtained from Equation 11.164 once $h_{11}^{(2)}(r)$ is known, so one does not need to go via Equation 11.165. In real space Equation 11.164 can be written as an OZ equation for the effective one-component system

$$h_{11}^{(2)}(r_{12}) = c_{11}^{(2)\rm eff}(r_{12}) + \int d\mathbf{r}_3\, c_{11}^{(2)\rm eff}(r_{13}) n_l^{\rm b} h_{11}^{(2)}(r_{32}). \qquad (11.168)$$

There also exist methods that combine simulation methods and integrals equation closures in order to extract $u_{11}^{\rm eff}(r)$ from $g_{11}^{(2)}(r)$, but we will not go into any details here.

11.6.2.2 General bulk systems

Let us now turn to the general case of a many-component bulk mixture of nonspherical particles with density $n_i^{\rm b}$ for species i. In the mixture, M_S species are considered as being solvent and M_D species are the solutes (subscript S for solvent and D for dissolved species), so we have a solution of the latter in the former. The same principles apply as in the binary case: starting from a treatment of all $M_D + M_S$ species on the same basis in the solution phase, one can obtain the effective pair interaction potential $u_{ij}^{\rm eff}(\mathbf{R}_1, \mathbf{R}_2)$ with implicit solvent for all pairs of solute species i and j. Thereby, $u_{ij}^{\rm eff}$ is to be determined such that the pair correlation functions g_{ij} for all solute species in the effective M_D-component system with pair potentials $u_{ij}^{\rm eff}$ is exactly the same as g_{ij} in the actual $(M_D + M_S)$-component system with the actual pair interactions

[33] A. P. Lyubartsev and A. Laaksonen, Phys. Rev. A **52** (1995) 3730 (https://doi.org/10.1103/PhysRevE.52.3730).

u_{ij}. The densities of the solute species in the effective system are the same as in the solution phase.

The derivation of the Adelman theory for the general bulk solution is presented in Appendix 11D. There it is shown that one can define an effective direct correlation function $c_{ij}^{(2)\text{eff}}$ for all pairs of solute species i and j in terms of $c_{i'j'}^{(2)}$ for all species in the solution phase in analogy to Equation 11.165 (see Equation 11.186 in the Appendix). This $c_{ij}^{(2)\text{eff}}$ satisfies an OZ equation for the effective M_D-component system (in analogy to Equation 11.168)

$$h_{ij}^{(2)}(\mathbf{R}_1, \mathbf{R}_2) = c_{ij}^{(2)\text{eff}}(\mathbf{R}_1, \mathbf{R}_2) + \sum_{l=1}^{M_D} \int d\mathbf{R}_3\, c_{il}^{(2)\text{eff}}(\mathbf{R}_1, \mathbf{R}_3) n_l^{\text{b}} h_{lj}^{(2)}(\mathbf{R}_3, \mathbf{R}_2),$$

where the sum is over the solute species alone (this equation follows from Equation 11.183 in the Appendix). Given $h_{ij}^{(2)}$ for the solute one can determine $c_{ij}^{(2)\text{eff}}$ by solving this equation.

Furthermore, the effective pair interaction potentials u_{ij}^{eff} for the solvated solute particles is related to the correlation functions for the effective M_D-component system in the same manner as before (cf. Equation 11.167)

$$g_{ij}^{(2)}(\mathbf{R}_1, \mathbf{R}_2) = e^{-\beta u_{ij}^{\text{eff}}(\mathbf{R}_1,\mathbf{R}_2) + h_{ij}^{(2)}(\mathbf{R}_1,\mathbf{R}_2) - c_{ij}^{(2)\text{eff}}(\mathbf{R}_1,\mathbf{R}_2) + b_{ij}^{(2)}(\mathbf{R}_1,\mathbf{R}_2;[\{u_{lm}^{\text{eff}}\}])},$$

where $b_{ij}^{(2)}(\mathbf{R}_1, \mathbf{R}_2; [\{u_{lm}^{\text{eff}}\}])$ is the bridge function for the effective system of solute particles and $\{u_{lm}^{\text{eff}}\}$ is the set of effective pair potentials (the indices l, m run over the M_D solute species).

In principle, one can obtain u_{ij}^{eff} for the solvated particles in an analogous manner as for the binary system of spherical particles discussed in the previous section. This is doable in practice at least for spherical solutes in a solvent of nonspherical molecules. One can then do a computer simulation for the bulk solution with molecular solvent in order to obtain $h_{ij}^{(2)}(r)$ for the solute and then use the same kind of procedures to determine $u_{ij}^{\text{eff}}(r)$ as mentioned in the previous section.

APPENDICES

Appendix 11A
The orientational variable ω

To describe the orientation of a nonspherical particle we introduce a "particle frame" $\mathfrak{F}^{\text{p}}$ associated with each particle and placed with the origin at the center of mass. This frame is attached to the particle and follows its translational and rotational motions. The orientation of the particle can be specified by comparing the directions of the particle frame axes with the axes of the laboratory frame $\mathfrak{F}^{\text{lab}}$ in the following way. We can imagine another frame $\mathfrak{F}$ centered at the particle (so its origin coincides that of the particle frame) but oriented so that its x, y and z axes are initially parallel to the corresponding axes of $\mathfrak{F}^{\rm lab}$. The finite rotation (or the sequence of rotations) that would bring $\mathfrak{F}$ to coincide with the particle frame $\mathfrak{F}^{\rm P}$ can be used to specify the orientation of the latter and thereby the particle orientation. The rotation(s) will be specified by a variable $\boldsymbol{\omega}$ and can, for example, be expressed in terms of the three Euler angles. These angles can be defined from the following set of three consecutive rotations:[34] A rotation through an angle ϕ around the z axis of $\mathfrak{F}$ followed by a rotation through θ around the new y axis of $\mathfrak{F}$. These two rotations make the rotated z axis of $\mathfrak{F}$ to coincide with the z axis of the particle frame $\mathfrak{F}^{\rm P}$ (the angles are the same as the spherical polar coordinate angles (ϕ, θ) for the z axis unit vector $\hat{z}^{\rm P}$ of the particle frame). A final third rotation of $\mathfrak{F}$ through the angle χ around the common z axis of $\mathfrak{F}^{\rm P}$ and $\mathfrak{F}$ makes the x and y axes of $\mathfrak{F}$ to coincide with those of $\mathfrak{F}^{\rm P}$. The particle frame and the rotated $\mathfrak{F}$ thereby fully coincide, so the set of angles (ϕ, θ, χ) specify the orientation of $\mathfrak{F}^{\rm P}$ and thereby that of the particle. We have $0 \leq \phi < 2\pi$, $0 \leq \theta \leq \pi$ and $0 \leq \chi < 2\pi$. For a linear molecule the last rotation through χ is redundant (the particle axis is then selected as the z axis of the particle frame) and for such particles the set (ϕ, θ) is sufficient to describe the orientation.

We shall, however, not select (ϕ, θ, χ) [or (ϕ, θ) for a linear molecule] to be our orientational variable $\boldsymbol{\omega}$. It is better to use $\boldsymbol{\omega}' = (\phi, \cos\theta, \chi)$ to describe the orientation of the particle. We have thereby selected $\cos\theta$ rather than θ because the differential then becomes $d\boldsymbol{\omega}' = d\phi\, d(\cos\theta)\, d\chi = -d\phi\, \sin\theta\, d\theta\, d\chi$, cf. the polar coordinate differential $d\phi\, \sin\theta\, d\theta$ that gives the differential solid angle. For a randomly oriented particle there is equal probability to have its $\hat{z}^{\rm P}$ axis within a certain solid angle around all different directions in space. For a linear molecule we have $\boldsymbol{\omega}' = (\phi, \cos\theta)$.

This choice of orientational variable has the property that $\int d\boldsymbol{\omega}' = \omega_\circ$, where $\omega_\circ = 8\pi^2$ except for linear molecules when we have $\omega_\circ = 4\pi$. It is more convenient to have a "normalized" variable $\boldsymbol{\omega}$ so that $\int d\boldsymbol{\omega} = 1$ in all cases. We therefore select $\boldsymbol{\omega} = (\frac{\phi}{2\pi}, \frac{\cos\theta}{2}, \frac{\chi}{2\pi})$ as the final choice of orientational variable, except for linear molecules when we take $\boldsymbol{\omega} = (\frac{\phi}{2\pi}, \frac{\cos\theta}{2})$. Note that $\boldsymbol{\omega}$ is dimensionless.

Appendix 11B
Formal definition of distribution functions for many-component systems

In this and the following Appendices we use a slightly different notation compared to the main text. At the end of the present Appendix it is explained how the two conventions are linked to each other.

Consider a fluid consisting of a mixture of M species. In the canonical ensemble at temperature T, volume V and for N_i particles of species

[34] There exist other conventions for the selection of which rotations are used to define the Euler angles. The one used here has the advantage that the first two Euler angles coincide with the angles for the z axis of $\mathfrak{F}^{\rm P}$ in spherical polar coordinates.

i, $1 \leq i \leq M$, the configurational probability density equals

$$\mathcal{P}_{\mathbf{N}}^{\{\mathbf{N}\}}(\mathbf{R}^{\mathbf{N}}) = \frac{1}{Z_{\mathbf{N}}} e^{-\beta \breve{U}_{\mathbf{N}}^{\mathrm{pot}}(\mathbf{R}^{\mathbf{N}})}, \tag{11.169}$$

where $\mathbf{N} = (N_1, \ldots, N_M)$ denote the set of number of particles of the various species,

$$\mathbf{R}^{\mathbf{N}} \equiv \mathbf{R}^{N_1}, \mathbf{R}^{N_2}, \ldots, \mathbf{R}^{N_M}$$

are the coordinates (i.e., $\mathbf{R}^{N_1}$ for species 1, $\mathbf{R}^{N_2}$ for species 2 etc.), $\{\mathbf{N}\} = \{N_1, \ldots, N_M\}$ is a list that specifies the number of variables for each species in $\mathcal{P}_{\mathbf{N}}^{\{\mathbf{N}\}}$, $\breve{U}_{\mathbf{N}}^{\mathrm{pot}}$ is the potential energy of the particles,

$$Z_{\mathbf{N}} = \int d\mathbf{R}^{\mathbf{N}} e^{-\beta \breve{U}_{N}^{\mathrm{pot}}(\mathbf{R}^{\mathbf{N}})} \tag{11.170}$$

is the configurational partition function and $d\mathbf{R}^{\mathbf{N}} = d\mathbf{R}^{N_1} \ldots d\mathbf{R}^{N_M}$. The canonical partition function is (cf. Equation 11.5)

$$Q_{\mathbf{N}} = \frac{Z_{\mathbf{N}}}{\mathbf{N}!} \prod_{i=1}^{M} \eta_i^{N_i}, \tag{11.171}$$

where $\mathbf{N}! = N_1! \ldots N_M!$ and $\eta_i = \mathfrak{q}_{\mathrm{r},i}/\Lambda_i^3$. The quantities η_i and $\mathfrak{q}_{\mathrm{r},i}$ are η and $\mathfrak{q}_{\mathrm{r}}$ for species i defined in Equation 11.6 and Λ_i is the thermal de Broglie wave length.

The l-point configurational probability density for $\mathbf{l} = (l_1, \ldots, l_M)$ particles of the various species with $l = \sum_i l_i$ equals

$$\begin{aligned} \mathcal{P}_{\mathbf{N}}^{\{\mathbf{l}\}}(\mathbf{R}^{\mathbf{l}}) &= \int d\mathbf{R}_{\mathbf{l}+1} \ldots d\mathbf{R}_{\mathbf{N}}\, \mathcal{P}_{\mathbf{N}}^{\{\mathbf{N}\}}(\mathbf{R}^{\mathbf{N}}) \\ &= \frac{1}{Z_{\mathbf{N}}} \int d\mathbf{R}_{\mathbf{l}+1} \ldots d\mathbf{R}_{\mathbf{N}}\, e^{-\beta \breve{U}_{\mathbf{N}}^{\mathrm{pot}}(\mathbf{R}^{\mathbf{N}})}, \end{aligned} \tag{11.172}$$

where integration is done over all variables in $\mathbf{R}^{\mathbf{N}}$ apart from those in $\mathbf{R}^{\mathbf{l}}$ and where $d\mathbf{R}_{\mathbf{l}+1} \ldots d\mathbf{R}_{\mathbf{N}} = \prod_{i=1}^{M} (d\mathbf{R}_{l_i+1} \ldots d\mathbf{R}_{N_i})$. The l-point density distribution function is

$$\mathfrak{n}_{\mathbf{N}}^{\{\mathbf{l}\}}(\mathbf{R}^{\mathbf{l}}) = \frac{\mathbf{N}!}{(\mathbf{N}-\mathbf{l})!} \mathcal{P}_{\mathbf{N}}^{\{\mathbf{l}\}}(\mathbf{R}^{\mathbf{l}}) \tag{11.173}$$

where $(\mathbf{N}-\mathbf{l})! = \prod_{i=1}^{N} (N_i - l_i)!$.

In the grand canonical ensemble for temperature T, volume V and chemical potentials μ_i of species i, $1 \leq i \leq M$, the probability that the system contains N_i particles of species i, $1 \leq i \leq M$, is

$$\mathscr{P}(\mathbf{N}) = \frac{Q_{\mathbf{N}} \exp\left(\beta \sum_{i=i}^{M} N_i \mu_i\right)}{\Xi} = \frac{\varsigma^{\mathbf{N}} Z_{\mathbf{N}}}{\mathbf{N}!\,\Xi},$$

where $\zeta^{\mathbf{N}} = \zeta_1^{N_1} \ldots \zeta_M^{N_M}$ with ζ_i = the activity of species i,

$$\Xi = \sum_{\mathbf{N}} \frac{\zeta^{\mathbf{N}} Z_{\mathbf{N}}}{\mathbf{N}!} = \sum_{\mathbf{N}} \frac{\zeta^{\mathbf{N}}}{\mathbf{N}!} \int d\mathbf{R}^{\mathbf{N}} e^{-\beta \check{U}^{\mathrm{pot}}_{\mathbf{N}}(\mathbf{R}^{\mathbf{N}})}, \tag{11.174}$$

is the grand canonical partition function and in the sum over $\mathbf{N}$, N_i goes between 0 and ∞ for all species: $\sum_{\mathbf{N}} = \sum_{N_1=0}^{\infty} \cdots \sum_{N_M=0}^{\infty}$. The l-point density distribution function is

$$\begin{aligned} \mathfrak{n}^{\{\mathbf{l}\}}(\mathbf{R}^{\mathbf{l}}) &= \sum_{\mathbf{N}=\mathbf{l}}^{\infty} \mathscr{P}(\mathbf{N}) \mathfrak{n}_{\mathbf{N}}^{\{\mathbf{l}\}}(\mathbf{R}^{\mathbf{l}}) \\ &= \frac{1}{\Xi} \sum_{\mathbf{N}=\mathbf{l}}^{\infty} \frac{\zeta^{\mathbf{N}}}{(\mathbf{N}-\mathbf{l})!} \int d\mathbf{R}_{\mathbf{l}+1} \ldots d\mathbf{R}_{\mathbf{N}}\, e^{-\beta \check{U}^{\mathrm{pot}}_{\mathbf{N}}(\mathbf{R}^{\mathbf{N}})}. \end{aligned} \tag{11.175}$$

In the sum, N_i goes from l_i to ∞ for all species. The usual convention for simplified notation is used, so whenever $N_i = l_i$ occurs in the terms of the rhs, the corresponding integral sign does not appear.

In $\mathfrak{n}^{\{\mathbf{l}\}}$ the species for the particle coordinates are specified by the order in the list $\{\mathbf{l}\} = \{l_1, \ldots, l_M\}$, that is, the first l_1 $\mathbf{R}$-coordinates are for species 1, the next l_2 species 2 etc. This is done instead of setting subscripts on the function as is commonly done. As an example, in the present notation the triplet density distribution function for a two-component system ($M = 2$) involving two particles of species 1 and one of species 2 is denoted $\mathfrak{n}^{\{2,1\}}(\mathbf{R}_1, \mathbf{R}_2, \mathbf{R}_3)$, where coordinates $\mathbf{R}_1$ and $\mathbf{R}_2$ are for species 1 and $\mathbf{R}_3$ for species 2. The usual notation is to denote this function as $\mathfrak{n}^{(3)}_{112}(\mathbf{R}_1, \mathbf{R}_2, \mathbf{R}_3)$, where the subscript show which components are involved. Thus we define $\mathfrak{n}^{(l)}$ (with ordinary parentheses in the superscript rather than braces)

$$\mathfrak{n}^{(l)}_{\mathtt{I}(\mathbf{l})}(\mathbf{R}^{\mathbf{l}}) \equiv \mathfrak{n}^{\{\mathbf{l}\}}(\mathbf{R}^{\mathbf{l}}),$$

where $l = \sum_i l_i$ and $\mathtt{I}(\mathbf{l})$ symbolizes a list of species indices of the particles. For $\mathfrak{n}^{\{\mathbf{l}\}}$ this means that each species index "i" of $\mathfrak{n}^{(l)}$ is repeated l_i times in $\mathtt{I}(\mathbf{l})$ like 112 above when $l_1 = 2$ and $l_2 = 1$. Other examples are the pair density distribution functions $\mathfrak{n}^{(2)}_{22}(\mathbf{R}_1, \mathbf{R}_2) = \mathfrak{n}^{\{0,2\}}(\mathbf{R}_1, \mathbf{R}_2)$ and $\mathfrak{n}^{(2)}_{12}(\mathbf{R}_1, \mathbf{R}_2) = \mathfrak{n}^{\{1,1\}}(\mathbf{R}_1, \mathbf{R}_2)$. Since the subscripts in the common notation tell which components are associated with the respective coordinates, we have $\mathfrak{n}^{(2)}_{ij}(\mathbf{R}_1, \mathbf{R}_2) = \mathfrak{n}^{(2)}_{ji}(\mathbf{R}_2, \mathbf{R}_1)$ and, for example,

$$\mathfrak{n}^{(3)}_{iij}(\mathbf{R}_1, \mathbf{R}_2, \mathbf{R}_3) = \mathfrak{n}^{(3)}_{iji}(\mathbf{R}_1, \mathbf{R}_3, \mathbf{R}_2) = \mathfrak{n}^{(3)}_{jii}(\mathbf{R}_3, \mathbf{R}_1, \mathbf{R}_2) = \mathfrak{n}^{(3)}_{jii}(\mathbf{R}_3, \mathbf{R}_2, \mathbf{R}_1).$$

These symmetries are a consequence of the fact that the total potential energy has the same symmetries.

Appendix 11C
Proofs of some equations of the McMillan-Mayer theory

In this Appendix we will derive Equations 11.153 and 11.157 of the McMillan-Mayer theory of solutions presented in Section 11.6.1. These equations will be given here in a more general forms than in the main text and we will use the notation of Appendix 11B and the results therein. For translation of notation to that in the main text see the end of that appendix.

The l-point density distribution function $\mathfrak{n}^{\{\mathbf{l}\}}$ at activities $\boldsymbol{\zeta} = (\zeta_1, \zeta_2, \ldots, \zeta_M)$ for the M components is given in Equation 11.175. By making the substitution $\mathbf{N} = \mathbf{l} + \mathbf{m}$ we can write it as

$$\mathfrak{n}^{\{\mathbf{l}\}}(\mathbf{R}^{\mathbf{l}};\boldsymbol{\zeta})\frac{\Xi(\boldsymbol{\zeta})}{\boldsymbol{\zeta}^{\mathbf{l}}} = \sum_{\mathbf{m}} \frac{\boldsymbol{\zeta}^{\mathbf{m}}}{\mathbf{m}!} \int d\mathbf{R}_{\mathbf{l}+1} \ldots d\mathbf{R}_{\mathbf{l}+\mathbf{m}}\, e^{-\beta \breve{U}^{\mathrm{pot}}_{\mathbf{l}+\mathbf{m}}(\mathbf{R}^{\mathbf{l}+\mathbf{m}})}, \tag{11.176}$$

whereby it is written in a manner that corresponds to Equation 11.125 for the one-component case. In the sum, m_i goes between 0 and ∞ for all species. Like in the one-component case, this is a Taylor expansion in the activities around zero, but in the present case it is an M-dimensional expansion. Our task is to make the corresponding expansion around another set of values of the activities $\boldsymbol{\zeta}' = (\zeta_1', \zeta_2', \ldots, \zeta_M')$ and we follow the same path as for the one-component case in Section 11.4.4.

For a function $f(\boldsymbol{\zeta}) = f(\zeta_1, \ldots, \zeta_M)$, we have according to Taylor's theorem the following expansion around $\boldsymbol{\zeta}'$, i.e., around ζ_i' for all i,

$$f(\boldsymbol{\zeta}) = \sum_{\boldsymbol{\nu}} \frac{1}{\boldsymbol{\nu}!} \left[\prod_{i=1}^{M} (\zeta_i - \zeta_i')^{\nu_i}\right] \left.\frac{\partial^{\nu} f(\boldsymbol{\zeta})}{\partial \zeta_1^{\nu_1} \ldots \partial \zeta_M^{\nu_M}}\right|_{\boldsymbol{\zeta}=\boldsymbol{\zeta}'}. \tag{11.177}$$

where $\boldsymbol{\nu} = (\nu_1, \ldots, \nu_M)$, $\nu = \sum_i \nu_i$ and in the sum ν_i goes between 0 and ∞ for all i. We take the function $f(\boldsymbol{\zeta})$ to be equal to the rhs of Equation 11.176 and obtain by differentiation

$$\begin{aligned}\frac{\partial^{\nu} f(\boldsymbol{\zeta})}{\partial \zeta_1^{\nu_1} \ldots \partial \zeta_M^{\nu_M}} &= \sum_{\mathbf{m}=\boldsymbol{\nu}}^{\infty} \frac{\boldsymbol{\zeta}^{\mathbf{m}-\boldsymbol{\nu}}}{(\mathbf{m}-\boldsymbol{\nu})!} \int d\mathbf{R}_{\mathbf{l}+1} \ldots d\mathbf{R}_{\mathbf{l}+\mathbf{m}} e^{-\beta \breve{U}^{\mathrm{pot}}_{\mathbf{l}+\mathbf{m}}(\mathbf{R}^{\mathbf{l}+\mathbf{m}})} \\ &= \sum_{\mathbf{t}} \frac{\boldsymbol{\zeta}^{\mathbf{t}}}{\mathbf{t}!} \int d\mathbf{R}_{\mathbf{l}+1} \ldots d\mathbf{R}_{\mathbf{l}+\boldsymbol{\nu}+\mathbf{t}}\, e^{-\beta \breve{U}^{\mathrm{pot}}_{\mathbf{l}+\boldsymbol{\nu}+\mathbf{t}}(\mathbf{R}^{\mathbf{l}+\boldsymbol{\nu}+\mathbf{t}})},\end{aligned}$$

where we have made the substitution $\mathbf{m} \to \boldsymbol{\nu} + \mathbf{t}$ to obtain the second line. By substituting $\mathbf{l} \to \mathbf{l} + \boldsymbol{\nu}$ and $\mathbf{m} \to \mathbf{t}$ in Equation 11.176 we see that

$$\mathfrak{n}^{\{\mathbf{l}+\boldsymbol{\nu}\}}(\mathbf{R}^{\mathbf{l}+\boldsymbol{\nu}};\boldsymbol{\zeta})\frac{\Xi(\boldsymbol{\zeta})}{\boldsymbol{\zeta}^{\mathbf{l}+\boldsymbol{\nu}}} = \sum_{\mathbf{t}} \frac{\boldsymbol{\zeta}^{\mathbf{t}}}{\mathbf{t}!} \int d\mathbf{R}_{\mathbf{l}+\boldsymbol{\nu}+1} \ldots d\mathbf{R}_{\mathbf{l}+\boldsymbol{\nu}+\mathbf{t}}\, e^{-\beta \breve{U}^{\mathrm{pot}}_{\mathbf{l}+\boldsymbol{\nu}+\mathbf{t}}(\mathbf{R}^{\mathbf{l}+\boldsymbol{\nu}+\mathbf{t}})}$$

and by inserting this into the previous equation we obtain

$$\frac{\partial^{\nu} f(\boldsymbol{\zeta})}{\partial \zeta_1^{\nu_1} \ldots \partial \zeta_M^{\nu_M}} = \int d\mathbf{R}_{\mathbf{l}+1} \ldots d\mathbf{R}_{\mathbf{l}+\boldsymbol{\nu}}\, \mathfrak{n}^{\{\mathbf{l}+\boldsymbol{\nu}\}}(\mathbf{R}^{\mathbf{l}+\boldsymbol{\nu}};\boldsymbol{\zeta})\frac{\Xi(\boldsymbol{\zeta})}{\boldsymbol{\zeta}^{\mathbf{l}+\boldsymbol{\nu}}}.$$

By inserting this result in Equation 11.177 and using the fact that $f(\zeta)$ is also equal to the lhs of Equation 11.176 we conclude that

$$\begin{aligned} \mathfrak{n}^{\{\mathbf{l}\}}(\mathbf{R}^{\mathbf{l}};\zeta)\frac{\Xi(\zeta)}{\zeta^{\mathbf{l}}} &= f(\zeta) = \sum_{\boldsymbol{\nu}} \frac{\Xi(\zeta')}{\boldsymbol{\nu}!} \prod_{i=1}^{M} \frac{(\zeta_i - \zeta_i')^{\nu_i}}{(\zeta_i')^{l_i+\nu_i}} \\ &\times \int d\mathbf{R}_{\mathbf{l}+1} \dots d\mathbf{R}_{\mathbf{l}+\boldsymbol{\nu}}\, \mathfrak{n}^{\{\mathbf{l}+\boldsymbol{\nu}\}}(\mathbf{R}^{\mathbf{l}+\boldsymbol{\nu}};\zeta'). \end{aligned} \quad (11.178)$$

Introducing the potential of mean force and the activity coefficient into this equation we obtain after substituting $\boldsymbol{\nu} \to \mathbf{m}$

$$\begin{aligned} \mathfrak{n}^{\{\mathbf{l}\}}(\mathbf{R}^{\mathbf{l}};\zeta)\frac{\Xi(\zeta)}{\zeta^{\mathbf{l}}} &= \Xi(\zeta') \sum_{\mathbf{m}} \frac{1}{\mathbf{m}!} \prod_{i=1}^{M} \frac{(\zeta_i - \zeta_i')^{m_i}}{(\gamma_i'^{\,\mathrm{b}})^{l_i+m_i}} \\ &\times \int d\mathbf{R}_{\mathbf{l}+1} \dots d\mathbf{R}_{\mathbf{l}+\mathbf{m}}\, e^{-\beta w^{\{\mathbf{l}+\mathbf{m}\}}(\mathbf{R}^{\mathbf{l}+\mathbf{m}};\zeta')}. \end{aligned} \quad (11.179)$$

The special case when all l_i are zero is

$$\frac{\Xi(\zeta)}{\Xi(\zeta')} = \sum_{\mathbf{m}} \frac{1}{\mathbf{m}!} \prod_{i=1}^{M} \left[\frac{\zeta_i - \zeta_i'}{\gamma_i'^{\,\mathrm{b}}}\right]^{m_i} \int d\mathbf{R}^{\mathbf{m}}\, e^{-\beta w^{\{\mathbf{m}\}}(\mathbf{R}^{\mathbf{m}};\zeta')}. \quad (11.180)$$

In the sums, m_i goes between 0 and ∞ for all species. These last two equations are the results for a general M-component system in the McMillan-Mayer theory.

Appendix 11D
Derivation of the Adelman theory for general bulk solutions

Since we deal with isotropic bulk systems we can write $h_{ij}^{(2)}(\mathbf{R}_1, \mathbf{R}_2) = h_{ij}^{(2)}(\mathbf{r}_{12}, \omega_1, \omega_2)$ because the pair functions depend only on the relative positions $\mathbf{r}_{12}$ of two particles and their orientations and the same for $c_{ij}^{(2)}$. The functions $H^{(2)}$ and $C^{(2)}$ for the solution phase are

$$\begin{aligned} H_{ij}^{(2)}(\mathbf{r}_{12}, \omega_1, \omega_2) &= n_i^{\mathrm{b}} \delta_{ij}^{(3)}(\mathbf{r}_{12}, \omega_1 - \omega_2) + n_i^{\mathrm{b}} n_j^{\mathrm{b}} h_{ij}^{(2)}(\mathbf{r}_{12}, \omega_1, \omega_2) \\ C_{ij}^{(2)}(\mathbf{r}_{12}, \omega_1, \omega_2) &= \frac{1}{n_i^{\mathrm{b}}} \delta_{ij}^{(3)}(\mathbf{r}_{12}, \omega_1 - \omega_2) - c_{ij}^{(2)}(\mathbf{r}_{12}, \omega_1, \omega_2) \end{aligned}$$

and they satisfy Equation 11.142[35]

$$\sum_l \int d\mathbf{r}_3 d\omega_3\, C_{il}^{(2)}(\mathbf{r}_{13}, \omega_1, \omega_3) H_{lj}^{(2)}(\mathbf{r}_{32}, \omega_3, \omega_2) = \delta_{ij}^{(3)}(\mathbf{r}_{12}, \omega_1 - \omega_2).$$

[35] See remark in footnote 25.

Since the $\mathbf{r}_3$ integration is a convolution one can eliminate that integration by going over to Fourier space, whereby the equation becomes

$$\sum_l \int d\omega_3\, \tilde{C}^{(2)}_{il}(\mathbf{k},\omega_1,\omega_3)\tilde{H}^{(2)}_{lj}(\mathbf{k},\omega_3,\omega_2) = \delta^{\omega}_{ij}(\omega_1-\omega_2), \qquad (11.181)$$

where $\delta^{\omega}_{ij}(\omega_1-\omega_2) = \delta_{ij}\delta^{\omega}(\omega_1-\omega_2)$.

Let us introduce the matrices $\mathbb{C}^{(2)} = \{\tilde{C}^{(2)}_{il}\}$ and $\mathbb{H}^{(2)} = \{\tilde{H}^{(2)}_{il}\}$, that is,

$$\mathbb{C}^{(2)} = \begin{bmatrix} \tilde{C}^{(2)}_{11} & \tilde{C}^{(2)}_{12} & \cdots & \tilde{C}^{(2)}_{1,M_D+M_S} \\ \tilde{C}^{(2)}_{21} & \tilde{C}^{(2)}_{22} & \cdots & \tilde{C}^{(2)}_{2,M_D+M_S} \\ \vdots & \vdots & \ddots & \vdots \\ \tilde{C}^{(2)}_{M_D+M_S,1} & \tilde{C}^{(2)}_{M_D+M_S,2} & \cdots & \tilde{C}^{(2)}_{M_D+M_S,M_D+M_S} \end{bmatrix}$$

and $\mathbb{H}^{(2)}$ likewise. We also introduce the diagonal matrix $\mathbb{I} = \{\delta^{\omega}_{ij}\}$. Then Equation 11.181 can be written

$$\int d\omega_3\, \mathbb{C}^{(2)}(\mathbf{k},\omega_1,\omega_3)\mathbb{H}^{(2)}(\mathbf{k},\omega_3,\omega_2) = \mathbb{I}(\omega_1-\omega_2),$$

where we have done a matrix multiplication of $\mathbb{C}^{(2)}$ and $\mathbb{H}^{(2)}$. To simplify notation, we will write this equation as

$$\mathbb{C}^{(2)} \star \mathbb{H}^{(2)} = \mathbb{I}. \qquad (11.182)$$

where $\star$ denotes matrix multiplication combined with ω_3 integration. The latter is, in fact, a continuous version of a matrix multiplication; instead of summing over a set of integers one integrates over a continuous variable ω_3.[36]

Let the indices $1,\ldots,M_D$ belong to the solute species and $M_D+1,\ldots,M_D+M_S$ belong to the solvent species. We can write $\mathbb{H}^{(2)}$ in terms of sub-matrices as

$$\mathbb{H}^{(2)} = \begin{bmatrix} \mathbb{H}^{(2)}_{DD} & \mathbb{H}^{(2)}_{DS} \\ \mathbb{H}^{(2)}_{SD} & \mathbb{H}^{(2)}_{SS} \end{bmatrix}$$

where, for example, the submatrix

$$\mathbb{H}^{(2)}_{DD} = \begin{bmatrix} \tilde{H}^{(2)}_{11} & \cdots & \tilde{H}^{(2)}_{1,M_D} \\ \vdots & \ddots & \vdots \\ \tilde{H}^{(2)}_{M_D,1} & \cdots & \tilde{H}^{(2)}_{M_D,M_D} \end{bmatrix}$$

[36] If one would discretize the ω_3 variable on a grid, the integral would become a finite sum. This means that the ω_3 integral of the product $\tilde{C}^{(2)}_{il}(\mathbf{k},\omega_1,\omega_3)\tilde{H}^{(2)}_{lj}(\mathbf{k},\omega_3,\omega_2)$ can be written like a matrix product with a sum over the grid points (multiplied by $d\omega_3$). When the grid is made denser and denser, the sum over the grid points goes over to an integral over ω_3 in the limit of infinitely dense grid.

is a $M_D \times M_D$ square matrix that involves only the solute species and

$$\mathbb{H}_{SD}^{(2)} = \begin{bmatrix} \tilde{H}_{M_D+1,1}^{(2)} & \cdots & \tilde{H}_{M_D+1,M_D}^{(2)} \\ \vdots & \ddots & \vdots \\ \tilde{H}_{M_D+M_S,1}^{(2)} & \cdots & \tilde{H}_{M_D+M_S,M_D}^{(2)} \end{bmatrix}$$

is a $M_S \times M_D$ rectangular matrix that contains correlation functions between the solute and solvent species. $\mathbb{C}^{(2)}$ can be written analogously.

Our first task is to show that we can identify an effective $C_{ij}^{(2)\text{eff}}$ for the effective M_D-component system that satisfies

$$\mathbb{C}_{DD}^{(2)\text{eff}} \star \mathbb{H}_{DD}^{(2)} = \mathbb{I}, \tag{11.183}$$

where $\mathbb{I} = \{\delta_{ij}^{\omega}\}$ is a diagonal $M_D \times M_D$ matrix (for simplicity we use the same notation $\mathbb{I}$ irrespectively of the dimension of matrix). To show this let us do the multiplication of $\mathbb{C}^{(2)}$ and $\mathbb{H}^{(2)}$ in Equation 11.182 in terms of their sub-matrices, whereby we have

$$\begin{bmatrix} \mathbb{C}_{DD}^{(2)} \star \mathbb{H}_{DD}^{(2)} + \mathbb{C}_{DS}^{(2)} \star \mathbb{H}_{SD}^{(2)} & \mathbb{C}_{DD}^{(2)} \star \mathbb{H}_{DS}^{(2)} + \mathbb{C}_{DS}^{(2)} \star \mathbb{H}_{SS}^{(2)} \\ \\ \mathbb{C}_{SD}^{(2)} \star \mathbb{H}_{DD}^{(2)} + \mathbb{C}_{SS}^{(2)} \star \mathbb{H}_{SD}^{(2)} & \mathbb{C}_{SD}^{(2)} \star \mathbb{H}_{DS}^{(2)} + \mathbb{C}_{SS}^{(2)} \star \mathbb{H}_{SS}^{(2)} \end{bmatrix} = \begin{bmatrix} \mathbb{I} & \mathbb{0}^{\intercal} \\ \mathbb{0} & \mathbb{I} \end{bmatrix},$$

where $\mathbb{0}$ is a rectangular matrix filled with zeros and $\mathbb{0}^{\intercal}$ is its transpose. The left column of sub-matrices on either side of the equation yield

$$\begin{aligned} \mathbb{C}_{DD}^{(2)} \star \mathbb{H}_{DD}^{(2)} + \mathbb{C}_{DS}^{(2)} \star \mathbb{H}_{SD}^{(2)} &= \mathbb{I} \\ \mathbb{C}_{SD}^{(2)} \star \mathbb{H}_{DD}^{(2)} + \mathbb{C}_{SS}^{(2)} \star \mathbb{H}_{SD}^{(2)} &= \mathbb{0}. \end{aligned} \tag{11.184}$$

The inverse $[\mathbb{C}_{SS}^{(2)}]^{-1}$ of $\mathbb{C}_{SS}^{(2)}$ satisfies (by definition) $\mathbb{C}_{SS}^{(2)} \star [\mathbb{C}_{SS}^{(2)}]^{-1} = \mathbb{I}$ and the corresponding relationship applies to the inverse $[\mathbb{H}_{DD}^{(2)}]^{-1}$.[37] The equation $\mathbb{C}_{SS}^{(2)} \star [\mathbb{C}_{SS}^{(2)}]^{-1} = \mathbb{I}$ is the same as the explicit expression

$$\sum_{l=M_D+1}^{M_D+M_S} \int d\omega_3\, \tilde{C}_{il}^{(2)}(\mathbf{k}, \omega_1, \omega_3)[\tilde{C}_{SS}^{(2)}]_{lj}^{-1}(\mathbf{k}, \omega_3, \omega_2) = \delta_{ij}^{\omega}(\omega_1 - \omega_2), \tag{11.185}$$

where l in the sum goes over the solvent species only. Given $\tilde{C}_{il}^{(2)}$ for all solvent species, this equation determines the elements $[\tilde{C}_{SS}^{(2)}]_{lj}^{-1}$ of the matrix $[\mathbb{C}_{SS}^{(2)}]^{-1}$ for these species. The corresponding arguments apply to $\mathbb{H}_{DD}^{(2)}$ and $[\mathbb{H}_{DD}^{(2)}]^{-1}$.

[37]Note that $[\mathbb{C}_{SS}^{(2)}]^{-1}$ and $[\mathbb{H}_{DD}^{(2)}]^{-1}$ are inverses of sub-matrices. They are different from $[\mathbb{C}^{(2)}]_{SS}^{-1}$ and $[\mathbb{H}^{(2)}]_{DD}^{-1}$, respectively, which satisfy $[\mathbb{C}^{(2)}]_{SS}^{-1} = \mathbb{H}_{SS}^{(2)}$ and $[\mathbb{H}^{(2)}]_{DD}^{-1} = \mathbb{C}_{DD}^{(2)}$ since they are sub-matrices of the inverse of the entire $\mathbb{C}^{(2)}$ and $\mathbb{H}^{(2)}$ matrices, which are inverses of each other.

When we multiply the first equation in 11.184 by $[\mathbb{H}_{DD}^{(2)}]^{-1}$ from the right it becomes

$$\mathbb{C}_{DD}^{(2)} + \mathbb{C}_{DS}^{(2)} \star \mathbb{H}_{SD}^{(2)} \star [\mathbb{H}_{DD}^{(2)}]^{-1} = [\mathbb{H}_{DD}^{(2)}]^{-1}$$

and by multiplying the second equation by the inverse $[\mathbb{C}_{SS}^{(2)}]^{-1}$ from the left and then by $[\mathbb{H}_{DD}^{(2)}]^{-1}$ from the right we obtain

$$[\mathbb{C}_{SS}^{(2)}]^{-1} \star \mathbb{C}_{SD}^{(2)} + \mathbb{H}_{SD}^{(2)} \star [\mathbb{H}_{DD}^{(2)}]^{-1} = \mathbb{0}.$$

If we combine these two equation we have

$$\mathbb{C}_{DD}^{(2)} - \mathbb{C}_{DS}^{(2)} \star [\mathbb{C}_{SS}^{(2)}]^{-1} \star \mathbb{C}_{SD}^{(2)} = [\mathbb{H}_{DD}^{(2)}]^{-1},$$

which implies Equation 11.183 if we identify

$$\mathbb{C}_{DD}^{(2)\text{eff}} = \mathbb{C}_{DD}^{(2)} - \mathbb{C}_{DS}^{(2)} \star [\mathbb{C}_{SS}^{(2)}]^{-1} \star \mathbb{C}_{SD}^{(2)} \tag{11.186}$$

and multiply by $\mathbb{H}_{DD}^{(2)}$ from the right. Thereby we have proven Equation 11.183.

Let us consider the case where the solvent consists of only one species labelled s. Then Equation 11.185 becomes in real space

$$\int d\mathbf{r}_3 d\omega_3\, C_{ss}^{(2)}(\mathbf{r}_{13}, \omega_1, \omega_3)[C_{SS}^{(2)}]_{ss}^{-1}(\mathbf{r}_{32}, \omega_3, \omega_2) = \delta^{(3)}(\mathbf{r}_{12}, \omega_1 - \omega_2). \tag{11.187}$$

By defining the function $\mathfrak{h}_{ss}^{(2)}(\mathbf{R}_1, \mathbf{R}_2) \equiv \mathfrak{h}_{ss}^{(2)}(\mathbf{r}_{12}, \omega_1, \omega_2)$ from the relationship

$$[C_{SS}^{(2)}]_{ss}^{-1}(\mathbf{r}_{12}, \omega_1, \omega_2) = n_s^{\text{b}} \delta^{(3)}(\mathbf{r}_{12}, \omega_1 - \omega_2) + (n_s^{\text{b}})^2 \mathfrak{h}_{ss}^{(2)}(\mathbf{r}_{12}, \omega_1, \omega_2), \tag{11.188}$$

we can write Equation 11.187 as

$$\mathfrak{h}_{ss}^{(2)}(\mathbf{R}_1, \mathbf{R}_2) = c_{ss}^{(2)}(\mathbf{R}_1, \mathbf{R}_2) + \int d\mathbf{R}_3\, c_{ss}^{(2)}(\mathbf{R}_1, \mathbf{R}_3) n_s^{\text{b}} \mathfrak{h}_{ss}^{(2)}(\mathbf{R}_3, \mathbf{R}_2), \tag{11.189}$$

which is an OZ-like equation. Given $c_{ss}^{(2)}$ for the solvent in the solution phase, this equation can be solved for $\mathfrak{h}_{ss}^{(2)}$. Note that $\mathfrak{h}_{ss}^{(2)}$ is different from $h_{ss}^{(2)}$ of the solution phase. Incidentally we may note that in the limit of zero solute concentration $\mathfrak{h}_{ss}^{(2)}$ approaches $h_{ss}^{(2)}$ of the bulk phase of pure solvent and Equation 11.189 becomes the OZ equation for the pure solvent.

For indices i and j of the solute species we have

$$C_{ij}^{(2)\text{eff}}(\mathbf{r}_{12}, \omega_1, \omega_2) = C_{ij}^{(2)\text{eff}}(\mathbf{R}_1, \mathbf{R}_2) = \frac{1}{n_j^{\text{b}}} \delta_{ij}^{(3)}(\mathbf{R}_1 - \mathbf{R}_2) - c_{ij}^{(2)\text{eff}}(\mathbf{R}_1, \mathbf{R}_2)$$

By inserting this, the definition of $C_{ij}^{(2)}$ and Equation 11.188 into Equation 11.186 transformed to real space, the latter can be written explicitly as

$$\begin{aligned} c_{ij}^{(2)\text{eff}}(\mathbf{R}_1, \mathbf{R}_2) &= c_{ij}^{(2)}(\mathbf{R}_1, \mathbf{R}_2) + \int d\mathbf{R}_3\, c_{is}^{(2)}(\mathbf{R}_1, \mathbf{R}_3) n_s^{\text{b}} c_{sj}^{(2)}(\mathbf{R}_3, \mathbf{R}_2) \quad (11.190) \\ &+ \int d\mathbf{R}_3 d\mathbf{R}_4\, c_{is}^{(2)}(\mathbf{R}_1, \mathbf{R}_3) n_s^{\text{b}} \mathfrak{h}_{ss}^{(2)}(\mathbf{R}_3, \mathbf{R}_4) n_s^{\text{b}} c_{sj}^{(2)}(\mathbf{R}_4, \mathbf{R}_2) \end{aligned}$$

for i and j of the solute species and where all functions in the rhs are for the solution phase. By inserting $\mathfrak{h}_{ss}^{(2)}$ from Equation 11.189 repeatedly in this equation, we obtain the generalization of Equation 11.166 to the present case, where the last integral in Equation 11.190 is written in terms of sums of all chains of direct correlations between the solvent molecules ($c_{ss}^{(2)}$ chains) in analogy to Equation 11.166.

IV

Electrolytes and polar fluids

CHAPTER 12

Electrolytes with Spherical Ions

This and the following chapters in Part IV of the book contain a unified exact treatment of electrolyte solutions, ionic liquids and polar fluids as well as approximate theories and applications.

In the present chapter we will study electrolytes consisting of spherically symmetric ions. They can be dissolved in a solvent modeled as a dielectric continuum or distributed in vacuum, a **classical plasma**. The latter case includes **molten salts** of simple ions. Later, in Chapter 13, we will treat nonspherical ions and electrolytes with molecular solvent.

We will first study inhomogeneous and then homogeneous electrolytes and investigate polarization response functions that form crucial ingredients in the general theory of electrolytes. This gives the basis for an analysis of the propagation of electrostatic interactions and screening in electrolytes as well as the identification of "dressed ions," which are the relevant entities for an understanding of the strength of the electrostatic interactions between ions. The dressed ions also give a base for a comprehension of the polarization response and nonlocal electrostatics.

The general concept of "unit screened Coulomb potential" is introduced and utilized in order to formulate the exact theory of electrolytes in terms of screened interactions between dressed ions – the Dressed Ion Theory (DIT). A key element in the analysis is the determination of the multiple decay modes that are in operation for the interactions, whereby each mode has its own decay length and, for oscillatory modes, its own wave length. The modes are solutions of an equation with multiple solutions derived from the dielectric function of the electrolyte. Furthermore, the strength of the interaction between ions is described in terms of effective ionic changes and effective dielectric permittivity; each mode have its own values for these entities. The phenomena of underscreening and overscreening in electrolytes are elegantly handled in terms of the effective charges of the ions. This way of formulating the exact

DOI: 10.1201/9781003286882-12

statistical mechanics gives a transparent representation of essential parts of the physics of electrolytes. Many different aspects of electrolyte systems are thereby dealt with, including various types of interparticle correlations and surface interactions.

Some of these topics were studied in Volume I of this treatise and for completeness we list below the relevant background material covered there. In the current volume we generalize the exact, general theory of both inhomogeneous and bulk electrolytes and give it a more thorough foundation. We will thereby recapitulate important results from Volume I on the way.

The classical approximations for electrolytes, the Poisson-Boltzmann (PB) approximation and its linearized version, the Debye-Hückel theory of electrolyte solutions, were treated in Chapters 6, 7, and 8. Section 6.1 dealt with bulk electrolytes, Section 7.1.1 with inhomogeneous electrolytes (electrical double layers) and Section 8.2 with surface interactions (electrical double layer interactions) in these approximations. Chapters 6 and 7 also contained a simple version of the exact, general DIT treatment of simple bulk electrolytes (spherical ions). In Section 6.2 the focus was the theory of electrostatic screening and the decay behavior of distribution functions, including explicit concrete examples, and Section 7.1.2 contained the exact treatment of screening in electrical double layers in simple cases. These chapters included an elementary treatment of polarization response functions and nonlocal electrostatics. The concepts of effective charges and dressed particles were introduced in both the PB approximation and the exact case and, for the latter case, effective dielectric permittivities were also dealt with.

12.1 BASIC DEFINITIONS AND FORMULAS

We will assume that each spherical ion of species i has a point charge q_i at its center and ions of species i and j interact with each other with a pair interaction potential

$$u_{ij}(r) = u_{ij}^{\mathrm{sh}}(r) + u_{ij}^{\mathrm{el}}(r), \tag{12.1}$$

where u_{ij}^{sh} is a short-range potential, u_{ij}^{el} is the electrostatic pair interaction for ions in vacuum

$$u_{ij}^{\mathrm{el}}(r) = q_i q_j \phi_{\mathrm{Coul}}(r) \tag{12.2}$$

and

$$\phi_{\mathrm{Coul}}(r) = \frac{1}{4\pi\varepsilon_0 r}, \tag{12.3}$$

where ε_0 is the permittivity of vacuum, is the **"unit" Coulomb potential** in vacuum. For an inhomogeneous electrolyte, the ions also interact with an external potential

$$v_i(\mathbf{r}) = v_i^{\mathrm{sh}}(\mathbf{r}) + q_i \Psi^{\mathrm{ext}}(\mathbf{r}),$$

where v_i^{sh} is short-ranged and Ψ^{ext} is the external electrostatic potential, that is, the potential from a charge distribution that is external to the system. The actual interaction potentials in real systems contain contributions from dispersion interactions, which means that $u_{ij}^{\rm sh}(r)$ decays like $1/r^6$ for large distances r, but unless we say otherwise we will here assume that $u_{ij}^{\rm sh}(r)$ has a finite range or decays to zero at least exponentially fast with increasing r. The same applies to $v_i^{\rm sh}(\mathbf{r})$ as function of distance from the body that gives rise to it.

The inclusion of dispersion interactions requires a special treatment; some aspects of this are dealt with in Sections 7.1.5 and 8.6 of Volume I. In the presence of such interactions between the ions, the pair correlation functions and ion-ion potential of mean force ultimately decays like $1/r^6$ for large distances r, see page 390 in Volume I. Therefore, the effects of such interactions always dominate for *very* large r because a power law function always decays slower than the exponentially screened electrostatic interactions. When $u_{ij}^{\rm sh}(r)$ decays faster than any power law, as assumed here, such long-range tails do not appear.

Dispersion interactions can have substantial effects for small distances between ions and, in particular, between ions and surfaces; an example is given in Figure 7.6 in Volume I. Van der Waals interactions have often large contributions for surface forces between macroscopic bodies at short separations. All short-range effects can be included in the interaction potentials $u_{ij}^{\rm sh}(r)$ and $v_i^{\rm sh}(\mathbf{r})$ in the present treatment, but the ultimate tails for very large separations due to dispersion interactions are not included.

In the **primitive model of electrolyte solutions**, the solvent is assumed to be dielectric continuum and we have instead of Equation 12.3

$$\phi_{\rm Coul}(r) = \frac{1}{4\pi\varepsilon_r^{\rm cont}\varepsilon_0 r} \qquad \text{(diel. continuum)}, \tag{12.4}$$

where $\varepsilon_r^{\rm cont}$ the dielectric constant of the continuum solvent (we have written superscript "cont" in order to distinguish $\varepsilon_r^{\rm cont}$ of the *continuum* from the dielectric constant ε_r of polar *molecular* fluids). In this model the short-range potential is a hard core potential $u_{ij}^{\rm sh} = u_{ij}^{\rm core}$, which is infinitely large when two ions overlap and zero otherwise. We will, however, derive the formulas for ions distributed in vacuum and for the case with a dielectric continuum one can replace ε_0 with $\varepsilon_r^{\rm cont}\varepsilon_0$.

The electrostatic potential ψ from a charge distribution ρ is

$$\psi(\mathbf{r}_3) = \int d\mathbf{r}_2 \, \frac{\rho(\mathbf{r}_2)}{4\pi\varepsilon_0 r_{23}} = \int d\mathbf{r}_2 \, \rho(\mathbf{r}_2)\phi_{\rm Coul}(r_{23}), \tag{12.5}$$

which we will refer to as the (ordinary) **Coulomb potential formula.**[1]

The number density of j-ions at $\mathbf{r}_2$ is denoted $n_j(\mathbf{r}_2)$ and the average charge–density distribution is

$$\rho(\mathbf{r}_2) = \sum_j q_j n_j(\mathbf{r}_2).$$

[1]There is also a screened version to be introduced later in Equation 12.64.

If we place an ion of species i with its center at the point $\mathbf{r}_1$ in the electrolyte, the density distribution for j-particles changes from $n_j(\mathbf{r}_2)$ to $n^{(1)}_{j;i}(\mathbf{r}_2|\mathbf{r}_1) \equiv n_j(\mathbf{r}_2)g^{(2)}_{ij}(\mathbf{r}_1,\mathbf{r}_2)$ and we have the charge distribution[2]

$$\rho_i(\mathbf{r}_2|\mathbf{r}_1) = \sum_j q_j n^{(1)}_{j;i}(\mathbf{r}_2|\mathbf{r}_1) = \sum_j q_j n_j(\mathbf{r}_2)g^{(2)}_{ij}(\mathbf{r}_1,\mathbf{r}_2) \tag{12.6}$$

in the surroundings of the ion. Thus, $\rho_i(\mathbf{r}_2|\mathbf{r}_1)$ is the charge density at $\mathbf{r}_2$ when an ion of species i is located at $\mathbf{r}_1$. The charge distribution that arises *because* we place the ion at $\mathbf{r}_1$ is accordingly

$$\Delta\rho_i(\mathbf{r}_2|\mathbf{r}_1) = \rho_i(\mathbf{r}_2|\mathbf{r}_1) - \rho(\mathbf{r}_2) = \sum_j q_j n_j(\mathbf{r}_2)h^{(2)}_{ij}(\mathbf{r}_1,\mathbf{r}_2). \tag{12.7}$$

One may say that the surroundings of the ion is polarized by the interactions with it and $\Delta\rho_i$ is the induced charge density. Note that this polarization is caused by all kinds of interactions, not only the electrostatic ones. For example, nonelectrostatic interactions that act differently on anions and cations give rise to charge separation and hence a charge density. It is common to denote $\Delta\rho_i(\mathbf{r}_2|\mathbf{r}_1)$ as the charge density of the **ion cloud** that surrounds the i-ion and we therefore write $\rho_i^{\mathrm{cloud}} = \Delta\rho_i$, that is,

$$\rho_i^{\mathrm{cloud}}(\mathbf{r}_2|\mathbf{r}_1) = \sum_j q_j n_j(\mathbf{r}_2)h^{(2)}_{ij}(\mathbf{r}_1,\mathbf{r}_2). \tag{12.8}$$

The *total* charge density of the ion and its ion cloud is

$$\begin{aligned}\rho_i^{\mathrm{tot}}(\mathbf{r}_2|\mathbf{r}_1) &= q_i\delta^{(3)}(\mathbf{r}_{12}) + \rho_i^{\mathrm{cloud}}(\mathbf{r}_2|\mathbf{r}_1) \\ &= q_i\delta^{(3)}(\mathbf{r}_{12}) + \sum_j q_j n_j(\mathbf{r}_2)h^{(2)}_{ij}(\mathbf{r}_1,\mathbf{r}_2),\end{aligned} \tag{12.9}$$

where $q_i\delta^{(3)}(r_{12})$ is the central point charge expressed as a charge density. This is the entire charge density that is associated with an i-ion located at $\mathbf{r}_1$. The electrostatic potential ψ_i from this charge density is obtained using the expression 12.5 for the Coulomb potential

$$\begin{aligned}\psi_i(\mathbf{r}_3|\mathbf{r}_1) &= \int d\mathbf{r}_2\, \rho_i^{\mathrm{tot}}(\mathbf{r}_2|\mathbf{r}_1)\,\phi_{\mathrm{Coul}}(r_{23}) \qquad (12.10)\\ &= q_i\phi_{\mathrm{Coul}}(r_{13}) + \int d\mathbf{r}_2\, \rho_i^{\mathrm{cloud}}(\mathbf{r}_2|\mathbf{r}_1)\,\phi_{\mathrm{Coul}}(r_{23})\end{aligned}$$

and is commonly known as the **mean electrostatic potential** due to the ion. It satisfies the Poisson equation

$$-\varepsilon_0\nabla_3^2\psi_i(\mathbf{r}_3|\mathbf{r}_1) = \rho_i^{\mathrm{tot}}(\mathbf{r}_3|\mathbf{r}_1), \tag{12.11}$$

[2]In an alternative notation $\rho_i(\mathbf{r}_2|\mathbf{r}_1) \equiv \rho(\mathbf{r}_2|\mathbf{r}_1;i) = \sum_j q_j n^{(1)}_j(\mathbf{r}_2|\mathbf{r}_1;i)$, cf. Equations 11.133 and 11.134.

where $\nabla_3^2 = \partial^2/\partial x_3^2 + \partial^2/\partial y_3^2 + \partial^2/\partial z_3^2$ is the Laplace operator for $\mathbf{r}_3$ and the boundary condition is $\psi_i(\mathbf{r}_3|\mathbf{r}_1) \to 0$ when r_{13} becomes very large (for a finite system we here assume that $\mathbf{r}_3$ remains inside the electrolyte). For the case with ions in a dielectric continuum we have the factor $\varepsilon_r^{\text{cont}}\varepsilon_0$ instead of ε_0 in Equation 12.11. The general formalism will, however, be derived and expressed for ions distributed in vacuum.

After these preliminaries we will consider an electrolyte where all ions are free to move, that is, we do not have any ion placed fixed at some point. The external charge density that gives rise to $\Psi^{\text{ext}}(\mathbf{r})$ is, however, always stationary and we will only consider equilibrium systems so there is no time dependence of the various properties of the electrolyte.

12.2 GENERAL FORMALISM FOR INHOMOGENEOUS ELECTROLYTES

12.2.1 Electrostatic polarization of electrolytes

We will first investigate how the electrolyte is polarized by variations in the external electrostatic field. Initially, the total electrostatic potential Ψ at point $\mathbf{r}_2$ is

$$\Psi(\mathbf{r}_2) = \Psi^{\text{ext}}(\mathbf{r}_2) + \int d\mathbf{r}_1\, \rho(\mathbf{r}_1)\phi_{\text{Coul}}(r_{12}).$$

If we change $\Psi^{\text{ext}}(\mathbf{r}_2)$ to $\Psi^{\text{ext}}(\mathbf{r}_2) + \delta\Psi^{\text{ext}}(\mathbf{r}_2)$ the number density is changed by, say, $\delta n_i(\mathbf{r}_1)$ and the average charge density by $\delta\rho(\mathbf{r}_1) = \sum_i q_i \delta n_i(\mathbf{r}_1)$, which we will denote as $\delta\rho^{\text{pol}}$. We use the subscript "pol" to specify that $\delta\rho^{\text{pol}}$ is the polarization charge density due solely to a change in the electrostatic field. This polarization gives rise to an additional electrostatic potential $\delta\Psi^{\text{pol}}$ that is given by Ψ

$$\delta\Psi^{\text{pol}}(\mathbf{r}_2) = \int d\mathbf{r}_1\, \delta\rho^{\text{pol}}(\mathbf{r}_1)\phi_{\text{Coul}}(r_{12}), \tag{12.12}$$

so the variation in total electrostatic potential $\delta\Psi$ caused by $\delta\Psi^{\text{ext}}$ is

$$\delta\Psi(\mathbf{r}_2) = \delta\Psi^{\text{ext}}(\mathbf{r}_2) + \delta\Psi^{\text{pol}}(\mathbf{r}_2) \tag{12.13}$$

and we will treat the case where $\delta\Psi^{\text{ext}}$ and hence $\delta\Psi$ are infinitesimally small.

Since $n_i(\mathbf{r}_1) = n_i^{\text{b}} e^{-\beta w_i(\mathbf{r}_1)}$ we have $\delta \ln n_i(\mathbf{r}_1) = \delta n_i(\mathbf{r}_1)/n_i(\mathbf{r}_1) = -\beta\delta w_i(\mathbf{r}_1)$, so if we can determine δw_i we can obtain δn_i and hence $\delta\rho^{\text{pol}}$. A pertinent question is: how can δw_i be determined from $\delta\Psi^{\text{ext}}$ or, in other words, what is the change in free energy of interaction, $\delta w_i(\mathbf{r}_1)$, for an i-ion at $\mathbf{r}_1$ when it interacts with the additional electrostatic potential $\delta\Psi^{\text{ext}}$? We will show that

$$\delta w_i(\mathbf{r}_1) = \int d\mathbf{r}_2\, \rho_i^{\text{tot}}(\mathbf{r}_2|\mathbf{r}_1)\delta\Psi^{\text{ext}}(\mathbf{r}_2), \tag{12.14}$$

which we can write as

$$\frac{\delta w_i(\mathbf{r}_1)}{\delta \Psi^{\mathrm{ext}}(\mathbf{r}_2)} = \rho_i^{\mathrm{tot}}(\mathbf{r}_2|\mathbf{r}_1). \tag{12.15}$$

Equation 12.14 implies that the change δw_i felt by an i-ion due to the variation $\delta\Psi^{\mathrm{ext}}$ in the external electrostatic potential is equal to the interaction energy between $\delta\Psi^{\mathrm{ext}}$ and total charge density ρ_i^{tot} associated with the ion, that is, the interaction of $\delta\Psi^{\mathrm{ext}}$ with the central ionic charge and the surrounding ion cloud of the i-ion.

This result can be shown as follows. The first Yvon equation for mixtures says that for any infinitesimal change $\delta v_j(\mathbf{r}_2)$ in external potential, the density distribution is changed by (cf. Equation 11.144)

$$\begin{aligned} \delta n_i(\mathbf{r}_1) &= -\beta \int d\mathbf{r}_2 \sum_j H_{ij}^{(2)}(\mathbf{r}_1,\mathbf{r}_2)\delta v_j(\mathbf{r}_2) \\ &= -\beta n_i(\mathbf{r}_1) \int d\mathbf{r}_2 \Bigg[\delta^{(3)}(\mathbf{r}_{12})\delta v_i(\mathbf{r}_2) \\ &+ \sum_j n_j(\mathbf{r}_2) h_{ij}^{(2)}(\mathbf{r}_1,\mathbf{r}_2)\delta v_j(\mathbf{r}_2) \Bigg], \end{aligned} \tag{12.16}$$

where we have inserted

$$H_{ij}^{(2)}(\mathbf{r}_1,\mathbf{r}_2) \equiv n_i(\mathbf{r}_1)\delta_{ij}^{(3)}(\mathbf{r}_{12}) + n_i(\mathbf{r}_1)n_j(\mathbf{r}_2)h_{ij}^{(2)}(\mathbf{r}_1,\mathbf{r}_2) \tag{12.17}$$

(cf. Equation 11.140) and $\delta_{ij}^{(3)}(\mathbf{r}) = \delta_{ij}\delta^{(3)}(\mathbf{r})$. In the current case $\delta v_j(\mathbf{r}_2) = q_j\delta\Psi^{\mathrm{ext}}(\mathbf{r}_2)$, so we have

$$\begin{aligned} \delta n_i(\mathbf{r}_1) &= -\beta n_i(\mathbf{r}_1) \int d\mathbf{r}_2 \Bigg[q_i \delta^{(3)}(\mathbf{r}_{12}) \\ &+ \sum_j q_j n_j(\mathbf{r}_2) h_{ij}^{(2)}(\mathbf{r}_1,\mathbf{r}_2) \Bigg] \delta\Psi^{\mathrm{ext}}(\mathbf{r}_2) \\ &= -\beta n_i(\mathbf{r}_1) \int d\mathbf{r}_2\, \rho_i^{\mathrm{tot}}(\mathbf{r}_2|\mathbf{r}_1)\, \delta\Psi^{\mathrm{ext}}(\mathbf{r}_2) \end{aligned} \tag{12.18}$$

Since $\delta n_i(\mathbf{r}_1) = -\beta n_i(\mathbf{r}_1)\delta w_i(\mathbf{r}_1)$ Equation 12.14 follows.

From Equation 12.18 it follows that

$$\delta\rho^{\mathrm{pol}}(\mathbf{r}_1) = \sum_i q_i \delta n_i(\mathbf{r}_1) = -\beta \int d\mathbf{r}_2 \sum_i q_i n_i(\mathbf{r}_1)\rho_i^{\mathrm{tot}}(\mathbf{r}_2|\mathbf{r}_1)\delta\Psi^{\mathrm{ext}}(\mathbf{r}_2).$$

We can write this as

$$\delta\rho^{\mathrm{pol}}(\mathbf{r}_1) = \int d\mathbf{r}_2\, \chi^{\rho}(\mathbf{r}_1,\mathbf{r}_2)\delta\Psi^{\mathrm{ext}}(\mathbf{r}_2), \tag{12.19}$$

where we have defined

$$\begin{aligned}\chi^{\rho}(\mathbf{r}_1,\mathbf{r}_2) &\equiv -\beta\sum_i q_i n_i(\mathbf{r}_1)\rho_i^{\mathrm{tot}}(\mathbf{r}_2|\mathbf{r}_1) \\ &= -\beta\sum_{ij} q_i q_j H_{ij}^{(2)}(\mathbf{r}_1,\mathbf{r}_2) = -\beta q_{\mathrm{e}}^2 H_{\mathrm{QQ}}(\mathbf{r}_1,\mathbf{r}_2).\end{aligned} \tag{12.20}$$

H_{QQ} is the **charge-charge correlation function** defined in Appendix 6C in Volume I (Equation I:6.216) and q_{e} is the elementary charge (the charge of a proton). We have[3]

$$\begin{aligned}H_{\mathrm{QQ}}(\mathbf{r}_1,\mathbf{r}_2) &\equiv \frac{1}{q_{\mathrm{e}}^2}\sum_{ij} q_i q_j H_{ij}^{(2)}(\mathbf{r}_1,\mathbf{r}_2), \\ &= \frac{1}{q_{\mathrm{e}}^2}\sum_i q_i n_i(\mathbf{r}_1)\left[q_i\delta^{(3)}(\mathbf{r}_{12}) + \sum_j q_j n_j(\mathbf{r}_2)h_{ij}^{(2)}(\mathbf{r}_1,\mathbf{r}_2)\right],\end{aligned} \tag{12.21}$$

where we have used Equation 12.17. The function χ^{ρ} is a **polarization response function**, which gives the polarization charge density at one point, $\mathbf{r}_1$, in terms of the variation in external electrostatic potential at another point, $\mathbf{r}_2$ (and vice versa). For bulk electrolytes we encountered this function in Section 6.2 of Volume I (Equations I:6.117 and I:6.118). It is a *linear response function*, which means that it gives the polarization caused by a small variation in potential and the response is therefore linear. Furthermore, χ^{ρ} is a *static* response function so it does not involve any time dependence.[4] As stated earlier, we are only dealing with equilibrium systems, the external potential is independent of time and no currents flow in the system. We can write Equation 12.19 as

$$\chi^{\rho}(\mathbf{r}_1,\mathbf{r}_2) = \frac{\delta\rho^{\mathrm{pol}}(\mathbf{r}_1)}{\delta\Psi^{\mathrm{ext}}(\mathbf{r}_2)} = \frac{\delta\rho(\mathbf{r}_1)}{\delta\Psi^{\mathrm{ext}}(\mathbf{r}_2)}, \tag{12.22}$$

where we have written $\delta\rho(\mathbf{r}_1)$ in the rhs because it is obvious that the variation in charge density is induced by the change in electrostatic potential. This relationship can serve as an alternative definition of χ^{ρ}.

The fact that the functions $\delta\Psi^{\mathrm{ext}}$, $\delta\rho^{\mathrm{pol}}$ and $\delta\Psi$ are linearly related to each other implies that there exists a function $\chi^{\star}(\mathbf{r}_1,\mathbf{r}_2)$ that expresses the linear relationship between $\delta\rho^{\mathrm{pol}}$ and $\delta\Psi$ as

$$\delta\rho^{\mathrm{pol}}(\mathbf{r}_1) = \int d\mathbf{r}_2\,\chi^{\star}(\mathbf{r}_1,\mathbf{r}_2)\delta\Psi(\mathbf{r}_2), \tag{12.23}$$

[3]For a binary bulk electrolyte this can be written as in Equation I:6.185.

[4]Frequency dependent response functions that are common in the study of polarization of electrolytes and polar fluids are therefore not included in the treatment.

which is similar to Equation 12.19. As we will show below, for a given χ^ρ the function $\chi^\star$ is the solution to the equation[5]

$$\chi^\rho(\mathbf{r}_1,\mathbf{r}_4) = \chi^\star(\mathbf{r}_1,\mathbf{r}_4) + \int d\mathbf{r}_2 d\mathbf{r}_3\, \chi^\star(\mathbf{r}_1,\mathbf{r}_2)\phi_{\text{Coul}}(r_{23})\chi^\rho(\mathbf{r}_3,\mathbf{r}_4), \qquad (12.24)$$

so it is known and can be calculated when χ^ρ is known. The function $\chi^\star$ will also be denoted as a "polarization response function" despite the fact that a true response function normally gives the response of a system due to an influence that we can completely control by external means. The total potential $\delta\Psi$ contains $\delta\Psi^{\text{pol}}$ from the polarization charge density that we cannot control directly, so $\chi^\star$ has a slightly different character than χ^ρ. Equation 12.23 can be written as

$$\chi^\star(\mathbf{r}_1,\mathbf{r}_2) = \frac{\delta\rho^{\text{pol}}(\mathbf{r}_1)}{\delta\Psi(\mathbf{r}_2)} = \frac{\delta\rho(\mathbf{r}_1)}{\delta\Psi(\mathbf{r}_2)}, \qquad (12.25)$$

which can serve as an alternative definition of $\chi^\star$. Note this relationship is valid for variations $\delta\Psi$ that are caused by changes $\delta\Psi^{\text{ext}}$ in the external potential. Both $\chi^\star(\mathbf{r}_1,\mathbf{r}_2)$ and $\chi^\rho(\mathbf{r}_1,\mathbf{r}_2)$ are symmetric functions in $\mathbf{r}_1$ and $\mathbf{r}_2$ when $1 \leftrightarrow 2$.

Equation 12.24 can easily be shown a follows. We have from Equation 12.23

$$\begin{aligned}
\delta\rho^{\text{pol}}(\mathbf{r}_1) &= \int d\mathbf{r}_2\, \chi^\star(\mathbf{r}_1,\mathbf{r}_2)[\delta\Psi^{\text{ext}}(\mathbf{r}_2) + \delta\Psi^{\text{pol}}(\mathbf{r}_2)] \\
&= \int d\mathbf{r}_4\, \chi^\star(\mathbf{r}_1,\mathbf{r}_4)\delta\Psi^{\text{ext}}(\mathbf{r}_4) \\
&+ \int d\mathbf{r}_2\, \chi^\star(\mathbf{r}_1,\mathbf{r}_2) \int d\mathbf{r}_3\, \delta\rho^{\text{pol}}(\mathbf{r}_3)\phi_{\text{Coul}}(r_{32}),
\end{aligned}$$

where we have changed the integration variable in the first integral on the rhs from $\mathbf{r}_2$ to $\mathbf{r}_4$ and inserted $\delta\Psi^{\text{pol}}$ from Equation 12.12. Let us now insert Equation 12.19 and obtain

$$\begin{aligned}
\delta\rho^{\text{pol}}(\mathbf{r}_1) &= \int d\mathbf{r}_4 \Big[\chi^\star(\mathbf{r}_1,\mathbf{r}_4) \\
&+ \int d\mathbf{r}_2 d\mathbf{r}_3\, \chi^\star(\mathbf{r}_1,\mathbf{r}_2)\chi^\rho(\mathbf{r}_3,\mathbf{r}_4)\phi_{\text{Coul}}(r_{32})\Big]\, \delta\Psi^{\text{ext}}(\mathbf{r}_4).
\end{aligned}$$

Since the left-hand side is equal to $\int d\mathbf{r}_4\, \chi^\rho(\mathbf{r}_1,\mathbf{r}_4)\delta\Psi^{\text{ext}}(\mathbf{r}_4)$ and $\delta\Psi^{\text{ext}}(\mathbf{r}_4)$ is arbitrary, Equation 12.24 follows.

From Equation 12.14 we saw that $\delta w_i(\mathbf{r}_1)$ due to the change $\delta\Psi^{\text{ext}}$ in the *external* electrostatic potential is equal to the interaction energy between $\delta\Psi^{\text{ext}}$ and total charge ρ_i^{tot} associated with the i-ion. We may now ask: how can

[5] As shown in Exercise 12.2 on page 297, Equation 12.24 implies that $\delta^{(3)}(\mathbf{r}_{23}) + \int d\mathbf{r}_4\, \chi^\rho(\mathbf{r}_4,\mathbf{r}_2)\phi_{\text{Coul}}(r_{43})$ and $\delta^{(3)}(\mathbf{r}_{12}) - \int d\mathbf{r}_5\, \chi^\star(\mathbf{r}_5,\mathbf{r}_1)\phi_{\text{Coul}}(r_{52})$ are functional inverses of each other.

$\delta w_i(\mathbf{r}_1)$ be expressed in terms of the change $\delta\Psi$ in *total* electrostatic potential? Since Ψ is the electrostatic potential that actually appear in the electrolyte, this is a very relevant question. Ψ^{ext} is the potential from the external source in the *absence* of the electrolyte and does not include the response of the electrolyte itself. We will show below that

$$\delta w_i(\mathbf{r}_1) = \int d\mathbf{r}_2\, \rho_i^\star(\mathbf{r}_2|\mathbf{r}_1)\, \delta\Psi(\mathbf{r}_2), \tag{12.26}$$

where $\rho_i^\star(\mathbf{r}_2|\mathbf{r}_1)$ is another charge density associated with the i-ion (defined below in Equation 12.29). This relationship is similar to Equation 12.14 and it implies that the change δw_i felt by an i-ion due to the change $\delta\Psi$ in the *total* electrostatic potential is equal to the interaction energy between $\delta\Psi$ and the charge density $\rho_i^\star$. The entity $\rho_i^\star(\mathbf{r}_2|\mathbf{r}_1)$ is the **dressed-particle charge density** of an i-ion, which is introduced for bulk electrolytes in Chapter 6 (in Volume I) and is called the "**dressed-ion charge density**." A **dressed ion** consists of the ion itself and its **dress**, which is a surrounding distribution of charge that is different from the ion cloud. We defined $\rho_i^\star$ first in the Poisson-Boltzmann approximation (Equations I:6.60-62) and then in the general case of bulk electrolytes (Equations I:6.129-131). Here, we reintroduce this concept in a somewhat different context and in what follows we will verify what we found in Volume I and investigate further aspects of dressed ions and, more generally, dressed particles. We will see that dressed particles and their charge density $\rho_i^\star$ play a major role for the properties of both electrolyte systems and polar fluids. This concept is therefore important for the understanding of such systems.

The charge density $\rho_i^\star$ is an essential ingredient in what is called **nonlocal electrostatics**, namely the fact that each ion experiences a potential of mean force that is determined by the electrostatic potential in the entire neighborhood of the ion, as expressed by Equation 12.26, and not only by the value of the potential at the place where the ion is located. In the Poisson-Boltzmann approximation it is assumed that the potential of mean force for an ion at $\mathbf{r}_1$ is given by the ionic charge q_i times the total electrostatic potential at the same point $\mathbf{r}_1$, so Equation 12.26 would read $\delta w_i(\mathbf{r}_1) = q_i\delta\Psi(\mathbf{r}_2)$. In other words, we have $\rho_i^\star(\mathbf{r}_2|\mathbf{r}_1) = q_i\delta^{(3)}(r_{12})$ in the PB approximation. Thus electrostatics is *strictly local* in this approximation, which is what the delta function expresses.[6] In reality the ion interacts with the potential $\delta\Psi$ via the charge density $\rho_i^\star(\mathbf{r}_2|\mathbf{r}_1)$ that is distributed around the ion. We will now return to the exact case.

[6]When considering the charge distribution around an ion in the PB approximation, the latter (the central ion) *does* have a dressed charge density, but the surrounding ions are assumed to interact with it via local electrostatics. This means that the potential of mean force acting on each of these ions is, as an approximation, given by its bare ionic charge times the mean electrostatic potential from the central ion, see Section 12.3.2.

In order to show Equation 12.26 and define $\rho_i^\star$ for inhomogeneous electrolytes, we use $\delta\Psi = \delta\Psi^{\mathrm{ext}} + \delta\Psi^{\mathrm{pol}}$ and write Equation 12.14 as

$$\delta w_i(\mathbf{r}_1) = \int d\mathbf{r}_2\, \rho_i^{\mathrm{tot}}(\mathbf{r}_2|\mathbf{r}_1)\, \delta\Psi(\mathbf{r}_2) - \int d\mathbf{r}_2\, \rho_i^{\mathrm{tot}}(\mathbf{r}_2|\mathbf{r}_1)\, \delta\Psi^{\mathrm{pol}}(\mathbf{r}_2). \quad (12.27)$$

Since

$$\begin{aligned} \delta\Psi^{\mathrm{pol}}(\mathbf{r}_2) &= \int d\mathbf{r}_3\, \delta\rho^{\mathrm{pol}}(\mathbf{r}_3)\phi_{\mathrm{Coul}}(r_{32}) \\ &= \int d\mathbf{r}_3\, d\mathbf{r}_2'\, \chi^\star(\mathbf{r}_3, \mathbf{r}_2')\delta\Psi(\mathbf{r}_2')\phi_{\mathrm{Coul}}(r_{32}), \end{aligned} \quad (12.28)$$

where we have used Equation 12.23, the last term in Equation 12.27 is

$$\begin{aligned} &\int d\mathbf{r}_2\, \rho_i^{\mathrm{tot}}(\mathbf{r}_2|\mathbf{r}_1)\, \delta\Psi^{\mathrm{pol}}(\mathbf{r}_2) \\ &= \int d\mathbf{r}_2' \left[\int d\mathbf{r}_2\, d\mathbf{r}_3\, \rho_i^{\mathrm{tot}}(\mathbf{r}_2|\mathbf{r}_1)\, \phi_{\mathrm{Coul}}(r_{32})\chi^\star(\mathbf{r}_3, \mathbf{r}_2')\right] \delta\Psi(\mathbf{r}_2') \\ &= \int d\mathbf{r}_2' \left[\int d\mathbf{r}_3\, \psi_i(\mathbf{r}_3|\mathbf{r}_1)\chi^\star(\mathbf{r}_3, \mathbf{r}_2')\right] \delta\Psi(\mathbf{r}_2'), \end{aligned}$$

where we have used Equation 12.10 to obtain the last line. If we define

$$\rho_i^\star(\mathbf{r}_2|\mathbf{r}_1) \equiv \rho_i^{\mathrm{tot}}(\mathbf{r}_2|\mathbf{r}_1) - \int d\mathbf{r}_3\, \psi_i(\mathbf{r}_3|\mathbf{r}_1)\chi^\star(\mathbf{r}_3, \mathbf{r}_2) \quad (12.29)$$

it follows that Equation 12.27 can be written as Equation 12.26, which thereby has been shown. The latter equation can be written as

$$\rho_i^\star(\mathbf{r}_2|\mathbf{r}_1) = \frac{\delta w_i(\mathbf{r}_1)}{\delta\Psi(\mathbf{r}_2)}, \quad (12.30)$$

which can be used as an alternative definition of $\rho_i^\star$. From Equation 12.29 we see that $\rho_i^\star$ is different from the charge density of the ion and its surrounding ion cloud, which is ρ_i^{tot}. Recall that $\rho_i^{\mathrm{tot}}(\mathbf{r}_2|\mathbf{r}_1)$ is the change in charge density when an i-ion is inserted at $\mathbf{r}_1$. It consists of the central ionic charge and the induced charge density in the surroundings. As we saw in Chapter 6 (in Volume I) and will confirm below, the last term in Equation 12.29 removes the linear part of the polarization response from the electrostatic field due to the i-ion. This linear part, the integral, is denoted ρ_i^{lin}.

In order to investigate the nature of $\rho_i^\star$, we first note that the electrostatic potential $\psi_i(\mathbf{r}_3|\mathbf{r}_1)$ due to the i-ion is the change in total electrostatic potential at $\mathbf{r}_3$ when the ion is inserted at $\mathbf{r}_1$. If $\psi_i(\mathbf{r}_3|\mathbf{r}_1)$ were weak for all $\mathbf{r}_3$, the polarization response would be linear and the electrostatically induced polarization charge density due to ψ_i would according to $\delta\rho^{\mathrm{pol}}(\mathbf{r}_2) = \int d\mathbf{r}_3\, \chi^\star(\mathbf{r}_2, \mathbf{r}_3)\delta\Psi(\mathbf{r}_3)$ (Equation 12.23) be equal to $\int d\mathbf{r}_3\, \chi^\star(\mathbf{r}_2, \mathbf{r}_3)\psi_i(\mathbf{r}_3|\mathbf{r}_1)$. However, since ψ_i is large close to the ion, there are also nonlinear contributions to the response and the integral gives only the linear part of this polarization response. By writing Equation 12.29 as

$$\rho_i^\star(\mathbf{r}_2|\mathbf{r}_1) = q_i\delta^{(3)}(\mathbf{r}_{12}) + \rho_i^{\mathrm{cloud}}(\mathbf{r}_2|\mathbf{r}_1) - \int d\mathbf{r}_3\, \chi^\star(\mathbf{r}_2,\mathbf{r}_3)\psi_i(\mathbf{r}_3|\mathbf{r}_1) \quad (12.31)$$

we see that $\rho_i^\star$ is equal to the sum of the charge densities of the central ionic charge (the $\delta^{(3)}$ term) and the surrounding ion cloud (ρ_i^{cloud}) apart from the *linear part* of the electrostatically induced polarization charge density (the integral, ρ_i^{lin}). Recall that the charge density $\rho_i^{\mathrm{cloud}} = \Delta\rho_i$ (defined in Equations 12.7 and 12.8) is induced by all kinds of interactions between the i-ion and the surrounding ions; some is induced by ψ_i and some by other interactions. Hence, the dressed charge density $\rho_i^\star$ consists of the central ionic charge and the contributions from *nonlinear* electrostatic response due to ψ_i together with nonelectrostatic contributions to ρ_i^{cloud}.[7] By comparing Equations 12.14 and 12.26 we see that *the dressed charge density $\rho_i^\star$ has the same role vis-à-vis the total potential $\delta\Psi$ as the total charge density ρ_i^{tot} has vis-à-vis the external potential $\delta\Psi^{\mathrm{ext}}$*.

Furthermore, since $\delta n_i(\mathbf{r}_1) = -\beta n_i(\mathbf{r}_1)\delta w_i(\mathbf{r}_1)$ we have

$$\begin{aligned} \delta\rho^{\mathrm{pol}}(\mathbf{r}_1) &= \sum_i q_i\delta n_i(\mathbf{r}_1) = -\beta\sum_i q_i n_i(\mathbf{r}_1)\delta w_i(\mathbf{r}_1) \\ &= -\beta\int d\mathbf{r}_2 \sum_i q_i n_i(\mathbf{r}_1)\rho_i^\star(\mathbf{r}_2|\mathbf{r}_1)\,\delta\Psi(\mathbf{r}_2). \end{aligned}$$

By comparing with Equation 12.23 we see that

$$\chi^\star(\mathbf{r}_1,\mathbf{r}_2) = -\beta\sum_i q_i n_i(\mathbf{r}_1)\rho_i^\star(\mathbf{r}_2|\mathbf{r}_1) \quad (12.32)$$

and it follows that $\rho_i^\star$ *has the same role vis-à-vis $\chi^\star$ as ρ_i^{tot} has vis-à-vis χ^ρ* in Equation 12.20. A summary of some important basic equations is given in Table 12.1.

12.2.2 The pair correlation functions $h_{ij}^{(2)}$, $h_{ij}^\star$ and $c_{ij}^\star$

In Equation 12.9 we saw how the charge density of an ion and its surrounding ion cloud, $\rho_i^{\mathrm{tot}}(\mathbf{r}_2|\mathbf{r}_1)$, is expressed in terms of the pair correlation functions $h_{ij}^{(2)}(\mathbf{r}_1,\mathbf{r}_2)$. Likewise, the dressed charge density $\rho_i^\star$ can be expressed in the same manner in terms of pair correlation functions $h_{ij}^\star(\mathbf{r}_1,\mathbf{r}_2)$, which are defined as

$$h_{ij}^\star(\mathbf{r}_1,\mathbf{r}_2) \equiv h_{ij}^{(2)}(\mathbf{r}_1,\mathbf{r}_2) + \beta\int d\mathbf{r}_3\, \psi_i(\mathbf{r}_3|\mathbf{r}_1)\rho_j^\star(\mathbf{r}_3|\mathbf{r}_2). \quad (12.33)$$

[7] Note that nonelectrostatic interactions that give rise to charge separations contribute to ψ_i, so weak effects from such interactions can be part of the linear electrostatic polarization response due to ψ_i and do not contribute to $\rho_i^\star$.

Table 12.1 Some important basic relationships in electrolyte theory. The equation numbers refer to the original ones in the text.

$\delta\rho^{\mathrm{pol}}(\mathbf{r}_1) = \int d\mathbf{r}_2\, \chi^{\rho}(\mathbf{r}_1,\mathbf{r}_2)\delta\Psi^{\mathrm{ext}}(\mathbf{r}_2)$	(12.19')
$= \int d\mathbf{r}_2\, \chi^{\star}(\mathbf{r}_1,\mathbf{r}_2)\delta\Psi(\mathbf{r}_2)$	(12.23')
$\delta w_i(\mathbf{r}_1) = \int d\mathbf{r}_2\, \rho_i^{\mathrm{tot}}(\mathbf{r}_2\vert\mathbf{r}_1)\delta\Psi^{\mathrm{ext}}(\mathbf{r}_2)$	(12.14')
$\delta w_i(\mathbf{r}_1) = \int d\mathbf{r}_2\, \rho_i^{\star}(\mathbf{r}_2\vert\mathbf{r}_1)\, \delta\Psi(\mathbf{r}_2)$	(12.26')
$\chi^{\rho}(\mathbf{r}_1,\mathbf{r}_2) = -\beta \sum_i q_i n_i(\mathbf{r}_1)\rho_i^{\mathrm{tot}}(\mathbf{r}_2\vert\mathbf{r}_1)$	
$= -\beta q_{\mathrm{e}}^2 H_{\mathrm{QQ}}(\mathbf{r}_1,\mathbf{r}_2)$	(12.20')
$\chi^{\star}(\mathbf{r}_1,\mathbf{r}_2) = -\beta \sum_i q_i n_i(\mathbf{r}_1)\rho_i^{\star}(\mathbf{r}_2\vert\mathbf{r}_1)$	(12.32')
$= -\beta q_{\mathrm{e}}^2 H_{\mathrm{QQ}}^{\star}(\mathbf{r}_1,\mathbf{r}_2)$	(12.42')
$\rho_i^{\star}(\mathbf{r}_2\vert\mathbf{r}_1) = \rho_i^{\mathrm{tot}}(\mathbf{r}_2\vert\mathbf{r}_1) - \int d\mathbf{r}_3\, \psi_i(\mathbf{r}_3\vert\mathbf{r}_1)\chi^{\star}(\mathbf{r}_3,\mathbf{r}_2)$	(12.29')
$\rho_i^{\mathrm{tot}}(\mathbf{r}_2\vert\mathbf{r}_1) = q_i\delta^{(3)}(\mathbf{r}_{12}) + \sum_j q_j n_j(\mathbf{r}_2)h_{ij}^{(2)}(\mathbf{r}_1,\mathbf{r}_2)$	(12.9')
$\rho_i^{\star}(\mathbf{r}_2\vert\mathbf{r}_1) = q_i\delta^{(3)}(\mathbf{r}_{12}) + \sum_j q_j n_j(\mathbf{r}_2)h_{ij}^{\star}(\mathbf{r}_1,\mathbf{r}_2)$	(12.34')

This follows because

$$\sum_j q_j n_j(\mathbf{r}_2)h_{ij}^{\star}(\mathbf{r}_1,\mathbf{r}_2) = \sum_j q_j n_j(\mathbf{r}_2)h_{ij}^{(2)}(\mathbf{r}_1,\mathbf{r}_2) + \\ + \beta\sum_j q_j n_j(\mathbf{r}_2)\int d\mathbf{r}_3\, \psi_i(\mathbf{r}_3\vert\mathbf{r}_1)\rho_j^{\star}(\mathbf{r}_3\vert\mathbf{r}_2),$$

where the last term is equal to $-\int d\mathbf{r}_3\, \psi_i(\mathbf{r}_3|\mathbf{r}_1)\chi^\star(\mathbf{r}_2, \mathbf{r}_3)$ so we obtain

$$\sum_j q_j n_j(\mathbf{r}_2) h^\star_{ij}(\mathbf{r}_1, \mathbf{r}_2) = \rho_i^{\rm tot}(\mathbf{r}_2|\mathbf{r}_1) - q_i\delta^{(3)}(\mathbf{r}_{12}) - \int d\mathbf{r}_3\, \psi_i(\mathbf{r}_3|\mathbf{r}_1)\chi^\star(\mathbf{r}_2, \mathbf{r}_3).$$

Using Equation 12.29 it hence follows that the charge density of the dressed ion is

$$\rho_i^\star(\mathbf{r}_2|\mathbf{r}_1) = q_i\delta^{(3)}(\mathbf{r}_{12}) + \sum_j q_j n_j(\mathbf{r}_2) h^\star_{ij}(\mathbf{r}_1, \mathbf{r}_2), \tag{12.34}$$

so $h^\star_{ij}$ *has the same role vis-à-vis* $\rho_i^\star$ *as* $h_{ij}^{(2)}$ *has vis-à-vis* $\rho_i^{\rm tot}$. We can write

$$\rho_i^\star(\mathbf{r}_2|\mathbf{r}_1) = q_i\delta^{(3)}(\mathbf{r}_{12}) + \rho_i^{\rm dress}(\mathbf{r}_2|\mathbf{r}_1), \tag{12.35}$$

where $\rho_i^{\rm dress}$ is the charge density of **the dress of the ion**

$$\rho_i^{\rm dress}(\mathbf{r}_2|\mathbf{r}_1) = \sum_j q_j n_j(\mathbf{r}_2) h^\star_{ij}(\mathbf{r}_1, \mathbf{r}_2), \tag{12.36}$$

and the pair correlation functions $h^\star_{ij}$ describe the ionic distributions in this dress (cf. Equation 12.9 for $\rho_i^{\rm tot}$ expressed in terms of $\rho_i^{\rm cloud}$ and $h_{ij}^{(2)}$). As shown in Exercise 12.1 $h^\star_{ij}$ is symmetric, $h^\star_{ij}(\mathbf{r}_1, \mathbf{r}_2) = h^\star_{ji}(\mathbf{r}_2, \mathbf{r}_1)$. It follows from Equation 12.31 that $\rho_i^{\rm dress}$ is the charge density of the ion cloud *apart from* the linear part of the polarization charge density around the i-ion.

{**Exercise 12.1:** Use the definition 12.10 of ψ_i and 12.29 of $\rho_j^\star$ to show that $h^\star_{ij}$ defined in Equation 12.33 is symmetric, $h^\star_{ij}(\mathbf{r}_1, \mathbf{r}_2) = h^\star_{ji}(\mathbf{r}_2, \mathbf{r}_1)$. Furthermore, show that $\chi^\star$ defined by Equation 12.32 satisfies Equation 12.24.}

The integral in the definition of $h^\star_{ij}$, Equation 12.33, can be given the following interpretation. As noticed earlier, the electrostatic potential $\psi_i(\mathbf{r}_3|\mathbf{r}_1)$ that appears in the integrand is the change in total electrostatic potential at $\mathbf{r}_3$ when an i-ion is inserted at $\mathbf{r}_1$. We know from Equation 12.26 that a small variation in total electrostatic potential $\delta\Psi$ at one location affects the potential of mean force for ions at other locations; for a j-ion at $\mathbf{r}_2$ the change is $\delta w_j(\mathbf{r}_2) = \int d\mathbf{r}_3\, \rho_j^\star(\mathbf{r}_3|\mathbf{r}_2)\, \delta\Psi(\mathbf{r}_3)$ due to a variation $\delta\Psi(\mathbf{r}_3)$ at $\mathbf{r}_3$. The integral gives the linear response $\delta w_j(\mathbf{r}_2)$ due to $\delta\Psi(\mathbf{r}_3)$. If $\delta\Psi(\mathbf{r}_3)$ is not small there are also nonlinear contributions to the change in $w(\mathbf{r}_2)$ and integral gives only the *linear part* of the entire response. We can apply this to the change $\psi_i(\mathbf{r}_3|\mathbf{r}_1)$ in total potential at $\mathbf{r}_3$ when an i-ion is inserted at $\mathbf{r}_1$ and in this case $\int d\mathbf{r}_3\, \rho_j^\star(\mathbf{r}_3|\mathbf{r}_2)\, \psi_i(\mathbf{r}_3|\mathbf{r}_1)$ is the linear part of the change in w_j for a j-ion at $\mathbf{r}_2$. More precisely, the latter is the linear contribution to the potential of mean force $w_{ij}^{(2)}(\mathbf{r}_1, \mathbf{r}_2)$ for the j-ion due to the electrostatic interactions with the i-ion at $\mathbf{r}_1$. This part will be denoted as $w_{ij}^{(2)\rm el}$, that is,

$$w_{ij}^{(2)\rm el}(\mathbf{r}_1, \mathbf{r}_2) \equiv \int d\mathbf{r}_3\, \psi_i(\mathbf{r}_3|\mathbf{r}_1)\rho_j^\star(\mathbf{r}_3|\mathbf{r}_2), \tag{12.37}$$

so $w_{ij}^{(2)\text{el}}$ is the *linear electrostatic part* of $w_{ij}^{(2)}$ due to the presence of one of the ions in the surroundings of the other. Thus, Equation 12.33 can be written

$$h_{ij}^{\star}(\mathbf{r}_1,\mathbf{r}_2) \equiv h_{ij}^{(2)}(\mathbf{r}_1,\mathbf{r}_2) + \beta w_{ij}^{(2)\text{el}}(\mathbf{r}_1,\mathbf{r}_2), \tag{12.38}$$

which emphasizes that $h_{ij}^{\star}$ and $h_{ij}^{(2)}$ differ by β time this linear part of $w_{ij}^{(2)}$. It follows that $h_{ij}^{\star}$ contains the nonlinear part of the response due to the electrostatic interactions of each ion with the other one. Accordingly, this is also contained in the dress of an ion and hence in $\rho_i^{\star}$ and $\rho_j^{\star}$ as we concluded in the analysis of Equation 12.31. The nonlinear parts will be examined in more detail later.

We have seen that the functions marked by a star like $h_{ij}^{\star}$ and $\rho_i^{\star}$ has similar roles as the functions $h_{ij}^{(2)}$ and ρ_i^{tot}. We will now see that there is a whole group of "starred" pair correlation functions $h_{ij}^{\star}$, $c_{ij}^{\star}$, $H_{ij}^{\star}$ and $C_{ij}^{\star}$ that satisfy relationships that are similar to those of $h_{ij}^{(2)}$, $c_{ij}^{(2)}$, $H_{ij}^{(2)}$ and $C_{ij}^{(2)}$. If we in analogy to the definition of $H_{ij}^{(2)}(\mathbf{r}_1,\mathbf{r}_2)$ introduce

$$H_{ij}^{\star}(\mathbf{r}_1,\mathbf{r}_2) \equiv n_i(\mathbf{r}_1)\delta_{ij}^{(3)}(\mathbf{r}_{12}) + n_i(\mathbf{r}_1)n_j(\mathbf{r}_2)h_{ij}^{\star}(\mathbf{r}_1,\mathbf{r}_2), \tag{12.39}$$

we can write Equation 12.32 as

$$\chi^{\star}(\mathbf{r}_1,\mathbf{r}_2) = -\beta \sum_i q_i q_j H_{ij}^{\star}(\mathbf{r}_1,\mathbf{r}_2), \tag{12.40}$$

which has the same form as $\chi^{\rho}(\mathbf{r}_1,\mathbf{r}_2)$ in Equation 12.20. By defining

$$H_{\text{QQ}}^{\star}(\mathbf{r}_1,\mathbf{r}_2) \equiv \frac{1}{q_{\text{e}}^2} \sum_{ij} q_i q_j H_{ij}^{\star}(\mathbf{r}_1,\mathbf{r}_2), \tag{12.41}$$

(cf. Equation 12.21) we can write

$$\chi^{\star}(\mathbf{r}_1,\mathbf{r}_2) = -\beta q_{\text{e}}^2 H_{\text{QQ}}^{\star}(\mathbf{r}_1,\mathbf{r}_2) \tag{12.42}$$

in analogy to $\chi^{\rho}(\mathbf{r}_1,\mathbf{r}_2) = -\beta q_{\text{e}}^2 H_{\text{QQ}}(\mathbf{r}_1,\mathbf{r}_2)$ (Equation 12.20).

Next, we introduce the function

$$\begin{aligned} c_{ij}^{\star}(\mathbf{r}_1,\mathbf{r}_2) &\equiv c_{ij}^{(2)}(\mathbf{r}_1,\mathbf{r}_2) + \beta u_{ij}^{\text{el}}(r_{12}) \\ &= c_{ij}^{(2)}(\mathbf{r}_1,\mathbf{r}_2) + \beta q_i q_j \phi_{\text{Coul}}(r_{12}), \end{aligned} \tag{12.43}$$

and define

$$C_{ij}^{\star}(\mathbf{r}_1,\mathbf{r}_2) \equiv \frac{1}{n_i(\mathbf{r}_2)}\delta_{ij}^{(3)}(\mathbf{r}_{12}) - c_{ij}^{\star}(\mathbf{r}_1,\mathbf{r}_2), \tag{12.44}$$

so $C_{ij}^{\star}$ is defined from $c_{ij}^{\star}$ in the same manner as $C_{ij}^{(2)}$ is defined from $c_{ij}^{(2)}$ (cf. Equation 11.141). Since $c_{ij}^{(2)}(\mathbf{r}_1,\mathbf{r}_2) \sim -\beta u_{ij}(r_{12}) \sim -\beta u_{ij}^{\text{el}}(r_{12})$ when

$r_{12} \to \infty$ it follows that $c^{\star}_{ij}(\mathbf{r}_1, \mathbf{r}_2)$ decays faster to zero than $c^{(2)}_{ij}(\mathbf{r}_1, \mathbf{r}_2)$ when $r_{12} \to \infty$, so we can designate $c^{\star}_{ij}$ as *the short-range part of the pair direct-correlation function* in the electrolyte.

As shown in the shaded text box below, $C^{\star}_{ij}$ and $H^{\star}_{ij}$ are functional inverses of each other, so we have

$$\sum_l \int d\mathbf{r}_3\, C^{\star}_{il}(\mathbf{r}_1, \mathbf{r}_3) H^{\star}_{lj}(\mathbf{r}_3, \mathbf{r}_2) = \delta^{(3)}_{ij}(\mathbf{r}_{12}) \tag{12.45}$$

in analogy to Equation 11.142 for $C^{(2)}_{ij}$ and $H^{(2)}_{ij}$. Note that this implies that $h^{\star}_{ij}$ and $c^{\star}_{ij}$ satisfy the Ornstein-Zernike equation

$$h^{\star}_{ij}(\mathbf{r}_1, \mathbf{r}_2) = c^{\star}_{ij}(\mathbf{r}_1, \mathbf{r}_2) + \sum_l \int d\mathbf{r}_3\, c^{\star}_{il}(\mathbf{r}_1, \mathbf{r}_3) n_l(\mathbf{r}_3) h^{\star}_{lj}(\mathbf{r}_3, \mathbf{r}_2). \tag{12.46}$$

Before we show that $H^{\star}_{ij}$ is the functional inverse of $C^{\star}_{ij}$ we will justify the definition 12.43 of $c^{\star}_{ij}$, which implies that $C^{\star}_{ij} = C^{(2)}_{ij} - \beta u^{\text{el}}_{ij}$. The second Yvon equation, which is the inverse of the first Yvon equation 12.16, is

$$\begin{aligned} \delta v_i(\mathbf{r}_1) &= -\frac{1}{\beta} \int d\mathbf{r}_2 \sum_j C^{(2)}_{ij}(\mathbf{r}_1, \mathbf{r}_2) \delta n_j(\mathbf{r}_2) \qquad (12.47) \\ &= -\frac{1}{\beta} \int d\mathbf{r}_2 \left[\frac{1}{n_i(\mathbf{r}_2)} \delta^{(3)}(\mathbf{r}_{12}) \delta n_i(\mathbf{r}_2) - \sum_j c^{(2)}_{ij}(\mathbf{r}_1, \mathbf{r}_2) \delta n_j(\mathbf{r}_2) \right] \end{aligned}$$

(cf. Equation 11.145). In the present case we apply it to the change $\delta\Psi^{\text{ext}}$ in external electrostatic potential, so we have $\delta v_i(\mathbf{r}_1) = q_i \delta\Psi^{\text{ext}}(\mathbf{r}_1)$.

Let us introduce $v^{\star}_i(\mathbf{r}_1) \equiv q_i \Psi(\mathbf{r}_1)$, i.e., the interaction energy of q_i with the total electrostatic potential. We have $\delta v^{\star}_i(\mathbf{r}_1) = q_i \delta\Psi(\mathbf{r}_1) = q_i \delta\Psi^{\text{ext}}(\mathbf{r}_1) + q_i \delta\Psi^{\text{pol}}(\mathbf{r}_1)$ and since

$$q_i \delta\Psi^{\text{pol}}(\mathbf{r}_1) = q_i \int d\mathbf{r}_2 \sum_j q_j \delta n_j(\mathbf{r}_2) \phi_{\text{Coul}}(r_{12}) = \int d\mathbf{r}_2 \sum_j \delta n_j(\mathbf{r}_2) u^{\text{el}}_{ij}(r_{12})$$

we get

$$\delta v^{\star}_i(\mathbf{r}_1) = \delta v_i(\mathbf{r}_1) + \int d\mathbf{r}_2 \sum_j \delta n_j(\mathbf{r}_2) u^{\text{el}}_{ij}(r_{12}).$$

By inserting the second Yvon equation into this relationship we obtain

$$\begin{aligned} \delta v^{\star}_i(\mathbf{r}_1) &= -\frac{1}{\beta} \int d\mathbf{r}_2 \left[\frac{1}{n_i(\mathbf{r}_2)} \delta^{(3)}(\mathbf{r}_{12}) \delta n_i(\mathbf{r}_2) \right. \\ &\quad \left. - \sum_j \left\{ c^{(2)}_{ij}(\mathbf{r}_1, \mathbf{r}_2) + \beta u^{\text{el}}_{ij}(r_{12}) \right\} \delta n_j(\mathbf{r}_2) \right] \end{aligned}$$

and therefore we can conclude that

$$\delta v_i^\star(\mathbf{r}_1) = -\frac{1}{\beta}\int d\mathbf{r}_2 C_{ij}^\star(\mathbf{r}_1,\mathbf{r}_2)\delta n_j(\mathbf{r}_2). \tag{12.48}$$

Note that δn_j in this relationship is the change in density due to the variation $\delta\Psi^{\text{Ext}}$ in external electrostatic potential (δn_j is not an arbitrary variation here). Equation 12.48 for $\delta v_i^\star$ corresponds to Equation 12.47 for δv_i, whereby $C_{ij}^\star$ in the former has a similar role as $C_{ij}^{(2)}$ in the latter.

Let us now show that the functional inverse to $C_{ij}^\star$ is equal to $H_{ij}^\star$ defined in Equation 12.39 with $h_{ij}^\star$ given in Equation 12.33. The latter equation implies that

$$h_{ij}^\star(\mathbf{r}_1,\mathbf{r}_2) = h_{ij}^{(2)}(\mathbf{r}_1,\mathbf{r}_2)+\beta\int d\mathbf{r}_3\, d\mathbf{r}_4\, \rho_i^{\text{tot}}(\mathbf{r}_4|\mathbf{r}_1)\,\phi_{\text{Coul}}(r_{43})\rho_j^\star(\mathbf{r}_3|\mathbf{r}_2). \tag{12.49}$$

By multiplying by $n_i(\mathbf{r}_1)n_j(\mathbf{r}_2)$ and adding $n_i(\mathbf{r}_1)\delta_{ij}^{(3)}(\mathbf{r}_{12})$ to both sides we can after simplification write this relationship as

$$\begin{aligned} H_{ij}^\star(\mathbf{r}_1,\mathbf{r}_2) &= H_{ij}^{(2)}(\mathbf{r}_1,\mathbf{r}_2) \\ &+ \beta\int d\mathbf{r}_3\, d\mathbf{r}_4 \sum_l q_l H_{il}^{(2)}(\mathbf{r}_1,\mathbf{r}_4)\phi_{\text{Coul}}(r_{43})\sum_m q_m H_{jm}^\star(\mathbf{r}_3,\mathbf{r}_2), \end{aligned} \tag{12.50}$$

where we have written ρ_i^{tot} and $\rho_j^\star$ in terms of the H functions. Note that this is equivalent to Equation 12.33.

Let us temporarily denote the inverse of $C_{ij}^\star$ as $H_{ij}^\sharp$. Our task is to show that $H_{ij}^\sharp$ is equal to $H_{ij}^\star$ that satisfies Equation 12.50. Since $H_{ij}^\sharp$ is the inverse of $C_{ij}^\star$ we have

$$\sum_m \int d\mathbf{r}_3\, C_{im}^\star(\mathbf{r}_1,\mathbf{r}_3)H_{mj}^\sharp(\mathbf{r}_3,\mathbf{r}_2) = \delta_{ij}^{(3)}(\mathbf{r}_{12})$$

and when we insert $C_{im}^\star = C_{im}^{(2)} - q_i q_m \phi_{\text{Coul}}$ we obtain

$$\begin{aligned} &\sum_m \int d\mathbf{r}_3 C_{im}^{(2)}(\mathbf{r}_1,\mathbf{r}_3)H_{mj}^\sharp(\mathbf{r}_3,\mathbf{r}_2) \\ &-\beta\sum_m \int d\mathbf{r}_3 q_i q_m \phi_{\text{Coul}}(r_{13})H_{mj}^\sharp(\mathbf{r}_3,\mathbf{r}_2) = \delta_{ij}^{(3)}(\mathbf{r}_{12}). \end{aligned}$$

By multiplying both sides with $H_{li}^{(2)}(\mathbf{r}_4,\mathbf{r}_1)$, summing over i and integrating over $\mathbf{r}_1$ we obtain after simplification, using the fact that $H_{li}^{(2)}$ and $C_{im}^{(2)}$ inverses of each other,

$$\begin{aligned} &H_{lj}^\sharp(\mathbf{r}_4,\mathbf{r}_2) - \beta\sum_{i,m}\int d\mathbf{r}_1 d\mathbf{r}_3 q_i H_{li}^{(2)}(\mathbf{r}_4,\mathbf{r}_1)\phi_{\text{Coul}}(r_{13})q_m H_{mj}^\sharp(\mathbf{r}_3,\mathbf{r}_2) \\ &= H_{lj}^{(2)}(\mathbf{r}_4,\mathbf{r}_2). \end{aligned}$$

By comparing with Equation 12.50 and adjusting the indices we can conclude that $H^{\sharp}_{ij}$ is equal to $H^{\star}_{ij}$.

In fact, the relationships derived so far in the present chapter can be obtained by *defining* $H^{\star}_{ij}$ as the functional inverse of $C^{\star}_{ij}$, which is defined as $C^{\star}_{ij} = C^{(2)}_{ij} - \beta u^{\mathrm{el}}_{ij}$. The entities $h^{\star}_{ij}$, $\rho^{\star}_i$ and $\chi^{\star}$ can then be defined from $H^{\star}_{ij}$ and the equations involving them can then be derived. The path followed above is, however, more transparent physically.

Earlier we saw that $c^{\star}_{ij}(\mathbf{r}_1, \mathbf{r}_2)$ is the short-range part of the pair direct-correlation function in the electrolyte. Its decay when $r_{12} \to \infty$ can be obtained as follows. Since

$$g^{(2)}_{ij} = h^{(2)}_{ij} + 1 = e^{-\beta u_{ij} + h^{(2)}_{ij} - c^{(2)}_{ij} + b^{(2)}_{ij}} = e^{-\beta u^{\mathrm{sh}}_{ij} + h^{(2)}_{ij} - c^{\star}_{ij} + b^{(2)}_{ij}} \tag{12.51}$$

we have

$$\begin{aligned} c^{\star}_{ij}(\mathbf{r}_1, \mathbf{r}_2) &= -\beta u^{\mathrm{sh}}_{ij}(r_{12}) - \ln\{h^{(2)}_{ij}(\mathbf{r}_1, \mathbf{r}_2) + 1\} + h^{(2)}_{ij}(\mathbf{r}_1, \mathbf{r}_2) + b^{(2)}_{ij}(\mathbf{r}_1, \mathbf{r}_2) \\ &\sim -\beta u^{\mathrm{sh}}_{ij}(r_{12}) + \tfrac{1}{2}[h^{(2)}_{ij}(\mathbf{r}_1, \mathbf{r}_2)]^2 + b^{(2)}_{ij}(\mathbf{r}_1, \mathbf{r}_2) \end{aligned} \tag{12.52}$$

when $r_{12} \to \infty$, so provided that u^{sh}_{ij} is a sufficiently short-range function (for example a hard-core potential) $c^{\star}_{ij}$ decays as $\frac{1}{2}[h^{(2)}_{ij}]^2 + b^{(2)}_{ij}$. The functions $b^{(2)}_{ij}$, $[h^{(2)}_{ij}]^2$ and the combination $\frac{1}{2}[h^{(2)}_{ij}]^2 + b^{(2)}_{ij}$ usually decay about equally fast to zero as functions of r_{12} when $r_{12} \to \infty$.[8]

12.2.3 Electric susceptibility and dielectric permittivity functions; the screened Coulomb potential

Let us further investigate the electrostatic potentials that arise from the polarization of an inhomogeneous electrolyte by the variation $\delta\Psi^{\mathrm{ext}}$ in external potential. Equation 12.28 says that

$$\begin{aligned} \delta\Psi^{\mathrm{pol}}(\mathbf{r}_2) &= \int d\mathbf{r}_3\, d\mathbf{r}_1\, \chi^{\star}(\mathbf{r}_3, \mathbf{r}_1)\delta\Psi(\mathbf{r}_1)\phi_{\mathrm{Coul}}(r_{32}) \\ &= -\int d\mathbf{r}_2\, \delta\Psi(\mathbf{r}_1)\chi^{\mathrm{el}}(\mathbf{r}_1; \mathbf{r}_2), \end{aligned} \tag{12.53}$$

where we have defined

$$\chi^{\mathrm{el}}(\mathbf{r}_1; \mathbf{r}_2) = -\int d\mathbf{r}_3\, \chi^{\star}(\mathbf{r}_1, \mathbf{r}_3)\phi_{\mathrm{Coul}}(r_{32}) \tag{12.54}$$

[8]It is not possible that $\frac{1}{2}[h^{(2)}]^2 + b^{(2)}$ decays slower than $b^{(2)}$ or $[h^{(2)}]^2$, but it can decay faster because of cancellations.

(recall that $\chi^{\star}$ is a symmetric function). The function χ^{el} will be called the **electric susceptibility function** and is also given by

$$\chi^{\mathrm{el}}(\mathbf{r}_1;\mathbf{r}_2) = -\frac{\delta\Psi^{\mathrm{pol}}(\mathbf{r}_2)}{\delta\Psi(\mathbf{r}_1)}, \tag{12.55}$$

as follows from Equation 12.53. The negative sign in the definition of χ^{el} is introduced in order to make it agree with the concept of electric susceptibility that is used in macroscopic electrostatics, as will be seen later.

Now, $\delta\Psi = \delta\Psi^{\mathrm{ext}} + \delta\Psi^{\mathrm{pol}}$ can be used to express $\delta\Psi$ and $\delta\Psi^{\mathrm{Ext}}$ in terms of each other. We obtain using Equation 12.53

$$\begin{aligned}\delta\Psi^{\mathrm{ext}}(\mathbf{r}_2) &= \delta\Psi(\mathbf{r}_2) + \int d\mathbf{r}_1\, \delta\Psi(\mathbf{r}_1)\chi^{\mathrm{el}}(\mathbf{r}_1;\mathbf{r}_2)\\ &= \int d\mathbf{r}_1\, \delta\Psi(\mathbf{r}_1)\left[\delta^{(3)}(\mathbf{r}_{12}) + \chi^{\mathrm{el}}(\mathbf{r}_1;\mathbf{r}_2)\right].\end{aligned}$$

If we define

$$\begin{aligned}\epsilon(\mathbf{r}_1;\mathbf{r}_2) &\equiv \delta^{(3)}(\mathbf{r}_{12}) + \chi^{\mathrm{el}}(\mathbf{r}_1;\mathbf{r}_2)\\ &= \delta^{(3)}(\mathbf{r}_{12}) - \int d\mathbf{r}_3\, \chi^{\star}(\mathbf{r}_1,\mathbf{r}_3)\phi_{\mathrm{Coul}}(r_{32})\end{aligned} \tag{12.56}$$

we can write this as

$$\delta\Psi^{\mathrm{ext}}(\mathbf{r}_2) = \int d\mathbf{r}_1\, \delta\Psi(\mathbf{r}_1)\epsilon(\mathbf{r}_1;\mathbf{r}_2) \tag{12.57}$$

or

$$\frac{\delta\Psi^{\mathrm{Ext}}(\mathbf{r}_2)}{\delta\Psi(\mathbf{r}_1)} = \epsilon(\mathbf{r}_1;\mathbf{r}_2). \tag{12.58}$$

The function $\epsilon(\mathbf{r}_1;\mathbf{r}_2)$ will be called the **dielectric permittivity function**. Note that $\chi^{\mathrm{el}}(\mathbf{r}_1;\mathbf{r}_2)$ and $\epsilon(\mathbf{r}_1;\mathbf{r}_2)$ are *not* symmetric functions of $\mathbf{r}_1$ and $\mathbf{r}_2$. The convention used here is that the variable before the semicolon (here $\mathbf{r}_1$) appears in the denominator of the derivative expression in the lhs of Equations 12.55 and 12.58, while the other variable (here $\mathbf{r}_2$) appears in the numerator. For isotropic homogeneous electrolytes, the functions depends solely on the distance $|\mathbf{r}_2 - \mathbf{r}_2|$, that is, $\chi^{\mathrm{el}} = \chi^{\mathrm{el}}(r_{12})$ and $\epsilon = \epsilon(r_{12})$.

It is important to note that *the nomenclature for the functions* ϵ *and* χ^{el} *is not universal* because in the literature there are other functions that have been given the names "electric susceptibility function" and "dielectric permittivity function." Furthermore, later in this chapter we will adopt the common terminology "the dielectric function" for $\tilde{\epsilon}(k)$, i.e. the Fourier transform of the function $\epsilon(r)$, which should not be confused with the dielectric permittivity function in ordinary space. In order to make a clearer distinction one may call $\epsilon(r)$ and $\epsilon(\mathbf{r}_1;\mathbf{r}_2)$ "spatial dielectric-permittivity functions," but this name will not be used here.

Another relationship between $\delta\Psi$ and $\delta\Psi^{\mathrm{Ext}}$ can be obtained by expressing $\delta\Psi^{\mathrm{pol}}$ in terms of $\delta\Psi^{\mathrm{Ext}}$ and χ^{ρ} (i.e., instead of expressing it in terms of $\delta\Psi$ and $\chi^{\star}$ or χ^{el} as in Equation 12.53). From Equations 12.12 and 12.19 we have

$$\delta\Psi^{\mathrm{pol}}(\mathbf{r}_3) = \int d\mathbf{r}_4\, d\mathbf{r}_2\, \chi^{\rho}(\mathbf{r}_4, \mathbf{r}_2)\delta\Psi^{\mathrm{ext}}(\mathbf{r}_2)\phi_{\mathrm{Coul}}(r_{43})$$

and from $\delta\Psi = \delta\Psi^{\mathrm{ext}} + \delta\Psi^{\mathrm{pol}}$ we therefore obtain

$$\delta\Psi(\mathbf{r}_3) = \int d\mathbf{r}_2 \left[\delta^{(3)}(\mathbf{r}_{23}) + \int d\mathbf{r}_4\, \chi^{\rho}(\mathbf{r}_4, \mathbf{r}_2)\phi_{\mathrm{Coul}}(r_{43})\right] \delta\Psi^{\mathrm{ext}}(\mathbf{r}_2), \quad (12.59)$$

which can be used as an alternative relationship between $\delta\Psi$ and $\delta\Psi^{\mathrm{ext}}$ that is equivalent to Equation 12.57. It can be written as

$$\frac{\delta\Psi(\mathbf{r}_3)}{\delta\Psi^{\mathrm{ext}}(\mathbf{r}_2)} = \delta^{(3)}(\mathbf{r}_{23}) + \int d\mathbf{r}_4\, \chi^{\rho}(\mathbf{r}_2, \mathbf{r}_4)\phi_{\mathrm{Coul}}(r_{43}) \equiv \epsilon^{\mathrm{R}}(\mathbf{r}_2; \mathbf{r}_3) \quad (12.60)$$

(recall that χ^{ρ} is a symmetric function). We will call $\epsilon^{\mathrm{R}}(\mathbf{r}_2; \mathbf{r}_3)$ the **reciprocal dielectric-permittivity function** and we have

$$\delta\Psi(\mathbf{r}_3) = \int d\mathbf{r}_2\, \delta\Psi^{\mathrm{ext}}(\mathbf{r}_2)\epsilon^{\mathrm{R}}(\mathbf{r}_2; \mathbf{r}_3). \quad (12.61)$$

From Equation 12.58 we see that $\epsilon^{\mathrm{R}}(\mathbf{r}_2; \mathbf{r}_3)$ is *the functional inverse of* $\epsilon(\mathbf{r}_1; \mathbf{r}_2)$. This fact is also a consequence of the relationship between χ^{ρ} and $\chi^{\star}$ given in Equation 12.24, as shown in the following Exercise. Hence, $\epsilon(\mathbf{r}_1; \mathbf{r}_2)$ can alternatively be obtained by inverting $\epsilon^{\mathrm{R}}(\mathbf{r}_2; \mathbf{r}_3)$ obtained from Equation 12.60.

{**Exercise 12.2:** Use Equation 12.24 to show that

$$\delta^{(3)}(\mathbf{r}_{23}) + \int d\mathbf{r}_4\, \chi^{\rho}(\mathbf{r}_2, \mathbf{r}_4)\phi_{\mathrm{Coul}}(r_{43}) \equiv \epsilon^{\mathrm{R}}(\mathbf{r}_2; \mathbf{r}_3)$$

is the functional inverse of

$$\delta^{(3)}(\mathbf{r}_{12}) - \int d\mathbf{r}_5\, \chi^{\star}(\mathbf{r}_1, \mathbf{r}_5)\phi_{\mathrm{Coul}}(r_{52}) \equiv \epsilon(\mathbf{r}_1; \mathbf{r}_2).$$

You can do this by calculating $\int d\mathbf{r}_2\, \epsilon^{\mathrm{R}}(\mathbf{r}_2; \mathbf{r}_3)\epsilon(\mathbf{r}_1; \mathbf{r}_2)$ from these two expressions and using Equation 12.24 to simplify the result in order to finally obtain $\delta^{(3)}(\mathbf{r}_{13})$.}

The functions ϵ^{R} and ϵ can be used to express ρ_i^{tot} in terms of $\rho_i^{\star}$ and vice versa, namely,

$$\rho_i^{\mathrm{tot}}(\mathbf{r}_2|\mathbf{r}_1) = \int d\mathbf{r}_4\, \epsilon^{\mathrm{R}}(\mathbf{r}_2; \mathbf{r}_4)\rho_i^{\star}(\mathbf{r}_4|\mathbf{r}_1) \quad (12.62)$$

$$\rho_j^{\star}(\mathbf{r}_3|\mathbf{r}_1) = \int d\mathbf{r}_2\, \epsilon(\mathbf{r}_3; \mathbf{r}_2)\rho_j^{\mathrm{tot}}(\mathbf{r}_2|\mathbf{r}_1). \quad (12.63)$$

Equation 12.62 can be derived from Equation 12.49 by multiplying both sides by $q_i n_i(\mathbf{r}_1)$, summing over i and adding $q_j \delta^{(3)}(\mathbf{r}_{12})$, whereby we obtain after simplification and rearrangement

$$\rho_j^{\text{tot}}(\mathbf{r}_1|\mathbf{r}_2) = \rho_j^\star(\mathbf{r}_1|\mathbf{r}_2) + \int d\mathbf{r}_3 \left[\int d\mathbf{r}_4\, \chi^\rho(\mathbf{r}_1,\mathbf{r}_4)\, \phi_{\text{Coul}}(r_{43})\right] \rho_j^\star(\mathbf{r}_3|\mathbf{r}_2).$$

We can write $\rho_j^\star(\mathbf{r}_1|\mathbf{r}_2) = \int d\mathbf{r}_3\, \delta^{(3)}(\mathbf{r}_{13})\rho_j^\star(\mathbf{r}_3|\mathbf{r}_2)$, so this term can be included in the large square bracket as a term $\delta^{(3)}(\mathbf{r}_{13})$. By swapping indices $1 \leftrightarrow 2$ and using the definition of ϵ^{R} in Equation 12.60 we obtain Equation 12.62. Equation 12.63 can be derived in a similar manner, but it is also a direct consequence of Equation 12.62; see the following Exercise.

{**Exercise 12.3:** Use the results in Exercise 12.2 to show that Equation 12.62 implies Equation 12.63.}

As we will see, the dielectric permittivity function $\epsilon(\mathbf{r}_1;\mathbf{r}_2)$ has a key role in the understanding of properties of electrolytes. It has an equally important role for polar fluids that will be treated in Chapter 14. For macroscopic homogeneous polar materials (with no mobile ions present) in the presence of a uniform electrostatic field, the dielectric permittivity and electric susceptibility functions reduce, as we will see later, to the usual dielectric constant ε_r and the macroscopic electric susceptibility (used in classical electrostatics), respectively, while the reciprocal dielectric permittivity function reduces to $1/\varepsilon_r$. Here, we are interested in the microscopic properties of fluids and the functions $\chi^{\text{el}}(\mathbf{r}_1;\mathbf{r}_2)$, $\epsilon(\mathbf{r}_1;\mathbf{r}_2)$ and $\epsilon^{\text{R}}(\mathbf{r}_1;\mathbf{r}_2)$ are the relevant entities. Likewise, the dressed-particle charge density $\rho_i^\star(\mathbf{r}_2|\mathbf{r}_1)$ and the starred correlation function $h_{ij}^\star(\mathbf{r}_1,\mathbf{r}_2)$ have important roles for the microscopic properties because they can be used to formulate the theory in a very utile manner. For instance, the mean electrostatic potential ψ_i can be written in terms of $\rho_i^\star$ instead of ρ_i^{tot} and the pair correlation function $h_{ij}^{(2)}$ can be expressed in a useful way in terms of $h_{ij}^\star$, $\rho_i^\star$ and $\rho_j^\star$ (some of the results that will be derived are shown in Table 12.2 on page 303).

Let us start with ψ_i and obtain a "screened version" of the Coulomb potential formula 12.5. The potential $\psi_i(\mathbf{r}_3|\mathbf{r}_1)$ is given be the first line in the following equation (which is the same as Equation 12.10) and by inserting ρ_i^{tot} from Equation 12.62 we obtain

$$\begin{aligned}
\psi_i(\mathbf{r}_3|\mathbf{r}_1) &= \int d\mathbf{r}_2\, \rho_i^{\text{tot}}(\mathbf{r}_2|\mathbf{r}_1)\phi_{\text{Coul}}(r_{23}) \\
&= \int d\mathbf{r}_2\, d\mathbf{r}_4\, \epsilon^{\text{R}}(\mathbf{r}_2;\mathbf{r}_4)\rho_i^\star(\mathbf{r}_4|\mathbf{r}_1)\phi_{\text{Coul}}(r_{23}) \\
&= \int d\mathbf{r}_4\, \rho_i^\star(\mathbf{r}_4|\mathbf{r}_1)\phi_{\text{Coul}}^\star(\mathbf{r}_4,\mathbf{r}_3),
\end{aligned} \qquad (12.64)$$

where we have defined the **unit screened Coulomb potential** $\phi^\star_{\rm Coul}$ for inhomogeneous fluids as

$$\begin{aligned}\phi^\star_{\rm Coul}(\mathbf{r}_4,\mathbf{r}_3) &\equiv \int d\mathbf{r}_2\, \epsilon^{\rm R}(\mathbf{r}_2;\mathbf{r}_4)\phi_{\rm Coul}(r_{23}) \qquad (12.65)\\ &= \phi_{\rm Coul}(r_{43}) + \int d\mathbf{r}_2\, d\mathbf{r}_1\, \chi^\rho(\mathbf{r}_1,\mathbf{r}_2)\phi_{\rm Coul}(r_{14})\phi_{\rm Coul}(r_{23})\end{aligned}$$

(the second line is obtained by inserting the definition of $\epsilon^{\rm R}$ and simplifying). We see that $\phi^\star_{\rm Coul}(\mathbf{r}_4,\mathbf{r}_3)$ is symmetric in $\mathbf{r}_4$ and $\mathbf{r}_3$. Note that $\rho^\star_i$ and $\phi^\star_{\rm Coul}$ have the same roles in the last line Equation 12.64 as $\rho^{\rm tot}_i$ and $\phi_{\rm Coul}$ have in the first line, which is simply the ordinary Coulomb potential formula applied to the charge distribution $\rho^{\rm tot}_i$. The last line constitute a *screened version of* this formula, which expresses the electrostatic potential in the same manner, but in terms of the screened Coulomb potential and the dressed particle charge distribution $\rho^\star_i$.

In Chapter 6 of Volume I we first encountered $\phi^\star_{\rm Coul}(r)$ for homogeneous bulk electrolytes in the Poisson-Boltzmann approximation (for ions in a dielectric continuum), where

$$\phi^\star_{\rm Coul}(r) = \phi_{\rm Coul}(r)e^{-\kappa_D r} = \frac{e^{-\kappa_D r}}{4\pi\varepsilon_r^{\rm cont}\varepsilon_0 r} \qquad \text{(PB, bulk electrolyte)}$$

(Equations I:6.68-69) and the decay parameter κ_D in the PB approximation is the **Debye parameter** defined from

$$\kappa_{\rm D}^2 = \frac{\beta}{\varepsilon_r^{\rm cont}\varepsilon_0}\sum_j n_j^{\rm b} q_j^2 \qquad \text{(diel. continuum)} \qquad (12.66)$$

($1/\kappa_{\rm D}$ is the **Debye length**). For ions distributed in vacuum (a classical plasma) we set $\varepsilon_r^{\rm cont} = 1$ in these formulas. We will continue to treat only the latter case unless we say otherwise. Later, in Section 12.3.2, it will be explained how the formulas for the primitive model with continuum solvent can obtained.

The general, exact theory for simple bulk electrolytes was presented in Section 6.2 of Volume I and we introduced $\phi^\star_{\rm Coul}(r)$ and the screened version of the Coulomb potential formula for this case (see Equations I:6.136-137; these equations are special cases of what follows). For inhomogeneous fluids $\phi^\star_{\rm Coul}(\mathbf{r},\mathbf{r}')$ depends on the absolute locations $\mathbf{r}$ and $\mathbf{r}'$ instead of the distance $|\mathbf{r}'-\mathbf{r}|$, because the screening of the electrostatic potential is different in different parts of the fluid. As we will see, $\phi^\star_{\rm Coul}$ has a central role in the exact theory.

We can interpret $\phi^\star_{\rm Coul}$ as follows. Let us expose the electrolyte to the electrostatic potential from a small point charge q that we place at $\mathbf{r}_3$, so the external potential changes by $\delta\Psi^{\rm ext}(\mathbf{r}_2)$ that is equal to $\delta\Psi^{\rm ext}_q(\mathbf{r}_2|\mathbf{r}_3) \equiv q\,\phi_{\rm Coul}(r_{23})$, where subscript q means that the potential originates from the charge q. The resulting total potential $\delta\Psi$ at $\mathbf{r}_4$ from the point charge is $\delta\Psi_q(\mathbf{r}_4|\mathbf{r}_3) \equiv \int d\mathbf{r}_2\, \epsilon^{\mathrm{R}}(\mathbf{r}_2;\mathbf{r}_4)\delta\Psi_q^{\mathrm{ext}}(\mathbf{r}_2|\mathbf{r}_3)$ (from Equation 12.61), that is,

$$\delta\Psi_q(\mathbf{r}_4|\mathbf{r}_3) = q\int d\mathbf{r}_2\, \epsilon^{\mathrm{R}}(\mathbf{r}_2;\mathbf{r}_4)\phi_{\mathrm{Coul}}(r_{23})$$

provided that q is sufficiently small so linear response is adequate (see the more detailed discussion below). From the first line in Equation 12.65 it follows that

$$\phi^{\star}_{\mathrm{Coul}}(\mathbf{r}_4,\mathbf{r}_3) = \lim_{q\to 0}\frac{\delta\Psi_q(\mathbf{r}_4|\mathbf{r}_3)}{q} \tag{12.67}$$

for $r_{43} > 0$. Thus, $q\phi^{\star}_{\mathrm{Coul}}$ gives the total potential from an infinitesimally small point charge q immersed in the electrolyte.

In more detail, we can analyze the second line of Equation 12.65 multiplied by q and we will show that $\phi^{\star}_{\mathrm{Coul}}$ contains the linear part of the electrostatic response of the electrolyte. If $q\phi_{\mathrm{Coul}}(r_{23})$ were small everywhere, the integral $q\int d\mathbf{r}_2\, \chi^{\rho}(\mathbf{r}_1,\mathbf{r}_2)\phi_{\mathrm{Coul}}(r_{23})$ would be the polarization response at $\mathbf{r}_1$ due to the external potential from the charge q placed at $\mathbf{r}_3$, that is, the charge density originating from this linear response. Since $q\phi_{\mathrm{Coul}}(r_{23})$ is large for small r_{23} (it is proportional to $1/r_{23}$), the integral is the linear part of the polarization response. However, in the limit $q \to 0$, as applied in Equation 12.67, linear response applies everywhere except in a small region around $r_{23} = 0$ that vanishes in this limit. The entire integral in the rhs of Equation 12.65 times q is the electrostatic potential at $\mathbf{r}_4$ from this polarization charge density. The first term, $q\phi_{\mathrm{Coul}}(r_{43})$, in the rhs gives the direct Coulomb potential between $\mathbf{r}_3$ and $\mathbf{r}_4$. Thus, in addition to the direct interaction, the screened Coulomb potential $\phi^{\star}_{\mathrm{Coul}}(\mathbf{r}_4,\mathbf{r}_3)$ contains effects of the linear electrostatic response of the electrolyte.

From the screened version of the Coulomb potential formula (the last line in Equation 12.64), we see that the dressed particle charge distribution $\rho_i^{\star}$ constitute the source of the potential ψ_i when $\phi^{\star}_{\mathrm{Coul}}$ is used instead of ϕ_{Coul}. In this equation $\phi^{\star}_{\mathrm{Coul}}$ provides the *linear* electrostatic contributions and $\rho_i^{\star}$ contains the *nonlinear* electrostatic contributions, as we saw earlier in the discussion of Equation 12.31 (cf. also the discussion of Equations 12.35 and 12.38).

In order to further illustrate the concept of the screened Coulomb potential $\phi^{\star}_{\mathrm{Coul}}$, let us consider the Poisson equation

$$-\varepsilon_0\nabla_3^2\psi_i(\mathbf{r}_3|\mathbf{r}_1) = \rho_i^{\mathrm{tot}}(\mathbf{r}_3|\mathbf{r}_1) \tag{12.68}$$

with the boundary condition $\psi_i(\mathbf{r}_3|\mathbf{r}_1) \to 0$ when $r_{13} \to \infty$ (we assume for simplicity that the system is infinitely large). Its solution is

$$\psi_i(\mathbf{r}_3|\mathbf{r}_1) = \int d\mathbf{r}_2\, \rho_i^{\mathrm{tot}}(\mathbf{r}_2|\mathbf{r}_1)\phi_{\mathrm{Coul}}(r_{23})$$

because $\phi_{\text{Coul}}(r_{23}) = \phi_{\text{Coul}}(|\mathbf{r}_3 - \mathbf{r}_2|)$ is the *Green's function*[9] associated with the differential operator $-\varepsilon_0 \nabla^2$, that is, ϕ_{Coul} satisfies the equation

$$-\varepsilon_0 \nabla_3^2 \phi_{\text{Coul}}(|\mathbf{r}_3 - \mathbf{r}_2|) = \delta^{(3)}(\mathbf{r}_3 - \mathbf{r}_2). \tag{12.69}$$

We will now see that $\phi^{\star}_{\text{Coul}}$ is *the Green's function associated with a modified Poisson equation,* namely the equation that is obtained by subtracting $\int d\mathbf{r}_4\, \psi_i(\mathbf{r}_4|\mathbf{r}_1)\chi^{\star}(\mathbf{r}_4, \mathbf{r}_3)$ from both sides of 12.68. The modified Poisson equation is accordingly

$$-\varepsilon_0 \nabla_3^2 \psi_i(\mathbf{r}_3|\mathbf{r}_1) - \int d\mathbf{r}_4\, \psi_i(\mathbf{r}_4|\mathbf{r}_1)\chi^{\star}(\mathbf{r}_4, \mathbf{r}_3)$$
$$= \rho_i^{\text{tot}}(\mathbf{r}_3|\mathbf{r}_1) - \int d\mathbf{r}_4\, \psi_i(\mathbf{r}_4|\mathbf{r}_1)\chi^{\star}(\mathbf{r}_4, \mathbf{r}_3),$$

which is the same as

$$-\varepsilon_0 \nabla_3^2 \psi_i(\mathbf{r}_3|\mathbf{r}_1) - \int d\mathbf{r}_4\, \psi_i(\mathbf{r}_4|\mathbf{r}_1)\chi^{\star}(\mathbf{r}_4, \mathbf{r}_3) = \rho_i^{\star}(\mathbf{r}_3|\mathbf{r}_1), \tag{12.70}$$

as follows from Equation 12.29. As we have seen earlier, the integral is the linear part of the polarization response from the electrostatic field due to the i-ion.

In Exercise 12.4 it is shown that the screened Coulomb potential $\phi^{\star}_{\text{Coul}}$ is the Green's function associated with Equation 12.70, that is, the solution to

$$-\varepsilon_0 \nabla_3^2 \phi^{\star}_{\text{Coul}}(\mathbf{r}_2, \mathbf{r}_3) - \int d\mathbf{r}_4\, \phi^{\star}_{\text{Coul}}(\mathbf{r}_2, \mathbf{r}_4)\chi^{\star}(\mathbf{r}_4, \mathbf{r}_3) = \delta^{(3)}(\mathbf{r}_{23}), \tag{12.71}$$

and hence Equation 12.70 has the solution

$$\psi_i(\mathbf{r}_3|\mathbf{r}_1) = \int d\mathbf{r}_2\, \rho_i^{\star}(\mathbf{r}_2|\mathbf{r}_1)\phi^{\star}_{\text{Coul}}(\mathbf{r}_3, \mathbf{r}_2),$$

which is the same as Equation 12.64. One can say that $\phi^{\star}_{\text{Coul}}(\mathbf{r}_3, \mathbf{r}_2)$ describes the spatial propagation of electrostatic interactions in the electrolyte between $\mathbf{r}_2$ and $\mathbf{r}_3$.

{**Exercise 12.4:** Use Equation 12.69 to show that Equation 12.65 implies that

$$-\varepsilon_0 \nabla_3^2 \phi^{\star}_{\text{Coul}}(\mathbf{r}_3, \mathbf{r}_2) = \delta^{(3)}(\mathbf{r}_3 - \mathbf{r}_2) + \int d\mathbf{r}_1\, \phi_{\text{Coul}}(r_{12})\chi^{\rho}(\mathbf{r}_1, \mathbf{r}_3).$$

[9]The concept of **Green's function** is treated in Chapter 6 of Volume I (see Example 6.2). Let us briefly summarize its main features. Consider an inhomogeneous partial differential equation $\mathscr{L}_{\mathbf{r}}F(\mathbf{r}) = f(\mathbf{r})$, where $\mathscr{L}_{\mathbf{r}}$ is a linear differential operator containing x, y, z derivatives of any order with constant coefficients and $f(\mathbf{r})$ is a given function. The Green's function $G(\mathbf{r}, \mathbf{r}')$ associated with $\mathscr{L}_{\mathbf{r}}$ is a solution to $\mathscr{L}_{\mathbf{r}}G(\mathbf{r}, \mathbf{r}') = \delta^{(3)}(\mathbf{r} - \mathbf{r}')$ and the solution to the original equation $\mathscr{L}_{\mathbf{r}}F(\mathbf{r}) = f(\mathbf{r})$ is given by $F(\mathbf{r}) = \int d\mathbf{r}'\, f(\mathbf{r}')G(\mathbf{r}, \mathbf{r}')$ as can be verified by inserting the rhs into the original equation and simplifying. $\mathscr{L}_{\mathbf{r}}$ can also contain terms that are not differentials operators, provided that $\mathscr{L}_{\mathbf{r}}F(\mathbf{r})$ is linear in F.

Then show that Equations 12.24 and 12.65 imply that

$$\int d\mathbf{r}_4\, \phi^{\star}_{\text{Coul}}(\mathbf{r}_4,\mathbf{r}_2)\chi^{\star}(\mathbf{r}_4,\mathbf{r}_3) = \int d\mathbf{r}_1\, \phi_{\text{Coul}}(r_{12})\chi^{\rho}(\mathbf{r}_1,\mathbf{r}_3).$$

Together these results give Equation 12.71.}

Finally. we will treat the pair correlation function $h^{(2)}_{ij}$ and write it is terms of starred entities. Using Equation 12.64 we can express $h^{(2)}_{ij}$ in Equation 12.33 as

$$\begin{aligned} h^{(2)}_{ij}(\mathbf{r}_1,\mathbf{r}_2) &= h^{\star}_{ij}(\mathbf{r}_1,\mathbf{r}_2) - \beta \int d\mathbf{r}_3\, d\mathbf{r}_4\, \rho^{\star}_i(\mathbf{r}_4|\mathbf{r}_1)\phi^{\star}_{\text{Coul}}(\mathbf{r}_4,\mathbf{r}_3)\rho^{\star}_j(\mathbf{r}_3|\mathbf{r}_2) \\ &= h^{\star}_{ij}(\mathbf{r}_1,\mathbf{r}_2) - \beta w^{(2)\text{el}}_{ij}(\mathbf{r}_1,\mathbf{r}_2) \end{aligned} \tag{12.72}$$

(cf. Equation 12.38), where $\rho^{\star}_i$ and $\rho^{\star}_j$ appear in a symmetric manner (cf. Equation 12.49 where ρ^{tot}_i and $\rho^{\star}_j$ have unsymmetric roles). Here, we have written the definition 12.37 of $w^{(2)\text{el}}_{ij}$ as

$$w^{(2)\text{el}}_{ij}(\mathbf{r}_1,\mathbf{r}_2) = \int d\mathbf{r}_3\, d\mathbf{r}_4\, \rho^{\star}_i(\mathbf{r}_4|\mathbf{r}_1)\phi^{\star}_{\text{Coul}}(\mathbf{r}_4,\mathbf{r}_3)\rho^{\star}_j(\mathbf{r}_3|\mathbf{r}_2). \tag{12.73}$$

Note that the integral has the form of an electrostatic interactions energy between the charge distributions $\rho^{\star}_i$ and $\rho^{\star}_j$ as mediated by $\phi^{\star}_{\text{Coul}}$. As we will see in the next section, this expression for $h^{(2)}_{ij}$ constitute a starting point for a general analysis of correlations in simple bulk electrolytes in the exact case.

It follows from these relationships that the expression $w^{(2)}_{ij} = u_{ij} - k_BT[h_{ij} - c_{ij} + b^{(2)}_{ij}]$ can be written as $w^{(2)}_{ij} = u^{\text{sh}}_{ij} + w^{(2)\text{el}}_{ij} - k_BT[h^{\star}_{ij} - c^{\star}_{ij} + b^{(2)}_{ij}]$, so we have

$$g^{(2)}_{ij} = h^{(2)}_{ij} + 1 = e^{-\beta[u^{\text{sh}}_{ij} + w^{(2)\text{el}}_{ij}] + h^{\star}_{ij} - c^{\star}_{ij} + b^{(2)}_{ij}} \tag{12.74}$$

(cf. Equation 12.51). Since $h^{\star}_{ij}$ and $c^{\star}_{ij}$ satisfy the OZ equation, the difference $h^{\star}_{ij} - c^{\star}_{ij}$ can be expressed as a sum of chains of $c^{\star}_{i'j'}$ functions like $c^{(2)}nc^{(2)}$, $c^{(2)}nc^{(2)}nc^{(2)}$ etc. in Equation 9.41.

Table 12.2 contains a summary of some important formulas for pair correlations and the Coulomb potential.

12.3 INTERACTIONS AND SCREENING IN SIMPLE BULK ELECTROLYTES, GENERAL ANALYSIS

12.3.1 Essential formulas for response functions, correlation functions and the dielectric function

The system under consideration is a bulk electrolyte consisting of simple spherical ions of density n^{b}_i for species i distributed in vacuum and total ionic density $n^{\text{b}}_{\text{tot}} = \sum_i n^{\text{b}}_i$. The theory will later be adapted to ions in a dielectric

Table 12.2 Important formulas for the pair correlation functions $h_{ij}^{(2)}$ and $h_{ij}^{\star}$, the linear electrostatic part of the pair potential of mean force $w_{ij}^{(2)\mathrm{el}}$ and the Screened Coulomb potential $\phi_{\mathrm{Coul}}^{\star}$. The equation numbers refer to the original ones in the text.

Splitting of the pair correlation function:

$$h_{ij}^{(2)}(\mathbf{r}_1,\mathbf{r}_2) = h_{ij}^{\star}(\mathbf{r}_1,\mathbf{r}_2) - \beta w_{ij}^{(2)\mathrm{el}}(\mathbf{r}_1,\mathbf{r}_2), \qquad \text{(12.38' and 12.72')}$$

where

$$w_{ij}^{(2)\mathrm{el}}(\mathbf{r}_1,\mathbf{r}_2) = \int d\mathbf{r}_3\, \psi_i(\mathbf{r}_3|\mathbf{r}_1)\rho_j^{\star}(\mathbf{r}_3|\mathbf{r}_2) \qquad \text{(12.37')}$$

$$= \int d\mathbf{r}_3\, d\mathbf{r}_4\, \rho_i^{\star}(\mathbf{r}_4|\mathbf{r}_1)\phi_{\mathrm{Coul}}^{\star}(\mathbf{r}_4,\mathbf{r}_3)\rho_j^{\star}(\mathbf{r}_3|\mathbf{r}_2) \qquad \text{(12.73')}$$

The (unit) Screened Coulomb potential:

$$\phi_{\mathrm{Coul}}^{\star}(\mathbf{r}_1,\mathbf{r}_2) = \int d\mathbf{r}_3\, \epsilon^{\mathrm{R}}(\mathbf{r}_3;\mathbf{r}_1)\phi_{\mathrm{Coul}}(r_{32}) \qquad \text{(12.65')}$$

Coulomb potential formulas (ordinary and screened):

$$\psi_i(\mathbf{r}_3|\mathbf{r}_1) = \int d\mathbf{r}_2\, \rho_i^{\mathrm{tot}}(\mathbf{r}_2|\mathbf{r}_1)\phi_{\mathrm{Coul}}(r_{23})$$

$$= \int d\mathbf{r}_4\, \rho_i^{\star}(\mathbf{r}_4|\mathbf{r}_1)\phi_{\mathrm{Coul}}^{\star}(\mathbf{r}_4,\mathbf{r}_3) \qquad \text{(12.64')}$$

continuum. We start by listing several relevant equations that we obtained for an inhomogeneous electrolyte in the previous sections of the present chapter and write them in the form they take for a bulk phase. Furthermore, we will define the *dielectric function* and derive some important new relationships.

The charge densities of the ion cloud and the dress of an i-ion are

$$\begin{aligned}\rho_i^{\text{cloud}}(\mathbf{r}_2|\mathbf{r}_1) &= \rho_i^{\text{cloud}}(r_{12}) \equiv \sum_j q_j n_j^{\text{b}} h_{ij}^{(2)}(r_{12}) \\ \rho_i^{\text{dress}}(\mathbf{r}_2|\mathbf{r}_1) &= \rho_i^{\text{dress}}(r_{12}) \equiv \sum_j q_j n_j^{\text{b}} h_{ij}^{\star}(r_{12}),\end{aligned}$$

respectively, and we have

$$\rho_i^{\text{tot}}(\mathbf{r}_2|\mathbf{r}_1) = \rho_i^{\text{tot}}(r_{12}) \equiv q_i\delta^{(3)}(r_{12}) + \rho_i^{\text{cloud}}(r_{12}) \qquad (12.75)$$

$$\rho_i^{\star}(\mathbf{r}_2|\mathbf{r}_1) = \rho_i^{\star}(r_{12}) \equiv q_i\delta^{(3)}(r_{12}) + \rho_i^{\text{dress}}(r_{12}) \qquad (12.76)$$

(cf. Table 12.1 on page 290). These latter entities include the central charge of the ion expressed as a charge density, i.e., $q_i\delta^{(3)}(r_{12})$. The entity $\rho_i^{\star}$ is the total charge density of a *dressed ion* of species i, which differs from ρ_i^{tot} by the linear part of the polarization response due to the i-ion as discussed in Section 12.2.1 (see Equations 12.29 and 12.31). The total charge of ρ_i^{tot} is zero according to the condition of **local electroneutrality**,

$$\int d\mathbf{r}_2\, \rho_i^{\text{tot}}(r_{12}) = q_i + \int d\mathbf{r}_2\, \rho_i^{\text{cloud}}(r_{12}) = 0,$$

while total charge $q_i^{\star}$ of $\rho_i^{\star}$,

$$q_i^{\star} \equiv \int d\mathbf{r}_2\, \rho_i^{\star}(r_{12}), \qquad (12.77)$$

is in general nonzero, $q_i^{\star} \neq 0$, because the linear part of the polarization response does not integrate to zero in general. The quantity $q_i^{\star}$ is the **charge of the dressed ion**. For bulk electrolytes we have

$$\rho_i(r_{12}) \equiv \sum_j q_j n_j^{\text{b}} g_{ij}^{(2)}(r_{12}) = \sum_j q_j n_j^{\text{b}} h_{ij}^{(2)}(r_{12})$$

because $\sum_j q_j n_j^{\text{b}} = 0$, so there is no difference between ρ_i and ρ_i^{cloud} (in inhomogeneous electrolytes they are different, cf. Equations 12.6 - 12.8). We will explicitly write the superscript "cloud" only when there is a some special reason to do so. In Appendix 12A it is described how to extract $\rho_i^{\star}(r)$ and $q_i^{\star}$ from $h_{ij}^{(2)}(r)$ or $\rho_i^{\text{cloud}}(r)$, which can be calculated from accurate approximate theories or by computer simulations.

If the bulk electrolyte is exposed to a weak external electrostatic potential $\delta\Psi^{\text{ext}}(\mathbf{r})$ that originates from some charge distribution that is not part of the system, the electrolyte is polarized and a polarization charge density $\delta\rho^{\text{pol}}(\mathbf{r}')$ arises. Note that in the present case $\delta\Psi^{\text{ext}}$ is the small variation in potential from 0, which is the value we adopt for the constant electrostatic potential of the bulk electrolyte. The charge density of the homogeneous electrolyte is originally zero (before $\delta\Psi^{\text{ext}}$ is turned on). The resulting total electrostatic potential is $\delta\Psi = \delta\Psi^{\rm ext} + \delta\Psi^{\rm pol}$, where $\delta\Psi^{\rm pol}$ arises from the induced $\delta\rho^{\rm pol}$ (Equation 12.12). Because of the presence of $\delta\Psi^{\rm ext}$, an i-ion experiences a potential of mean force δw_i that given by

$$\delta w_i(\mathbf{r}_1) = \int d\mathbf{r}_2\, \rho_i^{\rm tot}(r_{12})\delta\Psi^{\rm ext}(\mathbf{r}_2) = \int d\mathbf{r}_2\, \rho_i^\star(r_{12})\, \delta\Psi(\mathbf{r}_2) \qquad (12.78)$$

(see Table 12.1). The first equality says that δw_i is equal to the interaction energy between the weak potential $\delta\Psi^{\rm ext}$ and the ion together with its entire ion cloud, while the rhs shows that δw_i is also equal to the interaction energy between the weak total potential $\delta\Psi$ and the charge density of the dressed ion. These relationships are valid in the linear response domain, i.e., for infinitesimally small $\delta\Psi^{\rm ext}$ and $\delta\Psi$, and cannot be utilized for finite values of the potentials unless they are used as approximations.

The polarization response functions χ^ρ and $\chi^\star$ give the polarization charge density in terms of $\delta\Psi^{\rm ext}$ and $\delta\Psi$, respectively,

$$\delta\rho^{\rm pol}(\mathbf{r}_1) = \int d\mathbf{r}_2\, \chi^\rho(r_{12})\delta\Psi^{\rm ext}(\mathbf{r}_2) = \int d\mathbf{r}_2\, \chi^\star(r_{12})\delta\Psi(\mathbf{r}_2),$$

(see Table 12.1) and are given by

$$\chi^\rho(r_{12}) = -\beta\sum_i n_i^{\rm b} q_i \rho_i^{\rm tot}(r_{12}) = -\beta q_{\rm e}^2 H_{\rm QQ}(r_{12}) \qquad (12.79)$$

$$\chi^\star(r_{12}) = -\beta\sum_i n_i^{\rm b} q_i \rho_i^\star(r_{12}) = -\beta q_{\rm e}^2 H_{\rm QQ}^\star(r_{12}), \qquad (12.80)$$

where $H_{\rm QQ}(r_{12}) = \sum_{ij} q_i q_j H_{ij}^{(2)}(r_{12})/q_{\rm e}^2$ is the charge-charge correlation function and

$$H_{ij}^{(2)}(r_{12}) = n_i^{\rm b}\delta_{ij}^{(3)}(r_{12}) + n_i^{\rm b} n_j^{\rm b} h_{ij}^{(2)}(r_{12}) \qquad (12.81)$$

(cf. Equations 12.17 and 12.21). $H_{\rm QQ}^\star$ and $H_{ij}^\star$ are analogously defined. Equation 12.24 becomes

$$\chi^\rho(r_{14}) = \chi^\star(r_{14}) + \int d\mathbf{r}_2 d\mathbf{r}_3\, \chi^\star(r_{12})\phi_{\rm Coul}(r_{23})\chi^\rho(r_{34}) \qquad (12.82)$$

in the bulk phase, so $\chi^\star(r)$ can be determined when $H_{\rm QQ}(r)$ and hence $\chi^\rho(r)$ is known. This is most easily done in Fourier space because the equation then becomes $\tilde{\chi}^\rho(k) = \tilde{\chi}^\star(k) + \tilde{\chi}^\star(k)\tilde{\phi}_{\rm Coul}(k)\tilde{\chi}^\rho(k)$ since the integrals are convolutions and we obtain $\tilde{\chi}^\star(k) = \tilde{\chi}^\rho(k)/[1 + \tilde{\phi}_{\rm Coul}(k)\tilde{\chi}^\rho(k)]$.

The mean electrostatic potential ψ_i from an i-ion is given by the Coulomb potential formula

$$\psi_i(r_{13}) = \int d\mathbf{r}_2\, \rho_i^{\rm tot}(r_{12})\phi_{\rm Coul}(r_{23}) = \int d\mathbf{r}_4\, \rho_i^\star(r_{14})\phi_{\rm Coul}^\star(r_{43}), \qquad (12.83)$$

where the rhs is the screened form of this formula and the (unit) screened Coulomb potential $\phi^{\star}_{\mathrm{Coul}}$ is given by

$$\begin{aligned}\phi^{\star}_{\mathrm{Coul}}(r_{43}) &\equiv \int d\mathbf{r}_2\, \epsilon^{\mathrm{R}}(r_{42})\phi_{\mathrm{Coul}}(r_{23}) \qquad (12.84)\\ &= \phi_{\mathrm{Coul}}(r_{43}) + \int d\mathbf{r}_2\, d\mathbf{r}_1\, \chi^{\rho}(r_{12})\phi_{\mathrm{Coul}}(r_{14})\phi_{\mathrm{Coul}}(r_{23})\end{aligned}$$

(see Table 12.2 on page 303). The potential ψ_i is the solution to the Poisson equation

$$-\varepsilon_0\nabla_3^2\psi_i(r_{13}) = \rho_i^{\mathrm{tot}}(r_{13}) \qquad (12.85)$$

with the boundary condition $\psi_i(r_{13}) \to 0$ when $r_{13} \to \infty$. The expression for ψ_i in terms of $\phi^{\star}_{\mathrm{Coul}}$ is obtained as the solution to the modified Poisson equation

$$-\varepsilon_0\nabla_3^2\psi_i(r_{13}) - \int d\mathbf{r}_4\, \psi_i(r_{14})\chi^{\star}(r_{43}) = \rho_i^{\star}(r_{13}) \qquad (12.86)$$

with the same boundary condition. The two variants of the Poisson equation are completely equivalent because

$$\rho_i^{\star}(r_{13}) = \rho_i^{\mathrm{tot}}(r_{13}) - \int d\mathbf{r}_4\, \psi_i(r_{14})\chi^{\star}(r_{34}), \qquad (12.87)$$

see Equation 12.29. The Coulomb potential ϕ_{Coul} is the Green's function of the Poisson equation and its screened version, $\phi^{\star}_{\mathrm{Coul}}$, is the Green's function of the modified equation, that is, they are solutions to

$$-\varepsilon_0\nabla_3^2\phi_{\mathrm{Coul}}(r_{23}) = \delta^{(3)}(r_{23})$$

and

$$-\varepsilon_0\nabla_3^2\phi^{\star}_{\mathrm{Coul}}(r_{23}) - \int d\mathbf{r}_4\, \phi^{\star}_{\mathrm{Coul}}(r_{24})\chi^{\star}(r_{43}) = \delta^{(3)}(r_{23}), \qquad (12.88)$$

respectively.

The electric susceptibility function is

$$\chi^{\mathrm{el}}(r_{12}) = -\int d\mathbf{r}_3\, \chi^{\star}(r_{13})\phi_{\mathrm{Coul}}(r_{32}),$$

the dielectric permittivity function is

$$\epsilon(r_{12}) = \delta^{(3)}(r_{12}) + \chi^{\mathrm{el}}(r_{12}) = \delta^{(3)}(r_{12}) - \int d\mathbf{r}_3\, \chi^{\star}(r_{13})\phi_{\mathrm{Coul}}(r_{32}) \qquad (12.89)$$

and the reciprocal dielectric-permittivity function is

$$\epsilon^{\mathrm{R}}(r_{21}) = \delta^{(3)}(r_{21}) + \int d\mathbf{r}_3\, \chi^{\rho}(r_{23})\phi_{\mathrm{Coul}}(r_{31}). \qquad (12.90)$$

They satisfy

$$\begin{aligned}\delta\Psi^{\text{pol}}(\mathbf{r}_2) &= -\int d\mathbf{r}_2\,\delta\Psi(\mathbf{r}_1)\chi^{\text{el}}(r_{12})\\ \delta\Psi^{\text{ext}}(\mathbf{r}_2) &= \int d\mathbf{r}_1\,\delta\Psi(\mathbf{r}_1)\epsilon(r_{12})\\ \delta\Psi(\mathbf{r}_1) &= \int d\mathbf{r}_2\,\delta\Psi^{\text{ext}}(\mathbf{r}_2)\epsilon^{\text{R}}(r_{21})\end{aligned}$$

for any weak perturbation $\delta\Psi^{\text{Ext}}$ of the bulk phase and resulting weak $\delta\Psi^{\text{pol}}$ and $\delta\Psi$ (see Equations 12.53, 12.57 and 12.60). In Fourier space these equations become

$$\delta\tilde{\Psi}^{\text{pol}}(\mathbf{k}) = -\delta\tilde{\Psi}(\mathbf{k})\tilde{\chi}^{\text{el}}(k) \tag{12.91}$$

$$\delta\tilde{\Psi}^{\text{ext}}(\mathbf{k}) = \delta\tilde{\Psi}(\mathbf{k})\tilde{\epsilon}(k) \tag{12.92}$$

$$\delta\tilde{\Psi}(\mathbf{k}) = \delta\tilde{\Psi}^{\text{ext}}(\mathbf{k})\tilde{\epsilon}^{\text{R}}(k) \tag{12.93}$$

because the integrals are convolutions.[10] From these relationships it follows that

$$\tilde{\epsilon}(k) = 1 + \tilde{\chi}^{\text{el}}(k) \tag{12.94}$$

$$\tilde{\epsilon}^{\text{R}}(k) = \frac{1}{\tilde{\epsilon}(k)}. \tag{12.95}$$

The function $\tilde{\epsilon}(k)$ is called the **dielectric function,**[11] which should not be confused with the (spatial) dielectric permittivity function $\epsilon(r)$, i.e., its inverse Fourier transform.

By taking the Fourier transforms of Equations 12.89 and 12.90 and using Equation 12.95 we see that

$$\tilde{\epsilon}(k) = 1 - \tilde{\chi}^{\star}(k)\tilde{\phi}_{\text{Coul}}(k) = \frac{1}{1 + \tilde{\chi}^{\rho}(k)\tilde{\phi}_{\text{Coul}}(k)}. \tag{12.96}$$

The last equality can also be obtained from the Fourier transform of Equation 12.82.

We can write the first equality of Equation 12.96 as

$$\tilde{\epsilon}(k) = -\frac{\tilde{\chi}^{\star}(0)}{\varepsilon_0 k^2} + \left(1 - \frac{\tilde{\chi}^{\star}(k) - \tilde{\chi}^{\star}(0)}{\varepsilon_0 k^2}\right),$$

---

[10] Note that the corresponding integrals in inhomogeneous systems are *not* convolutions (in Equations 12.53, 12.57 and 12.60).

[11] When time dependence is involved, one deals with the dielectric function as function of both wave number k and angular frequency ω, that is, $\tilde{\epsilon}(k,\omega)$. Here, we only consider equilibrium systems, so we have $\omega = 0$. Hence $\tilde{\epsilon}(k) = \lim_{\omega\to 0}\tilde{\epsilon}(k,\omega)$ and $\tilde{\epsilon}(k)$ is often called the *static* dielectric function.

where we have used $\tilde{\phi}_{\text{Coul}}(k) = 1/(\varepsilon_0 k^2)$. The first term diverges like $1/k^2$ when $k \to 0$ while the rest remains finite for $k = 0$, as we will see. We therefore define the **singular part** of $\tilde{\epsilon}(k)$ as

$$\begin{aligned}\tilde{\epsilon}_{\text{sing}}(k) &\equiv -\frac{\tilde{\chi}^\star(0)}{\varepsilon_0 k^2} = -\frac{1}{\varepsilon_0 k^2}\int d\mathbf{r}\chi^\star(r)\\ &= \frac{\beta}{\varepsilon_0 k^2}\sum_i n_i^{\text{b}} q_i \int d\mathbf{r}\rho_i^\star(r) = \frac{\beta}{\varepsilon_0 k^2}\sum_i n_i^{\text{b}} q_i q_i^\star, \qquad (12.97)\end{aligned}$$

where we have used Equations 12.77 and 12.80. The **regular part** of $\tilde{\epsilon}(k)$ is defined as

$$\tilde{\epsilon}_{\text{reg}}(k) \equiv 1 - \frac{\tilde{\chi}^\star(k) - \tilde{\chi}^\star(0)}{\varepsilon_0 k^2} = 1 - \frac{1}{\varepsilon_0}\int d\mathbf{r}\, r^2 \left[\frac{\sin(kr) - kr}{(kr)^3}\right]\chi^\star(r), \quad (12.98)$$

which is finite when $k \to 0$

$$\mathcal{E}_r^{\star 0} \equiv \tilde{\epsilon}_{\text{reg}}(0) = 1 + \frac{1}{6\varepsilon_0}\int d\mathbf{r}\, r^2 \chi^\star(r). \qquad (12.99)$$

Thus, we have

$$\begin{aligned}\tilde{\epsilon}(k) &= \tilde{\epsilon}_{\text{sing}}(k) + \tilde{\epsilon}_{\text{reg}}(k)\\ &= \frac{\beta}{\varepsilon_0 k^2}\sum_i n_i^{\text{b}} q_i q_i^\star + \tilde{\epsilon}_{\text{reg}}(k). \qquad (12.100)\end{aligned}$$

This implies that when $k \to 0$ we have $\tilde{\epsilon}(k) \to \infty$ and hence $\tilde{\epsilon}^{\text{R}}(k) = 1/\tilde{\epsilon}(k) \to 0$ for an electrolyte. From Equation 12.93 we see that $\delta\tilde{\Psi}(\mathbf{k})$ therefore becomes zero when $k \to 0$ irrespectively of $\delta\tilde{\Psi}^{\text{ext}}(\mathbf{k})$ in the same limit. This reflects the fact that an electrolyte is a conductor because the limit $k \to 0$ describes the situation when the system is exposed to a uniform field and the total electric field $\delta\mathbf{E}(\mathbf{r}) = -\nabla\delta\Psi(\mathbf{r})$ inside the system then is zero.

Note that if the system does not contain any ions we have $q_i = 0$ for all i and $\tilde{\epsilon}(k) = \tilde{\epsilon}_{\text{reg}}(k)$, so $\tilde{\epsilon}(k)$ remains finite at $k = 0$ for nonelectrolytes. As we will see later, for a pure polar fluid (with no ions) we have $\lim_{k\to 0}\tilde{\epsilon}(k) \to \varepsilon_r$, that is, the dielectric constant, so for nonelectrolytes $\mathcal{E}_r^{\star 0} \equiv \tilde{\epsilon}_{\text{reg}}(0) = \varepsilon_r$.

We have

$$\frac{1}{\tilde{\epsilon}(k)} = 1 + \tilde{\chi}^\rho(k)\tilde{\phi}_{\text{Coul}}(k) = 1 - \frac{\beta}{\varepsilon_0 k^2}\sum_i n_i^{\text{b}} q_i \tilde{\rho}_i^{\text{tot}}(k) \qquad (12.101)$$

(from Equations 12.79 and 12.96). By using the Taylor expansion of $\sin(kr)/(kr)$, we have

$$\tilde{\rho}_i^{\text{tot}}(k) = \int d\mathbf{r}\frac{\sin(kr)}{kr}\rho_i^{\text{tot}}(r) = \int d\mathbf{r}\,\rho_i^{\text{tot}}(r) - \int d\mathbf{r}\frac{(kr)^2}{3!}\rho_i^{\text{tot}}(r) + \mathcal{O}(k^4). \qquad (12.102)$$

The first term on the rhs is zero due to local electroneutrality, so we obtain from Equation 12.101

$$\frac{1}{\tilde{\epsilon}(k)} = 1 + \frac{\beta}{6\varepsilon_0}\int d\mathbf{r}\, r^2 \sum_i n_i^{\mathrm{b}} q_i \rho_i^{\mathrm{tot}}(r) + \mathcal{O}(k^2).$$

Since $1/\tilde{\epsilon}(k) \to 0$ when $k \to 0$ for electrolytes, it follows that

$$\beta \int d\mathbf{r}\, r^2 \sum_i n_i^{\mathrm{b}} q_i \rho_i^{\mathrm{tot}}(r) = -6\varepsilon_0, \tag{12.103}$$

which is called the **Stillinger-Lovett condition**. By inserting the definition of $\rho_i^{\mathrm{tot}}(r)$ and noticing that the contribution from the term $q_i\delta^{(3)}(r)$ in $\rho_i^{\mathrm{tot}}(r)$ vanishes because of the factor r^2 in the integral, we obtain

$$\beta \sum_{ij} n_i^{\mathrm{b}} n_j^{\mathrm{b}} q_i q_j \int d\mathbf{r}\, r^2 h_{ij}^{(2)}(r) = -6\varepsilon_0, \tag{12.104}$$

which is a more common form of this condition. We can derive yet another form by writing Equation 12.101 as

$$\frac{1}{\tilde{\epsilon}(k)} = 1 - \beta \sum_i n_i^{\mathrm{b}} q_i \tilde{\rho}_i^{\mathrm{tot}}(k)\tilde{\phi}_{\mathrm{Coul}}(k) = 1 - \beta \sum_i n_i^{\mathrm{b}} q_i \tilde{\psi}_i(k),$$

where we have inserted the Fourier transform of the first equality in Equation 12.83. By setting $k = 0$ we obtain $1 - \beta\sum_i n_i^{\mathrm{b}} q_i \tilde{\psi}_i(0) = 0$, which yields

$$\beta \sum_i n_i^{\mathrm{b}} q_i \int d\mathbf{r}\, \psi_i(r) = 1. \tag{12.105}$$

All three forms 12.103, 12.104 and 12.105 of the Stillinger-Lovett condition are mathematically equivalent.

From the definition $\phi^{\star}_{\mathrm{Coul}}(r_{43}) \equiv \int d\mathbf{r}_2\, \epsilon^{\mathrm{R}}(r_{42})\phi_{\mathrm{Coul}}(r_{23})$ of the unit screened Coulomb potential (Equation 12.84), it follows that

$$\tilde{\phi}^{\star}_{\mathrm{Coul}}(k) = \frac{\tilde{\phi}_{\mathrm{Coul}}(k)}{\tilde{\epsilon}(k)} = \frac{1}{\varepsilon_0 k^2 \tilde{\epsilon}(k)}. \tag{12.106}$$

As shown in Exercise 13.5, this result can also be derived from Equation 12.88. From the Fourier transform of the screened Coulomb potential formula (the rhs in Equation 12.83) we obtain

$$\tilde{\psi}_i(k) = \tilde{\rho}_i^{\star}(k)\tilde{\phi}^{\star}_{\mathrm{Coul}}(k) = \frac{\tilde{\rho}_i^{\star}(k)\tilde{\phi}_{\mathrm{Coul}}(k)}{\tilde{\epsilon}(k)}, \tag{12.107}$$

which can also be obtained from the modified Poisson equation 12.86 as shown in the same exercise. Since $\tilde{\psi}_i(k) = \tilde{\rho}_i^{\mathrm{tot}}(k)\tilde{\phi}_{\mathrm{Coul}}(k)$ we can conclude that

$$\tilde{\rho}_i^{\mathrm{tot}}(k) = \frac{\tilde{\rho}_i^{\star}(k)}{\tilde{\epsilon}(k)}, \tag{12.108}$$

which can also be obtained from Equations 12.62 and 12.63 applied to bulk systems.

Since $\tilde{\varepsilon}(k) = \beta \sum_i n_i^{\mathrm{b}} q_i q_i^{\star}/[\varepsilon_0 k^2] + \tilde{\varepsilon}_{\mathrm{reg}}(k)$ (Equation 12.100) we can write Equation 12.106 and 12.107 as

$$\tilde{\phi}^{\star}_{\mathrm{Coul}}(k) = \frac{1}{\beta \sum_i n_i^{\mathrm{b}} q_i q_i^{\star} + \varepsilon_0 k^2 \tilde{\varepsilon}_{\mathrm{reg}}(k)} \tag{12.109}$$

and

$$\tilde{\psi}_i(k) = \frac{\tilde{\rho}_i^{\star}(k)}{\beta \sum_i n_i^{\mathrm{b}} q_i q_i^{\star} + \varepsilon_0 k^2 \tilde{\varepsilon}_{\mathrm{reg}}(k)}, \tag{12.110}$$

which will be very useful later.

{**Exercise 12.5:** Show that Equation 12.106 can alternatively be be obtained from the Fourier transform of Equation 12.88, whereby one can use Equation 9.249 in Appendix 9B to obtain the Fourier transform of $\nabla_3^2 \phi^{\star}_{\mathrm{Coul}}(r_{23})$. Show also that the modified Poisson equation 12.86 yields

$$\tilde{\psi}_i(k) = \frac{\tilde{\rho}_i^{\star}(k)}{\varepsilon_0 k^2 - \tilde{\chi}^{\star}(k)}$$

and that this expression can be written as Equation 12.107.}

As always, the pair correlation function $h_{ij}^{(2)}$ satisfies the Ornstein-Zernike equation

$$h_{ij}^{(2)}(r_{12}) = c_{ij}^{(2)}(r_{12}) + \sum_l n_l^{\mathrm{b}} \int d\mathbf{r}_3\, c_{il}^{(2)}(r_{13}) h_{lj}^{(2)}(r_{32})$$

and $h_{ij}^{\star}$ also satisfy an OZ equation

$$h_{ij}^{\star}(r_{12}) = c_{ij}^{\star}(r_{12}) + \sum_l n_l^{\mathrm{b}} \int d\mathbf{r}_3\, c_{il}^{\star}(r_{13}) h_{lj}^{\star}(r_{32}), \tag{12.111}$$

(Equation 12.46), where $c_{ij}^{\star} = c_{ij}^{(2)} + \beta u_{ij}^{\mathrm{el}} = c_{ij}^{(2)} + \beta q_i q_j \phi_{\mathrm{Coul}}$. Furthermore, the pair correlation function $h_{ij}^{(2)}$ can be expressed as

$$\begin{aligned} h_{ij}^{(2)}(r_{12}) &= h_{ij}^{\star}(r_{12}) - \beta w_{ij}^{\mathrm{el}}(r_{12}) \\ &= h_{ij}^{\star}(r_{12}) - \beta \int d\mathbf{r}_3\, d\mathbf{r}_4\, \rho_i^{\star}(r_{14}) \phi^{\star}_{\mathrm{Coul}}(r_{43}) \rho_j^{\star}(r_{23}) \end{aligned} \tag{12.112}$$

(see Table 12.2 on page 303). This can be written in Fourier space

$$\begin{aligned} \tilde{h}_{ij}^{(2)}(k) &= \tilde{h}_{ij}^{\star}(k) - \beta \tilde{\rho}_i^{\star}(k) \tilde{\phi}^{\star}_{\mathrm{Coul}}(k) \tilde{\rho}_j^{\star}(k) \qquad (12.113) \\ &= \tilde{h}_{ij}^{\star}(k) - \frac{\beta \tilde{\rho}_i^{\star}(k) \tilde{\rho}_j^{\star}(k)}{\varepsilon_0 k^2 \tilde{\varepsilon}(k)}, \end{aligned}$$

where we have inserted Equation 12.106. Using Equation Equations 12.100 we obtain

$$\tilde{h}_{ij}^{(2)}(k) = \tilde{h}_{ij}^{\star}(k) - \frac{\beta \tilde{\rho}_i^{\star}(k) \tilde{\rho}_j^{\star}(k)}{\beta \sum_i n_i^{\mathrm{b}} q_i q_i^{\star} + \varepsilon_0 k^2 \tilde{\varepsilon}_{\mathrm{reg}}(k)}. \tag{12.114}$$

These expressions will be very useful later.

Electrostatic fields and connection to macroscopic electrostatics

For completeness, let us consider the formalism in terms of electrostatic fields. The external electrostatic field $\mathbf{E}^{\mathrm{Ext}}$ is given in terms of its potential Ψ^{Ext} by $\mathbf{E}^{\mathrm{Ext}}(\mathbf{r}) = -\nabla\Psi^{\mathrm{Ext}}(\mathbf{r})$ and likewise the total electrostatic field is $\mathbf{E}(\mathbf{r}) = -\nabla\Psi(\mathbf{r})$. In Fourier space these relationships become $\tilde{\mathbf{E}}^{\mathrm{Ext}}(\mathbf{k}) = -i\mathbf{k}\tilde{\Psi}^{\mathrm{Ext}}(\mathbf{k})$ and $\tilde{\mathbf{E}}(\mathbf{k}) = -i\mathbf{k}\tilde{\Psi}(\mathbf{k})$ (see Appendix 9B). It is of particular interest to investigate the projection of the fields in the direction of $\mathbf{k}$, that is, the projection on the unit vector $\hat{\mathbf{k}}$, for example $\hat{\mathbf{k}}\cdot\tilde{\mathbf{E}}(\mathbf{k})$. This projection is called the *longitudinal component* of the field (along the direction of the planar wave[12] $\exp(i\mathbf{k}\cdot\mathbf{r})$). We have $\hat{\mathbf{k}}\cdot\tilde{\mathbf{E}}(\mathbf{k}) = -i\hat{\mathbf{k}}\cdot\mathbf{k}\tilde{\Psi}(\mathbf{k}) = -ik\tilde{\Psi}(\mathbf{k})$ and likewise for $\tilde{\mathbf{E}}^{\mathrm{Ext}}$. The transversal component (perpendicular to the wave) $\hat{\mathbf{k}}\times\tilde{\mathbf{E}}$, where $\times$ is the vector product, is uninteresting in electrostatics, since it is identically zero (since $\hat{\mathbf{k}}\times\mathbf{k} = 0$).[13] The fields are always directed along $\hat{\mathbf{k}}$ in this case.

In macroscopic electrostatics it is customary to consider the **polarization field** $\mathbf{P}(\mathbf{r})$ which is related to our Ψ^{pol} by $\mathbf{P}(\mathbf{r}) = -\varepsilon_0\mathbf{E}^{\mathrm{pol}}(\mathbf{r}) = \varepsilon_0\nabla\Psi^{\mathrm{pol}}(\mathbf{r})$. Furthermore, one uses the **displacement field** $\mathbf{D}(\mathbf{r})$, which is given by $\mathbf{D}(\mathbf{r}) = \varepsilon_0\mathbf{E}^{\mathrm{ext}}(\mathbf{r})$ in our notation. Note that $\mathbf{E}^{\mathrm{ext}}$ is the field from the external source charges alone, that is, in the absence of the fluid. The total electrostatic field, $\mathbf{E}$, is often called the **Maxwell field**. These matters will be discussed in more details later in Section 13.3.

In the present case we will only consider weak fields and the corresponding potentials. We have

$$\delta\tilde{\Psi}(\mathbf{k}) = \frac{\delta\tilde{\Psi}^{\mathrm{Ext}}(\mathbf{k})}{\tilde{\varepsilon}(k)}.$$

(Equation 12.92). Since $\hat{\mathbf{k}}\cdot\delta\tilde{\mathbf{E}}(\mathbf{k}) = -ik\delta\tilde{\Psi}(\mathbf{k})$ and $\hat{\mathbf{k}}\cdot\delta\tilde{\mathbf{E}}^{\mathrm{Ext}}(\mathbf{k}) = -ik\delta\tilde{\Psi}^{\mathrm{Ext}}(\mathbf{k})$ we can write this as

$$\hat{\mathbf{k}}\cdot\delta\tilde{\mathbf{E}}(\mathbf{k}) = \frac{\hat{\mathbf{k}}\cdot\delta\tilde{\mathbf{E}}^{\mathrm{Ext}}(\mathbf{k})}{\tilde{\epsilon}(k)},$$

which means that we are dealing with the longitudinal components of the fields. Our $\tilde{\varepsilon}(k)$ is therefore properly called the **static longitudinal dielectric function**, but we will for simplicity drop "static longitudinal" and simply call it the "dielectric function" as is often done when dealing with equilibrium systems.

[12]See the discussion at the beginning of Appendix 9B where it is mentioned that a Fourier transform can be regarded as a representation of a function by a superposition of planar waves.

[13]This is in agreement with Maxwell's equations which say that the curl of the electric field, $\nabla\times\mathbf{E}$, is zero for a static field. In Fourier space $\nabla\times\mathbf{E}$ corresponds to $i\hat{\mathbf{k}}\times\tilde{\mathbf{E}}$. The transversal component plays, however, an important role for dynamical phenomena in fluids.

12.3.2 The dielectric continuum solvent model and the Poisson-Boltzmann approximation

The results obtained so far in the current chapter, which apply for ions distributed in vacuum, can be generalized to ions in a dielectric continuum, for example electrolyte solutions in the primitive model, by replacing ε_0 by $\varepsilon_r^{\mathrm{cont}}\varepsilon_0$. There is, however, a further adjustment that has to be made. It concerns all relationships that involve the external electrostatic potential, the dielectric permittivity function, the dielectric function $\tilde{\epsilon}(k)$ and its reciprocal $\tilde{\epsilon}^{\mathrm{R}}(k) = 1/\tilde{\epsilon}(k)$. In order to see this we make the following observations.

The replacement of ε_0 by $\varepsilon_r^{\mathrm{cont}}\varepsilon_0$ makes all entities to be those of the system in the presence of the dielectric continuum. However, there is one entity that should be obtained for charges placed in vacuum, namely the external electrostatic potential used in the definition of the dielectric permittivity function. As we will see in Chapter 13, where electrolyte solutions with molecular solvent are treated, the external potential Ψ^{ext} in the definition of the dielectric permittivity function is the potential from the external charges in vacuum, that is, in the absence of the electrolyte solution consisting of the ions *and* the solvent molecules. In order to have the same meaning of $\epsilon(r)$ and $\tilde{\epsilon}(k)$ in the present case where we use a dielectric continuum model of the solvent, we need to use the same definition.

Hence, after the replacement of ε_0 by $\varepsilon_r^{\mathrm{cont}}\varepsilon_0$, the entity Ψ^{ext} is the potential obtained from the external charges in the *presence* of the continuum solvent, which is inappropriate to use in the definitions of $\epsilon(r)$ and $\tilde{\epsilon}(k)$. If Ψ^{ext} is instead the potential *in vacuum*, the potential in the presence of the continuum is $\Psi^{\mathrm{ext}}/\varepsilon_r^{\mathrm{cont}}$. Therefore, all occurrences of Ψ^{ext} should be replaced by $\Psi^{\mathrm{ext}}/\varepsilon_r^{\mathrm{cont}}$ in the formulas above. However, the relationship $\delta\tilde{\Psi}^{\mathrm{ext}}(\mathbf{k}) = \delta\tilde{\Psi}(\mathbf{k})\tilde{\epsilon}(k)$ must be maintained, so in order to ensure this after the replacement of Ψ^{ext} we must also replace $\tilde{\epsilon}(k)$ by $\tilde{\epsilon}(k)/\varepsilon_r^{\mathrm{cont}}$. This implies that $\tilde{\epsilon}(k)$ in all results we obtained above should be replaced by $\tilde{\epsilon}(k)/\varepsilon_r^{\mathrm{cont}}$ and likewise for $\tilde{\epsilon}_{\mathrm{sing}}(k)$ and $\tilde{\epsilon}_{\mathrm{reg}}(k)$. For the same reason $\tilde{\epsilon}^{\mathrm{R}}(k)$ should be replaced by $\tilde{\epsilon}^{\mathrm{R}}(k)\varepsilon_r^{\mathrm{cont}}$.

Thus, in the treatment of electrolytes in a dielectric continuum solvent, for example in the primitive model, we need to do the substitutions

$$\varepsilon_0 \to \varepsilon_r^{\mathrm{cont}}\varepsilon_0,\ \Psi^{\mathrm{ext}} \to \Psi^{\mathrm{ext}}/\varepsilon_r^{\mathrm{cont}},\ \tilde{\epsilon}(k) \to \tilde{\epsilon}(k)/\varepsilon_r^{\mathrm{cont}},\ \tilde{\epsilon}^{\mathrm{R}}(k) \to \tilde{\epsilon}^{\mathrm{R}}(k)\varepsilon_r^{\mathrm{cont}} \tag{12.115}$$

and the related ones

$$\tilde{\epsilon}_{\mathrm{sing}}(k) \to \tilde{\epsilon}_{\mathrm{sing}}(k)/\varepsilon_r^{\mathrm{cont}} \text{ and } \tilde{\epsilon}_{\mathrm{reg}}(k) \to \tilde{\epsilon}_{\mathrm{reg}}(k)/\varepsilon_r^{\mathrm{cont}} \tag{12.116}$$

This is the case irrespectively of the approximation used. All necessary substitutions for a dielectric continuum are made in the current section. (A couple of entities introduced in the next section[14] that are directly related to $\tilde{\varepsilon}(k)$ also need replacements, as discussed there.)

In the following we will deal with **asymptotic statements**, for example that a function $f_1(r)$ decays for large r like another function $f_2(r)$, which we write as $f_1(r) \sim f_2(r)$ when $r \to \infty$.

The following two statements are equivalent

$$f_1(r) \sim f_2(r),\ r \to \infty \quad \Leftrightarrow \quad \lim_{r\to\infty} \frac{f_1(r)}{f_2(r)} = 1.$$

Analogous statements apply for other limits, for instance $f_1(r) \sim f_2(r)$ when $r \to a$.

In an asymptotic expression the terms are placed in the order of their magnitude in the tail of the function for large r. The statement

$$F(r) \sim f_1(r) + f_2(r) + f_3(r),\ r \to \infty$$

means that

$$\lim_{r\to\infty} F(r)/f_1(r) = 1$$
$$\lim_{r\to\infty} [F(r) - f_1(r)]/f_2(r) = 1 \quad \text{and}$$
$$\lim_{r\to\infty} [F(r) - f_1(r) - f_2(r)]/f_3(r) = 1.$$

Let us recall some aspects of the **Poisson-Boltzmann approximation** for electrolytes in a dielectric continuum solvent that was presented in Chapter 6 of Volume I. This will also serve as an introduction to the general analysis of bulk electrolytes because during the recollection we will introduce some important techniques that will be used later to handle the general case. For simplicity the dielectric continuum is assumed to penetrate inside the ions, as is often done in the primitive model of electrolytes and in the PB approximation.

The basic approximation used in the PB approximation is that the potential of mean force $w_j(\mathbf{r})$ for a j-ion located at $\mathbf{r}$ is equal to the bare ionic charge q_i times total electrostatic potential at the same point $\mathbf{r}$

$$w_j(\mathbf{r}) = q_j \Psi(\mathbf{r}) \quad \text{(PB)}. \tag{12.117}$$

Thus it is assumed in this approximation that *electrostatics is local* (cf. the discussion of Equation 12.26). Likewise, the pair potential of mean force $w_{ij}^{(2)}$ between a j-ion and an i-ion is

$$w_{ij}^{(2)}(r) = q_j \psi_i(r), \quad r \geq d_{ij}^{\text{h}} \quad \text{(PB)},$$

[14]For the entities $\mathcal{E}_r^\star(\kappa)$ and $\mathcal{E}_r^{\text{eff}}(\kappa)$ that are introduced in Section 12.3.3, we must replace $\mathcal{E}_r^\star(\kappa) \to \mathcal{E}_r^\star(\kappa)/\varepsilon_r^{\text{cont}}$ and $\mathcal{E}_r^{\text{eff}}(\kappa) \to \mathcal{E}_r^{\text{eff}}(\kappa)/\varepsilon_r^{\text{cont}}$ because Equations 12.152 and 12.156 should remain valid as they stand after the substitution $\tilde{\varepsilon}_{\text{reg}}(k) \to \tilde{\varepsilon}_{\text{reg}}(k)/\varepsilon_r^{\text{cont}}$.

where ψ_i is the mean potential due to the i-ion and d^{h}_{ij} is the distance of closest approach of the ions. Thus, the density of j-ions around an i-ion located at $\mathbf{r}_2$ is

$$n^{(1)}_{j;i}(\mathbf{r}_1|\mathbf{r}_2) = n^{\mathrm{b}}_j g^{(2)}_{ij}(r_{12}) = n^{\mathrm{b}}_j e^{-\beta q_j \psi_i(r_{12})}, \quad r_{12} \geq d^{\mathrm{h}}_{ij} \quad \text{(PB)}. \tag{12.118}$$

Let us place the i-ion (the "central ion") at the origin and assume for simplicity that all ions have the diameter d^{h}. The charge density at distance r from the central i-ion is

$$\rho_i(r) = \sum_j q_j n^{\mathrm{b}}_j e^{-\beta q_j \psi_i(r)}, \quad r \geq d^{\mathrm{h}} \quad \text{(PB)}. \tag{12.119}$$

By inserting this into the Poisson equation 12.85 for ions in a dielectric continuum where $\varepsilon_r^{\mathrm{cont}} \neq 1$ (so ε_0 is replaced by $\varepsilon_r^{\mathrm{cont}}\varepsilon_0$), we obtain the (nonlinear) **Poisson-Boltzmann equation**

$$-\varepsilon_r^{\mathrm{cont}}\varepsilon_0 \nabla^2 \psi_i(r) = \begin{cases} q_i \delta^{(3)}(r), & r < d^{\mathrm{h}} \\ \sum_j q_j n^{\mathrm{b}}_j e^{-\beta q_j \psi_i(r)}, & r \geq d^{\mathrm{h}} \quad \text{(PB)} \end{cases} \tag{12.120}$$

(Equation I:6.29). In general, $g_{ij}(r) \neq g_{ji}(r)$ in the PB approximation, which is an inconsistency that arises because $\psi_i(r)/q_i \neq \psi_j(r)/q_j$ due to the nonlinearity in Equations 12.118 and 12.120, so in general we have $q_j\psi_i(r) \neq q_i\psi_j(r)$.[15]

Provided that $\beta q_j \psi_i(r)$ is sufficiently small we can linearize the exponential function in Equation 12.119 and obtain

$$\rho_i(r) = -\beta \sum_j n^{\mathrm{b}}_j q_j^2 \psi_i(r) = -\kappa_{\mathrm{D}}^2 \varepsilon_r^{\mathrm{cont}} \varepsilon_0 \psi_i(r), \quad r \geq d^{\mathrm{h}} \quad \text{(LPB)}, \tag{12.121}$$

where the Debye parameter κ_{D} is defined in Equation 12.66 and we have used the fact that $\sum_j q_j n^{\mathrm{b}}_j = 0$ due to electroneutrality. By inserting this into the Poisson equation we obtain the **Linearized Poisson-Boltzmann (LPB) equation** $\nabla^2\psi_i(r) = \kappa_{\mathrm{D}}^2 \psi_i(r)$ for $r \geq d^{\mathrm{h}}$. This equation is the basis for the Debye-Hückel theory of electrolyte solutions, which a quite good approximation for sufficiently dilute electrolytes. In the present case the LPB equation has the solution (obtained in Volume I)

$$\psi_i(r) = \frac{q_i^{\mathrm{eff}}}{\varepsilon_r^{\mathrm{cont}}\varepsilon_0} \cdot \frac{e^{-\kappa_{\mathrm{D}} r}}{4\pi r}, \quad r \geq d^{\mathrm{h}} \quad \text{(LPB)} \tag{12.122}$$

[15] This can easily be seen by expanding the exponential function in Equation 12.120 in a power series. For $r \geq d^{\mathrm{h}}$ we have $-\varepsilon_r^{\mathrm{cont}}\varepsilon_0\nabla^2\psi_i(r) = -\beta\sum_j q_j^2 n^{\mathrm{b}}_j \psi_i(r) + \beta^2 \sum_j q_j^3 n^{\mathrm{b}}_j \psi_i^2(r)/2 + \ldots$, which can be written $-\varepsilon_r^{\mathrm{cont}}\varepsilon_0\nabla^2 f_i(r) = -\beta\sum_j q_j^2 n^{\mathrm{b}}_j f_i(r) + q_i\beta^2\sum_j q_j^3 n^{\mathrm{b}}_j [f_i(r)]^2/2 + \ldots$, where $f_i(r) = \psi_i(r)/q_i$. The presence of the second order term makes $f_i(r)$ different from $f_l(r)$ for $l \neq i$ in the general case; for instance for a binary electrolyte we have $f_+(r) \neq f_-(r)$ when $q_+ \neq |q_-|$. Note that the second order term vanishes when $q_+ = |q_-|$.

(Equations I:6.33 and I:6.50), where q_i^{eff} is an effective charge given by

$$q_i^{\text{eff}} = q_i^{\text{eff}}(\kappa_D) = \frac{q_i e^{\kappa_D d^h}}{1 + \kappa_D d^h} \quad \text{(LPB)} \tag{12.123}$$

(Equation I:6.49). In the core region there are no charges apart from q_i at the center, so we *always* have

$$\psi_i(r) = q_i \phi_{\text{Coul}}(r) + \text{const} = \frac{q_i}{4\pi\varepsilon_r^{\text{cont}}\varepsilon_0}\left[\frac{1}{r} - \frac{1}{d^h}\right] + \psi_i(d^h), \quad r \leq d^h, \tag{12.124}$$

where the constant part originates from the spherically symmetric charge distribution outside the core and the terms in the square bracket in the rhs ensures that $\psi_i(r)|_{r=d^h} = \psi_i(d^h)$.[16] The charge density $\rho_i(r)$ is equal to zero for $r < d^h$ and from Equations 12.121 and 12.122 we obtain

$$\rho_i(r) = -\kappa_D^2 q_i^{\text{eff}} \cdot \frac{e^{-\kappa_D r}}{4\pi r}, \quad r \geq d^h \quad \text{(LPB)}. \tag{12.125}$$

We can also obtain this result by inserting Equation 12.122 into the Poisson equation and using the identity

$$\nabla^2 f(r) = \frac{1}{r} \cdot \frac{d^2[rf(r)]}{dr^2}. \tag{12.126}$$

The total charge density associated with the central ion is

$$\rho_i^{\text{tot}}(r) = \begin{cases} q_i \delta^{(3)}(r), & r < d^h \\ -\kappa_D^2 q_i^{\text{eff}} \dfrac{e^{-\kappa_D r}}{4\pi r}, & r \geq d^h \end{cases} \quad \text{(LPB)} \tag{12.127}$$

In the limit of zero ionic density, i.e., $n_j^b \to 0$ for all ionic species j, the Debye parameter $\kappa_D \to 0$ and the Debye length $1/\kappa_D \to \infty$. Furthermore, from Equation 12.123 it follows that $q_i^{\text{eff}} \to q_i$ in the same limit.

A function $e^{-\kappa r}/r$, where κ is some constant, is called a **Yukawa function** and its decay length is $1/\kappa$. Thus, the mean electrostatic potential and the charge density around an ion decays like a Yukawa function with decay parameter $\kappa = \kappa_D$ in the LPB approximation. In the limit of zero ionic density it is, in fact, a general exact result that $\psi_i(r)$ and $\rho_i(r)$ decays for large r like a Yukawa function $e^{-\kappa r}/r$ and, furthermore, that $\kappa/\kappa_D \to 1$ and $q_i^{\text{eff}} \to q_i$ in the same zero density limit. As we will see, a Yukawa function decay of $\psi_i(r)$ and $\rho_i(r)$ for large r is characteristic for electrolytes when the ion density is sufficiently low.

When a bulk electrolyte is exposed to a weak external potential $\delta\Psi^{\text{Ext}}$ that gives rise to the total potential $\delta\Psi$, we have $\delta w_j(\mathbf{r}_1) = q_j \delta\Psi(\mathbf{r}_1)$ in the

[16] $\psi_i(r)$ is continuous at $r = d^h$.

PB approximation (Equation 12.117). Since $\delta w_j(\mathbf{r}_1) = \int d\mathbf{r}_2\, \rho_j^\star(r_{12})\, \delta\Psi(\mathbf{r}_2)$ (Equation 12.78) it is seen that this implies the approximation $\rho_j^\star(r_{12}) = q_j\delta^{(3)}(r_{12})$. From $\rho_j^\star(r_{12}) = q_j\delta^{(3)}(r_{12}) + \rho_j^{\mathrm{dress}}(r_{12})$ we see that $\rho_j^{\mathrm{dress}} = 0$, that is, in the PB approximation an ion is treated as a bare ionic charge q_j and not as a dressed ion.[17] Since $\chi^\star(r_{12}) = -\beta\sum_j n_j^{\mathrm{b}} q_j \rho_j^\star(r_{12})$ (Equation 12.80) it follows that in the PB approximation we have

$$\chi^\star(r_{12}) = -\beta\sum_j n_j^{\mathrm{b}} q_j^2 \delta^{(3)}(r_{12}) = -\varepsilon_r^{\mathrm{cont}}\varepsilon_0\kappa_{\mathrm{D}}^2\delta^{(3)}(r_{12}) \quad \text{(PB)} \qquad (12.128)$$

(for ions distributed in vacuum one sets $\varepsilon_r^{\mathrm{cont}} = 1$). Thus $\tilde{\chi}^\star(k) = -\varepsilon_r^{\mathrm{cont}}\varepsilon_0\kappa_{\mathrm{D}}^2$, which is independent of k. From $\tilde{\epsilon}(k)/\varepsilon_r^{\mathrm{cont}} = 1 - \tilde{\chi}^\star(k)\tilde{\phi}_{\mathrm{Coul}}(k) = 1 - \tilde{\chi}^\star(k)/[\varepsilon_r^{\mathrm{cont}}\varepsilon_0 k^2]$ we obtain

$$\frac{\tilde{\epsilon}(k)}{\varepsilon_r^{\mathrm{cont}}} = 1 + \frac{\kappa_{\mathrm{D}}^2}{k^2} \qquad \text{(PB).} \qquad (12.129)$$

It is instructive to see that this result can also be obtained from $\tilde{\epsilon}(k) = \tilde{\epsilon}_{\mathrm{sing}}(k) + \tilde{\epsilon}_{\mathrm{reg}}(k)$ with the expression for $\tilde{\epsilon}_{\mathrm{reg}}(k)$ in a dielectric continuum

$$\tilde{\epsilon}_{\mathrm{reg}}(k) \equiv \varepsilon_r^{\mathrm{cont}} - \frac{\tilde{\chi}^\star(k) - \tilde{\chi}^\star(0)}{\varepsilon_0 k^2} \qquad \text{(diel. continuum)} \qquad (12.130)$$

(this can be obtained from the definition 12.98 of $\tilde{\epsilon}_{\mathrm{reg}}(k)$ with the substitutions 12.115 and 12.116). Since $\tilde{\chi}^\star(k) = \tilde{\chi}^\star(0)$ in the PB approximation ($\tilde{\chi}^\star(k)$ is independent of k) and $q_i^\star = q_i$ for ions without a dress we obtain

$$\tilde{\epsilon}_{\mathrm{reg}}(k) = \varepsilon_r^{\mathrm{cont}} \quad \text{and} \quad \tilde{\epsilon}_{\mathrm{sing}}(k) = \frac{\beta}{\varepsilon_0 k^2}\sum_i n_i^{\mathrm{b}} q_i^2 = \varepsilon_r^{\mathrm{cont}}\frac{\kappa_{\mathrm{D}}^2}{k^2} \qquad \text{(PB),}$$

which yields Equation 12.129.

By inserting Equation 12.129 into $\tilde{\phi}_{\mathrm{Coul}}^\star(k) = 1/[\varepsilon_0 k^2\tilde{\epsilon}(k)]$ (Equation 12.106) we obtain

$$\tilde{\phi}_{\mathrm{Coul}}^\star(k) = \frac{1}{\varepsilon_r^{\mathrm{cont}}\varepsilon_0(k^2 + \kappa_{\mathrm{D}}^2)} \qquad \text{(PB),} \qquad (12.131)$$

which implies that (cf. Equation I:6.69)

$$\phi_{\mathrm{Coul}}^\star(r) = \frac{1}{\varepsilon_r^{\mathrm{cont}}\varepsilon_0}\cdot\frac{e^{-\kappa_D r}}{4\pi r} = e^{-\kappa_D r}\phi_{\mathrm{Coul}}(r) \qquad \text{(PB)} \qquad (12.132)$$

because the inverse Fourier transform of $1/(k^2 + a^2)$ in three dimensions is equal to $e^{-ar}/(4\pi r)$.

[17] An exception is the central ion when dealing with pair distribution functions, see Equation 12.133.

In the PB approximation, which says

$$\rho_i(r) = \sum_j q_j n_j^{\mathrm{b}} e^{-\beta q_j \psi_i(r)} \qquad \text{for } r \geq d^{\mathrm{h}} \quad \text{(PB)}$$

(Equations 12.118 and 12.119), the central ion is, in fact, treated differently from the surrounding ions. As we have seen, the latter are approximated as being point ions with charge q_j without any dress and their interaction with the mean potential ψ_i from the central ion is given by $q_j\psi_i$. The central ion does, however, have a dress and its dressed-ion charge density is

$$\begin{aligned}\rho_i^{\star}(r_{13}) &= \rho_i^{\mathrm{tot}}(r_{13}) - \int d\mathbf{r}_4\, \psi_i(r_{14})\chi^{\star}(r_{34}) \\ &= q_i\delta^{(3)}(r_{13}) + \rho_i(r_{13}) + \varepsilon_r^{\mathrm{cont}}\varepsilon_0\kappa_{\mathrm{D}}^2\psi_i(r_{13}) \qquad \text{(PB)}, \qquad (12.133)\end{aligned}$$

(from Equation 12.87), where the linear part of the polarization response is subtracted from ρ_i^{tot}.

Let us expand the exponential function in the expression for $\rho_i(r)$ above in a Taylor series for $r \geq d^{\mathrm{h}}$

$$\begin{aligned}\rho_i(r) &= \sum_j q_j n_j^{\mathrm{b}}\left[1 - \beta q_j\psi_i(r) + \tfrac{1}{2}\left(\beta q_j \psi_i(r)\right)^2 - \tfrac{1}{3!}\left(\beta q_j\psi_i(r)\right)^3 + \ldots\right] \\ &= -\varepsilon_r^{\mathrm{cont}}\varepsilon_0\kappa_{\mathrm{D}}^2\psi_i(r) + \tfrac{1}{2}\sum_j q_j^3 n_j^{\mathrm{b}}\beta^2\psi_i^2(r) \\ &\quad - \tfrac{1}{3!}\sum_j q_j^4 n_j^{\mathrm{b}}\beta^3\psi_i^3(r) + \ldots \qquad (12.134)\end{aligned}$$

whereby Equation 12.133 yields

$$\rho_i^{\star}(r) = \begin{cases} q_i\delta^{(3)}(r) + \varepsilon_r^{\mathrm{cont}}\varepsilon_0\kappa_{\mathrm{D}}^2\psi_i(r), & r < d^{\mathrm{h}} \\ \frac{1}{2}\sum_j q_j^3 n_j^{\mathrm{b}}\beta^2\psi_i^2(r) - \frac{1}{3!}\sum_j q_j^4 n_j^{\mathrm{b}}\beta^3\psi_i^3(r) + \ldots, & r \geq d^{\mathrm{h}} \end{cases} \quad \text{(PB)} \qquad (12.135)$$

(cf. Equation I:6.63). Thus, outside the core $\rho_i^{\star}(r)$ is equal to the sum of the *nonlinear contributions* to $\rho_i(r)$ in the PB approximation. This is an explicit example of the previous conclusion that the dressed charge density $\rho_i^{\star}$ in the general case contains the central ionic charge and the nonlinear electrostatic contributions due to ψ_i.[18]

Furthermore, we see from the first line in Equation 12.135 that the linear term contributes inside the core, so the dress extends to the inside (this is also true in the general case). This part of the dress is, however, not a real charge distribution; it is a *nominal* charge distribution and is included in order to remove the polarization charge that the linear response to ψ_i *would have placed*

[18] In general there are also contributions from nonelectrostatic interactions with the central ion, but in the present case where u_{ij}^{sh} is equal to the hard core potential, such interactions are zero outside the core.

inside the central ion if the latter did not have a hard core. The contribution inside the ion is, in fact, vital in order to assure that the potential ψ_i is the same when it is calculated from $\rho_i^\star$ and $\phi_{\text{Coul}}^\star$ via the screened version of the Coulomb potential formula, i.e., the first equality in

$$\psi_i(r_{13}) = \int d\mathbf{r}_4\, \rho_i^\star(r_{14})\phi_{\text{Coul}}^\star(r_{43}) = \int d\mathbf{r}_2\, \rho_i^{\text{tot}}(r_{12})\phi_{\text{Coul}}(r_{23}) \qquad (12.136)$$

(Equation 12.83), as when it is calculated from ρ_i^{tot} and ϕ_{Coul} from the usual Coulomb potential formula, i.e., the rhs.

This fact is particularly clear in the LPB approximation. Then the nonlinear terms are absent so $\rho_i^\star(r) = 0$ for $r \geq d^{\text{h}}$, whereby the dressed ion and the bare ionic charge differ only by the nominal charge distribution inside the core

$$\rho_i^\star(r) = \begin{cases} q_i\delta^{(3)}(r) + \varepsilon_r^{\text{cont}}\varepsilon_0\kappa_{\text{D}}^2\psi_i(r), & r < d^{\text{h}} \\ 0, & r \geq d^{\text{h}} \end{cases} \quad \text{(LPB)}. \qquad (12.137)$$

When this $\rho_i^\star$ and $\phi_{\text{Coul}}^\star$ from Equation 12.132 are inserted into of Equation 12.136 one obtains $\psi_i(r)$ for all r,[19] that is, one obtains

$$\psi_i(r) = \frac{q_i^{\text{eff}}}{\varepsilon_r^{\text{cont}}\varepsilon_0} \cdot \frac{e^{-\kappa_{\text{D}} r}}{4\pi r}, \quad r \geq d^{\text{h}} \quad \text{(LPB)}$$

with $q_i^{\text{eff}} = q_i e^{\kappa_{\text{D}} d^{\text{h}}}/(1 + \kappa_{\text{D}} d^{\text{h}})$ (Equations 12.122 and 12.123) and for $r < d^{\text{h}}$ one obtains $\psi_i(r)$ given by Equation 12.124. This is, of course, also obtained when ρ_i^{tot} from Equation 12.127 is inserted in the rhs of Equation 12.136.

For the nonlinear PB case it was shown in Example 6.3 in Volume I that $\psi_i(r)$ decays like

$$\psi_i(r) \sim \frac{q_i^{\text{eff}}}{\varepsilon_r^{\text{cont}}\varepsilon_0} \cdot \frac{e^{-\kappa_{\text{D}} r}}{4\pi r} \sim q_i^{\text{eff}}\phi_{\text{Coul}}^\star(r), \quad r \to \infty \quad \text{(PB)} \qquad (12.138)$$

(Equation I:6.75), where

$$q_i^{\text{eff}} = q_i^{\text{eff}}(\kappa_{\text{D}}) = \int d\mathbf{r}'\, \rho_i^\star(r')\frac{\sinh(\kappa_{\text{D}} r')}{\kappa_{\text{D}} r'} \qquad (12.139)$$

(Equation I:6.76). These results were obtained from the screened Coulomb potential formula, Equation 12.136, with $\phi_{\text{Coul}}^\star(r) = e^{-\kappa_{\text{D}} r}/(4\pi\varepsilon_r^{\text{cont}}\varepsilon_0 r)$ inserted.[20] In Equation 12.139, $\rho_i^\star$ is taken from Equations 12.133 that is valid

[19] This fact is most easily shown in Fourier space.

[20] In the derivations in Example 6.3, the variable $\mathbf{r}'$ in the screened Coulomb potential formula $\psi_i(r) = \int d\mathbf{r}'\, \rho_i^\star(r')\phi_{\text{Coul}}^\star(|\mathbf{r} - \mathbf{r}'|)$ was expressed in spherical polar coordinates. The angular integration was done separately and we utilized the mathematical identity

$$\int d\hat{\mathbf{r}}'\, \frac{e^{-\kappa_D|\mathbf{r}-\mathbf{r}'|}}{|\mathbf{r} - \mathbf{r}'|} = \frac{e^{-\kappa_D r}}{r} \cdot \frac{4\pi \sinh(\kappa_D r')}{\kappa_D r'} \quad \text{when } r > r'$$

(Equation I:6.73), where $d\hat{\mathbf{r}}'$ denote the angular differential. The first factor in the rhs of this equation gives the r dependence in Equation 12.138 and the second factor appears in the integrand of Equation 12.139.

in the nonlinear PB approximation. The charge density $\rho_i(r)$ decays as

$$\rho_i(r) \sim -\kappa_D^2 q_i^{\text{eff}} \frac{e^{-\kappa_D r}}{4\pi r}, \quad r \to \infty, \quad \text{(PB)} \tag{12.140}$$

(from Equation 12.138 and the Poisson equation).

Equations 12.138 and 12.140 are asymptotic results that are valid for large r. As we will see later, there are other contributions to $\psi_i(r)$ and $\rho_i(r)$ that do not decay like Yukawa functions, but they decay faster to zero when $r \to \infty$ so the Yukawa function term shown in each case is the dominant contribution for large r. In the LPB approximation there is solely a Yukawa function term for $r \geq d^{\text{h}}$ and no other contributions.

In Exercise I:6.5 in Volume I it was explicitly shown in the LPB approximation that Equation 12.139 with $\rho_i^\star$ from Equation 12.137 yields $q_i^{\text{eff}} = q_i e^{\kappa_D d^{\text{h}}}/(1 + \kappa_D d^{\text{h}})$ in agreement with Equation 12.123. The PB approximation gives a different q_i^{eff} value since $\rho_i^\star$ are different in the two cases; both values are approximations to the actual q_i^{eff} of the system. The PB value is approximately equal to the LPB value when $\beta q_j \psi_i(r)$ is small for all $r \geq d^{\text{h}}$ and the nonlinear terms in Equation 12.135 therefore are unimportant. In the limit of zero ionic density $q_i^{\text{eff}} \to q_i$ in all cases and Equations 12.138 and 12.140 with $q_i^{\text{eff}} = q_i$ are, in fact, general exact results in this limit.

We see that it is important to distinguish between the ion distribution of the cloud and that of the dress. The latter form $\rho_i^\star$ together with the central charge of the ion and it is this entity that is used to calculate the effective charge of an ion. One may think that the effective charge is a measure of the charge of the ion and that in the cloud close to the ion, but this is not the case. It is the dressed-ion charge density $\rho_i^\star$ that matters.

There exist a mathematically elegant method to obtain asymptotic formulas like Equation 12.138 with coefficient given by Equation 12.139 from the Fourier transforms of the functions. This is explained for the general case in the following shaded text box and we will apply this method to obtain Equation 12.138 from

$$\tilde{\psi}_i(k) = \tilde{\rho}_i^\star(k)\tilde{\phi}_{\text{Coul}}^\star(k) = \frac{\tilde{\rho}_i^\star(k)}{\varepsilon_0 k^2 \tilde{\epsilon}(k)} = \frac{\tilde{\rho}_i^\star(k)}{\varepsilon_r^{\text{cont}} \varepsilon_0 (k^2 + \kappa_D^2)} \quad \text{(PB)}, \tag{12.141}$$

which follows from Equations 12.106, 12.107 and 12.131. The method will first be explained in relatively simple terms and then we will show a general manner to derive it. It is important in the general analysis of pair correlations and screening in electrolytes.

Asymptotic decay determined by analysis in complex k-space.

In this text box we will investigate how one can determine that a radial function $\mathsf{F}(r)$ decays like a Yukawa function $e^{-ar}/(4\pi r)$ when $r \to \infty$ by analyzing the three-dimensional Fourier transform $\tilde{\mathsf{F}}(k)$ of $\mathsf{F}(r)$.

When we obtained Equation 12.132 we used the fact that the Fourier transform of $e^{-ar}/(4\pi r)$ is $1/(k^2+a^2)$. We will now investigate other functions that decay in the same manner for large r. Let us denote $e^{-ar}/(4\pi r) = \mathsf{f}_0(r)$ for $r \geq 0$ and consider the function

$$\mathsf{f}_d(r) = \begin{cases} 0, & r < d \\ \dfrac{e^{-ar}}{4\pi r} & r \geq d, \end{cases}$$

which has the Fourier transform

$$\tilde{\mathsf{f}}_d(k) = \frac{e^{-ad}}{k^2+a^2}\left[\cos(dk) + ad\frac{\sin(dk)}{dk}\right].$$

Note that $\tilde{\mathsf{f}}_d(k)$ has k^2+a^2 in the denominator and the same applies to $\tilde{\mathsf{f}}_0(k) = 1/(k^2+a^2)$. Since $k^2+a^2 = 0$ when $k = \pm ia$, where i is the imaginary unit, both functions have simple poles for $k = \pm ia$.

Let us investigate $\tilde{\mathsf{f}}_d(k)$ when $k \to \pm ia$. The square bracket in $\tilde{\mathsf{f}}_d(k)$ is equal to e^{ad} for $k = \pm ia$ so $\tilde{\mathsf{f}}_d(k)$ diverges like

$$\tilde{\mathsf{f}}_d(k) \sim \frac{1}{k^2+a^2} \quad \text{when } k \to \pm ia.$$

Thus $1/(k^2+a^2)$ is the dominant contribution to $\tilde{\mathsf{f}}_d(k)$ when k approaches any of the poles $\pm ia$. Note that this is a property of $\tilde{\mathsf{f}}_d(k)$ that is independent of the value of $d \geq 0$. The appearance of the contribution $1/(k^2+a^2)$ is thus connected to the ultimate tail of $\mathsf{f}_d(r)$ when $r \to \infty$.

The same applies to any function that decays like $e^{-ar}/(4\pi r)$ when $r \to \infty$. Consider a function $\mathsf{F}(r)$ that decays in this manner when $r \to \infty$ and define $\mathsf{G}(r) \equiv \mathsf{F}(r) - \mathsf{f}_d(r)$ for a large, but arbitrary d value. We have $\mathsf{G}(r) = \mathsf{F}(r)$ for $r < d$, but the tails of $\mathsf{F}(r)$ and $\mathsf{f}_d(r)$ for very large r cancel each other and we have $\mathsf{G}(r)/[e^{-ar}/(4\pi r)] \to 0$ when $r \to \infty$. Thus $|\mathsf{G}(r)|$ decays faster to zero with increased r than $e^{-ar}/(4\pi r)$ in this limit. This is true irrespectively of how large d is. We have $\tilde{\mathsf{G}}(k) = \tilde{\mathsf{F}}(k) - \tilde{\mathsf{f}}_d(k)$, where the second term gives the contribution $-1/(k^2+a^2)$ when $k \to \pm ia$ independently of the value of d. $\tilde{\mathsf{F}}(k)$ has a contribution $1/(k^2+a^2)$ from the ultimate tail of $\mathsf{F}(r)$ that cancels this $-1/(k^2+a^2)$ and $\tilde{\mathsf{G}}(k)$ stays finite when $k \to \pm ia$.

Any function $\mathsf{F}(r)$ that decays like

$$\mathsf{F}(r) \sim \frac{\mathsf{C}e^{-ar}}{4\pi r} \quad \text{when } r \to \infty \tag{12.142}$$

with some constant C has a Fourier transform $\tilde{\mathsf{F}}(k)$ with simple poles at $k = \pm ia$ and we have $\tilde{\mathsf{F}}(k) \sim \mathsf{C}/(k^2+a^2)$ when $k \to \pm ia$. Vice versa, any $\tilde{\mathsf{F}}(k)$ with simple poles at $k = \pm ia$ has a contribution to $\mathsf{F}(r)$ that decays like $\mathsf{C}e^{-ar}/(4\pi r)$ for some constant $\mathsf{C} \neq 0$ when $r \to \infty$.

Note that $\tilde{\mathsf{F}}(-k) = \tilde{\mathsf{F}}(k)$ (Equation 9.241), so we only need to consider one of the poles, $k = \mathrm{i}a$, because $k = -\mathrm{i}a$ must also be a pole. We can write $\tilde{\mathsf{F}}$ as

$$\tilde{\mathsf{F}}(k) = \mathsf{s}(k)/\mathsf{t}(k), \tag{12.143}$$

where $\mathsf{t}(k)$ has a simple zero at $k = \mathrm{i}a$ and $\mathsf{s}(k)$ is finite and nonzero there. Since

$$\lim_{k \to \mathrm{i}a} \frac{\mathsf{t}(k)}{k^2 + a^2} = \left. \frac{\mathsf{t}'(k)}{2k} \right|_{k=\mathrm{i}a},$$

where we have used l'Hospital's rule, we have

$$\tilde{\mathsf{F}}(k) = \frac{\mathsf{s}(k)}{\mathsf{t}(k)} \sim \left. \frac{\mathsf{s}(k)}{\mathsf{t}'(k)/(2k)} \right|_{k=\mathrm{i}a} \frac{1}{k^2 + a^2} \quad \text{when} \quad k \to \mathrm{i}a$$

and hence

$$\mathsf{C} = \frac{\mathsf{s}(\mathrm{i}a)}{\mathsf{t}'(\mathrm{i}a)/(2\mathrm{i}a)}, \tag{12.144}$$

which gives the coefficient C in terms of $\mathsf{s}(k)$ and $\mathsf{t}(k)$.

For completeness, we will derive these results from pole analysis in a more elaborate manner by using Cauchy's residue theorem in complex analysis. Our task is to evaluate the inverse Fourier transform

$$\mathsf{F}(r) = \frac{1}{2\pi^2} \int_0^\infty dk\, \tilde{\mathsf{F}}(k) k \left[\frac{e^{\mathrm{i}kr} - e^{-\mathrm{i}kr}}{2\mathrm{i}r} \right]$$

(Equation 9.244), where we have inserted $\sin(kr) = (e^{\mathrm{i}kr} - e^{-\mathrm{i}kr})/2\mathrm{i}$. Since $\tilde{\mathsf{F}}(-k) = \tilde{\mathsf{F}}(k)$ and $\tilde{\mathsf{F}}(k) = \mathsf{s}(k)/\mathsf{t}(k)$ we can write this as

$$\mathsf{F}(r) = \frac{1}{(2\pi)^2} \int_{-\infty}^{\infty} dk\, \frac{\mathsf{s}(k)k}{\mathsf{t}(k)} \cdot \frac{e^{\mathrm{i}kr}}{\mathrm{i}r}.$$

Assume for simplicity that $\mathsf{s}(k)$ and $\mathsf{t}(k)$ are analytic functions in complex k space and that $\mathsf{t}(k)$ has a number of simple zeros $\{\pm \mathrm{i}a_1, \pm \mathrm{i}a_2, \pm \mathrm{i}a_3, \dots\}$ while $\mathsf{s}(k) \neq 0$ there, so $\tilde{\mathsf{F}}(k)$ has poles for these k values.[21] The integral over k can be evaluated by closing the contour integral around the upper half plane in complex k space and we have from the residue theorem[22]

$$\mathsf{F}(r) = \frac{1}{(2\pi)^2} \cdot 2\pi\mathrm{i} \sum_l \operatorname*{Res}_{k=\mathrm{i}a_l} \left(\frac{\mathsf{s}(k)k}{\mathsf{t}(k)} \cdot \frac{e^{\mathrm{i}kr}}{\mathrm{i}r} \right) = 2 \sum_l \frac{\mathsf{s}(\mathrm{i}a_l)\mathrm{i}a_l}{\mathsf{t}'(\mathrm{i}a_l)} \cdot \frac{e^{-a_l r}}{4\pi r},$$

---

[21] In general, $\mathsf{s}(k)$ and $\mathsf{t}(k)$ are not analytic functions and $\tilde{\mathsf{F}}(k)$ has other singularities apart from simple poles. Here, we only treat contributions from poles and we will consider some other kinds of singularities later.

[22] We here assume that in the integrand of $\mathsf{F}(r)$ the function $\mathsf{s}(k)ke^{\mathrm{i}kr}/\mathsf{t}(k)$ with $e^{\mathrm{i}kr} = e^{\mathrm{i}\Re(k)r}e^{-\Im(k)r}$ goes to zero when $\Im(k) \to \infty$ so that the contribution from the contour integral around the upper half plane in complex k space gives zero.

where $\sum_l \operatorname*{Res}_{k=\mathrm{i}a_l}(\cdot)$ is the sum of the residues[23] at the simple poles $k = \mathrm{i}a_l$ in the upper half plane of complex k space. Thus each pole $k = \mathrm{i}a_l$ gives rise to a contribution $\mathsf{C}_l e^{-a_l r}/(4\pi r)$ to $\mathsf{F}(r)$ with $\mathsf{C}_l = 2\mathsf{s}(\mathrm{i}a_l)\mathrm{i}a_l/\mathsf{t}'(\mathrm{i}a_l)$ in agreement with Equation 12.144 and we obtain

$$\mathsf{F}(r) = \sum_l \frac{\mathsf{C}_l e^{-a_l r}}{4\pi r}. \tag{12.145}$$

We have $\mathsf{F}(r) \sim \mathsf{C}_1 e^{-a_1 r}/(4\pi r)$ when $r \to \infty$, where $\mathrm{i}a_1$ is the pole with smallest value of $a_l > 0$, that is, the pole that lies closest to the real axis.

Equation 12.145 can also be applied for complex-valued a_l with $\Re(a_l) > 0$, where $\Re(\cdot)$ is the real part. In our applications of these results, $\mathrm{i}\overline{a_l}$ is also a pole as will be explained later ($\overline{a_l}$ is the complex conjugate of a_l). The pair of poles give a contribution to $\mathsf{F}(r)$ that is exponentially damped oscillatory with decay length $1/\Re(a_l)$ and wave length $2\pi/\Im(a_l)$, where $\Im(\cdot)$ is the imaginary part. This will be explored in more detail later (cf. Equations 12.173 and 12.174).

In the LPB approximation $\rho_i(r)$ is given by Equation 12.125 and we have

$$\tilde{\rho}_i(k) = -\frac{\kappa_{\mathrm{D}}^2 q_i^{\mathrm{eff}}}{k^2 + \kappa_{\mathrm{D}}^2} \quad \text{(LPB)}.$$

There are only the purely imaginary poles $k = \pm\mathrm{i}\kappa_{\mathrm{D}}$, so if we apply Equation 12.145 to $\mathsf{F}(r) = \rho_i(r)$, there is only one term in the sum and it has $a_1 = \kappa_{\mathrm{D}}$. In the nonlinear PB approximation $\tilde{\rho}_i(k)$ has the same poles, $k = \pm\mathrm{i}\kappa_{\mathrm{D}}$, and there is only one term that decays like $e^{-\kappa_{\mathrm{D}} r}/(4\pi r)$, but in this case there are other contributions as well. Let us investigate the decay of $\rho_i(r)$ in this latter approximation. From the Taylor expansion of $\rho_i(r)$ in Equation 12.134 and the decay of $\psi_i(r)$ in Equation 12.138 it follows that[24]

$$\begin{aligned}\rho_i(r) &\sim -\frac{\kappa_{\mathrm{D}}^2 q_i^{\mathrm{eff}}}{4\pi} \cdot \frac{e^{-\kappa_{\mathrm{D}} r}}{r} + \mathsf{B}_i \frac{e^{-2\kappa_{\mathrm{D}} r}}{r^2} \\ &+ \text{terms proportional to } \frac{e^{-m\kappa_{\mathrm{D}} r}}{r^\ell} \text{ with } m > 2, \ell \geq 2 \qquad \text{(PB)},\end{aligned} \tag{12.146}$$

when $r \to \infty$, where

$$\mathsf{B}_i = \tfrac{1}{2} \sum_j q_j^3 n_j^{\mathrm{b}} \beta^2 \left[\frac{q_i^{\mathrm{eff}}}{4\pi\varepsilon_r^{\mathrm{cont}}\varepsilon_0}\right]^2.$$

[23]The residue of a function $f(k)$ at a simple pole $k = b$ is $\operatorname*{Res}_{k=b}(f(k)) = \lim_{k\to b}(k-b)f(k)$. If $f(k) = f_1(k)/f_2(k)$, where $f_2(k)$ has a simple zero for $k = b$ and $f_1(b) \neq 0$, we have $\lim_{k\to b}(k-b)f(k) = f_1(b)/f_2'(b)$.

[24]One also has to use the PB equation to assess the coefficients of the terms in the second line of Equation 12.146.

The second term decays faster with increasing r than the first one; its decay length is half a Debye length. The decay of all terms can be analyzed in complex k space and for each term there is a different kind of singularity in complex k space. The singularity determines the mathematical form of the decay of each term and vice versa. The analysis is done for the second term in the following paragraph, which can be skipped in the first reading (it is included here for later reference).

★ Equation 12.145 is valid for functions that have no other singularities than simples poles in Fourier space. In general, $\tilde{\mathsf{F}}(k)$ also has other singularities and there are terms in $\mathsf{F}(r)$ with different mathematical forms than $e^{-ar}/(4\pi r)$, that is, in addition to those in Equation 12.145. If $\mathrm{i}a_1$ is the simple pole that lies nearest the real axis in complex k space and all other singularities lie further away, we have $\mathsf{F}(r) \sim \mathsf{C}_1 e^{-a_1 r}/(4\pi r)$ when $r \to \infty$ and the other terms decay faster with increasing r. A concrete example is $\rho_i(r)$ in the nonlinear PB approximation. In complex k space the first term in 12.146 gives the simple poles at $k = \pm\mathrm{i}\kappa_\mathrm{D}$. The second term has another type of singularity at $k = \pm\mathrm{i}2\kappa_\mathrm{D}$ because the Fourier transform of

$$f(r) = \frac{e^{-2\kappa_\mathrm{D} r}}{4\pi r^2}$$

is equal to $\tilde{f}(k) = \arctan(k/(2\kappa_\mathrm{D}))/k$, which has a logarithmic branch point singularity at $k = \pm\mathrm{i}2\kappa_\mathrm{D}$. This can be seen by setting $k = \mathrm{i}y$, which gives

$$\tilde{f}(\mathrm{i}y) = \frac{\arctan(\mathrm{i}y/(2\kappa_\mathrm{D}))}{\mathrm{i}y} = \frac{\ln\left[(2\kappa_\mathrm{D} + y)/(2\kappa_\mathrm{D} - y)\right]}{2y}, \qquad (12.147)$$

so $\tilde{f}(k)$ diverges logarithmically at $k = \mathrm{i}y = \pm\mathrm{i}2\kappa_\mathrm{D}$. The higher order terms in Equation 12.146 have yet other types of singularities for $k = \pm\mathrm{i}\alpha\kappa_\mathrm{D}$ with $\alpha \geq 2$ and decay even faster with increasing r.

We are now ready to demonstrate how the decay behavior of $\psi_i(r)$ shown in Equation 12.138 with coefficient q_i^{eff} given by Equation 12.139 can easily be obtained from $\tilde{\psi}_i(k)$ in Equation 12.141. We see that $\tilde{\psi}_i(k)$ has the form (cf. Equation 12.143)

$$\tilde{\psi}_i(k) = \mathsf{s}(k)/\mathsf{t}(k) = \tilde{\mathsf{F}}(k)$$

with

$$\mathsf{s}(k) = \tilde{\rho}_i^{\star}(k) \quad \text{and} \quad \mathsf{t}(k) = \varepsilon_0 k^2 \tilde{\epsilon}(k) = \varepsilon_r^{\mathrm{cont}} \varepsilon_0 (k^2 + \kappa_\mathrm{D}^2).$$

The function $\mathsf{t}(k)$ is zero when $k^2\tilde{\epsilon}(k) = k^2 + \kappa_\mathrm{D}^2$ is zero, that is, for $k = \pm\mathrm{i}\kappa_\mathrm{D}$, so $\tilde{\psi}_i(k)$ has poles for these k values. From Equations 12.142–12.144 and the fact that $\mathsf{t}'(\mathrm{i}\kappa_\mathrm{D})/(2\mathrm{i}\kappa_\mathrm{D}) = \varepsilon_r^{\mathrm{cont}}\varepsilon_0$ it therefore follows that[25]

$$\psi_i(r) \sim \frac{\tilde{\rho}_i^\star(\mathrm{i}\kappa_\mathrm{D})}{\varepsilon_r^{\mathrm{cont}}\varepsilon_0} \cdot \frac{e^{-\kappa_\mathrm{D} r}}{4\pi r},$$

so we can designate

$$q_i^{\mathrm{eff}}(\kappa_\mathrm{D}) = \tilde{\rho}_i^\star(\mathrm{i}\kappa_\mathrm{D}).$$

Since

$$\tilde{\rho}_i^\star(\mathrm{i}\kappa_\mathrm{D}) = \int d\mathbf{r}\, \rho_i^\star(r)\frac{\sinh(\kappa_\mathrm{D} r)}{\kappa_\mathrm{D} r},$$

(from Equation 9.242) the result agrees with Equation 12.139.

12.3.3 Electrostatic screening and decay modes in the general case

We are now ready to do an exact analysis of ionic correlations and screening in bulk electrolytes consisting of simple ions distributed in vacuum (a classical plasma). The case of ions in a dielectric continuum solvent can be obtained by using the substitutions in Equations 12.115, 12.116 and footnote 14. A more elementary, but nevertheless exact treatment of the latter case was done in Section 6.2 in Volume I. The theory is formally exact and expressed in a manner that helps the physical interpretation of the relationships. In order to obtain numerical results, we need, however, to introduce approximations in most cases.

As mentioned earlier, for low ionic densities the electrostatic potential $\psi_i(r)$ around an ion decays like a Yukawa function when $r \to \infty$. The decay is like in the PB approximation, Equation 12.138, but as we will see, the values of the decay length and the prefactor are in general different. It is suitable to simultaneously investigate $\psi_i(r)$ and the unit screened Coulomb potential $\phi_{\mathrm{Coul}}^\star(r)$, which are given in Fourier space by

$$\tilde{\phi}_{\mathrm{Coul}}^\star(k) = \frac{1}{\beta\sum_i n_i^{\mathrm{b}} q_i q_i^\star + \varepsilon_0 k^2 \tilde{\epsilon}_{\mathrm{reg}}(k)}$$

and

$$\tilde{\psi}_i(k) = \tilde{\rho}_i^\star(k)\tilde{\phi}_{\mathrm{Coul}}^\star(k) = \frac{\tilde{\rho}_i^\star(k)}{\beta\sum_i n_i^{\mathrm{b}} q_i q_i^\star + \varepsilon_0 k^2 \tilde{\epsilon}_{\mathrm{reg}}(k)}$$

(Equations 12.109 and 12.110). Both $\tilde{\phi}_{\mathrm{Coul}}^\star(k)$ and $\tilde{\psi}_i(k)$ have the form $\tilde{\mathsf{F}}(k) = \mathsf{s}(k)/\mathsf{t}(k)$ with

$$\mathsf{t}(k) = \beta\sum_i n_i^{\mathrm{b}} q_i q_i^\star + \varepsilon_0 k^2 \tilde{\epsilon}_{\mathrm{reg}}(k) \tag{12.148}$$

[25] Since $\rho_i^\star(r)$ decays like $[\psi_i(r)]^2$ when $r \to \infty$ (Equation 12.135), it decays faster to zero than $\psi_i(r)$. The singularity of $\tilde{\rho}_i^\star(k)$ therefore lies further away from the real axis than $k = \mathrm{i}\kappa_\mathrm{D}$.

and $\mathsf{s}(k) = 1$ and $\mathsf{s}(k) = \tilde{\rho}_i^\star(k)$, respectively. They have poles when $\mathsf{t}(k) = 0$. From the discussion in the shaded text box on page 319 ff we see that a simple pole at $k = \mathrm{i}\kappa$ corresponds to a contribution to the respective function that decays like $e^{-\kappa r}/(4\pi r)$ and we see that the **decay parameter** κ is a solution to the equation

$$\mathsf{t}(\mathrm{i}\kappa) = \beta \sum_i n_i^{\mathrm{b}} q_i q_i^\star - \varepsilon_0 \kappa^2 \tilde{\epsilon}_{\mathrm{reg}}(\mathrm{i}\kappa) = 0, \tag{12.149}$$

which is the same as

$$\tilde{\epsilon}(\mathrm{i}\kappa) = 0 \tag{12.150}$$

as follows from Equation 12.100. We are initially interested in the contribution with the longest decay length because it will dominate when $r \to \infty$, which means that $k = \mathrm{i}\kappa$ is the pole that lies closest to the real axis in complex k space.[26] Equation 12.149 can be written as

$$\kappa^2 = \frac{\beta}{\mathcal{E}_r^\star(\kappa)\varepsilon_0} \sum_i n_i^{\mathrm{b}} q_i q_i^\star, \tag{12.151}$$

where we have defined

$$\mathcal{E}_r^\star(\kappa) \equiv \tilde{\epsilon}_{\mathrm{reg}}(\mathrm{i}\kappa). \tag{12.152}$$

$\mathcal{E}_r^\star(\kappa)$ will be denoted the **dielectric factor**. From the definition 12.98 of $\tilde{\epsilon}_{\mathrm{reg}}(k)$ it follows that

$$\mathcal{E}_r^\star(\kappa) = 1 + \frac{1}{\varepsilon_0} \int d\mathbf{r}\, r^2 \left[\frac{\sinh(\kappa r) - \kappa r}{(\kappa r)^3}\right] \chi^\star(r), \tag{12.153}$$

where we have used $\sin(\mathrm{i}x)/(\mathrm{i}x) = \sinh(x)/x$. For ions in a dielectric continuum solvent we have instead (cf. Equation I:6.164)

$$\mathcal{E}_r^\star(\kappa) = \varepsilon_r^{\mathrm{cont}} + \frac{1}{\varepsilon_0} \int d\mathbf{r}\, r^2 \left[\frac{\sinh(\kappa r) - \kappa r}{(\kappa r)^3}\right] \chi^\star(r) \quad \text{(diel. continuum)}$$

(from Equation 12.130), but Equation 12.151 is valid as it stands. Note that when we do the replacements in Equation 12.116 we also need to replace $\mathcal{E}_r^\star(\kappa) \to \mathcal{E}_r^\star(\kappa)/\varepsilon_r^{\mathrm{cont}}$ in order to maintain Equation 12.152 as it is (cf. footnote 14).

The exact expression 12.151 for the decay parameter κ replaces the expression for the Debye parameter κ_{D} in

$$\kappa_{\mathrm{D}}^2 = \frac{\beta}{\varepsilon_r^{\mathrm{cont}}\varepsilon_0} \sum_i n_i^{\mathrm{b}} q_i^2 \quad \text{(diel. continuum)} \tag{12.154}$$

[26] At least when the ionic density is sufficiently low, $\rho_j^\star(r)$ decays faster to zero with increasing r than $\rho_j(r)$ and $\psi_j(r)$ and therefore it follows that any singularity of $\tilde{\rho}_j^\star(k)$ in complex k space and therefore of $\chi^\star(r)$ and $\tilde{\epsilon}_{\mathrm{reg}}(k)$ occur further away from the real axis than $\mathrm{i}\kappa$. As we saw from Equation 12.135, in the PB approximation $\rho_j^\star(r)$ decays like $\psi_j^2(r)$ or faster with increasing r which is in accordance with this.

(Equation 12.66), which gives an approximation, $\kappa \approx \kappa_D$, for the actual decay parameter κ in electrolytes. The two expressions differ in two places only: one of the factors q_i in the Debye parameter expression is replaced by $q_i^\star$ in the exact one and $\varepsilon_r^{\mathrm{cont}}$ is replaced by $\mathcal{E}_r^\star(\kappa)$. The second term in $\mathcal{E}_r^\star(\kappa)$ above gives the contribution from the ionic correlations in the electrolyte. For ions in a dielectric continuum solvent, $\varepsilon_r^{\mathrm{cont}}$ is equal to the dielectric constant of the *pure solvent* (without ions) and we see that the appropriate dielectric factor is not $\varepsilon_r^{\mathrm{cont}}$ as in the Debye parameter expression, but $\mathcal{E}_r^\star(\kappa)$ that depends on the presence of ions. It is a kind of renormalized "dielectric constant" for the *electrolyte solution*. When the ion concentration goes to zero, the decay parameter $\kappa \to 0$ and we have $q_i^\star \to q_i$, $\mathcal{E}_r^\star(\kappa) \to \varepsilon_r^{\mathrm{cont}}$ (because $\chi^\star \to 0$) and $\kappa/\kappa_D \to 1$, so the Debye parameter expression 12.154 becomes more and more accurate.

From Equations 12.142 – 12.144 with $\tilde{\mathsf{F}}(k) = \tilde{\phi}^\star_{\mathrm{Coul}}(k)$, $\mathsf{s}(k) = 1$ and $\mathsf{t}(k)$ given by Equation 12.148, it follows that

$$\phi^\star_{\mathrm{Coul}}(r) \sim \frac{1}{\mathcal{E}_r^{\mathrm{eff}}(\kappa)\varepsilon_0} \cdot \frac{e^{-\kappa r}}{4\pi r} \quad \text{when } r \to \infty \tag{12.155}$$

where $\mathcal{E}_r^{\mathrm{eff}}(\kappa)\varepsilon_0 = \mathsf{t}'(\mathrm{i}\kappa)/[2\mathrm{i}\kappa]$, that is,

$$\mathcal{E}_r^{\mathrm{eff}}(\kappa) = \tilde{\epsilon}_{\mathrm{reg}}(\mathrm{i}\kappa) + \frac{\mathrm{i}\kappa}{2}\tilde{\epsilon}'_{\mathrm{reg}}(\mathrm{i}\kappa) = \mathcal{E}_r^\star(\kappa) + \frac{\mathrm{i}\kappa}{2}\tilde{\epsilon}'_{\mathrm{reg}}(\mathrm{i}\kappa), \tag{12.156}$$

which will be denoted as the **effective dielectric permittivity**. Note that $\mathcal{E}_r^{\mathrm{eff}}(\kappa)$ occurs in $\phi^\star_{\mathrm{Coul}}(r)$ and $\mathcal{E}_r^\star(\kappa)$ in Equation 12.151, while $\varepsilon_r^{\mathrm{cont}}$ occurs at both places in the PB approximation. When $\kappa \to 0$ we have $\mathcal{E}_r^\star(\kappa) \to \mathcal{E}_r^{\star 0} \equiv \tilde{\epsilon}_{\mathrm{reg}}(0)$ (see Equation 12.99) and $\mathcal{E}_r^{\mathrm{eff}}(\kappa) \to \mathcal{E}_r^{\star 0}$ as well. When $\kappa \neq 0$ the values of $\mathcal{E}_r^{\mathrm{eff}}(\kappa)$ and $\mathcal{E}_r^\star(\kappa)$ are different due to the second term in Equation 12.156.

From these results it follows that we can write

$$\varepsilon_0 k^2 \tilde{\epsilon}(k) = \beta \sum_i n_i^{\mathrm{b}} q_i q_i^\star + \varepsilon_0 k^2 \tilde{\epsilon}_{\mathrm{reg}}(k) = (\kappa^2 + k^2)\mathsf{E}_\kappa(k), \tag{12.157}$$

where $\mathsf{E}_\kappa(k)$ is a function with the property $\mathsf{E}_\kappa(\mathrm{i}\kappa) = \mathcal{E}_r^{\mathrm{eff}}(\kappa)\varepsilon_0$, which can be concluded from

$$\lim_{k \to \mathrm{i}\kappa} \mathsf{E}_\kappa(k) = \lim_{k \to \mathrm{i}\kappa} \frac{\varepsilon_0 k^2 \tilde{\epsilon}(k)}{\kappa^2 + k^2} = \lim_{k \to \mathrm{i}\kappa} \frac{\mathsf{t}(k)}{\kappa^2 + k^2} = \frac{\mathsf{t}'(\mathrm{i}\kappa)}{2\mathrm{i}\kappa} = \mathcal{E}_r^{\mathrm{eff}}(\kappa)\varepsilon_0$$

and we have used 'Hospital's rule to obtain the last equality. Equation 12.156 can alternatively be written as

$$\mathcal{E}_r^{\mathrm{eff}}(\kappa) = \left[\frac{1}{2k} \cdot \frac{d(k^2 \tilde{\epsilon}_{\mathrm{reg}}(k))}{dk}\right]_{k=\mathrm{i}\kappa} = \left[\frac{1}{2k} \cdot \frac{d(k^2 \tilde{\epsilon}(k))}{dk}\right]_{k=\mathrm{i}\kappa} = \frac{\mathrm{i}\kappa}{2}\tilde{\epsilon}'(\mathrm{i}\kappa), \tag{12.158}$$

where we have used $\tilde{\epsilon}(\mathrm{i}\kappa) = 0$ to obtain the last equality.

Yet another expression for $\mathcal{E}_r^{\text{eff}}(\kappa)$ can be derived. Since $\varepsilon_0 k^2 \tilde{\epsilon}_{\text{reg}}(k) = \varepsilon_0 k^2 - \tilde{\chi}^\star(k) + \tilde{\chi}^\star(0)$, as follows from the definition 12.98 of $\tilde{\epsilon}_{\text{reg}}(k)$, we have $\mathsf{t}(k) = \beta \sum_i n_i^{\text{b}} q_i q_i^\star + \varepsilon_0 k^2 - \tilde{\chi}^\star(k) + \tilde{\chi}^\star(0)$. Therefore, $\mathcal{E}_r^{\text{eff}}(\kappa)\varepsilon_0 = \mathsf{t}'(\text{i}\kappa)/[2\text{i}\kappa] = [\varepsilon_0 2\text{i}\kappa - \tilde{\chi}^{\star\prime}(\text{i}\kappa)]/[2\text{i}\kappa]$, so we have

$$\mathcal{E}_r^{\text{eff}}(\kappa) = 1 - \frac{\tilde{\chi}^{\star\prime}(\text{i}\kappa)}{2\varepsilon_0 \text{i}\kappa} = 1 + \frac{1}{2\varepsilon_0}\int d\mathbf{r}\, r^2 \left[\frac{\kappa r \cosh(\kappa r) - \sinh(\kappa r)}{(\kappa r)^3}\right] \chi^\star(r), \tag{12.159}$$

which can be compared with Equation 12.153. (Equation 12.159 was quoted without proof in Volume I, Equation I:6.165, for the case of dielectric continuum solvent where the initial 1 in the rhs is replaced by $\varepsilon_r^{\text{cont}}$.)

When $\tilde{\mathsf{F}}(k) = \tilde{\psi}_i(k)$ and $\mathsf{s}(k) = \tilde{\rho}_i^\star(k)$ we obtain in the same manner

$$\psi_i(r) \sim \frac{q_i^{\text{eff}}(\kappa)}{\mathcal{E}_r^{\text{eff}}(\kappa)\varepsilon_0} \cdot \frac{e^{-\kappa r}}{4\pi r} \quad \text{when } r \to \infty \tag{12.160}$$

where

$$q_i^{\text{eff}}(\kappa) = \tilde{\rho}_i^\star(\text{i}\kappa) = \int d\mathbf{r}'\, \rho_i^\star(r') \frac{\sinh(\kappa r')}{\kappa r'} \tag{12.161}$$

is the **effective charge** of an i-ion, which is a function of κ. Also in the LPB and PB approximations, Equation 12.123 and 12.139, q_i^{eff} is a function of the decay parameter, $q_i^{\text{eff}} = q_i^{\text{eff}}(\kappa_{\text{D}})$. In order to interpret the κ dependence of the effective charge we note that[27]

$$\int d\mathbf{r}'\, e^{-\kappa z'} \rho_i^\star(r') = \int d\mathbf{r}'\, \frac{\sinh(\kappa r')}{\kappa r'} \rho_i^\star(r') = q_i^{\text{eff}}(\kappa), \tag{12.162}$$

so $q_i^{\text{eff}}(\kappa)$ is the projection of $\rho_i^\star(r')$ on an exponentially decaying function $e^{-\kappa z'}$ in the z'-direction (or any other direction in space).

There is another exact equation for κ that is equivalent to Equations 12.150 and 12.151, but it is valid only for spherical ions while the other two are valid for all cases. It can be obtained from $\tilde{\epsilon}(k) = 1 - \tilde{\chi}^\star(k)\tilde{\phi}_{\text{Coul}}(k)$ (Equation 12.96) and $\tilde{\chi}^\star(k) = -\beta \sum_i n_i^{\text{b}} q_i \tilde{\rho}_i^\star(k)$ (the Fourier transform of Equation 12.80), which imply

$$\tilde{\epsilon}(k) = 1 + \frac{\beta \sum_i n_i^{\text{b}} q_i \tilde{\rho}_i^\star(k)}{\varepsilon_0 k^2} \quad \text{(spherical ions).} \tag{12.163}$$

---

[27] The integral can be evaluated by using spherical polar coordinates with the polar angle θ' measured from the z axis

$$\begin{aligned}\int d\mathbf{r}'\, e^{-\kappa z'} \rho_i^\star(r') &= 2\pi \int_0^\infty dr'\, r'^2 \left[\int_0^\pi d\theta' \sin\theta'\, e^{-\kappa r' \cos\theta'}\right] \rho_i^\star(r') \\ &= \int d\mathbf{r}'\, \frac{\sinh(\kappa r')}{\kappa r'} \rho_i^\star(r'),\end{aligned}$$

where we have inserted $z' = r' \cos\theta'$.

Since $\tilde{\epsilon}(i\kappa) = 0$ and $\tilde{\rho}_i^\star(i\kappa) = q_i^{\mathrm{eff}}(\kappa)$ it follows that the decay parameter κ is also a solution to

$$\kappa^2 = \frac{\beta}{\varepsilon_0} \sum_i n_i^{\mathrm{b}} q_i q_i^{\mathrm{eff}}(\kappa) \quad \text{(spherical ions).} \tag{12.164}$$

It may appear from a comparison of Equations 12.151 and 12.164 that $q_i^\star/\mathcal{E}_r^\star(\kappa)$ is equal to $q_i^{\mathrm{eff}}(\kappa)$, but this is not the case in general. The various expressions for the decay parameter κ are listed in Table 12.3.

For ions in a dielectric continuum solvent Equations 12.155 - 12.161 are valid as they stand, but

$$\tilde{\epsilon}(k) = \varepsilon_r^{\mathrm{cont}} + \frac{\beta \sum_i n_i^{\mathrm{b}} q_i \tilde{\rho}_i^\star(k)}{\varepsilon_0 k^2} \quad \text{(spherical ions, diel. continuum)} \tag{12.165}$$

and

$$\kappa^2 = \frac{\beta}{\varepsilon_r^{\mathrm{cont}} \varepsilon_0} \sum_i n_i^{\mathrm{b}} q_i q_i^{\mathrm{eff}}(\kappa) \quad \text{(spherical ions, diel. continuum).} \tag{12.166}$$

Note that $\tilde{\epsilon}(k)$, $\mathcal{E}_r^\star(\kappa)$ and hence $\mathcal{E}_r^{\mathrm{eff}}(\kappa)$ in this case contain a term that is equal to $\varepsilon_r^{\mathrm{cont}}$.

As we will see in Chapter 13, for ions of arbitrary shape the expressions corresponding to Equations 12.164 and 12.166 are more complicated because $\tilde{\rho}_i^\star$ depends on the orientation of the i-ion. In these expressions the effective charge q_i^{eff} is replaced by a "multipolar effective charge" $\mathrm{Q}_i^{\mathrm{eff}}$ that is orientation dependent (see the last equation in Table 12.3). The entity $\mathrm{Q}_i^{\mathrm{eff}}$, which is relevant also for the mean potential from a nonspherical particle in the PB approximation, was introduced in Chapter 6 in Volume I. Here, we will continue with the spherical case. The nonspherical case is dealt with in Chapter 13.

The expression for the Debye parameter κ_{D} in Equation 12.154 differs from the exact Equation 12.166 only by the replacement of one of the factors q_i in the former by $q_i^{\mathrm{eff}}(\kappa)$ in the latter. The appearance of $\sum_i n_i^{\mathrm{b}} q_i^2$ in the definition of the Debye parameter originates from the Poisson-Boltzmann equation $-\varepsilon_r^{\mathrm{cont}} \varepsilon_0 \nabla^2 \psi_j(r) = \sum_i n_i^{\mathrm{b}} q_i e^{-\beta q_i \psi_j(r)}$ (Equation 12.120), where the approximation $w_{ij}^{(2)}(r) = q_i \psi_j(r)$ has been used in $e^{-\beta w_{ij}^{(2)}(r)}$. Linearization of the rhs of the PB equation yields $-\sum_i n_i^{\mathrm{b}} q_i \times \beta q_i \psi_j(r) = -\varepsilon_r^{\mathrm{cont}} \varepsilon_0 \kappa_{\mathrm{D}}^2 \psi_j(r)$, where the second factor of q_i stems from $\beta q_i \psi_j(r)$ in the exponent. Recall that $w_{ij}^{(2)}(r) = q_i \psi_j(r)$, which is a key assumption in the PB approximation, means that all ions in the surroundings of the central ion are treated like bare point charges without dress. Only the central ion has a dress in the PB

Table 12.3 Definition of the Debye parameter $\kappa_{\rm D}$ and a list with some exact equations for the decay parameter κ that are mathematically equivalent. Equations 12.150', 12.151' and 13.86' are valid in general, while Equation 12.164' and 12.166' are only valid for spherical ions. Each solution to these equations constitute a ***decay mode*** for the electrostatic interactions. The definitions of $q_i^\star$, $q_i^{\rm eff}(\kappa)$ and $\mathcal{E}_r^\star(\kappa)$ are also given. The equation numbers refer to the original ones in the text.

$$\kappa_{\rm D}^2 = \frac{\beta}{\varepsilon_r^{\rm cont}\varepsilon_0}\sum_i n_i^{\rm b} q_i^2 \qquad (12.66', 12.154')$$

(for ions in dielectric continuum).

Mathematically equivalent equations for κ:

$$\tilde{\epsilon}(\mathrm{i}\kappa) = 0 \qquad (12.150')$$

$$\kappa^2 = \frac{\beta}{\mathcal{E}_r^\star(\kappa)\varepsilon_0}\sum_i n_i^{\rm b} q_i q_i^\star \qquad (12.151')$$

$$\kappa^2 = \frac{\beta}{\varepsilon_0}\sum_i n_i^{\rm b} q_i q_i^{\rm eff}(\kappa) \qquad (12.164')$$

(for spherical ions)

$$\kappa^2 = \frac{\beta}{\varepsilon_r^{\rm cont}\varepsilon_0}\sum_i n_i^{\rm b} q_i q_i^{\rm eff}(\kappa) \qquad (12.166')$$

(for spherical ions in dielectric continuum)

$$q_i^\star \equiv \tilde{\rho}_i^\star(0) = \int d\mathbf{r}\,\rho_i^\star(r), \quad q_i^{\rm eff}(\kappa) \equiv \tilde{\rho}_i^\star(\mathrm{i}\kappa) \qquad (12.77'), (12.161')$$

$$\mathcal{E}_r^\star(\kappa) \equiv \tilde{\epsilon}_{\rm reg}(\mathrm{i}\kappa) \qquad (12.152')$$

$$\kappa^2 = \frac{\beta}{\varepsilon_0}\sum_i n_i^{\rm b}\left\langle \mathrm{Q}_i(-\hat{\mathbf{r}},\omega,\kappa)\mathrm{Q}_i^{\rm eff}(\hat{\mathbf{r}},\omega,\kappa)\right\rangle_\omega \qquad (13.86')$$

(for nonspherical ions/particles).[28]

[28] $\mathrm{Q}_i^{\rm eff}$ is the "multipolar effective charge" of a nonspherical i-particles defined in Equation 13.75 and Q_i is defined analogously from the bare charge-distribution inside the particle, see Equation 13.85. The average over orientations ω makes the rhs to be independent of the direction of $\hat{\mathbf{r}}$.

approximation and therefore it has an effective charge $q_j^{\text{eff}} \neq q_j$. In a correct treatment *all ions must treated on an equal basis*, so all ions have a dress and an effective charge $q_i^{\text{eff}} \neq q_i$. Then $w_{ij}^{(2)}(r) \sim q_i^{\text{eff}}\psi_j(r)$ for large r and we obtain $\sum_i n_i^{\text{b}} q_i q_i^{\text{eff}}$ as in Equation 12.166 when $e^{-\beta w_{ij}^{(2)}(r)}$ is linearized. To replace q_i by $q_i^{\text{eff}}(\kappa)$ may seem to be a very small change, but it has quite dramatic consequences, for example the appearance of more than one decay length for the electrostatic interactions and the possibility of oscillatory decay. Some of these consequences were explored in Sections 6.1.5 ff in Volume I. Here, we will investigate this topic in more detail and later we will give an explicit example (Example 12:A) that extends the approximate treatment presented in Section 6.1.5.

In the special case of a binary electrolyte, the total ionic density is $n_{\text{tot}}^{\text{b}} = n_+^{\text{b}} + n_-^{\text{b}}$ and we have

$$q_+ n_+^{\text{b}} = -q_- n_-^{\text{b}} = q_\pm n_{\text{tot}}^{\text{b}}/2 \tag{12.167}$$

where

$$q_\pm \equiv \mathsf{x}_+^{\text{b}} q_+ + \mathsf{x}_-^{\text{b}} |q_-| = \frac{2q_+|q_-|}{q_+ + |q_-|} \tag{12.168}$$

and $\mathsf{x}_i^{\text{b}} = n_i^{\text{b}}/n_{\text{tot}}^{\text{b}}$ is a mole fraction that only involves the ionic species. The definition 12.154 of κ_{D} with $\varepsilon_r^{\text{cont}} = 1$ can be written as

$$\kappa_{\text{D}}^2 = \frac{\beta q_\pm n_{\text{tot}}^{\text{b}}}{2\varepsilon_0}\left[q_+ - q_-\right] = \frac{\beta q_\pm n_{\text{tot}}^{\text{b}}}{2\varepsilon_0}\left[q_+ + |q_-|\right]$$

and Equation 12.164 as

$$\kappa^2 = \frac{\beta q_\pm n_{\text{tot}}^{\text{b}}}{2\varepsilon_0}\left[q_+^{\text{eff}}(\kappa) - q_-^{\text{eff}}(\kappa)\right].$$

Hence

$$\left(\frac{\kappa}{\kappa_{\text{D}}}\right)^2 = \frac{q_+^{\text{eff}}(\kappa) - q_-^{\text{eff}}(\kappa)}{q_+ + |q_-|} \quad \text{(spherical ions).} \tag{12.169}$$

For low ionic densities $q_+^{\text{eff}}(\kappa)$ is positive and $q_-^{\text{eff}}(\kappa)$ negative, but for high densities it is not necessarily the case due to the contributions from the dresses of the ions.

For a symmetric electrolyte, where $q_+ = -q_-$ and $n_+^{\text{b}} = n_-^{\text{b}}$, we have $q_\pm = q_+ = -q_- \equiv q$. In the **restricted primitive model (RPM)**, where all ions have the same diameter, the anions and cations differ only by the signs of their charges. In this case Equation 12.169 becomes

$$\left(\frac{\kappa}{\kappa_{\text{D}}}\right)^2 = \frac{q^{\text{eff}}(\kappa)}{q} \quad \text{(RPM)} \tag{12.170}$$

because $q_+^{\text{eff}}(\kappa) = -q_-^{\text{eff}}(\kappa) \equiv q^{\text{eff}}(\kappa)$.

The Debye parameter κ_D is expressed in Equation 12.154 solely in terms of the system parameters β, n_i^b, q_i and ε_r^{cont}, so it is uniquely defined from these parameters. On the contrary, Equation 12.166 is *an equation for* κ that has several solutions that we may call κ, κ' etc. The same is true for

$$\tilde{\epsilon}(i\kappa) = 0 \qquad \text{and} \qquad \kappa^2 = \frac{\beta}{\mathcal{E}_r^\star(\kappa)\varepsilon_0}\sum_i n_i^b q_i q_i^\star$$

(Equations 12.150 and 12.151) and they have the same solutions κ, κ' etc. Thus we have the three equivalent equations for κ'

$$\begin{aligned}\tilde{\epsilon}(i\kappa') &= 0\\ (\kappa')^2 &= \frac{\beta}{\mathcal{E}_r^\star(\kappa')\varepsilon_0}\sum_i n_i^b q_i q_i^\star\\ (\kappa')^2 &= \frac{\beta}{\varepsilon_0}\sum_j n_j^b q_j q_j^{eff}(\kappa') \quad \text{(spherical ions)}\end{aligned}$$

(cf. Table 12.3) and likewise for any other solution κ'', κ''' etc. Each solution corresponds to a pole $k = i\kappa$, $k = i\kappa'$ etc. of $\tilde{\phi}^\star_{Coul}(k)$ and $\tilde{\psi}_i(k)$ and gives a contribution to $\phi^\star_{Coul}(r)$ and $\psi_i(r)$ with a Yukawa function decay, so we have for sufficiently large r values[29]

$$\begin{aligned}\phi^\star_{Coul}(r) &= \frac{1}{\mathcal{E}_r^{eff}(\kappa)\varepsilon_0}\cdot\frac{e^{-\kappa r}}{4\pi r} + \frac{1}{\mathcal{E}_r^{eff}(\kappa')\varepsilon_0}\cdot\frac{e^{-\kappa' r}}{4\pi r} + \ldots \qquad (12.171)\\ \psi_i(r) &= \frac{q_i^{eff}(\kappa)}{\mathcal{E}_r^{eff}(\kappa)\varepsilon_0}\cdot\frac{e^{-\kappa r}}{4\pi r} + \frac{q_i^{eff}(\kappa')}{\mathcal{E}_r^{eff}(\kappa')\varepsilon_0}\cdot\frac{e^{-\kappa' r}}{4\pi r} + \ldots \quad \text{(large } r\text{)},\end{aligned}$$

where we have explicitly shown only the first two Yukawa function terms. From the last equation and the Poisson equation $-\varepsilon_0\nabla^2\psi_i(r) = \rho_i^{tot}(r)$ it follows that

$$\rho_i(r) = -\frac{\kappa^2 q_i^{eff}(\kappa)}{\mathcal{E}_r^{eff}(\kappa)}\cdot\frac{e^{-\kappa r}}{4\pi r} - \frac{(\kappa')^2 q_i^{eff}(\kappa')}{\mathcal{E}_r^{eff}(\kappa')}\cdot\frac{e^{-\kappa' r}}{4\pi r} - \ldots \quad \text{(large } r\text{)}. \qquad (12.172)$$

Thus, there are multiple decay parameters for an electrolyte and not only one as predicted by the PB approximation. Each solution of $\tilde{\epsilon}(i\kappa) = 0$ constitutes a **decay mode** for the electrostatic interactions and each mode has its own value of the effective charge $\{q_i^{eff}(\kappa), q_i^{eff}(\kappa')\ldots\}$ and the effective dielectric permittivity $\{\mathcal{E}_r^{eff}(\kappa), \mathcal{E}_r^{eff}(\kappa')\ldots\}$. As we saw in Equation 12.162, q_i^{eff} is the projection of $\rho_i^\star(r)$ on an exponentially decaying function with decay length $1/\kappa$, $1/\kappa'$ etc. Likewise each mode has its own value of the dielectric factor $\{\mathcal{E}_r^\star(\kappa), \mathcal{E}_r^\star(\kappa')\ldots\}$ in Equation 12.151. We may denote the various modes as the κ mode, κ' mode etc.

[29] Like in the PB approximation there are additional contributions with other functional form, cf. Equation 12.146, but at least for low electrolyte densities the first term dominates for large r. Other types of terms will be treated later in the shaded text box on page 351.

In fact, the solutions to the equations for κ in Table 12.3 can be complex-valued, $\kappa = \kappa_{\Re} + \mathrm{i}\kappa_{\Im}$, where $\kappa_{\Re}$ and $\kappa_{\Im}$ are the real and imaginary parts, respectively. Then, the there is always another solution κ' that is the complex conjugate $\kappa' = \kappa_{\Re} - \mathrm{i}\kappa_{\Im} = \underline{\kappa}$ (underbar means complex conjugate). Thus we have two decay parameters that are complex conjugates of each other. Since the sum of a complex number A and its complex conjugate $\underline{A}$ is given by $A + \underline{A} = 2\Re(A)$, where $\Re(A)$ stands for the real part of A, we have

$$\frac{1}{\mathcal{E}_r^{\mathrm{eff}}(\kappa)\varepsilon_0} \cdot \frac{e^{-\kappa r}}{4\pi r}\phi^{\star}_{\mathrm{Coul}}(r) + \frac{1}{\mathcal{E}_r^{\mathrm{eff}}(\underline{\kappa})\varepsilon_0} \cdot \frac{e^{-\underline{\kappa} r}}{4\pi r} = 2\Re\left[\frac{e^{-(\kappa_{\Re}+\mathrm{i}\kappa_{\Im})r}}{4\pi|\mathcal{E}_r^{\mathrm{eff}}|e^{-\mathrm{i}\vartheta_{\mathcal{E}}}\varepsilon_0 r}\right],$$

where we have written $\mathcal{E}_r^{\mathrm{eff}}(\kappa_{\Re} + \mathrm{i}\kappa_{\Im}) = |\mathcal{E}_r^{\mathrm{eff}}|\exp(-\mathrm{i}\vartheta_{\mathcal{E}})$.[30] Thus[31]

$$\phi^{*}_{\mathrm{Coul}}(r) \sim \frac{1}{2\pi|\mathcal{E}_r^{\mathrm{eff}}|\varepsilon_0} \cdot \frac{e^{-\kappa_{\Re} r}}{r}\cos(\kappa_{\Im} r - \vartheta_{\mathcal{E}}) \quad \text{when } r \to \infty, \tag{12.173}$$

where $\vartheta_{\mathcal{E}}$ is a phase shift, $1/\kappa_{\Re}$ is the decay length and $2\pi/\kappa_{\Im}$ is the wave length. In the same manner we obtain

$$\psi_i(r) \sim \frac{|q_i^{\mathrm{eff}}|}{2\pi|\mathcal{E}_r^{\mathrm{eff}}|\varepsilon_0} \cdot \frac{e^{-\kappa_{\Re} r}}{r}\cos(\kappa_{\Im} r + \vartheta_i - \vartheta_{\mathcal{E}}) \quad \text{when } r \to \infty, \tag{12.174}$$

where $q_i^{\mathrm{eff}}(\kappa_{\Re} + \mathrm{i}\kappa_{\Im}) = |q_i^{\mathrm{eff}}|\exp(-\mathrm{i}\vartheta_i)$.[32] We will denote a function $e^{-\mathfrak{a}r}\cos(\mathfrak{b}r + \mathfrak{c})/r$ with real $\mathfrak{a}$, $\mathfrak{b}$ and $\mathfrak{c}$ as an **oscillatory Yukawa function**. In the PB approximation κ_{D} is always real-valued, so there cannot be any oscillations in $\phi^{*}_{\mathrm{Coul}}(r)$ and $\psi_i(r)$. These functions always decay monotonically as an ordinary Yukawa function when $r \to \infty$ in this approximation and there is only one decay mode.

It is illustrative to derive Equation 12.171 in another manner from the fact that $\tilde{\epsilon}(k) = 0$ have several solutions $k = \pm\mathrm{i}\kappa$, $k = \pm\mathrm{i}\kappa'$, $k = \pm\mathrm{i}\kappa''$ etc. and we assume that they are simple zeros of $\tilde{\epsilon}(k)$ so they are simple poles of $\tilde{\phi}^{*}_{\mathrm{Coul}}(k)$. From Equation 12.157 it follows that

$$\tilde{\phi}^{*}_{\mathrm{Coul}}(k) = \frac{1}{(\kappa^2 + k^2)\mathsf{E}_\kappa(k)} = \frac{1}{(\kappa^2 + k^2)\mathsf{E}_\kappa(\mathrm{i}\kappa)} + F_1(k),$$

where the diverging part of $\tilde{\phi}^{*}_{\mathrm{Coul}}(k)$ when $k \to \mathrm{i}\kappa$ is given by the first term on the rhs and where $F_1(k)$ is a function that is finite for $k = \mathrm{i}\kappa$ but has poles for $k = \mathrm{i}\kappa'$, $k = \mathrm{i}\kappa''$ etc. Recall that $\mathsf{E}_\kappa(\mathrm{i}\kappa) = \mathcal{E}_r^{\mathrm{eff}}(\kappa)\varepsilon_0$. By repeating the same arguments for $F_1(k)$ we can conclude that we can write

$$\tilde{\phi}^{*}_{\mathrm{Coul}}(k) = \frac{1}{(\kappa^2 + k^2)\mathsf{E}_\kappa(\mathrm{i}\kappa)} + \frac{1}{(\kappa^2 + k'^2)\mathsf{E}_{\kappa'}(\mathrm{i}\kappa')} + F_2(k),$$

[30] We may always select a phase ϑ in the interval $-\pi < \vartheta \le \pi$ for it to be definite.

[31] This asymptotic formula holds when $k = \mathrm{i}\kappa$ is the singularity closest to the real axis in complex k space.

[32] Note that any difference in sign for $\psi_+(r)$ and $\psi_-(r)$ for a binary electrolyte is incorporated in the phase ϑ_i. For instance, if $\psi_+(r) \approx -\psi_+(r)$ for a range of r values we normally have $\vartheta_+ - \vartheta_- \approx \pi$.

where $\mathsf{E}_{\kappa'}(\mathrm{i}\kappa') = \mathcal{E}_r^{\text{eff}}(\kappa')\varepsilon_0$ and $F_2(k)$ is a function that is finite for $k = \mathrm{i}\kappa$ and $k = \mathrm{i}\kappa'$ but has poles for $k = \mathrm{i}\kappa''$ etc. By continuing in the same manner we can conclude that[33]

$$\begin{aligned}\tilde{\phi}^*_{\text{Coul}}(k) &= \frac{1}{(\kappa^2 + k^2)\mathcal{E}_r^{\text{eff}}(\kappa)\varepsilon_0} + \frac{1}{(\kappa'^2 + k^2)\mathcal{E}_r^{\text{eff}}(\kappa')\varepsilon_0} \\ &+ \frac{1}{(\kappa''^2 + k^2)\mathcal{E}_r^{\text{eff}}(\kappa'')\varepsilon_0} + \ldots, \end{aligned} \tag{12.175}$$

which is the Fourier transform of Equation 12.171. Thus we have derived the latter in another manner than before.

Likewise it is illustrative to view the expression for $\psi_i(r)$ in Equation 12.171 from another perspective. To do this, let us start with

$$\begin{aligned}\psi_i(r) &= \int d\mathbf{r}'\, \rho_i^\star(r')\phi^\star_{\text{Coul}}(|\mathbf{r} - \mathbf{r}'|) \\ &\sim \frac{1}{\mathcal{E}_r^{\text{eff}}(\kappa)\varepsilon_0}\int d\mathbf{r}'\, \rho_i^\star(r')\frac{e^{-\kappa|\mathbf{r}-\mathbf{r}'|}}{4\pi|\mathbf{r} - \mathbf{r}'|} \quad \text{when } r \to \infty, \end{aligned} \tag{12.176}$$

where we have inserted the first term of $\phi^\star_{\text{Coul}}$ in Equation 12.171. The integral in the second line can be written as

$$\int\limits_0^\infty dr'\, r'^2\rho_i^\star(r')\left[\int d\hat{\mathbf{r}}'\frac{e^{-\kappa|\mathbf{r}-\mathbf{r}'|}}{4\pi|\mathbf{r} - \mathbf{r}'|}\right] = \int\limits_0^\infty dr'\, r'^2\rho_i^\star(r')\left[\frac{e^{-\kappa|r-r'|} - e^{-\kappa|r+r'|}}{2\kappa r r'}\right],$$

where $\hat{\mathbf{r}}' = \mathbf{r}'/r'$ is the unit vector with polar angles (φ'_r, θ'_r) and $d\hat{\mathbf{r}}' = \sin(\theta'_r)d\theta'_r d\varphi'_r$ stands for the angle differential. The angular integration in the square brackets has been performed analytically to obtain the rhs.[34] If $r > r'$ we have $|r \pm r'| = r \pm r'$, whereby this becomes

$$\int\limits_0^\infty dr'\, r'^2\rho_i^\star(r')\frac{e^{-\kappa r}}{r}\cdot\frac{\sinh(\kappa r')}{\kappa r'} = \frac{e^{-\kappa r}}{4\pi r}\int d\mathbf{r}'\, \rho_i^\star(r')\frac{\sinh(\kappa r')}{\kappa r'} = \frac{q_i^{\text{eff}}(\kappa)e^{-\kappa r}}{4\pi r}.$$

When $\rho_i^\star(r')$ decays to zero when $r \to \infty$ faster than $e^{-\kappa r}/r$ this can be used for large r whereby we can conclude that

$$\psi_i(r) \sim \frac{q_i^{\text{eff}}(\kappa)}{\mathcal{E}_r^{\text{eff}}(\kappa)\varepsilon_0}\cdot\frac{e^{-\kappa r}}{4\pi r} \quad \text{when } r \to \infty.$$

---

[33]The additional contributions with other functional forms mentioned in footnote 29 are contained in $F_1(k)$, $F_2(k)$ etc. and are not explicitly shown here.

[34]By selecting the z axis in the direction of $\mathbf{r}$ and using the law of cosines in trigonometry $|\mathbf{r}-\mathbf{r}'|^2 = r^2+(r')^2-2rr'\cos\theta'_r$, where θ'_r is the angle between the vectors $\mathbf{r}$ and $\mathbf{r}'$, one can easily perform the angular integration after the change of variable $t^2 = r^2+(r')^2-2rr'\cos\theta'_r$, whereby $t\,dt = rr'\sin(\theta'_r)d\theta'_r$ and the limits $0 \le \theta'_r \le \pi$ correspond to $|r-r'| \le t \le |r+r'|$. The integration over φ'_r gives a factor 2π.

This is the first term in the expression for $\psi_i(r)$ in Equation 12.171. The term from the κ' mode can be treated in the same manner.

Next, we will present an example of how this formalism can be used, namely an approximation with two decay modes that works very well for aqueous $1:-1$ electrolyte solutions in the primitive model and for classical plasmas at sufficiently high temperatures. It includes two monotonic modes with decay parameters κ and κ' as shown in Equations 12.171 and 12.172 and it predicts the occurrence complex-valued decay parameters, $\kappa_{\Re} \pm i\kappa_{\Im}$, which give rise to oscillations as in Equations 12.173 and 12.174. Furthermore, the approximation provides explicit expressions for the effective charges $q_i^{\text{eff}}(\kappa)$, $q_i^{\text{eff}}(\kappa')$ and the effective dielectric permittivities $\mathcal{E}_r^{\text{eff}}(\kappa)$ and $\mathcal{E}_r^{\text{eff}}(\kappa')$ for these cases. The example has a particular pedagogical value because it demonstrates in an explicit manner the importance of the different decay modes in operation in electrolytes and, especially, the appearance of exponentially damped oscillatory decay of the charge distribution and the electrostatic interactions when the ion density is increased beyond a certain threshold. It also illustrates the meaning of the effective dielectric permittivities and shows how they behave. The approximation used does, however, not work for higher valencies, for instance divalent ions in water, and for lower temperatures, so more refined approximations are needed in these cases.

Example 12:A. A simple, accurate approximation with two decay modes[35]

In the LPB approximation (Debye-Hückel theory) the central ion has an effective charge

$$q_i^{\text{eff}}(\kappa) = \frac{q_i e^{\kappa d^{\text{h}}}}{1 + \kappa d^{\text{h}}}$$

with $\kappa = \kappa_{\text{D}}$ (Equation 12.123), while the surrounding ions are treated as bare point charges with charge q_i that do not correlate with each other. They correlate with the central ion, which is the reason why the latter has an effective charge $q_i^{\text{eff}} \neq q_i$. In a correct treatment, all ions are treated in the same manner and all correlate with each other, which means that *all ions* have effective charges different from the bare charges. A simple way to improve the LPB approximation is to assume that the effective charges are given by the same expression as the central ion above, but we do not assume that $\kappa = \kappa_{\text{D}}$.

[35]This example is based on the publication R. Kjellander, Phys. Chem. Chem. Phys. **22** (2020) 23952 (https://doi.org/10.1039/D0CP02742A).

Let us consider an electrolyte solution in the restricted primitive model, where we have $\varepsilon_r^{\text{cont}} \neq 1$ and $q_+^{\text{eff}}(\kappa) = -q_-^{\text{eff}}(\kappa) = q^{\text{eff}}(\kappa)$. By inserting the expression for $q_i^{\text{eff}}(\kappa)$ into Equation 12.170 we obtain

$$\left(\frac{\kappa}{\kappa_{\text{D}}}\right)^2 = \frac{e^{\kappa d^{\text{h}}}}{1+\kappa d^{\text{h}}}. \tag{12.177}$$

This is an equation for the decay parameter κ that gives κ as a function of κ_{D}, which is determined by the ion density, temperature etc. Note that κ_{D} is simply a variable here, not a decay parameter. When the ion density is nonzero and therefore $\kappa > 0$, the rhs $\neq 1$, which implies that $\kappa \neq \kappa_{\text{D}}$. This is a consequence of the equal treatment of all ions.

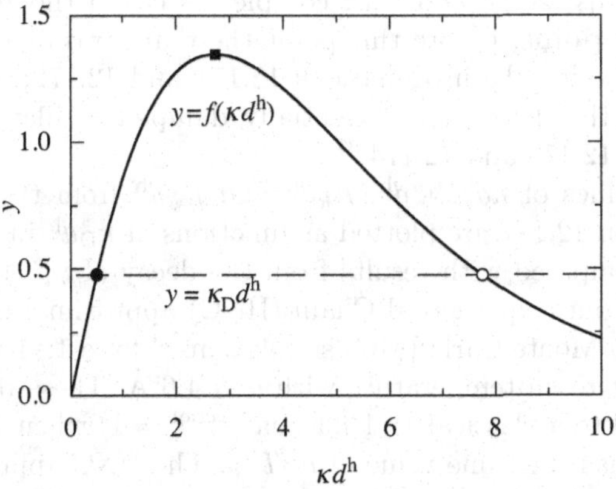

Figure 12.1 The full curve shows a plot of the function $f(\kappa d^{\text{h}})$, where $f(x) = [x^2(1+x)/e^x]^{1/2}$. The horizontal dashed line shows $y = \kappa_{\text{D}} d^{\text{h}}$ with $\kappa_{\text{D}} d^{\text{h}} = 0.477$. It has two intersections with the full curve that are shown as a full circle and an open circle, respectively. They correspond to the two solutions κ and κ' of Equation 12.177 when $\kappa_{\text{D}} d^{\text{h}} = 0.477$. The maximum of the curve is marked with a full square.

Equation 12.177 can easily be solved for κ numerically. In order to illustrate the solutions, we write the equation as $f(\kappa d^{\text{h}}) = \kappa_{\text{D}} d^{\text{h}}$, where $f(x) = [x^2(1+x)/e^x]^{1/2}$, and plot $y = f(\kappa d^{\text{h}})$ as function of $x = \kappa d^{\text{h}}$. This is done in Figure 12.1 and a solution is given by a point where the function $f(\kappa d^{\text{h}})$ has the value $\kappa_{\text{D}} d^{\text{h}}$, that is, at the intersection with the horizontal line $y = \kappa_{\text{D}} d^{\text{h}}$; the figure shows an example with $\kappa_{\text{D}} d^{\text{h}} = 0.477$ and in this case there is an intersection (full circle) at $\kappa d^{\text{h}} = 0.500$. There is also a second intersection, which is marked by an open circle in the figure and occurs at $\kappa d^{\text{h}} = 7.75$. Hence there are two solutions to Equation 12.177, which we denote by κ and κ'; for $\kappa_D d^h = 0.477$ the solutions are given by $\kappa d^h = 0.500$ and $\kappa' d^h = 7.75$.

The curve has a maximum marked by a full square in the figure. Its value is equal to 1.346 and it occurs for $\kappa d^h = 1 + \sqrt{3} \approx 2.732$. For any value of $\kappa_D d^h$ below the maximum, that is, for $\kappa_D d^h < 1.346$, there are two intersections and hence two solutions. Since $f(x)$ goes to zero when $x \to \infty$, the second solution κ' goes to ∞ when $\kappa_D d^h \to 0$, while the first solution κ goes to zero and we have $\kappa/\kappa_D \to 1$ in this limit as follows from Equation 12.177. When $\kappa_D d^h$ is increased from zero the two solutions κ and κ' approach each other and for $\kappa_D d^h = 1.346$ they merge. When $\kappa_D d^h > 1.346$ the two solutions turn complex-valued, $\kappa = \kappa_\Re + i\kappa_\Im$ and $\kappa' = \kappa_\Re - i\kappa_\Im = \underline{\kappa}$.[36] The point where the two real solutions merge and turn complex is called the **Kirkwood crossover point**. Before this point there are two monotonically decaying modes like in Equations 12.171 and 12.172, while after this point the decay is exponentially damped oscillatory like in Equations 12.173 and 12.174.[37]

The values of κd^h, $\kappa' d^h$, $\kappa_\Re d^h$ and $\kappa_\Im d^h$ from the solutions to Equation 12.177 are plotted as functions of $\kappa_D d^h$ in Fig. 12.2a and are compared with results from the decay of $\psi_i(r)$ and $\rho_i(r)$ obtained from Hypernetted Chain (HNC) approximation calculations[38] and Monte Carlo (MC) simulations[39] for a 1:–1 electrolyte in water at room temperature with $a = 4.6$ Å. These results also correspond to a classical 1:–1 plasma ($\varepsilon_r^{\text{cont}} = 1$) when $T = 23400$ K, which has the same value of $k_B T \varepsilon_r$. The HNC approximation is very accurate for these systems.

We see that the value of κ is close to κ_D for low ionic densities and increases with the density faster than κ_D, while κ' decreases from a large value. At the Kirkwood crossover point, which is marked by a filled symbol in the figure, κ and κ' merge with each other and thereafter $\psi_i(r)$ and $\rho_i(r)$ decay as oscillatory Yukawa functions with decay length $1/\kappa_\Re$ and wave length $2\pi/\kappa_\Im$. At the crossover point the wave length is infinitely large ($\kappa_\Im = 0$) and then decreases with increasing density. It is seen in the figure that the predictions from the present theory, Equation 12.177, agree

[36] This is similar to the solutions to a polynomial equation, for example $x^2 = a$ when a decreases from positive values and passes $a = 0$, whereby the two solutions become complex.

[37] There also exist other complex-valued solutions to Equation 12.177, but here we are only concerned with those two that are connected continuously with the two real solutions κ and κ'.

[38] J. Ennis, 1993, *Electrostatic and steric interactions in surfactant systems*, Ph. D. thesis, Australian National University, Canberra, Australia; J. Ennis, R. Kjellander, and D. J. Mitchell, J. Chem. Phys. **102** (1995) 975 (https://doi.org/10.1063/1.469166).

[39] J. Ulander, and R. Kjellander, J. Chem. Phys. **114** (2001) 4893 (https://doi.org/10.1063/1.1350449).

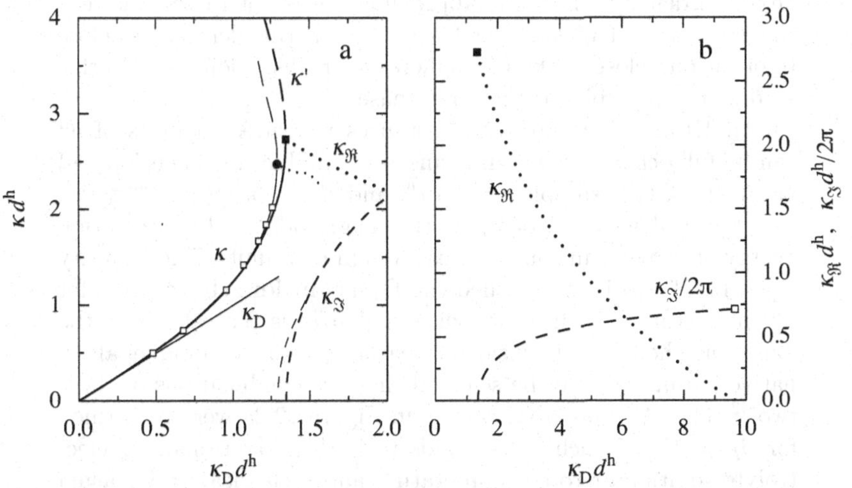

Figure 12.2 (a) Decay parameters times the ionic diameter d^{h} for a 1:–1 aqueous electrolyte solution at room temperature plotted as functions of $\kappa_{\mathrm{D}}d^{\mathrm{h}}$, which is proportional to the square root of the ion density. The same results apply to a classical 1:–1 plasma at T = 23400 K (no solvent). The thick curves are obtained from Equation 12.177, the thin curves show HNC results[38] and the open symbols show MC simulation results.[39] The straight full line shows the Debye parameter plotted as $\kappa_{\mathrm{D}}d^{\mathrm{h}}$. The Kirkwood crossover point is shown as a filled symbol for the results from HNC and from Equation 12.177. (b) The real and imaginary parts of κ from Equation 12.177 plotted as $\kappa_{\Re}d^{\mathrm{h}}$ and $\kappa_{\Im}d^{\mathrm{h}}/(2\pi)$, which are proportional to the inverse decay and wave lengths, respectively. The open square shows the point of the $\kappa_{\Im}$ curve where $\kappa_{\Re} = 0$.

very well with the MC and HNC results, including the crossover point and beyond.[40] The HNC approximation gives the crossover point at $\kappa_{\mathrm{D}}d^{\mathrm{h}} \approx 1.29$ while the present, very simple approximation gives $\kappa_{\mathrm{D}}d^{\mathrm{h}} = 1.35$.[41] Fig. 12.2b shows $\kappa_{\Re}d^{\mathrm{h}}$ and $\kappa_{\Im}d^{\mathrm{h}}/(2\pi)$ obtained from Equation 12.177 beyond the Kirkwood crossover point. The parameter $\kappa_{\Re}$ decreases with increasing density and hence the decay length $1/\kappa_{\Re}$ of the oscillatory mode becomes longer; the latter becomes longer than the wave length $2\pi/\kappa_{\Im}$ at $\kappa_{\mathrm{D}}d^{\mathrm{h}} = 6.1$, where the two curves cross each other. The present theory predicts that $\kappa_{\Re}$ becomes equal to zero and the decay length hence infinitely large for $\kappa_{\mathrm{D}}d^{\mathrm{h}} = 9.64$, but this corresponds to a very

[40]The HNC data for $\kappa_{\mathrm{D}}d^{\mathrm{h}} \gtrsim 1.5$ in the reference cited in footnote 38 were not reliable and have been ignored in the plot.

[41]The Mean Spherical approximation, which is considerably more complicated than the present approximation, gives results that are very similar to those of Equation 12.177 and has a Kirkwood crossover point at $\kappa_{\mathrm{D}}d^{\mathrm{h}} \approx 1.23$.

large ion density where the approximation is not reasonable. For the classical 1:–1 plasma mentioned earlier, this density lies close to or beyond close-packing of spheres for realistic ion sizes, so this state is not possible for the fluid phase.

An RPM electrolyte with distances measured in units of d^{h} can be fully characterized in terms of two reduced (dimensionless) parameters, for example $\tau = \kappa_{\mathrm{D}} d^{\mathrm{h}}$ and $\beta_R = \beta q^2/(4\pi\varepsilon_r^{\mathrm{cont}}\varepsilon_0 d^{\mathrm{h}})$. The physical meaning of β_R is the magnitude of the interaction energy of two ions in contact measured in $k_B T$ units. The quantity $\beta q_{\mathrm{e}}^2/(4\pi\varepsilon_r^{\mathrm{cont}}\varepsilon_0)$ is often called the **Bjerrum length**, so β_R differ by the factor z^2/d^{h} from it, where $z = q/q_{\mathrm{e}}$ is the valency of the ions. The choice of dimensionless parameters is not unique; alternative parameters can be selected that are combinations of these two.[42] The MC and HNC results in Figure 12.2a were performed for $\beta_R = 1.55$, which corresponds to both an aqueous 1:–1 electrolyte solution at room temperature and a classical 1:–1 plasma at $T = 23400$ K (no solvent). The current approximation does not work for higher values of β_R, for example for 2:–2 electrolytes at the same temperature (cf. Figure 6.7a in Volume I and Figure 12.8, where one set of data shows the case $\beta_R = 6.22$ that corresponds to divalent ions in water).

As a simple approximation for the charge density $\rho_i(r)$ around an i-ion we take the first two terms in Equation 12.172, where we have made the substitution $\mathcal{E}_r^{\mathrm{eff}}(\kappa) \to \mathcal{E}_r^{\mathrm{eff}}(\kappa)/\varepsilon_r^{\mathrm{cont}}$, and assume that they are valid for all $r \geq d^{\mathrm{h}}$

$$\rho_i(r) = -\left[\frac{\kappa^2 q_i^{\mathrm{eff}}}{\mathcal{E}_r^{\mathrm{eff}}/\varepsilon_r^{\mathrm{cont}}} \cdot \frac{e^{-\kappa r}}{4\pi r} + \frac{\kappa'^2 q_i'^{\mathrm{eff}}}{\mathcal{E}_r'^{\mathrm{eff}}/\varepsilon_r^{\mathrm{cont}}} \cdot \frac{e^{-\kappa' r}}{4\pi r}\right] \quad \text{for } r \geq d^{\mathrm{h}}, \tag{12.178}$$

where $q_i^{\mathrm{eff}} = q_i^{\mathrm{eff}}(\kappa)$, $q_i'^{\mathrm{eff}} = q_i^{\mathrm{eff}}(\kappa')$, $\mathcal{E}_r^{\mathrm{eff}} = \mathcal{E}_r^{\mathrm{eff}}(\kappa)$ and $\mathcal{E}_r'^{\mathrm{eff}} = \mathcal{E}_r^{\mathrm{eff}}(\kappa')$. As always, we have $\rho_i(r) = 0$ for $r < d^{\mathrm{h}}$. Thus we assume that there are two decay modes with decay parameters κ and κ' given by the solutions to Equation 12.177 and that no contributions with other functional forms than Yukawa functions exist in $\rho_i(r)$ and neither in $\psi_i(r)$. This is, of course, not true in general, but as we will see it works surprisingly well in many cases when β_R is of the order 2 or less. The mean electrostatic potential $\psi_i(r)$ for $r \geq d^{\mathrm{h}}$ and the screened Coulomb potential $\phi^{\star}_{\mathrm{Coul}}(r)$ in this approximation are likewise given by the first two terms in Equation 12.171 so they have the same decay modes.

[42] Alternative dimensionless parameters that can be used to characterize an RPM electrolyte can be written as combinations of τ and β_R, for example, the reduced density $n_R^{\mathrm{b}} = n^{\mathrm{b}}(d^{\mathrm{h}})^3 = \tau^2/(8\pi\beta_R)$ and the so-called plasma parameter $(\beta_R\tau)^{-1}$, which is often denoted Λ (which different from the thermal de Broglie wave length Λ in this treatise).

Since $q^{\text{eff}}(\kappa)/q = (\kappa/\kappa_{\text{D}})^2 = e^{\kappa d^{\text{h}}}/(1+\kappa d^{\text{h}})$, we have

$$q_i^{\text{eff}}(\kappa) = \frac{q_i e^{\kappa d^{\text{h}}}}{1+\kappa d^{\text{h}}}, \tag{12.179}$$

which was our starting point. The parameters $\mathcal{E}_r^{\text{eff}}$ and $\mathcal{E}_r'^{\text{eff}}$ can be determined from the necessary requirements: local electroneutrality $\int d\mathbf{r}\,\rho_i(r) + q_i = 0$ and the Stillinger-Lovett condition, which can be written as

$$\beta \int d\mathbf{r}\, r^2 \sum_i n_i^{\text{b}} q_i \rho_i(r) = -6\varepsilon_r^{\text{cont}}\varepsilon_0 \quad \text{(diel. continuum)} \tag{12.180}$$

(Equation 12.103). The former requirement applied to Equation 12.178 yields[43]

$$\frac{\varepsilon_r^{\text{cont}}}{\mathcal{E}_r^{\text{eff}}} + \frac{\varepsilon_r^{\text{cont}}}{\mathcal{E}_r'^{\text{eff}}} = 1. \tag{12.181}$$

and the latter[44]

$$\frac{\varepsilon_r^{\text{cont}}}{\mathcal{E}_r^{\text{eff}}} e^{-\kappa d^{\text{h}}} \exp_3(\kappa d^{\text{h}}) + \frac{\varepsilon_r^{\text{cont}}}{\mathcal{E}_r'^{\text{eff}}} e^{-\kappa' d^{\text{h}}} \exp_3(\kappa' d^{\text{h}}) = 1, \tag{12.182}$$

where we have introduced the "exponential sum function"

$$\exp_\ell(x) = \sum_{\nu=0}^{\ell} \frac{x^\nu}{\nu!} \tag{12.183}$$

(note that $\exp_\infty(x) = \exp(x) = e^x$). Together, these two equations for $\mathcal{E}_r^{\text{eff}}$ and $\mathcal{E}_r'^{\text{eff}}$ give

$$\frac{\mathcal{E}_r^{\text{eff}}}{\varepsilon_r^{\text{cont}}} = \frac{e^{-\kappa d^{\text{h}}} \exp_3(\kappa d^{\text{h}}) - e^{-\kappa' d^{\text{h}}} \exp_3(\kappa' d^{\text{h}})}{1 - e^{-\kappa' d^{\text{h}}} \exp_3(\kappa' d^{\text{h}})} \tag{12.184}$$

$$\frac{\mathcal{E}_r'^{\text{eff}}}{\varepsilon_r^{\text{cont}}} = -\frac{e^{-\kappa d^{\text{h}}} \exp_3(\kappa d^{\text{h}}) - e^{-\kappa' d^{\text{h}}} \exp_3(\kappa' d^{\text{h}})}{1 - e^{-\kappa d^{\text{h}}} \exp_3(\kappa d^{\text{h}})}. \tag{12.185}$$

In Figure 12.3a the values of $\mathcal{E}_r^{\text{eff}}/\varepsilon_r^{\text{cont}}$ from Equation 12.184 before the Kirkwood crossover point are compared to results from MC simulations and the HNC approximation for the same system as in Figure 12.2.[45] The agreement is good; the curve from the present theory is somewhat shifted horizontally relative to that from HNC due to the slight difference in crossover point as seen in Figure 12.2. $\mathcal{E}_r'^{\text{eff}}/\varepsilon_r^{\text{cont}}$ is plotted in Figure 12.3b and we see that it increases quickly toward zero from large negative values.

[43]When $q_i^{\text{eff}}(\kappa)$ is given by Equation 12.179 we have $\int_{r=d^{\text{h}}}^{r=\infty} d\mathbf{r}\,\kappa^2 q_i^{\text{eff}}(\kappa) e^{-\kappa r}/(4\pi r) = q_i$.

[44]When Equation 12.178 is inserted into the Stillinger-Lovett condition 12.180, we can simplify the resulting expression by using Equation 12.166 and the fact that $\int_{r=d^{\text{h}}}^{r=\infty} d\mathbf{r}\, r^2 e^{-\kappa r}/(4\pi r) = 6e^{-\kappa d^{\text{h}}} \exp_3(\kappa d^{\text{h}})/\kappa^4$.

[45]MC and HNC data are taken from the reference in footnote 39.

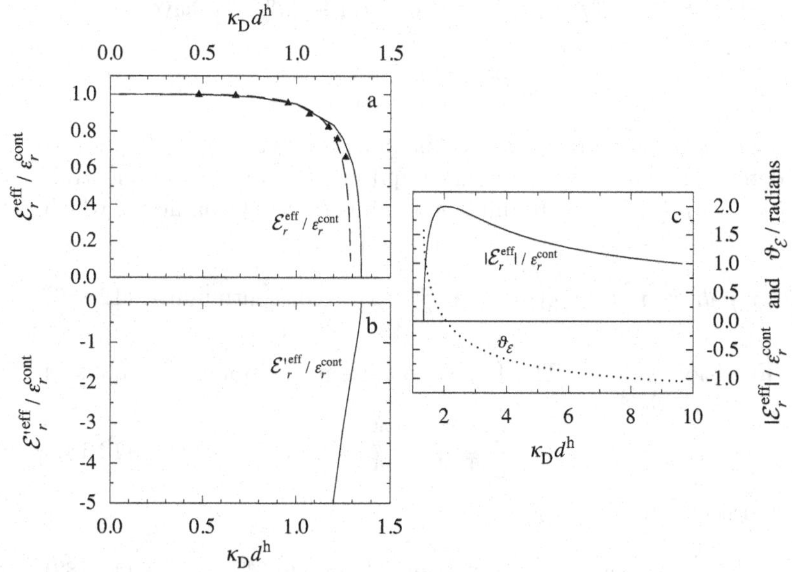

Figure 12.3 (a) The ratio $\mathcal{E}_r^{\mathrm{eff}}/\varepsilon_r^{\mathrm{cont}}$ and (b) $\mathcal{E}_r^{\prime\,\mathrm{eff}}/\varepsilon_r^{\mathrm{cont}}$ as functions of $\kappa_\mathrm{D} d^\mathrm{h}$. The full curves show the result from Equations 12.184 and 12.185, the dashed curve show HNC results and the symbols show MC simulation results.[45] The HNC and MC data are for the same system as in Figure 12.2. Note the difference in ordinate scales of frames (a) and (b). Frame (c) shows the ratio $|\mathcal{E}_r^{\mathrm{eff}}|/\varepsilon_r^{\mathrm{cont}}$ and the phase angle $\vartheta_\mathcal{E}$ of $\mathcal{E}_r^{\mathrm{eff}} = |\mathcal{E}_r^{\mathrm{eff}}|\, e^{-i\vartheta_\mathcal{E}}$ from Equations 12.188 and 12.189 as functions of $\kappa_\mathrm{D} d^\mathrm{h}$ beyond the Kirkwood crossover point where κ is complex-valued.

Equation 12.184 implies that $\varepsilon_r^{\mathrm{cont}} \geq \mathcal{E}_r^{\mathrm{eff}} \geq 0$ and Equation 12.185 that $-\infty \leq \mathcal{E}_r^{\prime\,\mathrm{eff}} \leq 0$ (provided that κ and κ' are real-valued), so $\mathcal{E}_r^{\mathrm{eff}}$ is positive and $\mathcal{E}_r^{\prime\,\mathrm{eff}}$ is negative before the crossover point as seen in the figure. The value zero occurs at that point, i.e.,

$$\kappa = \kappa' \quad \Rightarrow \quad \mathcal{E}_r^{\mathrm{eff}} = 0, \quad \mathcal{E}_r^{\prime\,\mathrm{eff}} = 0. \tag{12.186}$$

This means that each of the terms of $\rho_i(r)$ in the rhs of Equation 12.178 becomes infinite there, but their sum remains, however, finite since the two infinities from $1/\mathcal{E}_r^{\mathrm{eff}}$ and $1/\mathcal{E}_r^{\prime\,\mathrm{eff}}$ in the respective term have different signs and cancel each other. In the limit of infinite dilution where $\mathcal{E}_r^{\mathrm{eff}} \to \varepsilon_r$ and $\mathcal{E}_r^{\prime\,\mathrm{eff}} \to -\infty$, the second term in Equation (12.178) vanishes, but close to the Kirkwood crossover point this second term is about equally important as the first term.

Equation 12.186 is, in fact, *valid in the general case* because $i\kappa$ and $i\kappa'$ are two consecutive solutions of $\tilde{\epsilon}(k) = 0$ and we have

$\mathcal{E}_r^{\text{eff}}(\kappa) = \mathrm{i}\kappa\tilde{\varepsilon}'(\mathrm{i}\kappa)/2$ (Equation 12.158). Before the crossover point the signs of $\tilde{\varepsilon}'(\mathrm{i}\kappa)$ and $\tilde{\varepsilon}'(\mathrm{i}\kappa')$ must be different because the derivative of $\tilde{\varepsilon}(k)$ at the two consecutive zeros $k = \mathrm{i}\kappa$ and $k = \mathrm{i}\kappa'$ must have different signs.[46] $\mathcal{E}_r^{\text{eff}}(\kappa)$ must be positive because it goes to $\varepsilon_r^{\text{cont}}$ in the limit of infinite dilution, so $\mathcal{E}_r^{\text{eff}}(\kappa')$ is negative. When κ and κ' merge the derivative $\tilde{\varepsilon}'(k)$ becomes zero so both $\mathcal{E}_r^{\text{eff}}(\kappa)$ and $\mathcal{E}_r^{\text{eff}}(\kappa')$ are zero at the crossover point.

In Figures 12.4a and b the function $r|\rho(r)|$ is plotted as function of r on a semilogarithmic scale for the cases $\kappa_{\text{D}}d^{\text{h}} = 1.07$ and 1.27, where κ and κ' are real-valued. This corresponds to 0.5 and 0.7 M aqueous electrolytes at room temperature. We see in Figure 12.4a that the MC[47] and HNC results[48] virtually coincide with each other, so the HNC approximation is very accurate for the monovalent electrolyte. The prediction from Equation 12.178 (dotted curve) agrees very well with the MC and HNC results for all r except in a very short interval just outside the core region, $1 \leq r \lesssim 1.3d^{\text{h}}$. The curves virtually coincide for $r \gtrsim 2d^{\text{h}}$. In Fig. 12.4b the agreement is equally good.

The dashed curves, which are straight lines, are calculated from the first term in Equation 12.178, which is leading for large r (a Yukawa function becomes a straight line in a semilogarithmic plot of this kind). In Figure 12.4b this line coincides with the accurate results $r \gtrsim 2.5d^{\text{h}}$, but it deviates substantially from the dotted curve for smaller r values. This implies that the contribution from the second Yukawa term with decay parameter κ' is important for the 0.7 M case. For 0.5 M it is less important, but it still gives a noticeable contribution that makes the dotted curve to deviate from the dashed one. The difference in importance of the second term for the 0.7 and 0.5 M cases is due to the fact that the magnitude of $\mathcal{E}_r'^{\text{eff}}$ in the denominator of this term increases rapidly when the concentration is decreased, see Figure 12.3. We see that Equation 12.178 gives very good results for the 1:–1 electrolyte systems; the deviation for small r has little consequence in most cases and is caused by terms in $\rho(r)$ with shorter decay lengths that are not included in this approximation.

We can conclude that close to the Kirkwood crossover point it is very important to have the two decay modes present in the theory in order to describe the electrolyte in an appropriate manner. This conclusion is further emphasized beyond this point where κ and κ' are complex-valued, $\kappa = \kappa_{\Re} + \mathrm{i}\kappa_{\Im}$. We then have $q_i^{\text{eff}} = |q_i^{\text{eff}}|\, e^{-\mathrm{i}\vartheta_i}$ and $q_i'^{\text{eff}} = \underline{q_i^{\text{eff}}} = |q_i^{\text{eff}}|\, e^{\mathrm{i}\vartheta_i}$, where the phase ϑ_i is a real number.

[46] This is like the derivative of $f(x) = (x - a)(x - b)$ at $x = a$ and $x = b$, where $f(x) = 0$.

[47] MC data are taken from the reference in footnote 39.

[48] HNC data are taken from the references in footnote 38.

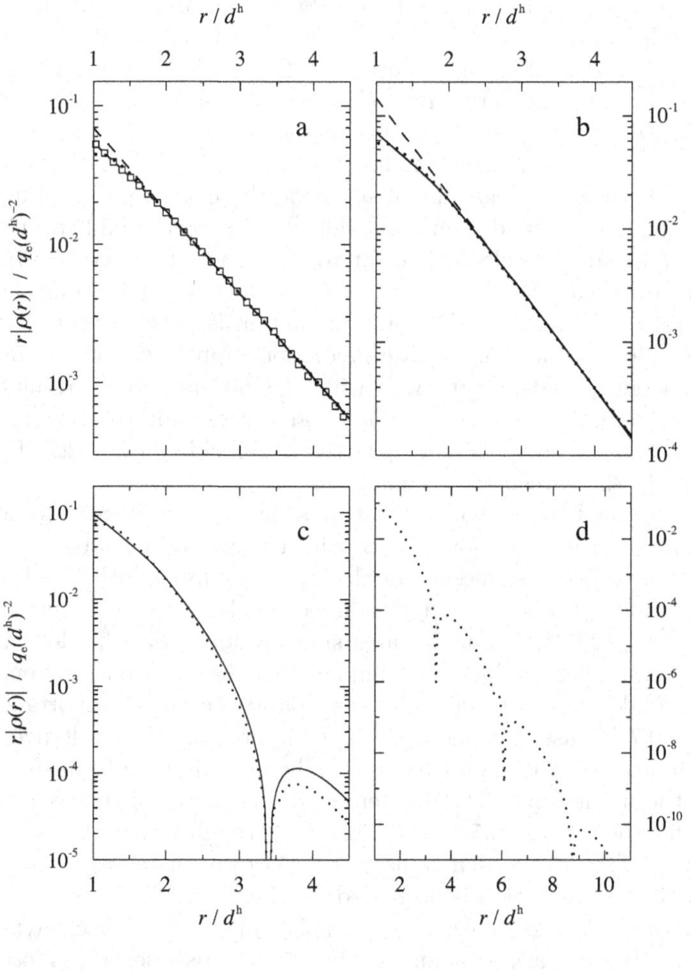

Figure 12.4 The radial charge distribution $\rho(r)$ for a 1:–1 aqueous electrolyte solution at room temperature plotted as $r\rho(r)$ as a function of r on a semilogarithmic scale. The ion diameter is $d^{\mathrm{h}} = 4.6$ Å and the concentration is 0.5 M in frame (a), 0.7 M in frame (b) and 1.0 M in frame (c), which correspond to $\kappa_{\mathrm{D}} d^{\mathrm{h}} = 1.07$, 1.27 and 1.51, respectively. The open squares in frame (a) show results from MC simulations.[47] In frames (a)–(c) the full curves show HNC results[48] and the dotted curves show results from the present theory, Equations 12.178 and 12.187. The dashed curves in frames (a) and (b) show results from solely the first (leading) term in the former equation. Frame (d) shows the same dotted curve as in frame (c), but in a wider range. The unit for $r\rho(r)$ is $q_e(d^{\mathrm{h}})^{-2}$.

Likewise, $\mathcal{E}_r^{\mathrm{eff}} = |\mathcal{E}_r^{\mathrm{eff}}|\, e^{-i\vartheta_{\mathcal{E}}}$, $\mathcal{E}_r'^{\,\mathrm{eff}} = \overline{\mathcal{E}_r^{\mathrm{eff}}} = |\mathcal{E}_r^{\mathrm{eff}}|\, e^{i\vartheta_{\mathcal{E}}}$ and $\kappa = |\kappa|\, e^{-i\vartheta_\kappa}$. The radial charge distribution is exponentially damped oscillatory and Equation 12.178 can be written as

$$\rho_i(r) = \frac{|\kappa|^2 |q_i^{\mathrm{eff}}|\, \varepsilon_r^{\mathrm{cont}}}{2\pi |\mathcal{E}_r^{\mathrm{eff}}|} \cdot \frac{e^{-\kappa_{\Re} r}}{r} \cos(\kappa_{\Im} r + \alpha_i) \quad \text{when } r \geq d^{\mathrm{h}}, \tag{12.187}$$

where $\alpha_i = \vartheta_i + 2\vartheta_\kappa - \vartheta_{\mathcal{E}}$. Note that $\vartheta_\kappa = -\arctan(\kappa_{\Im}/\kappa_{\Re})$.

Equation 12.181 yields

$$\frac{|\mathcal{E}_r^{\mathrm{eff}}|}{\varepsilon_r^{\mathrm{cont}}} = 2\cos\vartheta_{\mathcal{E}} \tag{12.188}$$

and one can derive from Equation 12.184 that

$$\tan\vartheta_{\mathcal{E}} = \frac{e^{\kappa_{\Re} d^{\mathrm{h}}} - \mathsf{a}\cos(\kappa_{\Im} d^{\mathrm{h}}) - \mathsf{b}\sin(\kappa_{\Im} d^{\mathrm{h}})}{\mathsf{a}\sin(\kappa_{\Im} d^{\mathrm{h}}) - \mathsf{b}\cos(\kappa_{\Im} d^{\mathrm{h}})}, \tag{12.189}$$

where

$$\begin{aligned} \mathsf{a} &= \mathsf{a}(\kappa_{\Re}, \kappa_{\Im}) = \exp_3(\kappa_{\Re} d^{\mathrm{h}}) - (\kappa_{\Im} d^{\mathrm{h}})^2 \frac{1 + \kappa_{\Re} d^{\mathrm{h}}}{2} \\ \mathsf{b} &= \mathsf{b}(\kappa_{\Re}, \kappa_{\Im}) = \kappa_{\Im} d^{\mathrm{h}} \left[\exp_2(\kappa_{\Re} d^{\mathrm{h}}) - \frac{(\kappa_{\Im} d^{\mathrm{h}})^2}{6} \right]. \end{aligned}$$

In Figure 12.3c we have plotted $|\mathcal{E}_r^{\mathrm{eff}}|/\varepsilon_r^{\mathrm{cont}}$ and $\vartheta_{\mathcal{E}}$ as functions of $\kappa_{\mathrm{D}} d^{\mathrm{h}}$. At the Kirkwood crossover point, to the left in the figure, we have $|\mathcal{E}_r^{\mathrm{eff}}| = 0$ and $\vartheta_{\mathcal{E}} = \pi/2$.

In Figures 12.4c and d the function $r|\rho(r)|$ is plotted for $\kappa_{\mathrm{D}} d^{\mathrm{h}} = 1.51$, which corresponds to 1.0 M aqueous electrolytes at room temperature. The agreement with the accurate HNC curve is about equally good for $r \lesssim 3.5 d^{\mathrm{h}}$ as the results for the other concentrations. For $r \gtrsim 3.5 d^{\mathrm{h}}$ Equation 12.187 underestimates the magnitude of the charge density a little bit, but the values are very small there, $r|\rho(r)| \lesssim 10^{-4}$, so this has little consequences. In Figure 12.4d the function $r|\rho(r)|$ is plotted for a wider range of r values in order to show some oscillations in the tail for large r. Whenever $\rho(r)$ passes zero and changes sign the logarithm $\ln(r|\rho(r)|)$ becomes infinite, which causes the cusps seen in the figure (they are truncated at finite values in the plot).

Thus, the two-mode approximation used in Equations 12.178 together with Equation 12.177 works very well for the systems investigated in this Example.[49] It is restricted to systems where β_R is of the order 2 or less because the expression for $\rho_i(r)$ is

[49] This approach works very well also for the calculation of thermodynamic data, like chemical potentials and the osmotic coefficient, see the reference cited in footnote 35.

linear in the ionic charges, so nonlinear effects that are important for stronger electrostatic coupling are neglected. Furthermore, we have neglected effects of packing of the hard core ions that are important at high ionic densities. The latter will be considered later in Example 12:C.

12.3.4 Pair correlations in the main, general case

Next, we turn to the pair correlation functions and we will treat ions distributed in vacuum ($\varepsilon_r^{\mathrm{cont}} = 1$). We have

$$h_{ij}^{(2)}(r_{12}) = h_{ij}^{\star}(r_{12}) - \beta w_{ij}^{(2)\mathrm{el}}(r_{12}),$$

where

$$w_{ij}^{(2)\mathrm{el}}(r_{12}) = \int d\mathbf{r}_3\, d\mathbf{r}_4\, \rho_i^{\star}(r_{14})\phi_{\mathrm{Coul}}^{\star}(r_{43})\rho_j^{\star}(r_{23})$$

(see Table 12.2 on page 303 and Equation 12.112). In Fourier space

$$\tilde{h}_{ij}^{(2)}(k) = \tilde{h}_{ij}^{\star}(k) - \frac{\beta\tilde{\rho}_i^{\star}(k)\tilde{\rho}_j^{\star}(k)}{\varepsilon_0 k^2 \tilde{\epsilon}(k)} \qquad (12.190)$$

because

$$\tilde{w}_{ij}^{(2)\mathrm{el}}(k) = \tilde{\rho}_i^{\star}(k)\tilde{\phi}_{\mathrm{Coul}}^{\star}(k)\tilde{\rho}_j^{\star}(k) = \frac{\tilde{\rho}_i^{\star}(k)\tilde{\rho}_j^{\star}(k)}{\varepsilon_0 k^2 \tilde{\epsilon}(k)}$$

(cf. Equation 12.113). Note that $\tilde{w}_{ij}^{(2)\mathrm{el}}(k) = \tilde{\rho}_i^{\star}(k)\tilde{\psi}_j(k) = \tilde{\psi}_i(k)\tilde{\rho}_j^{\star}(k)$ so the pole $k = i\kappa$ given by $\tilde{\epsilon}(i\kappa) = 0$ gives rise to the leading decay mode of $w_{ij}^{(2)\mathrm{el}}(r)$

$$w_{ij}^{(2)\mathrm{el}}(r) \sim \frac{q_i^{\mathrm{eff}}(\kappa)q_j^{\mathrm{eff}}(\kappa)}{\mathcal{E}_r^{\mathrm{eff}}(\kappa)\varepsilon_0} \cdot \frac{e^{-\kappa r}}{4\pi r} \quad \text{when } r \to \infty \qquad (12.191)$$

when κ is a real number, for example for sufficiently low ionic densities.

We will now show a very important fact:

> In the vast majority of cases *all poles of* $\tilde{h}_{ij}^{(2)}(k)$ *are given by the zeros of* $\tilde{\epsilon}(k)$ and are therefore the same as those of $\tilde{\phi}_{\mathrm{Coul}}^{\star}(k)$, $\tilde{\psi}_i(k)$ and $\tilde{\rho}_i(k)$. The decay modes of these functions are therefore the same, so $h_{ij}^{(2)}(r)$ has the same Yukawa function decays *as* $\phi_{\mathrm{Coul}}^{\star}(r)$, $\psi_i(r)$ and $\rho_i(r)$ with the ***same set of decay parameters***. The same applies to $H_{\mathrm{NN}}(r)$, $H_{\mathrm{NQ}}(r)$ and $H_{\mathrm{QQ}}(r)$.

For sufficiently large r values we have[50]

$$h_{ij}^{(2)}(r) = -\frac{\beta q_i^{\text{eff}}(\kappa) q_j^{\text{eff}}(\kappa)}{\mathcal{E}_r^{\text{eff}}(\kappa)\varepsilon_0} \cdot \frac{e^{-\kappa r}}{4\pi r} - \frac{\beta q_i^{\text{eff}}(\kappa') q_j^{\text{eff}}(\kappa')}{\mathcal{E}_r^{\text{eff}}(\kappa')\varepsilon_0} \cdot \frac{e^{-\kappa' r}}{4\pi r} - \ldots \quad (\text{large } r) \tag{12.192}$$

where only the first two Yukawa function contributions are explicitly shown (cf. Equations 12.171 and 12.172). Note that we can obtain the corresponding expression for $\rho_i(r)$, Equation 12.172, by inserting Equation 12.192 into $\rho_i(r) = \sum_j n_j^{\text{b}} q_j h_{ij}^{(2)}(r)$ and using Equation 12.164 where κ^2 is expressed in terms of the effective charges.

For oscillatory decay modes, where the decay parameters κ and κ' are complex-valued, the two terms in Equation 12.192 can be replaced by an exponentially damped, oscillatory function,

$$h_{ij}^{(2)}(r) = -\frac{\beta |q_i^{\text{eff}} q_j^{\text{eff}}|}{|\mathcal{E}_r^{\text{eff}}|\varepsilon_0} \cdot \frac{e^{-\kappa_{\Re} r}}{2\pi r} \cos(\kappa_{\Im} r + \vartheta_i + \vartheta_j - \vartheta_{\mathcal{E}}) - \ldots \quad (\text{large } r), \tag{12.193}$$

where, as before, the phase shifts $\vartheta_{\mathcal{E}}$ and ϑ_i are defined from $\mathcal{E}_r^{\text{eff}}(\kappa_{\Re} + i\kappa_{\Im}) = |\mathcal{E}_r^{\text{eff}}| \exp(-i\vartheta_{\mathcal{E}})$ and $q_i^{\text{eff}}(\kappa_{\Re} + i\kappa_{\Im}) = |q_i^{\text{eff}}| \exp(-i\vartheta_i)$, cf. Equations 12.173 and 12.174. As we have seen in Figure 12.4 above and also in Figures 6.9–6.11 of Volume I, the value of r does not necessarily need to be particularly large for these types of Yukawa function decays to be dominating.

It is, however, far from self-evident that $\tilde{h}_{ij}^{(2)}(k)$ has poles only when $\tilde{\epsilon}(k)$ is zero because the first term $\tilde{h}_{ij}^{\star}(k)$ in Equation 12.190 has poles that are not zeros of $\tilde{\epsilon}(k)$. There are, in fact, some exceptional cases where the poles of $\tilde{h}_{ij}^{(2)}(k)$ originate from both the zeros of $\tilde{\epsilon}(k)$ and the poles of $\tilde{h}_{ij}^{\star}(k)$; the most notable example is the restricted primitive model where anions and cations only differ in their signs. We will deal with these exceptions later and treat the main cases first. What normally takes place is that any simple pole of $\tilde{h}_{ij}^{\star}(k)$ is cancelled by a corresponding pole of the second term in the rhs of Equation 12.190, which is equal to $-\beta\tilde{w}_{ij}^{(2)\text{el}}(k)$. The only simple poles that remain in the rhs are therefore the zeros of $\tilde{\epsilon}(k)$. In order to show this, let us first analyze $\tilde{h}_{ij}^{\star}(k)$, then $\tilde{w}_{ij}^{(2)\text{el}}(k)$ and finally $\tilde{h}_{ij}^{(2)}(k)$. For simplicity we will do the derivation for binary electrolytes; the general case can be done in a similar manner.[51]

Proof: The poles of $\tilde{h}_{ij}^{(2)}(k)$ and $\tilde{\phi}_{\text{Coul}}^{\star}(k)$ are the same and are given by $\tilde{\epsilon}(k) = 0$.

[50]There are additional contributions with other functional forms, cf. footnote 29, but at least for low electrolyte densities the first term dominates for large r. We will deal with the other types of terms later in the shaded text box on page 351, see Equation 12.203.

[51]A more general derivation is done in Section 13.3.3 for nonspherical particles. It is, of course, valid in the spherical case too.

The functions $h^{\star}_{ij}(r)$ and $c^{\star}_{ij}(r)$ satisfy the Ornstein Zernike equation 12.111, which can be written as

$$\sum_l \int d\mathbf{r}_3\, C^{\star}_{il}(r_{13}) H^{\star}_{lj}(r_{32}) = \delta^{(3)}_{ij}(r_{12})$$

(cf. Equation 12.45), where $H^{\star}_{ij}(r) \equiv n^{\mathrm{b}}_i \delta^{(3)}_{ij}(r) + n^{\mathrm{b}}_i n^{\mathrm{b}}_j h^{\star}_{ij}(r)$ and $C^{\star}_{ij}(r) \equiv \delta^{(3)}_{ij}(r)/n^{\mathrm{b}}_i - c^{\star}_{ij}(r)$, cf. Equations 12.39 and 12.44. In Fourier space, we have

$$\sum_l \tilde{C}^{\star}_{il}(k) \tilde{H}^{\star}_{lj}(k) = \delta_{ij}.$$

We can write this system of equations in matrix form as

$$\mathbb{C}^{\star}(k)\mathbb{H}^{\star}(k) = \mathbb{I},$$

where for the present case of a binary electrolyte

$$\mathbb{C}^{\star}(k) = \begin{bmatrix} \tilde{C}^{\star}_{++}(k) & \tilde{C}^{\star}_{+-}(k) \\ \tilde{C}^{\star}_{-+}(k) & \tilde{C}^{\star}_{--}(k) \end{bmatrix}, \qquad \mathbb{H}^{\star}(k) = \begin{bmatrix} \tilde{H}^{\star}_{++}(k) & \tilde{H}^{\star}_{+-}(k) \\ \tilde{H}^{\star}_{-+}(k) & H^{\star}_{--}(k) \end{bmatrix}$$

and $\mathbb{I}$ is the unit matrix. Note that $\tilde{C}^{\star}_{-+} = \tilde{C}^{\star}_{+-}$ and $\tilde{H}^{\star}_{-+} = \tilde{H}^{\star}_{+-}$ so the matrices are symmetric. $\mathbb{H}^{\star}$ is the matrix inverse of $\mathbb{C}^{\star}$ and using the fact that an inverse matrix is equal to the adjugate matrix divided by the determinant[52] we obtain

$$\mathbb{H}^{\star}(k) = [\mathbb{C}^{\star}(k)]^{-1} = \frac{\operatorname{Adj}\mathbb{C}^{\star}(k)}{D^{\star}(k)}, \tag{12.194}$$

where $D^{\star}(k) = \det \mathbb{C}^{\star}(k)$ is the determinant and $\operatorname{Adj}\mathbb{C}^{\star}(k)$ is the adjugate matrix of $\mathbb{C}^{\star}(k)$. In the present case

$$\operatorname{Adj}\mathbb{C}^{\star}(k) = \begin{bmatrix} \tilde{C}^{\star}_{--}(k) & -\tilde{C}^{\star}_{+-}(k) \\ -\tilde{C}^{\star}_{+-}(k) & \tilde{C}^{\star}_{++}(k) \end{bmatrix}.$$

Each element of $\mathbb{H}^{\star}(k)$ has a (simple) pole for every k value where $\det \mathbb{C}^{\star}(k)$ has a (simple) zero

$$D^{\star}(k) = \det \mathbb{C}^{\star}(k) = \tilde{C}^{\star}_{++}(k)\tilde{C}^{\star}_{--}(k) - [\tilde{C}^{\star}_{+-}(k)]^2 = 0 \tag{12.195}$$

[52]The inverse of a square matrix $\mathbb{A}$ is equal to $[\det\mathbb{A}]^{-1}\operatorname{Adj}\mathbb{A}$, where $\det\mathbb{A}$ is the determinant and $\operatorname{Adj}\mathbb{A}$ is the adjugate matrix of $\mathbb{A}$. $\operatorname{Adj}\mathbb{A}$ is the transpose of the matrix of cofactors $\mathrm{cof}_{ij}\mathbb{A}$. One obtains $\mathrm{cof}_{ij}\mathbb{A}$ by taking the determinant of the matrix obtained by striking the i'th row and the j'th column of $\mathbb{A}$ and choosing a positive/negative sign if $i+j$ is even/odd.

provided that the corresponding element of $\mathrm{Adj}\,\mathbb{C}^\star(k)$ is nonzero. Say that the determinant is zero for $k = ib$, where b may be a real or a complex number, so $\tilde{H}^\star_{ij}(k)$ has a pole for $k = ib$. Let us select the parameters α_+ and α_- such that $\tilde{C}^\star_{++}(ib) = \varsigma\alpha_-^2$ and $\tilde{C}^\star_{--}(ib) = \varsigma\alpha_+^2$, where ς is either $+1$ or -1 (this is to be determined later) and where the other quantities may be complex numbers. Since $D^\star(ib) = 0$ we have

$$[\tilde{C}^\star_{+-}(ib)]^2 = \tilde{C}^\star_{++}(ib)\tilde{C}^\star_{--}(ib) = \varsigma^2\alpha_-^2\alpha_+^2 = \alpha_-^2\alpha_+^2$$

so $\tilde{C}^\star_{+-}(ib) = \pm\alpha_+\alpha_-$, where one of the signs applies. We can always select ς such that $-\tilde{C}^\star_{+-}(ib) = \varsigma\alpha_+\alpha_-$. Thus we have

$$\mathrm{Adj}\,\mathbb{C}^\star(ib) = \varsigma\begin{bmatrix} \alpha_+^2 & \alpha_+\alpha_- \\ \alpha_+\alpha_- & \alpha_-^2 \end{bmatrix}$$

so the i, j element of this matrix factorizes like $\varsigma\alpha_i\alpha_j$. From this result and Equation 12.194 it follows that $\tilde{H}^\star_{ij}(k)$ is dominated for k values near ib by the diverging contribution

$$\tilde{H}^\star_{ij}(k) \sim \frac{\varsigma\alpha_i(b)\alpha_j(b)}{D^\star(k)} \sim \frac{\alpha_i(b)\alpha_j(b)}{(k^2+b^2)\mathcal{K}^\star(b)} \quad \text{when } k \to ib, \tag{12.196}$$

where $\mathcal{K}^\star(b)$ is a constant and we have explicitly shown that α_i depends on b. Since $\tilde{H}^\star_{ij}(k) = n_i^{\mathrm{b}}\delta_{ij} + n_i^{\mathrm{b}}n_j^{\mathrm{b}}\tilde{h}^\star_{ij}(k)$, the same contribution dominates in $n_i^{\mathrm{b}}n_j^{\mathrm{b}}\tilde{h}^\star_{ij}(k)$ at the pole $k = ib$. Furthermore, since $\tilde{\rho}_i^\star(k) = \sum_j q_j\tilde{H}^\star_{ij}(k)/n_i^{\mathrm{b}}$ and $\tilde{\varepsilon}(k) = 1+\beta\sum_{ij} q_iq_j\tilde{H}^\star_{ij}(k)\tilde{\phi}_{\mathrm{Coul}}(k)$ (as follows from Equation 12.76, 12.80 and 12.96), we see that when $k \to ib$

$$\begin{aligned} n_i^{\mathrm{b}}\tilde{\rho}_i^\star(k) &\sim \frac{\alpha_i(b)\mathcal{F}(b)}{(k^2+b^2)\mathcal{K}^\star(b)} \\ \varepsilon_0 k^2\tilde{\varepsilon}(k) &\sim \frac{\beta\mathcal{F}^2(b)}{(k^2+b^2)\mathcal{K}^\star(b)}, \end{aligned} \tag{12.197}$$

where $\mathcal{F}(b) = \sum_j q_j\alpha_j(b)$. This expression is valid provided that $\mathcal{F}(b)$ is nonzero.

Next, we investigate $n_i^{\mathrm{b}}n_j^{\mathrm{b}}\beta\tilde{w}^{(2)\mathrm{el}}_{ij}(k)$, which is equal to the second term in the rhs of Equation 12.190 times $n_i^{\mathrm{b}}n_j^{\mathrm{b}}$. It is dominated when $k \to ib$ by the diverging contribution from the pole

$$\begin{aligned} n_i^{\mathrm{b}}n_j^{\mathrm{b}}\beta\tilde{w}^{(2)\mathrm{el}}_{ij}(k) \equiv n_i^{\mathrm{b}}n_j^{\mathrm{b}}\frac{\beta\tilde{\rho}_i^\star(k)\tilde{\rho}_j^\star(k)}{\varepsilon_r^{\mathrm{cont}}\varepsilon_0 k^2\tilde{\varepsilon}(k)} &\sim \frac{\beta\dfrac{\alpha_i(b)\mathcal{F}(b)}{(k^2+b^2)\mathcal{K}^\star(b)}\cdot\dfrac{\alpha_j(b)\mathcal{F}(b)}{(k^2+b^2)\mathcal{K}^\star(b)}}{\dfrac{\beta\mathcal{F}^2(b)}{(k^2+b^2)\mathcal{K}^\star(b)}} \\ &= \frac{\alpha_i(b)\alpha_j(b)}{(k^2+b^2)\mathcal{K}^\star(b)} \end{aligned}$$

which is the same as the contribution from the pole of $n_i^{\rm b} n_j^{\rm b} \tilde{h}_{ij}^{\star}(k)$. Thus, in $\tilde{h}_{ij}^{(2)}(k)$ as given by Equation 12.190, the pole at $k = ib$ for $\tilde{h}_{ij}^{\star}(k)$ is cancelled by the pole from the second term in the rhs. We have thereby shown that provided $\mathcal{F}(b) \neq 0$, the only poles that remain in the rhs and hence in $\tilde{h}_{ij}^{(2)}(k)$ are those from the zeros of $\tilde{\varepsilon}(k)$. Thus $\tilde{h}_{ij}^{(2)}(k)$ has the same poles as $\tilde{\phi}_{\rm Coul}^{\star}(k)$.

The pole of $\tilde{h}_{ij}^{\star}(k)$ at $k = ib$ gives a contribution to $h_{ij}^{\star}(r)$ that decays like a Yukawa function and when it is the singularity of $\tilde{h}_{ij}^{\star}(k)$ that lies closest to the real axis in complex k-space it follows from Equation 12.196 that

$$n_i^{\rm b} n_j^{\rm b} h_{ij}^{\star}(r) \sim \frac{\alpha_i(b)\alpha_j(b)}{\mathcal{K}^{\star}(b)} \cdot \frac{e^{-br}}{4\pi r} \qquad \text{when } r \to \infty. \tag{12.198}$$

Note that $\varsigma\mathcal{K}^{\star}(b) = [dD^{\star}(k)/dk]_{k=ib}/(2ib)$, so Equation 12.198 can also be obtained by application of Equation 12.144.

We have derived Equations 12.196 and 12.198 for the binary case, but they are valid any number of components and we have in the general case the factorization[53]

$$[\mathrm{Adj}\, \mathbb{C}^{\star}(ib)]_{ij} = \varsigma\alpha_i(b)\alpha_j(b)$$

when $\det \mathbb{C}^{\star}(ib) = 0$.

We have thus shown that $\tilde{h}_{ij}^{(2)}(k)$, $\tilde{\phi}_{\rm Coul}^{\star}(k)$, $\tilde{\rho}_i(k)$ and $\tilde{\psi}_i(k)$ normally have the same poles, which are different from those of $\tilde{h}_{ij}^{\star}(k)$. It follows that $h_{ij}^{(2)}(r)$, $\psi_i(r)$, $\rho_i(r)$ *and* $\phi_{\rm Coul}^{\star}(r)$ *have the same decay modes with the same set of decay parameters* that are all determined by the solutions of the equation $\tilde{\varepsilon}(i\kappa) = 0$, which is equivalent to Equations 12.151 and 12.166. This is our main case.

However, these findings *do not apply to electrolytes in the restricted primitive model* because in this case $c_{++}^{\star}(r) \equiv c_{--}^{\star}(r)$, so $\alpha_+^2(b) = \alpha_-^2(b)$ and there are two different groups of poles for $\tilde{h}_{ij}^{\star}(k)$, those that have $\alpha_+(b) = \alpha_-(b)$ and those that have $\alpha_+(b) = -\alpha_-(b)$. For the poles $k = ib$ that belong to the first group we have $\mathcal{F}(b) = q[\alpha_+(b) - \alpha_-(b)] = 0$. Then the second term in the rhs of Equation 12.190 does *not* have poles for the same k values as $\tilde{h}_{ij}^{\star}(k)$ and hence $k = ib$ is also a pole of $\tilde{h}_{ij}^{(2)}(k)$ (but not zeros of $\tilde{\varepsilon}(k)$). For each of them there is a term in $h_{ij}^{(2)}(r)$ that decays like in Equation 12.198. The poles of $\tilde{h}_{ij}^{\star}(k)$ that belong to the second group have $\mathcal{F}(b) \neq 0$ so they cannot be poles of $\tilde{h}_{ij}^{(2)}(k)$ (like in our main case).

[53] The factorization of the numerator of the coefficient in Equations 12.196 and 12.198 is a consequence of the fact that elements of the adjugate matrix of any square matrix $\mathbb{C}$ that has zero determinant, $\det \mathbb{C} = 0$, factorize so that $[\mathrm{Adj}\, \mathbb{C}]_{ij} = \varsigma\alpha_i\alpha_j$ for some α_l, $l = i, j$. This is the basis for a generalization of the current proof to electrolytes with any number of components.

The same is true for any **charge-inversion invariant system**, that is, any system that stays the same when all positive and negative charges reverse their signs, which implies that $c^{\star}_{++}(r) \equiv c^{\star}_{--}(r)$. In these exceptional cases ***the statements in the box on page 344 do not apply*** and $\tilde{h}^{(2)}_{ij}(k)$ has instead two distinct sets of poles: those that are also poles of $\tilde{h}^{\star}_{ij}(k)$ (zeros of $D^{\star}(k)$) and those that are due to the zeros of $\tilde{\varepsilon}(k)$. The members of the latter set are poles of $\tilde{\phi}^{\star}_{\text{Coul}}(k)$, $\tilde{\rho}_i(k)$ and $\tilde{\psi}_i(k)$ but the members of the former set are *not* poles of these functions.

Anything that breaks the charge-inversion symmetry makes the main case to apply, for example a slight difference in size for anions and cations with the same absolute charge (i.e., a slight deviation from RPM). Then all poles of $\tilde{h}^{(2)}_{ij}(k)$ are given by the zeros of $\tilde{\varepsilon}(k)$, but one can nevertheless still distinguish two sets of poles: those that lie close to some pole of $\tilde{h}^{\star}_{ij}(k)$ and those that do not. This difference between the two sets has important consequences for the behavior of $h^{(2)}_{ij}(r)$. As will be show later, when the size difference between anions and cations is made to vanish, the former set merges with the poles of $\tilde{h}^{\star}_{ij}(k)$ and ceases to be zeros of $\tilde{\varepsilon}(k)$.

In order to further illuminate the difference between the main case and the exceptions let us investigate the poles of $\tilde{h}^{\star}_{ij}(k)$ and $\tilde{h}^{(2)}_{ij}(k)$ a bit more and for simplicity we continue to do the explicit derivations for binary electrolytes. As we have seen, the former poles occurs when the determinant $D^{\star}(k)$ in Equation 12.194 is zero, Equation 12.195. In the same manner as the derivation of Equation 12.194, we obtain

$$\mathbb{H}^{(2)}(k) = [\mathbb{C}^{(2)}(k)]^{-1} = \frac{\text{Adj}\,\mathbb{C}^{(2)}(k)}{D^{(2)}(k)}, \tag{12.199}$$

where the matrix elements of $\mathbb{H}^{(2)}$ and $\mathbb{C}^{(2)}$ are $\tilde{H}^{(2)}_{ij}(k) \equiv n^{\text{b}}_i\delta_{ij} + n^{\text{b}}_i n^{\text{b}}_j \tilde{h}^{(2)}_{ij}(k)$ and $\tilde{C}^{(2)}_{ij}(k) \equiv \delta_{ij}/n^{\text{b}}_i - \tilde{c}^{(2)}_{ij}(k)$. Thus, the poles of $\tilde{h}^{(2)}_{ij}(k)$ occur for k values where the determinant $D^{(2)}(k) = \tilde{C}^{(2)}_{++}(k)\tilde{C}^{(2)}_{--}(k) - [\tilde{C}^{(2)}_{+-}(k)]^2$ is zero. The definition of $c^{\star}_{ij}$ in Equation 12.43 implies that $\tilde{C}^{(2)}_{ij}(k) = \tilde{C}^{\star}_{ij}(k) + \beta q_i q_j \tilde{\phi}_{\text{Coul}}(k)$ and by inserting this into $D^{(2)}(k)$ and simplifying we obtain[54]

$$\begin{aligned} D^{(2)}(k) &= D^{\star}(k) + \beta\left[q_+^2 \tilde{C}^{\star}_{--}(k) + q_-^2 \tilde{C}^{\star}_{++}(k) - 2q_+q_-\tilde{C}^{\star}_{+-}(k)\right]\tilde{\phi}_{\text{Coul}}(k) \\ &= D^{\star}(k) + \beta\sum_{ij} q_i q_j [\text{Adj}\,\mathbb{C}^{\star}(k)]_{ij}\,\tilde{\phi}_{\text{Coul}}(k). \end{aligned} \tag{12.200}$$

---

[54]This result is in the general case a consequence of the matrix identity $\text{Det}(\mathbb{A} + \mathbb{a}\mathbb{b}^{\mathsf{T}}) = \text{Det}(\mathbb{A}) + \mathbb{b}^{\mathsf{T}}\text{Adj}(\mathbb{A})\mathbb{a}$, where $\mathbb{A}$ is a square matrix and $\mathbb{a}$ and $\mathbb{b}$ are column matrices. Equation 12.200 follows if we set $\mathbb{A} = \mathbb{C}^{\star}$, $\mathbb{b}^{\mathsf{T}} = (q_1, q_2, \ldots q_M)$, where M is the number of ionic components, and $\mathbb{a} = \beta\tilde{\phi}_{\text{Coul}}\,\mathbb{b}$, whereby $\mathbb{A} + \mathbb{a}\mathbb{b}^{\mathsf{T}} = \mathbb{C}^{(2)}$.

From this result it follows that $D^{(2)}(k)$ and $D^{\star}(k)$ cannot both be equal to zero for the same k value unless $\sum_{ij} q_i q_j [\mathrm{Adj}\,\mathbb{C}^{\star}(k)]_{ij} = 0$ for this k. This is in line with our findings in the shaded text box above because when $D^{\star}(k)$ is zero, say for $k = \mathrm{i}b$ as investigated there, we have $[\mathrm{Adj}\,\mathbb{C}^{\star}(\mathrm{i}b)]_{ij} = \varsigma\alpha_i(b)\alpha_j(b)$ so $\sum_{ij} q_i q_j [\mathrm{Adj}\,\mathbb{C}^{\star}(k)]_{ij} = \varsigma\mathcal{F}^2(b)$, where $\mathcal{F}(b) = \sum_j q_j \alpha_j(b)$, so we have

$$D^{(2)}(\mathrm{i}b) = D^{\star}(\mathrm{i}b) + \beta\varsigma\mathcal{F}^2(b)\tilde{\phi}_{\mathrm{Coul}}(\mathrm{i}b) = -\frac{\beta\varsigma\mathcal{F}^2(b)}{\varepsilon_0 b^2} \qquad (12.201)$$

because $D^{\star}(\mathrm{i}b) = 0$. $\mathcal{F}(b)$ is normally nonzero, so $D^{(2)}(k)$ and $D^{\star}(k)$ are not both equal to zero for $k = \mathrm{i}b$. It follows that the poles of $\tilde{h}^{\star}_{ij}(k)$ and of $\tilde{h}^{(2)}_{ij}(k)$ cannot coincide. For charge-inversion invariant systems, however, $\mathcal{F}(b)$ is zero for some zeros of $D^{\star}(k)$; for binary systems this occurs when $\alpha_+(b) = \alpha_-(b)$. In this case $D^{(2)}(\mathrm{i}b)$ is zero when $D^{\star}(\mathrm{i}b)$ is zero so $k = \mathrm{i}b$ is a coinciding pole. It is, of course, possible in the general case that $\mathcal{F}(b)$ *happen* to be zero for some system parameter values and then a pole of $\tilde{h}^{\star}_{ij}(k)$ and of $\tilde{h}^{(2)}_{ij}(k)$ can coincide, but this is a rare exception.

We can take the argument a step further because from Equation 12.200 and $\mathrm{Adj}\,\mathbb{C}^{\star}(k) = D^{\star}(k)\mathbb{H}^{\star}(k)$ (Equation 12.194) it follows that

$$D^{(2)}(k) = D^{\star}(k)\left[1 + \beta\sum_{ij} q_i q_j \tilde{H}^{\star}_{ij}(k)\tilde{\phi}_{\mathrm{Coul}}(k)\right] = D^{\star}(k)\tilde{\epsilon}(k). \qquad (12.202)$$

For charge-inversion invariant systems it is possible to have $D^{(2)}(k) = 0$ when either $D^{*}(k) = 0$ or $\tilde{\epsilon}(k) = 0$, but in general, $D^{(2)}(k)$ is zero *only* when $\tilde{\epsilon}(k)$ is zero because $D^{(2)}(k)$ and $D^{\star}(k)$ cannot simultaneously be zero.

Let us now return to the general case. As pointed out earlier there are, of course, contributions to the radial distribution functions like $\rho_i(r)$ and $h^{(2)}_{ij}(r)$ that do not decay like Yukawa functions. In the (nonlinear) PB approximation we saw in Equation 12.146 that there are terms that decay like $e^{-m\kappa_{\mathrm{D}} r}/r^{\ell}$ with integers $m \geq 2$ and $\ell \geq 2$. They arise because distribution functions like $\rho_i(r)$ and $g_{ij}(r) = e^{-\beta w^{(2)}_{ij}(r)}$ are nonlinear functions of $w^{(2)}_{ij}(r)$, so a Yukawa term $e^{-\kappa_{\mathrm{D}} r}/r$ in $w^{(2)}_{ij}(r)$ gives rise to a series of exponential terms with decay parameters $2\kappa_{\mathrm{D}}$, $3\kappa_{\mathrm{D}}$ etc. and with r^2, r^3 etc. in the denominator. They all decay faster with increasing r than the Yukawa function term and in complex k space they have singularities at $k = \mathrm{i}m\kappa_{\mathrm{D}}$ for $m \geq 2$ that are not poles. For example, $e^{-2\kappa_{\mathrm{D}} r}/r^2$ has a logarithmic branch point singularity at $k = \mathrm{i}2\kappa_{\mathrm{D}}$ as shown in Equation 12.147. All these singularities lie further away from the real axis than the pole at $k = \mathrm{i}\kappa_{\mathrm{D}}$.

Likewise, in the general case a pole at $k = \mathrm{i}\kappa$ generates other types of singularities like branch points in k-space. They occur at $k = \mathrm{i}m\kappa$ for $m \geq 2$. Each singularity corresponds in real space to a non-Yukawa term that contains a factor $e^{-m\kappa r}/r^{\ell}$ with $m \geq 2$ and $\ell \geq 2$. Any other pole, say $k = \mathrm{i}\kappa'$, gives also rise to such terms and furthermore there are cross-terms with decay parameters such as $m\kappa + m'\kappa'$ with $m + m' \geq 2$. In many cases it is sufficient to include a few Yukawa function contributions and sometimes also a branch point contribution in order to describe a radial distribution function quite accurately for most r.[55] Here, we will only briefly touch upon this subject.

Higher order terms generated by Yukawa terms
Equation 12.74 says

$$h_{ij}^{(2)} + 1 = e^{-\beta w_{ij}^{(2)}} = e^{-\beta[u_{ij}^{\mathrm{sh}} + w_{ij}^{(2)\mathrm{el}}] + h_{ij}^{\star} - c_{ij}^{\star} + b_{ij}^{(2)}}$$

and from Equation 12.52 it follows that

$$\begin{aligned} c_{ij}^{\star}(r) &= -\beta u_{ij}^{\mathrm{sh}}(r) + b_{ij}^{(2)}(r) - \ln\{h_{ij}^{(2)}(r) + 1\} + h_{ij}^{(2)}(r) \\ &= -\beta u_{ij}^{\mathrm{sh}}(r) + b_{ij}^{(2)}(r) + \frac{[h_{ij}^{(2)}(r)]^2}{2} - \frac{[h_{ij}^{(2)}(r)]^3}{3} + \frac{[h_{ij}^{(2)}(r)]^4}{4} + \ldots, \end{aligned}$$

which is valid when $-1 < h_{ij}^{(2)}(r) \leq 1$. We assume that $u_{ij}^{\mathrm{sh}}(r)$ is a short-range function that decay to zero faster than the other functions. The Yukawa function contributions to $h_{ij}^{(2)}$ inside the square and cubic terms etc. give rise to terms in $c_{ij}^{\star}(r)$ of the kind just mentioned. The leading term of this kind for large r is proportional to $e^{-2\kappa r}/r^2$ when κ is real. However, the bridge function $b_{ij}^{(2)}(r)$ also have terms that decay exponentially in a similar fashion. They contain a factor $e^{-2\kappa r}/r^2$ but have also other r-dependent factors.[56] For example, the first term in the bridge function given by Equation 9.161 decays like $\ln(r)e^{-2\kappa r}/r^2$ for large r. The (infinite) sum of all contributions to $c_{ij}^{\star}(r)$ with decay parameter 2κ gives a term that decays like $\mathscr{F}_1(r)e^{-2\kappa r}/r^2$, where $\mathscr{F}_1(r)$ is a function[57] that varies slowly with r. One can show that $\mathscr{F}_1(r) \sim \mathrm{const}/\ln^2(r)$ when $r \to \infty$ and at least for low ionic densities one has $\mathscr{F}_1(r) \approx 1/2$ up to rather large values of r. In several cases it is a good approximation to set $\mathscr{F}_1(r) = 1/2$ for all r because $\mathscr{F}_1(r)e^{-2\kappa r}/r^2$ is small for large r where $\mathscr{F}_1(r)$ deviates from 1/2. In complex k space this term has a branch point singularity at $k = \mathrm{i}2\kappa$.

One can obtain the corresponding contribution to $h_{ij}^{\star}(r)$ by using the OZ equation, which is fulfilled by $c_{ij}^{\star}(r)$ and $h_{ij}^{\star}(r)$, and one finds that it decays

[55] A. Härtel, M. Bültmann, and F. Coupette, Phys. Rev. Letters **130** (2023) 108202 (https://doi.org/10.1103/PhysRevLett.130.108202), in particular the Supplemental Material at http://link.aps.org/supplemental/10.1103/PhysRevLett.130.108202. See also Figures 6.9 and 6.10 in Volume I of the present treatise and the original work cited there.

[56] R. Kjellander and D. J. Mitchell, J. Chem. Phys. **101** (1994) 603 (https://doi.org/10.1063/1.468116).

[57] $\mathscr{F}_1(r) = L^2\vartheta(4\kappa r/e^L)/2$ where $\vartheta(x) = \int_0^\infty dt\, x e^{-tx}/[\ln^2(t) + \pi^2]$ and L is a constant that depends on the system's state.

like $\mathscr{F}_2(r)e^{-2\kappa r}/r^2$, where $\mathscr{F}_2(r)$ has the same properties as $\mathscr{F}_1(r)$.[58] As we have seen, there are also Yukawa function terms in $h^{\star}_{ij}(r)$ (contributions from the poles of $\hat{h}^{\star}_{ij}(k)$), but here we are mainly interested in contributions from the other kinds of singularities because we already know how to handle the poles.

Since $\hat{\rho}^{\star}_i(k)$ and $\tilde{\epsilon}(k)$ are sums of $\hat{h}^{\star}_{ij}(k)$ they also have a branch point singularity at $k = \mathrm{i}2\kappa$. As a consequence, $\tilde{h}^{(2)}_{ij}(k)$ also has such a singularity and hence $h^{(2)}_{ij}(r)$ has a term with a similar decay behavior, $\mathscr{F}_3(r)e^{-2\kappa r}/r^2$, where $\mathscr{F}_3(r)$ is yet another slowly varying function. When $2\kappa > \kappa'$ this term is the third leading one in the asymptotic decay behavior[59] of $h^{(2)}_{ij}(r)$ for large r, so we have when $r \to \infty$

$$h^{(2)}_{ij}(r) \sim -\frac{\beta q^{\mathrm{eff}}_i q^{\mathrm{eff}}_j}{\mathcal{E}^{\mathrm{eff}}_r \varepsilon_0} \cdot \frac{e^{-\kappa r}}{4\pi r} - \frac{\beta q'^{\mathrm{eff}}_i q'^{\mathrm{eff}}_j}{\mathcal{E}'^{\mathrm{eff}}_r \varepsilon_0} \cdot \frac{e^{-\kappa' r}}{4\pi r} + \frac{t_i t_j \mathscr{F}_3(r) e^{-2\kappa r}}{r^2}, \quad (12.203)$$

where t_i and t_j are constants. This is the case for 1:–2 aqueous electrolytes at low ionic densities, see below. If $2\kappa < \kappa'$ the order of the second and third terms are reversed in this asymptotic expression. This occurs, for example, for 1:–1 aqueous electrolytes at low densities.

The cubic term in $c^{\star}_{ij}(r)$ gives rise to a contribution that is proportional to $e^{-3\kappa r}/r^3$ and the bridge function also contain contributions that decay exponentially with decay parameter 3κ. Their sum is proportional to $e^{-3\kappa r}/r^3$ times a slowly varying function and lead to a similar term in $h^{(2)}_{ij}(r)$ with decay parameter 3κ. The corresponding is true for the higher order terms in $c^{\star}_{ij}(r)$. Thus $h^{(2)}_{ij}(r)$ contains many contributions with shorter range than the corresponding Yukawa terms. *They are all consequences of the existence of the fundamental decay modes* (the Yukawa terms), which arise due to the zeros of $\epsilon(k)$ in our main case. For charge reversal invariant systems like the RPM, the Yukawa function terms arise due to either $D^*(k) = 0$ or $\tilde{\epsilon}(k) = 0$. This latter case will be treated in more detail later.

For 1:–2 aqueous electrolytes below the Kirkwood cross-over point, we have $2\kappa > \kappa'$ so the asymptotic decay of $h^{(2)}_{ij}(r)$ is given by Equation 12.203. In this case $\kappa' \sim 2\kappa$ when $n^{\mathrm{b}}_{\mathrm{tot}} \to 0$ as shown in Figure 12.5, which show data from MC simulations and HNC calculations in the primitive model.[60] For low ionic densities $\mathcal{E}'^{\mathrm{eff}}_r$ is a very large negative number so the second term in Equation 12.203 is very small and the third term becomes *in practice* the second leading term in $h^{(2)}_{ij}(r)$ for large r.

In the case of 1:–1 aqueous electrolytes the third term of Equation 12.203 *constitute* the second leading term because $2\kappa < \kappa'$ and for low ionic densities

---

[58] $\mathscr{F}_2(r)$ differs from $\mathscr{F}_1(r)$ only in the value of L in footnote 57.

[59] Recall that in an asymptotic expression the terms are placed in the order of their magnitude in the tail of the function for large r, see the box on page 313.

[60] The data are taken from the reference in footnote 39.

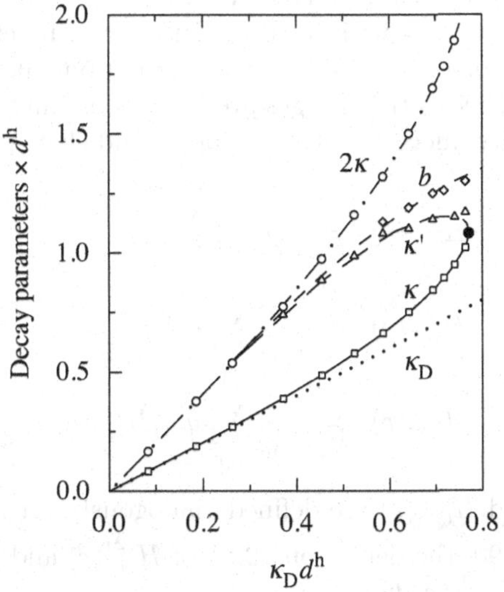

Figure 12.5 Decay parameters for 1:–2 aqueous electrolytes at room temperature with $d^h = 4.6$ Å plotted as functions of $\kappa_D d^h$. The curves show results from HNC calculations and open symbols MC simulation results.[60] Full line and open squares show κd^h, long dashes and open triangles $\kappa' d^h$, dashed-dotted line and open circles $2\kappa d^h$, short dashes and diamonds bd^h and dotted straight line $\kappa_D d^h$. The Kirkwood crossover point for the HNC data is shown as a filled circle. No data is shown beyond this point.

the remaining term is small because $\mathcal{E}_r'^{\text{eff}}$ is a very large negative number. In both cases the same two terms are leading in the tail of $h_{ij}^{(2)}(r)$ for low densities when $r \to \infty$. In the limit $n_{\text{tot}}^{\text{b}} \to 0$ the function $\mathscr{F}_3(r)$ approaches its PB value 1/2 for any finite r and the two leading terms approach the PB prediction.

Note that the 1:–1 and 1:–2 electrolytes have very different behavior for κ' when $n_{\text{tot}}^{\text{b}} \to 0$; in the 1:–1 case κ' becomes very large for small densities as we have seen in Figure 12.2, while in the 1:–2 case $\kappa' \sim 2\kappa$. Figure 12.5 also shows the decay parameter b of the leading term of $h_{ij}^{\star}(r)$ in Equation 12.198. The decay parameter b approaches 2κ when the density goes to zero, which is the same behavior as κ'. Also for 1:–1 electrolytes (not shown) b approaches 2κ. An example of pair distribution functions for the system in Figure 12.5 is shown in Figure 6.9 in Volume I. This example is discussed in connection to Equation I:6.183 ff.

Our findings for the decay of $h_{ij}^{(2)}(r)$ have important consequences for the **density-density, charge-charge and density-charge correlation functions,** $H_{\text{NN}}(r_{12})$, $H_{\text{QQ}}(r_{12})$, $H_{\text{NQ}}(r_{12}) \equiv H_{\text{QN}}(r_{12})$, respectively, which were defined in Appendix 6C of Volume I (also defined below). $H_{\text{NN}}(r_{12})$ describe correlations in density fluctuations at $\mathbf{r}_1$ with those at $\mathbf{r}_2$, including self-correlations of each particle with itself (a Dirac delta function contribution). $H_{\mathrm{QQ}}(r_{12})$ describe correlations in charge fluctuations at these points, irrespectively of which species the charges are located on, and $H_{\mathrm{NQ}}(r_{12})$ describe correlations in density fluctuations at one point and charge fluctuations at the other. We have

$$\begin{aligned} H_{\mathrm{NN}}(r) &= \sum_{ij} H_{ij}^{(2)}(r) \\ H_{\mathrm{NQ}}(r) &= \frac{1}{q_{\mathrm{e}}} \sum_{ij} q_i H_{ij}^{(2)}(r). \\ H_{\mathrm{QQ}}(r) &= \frac{1}{q_{\mathrm{e}}^2} \sum_{ij} q_i q_j H_{ij}^{(2)}(r) \end{aligned} \tag{12.204}$$

$H_{\mathrm{NN}}^{\star}(r)$, $H_{\mathrm{NQ}}^{\star}(r)$ and $H_{\mathrm{QQ}}^{\star}(r)$ are defined analogously with $H_{ij}^{\star}(r)$ in the rhs. From Equation 12.190, the definition 12.81 of $H_{ij}^{(2)}(r)$ and the analogous definition of $H_{ij}^{\star}(r)$ it follows that

$$\begin{aligned} \tilde{H}_{\mathrm{NN}}(k) &= \tilde{H}_{\mathrm{NN}}^{\star}(k) - \beta(n_{\mathrm{tot}}^{\mathrm{b}})^2 \frac{[\tilde{\rho}_{\mathrm{n}}^{\star}(k)]^2}{\varepsilon_0 k^2 \tilde{\epsilon}(k)} \\ \tilde{H}_{\mathrm{NQ}}(k) &= \tilde{H}_{\mathrm{NQ}}^{\star}(k) - \beta(n_{\mathrm{tot}}^{\mathrm{b}})^2 \cdot \frac{\tilde{\rho}_{\mathrm{n}}^{\star}(k)\tilde{\rho}_{\mathrm{q}}^{\star}(k)}{\varepsilon_0 k^2 \tilde{\epsilon}(k)} \\ \tilde{H}_{\mathrm{QQ}}(k) &= \tilde{H}_{\mathrm{QQ}}^{\star}(k) - \beta(n_{\mathrm{tot}}^{\mathrm{b}})^2 \cdot \frac{[\tilde{\rho}_{\mathrm{q}}^{\star}(k)]^2}{\varepsilon_0 k^2 \tilde{\epsilon}(k)} \end{aligned} \tag{12.205}$$

where

$$\begin{aligned} \tilde{\rho}_{\mathrm{n}}^{\star}(k) &= \frac{1}{n_{\mathrm{tot}}^{\mathrm{b}}} \sum_i n_i^{\mathrm{b}} \tilde{\rho}_i^{\star}(k) = \sum_i \mathrm{x}_i^{\mathrm{b}} \tilde{\rho}_i^{\star}(k) \\ \tilde{\rho}_{\mathrm{q}}^{\star}(k) &= \frac{1}{n_{\mathrm{tot}}^{\mathrm{b}} q_{\mathrm{e}}} \sum_i n_i^{\mathrm{b}} q_i \tilde{\rho}_i^{\star}(k) = \sum_i \mathrm{x}_i^{\mathrm{b}} \mathrm{z}_i \tilde{\rho}_i^{\star}(k) \end{aligned}$$

and $\mathrm{z}_i = q_i/q_{\mathrm{e}}$ is the valency of an i-ion. We likewise define[61]

$$\rho_{\mathrm{n}}(r) = \sum_i \mathrm{x}_i^{\mathrm{b}} \rho_i(r) \qquad \text{and} \qquad \rho_{\mathrm{q}}(r) = \sum_i \mathrm{x}_i^{\mathrm{b}} \mathrm{z}_i \rho_i(r).$$

[61]There is a slight difference in notation compared to Chapter 6 in Volume I where we treated $H_{\mathrm{NN}}(r)$, $H_{\mathrm{QQ}}(r)$ and $H_{\mathrm{QN}}(r)$. For a binary system we used the charge density $\rho_{\mathrm{Q}}(r) = [\rho_+(r) - \rho_-(r)]/2$ instead of $\rho_{\mathrm{q}}(r) = \sum_i \mathrm{x}_i^{\mathrm{b}} \mathrm{z}_i \rho_i(r)$ that we use here (for a general system $\rho_{\mathrm{Q}}(r) = \sum_i \mathrm{x}_i^{\mathrm{b}} q_i \rho_i(r)/q_\pm$, where $q_\pm = \sum_i \mathrm{x}_i^{\mathrm{b}} |q_i|$). The two functions differ just by the factor $q_\pm/q_{\mathrm{e}}$ and we have $\rho_{\mathrm{q}}(r) = \rho_{\mathrm{Q}}(r) q_\pm / q_{\mathrm{e}}$. Likewise we used $\rho_{\mathrm{Q}}^{\star}(r)$ instead of $\rho_{\mathrm{q}}^{\star}(r)$ and we have $q_{\mathrm{q}}^{\mathrm{eff}}(\kappa) = q_{\mathrm{Q}}^{\mathrm{eff}}(\kappa) q_\pm / q_{\mathrm{e}}$. The reason why $\rho_{\mathrm{q}}^{\star}(r)$ and $\rho_{\mathrm{q}}(r)$ are preferred here is that they are more suitable for a generalization to nonspherical particles, including electroneutral ones. Note that the function $\rho_{\mathrm{n}}(r)$ used here is identical to $\rho_{\mathrm{N}}(r)$ in Chapter 6. We have $H_{\mathrm{QQ}}(r) = n_{\mathrm{tot}}^{\mathrm{b}} [\rho_{\mathrm{q}}(r) + q_{\mathrm{q}} \delta^{(3)}(r)]/q_{\mathrm{e}}$ (cf. Equation I:6.185), where $q_{\mathrm{q}} = \sum_i \mathrm{x}_i^{\mathrm{b}} \mathrm{z}_i q_i$, and $H_{\mathrm{NQ}}(r) = n_{\mathrm{tot}}^{\mathrm{b}} \rho_{\mathrm{n}}(r)/q_{\mathrm{e}}$ (the same as Equation I:6.191).

From the definition of $H^\star_{ij}(r)$ it follows that $H^\star_{\rm QQ}(r) = n^{\rm b}_{\rm tot}\rho^\star_{\rm q}(r)/q_{\rm e}$ and $H^\star_{\rm NQ}(r) = n^{\rm b}_{\rm tot}\rho^\star_{\rm n}(r)/q_{\rm e}$.

{**Exercise 12.6:** Use Equation 12.108 to show that

$$\tilde{H}_{\rm QQ}(k) = \frac{\tilde{H}^\star_{\rm QQ}(k)}{\tilde{\epsilon}(k)} \quad \text{and} \quad \tilde{H}_{\rm NQ}(k) = \frac{\tilde{H}^\star_{\rm NQ}(k)}{\tilde{\epsilon}(k)}. \tag{12.206}$$

Then, show that the last two lines in Equation 12.205 can be simplified to give these results (the first line cannot be simplified in a similar manner).}

In our main case, the poles of $\tilde{H}^\star_{IJ}(k)$ are different from those of $\tilde{H}_{IJ}(k)$ for $I, J = N$ or Q, because these functions cannot have any other poles than $\tilde{h}^\star_{ij}(k)$ and $\tilde{h}^{(2)}_{ij}(k)$, respectively. It therefore follows from Equation 12.205 that all poles of $\tilde{H}_{\rm NN}(k)$, $\tilde{H}_{\rm NQ}(k)$ and $\tilde{H}_{\rm QQ}(k)$ are given by zeros of $\tilde{\epsilon}(k)$ and hence that $H_{\rm NN}(r)$, $H_{\rm NQ}(r)$ and $H_{\rm QQ}(r)$ have the same decay modes and the same decay parameters as $h^{(2)}_{ij}(r)$, $\psi_i(r)$, $\rho_i(r)$ and $\phi^\star_{\rm Coul}(r)$. Thus we have[62] for large r

$$\begin{aligned}
H_{\rm NN}(r) &= -\frac{\left[q^{\rm eff}_{\rm n}(\kappa)\right]^2}{\mathcal{B}(\kappa)}\cdot\frac{e^{-\kappa r}}{4\pi r} - \frac{\left[q^{\rm eff}_{\rm n}(\kappa')\right]^2}{\mathcal{B}(\kappa')}\cdot\frac{e^{-\kappa' r}}{4\pi r} - \ldots \\
H_{\rm NQ}(r) &= -\frac{q^{\rm eff}_{\rm n}(\kappa)q^{\rm eff}_{\rm q}(\kappa)}{\mathcal{B}(\kappa)}\cdot\frac{e^{-\kappa r}}{4\pi r} - \frac{q^{\rm eff}_{\rm n}(\kappa')q^{\rm eff}_{\rm q}(\kappa')}{\mathcal{B}(\kappa')}\cdot\frac{e^{-\kappa' r}}{4\pi r} - \ldots \\
H_{\rm QQ}(r) &= -\frac{\left[q^{\rm eff}_{\rm q}(\kappa)\right]^2}{\mathcal{B}(\kappa)}\cdot\frac{e^{-\kappa r}}{4\pi r} - \frac{\left[q^{\rm eff}_{\rm q}(\kappa')\right]^2}{\mathcal{B}(\kappa')}\cdot\frac{e^{-\kappa' r}}{4\pi r} - \ldots
\end{aligned} \tag{12.207}$$

(cf. Equation I:6.198), where we have only shown the first two Yukawa function terms and where

$$q^{\rm eff}_{\rm n}(\kappa) \equiv \tilde{\rho}^\star_{\rm n}(i\kappa) = \sum_i {\rm x}^{\rm b}_i q^{\rm eff}_i(\kappa)$$

$$q^{\rm eff}_{\rm q}(\kappa) \equiv \tilde{\rho}^\star_{\rm q}(i\kappa) = \sum_i {\rm x}^{\rm b}_i {\rm z}_i q^{\rm eff}_i(\kappa)$$

and

$$\mathcal{B}(\kappa) = \frac{\mathcal{E}^{\rm eff}_r(\kappa)\varepsilon_0}{\beta(n^{\rm b}_{\rm tot})^2}.$$

For comparison we can write the Yukawa terms of $h^{(2)}_{ij}(r)$ for large r as

$$(n^{\rm b}_{\rm tot})^2 h^{(2)}_{ij}(r) = -\frac{q^{\rm eff}_i(\kappa)q^{\rm eff}_j(\kappa)}{\mathcal{B}(\kappa)}\cdot\frac{e^{-\kappa r}}{4\pi r} - \frac{q^{\rm eff}_i(\kappa')q^{\rm eff}_j(\kappa')}{\mathcal{B}(\kappa')}\cdot\frac{e^{-\kappa' r}}{4\pi r} - \ldots$$

---

[62] As in the case of Equation 12.192 there are additional contributions with other functional forms, which we treated in the shaded text box on page 351, see Equation 12.203. The first term in Equation 12.207 dominates for large r at least for low electrolyte densities.

Table 12.4 A summary of the Yukawa terms that appear in the charge-charge, density-charge and density-density correlation functions $H_{\rm NN}(r)$, $H_{\rm NQ}(r)$ and $H_{\rm QQ}(r)$ in bulk electrolytes. It applies in general, both for spherical ions and systems consisting nonspherical particles, which are treated in Section 13.3.3. Each decay mode with decay parameter κ, κ' etc. has one Yukawa term (the parameters are solutions to $\tilde{\varepsilon}(\mathrm{i}\kappa) = 0$, see Table 12.3 on page 329). The function $H^{\star}_{\rm NN}(r)$ has decay parameters b, b' etc. When some decay parameters are complex-valued, a pair of terms forms an oscillatory Yukawa function contribution. The equation numbers refer to the original ones in the text.

Kind of system	Yukawa terms
Normal case	
$H_{\rm NN}(r)/(n^{\rm b}_{\rm tot})^2$	$-\dfrac{\beta}{\mathcal{E}^{\rm eff}_r(\kappa)\varepsilon_0}\left[q^{\rm eff}_n(\kappa)\right]^2\dfrac{e^{-\kappa r}}{4\pi r}$
$H_{\rm QQ}(r)/(n^{\rm b}_{\rm tot})^2$	$-\dfrac{\beta}{\mathcal{E}^{\rm eff}_r(\kappa)\varepsilon_0}\left[q^{\rm eff}_{\rm q}(\kappa)\right]^2\dfrac{e^{-\kappa r}}{4\pi r}$
$H_{\rm NQ}(r)/(n^{\rm b}_{\rm tot})^2$	$-\dfrac{\beta}{\mathcal{E}^{\rm eff}_r(\kappa)\varepsilon_0}\left[q^{\rm eff}_{\rm n}(\kappa)q^{\rm eff}_{\rm q}(\kappa)\right]\dfrac{e^{-\kappa r}}{4\pi r}$
	(12.207') and (13.100')
Charge-inversion invariant system	
$H_{\rm NN}(r) \equiv H^{\star}_{\rm NN}(r)$	$\dfrac{[\mathcal{C}_{\rm n}(b)]^2}{\mathcal{K}^{\star}(b)}\cdot\dfrac{e^{-br}}{4\pi r}$
$H_{\rm QQ}(r)/(n^{\rm b}_{\rm tot})^2$	$-\dfrac{\beta}{\mathcal{E}^{\rm eff}_r(\kappa)\varepsilon_0}\left[q^{\rm eff}_{\rm q}(\kappa)\right]^2\dfrac{e^{-\kappa r}}{4\pi r}$
$H_{\rm NQ}(r)/(n^{\rm b}_{\rm tot})^2 \equiv 0$	
	(12.211'), (13.107') and (13.108')
In all cases we have	$q^{\rm eff}_{\rm q}(\kappa) = \dfrac{\kappa^2\varepsilon_0}{\beta q_{\rm e} n^{\rm b}_{\rm tot}}$
	$q^{\rm eff}_{\rm n}(\kappa) = \sum_i \mathrm{x}^{\rm b}_i q^{\rm eff}_i(\kappa)$

(Equation 12.192), which has the same structure. As usual, a pair of complex decay parameters give an oscillatory Yukawa function term in each of $h_{ij}^{(2)}(r)$, $H_{\mathrm{NN}}(r)$, $H_{\mathrm{NQ}}(r)$ and $H_{\mathrm{QQ}}(r)$.

The explicit expression for $H_{\mathrm{QQ}}(r)$ is

$$H_{\mathrm{QQ}}(r) = -\frac{\beta\left[n_{\mathrm{tot}}^{\mathrm{b}} q_{\mathrm{q}}^{\mathrm{eff}}(\kappa)\right]^2}{\mathcal{E}_r^{\mathrm{eff}}(\kappa)\varepsilon_0} \cdot \frac{e^{-\kappa r}}{4\pi r} - \frac{\beta\left[n_{\mathrm{tot}}^{\mathrm{b}} q_{\mathrm{q}}^{\mathrm{eff}}(\kappa')\right]^2}{\mathcal{E}_r^{\mathrm{eff}}(\kappa')\varepsilon_0} \cdot \frac{e^{-\kappa' r}}{4\pi r} - \ldots \quad (12.208)$$

for large r and analogously for the other functions. Note that

$$n_{\mathrm{tot}}^{\mathrm{b}} q_{\mathrm{q}}^{\mathrm{eff}}(\kappa) = \frac{1}{q_{\mathrm{e}}} \sum_i n_i^{\mathrm{b}} q_i q_i^{\mathrm{eff}}(\kappa) = \frac{\kappa^2 \varepsilon_0}{\beta q_{\mathrm{e}}},$$

so we have for large r

$$H_{\mathrm{QQ}}(r) = -\frac{\kappa^4 \varepsilon_0}{\mathcal{E}_r^{\mathrm{eff}}(\kappa)\beta q_{\mathrm{e}}^2} \cdot \frac{e^{-\kappa r}}{4\pi r} - \frac{(\kappa')^4 \varepsilon_0}{\mathcal{E}_r^{\mathrm{eff}}(\kappa')\beta q_{\mathrm{e}}^2} \cdot \frac{e^{-\kappa' r}}{4\pi r} - \ldots \quad (12.209)$$

The corresponding coefficients for $H_{\mathrm{NQ}}(r)$ contain the factor $n_{\mathrm{tot}}^{\mathrm{b}} q_{\mathrm{q}}^{\mathrm{eff}}$ so they are proportional to κ^2 and $(\kappa')^2$, respectively. The coefficients for $H_{\mathrm{NQ}}(r)$ contain also the factor $n_{\mathrm{tot}}^{\mathrm{b}} q_{\mathrm{n}}^{\mathrm{eff}}$ and those for $H_{\mathrm{NN}}(r)$ contain the factor $[n_{\mathrm{tot}}^{\mathrm{b}} q_{\mathrm{n}}^{\mathrm{eff}}]^2$ that does not have such a direct relationship to the decay parameter κ.

{**Exercise 12.7:** Do a pole analysis of Equation 12.206 in accordance with Equations 12.142-12.144 to show that the pole at $k = i\kappa$ gives rise to the Yukawa contributions

$$-\frac{\kappa^2 n_{\mathrm{tot}}^{\mathrm{b}} q_{\mathrm{q}}^{\mathrm{eff}}(\kappa)}{\mathcal{E}_r^{\mathrm{eff}}(\kappa) q_{\mathrm{e}}} \cdot \frac{e^{-\kappa r}}{4\pi r} \text{ in } H_{\mathrm{QQ}}(r) \quad \text{and} \quad -\frac{\kappa^2 n_{\mathrm{tot}}^{\mathrm{b}} q_{\mathrm{n}}^{\mathrm{eff}}(\kappa)}{\mathcal{E}_r^{\mathrm{eff}}(\kappa) q_{\mathrm{e}}} \cdot \frac{e^{-\kappa r}}{4\pi r} \text{ in } H_{\mathrm{NQ}}(r).$$

Note that this result for $H_{\mathrm{QQ}}(r)$ agrees with Equations 12.208 and 12.209. **Hint:** Multiply the numerators and denominators in Equation 12.206 by k^2 before you start the analysis.}

A summary of the decay modes for $H_{\mathrm{NN}}(r)$, $H_{\mathrm{NQ}}(r)$ and $H_{\mathrm{QQ}}(r)$ is given in Table 12.4 on the preceding page and applies in general, both for spherical electrolytes and systems consisting nonspherical particles, which are treated in Section 13.3.3.3.

Example 12:B. Two examples of dominance of $H_{\mathrm{QQ}}(r)$ or $H_{\mathrm{NN}}(r)$.

As examples of the use of Equation 12.207 we will consider two cases where distinct behaviors of $q_{\mathrm{n}}^{\mathrm{eff}}$ and $q_{\mathrm{q}}^{\mathrm{eff}}$ are important. In the limit of low ionic density, we have $q_i^{\mathrm{eff}}(\kappa) \to q_i$ for the leading decay mode. We can express this as $q_i^{\text{eff}}(\kappa) = q_i + \tau_i(\kappa)$,[63] where $\tau_i(\kappa) \to 0$ when $n_{\text{tot}}^{\text{b}} \to 0$. Hence

$$
\begin{aligned}
q_{\text{n}}^{\text{eff}}(\kappa) &= \sum_i \mathrm{x}_i^{\text{b}} q_i^{\text{eff}}(\kappa) = \sum_i \mathrm{x}_i^{\text{b}} q_i + \sum_i \mathrm{x}_i^{\text{b}} \tau_i(\kappa) = \sum_i \mathrm{x}_i^{\text{b}} \tau_i(\kappa) \\
q_{\text{q}}^{\text{eff}}(\kappa) &= \sum_i \mathrm{x}_i^{\text{b}} q_i q_i^{\text{eff}}(\kappa) = \sum_i \mathrm{x}_i^{\text{b}} q_i^2 + \sum_i \mathrm{x}_i^{\text{b}} q_i \tau_i(\kappa)
\end{aligned}
$$

and we can conclude that $q_{\text{n}}^{\text{eff}}(\kappa) \to 0$ while $q_{\text{q}}^{\text{eff}}(\kappa)$ stays finite when $n_{\text{tot}}^{\text{b}} \to 0$. This implies that the leading decay mode of $H_{\text{NN}}(r)$ and $H_{\text{NQ}}(r)$ are much smaller in magnitude for low ionic densities than that of $H_{\text{QQ}}(r)$, which hence dominates. In our main case, all three correlation functions have equal decay length, but at sufficiently low ionic densities $H_{\text{QQ}}(r)$ dominates for large r because it has the largest magnitude.

The other example concerns conditions close to the critical point of an electrolyte, where long-range density-density correlations appear (the decay length for $H_{\text{NN}}(r)$ diverges when the critical point is approached). An example of this behavior is shown in Figure 6.12 of Volume I, which presents a case where one decay mode (with decay parameter κ'') has a decay length $1/\kappa''$ that becomes large near the critical point, while the decay lengths of the other modes remain relatively small. The $1/\kappa''$ mode then dominates in $H_{\text{NN}}(r)$ for large r because it has the longest decay length. In our main case, $H_{\text{QQ}}(r)$ and $H_{\text{NQ}}(r)$ have contributions with the same decay length so these correlations are equally long-ranged. However, because of the appearance of the factors $(\kappa'')^4$ and $(\kappa'')^2$ in the respective coefficients, the magnitudes of these correlations vanish in the limit $\kappa'' \to 0$ where the decay length diverges. Near the critical point $q_{\text{n}}^{\text{eff}}(\kappa'')$ is much larger than $q_{\text{q}}^{\text{eff}}(\kappa'')$, so $H_{\text{NN}}(r)$ dominates over the other two correlation functions because its magnitude is larger.

These two examples are in a way each other's opposite because in the first $H_{\text{QQ}}(r)$ dominates in magnitude over $H_{\text{NN}}(r)$ and in the second it is the other way around. $H_{\text{NQ}}(r)$ lies between these two.

12.3.5 Charge-inversion invariant electrolytes and borderline cases

Next, we treat the case of charge-inversion invariant systems like the RPM and we will see that $H_{\text{NN}}(r)$, $H_{\text{QQ}}(r)$ and $H_{\text{NQ}}(r)$ are quite different from the

[63] From $\tilde{\rho}_i^\star(k) = q_i + \sum_j q_j n_j^{\text{b}} \tilde{h}_{ij}^\star(k)$ it follows that $\tau_i(\kappa) = \sum_j q_j n_j^{\text{b}} \tilde{h}_{ij}^\star(\text{i}\kappa)$.

main case that we have considered so far. First of all, $H_{\mathrm{NQ}}(r)$ is identically equal to zero for all r. This follows because we have $H_{\mathrm{NQ}}(r) = \sum_j n_j^{\mathrm{b}} \rho_j(r)/q_{\mathrm{e}}$, so if we invert the sign of all charges $H_{\mathrm{NQ}}(r)$ changes sign, but if the system is invariant under such an inversion $H_{\mathrm{NQ}}(r)$ must remain the same and the only possibility is that $H_{\mathrm{NQ}}(r) = 0$. The same is true for $n_{\mathrm{tot}}^{\mathrm{b}} \tilde{\rho}_{\mathrm{n}}^{\star}(k) = \sum_j n_j^{\mathrm{b}} \tilde{\rho}_j^{\star}(k)$, so $\tilde{\rho}_{\mathrm{n}}^{\star}(k) \equiv 0$ and Equation 12.205 yields

$$
\begin{aligned}
\tilde{H}_{\mathrm{NN}}(k) &= \tilde{H}_{\mathrm{NN}}^{\star}(k) \\
\tilde{H}_{\mathrm{QQ}}(k) &= \tilde{H}_{\mathrm{QQ}}^{\star}(k) - \beta (n_{\mathrm{tot}}^{\mathrm{b}})^2 \cdot \frac{\left[\tilde{\rho}_{\mathrm{q}}^{\star}(k)\right]^2}{\varepsilon_0 k^2 \tilde{\epsilon}(k)} \\
H_{\mathrm{NQ}}(k) &\equiv 0 \qquad \text{(charge-inv. invar.).} \qquad (12.210)
\end{aligned}
$$

It follows immediately that the zeros of $\tilde{\epsilon}(k)$ determine the Yukawa decay modes of $H_{\mathrm{QQ}}(r)$ but not the Yukawa terms of $H_{\mathrm{NN}}(r)$. The latter are instead determined by the zeros of $D^{\star}(k)$ that give rise to poles of $\tilde{H}_{\mathrm{NN}}^{\star}(k)$. We know from Equation 12.201 that a zero $k = ib$ of $D^{\star}(k)$ gives rise to a pole of $\tilde{h}_{ij}^{(2)}(k)$ provided that $\mathcal{F}(b) = \sum_j q_j \alpha_j(b) = 0$. Hence, it is only the set of zeros that satisfy the condition $\mathcal{F}(b) = 0$ that can give the Yukawa terms of $H_{\mathrm{NN}}(r)$. Since

$$
\tilde{H}_{ij}^{\star}(k) \sim \frac{\alpha_i(b)\alpha_j(b)}{(k^2 + b^2)\mathcal{K}^{\star}(b)} \qquad \text{when } k \to ib,
$$

(Equation 12.196) we obtain

$$
\tilde{H}_{\mathrm{NN}}(k) = \tilde{H}_{\mathrm{NN}}^{\star}(k) = \sum_{ij} H_{ij}^{\star}(r) \sim \frac{[\sum_i \alpha_i(b)]^2}{(k^2 + b^2)\mathcal{K}^{\star}(b)} \qquad \text{when } k \to ib.
$$

Note that this set of zeros of $D^{\star}(k)$ does not give any poles of $\tilde{H}_{\mathrm{QQ}}^{\star}(k) = \sum_{ij} q_i q_j H_{ij}^{(2)}(r)/q_{\mathrm{e}}^2$ because the term from $\tilde{H}_{ij}^{\star}(k)$ that diverge at $k = ib$ vanish from $\tilde{H}_{\mathrm{QQ}}^{\star}(k)$ since the sum contains the factor $[\sum_j q_j \alpha_j(b)]^2 = \mathcal{F}^2(b) = 0$. Only the set of zeros $k = ib$ of $D^{\star}(k)$ that satisfy $\mathcal{F}(b) \neq 0$ gives rise to poles of $\tilde{H}_{\mathrm{QQ}}^{\star}(k)$.

When $k = ib$ is the singularity of $\tilde{H}_{\mathrm{NN}}(k)$ that lies closest to the real axis in complex k space and $k = i\kappa$ is the singularity of $\tilde{H}_{\mathrm{QQ}}(k)$ that lies closest to this axis, we have

$$
\begin{aligned}
H_{\mathrm{NN}}(r) &\sim \frac{[\mathcal{C}_{\mathrm{n}}(b)]^2}{\mathcal{K}^{\star}(b)} \cdot \frac{e^{-br}}{4\pi r} \\
H_{\mathrm{QQ}}(r) &\sim -\frac{\beta \left[n_{\mathrm{tot}}^{\mathrm{b}} q_{\mathrm{q}}^{\mathrm{eff}}(\kappa)\right]^2}{\mathcal{E}_r^{\mathrm{eff}}(\kappa)\varepsilon_0} \cdot \frac{e^{-\kappa r}}{4\pi r} \qquad \text{when } r \to \infty. \qquad (12.211)
\end{aligned}
$$

where $\mathcal{C}_{\mathrm{n}}(b) = \sum_i \alpha_i(b)$. The other zeros like $\tilde{\epsilon}(i\kappa') = 0$ and $D^{\star}(ib') = 0$ subject to $\mathcal{F}(b') = 0$ give rise to the same kind of terms. A pair of complex zeros give an oscillatory Yukawa function term as usual.

It follows that the decay modes and hence the decay parameters fall into *three distinct sets*: the solutions of

(i) $\tilde{\epsilon}(\mathrm{i}\kappa) = 0$ (Yukawa terms of H_{QQ}),
(ii) $D^{\star}(\mathrm{i}b) = 0$ with $\sum_j q_j \alpha_j(b) = 0$ (Yukawa terms of $H_{\mathrm{NN}} = H^{\star}_{\mathrm{NN}}$),
(iii) $D^{\star}(\mathrm{i}b) = 0$ with $\sum_j q_j \alpha_j(b) \neq 0$ (Yukawa terms of $H^{\star}_{\mathrm{QQ}}$).

The Yukawa function contributions of $h^{(2)}_{ij}(r)$ arise from the *first two sets* because $D^{(2)}(k) = D^{\star}(k)\tilde{\epsilon}(k) = 0$ for both $k = \mathrm{i}\kappa$ and $k = \mathrm{i}b$ (with $\mathcal{F}(b) = 0$). We have for sufficiently large r

$$\begin{aligned} &h^{(2)}_{ij}(r) \\ &= -\frac{\beta q^{\mathrm{eff}}_i(\kappa) q^{\mathrm{eff}}_j(\kappa)}{\mathcal{E}^{\mathrm{eff}}_r(\kappa)\varepsilon_0} \cdot \frac{e^{-\kappa r}}{4\pi r} - \frac{\beta q^{\mathrm{eff}}_i(\kappa') q^{\mathrm{eff}}_j(\kappa')}{\mathcal{E}^{\mathrm{eff}}_r(\kappa')\varepsilon_0} \cdot \frac{e^{-\kappa' r}}{4\pi r} - \dots \\ &+ \frac{a_i(b) a_j(b)}{\mathcal{K}^{\star}(b)} \cdot \frac{e^{-br}}{4\pi r} + \frac{a_i(b') a_j(b')}{\mathcal{K}^{\star}(b')} \cdot \frac{e^{-b' r}}{4\pi r} + \dots \quad \text{(charge-inv. invar.)} \end{aligned} \tag{12.212}$$

where $a_i(b) = \alpha_i(b)/n^{\mathrm{b}}_i$ and we have only shown the first two Yukawa function contributions of each kind. Note that $\sum_i n^{\mathrm{b}}_i q_i a_i(b) = 0$ for these coefficients due to the condition $\mathcal{F}(b) = 0$. Like before there are short-range terms with other functional forms as well.

The charge distribution $\rho_i(r)$ contains only the decay modes of $H_{\mathrm{QQ}}(r)$ because the decay parameters of both functions are determined by the zeros of $\tilde{\epsilon}(k)$. This can also be seen from the fact that $\rho_i(r) = \sum_j n^{\mathrm{b}}_j q_j h^{(2)}_{ij}(r)$ so the terms from the second line of Equation 12.212 vanish because $\sum_j n^{\mathrm{b}}_j q_j a_j(b) = 0$ and the same for the term with b'. The same is true for $\psi_i(r)$ and $\phi^{\star}_{\mathrm{Coul}}(r)$ that contain only the $H_{\mathrm{QQ}}(r)$ modes (like in Equation 12.171). The total density distribution of ions around an i-ion, $\sum_j n^{\mathrm{b}}_j g^{(2)}_{ij}(r)$, contains only contributions from the second line because $\sum_j n^{\mathrm{b}}_j q^{\mathrm{eff}}_j(\kappa) = n^{\mathrm{b}}_{\mathrm{tot}} q^{\mathrm{eff}}_{\mathrm{n}}(\kappa) = n^{\mathrm{b}}_{\mathrm{tot}} \tilde{\rho}^{\star}_{\mathrm{n}}(\mathrm{i}\kappa) = 0$ for charge-inversion invariant systems.

For binary electrolytes we have $q_+ = -q_- \equiv q$, $n^{\mathrm{b}}_+ = n^{\mathrm{b}}_- = n^{\mathrm{b}}_{\mathrm{tot}}/2$ and $h^{(2)}_{++}(r) = h^{(2)}_{--}(r)$. Furthermore,

$$\begin{aligned} H_{\mathrm{NN}}(r) &= n^{\mathrm{b}}_{\mathrm{tot}} \left[\delta^{(3)}(r) + n^{\mathrm{b}}_{\mathrm{tot}} h^{(2)}_{\mathrm{S}}(r)\right] \\ H_{\mathrm{QQ}}(r) &= \mathsf{z}^2 n^{\mathrm{b}}_{\mathrm{tot}} \left[\delta^{(3)}(r) + n^{\mathrm{b}}_{\mathrm{tot}} h^{(2)}_{\mathrm{D}}(r)\right] \end{aligned}$$

where $\mathsf{z} = q/q_{\mathrm{e}}$ and we have defined the "sum" and "difference" correlation functions

$$h^{(2)}_{\mathrm{S}} = \tfrac{1}{2}[h^{(2)}_{++} + h^{(2)}_{+-}] \quad \text{and} \quad h^{(2)}_{\mathrm{D}} = \tfrac{1}{2}[h^{(2)}_{++} - h^{(2)}_{+-}]. \tag{12.213}$$

The functions $h_S^\star$ and $h_D^\star$ are defined analogously and we have

$$\begin{aligned} h_{+\pm}^{(2)}(r) &= h_S^{(2)}(r) \pm h_D^{(2)}(r) \\ h_S^{(2)}(r) &= h_S^\star(r) \\ h_D^{(2)}(r) &= h_D^\star(r) - \frac{\beta\left[\tilde{\rho}^\star(k)\right]^2}{\varepsilon_0 k^2 \tilde{\epsilon}(k)}, \end{aligned}$$

where $\tilde{\rho}^\star(k) = \tilde{\rho}_+^\star(k) = -\tilde{\rho}_-^\star(k)$. We also have $q_+^{\text{eff}} = -q_-^{\text{eff}} \equiv q^{\text{eff}}$ and $a_+ = a_- \equiv a$ and Equation 12.212 yields

$$\begin{aligned} h_S^{(2)}(r) &= \frac{a^2(b)}{\mathcal{K}^\star(b)} \cdot \frac{e^{-br}}{4\pi r} + \frac{a^2(b')}{\mathcal{K}^\star(b')} \cdot \frac{e^{-b'r}}{4\pi r} + \ldots \qquad \text{(large } r\text{)} \qquad (12.214) \\ h_D^{(2)}(r) &= -\frac{\beta[q^{\text{eff}}(\kappa)]^2}{\mathcal{E}_r^{\text{eff}}(\kappa)\varepsilon_0} \cdot \frac{e^{-\kappa r}}{4\pi r} - \frac{\beta[q^{\text{eff}}(\kappa')]^2}{\mathcal{E}_r^{\text{eff}}(\kappa')\varepsilon_0} \cdot \frac{e^{-\kappa' r}}{4\pi r} - \ldots. \end{aligned}$$

Again we have only shown the first two Yukawa function contributions of each kind and there are also short-range terms with other functional forms. Note that in the present case we have $\sum_i \alpha_i(b) = a(b) n_{\text{tot}}^{\text{b}}$ and $q_{\text{q}}^{\text{eff}}(\kappa) = \mathsf{z} q^{\text{eff}}(\kappa)$, so these results agree with Equation 12.211.

In order to illustrate these formal results, we will give an explicit example of $h_{ij}^{(2)}(r)$ for an aqueous electrolyte solution in the restricted primitive model. It is based on a similar approximation as in Example 12:A, but here we introduce some important improvements that make the results accurate also for high ionic densities. As in Example 12:A, the validity of the approximation is limited to values of β_R of the order 2 or less, that is, for sufficiently high temperatures and/or high values of $\varepsilon_r^{\text{cont}}$.

Example 12:C. A simple approximation for high ionic densities[64]

In Example 12:A we saw that the two-mode approximation 12.178 for the charge distribution $\rho_i(r)$ is very successful for electrolytes when β_R is of the order 2 or less and the ionic densities are rather low. Here, we are going to extend this approach to higher densities, whereby we need to consider not only the electrostatic interactions between the ions but also the core-core interactions and we assume that all ions have the same diameter d^{h}. We make an Ansatz where the electrostatic correlations between the ions are considered by means of a two-mode approximation for $w_{ij}^{(2)\text{el}}(r)$

[64] This example is based on the reference cited in footnote 35.

and the core-core correlations are included by means of an approximate potential of mean force $w_{ij}^{(2)\mathrm{core}}(r)$ for their interaction. From Equation 12.191 we know that

$$w_{ij}^{(2)\mathrm{el}}(r) \sim \frac{q_i^{\mathrm{eff}}(\kappa)q_j^{\mathrm{eff}}(\kappa)}{\mathcal{E}_r^{\mathrm{eff}}(\kappa)\varepsilon_0} \cdot \frac{e^{-\kappa r}}{4\pi r} \quad \text{when } r \to \infty.$$

Let us include the first two decay modes and assume as an approximation that

$$w_{ij}^{(2)\mathrm{el}}(r) = \frac{q_i^{\mathrm{eff}}(\kappa)q_j^{\mathrm{eff}}(\kappa)}{\mathcal{E}_r^{\mathrm{eff}}(\kappa)\varepsilon_0} \cdot \frac{e^{-\kappa r}}{4\pi r} + \frac{q_i^{\mathrm{eff}}(\kappa')q_j^{\mathrm{eff}}(\kappa')}{\mathcal{E}_r^{\mathrm{eff}}(\kappa')\varepsilon_0} \cdot \frac{e^{-\kappa' r}}{4\pi r}$$

is valid for all $r \geq d^{\mathrm{h}}$. We furthermore assume that

$$g_{ij}^{(2)}(r) = e^{-\beta[w_{ij}^{(2)\mathrm{el}}(r) + w_{ij}^{(2)\mathrm{core}}(r)]},$$

where $w_{ij}^{(2)\mathrm{core}}(r) = w^{(2)\mathrm{hs}}(r)$ is taken from the pair distribution function $g^{(2)\mathrm{hs}}(r)$ for a fluid of hard spheres of diameter d^{h}, so $g^{(2)\mathrm{hs}}(r) = e^{-\beta w^{(2)\mathrm{hs}}(r)}$. The hard sphere density is set equal to the total ionic density $n_{\mathrm{tot}}^{\mathrm{b}}$.

We will implement this approximation for the restricted primitive model, so we have

$$g_{+\pm}^{(2)}(r) = e^{-\beta[\pm w^{(2)\mathrm{el}}(r) + w^{(2)\mathrm{hs}}(r)]} = e^{\mp\beta w^{(2)\mathrm{el}}(r)} g^{(2)\mathrm{hs}}(r) \tag{12.215}$$

where

$$w^{(2)\mathrm{el}}(r) = \frac{\left[q^{\mathrm{eff}}\right]^2}{\mathcal{E}_r^{\mathrm{eff}}\varepsilon_0} \cdot \frac{e^{-\kappa r}}{4\pi r} + \frac{\left[q'^{\mathrm{eff}}\right]^2}{\mathcal{E}_r'^{\mathrm{eff}}\varepsilon_0} \cdot \frac{e^{-\kappa' r}}{4\pi r}. \tag{12.216}$$

The electrolyte is in a dielectric continuum so $\varepsilon_r^{\mathrm{cont}} \neq 1$ in the current example.

Let us introduce $g_{\mathrm{S}}^{(2)}(r)$ and $g_{\mathrm{D}}^{(2)}(r)$ in analogy to $h_{\mathrm{S}}^{(2)}(r)$ and $h_{\mathrm{D}}^{(2)}(r)$ in Equation 12.213, so we have $g_{+\pm}^{(2)}(r) = g_{\mathrm{S}}^{(2)}(r) \pm g_{\mathrm{D}}^{(2)}(r)$ and $g_{\mathrm{S}}^{(2)}(r) = h_{\mathrm{S}}^{(2)}(r) + 1$ and $g_{\mathrm{D}}^{(2)}(r) = h_{\mathrm{D}}^{(2)}(r)$. Equation 12.215 implies

$$\begin{aligned} g_{\mathrm{S}}^{(2)}(r) &= \cosh\left[\beta w^{(2)\mathrm{el}}(r)\right] g^{(2)\mathrm{hs}}(r) \\ g_{\mathrm{D}}^{(2)}(r) &= -\sinh\left[\beta w^{(2)\mathrm{el}}(r)\right] g^{(2)\mathrm{hs}}(r). \end{aligned} \tag{12.217}$$

When the electrostatic coupling is sufficiently weak, so $\beta w^{(2)\mathrm{el}}(r)$ is small, we can use the approximations $\cosh(x) \approx 1 + x^2/2$ and

$\sinh(x) \approx x$ in these expressions and obtain as a good approximation

$$\begin{aligned} g_{\rm S}^{(2)}(r) &= \left(1 + \tfrac{1}{2}\left[\beta w^{(2)\rm el}(r)\right]^2\right) g^{(2)\rm hs}(r) \\ g_{\rm D}^{(2)}(r) &= -\beta w^{(2)\rm el}(r) g^{(2)\rm hs}(r). \end{aligned} \quad (12.218)$$

The square term in $g_{\rm S}^{(2)}(r)$ has a special role because Equation 12.218 implies that the pair distribution function for equally charged ions is

$$\begin{aligned} g_{++}^{(2)}(r) &= g_{\rm S}^{(2)}(r) + g_{\rm D}^{(2)}(r) \\ &= \left(1 - \beta w^{(2)\rm el}(r) + \tfrac{1}{2}\left[\beta w^{(2)\rm el}(r)\right]^2\right) g^{(2)\rm hs}(r), \end{aligned}$$

which is guaranteed to be positive. If the square term is left out we have instead $g_{++}^{(2)}(r) = \left(1 - \beta w^{(2)\rm el}(r)\right) g^{(2)\rm hs}(r)$, which can be negative. This is unphysical because $n_+^{\rm b} g_{++}^{(2)}(r)$ is the density of positive ions around a positive ion. To avoid this, the power series of $\cosh(x)$ above should include one term more than that of $\sinh(x)$.

The charge density around a positive or negative ion is $\rho_\pm(r) = \pm q n_{\rm tot}^{\rm b} g_{\rm D}^{(2)}(r)$, so we have

$$\begin{aligned} \rho_\pm(r) &= \mp \beta q n_{\rm tot}^{\rm b} \left[\frac{\left[q^{\rm eff}\right]^2}{\mathcal{E}_r^{\rm eff} \varepsilon_0} \cdot \frac{e^{-\kappa r}}{4\pi r} + \frac{\left[q'^{\rm eff}\right]^2}{\mathcal{E}_r'^{\rm eff} \varepsilon_0} \cdot \frac{e^{-\kappa' r}}{4\pi r} \right] g^{(2)\rm hs}(r) \\ &= \mp \left[\frac{\kappa^2 q^{\rm eff}}{\mathcal{E}_r^{\rm eff}/\varepsilon_r^{\rm cont}} \cdot \frac{e^{-\kappa r}}{4\pi r} + \frac{\kappa'^2 q'^{\rm eff}}{\mathcal{E}_r'^{\rm eff}/\varepsilon_r^{\rm cont}} \cdot \frac{e^{-\kappa' r}}{4\pi r} \right] g^{(2)\rm hs}(r), \end{aligned}$$

where we have used $\kappa^2 = \beta n_{\rm tot}^{\rm b} q q^{\rm eff}/(\varepsilon_r^{\rm cont}\varepsilon_0)$, as follows from Equation 12.166. If we neglect the core-core correlations and set $g^{(2)\rm hs}(r) = 1$ we obtain the Ansatz 12.178 that we used in Example 12:A.

Expressed in terms of $h_{\rm S}^{(2)}$, $h_{\rm D}^{(2)}$ and $h^{(2)\rm hs} = g^{(2)\rm hs} - 1$, Equation 12.218 can be written as

$$\begin{aligned} h_{\rm S}^{(2)}(r) &= h^{(2)\rm hs}(r) + \tfrac{1}{2}\left[\beta w^{(2)\rm el}(r)\right]^2 + \tfrac{1}{2}\left[\beta w^{(2)\rm el}(r)\right]^2 h^{(2)\rm hs}(r) \\ h_{\rm D}^{(2)}(r) &= -\beta \left[w^{(2)\rm el}(r) + w^{(2)\rm el}(r) h^{(2)\rm hs}(r)\right]. \end{aligned} \quad (12.219)$$

Let us see what these contributions represent. The pair correlation function $h^{(2)\rm hs}(r)$ for hard spheres contains oscillatory Yukawa function terms that originates from the complex poles of $\tilde{h}^{(2)\rm hs}(k)$, that is, the zeros of denominator $1 - n^{\rm b}\tilde{c}^{(2)\rm hs}(k)$ in Equation 9.77

applied to a hard sphere fluid. Recall that $h^{(2)\mathrm{hs}}(r)$ from the Percus-Yevick approximation can be written as a superposition of such functions as mentioned in connection to Equation 10.2. The Yukawa function terms of $h_{\mathrm{S}}^{(2)}(r)$ in Equation 12.214 are accordingly approximated by such terms for hard spheres.

The remaining contributions to $h^{(2)\mathrm{hs}}(r)$ have other functional forms than Yukawa functions. As we have seen, such contributions are always present in nonlinear theories. The second term for $h_{\mathrm{S}}^{(2)}(r)$ in Equation 12.219 with $w^{(2)\mathrm{el}}(r)$ from Equation 12.216 give three terms proportional to $e^{-2\kappa r}/r^2$, $e^{-2\kappa' r}/r^2$ and $e^{-2(\kappa+\kappa')r}/r^2$, respectively, which all correspond to logarithmic branch points of $\tilde{h}_{\mathrm{S}}^{(2)}(k)$ in complex k space (cf. the shaded text box on page 351). The third term in the first line contains cross-terms between electrostatic and core-core correlations. It always decays faster to zero with increasing r than the leading term of $h_{\mathrm{S}}^{(2)}(r)$.

In $h_{\mathrm{D}}^{(2)}(r)$ of Equation 12.219, the first term in the square bracket gives the two first Yukawa terms of $h_{\mathrm{D}}^{(2)}(r)$ in Equation 12.214, while the second term contains cross-terms between electrostatic and core-core correlations. The latter always decays faster than the leading term of $h_{\mathrm{D}}^{(2)}(r)$.

The decay parameters κ and κ' used in this approximation are the same as in Example 12:A and are plotted in Figure 12.2. Thus, we use the relationship

$$\frac{q^{\mathrm{eff}}(\kappa)}{q} = \left[\frac{\kappa}{\kappa_{\mathrm{D}}}\right]^2 = \frac{e^{\kappa d^{\mathrm{h}}}}{1+\kappa d^{\mathrm{h}}},$$

where the last equality constitute the approximate Equation 12.177 for κ and κ'.

$\mathcal{E}_r^{\mathrm{eff}}/\varepsilon_r^{\mathrm{cont}}$ and $\mathcal{E}_r'^{\mathrm{eff}}/\varepsilon_r^{\mathrm{cont}}$ are determined from the requirement that both the local electroneutrality and the Stillinger-Lovett conditions are fulfilled. In the current approximation this leads to expressions[65] for $\mathcal{E}_r^{\mathrm{eff}}/\varepsilon_r^{\mathrm{cont}}$ and $\mathcal{E}_r'^{\mathrm{eff}}/\varepsilon_r^{\mathrm{cont}}$ that involve $g^{(2)\mathrm{hs}}(r)$, κ and κ'. They have to be evaluated numerically. The pair distribution function $g^{(2)\mathrm{hs}}(r)$ for hard sphere fluids that is used here can be calculated from a very accurate semi-empirical formula due to Verlet and Weis.[66]

[65] The expressions for $\mathcal{E}_r^{\mathrm{eff}}/\varepsilon_r^{\mathrm{cont}}$ and $\mathcal{E}_r'^{\mathrm{eff}}/\varepsilon_r^{\mathrm{cont}}$ can be found in Appendix A of the reference cited in footnote 35.

[66] L. Verlet and J.-J. Weis, Phys. Rev. A, **5** (1972) 939 (https://doi.org/10.1103/PhysRevA.5.939). A Fortran code for the Verlet-Weis semi-empirical formula for $g^{(2)\mathrm{hs}}(r)$ is available in: W. R. Smith, D. J. Henderson, P. J. Leonard, J. A. Barker, and E. W. Grundke, Molec. Phys., **106** (2008) 3 (https://doi.org/10.1080/00268970701628423).

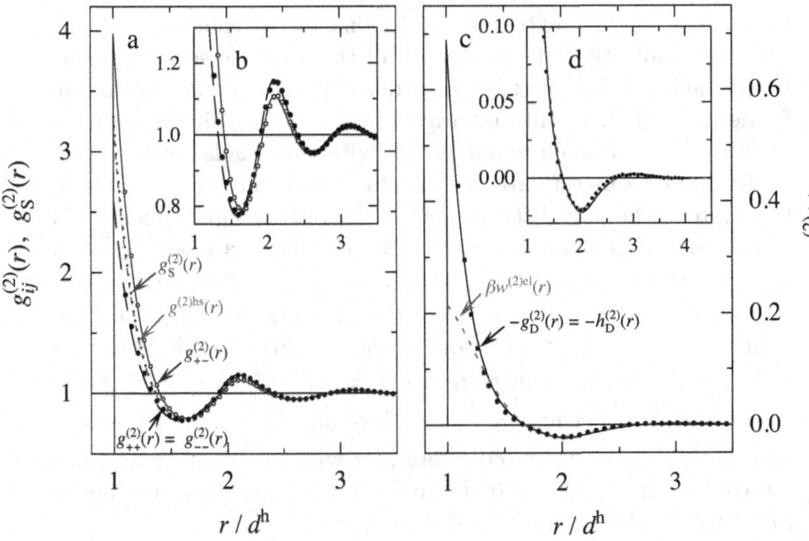

Figure 12.6 **(See color insert).** (a) The black and blue curves show radial distribution function $g_{ij}^{(2)}(r)$ for a 2 M 1:–1 aqueous electrolyte solution with ions of diameter $d^h = 6.6$ Å ($\kappa_D d^h = 3.11$ and $\beta_R = 1.10$) calculated from Equation 12.215 with the two-mode expression for $w^{(2)el}(r)$ from Equation 12.216: the black long dashes show $g_{++}^{(2)}(r) = g_{--}^{(2)}(r)$ and the blue full curve show $g_{+-}^{(2)}(r)$. They are compared with the same functions from MC simulations[67] (black and blue symbols). The green curve (dots) is $g_S^{(2)}(r) = h_S^{(2)}(r) + 1$ from Equation 12.219 and we have $g_{+\pm}^{(2)}(r) = g_S^{(2)}(r) \pm g_D^{(2)}(r)$. The red curve (short dashes) shows the pair distribution $g^{(2)hs}(r)$ for a hard sphere fluid of the same density and sphere diameter (packing fraction $\eta^{hs} = 0.37$). It nearly coincides with the green dotted curve. Insert (b) shows a magnified view of the $g_{ij}^{(2)}(r)$ curves in frame (a). (c) The function $-g_D^{(2)}(r) \equiv -h_D^{(2)}(r)$ is proportional to the radial charge-density distribution $\rho_-(r)$ around a negative ion. The full curve is calculated from Equation 12.219 and the symbols are the MC results. The red curve (short dashes) shows $\beta w^{(2)el}(r)$ from Equation 12.216, which is the electrostatic part of $-g_D^{(2)}(r)$ in the present approximation. The insert (d) shows a magnified view of the curves in frame (c). Note the large differences in ordinate scales of the various frames.

An example of the results from this theory is displayed in Figure 12.6, which shows the pair distributions for an electrolyte with rather large ions, $d^h = 6.6$ Å, and high ion concentration. In this system the core-core correlations are important and the charge density distribution $\rho(r)$ is oscillatory, so the two decay parameters κ and κ' are complex conjugates of each other $\kappa_{\Re} \pm i\kappa_{\Im}$. The pair distributions are compared in the figure with MC simulations data by Jos W. Zwanikken and coworkers[67] and we can see that the present theory gives excellent results that nearly coincide with the MC ones. Note that the resolution of the ordinate scale in frame (d) of the figure is more than an order of magnitude larger than in frame (a) and that quite intricate details of the pair distribution functions are obtained in agreement with the simulations.

In order to investigate the effects of core-core correlations in this system, the pair distribution $g^{(2)\text{hs}}(r)$ of the hard sphere fluid is plotted in Figure 12.6a and it is seen that the main oscillation of $g_{ij}^{(2)}(r)$ with a wave length $\approx d^{\text{h}}$ is caused by these correlations. It is also seen in frames (a) and (b) of the figure that there is another oscillatory component with longer wave length superimposed on the main oscillation. This component is $-\beta w^{(2)\text{el}}(r)$ in $h_{\text{D}}^{(2)}(r)$ and it is oscillatory because κ is complex-valued. It makes the $g_{++}^{(2)}$ and $g_{+-}^{(2)}$ curves to repeatedly intersect with one another and has a wave length $2\pi/\kappa_{\Im} = 2.06d^{\text{h}}$. In frames (c) and (d) of the figure $\beta w^{(2)\text{el}}(r)$ is plotted as the red dashed curve.

It is of interest to investigate the importance of the various contributions to Equation 12.219. In frame (a) of the figure we see that $g^{(2)\text{hs}}(r)$ and $g_{\text{S}}^{(2)}(r) = h_{\text{S}}^{(2)}(r) + 1$ nearly coincide, which means that the two last contributions to $h_{\text{S}}^{(2)}(r)$ in the equation are unimportant in the present case. For $h_{\text{D}}^{(2)}(r)$ the situation is different because in frame (c) of the figure we see that $\beta w^{(2)\text{el}}(r)$ gives only about 1/3 of $-h_{\text{D}}^{(2)}(r)$ for $r = d^{\text{h}}$. Thus, the second term in the square bracket, i.e., the cross term between electrostatic and core-core correlation, is very important for small r. However, for $r \gtrsim 1.3d^{\text{h}}$ $\beta w^{(2)\text{el}}(r)$ gives nearly the entire $-h_{\text{D}}^{(2)}(r)$ as seen in frames (c) and (d).

To conclude we have seen that the pair correlations of a rather dense electrolyte can accurately be obtained in an approximation with two electrostatic decay modes together with non-electrostatic contributions taken from a hard sphere fluid, provided that the temperature and/or $\varepsilon_r^{\text{cont}}$ are sufficiently high (β_R of the order 2 or less; in this case β_R =1.10). In the present example the two modes are combined into an oscillatory decay mode. Due to the inherent nonlinear nature of the pair distributions in terms of the potential of mean force, the Yukawa function contributions are accompanied by cross-terms involving electrostatic decay modes between themselves and in combination with the core-core contributions. These cross-terms constitute short-range contributions to the pair distributions.

---

[67] J. W. Zwanikken, P. K. Jha and M. Olvera de la Cruz, J. Chem. Phys., **135** (2011) 064106 (https://doi.org/10.1063/1.3624809).

Before we proceed, let us recapitulate some basic facts. The charge-inversion invariant electrolyte systems stand out from all other electrolytes because they have two distinct sets of decay parameters for $h_{ij}^{(2)}(r)$, namely the solutions of $\tilde{\varepsilon}(i\kappa) = 0$ that govern the decay of $H_{\text{QQ}}(r)$ and the solutions of $D^{\star}(ib) = 0$ that govern the decay of $H_{\text{NN}}(r) = H^{\star}_{\text{NN}}(r)$. The decay modes of $\rho_i(r)$, $\psi_i(r)$ and $\phi^{\star}_{\text{Coul}}(r)$ are the same as those of $H_{\text{QQ}}(r)$. The reason for the two distinct sets is that such systems are models with an extreme symmetry that makes the density-charge correlation function to be identically equal to zero, $H_{\text{NQ}}(r) \equiv 0$. This means that the charge density and the number density distributions are decoupled on the linear level. There is, however, an interdependency of these distributions on the nonlinear level. As we saw in Example 12:C, $h_{\text{S}}^{(2)}(r)$ in that case is nearly the same as for a hard sphere fluid, but not exactly the same because there are contributions from the cross-terms in Equation 12.219 between the electrostatic and core-core correlations. The cross-terms arise from the non-linearity of the theory and we saw that such a term is very important for $h_{\text{D}}^{(2)}(r)$ for small r in that case.

Systems with charge-inversion invariance are idealized models. In reality, the anions and cations differ from each other in other ways than the sign of their charges and then $H_{\text{NQ}}(r) \neq 0$. This is our main case and there is only one set of decay parameters for $h_{ij}^{(2)}(r)$, namely the solutions of $\tilde{\varepsilon}(i\kappa) = 0$ that govern the decay of $H_{\text{QQ}}(r)$, $H_{\text{NN}}(r)$, $H_{\text{NQ}}(r)$ and all other functions just mentioned apart from $H^{\star}_{\text{NN}}(r)$, which has a different set of decay parameters. It does not matter how little the anions and cations are different from each other apart from the sign of their charges; as soon as the symmetry is broken, the main case applies. The density-charge correlations described by $H_{\text{NQ}}(r)$ make the density-density and charge-charge correlations to be coupled also on the linear level, whereby they acquire the same modes instead of distinct ones.

A pertinent question is exactly what happens when the symmetry is broken. The properties of a charge-inversion invariant system must be virtually the same as those of a system that deviates very slightly from it, i.e., **borderline cases** that are nearly charge-inversion invariant but nevertheless behave as in our main case. An example of such a borderline case is a binary electrolyte with slightly different sizes of anions and cations and with $q_+ = -q_-$. $H_{\text{NN}}(r)$ in the invariant system must be virtually the same as in the borderline system despite that in the former case the decay parameters are given by zeros of $D^{\star}(k)$ and not zeros of $\tilde{\varepsilon}(k)$, but the other way around in the latter: they are given by zeros of $\tilde{\varepsilon}(k)$ and not of $D^{\star}(k)$.

A way to understand what is going on is to follow the "trajectories" of the poles of $\tilde{H}_{\text{NN}}(k)$ and $\tilde{H}^{\star}_{\text{NN}}(k)$ in complex k space as functions of the system parameters like ion sizes, that is, we follow the trajectories of the corresponding zeros of $\tilde{\varepsilon}(k)$ and $D^{\star}(k)$, respectively. For a system that is nearly charge-inversion invariant we will show below that a set of poles of $\tilde{H}_{\text{NN}}(k)$ lie close to poles of $\tilde{H}^{\star}_{\text{NN}}(k)$; let us call them set A. The trajectories of any of these poles of $\tilde{H}_{\mathrm{NN}}(k)$ and the corresponding pole of $\tilde{H}^{\star}_{\mathrm{NN}}(k)$ cross each other at the point where the system become invariant. The poles coincide at the crossing point, whereby the pole of $\tilde{H}_{\mathrm{NN}}(k)$ ceases to be a zero of $\tilde{\epsilon}(k)$ and becomes instead a zero of $D^{\star}(k)$ (i.e., a pole of both $\tilde{H}^{\star}_{\mathrm{NN}}(k)$ and $\tilde{H}_{\mathrm{NN}}(k)$ that become equal at this point).

The poles of $\tilde{H}_{\mathrm{NN}}(k)$ that do not belong to set A will be called set B. For any of these poles, say $k = \mathrm{i}\kappa'$, the prefactor $q^{\mathrm{eff}}_{\mathrm{n}}(\kappa')$ of its Yukawa function contribution in $H_{\mathrm{NN}}(r)$ (Equation 12.207) becomes zero when the system becomes invariant, so this contribution vanish. Recall that $q^{\mathrm{eff}}_{\mathrm{n}}(\kappa') \equiv \tilde{\rho}^{\star}_{\mathrm{n}}(\mathrm{i}\kappa')$, which is zero for such a system.

We will now show that the poles in set A behave in the manner indicated. Equation 12.197 implies that

$$\tilde{\epsilon}(k) = \frac{\beta \mathcal{F}^2(b)}{\varepsilon_0 k^2 \left[k^2 + b^2\right] \mathcal{K}^{\star}(b)} + \tilde{\epsilon}^{\mathrm{rest}}(k), \tag{12.220}$$

where $\tilde{\epsilon}^{\mathrm{rest}}(k)$ is a function that remains finite when $k \to ib$. Consider one of the zeros of $\tilde{\epsilon}(k)$, say $k = \mathrm{i}\kappa$. Since $\tilde{\epsilon}(\mathrm{i}\kappa) = 0$ we have

$$\frac{-\beta \mathcal{F}^2(b)}{\varepsilon_0 \kappa^2 \left[b^2 - \kappa^2\right] \mathcal{K}^{\star}(b)} + \tilde{\epsilon}^{\mathrm{rest}}(\mathrm{i}\kappa) = 0 \tag{12.221}$$

and provided that $\mathcal{F}(b) \neq 0$ we can write this as

$$\tilde{\epsilon}^{\mathrm{rest}}(\mathrm{i}\kappa)\kappa^2 \left[b^2 - \kappa^2\right] = \frac{\beta \mathcal{F}^2(b)}{\varepsilon_0 \mathcal{K}^{\star}(b)}. \tag{12.222}$$

When $\mathcal{F}(b) \approx 0$ this equation has a solution $\kappa \approx \pm b$ with the property $\kappa \to \pm b$ when $\mathcal{F}(b) \to 0$. This is the solution that we are looking for. Its existence can also be inferred directly from Equation 12.221 because $\mathcal{F}^2(b)/[b^2 - \kappa^2]$ must remain finite in this limit since $\tilde{\epsilon}^{\mathrm{rest}}(\mathrm{i}\kappa)$ stays finite there. Likewise, there exist solutions of $\tilde{\epsilon}(k) = 0$ that are connected in the same manner to other poles of $\tilde{H}^{\star}_{\mathrm{NN}}(k)$. Any remaining solutions κ of Equation 12.222 are not interesting here because in the same limit they go to the solutions of $\tilde{\epsilon}^{\mathrm{rest}}(\mathrm{i}\kappa) = 0$, which is the same as $\tilde{\epsilon}(\mathrm{i}\kappa) = 0$ as seen from from Equation 12.220 with $\mathcal{F}(b) = 0$. Thus they belong to set B.

Thus, for a nearly charge-inversion invariant system the zeros of $\tilde{\epsilon}(k)$ belong to either of two sets. Set A: each zero in this set lies close to some zero of $D^{\star}(k)$ and will coincide with it when the system becomes charge-inversion invariant. Set B consists of all other zeros. The poles of $\tilde{H}_{\mathrm{NN}}(k)$ due to the zeros in set A will remain as poles of $\tilde{H}_{\mathrm{NN}}(k)$ at the crossing point, while the contributions to $H_{\mathrm{NN}}(r)$ from the zeros in set B will vanish when at the system becomes invariant as noticed earlier. The latter zeros are poles of $\tilde{H}_{\mathrm{QQ}}(k)$ throughout.

The zeros in set A, however, do not give any contributions to $H_{QQ}(r)$ at the crossing point. The reason is that $\mathcal{E}_r^{\text{eff}}(\kappa)$ diverges there for this set, whereby the Yukawa function contribution to $H_{QQ}(r)$ in Equation 12.209 vanishes. This can be shown as follows. The derivative of Equation 12.220 is

$$\tilde{\epsilon}'(k) = -\frac{2\left[2k^2+b^2\right]\beta\mathcal{F}^2(b)}{\varepsilon_0 k^3\left[k^2+b^2\right]^2\mathcal{K}^\star(b)} + \tilde{\epsilon}'^{\,\text{rest}}(k)$$

so we have from Equation 12.158

$$\mathcal{E}_r^{\text{eff}}(\kappa) = \frac{\left[b^2-2\kappa^2\right]\beta\mathcal{F}^2(b)}{\varepsilon_0\kappa^2\left[b^2-\kappa^2\right]^2\mathcal{K}^\star(b)} + \frac{\text{i}\kappa}{2}\tilde{\epsilon}'^{\,\text{rest}}(\text{i}\kappa).$$

Since $\mathcal{F}^2(b)/[b^2-\kappa^2]$ remains finite when $\kappa\to b$ the ratio $\mathcal{F}^2(b)/[b^2-\kappa^2]^2$ diverges like $1/[b^2-\kappa^2]$, so it follows that $\mathcal{E}_r^{\text{eff}}(\kappa)$ diverges for decay modes in set A. The contribution from such a decay mode in $H_{QQ}(r)$, $\rho_i(r)$, $\psi_i(r)$ and $\phi^\star_{\text{Coul}}(r)$ vanishes when $\kappa\to b$ because of this divergence. However, the combination $q_i^{\text{eff}}(\kappa)q_j^{\text{eff}}(\kappa)/\mathcal{E}_r^{\text{eff}}(\kappa)$ in $h_{ij}^{(2)}(r)$ stays finite when $\kappa\to b$ because the divergence in $\mathcal{E}_r^{\text{eff}}(\kappa)$ is cancelled[68] by a divergence in $q_l^{\text{eff}}(\kappa)$ for $l=i,j$. We have $-\beta q_i^{\text{eff}}(\kappa)q_j^{\text{eff}}(\kappa)/[\mathcal{E}_r^{\text{eff}}(\kappa)\varepsilon_0]\to a_i(b)a_j(b)/\mathcal{K}^\star(b)$. The same is true for $[q_{\text{n}}^{\text{eff}}(\kappa)]^2/\mathcal{E}_r^{\text{eff}}(\kappa)$ in the coefficient of $H_{NN}(r)$ and the entire coefficient goes to $[\sum_i \alpha_i(b)]^2/\mathcal{K}^\star(b)$ in Equation 12.211 when $\kappa\to b$. The effective charges $q_i^{\text{eff}}(\kappa)$ of the decay modes in set A loose their physical meaning when $\kappa\to b$ so they do not have any role for the invariant system. However, the combinations $q_i^{\text{eff}}(\kappa)/\mathcal{E}_r^{\text{eff}}(\kappa)$, which goes to zero, and $q_i^{\text{eff}}(\kappa)q_j^{\text{eff}}(\kappa)/\mathcal{E}_r^{\text{eff}}(\kappa)$, which stays finite, maintain physical roles in this limit.

The pair correlation function $h_{ij}^{(2)}(r)$ for a borderline system contains contribution from all decay modes of the main case, Equation 12.192. There are no Yukawa function contributions other than those with decay parameters from the solutions of $\tilde{\epsilon}(\text{i}\kappa)=0$. $H_{QQ}(r)$, $H_{NN}(r)$ and $H_{NQ}(r)$ contain contributions from all these modes, but for modes that belong to set A the terms in $H_{QQ}(r)$ are small because they are zero in the invariant case and likewise the terms in $H_{NN}(r)$ are small for modes that belong to set B. All modes give small contributions to $H_{NQ}(r)$. How does this relate to the contributions in $h_{ij}^{(2)}(r)$ from the two types of modes?

---

68 From Equation 12.197 it follows that $q_i^{\text{eff}}(\kappa)\equiv\tilde{\rho}_i^\star(\text{i}\kappa)$ is proportional to $\alpha_i(b)\mathcal{F}(b)/(b^2-\kappa^2)$, which diverges like $\mathcal{F}(b)/(b^2-\kappa^2)$ when $\kappa\to b$. This implies that the combination $q_i^{\text{eff}}(\kappa)q_j^{\text{eff}}(\kappa)/\mathcal{E}_r^{\text{eff}}(\kappa)$ stays finite in this limit. However, $q_i^{\text{eff}}(\kappa)/\mathcal{E}_r^{\text{eff}}(\kappa)$ that occurs in $\rho_i(r)$ and $\psi_i(r)$ goes to zero.

Let us take a binary electrolyte as an example, so we have $q_+ = -q_- \equiv q$ and $n_+^{\mathrm{b}} = n_-^{\mathrm{b}} = n_{\mathrm{tot}}^{\mathrm{b}}/2$. In this case we have

$$q_{\mathrm{n}}^{\mathrm{eff}}(\kappa) = \sum_i \mathrm{x}_i^{\mathrm{b}} \left[q_i^{\mathrm{eff}}(\kappa)\right] = \frac{1}{2}\left[q_+^{\mathrm{eff}}(\kappa) + q_-^{\mathrm{eff}}(\kappa)\right]$$

$$q_{\mathrm{q}}^{\mathrm{eff}}(\kappa) = \sum_i \mathrm{x}_i^{\mathrm{b}} \mathrm{z}_i q_i^{\mathrm{eff}}(\kappa) = \frac{\mathrm{z}}{2}\left[q_+^{\mathrm{eff}}(\kappa) - q_-^{\mathrm{eff}}(\kappa)\right]$$

and the coefficient for a Yukawa function term of H_{IJ} in Equation 12.207 contains the factor

$$H_{\mathrm{NN}}: \quad \frac{\left[q_{\mathrm{n}}^{\mathrm{eff}}(\kappa)\right]^2}{\mathcal{E}_r^{\mathrm{eff}}(\kappa)} = \frac{\left[q_+^{\mathrm{eff}}(\kappa) + q_-^{\mathrm{eff}}(\kappa)\right]^2}{\mathcal{E}_r^{\mathrm{eff}}(\kappa)}$$

$$H_{\mathrm{QQ}}: \quad \frac{\left[q_{\mathrm{q}}^{\mathrm{eff}}(\kappa)\right]^2}{\mathcal{E}_r^{\mathrm{eff}}(\kappa)} = \frac{\left[q_+^{\mathrm{eff}}(\kappa) - q_-^{\mathrm{eff}}(\kappa)\right]^2}{\mathcal{E}_r^{\mathrm{eff}}(\kappa)}$$

$$H_{\mathrm{NQ}}: \quad \frac{q_{\mathrm{n}}^{\mathrm{eff}}(\kappa) q_{\mathrm{q}}^{\mathrm{eff}}(\kappa)}{\mathcal{E}_r^{\mathrm{eff}}(\kappa)} = \frac{\left[q_+^{\mathrm{eff}}(\kappa)\right]^2 - \left[q_-^{\mathrm{eff}}(\kappa)\right]^2}{\mathcal{E}_r^{\mathrm{eff}}(\kappa)}.$$

For borderline cases $h_{++}^{(2)}(r) \approx h_{--}^{(2)}(r)$.

We start with the modes in set B, which are straightforward to treat. In this case $q_+^{\mathrm{eff}}(\kappa) \approx -q_-^{\mathrm{eff}}(\kappa)$ and the contributions from such a mode in $H_{\mathrm{NN}}(r)$ and $H_{\mathrm{NQ}}(r)$ are small because $q_+^{\mathrm{eff}}(\kappa) + q_-^{\mathrm{eff}}(\kappa) \approx 0$ and $\left[q_+^{\mathrm{eff}}(\kappa)\right]^2 \approx \left[q_-^{\mathrm{eff}}(\kappa)\right]^2$, respectively. The effective charges can change sign so it is possible to have $q_+^{\mathrm{eff}}(\kappa) < 0$ and $q_-^{\mathrm{eff}}(\kappa) > 0$ when κ is real.

The modes in set A require a bit more thought. For an invariant system we have $\alpha_+(b) = \alpha_-(b)$ because $\mathcal{F}(b) = \sum_j q_j \alpha_j(b) = q[\alpha_+(b) - \alpha_-(b)] = 0$. The Yukawa function contribution to $h_{++}^{(2)}(r)$ and $h_{+-}^{(2)}(r)$ from this kind of mode is therefore equal and for continuity reasons this must be approximately true for the borderline cases. We can therefore conclude that $q_+^{\mathrm{eff}}(\kappa) \approx q_-^{\mathrm{eff}}(\kappa)$.[69] Note that $q_+^{\mathrm{eff}}(\kappa) - q_-^{\mathrm{eff}}(\kappa)$ stays finite and nonzero throughout[70] as it must because $\kappa^2 = \beta n_{\mathrm{tot}}^{\mathrm{b}} q \left[q_+^{\mathrm{eff}}(\kappa) - q_-^{\mathrm{eff}}(\kappa)\right]/(2\varepsilon_0)$. The reason why the contribution to $H_{\mathrm{QQ}}(r)$ is small is, as we saw earlier, that $\mathcal{E}_r^{\mathrm{eff}}(\kappa)$ diverges when $\kappa \to b$ so $|\mathcal{E}_r^{\mathrm{eff}}(\kappa)|$ must be large for a borderline system. The contribution to $H_{\mathrm{NQ}}(r)$ is small for the same reason as before.

We can conclude that for a borderline system the Yukawa function contributions to $h_{++}^{(2)}(r)$ and $h_{+-}^{(2)}(r)$ from a mode in set A decay to zero when $r \to \infty$ *with the same sign*, either positive or negative.[71] This illustrated

---

[69] This also follows from Equation 12.197 that implies that $q_+^{\mathrm{eff}}(\kappa)/q_-^{\mathrm{eff}}(\kappa) \approx \alpha_+(b)/\alpha_-(b) \approx 1$ when $\kappa \approx b$ for modes in set A. Recall that $q_i^{\mathrm{eff}}(\kappa) \equiv \tilde{\rho}_i^{\star}(\mathrm{i}\kappa)$.

[70] From footnote 68 we see that $q_+^{\mathrm{eff}}(\kappa) - q_-^{\mathrm{eff}}(\kappa) = [q_+ q_+^{\mathrm{eff}}(\kappa) + q_- q_-^{\mathrm{eff}}(\kappa)]/q$ is proportional to $[q_+\alpha_+(b) + q_-\alpha_-(b)]\mathcal{F}(b)/(b^2 - \kappa^2) = \mathcal{F}^2(b)/(b^2 - \kappa^2)$, which remains finite when $\kappa \to b$.

[71] This holds provided that the charge-inversion invariant system has a non-zero coefficient for its Yukawa function contribution with decay parameter $b \approx \kappa$ and that the borderline system does not deviate too much from the invariant one.

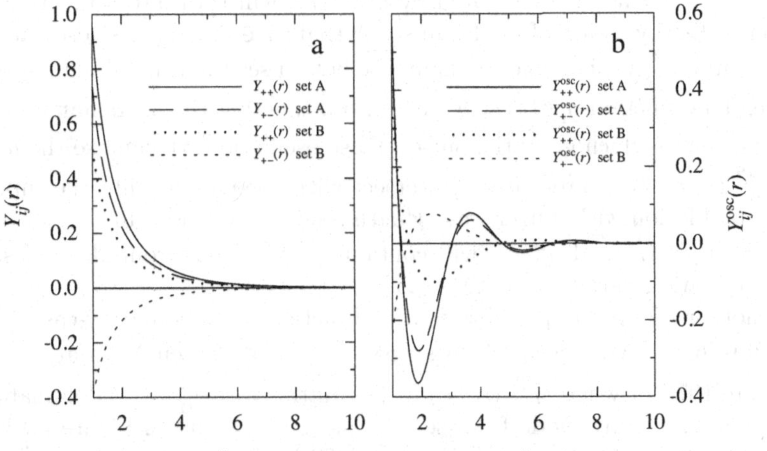

Figure 12.7 (a) Examples of Yukawa functions $Y_{ij}(r) = q_i^{\text{eff}} q_j^{\text{eff}} e^{-\kappa r}/(\mathcal{E}_r^{\text{eff}} r)$ that appear in the various decay modes in $h_{ij}^{(2)}(r)$ for a binary electrolyte in borderline cases where the electrolyte is nearly charge-inversion invariant. The curves show the functions for different types of decay modes: set A were $q_+^{\text{eff}}(\kappa) \approx q_-^{\text{eff}}(\kappa)$ and set B where $q_+^{\text{eff}}(\kappa) \approx -q_-^{\text{eff}}(\kappa)$. (b) Corresponding examples for oscillatory Yukawa functions $Y_{ij}^{\text{osc}}(r) = |q_i^{\text{eff}} q_j^{\text{eff}}| \, e^{-\kappa_{\Re} r} \cos(\kappa_{\Im} r + \vartheta_{ij})/(|\mathcal{E}_r^{\text{eff}}| r)$ in $h_{ij}^{(2)}(r)$ and the curves show the two different types of decay modes: set A where the oscillations are in phase and set B where they are out of phase. The electrostatically dominated contributions to $h_{ij}^{(2)}(r)$ belong to set B (both monotonic and oscillatory, for example before and after the Kirkwood cross-over) and those that are dominated by core-core packing effects belong to set A (oscillatory) and also the long-range density-density correlations that appear near a critical point (monotonic).

in Figure 12.7a where examples of $Y_{ij}(r) = q_i^{\text{eff}} q_j^{\text{eff}} e^{-\kappa r}/(\mathcal{E}_r^{\text{eff}} r)$ are plotted and both $Y_{++}(r)$ and $Y_{+-}(r)$ for set A decay to zero from above. Oscillatory Yukawa function contributions from set A to each of $h_{++}^{(2)}(r)$ and $h_{+-}^{(2)}(r)$ oscillate *in phase* when $r \to \infty$, see Figure 12.7b where $Y_{ij}^{\text{osc}}(r) = |q_i^{\text{eff}} q_j^{\text{eff}}| \, e^{-\kappa_{\Re} r} \cos(\kappa_{\Im} r + \vartheta_{ij})/(|\mathcal{E}_r^{\text{eff}}| r)$ is plotted. The corresponding contributions from set B oscillate *out of phase* as also shown the figure. For set B the (monotonic) Yukawa function contributions decay to zero when $r \to \infty$ from opposite sides, one negative and the other positive, see $Y_{++}(r)$ and $Y_{+-}(r)$ for set B in Figure 12.7a. This is typical for a mode that is electrostatically dominated and gives the behavior of $h_{ij}^{(2)}(r)$ in, for example, the Poisson-Boltzmann approximation where $q_+^{\text{eff}}(\kappa_D) > 0$ and $q_-^{\text{eff}}(\kappa_D) < 0$.

As an example of a borderline case we can take the system in Figure 12.6 of Example 12:C and make the anion and cation sizes slightly different.

Then $g_{\rm S}^{(2)}(r)$, which is dominated by $g^{(2)\rm hs}(r)$, will turn into a function that is dominated by a sum of modes in set A that give the in-phase oscillation in $g_{i+}^{(2)}(r)$ and $g_{i-}^{(2)}(r)$, i.e., like the main oscillation seen in frames a and b of the figure. The function $g_{\rm D}^{(2)}(r)$ will turn into a function that is dominated by a mode in set B which gives the out-of-phase oscillations that make the $g_{i+}^{(2)}(r)$ and $g_{i-}^{(2)}(r)$ curves to repeatedly intersect with one another, like the superimposed oscillation with longer wave length seen in the figure. In the borderline case both $g_{\rm S}^{(2)}(r)$ and $g_{\rm D}^{(2)}(r)$ will contain contributions from all modes, but the dominating parts will be those just described.

Another example of pair distribution functions for a boundary-case system is Figure 6.8 in Volume I, which shows $g_{ij}^{(2)}(r)$ for molten NaCl at the temperature 1073 K, where $g_{++}^{(2)}(r) \approx g_{--}^{(2)}(r)$. In this case $g_{ij}^{(2)}(r)$ is dominated by an oscillatory decay mode from set B. Another example is Figure 6.12 that shows the decay parameters for NaCl at 3000 K. This case has one monotonic decay mode from set A and one oscillatory mode from set B, see the shaded text box on page 335 in Volume I.[72] The former mode has a decay length that is longer than the latter at rather low densities, so the former dominates in $h_{ij}^{(2)}(r)$ for large r, but for high densities the situation is the reverse. At very low densities, not shown in that figure, $h_{ij}^{(2)}(r)$ is dominated by a monotonically decaying mode that belongs to set B and is well described by the PB approximation.

12.4 INHOMOGENEOUS ELECTROLYTES ON THE SINGLET LEVEL; SURFACE INTERACTIONS

In Sections 12.2.1–12.2.3 we introduced the basic concepts and set up the machinery for treating inhomogeneous electrolytes on the pair correlation level. We applied the resulting general, exact theory to homogeneous bulk systems in Section 12.3. Here, we will use the general results for bulk electrolytes in order to formulate the corresponding exact theory for inhomogeneous electrolytes on the singlet level. We will thereby both threat interactions for surfaces with uniformly smeared-out surface charges and those that have a lateral structure, for example crystalline surfaces in contact with electrolytes.

Initially we will use the trick presented in Section 10.1.3 in connection to integral equations, namely to transform a bulk fluid mixture consisting small and large particles into an inhomogeneous fluid of small particles in the presence of a macroscopic body that originally was a large particle that has grown in size. Since there is no approximation involved in this trick, the resulting theory is still exact and general. By having two large particles we will likewise formulate an exact theory for surface interactions in electrolytes.

[72] There is a slight difference in notation in Volume I for functions like $\rho_{\rm Q}(r)$ and $\rho_{\rm N}(r)$ that appear there, see footnote 61 for an explanation. In the current notation have $\rho_{\rm q}(r) = \rho_{\rm Q}(r)q_{\pm}/q_{\rm e}$ and $\rho_{\rm n}(r) = \rho_{\rm N}(r)$.

We therefore study an electrolyte system that contains one species of large charged particles and where the rest of the ions are as before, but there is an excess of those ions that serve as counterions to the large particles so electroneutrality is maintained. The large species (L) has density $n_{\mathrm{L}}^{\mathrm{b}}$ and consists of spheres of diameter d_{L} with surface charge density $\sigma^{\mathcal{S}}$ that is uniformly smeared out (superscript $\mathcal{S}$ stands for surface). Since the electrostatic potential outside the particle from such a surface charge is identical to the potential from a charge $q_{\mathrm{L}} = \pi d_{\mathrm{L}}^2 \sigma^{\mathcal{S}}$ placed at the sphere center, we will initially treat these particles as if they have a central charge q_{L} instead of the surface charge. The theory in the previous parts of this chapter is valid for any kinds of spherical ions, irrespectively of the sizes and short-range pair potential $u_{ij}^{\mathrm{sh}}(r)$,[73] so we can use the results that we have obtained.

Thus, the decay modes of the distribution functions are determined by the solutions of $\tilde{\epsilon}(i\kappa) = 0$ and they give the Yukawa function contributions to $\psi_j(r)$ and $h_{ij}^{(2)}(r)$ given by Equations 12.171 and 12.192, respectively. In particular, for the case $j = \mathrm{L}$ the mean potential $\psi_{\mathrm{L}}(r)$ outside an L-particle is given by

$$\psi_{\mathrm{L}}(r) = \frac{q_{\mathrm{L}}^{\mathrm{eff}}(\kappa)}{\mathcal{E}_r^{\mathrm{eff}}(\kappa)\varepsilon_0} \cdot \frac{e^{-\kappa r}}{4\pi r} + \frac{q_{\mathrm{L}}^{\mathrm{eff}}(\kappa')}{\mathcal{E}_r^{\mathrm{eff}}(\kappa')\varepsilon_0} \cdot \frac{e^{-\kappa' r}}{4\pi r} - \ldots \quad \text{(large } r\text{)} \tag{12.223}$$

where we have only shown the terms from the first two decay modes. Since $h_{i\mathrm{L}}^{(2)}(r) = e^{-\beta w_{i\mathrm{L}}(r)} - 1 \sim -\beta w_{i\mathrm{L}}(r)$ the potential of mean force for i-ions outside an L-particle contains the contributions

$$w_{i\mathrm{L}}(r) = \frac{q_i^{\mathrm{eff}}(\kappa) q_{\mathrm{L}}^{\mathrm{eff}}(\kappa)}{\mathcal{E}_r^{\mathrm{eff}}(\kappa)\varepsilon_0} \cdot \frac{e^{-\kappa r}}{4\pi r} + \frac{q_i^{\mathrm{eff}}(\kappa') q_{\mathrm{L}}^{\mathrm{eff}}(\kappa')}{\mathcal{E}_r^{\mathrm{eff}}(\kappa')\varepsilon_0} \cdot \frac{e^{-\kappa' r}}{4\pi r} - \ldots \quad \text{(large } r\text{)} \tag{12.224}$$

from the decay modes. These formulas apply also when there is only one single L-particle present and $n_{\mathrm{L}}^{\mathrm{b}} = 0$. As before, there are also short-range contributions with other functional forms and with decay parameters 2κ, $2\kappa'$ etc. (including cross-terms).

When κ is real and $k = i\kappa$ is the singularity that lies closest to the real axis in complex k space, we have

$$\psi_{\mathrm{L}}(r) \sim \frac{q_{\mathrm{L}}^{\mathrm{eff}}(\kappa)}{\mathcal{E}_r^{\mathrm{eff}}(\kappa)\varepsilon_0} \cdot \frac{e^{-\kappa r}}{4\pi r} \quad \text{and} \quad w_{i\mathrm{L}}(r) \sim \frac{q_i^{\mathrm{eff}}(\kappa) q_{\mathrm{L}}^{\mathrm{eff}}(\kappa)}{\mathcal{E}_r^{\mathrm{eff}}(\kappa)\varepsilon_0} \cdot \frac{e^{-\kappa r}}{4\pi r}$$

when $r \to \infty$. We will increase the diameter d_{L} of the L-particle, but before we do this we will change variable and use the distance $\ell = r - d_{\mathrm{L}}/2$ from the sphere surface instead of the distance r (measured from the sphere center). Let us do this in the factor

$$\frac{q_{\mathrm{L}}^{\mathrm{eff}}(\kappa) e^{-\kappa r}}{4\pi r} = \frac{q_{\mathrm{L}}^{\mathrm{eff}}(\kappa) e^{-\kappa d_{\mathrm{L}}/2}}{4\pi(\ell + d_{\mathrm{L}}/2)} \cdot e^{-\kappa \ell} \tag{12.225}$$

[73]The short-range pair potential $u_{ij}^{\mathrm{sh}}(r)$ is subject to the condition that it decays to zero with increasing r at least exponentially fast, as pointed out earlier.

that appears in the rhs. From the definition of $q_{\mathrm{L}}^{\mathrm{eff}}$, Equation 12.161, we have

$$
\begin{aligned}
q_{\mathrm{L}}^{\mathrm{eff}}(\kappa) &= 4\pi \int_0^\infty dr'\, r'^2 \rho_{\mathrm{L}}^\star(r') \frac{\sinh(\kappa r')}{\kappa r'} \\
&= \frac{4\pi}{\kappa} \int_{-d_{\mathrm{L}}/2}^{\infty} d\ell'\, (\ell' + d_{\mathrm{L}}/2)\rho_{\mathrm{L}}^\star(\ell' + d_{\mathrm{L}}/2) \sinh\left(\kappa[\ell' + d_{\mathrm{L}}/2]\right).
\end{aligned}
$$

Since $\sinh\left(\kappa[\ell' + d_{\mathrm{L}}/2]\right) = [e^{\kappa(\ell'+d_{\mathrm{L}}/2)} - e^{-\kappa(\ell'+d_{\mathrm{L}}/2)}]/2$ we can write coefficient in the rhs of Equation 12.225 as

$$
\frac{q_{\mathrm{L}}^{\mathrm{eff}}(\kappa)e^{-\kappa d_{\mathrm{L}}/2}}{4\pi(\ell + d_{\mathrm{L}}/2)} = \frac{1}{2\kappa} \int_{-d_{\mathrm{L}}/2}^{\infty} d\ell' \, \frac{(\ell' + d_{\mathrm{L}}/2)\left[e^{\kappa\ell'} - e^{-\kappa(\ell'+d_{\mathrm{L}})}\right]}{\ell + d_{\mathrm{L}}/2} \rho_{\mathrm{L}}^\star(\ell' + d_{\mathrm{L}}/2). \tag{12.226}
$$

Let us define $\rho_{\mathrm{w}}^\star(\ell) \equiv \rho_{\mathrm{L}}^\star(\ell + d_{\mathrm{L}}/2)$, $\psi_{\mathrm{w}}(\ell) = \psi_{\mathrm{L}}(\ell + d_{\mathrm{L}}/2)$ and $w_{i;\mathrm{w}}(\ell) \equiv w_{i\mathrm{L}}(\ell + d_{\mathrm{L}}/2)$, so all functions are expressed in terms of the distance from the surface. Recall that the L-particle has a surface charge density σ^S rather than a central charge q_{L}. From now on we will denote this surface charge density as σ_{w}^S, so the bare charge distribution of the L-particle is $\sigma_{\mathrm{w}}^S \delta(\ell)$.

By taking the limit $d_{\mathrm{L}} \to \infty$ while maintaining σ_{w}^S constant, the L-particle is transformed into a planar wall, w, with surface charge density σ_{w}^S and dressed charge distribution $\rho_{\mathrm{w}}^\star(\ell)$. Thereby

$$
\frac{(\ell' + d_{\mathrm{L}}/2)\left[e^{\kappa\ell'} - e^{-\kappa(\ell'+d_{\mathrm{L}})}\right]}{\ell + d_{\mathrm{L}}/2} \to e^{\kappa\ell'}
$$

and from Equations 12.225 and 12.226 it follows that

$$
\frac{q_{\mathrm{L}}^{\mathrm{eff}}(\kappa)e^{-\kappa r}}{4\pi r} \to \frac{e^{-\kappa\ell}}{2\kappa} \int_{-\infty}^{\infty} d\ell'\, e^{\kappa\ell'} \rho_w^\star(\ell').
$$

The integral converges because $\rho_{\mathrm{w}}^\star(\ell')$ decays to zero faster than $e^{-\kappa\ell'}$ when $\ell' \to \infty$.

We now place the origin of the coordinate system at the wall surface with the z axis in the perpendicular direction out from the wall. Thereby we can replace the distance ℓ by z, so we consider the functions $\psi_{\mathrm{w}}(z)$ and $w_{i;\mathrm{w}}(z)$. It follows from our results above that when $z \to \infty$

$$
\psi_{\mathrm{w}}(z) \sim \frac{\sigma_{\mathrm{w}}^{S\mathrm{eff}}(\kappa)}{\mathcal{E}_r^{\mathrm{eff}}(\kappa)\varepsilon_0} \cdot \frac{e^{-\kappa z}}{\kappa} \quad \text{and} \quad w_{i;\mathrm{w}}(z) \sim \frac{q_i^{\mathrm{eff}}(\kappa)\sigma_{\mathrm{w}}^{S\mathrm{eff}}(\kappa)}{\mathcal{E}_r^{\mathrm{eff}}(\kappa)\varepsilon_0} \cdot \frac{e^{-\kappa z}}{\kappa}, \tag{12.227}
$$

where

$$
\sigma_{\mathrm{w}}^{S\mathrm{eff}}(\kappa) = \frac{1}{2} \int_{-\infty}^{\infty} d\ell'\, e^{\kappa\ell'} \rho_{\mathrm{w}}^\star(\ell') \tag{12.228}
$$

is the **effective surface charge density** of the wall. These results are the same as those we obtained in Volume I for $\psi_{\mathtt{w}}(z)$ in the Poisson-Boltzmann approximation, Equations I:7.31 and I:7.32, and in the general exact case, Equations I:7.34 and I:7.35. They were derived in a different manner in Section 7.1 that we will recapitulate and generalize later.

We conclude that when particle L is grown in order to become the planar wall w, the consequence is that $q_{\mathtt{L}}^{\mathrm{eff}}(\kappa)e^{-\kappa r}/(4\pi r)$ is replaced by $\sigma_{\mathtt{w}}^{\mathcal{S}\mathrm{eff}}(\kappa)e^{-\kappa z}/\kappa$, whereby the Yukawa function decay is replaced by a plain exponential decay. Note that the decay parameter κ is that of *the bulk electrolyte in absence of the* L*-particle*, because the density of such particles is zero, $n_{\mathtt{L}}^{\mathrm{b}} = 0$. The same applies to the other decay parameters like κ'. Recall from Section 10.1.3 that, for example, the OZ equation and the other equations for all species apart from L are those of the bulk fluid with $n_{\mathtt{L}}^{\mathrm{b}} = 0$.

From Equations 12.223 and 12.224 we obtain for large z

$$\begin{aligned}\psi_{\mathtt{w}}(z) &= \frac{\sigma_{\mathtt{w}}^{\mathcal{S}\mathrm{eff}}(\kappa)}{\mathcal{E}_r^{\mathrm{eff}}(\kappa)\varepsilon_0}\cdot\frac{e^{-\kappa z}}{\kappa} + \frac{\sigma_{\mathtt{w}}^{\mathcal{S}\mathrm{eff}}(\kappa')}{\mathcal{E}_r^{\mathrm{eff}}(\kappa')\varepsilon_0}\cdot\frac{e^{-\kappa' z}}{\kappa'} - \ldots \qquad (12.229)\\ w_{i;\mathtt{w}}(z) &= \frac{q_i^{\mathrm{eff}}(\kappa)\sigma_{\mathtt{w}}^{\mathcal{S}\mathrm{eff}}(\kappa)}{\mathcal{E}_r^{\mathrm{eff}}(\kappa)\varepsilon_0}\cdot\frac{e^{-\kappa z}}{\kappa} + \frac{q_i^{\mathrm{eff}}(\kappa')\sigma_{\mathtt{w}}^{\mathcal{S}\mathrm{eff}}(\kappa')}{\kappa'\mathcal{E}_r^{\mathrm{eff}}(\kappa')\varepsilon_0}\cdot\frac{e^{-\kappa' z}}{\kappa'} + \ldots\end{aligned}$$

The density distribution $n_{i;\mathtt{w}}(z) = n_i^{\mathrm{b}}[h_{i\mathtt{w}}^{(2)}(z) + 1]$ of i-ions outside the wall w is given by

$$\frac{n_{i;\mathtt{w}}(z)}{n_i^{\mathrm{b}}} = 1 - \frac{\beta q_i^{\mathrm{eff}}(\kappa)\sigma_{\mathtt{w}}^{\mathcal{S}\mathrm{eff}}(\kappa)}{\mathcal{E}_r^{\mathrm{eff}}(\kappa)\varepsilon_0}\cdot\frac{e^{-\kappa z}}{\kappa} - \frac{\beta q_i^{\mathrm{eff}}(\kappa')\sigma_{\mathtt{w}}^{\mathcal{S}\mathrm{eff}}(\kappa')}{\mathcal{E}_r^{\mathrm{eff}}(\kappa')\varepsilon_0}\cdot\frac{e^{-\kappa' z}}{\kappa} - \ldots,$$

as follows from the expression for $w_{i;\mathtt{w}}(z)$. Since the decay parameters are those of the bulk electrolyte that is in contact with the wall, the decay lengths and, for the oscillatory modes, the wave lengths are the same as in bulk. The wall affects only the amplitudes of $\psi_{\mathtt{w}}$ and $w_{i;\mathtt{w}}$ and, for the oscillatory modes, the phase of the oscillations. The influence occurs via the prefactors $\sigma_{\mathtt{w}}^{\mathcal{S}\mathrm{eff}}$. As before there are short-range terms with other functional forms. In addition, the surface may induce some additional short-range contribution, but only contributions that are commensurate with the decay modes of the bulk survive with increasing z.

Let us now introduce a second wall into the system and for this purpose we will further investigate the potential of mean force $w_{i;\mathtt{w}}(z)$ between an i-ion with its center at z and the planar wall. Since the theory is valid irrespectively of the sizes of the ions, we can introduce a charged particle of another species, $i = \mathtt{L}_2$, and let is grow to form another wall $\mathtt{w}_2$ at, say, distance $D^{\mathcal{S}}$ from the first one as measured between their surfaces. However, it is easier to instead place a planar particle with a large, but finite surface area A at distance $D^{\mathcal{S}}$ from the first wall. We then increase A to infinity, whereby the end result will be the same as when we grow a sphere $\mathtt{L}_2$, namely two infinite planar walls with separation $D^{\mathcal{S}}$. The boundary effects at the edges of the second planar wall do not matter in the limit $\mathsf{A} \to \infty$. The wall $\mathtt{w}_2$ has a uniform surface charge density $\sigma^S_{\mathtt{w}_2}$ and, in general, a different short-range non-electrostatic interaction with the other ions than the other one. From now on, the first wall will be called $\mathtt{w}_1$ instead of $\mathtt{w}$.

In order to deal with the second wall, let us first see what happens with the factor q_i^{eff} in $w_{i;\mathtt{w}_1}$ for $i = \mathrm{L}_2$ when an L_2-particle becomes large. From this we can figure out how to treat the planar wall with area A. We can get a clue from the fact that the expressions for $\psi_{\mathtt{w}_1}(z)$ and $w_{i;\mathtt{w}_1}(z)$ differ only by the factor q_i^{eff} in each term. If the term for each decay mode in $\psi_{\mathtt{w}_1}(z)$ is multiplied by q_i^{eff} we obtain the corresponding term in $w_{i;\mathtt{w}_1}(z)$, which is the interaction free energy of the i-ion with wall $\mathtt{w}_1$. In fact, the factor q_i^{eff} for each mode can be obtained from the interaction free energy between $\psi_{\mathtt{w}_1}(z)$ from the wall and the dressed-ion charge density $\rho_i^\star$ for an ion placed at z, that is, $\int d\mathbf{r}'\, \psi_{\mathtt{w}_1}(z+z')\rho_i^\star(r')$. For the κ-mode in $\psi_{\mathtt{w}_1}(z)$ (the first term in Equation 12.229) we get the integral

$$\int d\mathbf{r}'\, e^{-\kappa(z+z')}\rho_i^\star(r') = e^{-\kappa z}\int d\mathbf{r}'\, \frac{\sinh(\kappa r')}{\kappa r'}\rho_i^\star(r') = e^{-\kappa z} q_i^{\mathrm{eff}}(\kappa),$$

where we have used Equation 12.162. This shows that the factor $q_i^{\mathrm{eff}}(\kappa)$ is obtained as asserted. For a huge sphere of species $i = \mathrm{L}_2$ the charge distribution $\rho_i^\star(r')$ near each point on the surface has virtually planar geometry. Thus, for the planar wall we will investigate the same integral, but with the dressed charge distribution of a planar wall inserted instead of that of the large sphere of species L_2.

Let us now treat a planar wall $\mathtt{w}_2$ with dressed charge distribution $\rho^\star_{\mathtt{w}_2}(-z')$ and area A, which we will let go to infinity. There is a minus sign in the argument of $\rho^\star_{\mathtt{w}_2}(-z')$ because wall $\mathtt{w}_2$ is facing wall $\mathtt{w}_1$ so it has the opposite orientation (its normal points in the negative z direction). For a planar wall we consider the free energy *per unit area*. From the first term of $\psi_{\mathtt{w}_1}(z)$ in Equation 12.229 we get the integral

$$\lim_{\mathsf{A}\to\infty}\frac{1}{\mathsf{A}}\int d\mathbf{r}'\, e^{-\kappa(z+z')}\rho^\star_{\mathtt{w}_2}(-z') = e^{-\kappa z}\int_{-\infty}^{\infty} dz'\, e^{-\kappa z'}\rho^\star_{\mathtt{w}_2}(-z') = 2e^{-\kappa z}\sigma^{S\mathrm{eff}}_{\mathtt{w}_2}(\kappa).$$

Thus, instead of the factor q_i^{eff} in $w_{i;\mathtt{w}_1}$ for $i = \mathrm{L}_2$ we have the factor $2\sigma^{S\mathrm{eff}}_{\mathtt{w}_2}$ in the **free energy of surface interaction per unit area** (potential of mean force per unit area) $\mathscr{W}^S_{\mathtt{w}_1,\mathtt{w}_2}$ between the walls $\mathtt{w}_1$ and $\mathtt{w}_2$. When the two walls are at distance D^S from each other we therefore have

$$\mathscr{W}^S_{\mathtt{w}_1,\mathtt{w}_2}(D^S) = \frac{2\sigma^{S\mathrm{eff}}_{\mathtt{w}_1}(\kappa)\sigma^{S\mathrm{eff}}_{\mathtt{w}_2}(\kappa)}{\mathcal{E}_r^{\mathrm{eff}}(\kappa)\varepsilon_0}\cdot\frac{e^{-\kappa D^S}}{\kappa} + \frac{2\sigma^{S\mathrm{eff}}_{\mathtt{w}_1}(\kappa')\sigma^{S\mathrm{eff}}_{\mathtt{w}_2}(\kappa')}{\mathcal{E}_r^{\mathrm{eff}}(\kappa')\varepsilon_0}\cdot\frac{e^{-\kappa' D^S}}{\kappa'} + \dots \tag{12.230}$$

for sufficiently large D^S (cf. Equation I:7.43). The electrolyte between the walls is in equilibrium with a bulk electrolyte with the composition of the original one (with $n^{\rm b}_{\rm L_1} = n^{\rm b}_{\rm L_2} = 0$), which has the decay parameters κ, κ' etc. It follows that the *decay modes of the surface interactions in electrolytes are governed by the zeros of* $\tilde{\varepsilon}(k)$ *for the bulk phase* that is in equilibrium with the system. Again there are terms with other functional forms that are more short-range than the first one and there may be some additional short-range contribution induced by the surfaces. Only contributions that are commensurate with the decay modes of the bulk survive with increasing $D^{\mathcal{S}}$.

In Equation 12.230 the influence of each wall occurs only via the prefactors $\sigma^{\mathcal{S}\rm eff}_{\rm w_\nu}(\kappa)$, $\nu = 1, 2$. Everything else is determined by the bulk phase. This means that the *amplitude* of each decay mode of $\mathscr{W}^{\mathcal{S}}_{\rm w_1,w_2}$ depends on the walls (for oscillatory modes the phase of the oscillations are also affected by the walls), but nothing else. For some surfaces, one decay mode can have larger amplitude than the other modes, while for other surfaces immersed in the same electrolyte some other mode can have the largest amplitude. All this depends on the relative magnitudes of the prefactors $\sigma^{\mathcal{S}\rm eff}_{\rm w_\nu}(\kappa)$. When the surface separation is increased, the mode with the longest decay length will eventually dominate even if it has a relatively small amplitude compared to the other ones, because the latter modes decay to zero more rapidly with distance.

For highly charged surfaces, it is likely that the modes with large contributions in $H_{\rm QQ}(r)$ in bulk have larger amplitude in $\mathscr{W}^{\mathcal{S}}_{\rm w_1,w_2}$ than the modes with large contributions only in $H_{\rm NN}(r)$. For latter kind of mode we have seen that in borderline binary electrolytes the corresponding terms in h_{++} and h_{+-} decay to zero with the same sign (cf. the curves for set A in Figure 12.7). Therefore, their contribution to the charge density $n^{\rm b}_{\rm tot} q[h_{++} - h_{+-}]/2$ carries little charge and interacts less with the surface charges than those from the former modes, which carries much more charge because h_{++} and h_{+-} decay to zero with opposite signs (cf. set B in Figure 12.7). It is likely that larger charge-charge correlations give a stronger coupling with the surface charges leading to a larger amplitude.

So far we have obtained results for electrolytes in contact with smooth surfaces with uniform surface-charge densities. We will now generalize the treatment of inhomogeneous electrolytes in planar geometry and thereby include the case of planar walls with an *internal* charge distribution $\sigma_{\rm w}(\mathbf{r})$ that may vary in the lateral direction along the surface. The short-range non-electrostatic interaction potential between the ions and the wall may also vary in this direction. Systems with uniform surface charge densities constitute the special case where $\sigma_{\rm w}(\mathbf{r}) = \sigma^{\mathcal{S}}_{\rm w}\delta(z)$. We will do the derivations using a different approach than previously in the current section; it is a generalization of the approach used in Chapter 7 of Volume I. As before we have the coordinate system with the origin at the wall surface with the z axis in the perpendicular direction out from the wall and $\sigma(\mathbf{r}) = 0$ for $z > 0$.

The charge distribution $\rho_{\rm w}(\mathbf{r})$ of the electrolyte outside a wall w and the associated total charge distribution $\rho^{\rm tot}_{\rm w}(\mathbf{r})$ are

$$\rho_{\rm w}(\mathbf{r}) = \sum_j q_j n_{j;\rm w}(\mathbf{r}) \quad \text{and} \quad \rho^{\rm tot}_{\rm w}(\mathbf{r}) = \sigma_{\rm w}(\mathbf{r}) + \rho_{\rm w}(\mathbf{r}).$$

The dressed charge distribution $\rho^\star_{\mathrm{w}}(\mathbf{r})$ of the wall is given by

$$\rho^\star_{\mathrm{w}}(\mathbf{r}) = \rho^{\mathrm{tot}}_{\mathrm{w}}(\mathbf{r}) - \int d\mathbf{r}'\, \psi_{\mathrm{w}}(\mathbf{r}')\chi^\star(|\mathbf{r}-\mathbf{r}'|),$$

in analogy to the case of a particle immersed in the electrolyte (Equation 12.87). Here, $\chi^\star(r)$ is the polarization response function of the bulk electrolyte that is in contact with the wall surface. The mean potential ψ_{w} can be obtained from the screened Coulomb potential formula

$$\psi_{\mathrm{w}}(\mathbf{r}) = \int d\mathbf{r}'\, \rho^\star_{\mathrm{w}}(\mathbf{r}')\phi^\star_{\mathrm{Coul}}(|\mathbf{r}-\mathbf{r}'|)$$

(cf. Equation 12.83), where $\phi^\star_{\mathrm{Coul}}(r)$ is the unit screened Coulomb potential for the bulk phase. Recall that

$$\phi^\star_{\mathrm{Coul}}(r) = \frac{1}{\mathcal{E}^{\mathrm{eff}}_r(\kappa)\varepsilon_0}\cdot\frac{e^{-\kappa r}}{4\pi r} + \frac{1}{\mathcal{E}^{\mathrm{eff}}_r(\kappa')\varepsilon_0}\cdot\frac{e^{-\kappa' r}}{4\pi r} + \ldots \quad \text{(large } r\text{)}$$

(Equation 12.171), where only the two first Yukawa function contributions are shown.

Let us first treat the case with a uniform surface charge density, $\sigma_{\mathrm{w}}(\mathbf{r}) = \sigma^S_{\mathrm{w}}\delta(z)$, whereby $\psi_{\mathrm{w}}(z)$ and $\rho^\star_{\mathrm{w}}(z)$ depend on z only. When κ is real and $k = i\kappa$ is the singularity of $\tilde{\phi}^\star_{\mathrm{Coul}}(k)$ closest to the real axis in complex k space, we have

$$\begin{aligned}\psi_{\mathrm{w}}(z) &\sim \frac{1}{\mathcal{E}^{\mathrm{eff}}_r(\kappa)\varepsilon_0}\int d\mathbf{r}'\, \rho^\star_{\mathrm{w}}(z')\frac{e^{-\kappa|\mathbf{r}-\mathbf{r}'|}}{4\pi|\mathbf{r}-\mathbf{r}'|} \quad \text{when } z\to\infty \\ &= \frac{1}{\mathcal{E}^{\mathrm{eff}}_r(\kappa)\varepsilon_0}\int_{-\infty}^{\infty} dz'\rho^\star_{\mathrm{w}}(z')\frac{e^{-\kappa|z-z'|}}{2\kappa},\end{aligned} \tag{12.231}$$

where the second line is obtained after some elementary steps.[74] Since $\rho^\star_{\mathrm{w}}(z')$ decays to zero faster than $e^{-\kappa z'}$ when z' increases, we can take $z' < z$ and in the limit $z\to\infty$ we obtain

$$\psi_{\mathrm{w}}(z) \sim \frac{e^{-\kappa z}}{\kappa\mathcal{E}^{\mathrm{eff}}_r(\kappa)\varepsilon_0}\cdot\frac{1}{2}\int_{-\infty}^{\infty} dz'\rho^\star_{\mathrm{w}}(z')e^{\kappa z'} = \frac{\sigma^{S\mathrm{eff}}_{\mathrm{w}}(\kappa)}{\mathcal{E}^{\mathrm{eff}}_r(\kappa)\varepsilon_0}\cdot\frac{e^{-\kappa z}}{\kappa}.$$

This result is the same as before.

---

[74]The second line in Equation 12.231 can be obtained as follows. We have

$$\int d\mathbf{r}'\, \rho^\star_{\mathrm{w}}(z')\frac{e^{-\kappa|\mathbf{r}-\mathbf{r}'|}}{4\pi|\mathbf{r}-\mathbf{r}'|} = \frac{1}{4\pi}\int_{-\infty}^{\infty} dz'\rho^\star_{\mathrm{w}}(z')\int d\mathbf{s}\,\frac{e^{-\kappa[s^2+(z-z')^2]^{1/2}}}{[s^2+(z-z')^2]^{1/2}},$$

where $\mathbf{s} = (x-x', y-y')$, $s = |\mathbf{s}|$ and $d\mathbf{s}$ can be replaced by $2\pi s\, ds$. We obtain the result in Equation 12.231 by using the identity

$$\int_0^{\infty} ds\, s\,\frac{e^{-\kappa[s^2+a^2]^{1/2}}}{[s^2+a^2]^{1/2}} = \frac{e^{-\kappa|a|}}{\kappa}$$

that can easily be shown by making the variable substitution $s^2+a^2=t^2$ in the integral.

Next, we proceed with the cases with a *nonuniform surface charge density* or an internal charge distribution $\sigma_{\mathrm{w}}(\mathbf{r})$ of the wall that depends also on x and y. The short-range non-electrostatic interaction potential between the ions and the wall also depend on x and y in general. The same is the case for $\rho_{\mathrm{w}}^*(\mathbf{r})$ and $\psi_{\mathrm{w}}(\mathbf{r})$ and we have

$$\psi_{\mathrm{w}}(\mathbf{r}) \sim \frac{1}{\mathcal{E}_r^{\mathrm{eff}}(\kappa)\varepsilon_0} \int d\mathbf{r}' \rho_{\mathrm{w}}^*(\mathbf{r}') \frac{e^{-\kappa|\mathbf{r}-\mathbf{r}'|}}{4\pi|\mathbf{r}-\mathbf{r}'|} \quad \text{when } z \to \infty. \tag{12.232}$$

In the shaded text box on the next page it is shown that the x and y dependences of $\psi_{\mathrm{w}}(\mathbf{r})$ vanish when $z \to \infty$ more quickly than the z dependence, which still goes like $\exp(-\kappa z)$. Thus, $\psi_{\mathrm{w}}(\mathbf{r})$ ultimately depends only on z for large z. This means that the potential still decays like

$$\psi_{\mathrm{w}}(\mathbf{r}) \sim \frac{\sigma_{\mathrm{w}}^{\mathcal{S}\mathrm{eff}}(\kappa)}{\mathcal{E}_r^{\mathrm{eff}}(\kappa)\varepsilon_0} \cdot \frac{e^{-\kappa z}}{\kappa} \quad \text{when } z \to \infty, \tag{12.233}$$

where in this case the effective surface charge density is given by

$$\sigma_{\mathrm{w}}^{\mathcal{S}\mathrm{eff}}(\kappa) = \lim_{\mathsf{A}\to\infty} \frac{1}{2\mathsf{A}} \int d\mathbf{r}' \rho_{\mathrm{w}}^*(\mathbf{r}') e^{\kappa z'}. \tag{12.234}$$

(this follows from Equation 12.248).

The entire contribution to $\psi_{\mathrm{w}}(\mathbf{r})$ from the κ mode is

$$\psi_{\mathrm{w}}(\mathbf{r}) \sim \frac{\sigma_{\mathrm{w}}^{\mathcal{S}\mathrm{eff}}(\kappa) + \mathscr{S}_{\mathrm{wR}}(\mathbf{r},\kappa)}{\mathcal{E}_r^{\mathrm{eff}}(\kappa)\varepsilon_0} \cdot \frac{e^{-\kappa z}}{\kappa} \quad \text{when } z \to \infty, \tag{12.235}$$

where $\mathscr{S}_{\mathrm{wR}}(\mathbf{r},\kappa)$ is defined in Equation 12.249. The x and y dependence is carried by the function $\mathscr{S}_{\mathrm{wR}}(\mathbf{r},\kappa)$, which goes to zero when $z \to \infty$, so this term in the numerator gradually vanishes when z is increased. This means that the x and y dependence of $\psi_{\mathrm{w}}(\mathbf{r})$ fades away when z increases. In the shaded text box below we analyze the x and y dependence of $\mathscr{S}_{\mathrm{wR}}(\mathbf{r},\kappa)$ in more detail for the case of a periodic structure of $\sigma_{\mathrm{w}}(\mathbf{r})$, for instance when the wall consists of a crystal. The leading contribution to $e^{-\kappa z}\mathscr{S}_{\mathrm{wR}}(\mathbf{r},\kappa)$ then decreases like $\exp(-[\kappa^2 + \alpha^2]^{1/2} z)$ when $z \to \infty$, where α is determined by the periodicity of the structure.[75]

Likewise, when κ is real and $k = i\kappa$ is the singularity of $\tilde{w}_{ij}(k)$ closest to the real axis, the potential of mean force for an i-ion outside the wall w decays like

$$w_{i;\mathrm{w}}(\mathbf{r}) \sim \frac{q_i^{\mathrm{eff}}(\kappa)\left[\sigma_{\mathrm{w}}^{\mathcal{S}\mathrm{eff}}(\kappa) + \mathscr{S}_{\mathrm{wR}}(\mathbf{r},\kappa)\right]}{\mathcal{E}_r^{\mathrm{eff}}(\kappa)\varepsilon_0} \cdot \frac{e^{-\kappa z}}{\kappa} \quad \text{when } z \to \infty$$

[75]The value of α is determined by the longest wave length of the two-dimensional structure (the reciprocal lattice point that lies closest to the origin), see the discussion in connection to Equation 12.242.

(cf. Equation 12.229). For complex κ one can do the corresponding analysis for the contribution with decay parameter $\kappa' = \underline{\kappa}$ and add the two terms in order to obtain the exponentially decaying oscillatory term in $\psi_{\mathrm{w}}(\mathbf{r})$ and $w_{i;\mathrm{w}}(\mathbf{r})$. This will not be done here.

Electrostatic potential from a wall with a nonuniform surface charge density and/or any internal charge distribution in contact with an electrolyte.

We start by calculating the screened electrostatic potential from solely the internal charge distribution $\sigma_{\mathrm{w}}(\mathbf{r})$ of the wall, more precisely the contribution from the κ mode of the screened potential

$$\psi_\sigma(\mathbf{r}) = \frac{1}{\mathcal{E}_r^{\mathrm{eff}}\varepsilon_0}\int d\mathbf{r}'\,\sigma_{\mathrm{w}}(\mathbf{r}')\frac{e^{-\kappa|\mathbf{r}-\mathbf{r}'|}}{4\pi|\mathbf{r}-\mathbf{r}'|}.$$

Since the rhs is a convolution it can be written in Fourier space as a product and we obtain

$$\psi_\sigma(\mathbf{r}) = \frac{1}{\mathcal{E}_r^{\mathrm{eff}}\varepsilon_0(2\pi)^3}\int d\mathbf{k}\frac{\tilde{\sigma}_{\mathrm{w}}(\mathbf{k})}{\kappa^2+k^2}e^{i\mathbf{k}\cdot\mathbf{r}}, \tag{12.236}$$

where $1/(\kappa^2+k^2)$ is the Fourier transform of $\exp(-\kappa r)/(4\pi r)$ and we have made an inverse Fourier transform (defined in Equation 9.243). Alternatively, this expression can be obtained from the inverse Fourier transform of $\tilde{\psi}_\sigma(\mathbf{k}) = \tilde{\sigma}_{\mathrm{w}}(\mathbf{k})\tilde{\phi}^*_{\mathrm{Coul}}(k)$ with the first term of Equation 12.175 inserted instead of $\tilde{\phi}^*_{\mathrm{Coul}}(k)$.

Consider first the special case of the potential from a nonuniform surface charge density located at $z=0$, namely $\sigma_{\mathrm{w}}(\mathbf{r}) = \delta(z)\sigma^S(x,y) = \delta(z)\sigma^S(\mathbf{r}_{xy})$ where $\mathbf{r}_{xy} = (x,y)$. We will denote the resulting potential as $\psi_S(\mathbf{r})$. In this case $\tilde{\sigma}_{\mathrm{w}}(\mathbf{k}) = \underset{\sim}{\sigma}^S(\mathbf{k}_{xy})$, where $\mathbf{k}_{xy} = (k_x,k_y)$ and $\underset{\sim}{\sigma}^S(\mathbf{k}_{xy})$ is the two-dimensional Fourier transform of $\sigma^S(\mathbf{r}_{xy})$. We will split σ^S in two parts

$$\sigma^S(\mathbf{r}_{xy}) = \sigma^S_{\mathrm{Av}} + \sigma^S_{\mathrm{R}}(\mathbf{r}_{xy}), \tag{12.237}$$

where σ^S_{Av} is the average surface charge density defined as

$$\sigma^S_{\mathrm{Av}} = \lim_{\mathsf{A}\to\infty}\mathsf{A}^{-1}\int_{\mathsf{A}} d\mathbf{r}_{xy}\,\sigma^S(\mathbf{r}_{xy})$$

and $\sigma^S_{\mathrm{R}}(\mathbf{r}_{xy})$ is the rest of surface charge distribution that is net electroneutral. Since σ^S_{Av} is a uniform surface charge density we have in Fourier space

$$\underset{\sim}{\sigma}^S(\mathbf{k}_{xy}) = \sigma^S_{\mathrm{Av}}4\pi^2\delta^{(2)}(\mathbf{k}_{xy}) + \underset{\sim}{\sigma}^S_{\mathrm{R}}(\mathbf{k}_{xy}), \tag{12.238}$$

where $\delta^{(2)}$ is the two-dimensional Dirac function. Note that $\underset{\sim}{\sigma}^S_{\mathrm{R}}(\mathbf{0}) = 0$ since σ^S_{R} is electroneutral.

By inserting $\widetilde{\sigma}_w(\mathbf{k}) = \underset{\sim}{\sigma}^S(\mathbf{k}_{xy})$ into Equation (12.236), we obtain

$$\begin{aligned}\psi_S(\mathbf{r}) &= \frac{1}{\mathcal{E}_r^{\rm eff}\varepsilon_0(2\pi)^3}\int d\mathbf{k}_{xy}\int_{-\infty}^{\infty} dk_z \frac{\underset{\sim}{\sigma}^S(\mathbf{k}_{xy})}{\kappa^2 + k_{xy}^2 + k_z^2} e^{i(\mathbf{k}_{xy}\cdot\mathbf{r}_{xy}+k_z z)} \\ &= \frac{1}{\mathcal{E}_r^{\rm eff}\varepsilon_0 8\pi^2}\int d\mathbf{k}_{xy}\,\underset{\sim}{\sigma}^S(\mathbf{k}_{xy})\frac{e^{-z\,[\kappa^2+k_{xy}^2]^{1/2}}}{[\kappa^2 + k_{xy}^2]^{1/2}} e^{i\mathbf{k}_{xy}\cdot\mathbf{r}_{xy}}, \qquad (12.239)\end{aligned}$$

where $k_{xy} = |\mathbf{k}_{xy}|$ and the last equality follows from the following inverse one-dimensional FT

$$\frac{1}{\pi}\int_{-\infty}^{\infty} dk_z \frac{e^{ik_z z}}{a^2 + k_z^2} = \frac{e^{-a|z|}}{a}$$

with $a > 0$, which we apply for the case $a^2 = \kappa^2 + k_{xy}^2$ and with $z > 0$. By breaking out a factor $\exp(-\kappa z)/\kappa$ to the outside of the integral we obtain from Equation 12.239

$$\psi_S(\mathbf{r}) = \frac{1}{\mathcal{E}_r^{\rm eff}\varepsilon_0 8\pi^2}\cdot\frac{e^{-\kappa z}}{\kappa}\int d\mathbf{k}_{xy}\,\underset{\sim}{\sigma}^S(\mathbf{k}_{xy})\mathcal{F}(z, k_{xy})e^{i\mathbf{k}_{xy}\cdot\mathbf{r}_{xy}},$$

where

$$\mathcal{F}(z, k_{xy}) = \frac{\exp\left(-\kappa z\,\left\{[1 + k_{xy}^2/\kappa^2]^{1/2} - 1\right\}\right)}{[1 + k_{xy}^2/\kappa^2]^{1/2}}.$$

Let us now consider the contributions to $\psi_S(\mathbf{r})$ from the two terms of $\underset{\sim}{\sigma}^S(\mathbf{k}_{xy})$ in Equation 12.238. From the first one we obtain

$$\begin{aligned}\psi_{S_{\rm Av}}(\mathbf{r}) &\equiv \frac{\sigma_{\rm Av}^S}{\mathcal{E}_r^{\rm eff}\varepsilon_0 8\pi^2}\cdot\frac{e^{-\kappa z}}{\kappa}\int d\mathbf{k}_{xy}\,4\pi^2\delta^{(2)}(\mathbf{k}_{xy})\mathcal{F}(z, k_{xy})e^{i\mathbf{k}_{xy}\cdot\mathbf{r}_{xy}} \\ &= \frac{\sigma_{\rm Av}^S}{2\mathcal{E}_r^{\rm eff}\varepsilon_0}\cdot\frac{e^{-\kappa z}}{\kappa}, \qquad (12.240)\end{aligned}$$

because $F(z, 0) = 1$, and from the second

$$\psi_{S_{\rm R}}(\mathbf{r}) \equiv \frac{1}{\mathcal{E}_r^{\rm eff}\varepsilon_0 8\pi^2}\cdot\frac{e^{-\kappa z}}{\kappa}\int d\mathbf{k}_{xy}\,\underset{\sim}{\sigma}_{\rm R}^S(\mathbf{k}_{xy})\mathcal{F}(z, k_{xy})e^{i\mathbf{k}_{xy}\cdot\mathbf{r}_{xy}}. \qquad (12.241)$$

Note that the x and y dependence of $\psi_S(\mathbf{r})$ originates solely from $\psi_{S_{\rm R}}(\mathbf{r})$. When $k_{xy} \neq 0$ we have $F(z, k_{xy}) \to 0$ exponentially fast when $z \to \infty$. Since $\underset{\sim}{\sigma}_{\rm R}^S(\mathbf{0}) = 0$, the integral in Equation 12.241 goes to zero in this limit, which means that $\psi_{S_{\rm R}}(\mathbf{r})$ goes to zero faster than $\psi_{S_{\rm Av}}(\mathbf{r})$, so the x and y dependence vanishes for very large z where $\psi_{S_{\rm Av}}(\mathbf{r})$ dominates.

As an example, take a surface charge density $\sigma^S(x, y)$ that is periodic, for instance a lattice of charged particles at the wall like a two-dimensional crystal. Then, $\tilde{\sigma}^S_{\rm R}(\mathbf{k}_{xy})$ is equal to an infinite sum of Dirac delta functions located at the lattice points of the reciprocal lattice of this structure and

$$\int d\mathbf{k}_{xy}\, \tilde{\sigma}^S_{\rm R}(\mathbf{k}_{xy})\mathcal{F}(z,k_{xy})e^{\mathrm{i}\mathbf{k}_{xy}\cdot\mathbf{r}_{xy}} = \sum_{\substack{\text{reciprocal}\\ \text{lattice points } \boldsymbol{\tau}_{xy}}} a(\boldsymbol{\tau}_{xy})\mathcal{F}(z,|\boldsymbol{\tau}_{xy}|)e^{\mathrm{i}\boldsymbol{\tau}_{xy}\cdot\mathbf{r}_{xy}}, \tag{12.242}$$

where $a(\boldsymbol{\tau}_{xy})$ is the coefficient for the delta function contribution in $\tilde{\sigma}^S_{\rm R}$ and no contribution from $\mathbf{k}_{xy} = 0$ is present because $\tilde{\sigma}^S_{\rm R}(\mathbf{0}) = 0$. The rhs is a two-dimensional Fourier series in the variables (x, y) and the wave lengths of its terms are determined by $\boldsymbol{\tau}_{xy}$.

Since $e^{-\kappa z}\mathcal{F}(z, |\boldsymbol{\tau}_{xy}|)$ is proportional to $e^{-z\,[\kappa^2+|\boldsymbol{\tau}_{xy}|^2]^{1/2}}$ the leading decay of the function $\psi_{S_{\rm R}}(\mathbf{r})$ as function of z is determined by the smallest value of $|\boldsymbol{\tau}_{xy}|$, that is, from the lattice point(s) closest to the origin in the reciprocal lattice. The leading term decays like $e^{-z\,[\kappa^2+\alpha^2]^{1/2}}$, where α is the smallest $|\boldsymbol{\tau}_{xy}|$. Thus the corresponding decay length is only a fraction of $1/\kappa$, so the x and y dependence vanishes quite quickly when z is increased. The entire x and y dependence of $\psi_{S_{\rm R}}(\mathbf{r})$ is given by the sum over all reciprocal lattice points and the contributions from the points far from the origin (large $|\boldsymbol{\tau}_{xy}|$) fade away more rapidly when z increases than from those closer to it. Since a large $|\boldsymbol{\tau}_{xy}|$ means a small spatial wave length in the x, y space for the terms in the Fourier series, the sharper features of the x and y dependence of $\psi_{S_{\rm R}}(\mathbf{r})$ vanishes first, so the potential becomes more and more smoothed out in the x and y directions when z increases.

For a non-periodic surface charge distribution $\sigma^S(x, y)$ there is not a distinct set of decay lengths in the z-dependence of $\psi_{S_{\rm R}}(\mathbf{r})$, but instead a continuum of them and all are smaller than $1/\kappa$. For small z-values the contributions to $\psi_{S_{\rm R}}(\mathbf{r})$ with large values of $\tilde{\sigma}^S_{\rm R}(\mathbf{k}_{xy})$ will dominate, but if they have a large $|\mathbf{k}_{xy}|$ they will vanish quickly with increased z and other contributions with smaller $|\mathbf{k}_{xy}|$ will dominate instead.

Next, we will calculate $\psi_\sigma(\mathbf{r})$ from entire internal charge-distribution $\sigma_{\rm w}(\mathbf{r}')$ of the wall. For this task we can use the results in Equations 12.240 and 12.241 because $\sigma_{\rm w}(\mathbf{r}')$ inside the wall can be regarded as a set of two-dimensional distributions with different z' coordinates, whereby the two-dimensional charge density in the interval $(z', z' + dz')$ is equal to $\sigma_w(\mathbf{r}'_{xy}, z')dz'$ for an infinitesimally small dz'. The two-dimensional Fourier transform of this distribution is $\tilde{\sigma}_{\rm w}(\mathbf{k}_{xy}, z')dz'$. Like we did for σ^S in Equation 12.238, we write

$$\tilde{\sigma}_{\rm w}(\mathbf{k}_{xy}, z') = \sigma_{\rm wAv}(z')4\pi^2\delta^{(2)}(\mathbf{k}_{xy}) + \tilde{\sigma}_{\rm wR}(\mathbf{k}_{xy}, z'), \tag{12.243}$$

where

$$\sigma_{\rm wAv}(z') \equiv \lim_{\mathsf{A}\to\infty} \mathsf{A}^{-1} \int\limits_{\mathsf{A}} d\mathbf{r}'_{xy}\, \sigma_{\rm w}(\mathbf{r}'_{xy}, z') \tag{12.244}$$

($\sigma_{\text{wAv}}(z')dz'$ is the average two-dimensional density) and ϱ_{wR} is the remaining part of ϱ_{w}.

The electrostatic potential from the charge distribution $\sigma_{\text{w}}(\mathbf{r}') = \sigma_{\text{w}}(\mathbf{r}'_{xy}, z')$ in the plane at z' is obtained from Equations 12.240 and 12.241 by replacing $\sigma^{\mathcal{S}}_{\text{Av}}$ and $\varrho^{\mathcal{S}}_{\text{R}}(\mathbf{k}_{xy})$ by $\sigma_{\text{wAv}}(z')dz'$ and $\varrho_{\text{wR}}(\mathbf{k}_{xy}, z')dz'$, respectively, and replacing z by $z - z'$ in the formulas. The total $\psi_\sigma(\mathbf{r})$ is then obtained by summing over all $z' < 0$, whereby we get the part $\psi_{\sigma_{\text{Av}}}$ from the average σ_{wAv}

$$\psi_{\sigma_{\text{Av}}}(\mathbf{r}) \equiv \int_{-\infty}^{0} dz' \, \frac{\sigma_{\text{wAv}}(z')}{2\mathcal{E}_r^{\text{eff}}\varepsilon_0} \cdot \frac{e^{-\kappa(z-z')}}{\kappa}.$$

We can write this as

$$\psi_{\sigma_{\text{Av}}}(z) = \frac{e^{-\kappa z}}{2\mathcal{E}_r^{\text{eff}}\varepsilon_0 \kappa} \int_{-\infty}^{0} dz' \, \sigma_{\text{wAv}}(z') e^{\kappa z'} = \frac{\sigma^{\mathcal{S}\text{tot}}_{\text{wAv}}}{\mathcal{E}_r^{\text{eff}}\varepsilon_0} \cdot \frac{e^{-\kappa z}}{\kappa} \tag{12.245}$$

where

$$\sigma^{\mathcal{S}\text{tot}}_{\text{wAv}} = \frac{1}{2} \int_{-\infty}^{0} dz' \, \sigma^{\mathcal{S}}_{\text{wAv}}(z') e^{\kappa z'} \tag{12.246}$$

and we have explicitly shown that $\psi_{\sigma_{\text{Av}}}$ only depends on z. Likewise, we obtain the part of $\psi_\sigma(\mathbf{r})$ from ϱ_{wR}

$$\begin{aligned}\psi_{\sigma\text{R}}(\mathbf{r}) \quad &\equiv \quad \frac{1}{\mathcal{E}_r^{\text{eff}}\varepsilon_0 8\pi^2} \cdot \int_{-\infty}^{0} dz' \, \frac{e^{-\kappa(z-z')}}{\kappa} \\ &\times \quad \int d\mathbf{k}_{xy} \, \varrho_{\text{wR}}(\mathbf{k}_{xy}, z') \mathcal{F}(z - z', k_{xy}) e^{i\mathbf{k}_{xy}\cdot\mathbf{r}_{xy}}\end{aligned}$$

(cf. Equation 12.241), which we can write as

$$\psi_{\sigma\text{R}}(\mathbf{r}) = \frac{\mathscr{S}_{\sigma\text{R}}(\mathbf{r}, \kappa)}{\mathcal{E}_r^{\text{eff}}\varepsilon_0} \cdot \frac{e^{-\kappa z}}{\kappa},$$

where

$$\mathscr{S}_{\sigma\text{R}}(\mathbf{r}, \kappa) = \frac{1}{8\pi^2} \int_{-\infty}^{0} dz' \, e^{\kappa z'} \int d\mathbf{k}_{xy} \, \varrho_{\text{wR}}(\mathbf{k}_{xy}, z') \mathcal{F}(z - z', k_{xy}) e^{i\mathbf{k}_{xy}\cdot\mathbf{r}_{xy}}. \tag{12.247}$$

In total we have $\psi_\sigma(\mathbf{r}) = \psi_{\sigma_{\text{Av}}}(z) + \psi_{\sigma\text{R}}(\mathbf{r})$. For the same reasons as before, $\psi_{\sigma\text{R}}(\mathbf{r})$ decays to zero faster than $e^{-\kappa z}$ when $z \to \infty$ and we have $\psi_\sigma(\mathbf{r}) \sim \psi_{\sigma_{\text{Av}}}(z)$.

The decay behavior of the potential $\psi_{\rm w}(\mathbf{r})$ as calculated from Equation 12.232 can be obtained in the same manner because $\rho^*_{\rm w}(\mathbf{r}')$ can also be regarded as a set of two-dimensional distributions with different z' coordinates. In this case, however, the charge density is not limited to $z' < 0$, but when the density $\rho^*_{\rm w}(\mathbf{r}')$ decays to zero faster than $e^{-\kappa z'}$ when z' increases we can obtain the leading term in $\psi_{\rm w}(\mathbf{r})$ as before. Thus, in analogy to Equation 12.245 together with Equations 12.244 and 12.246 we get

$$\psi_{\rm w}(\mathbf{r}) \sim \frac{\sigma_{\rm w}^{\mathcal{S}\rm eff}(\kappa)}{\mathcal{E}_r^{\rm eff}(\kappa)\varepsilon_0} \cdot \frac{e^{-\kappa z}}{\kappa} \qquad \text{when } z \to \infty,$$

where

$$\sigma_{\rm w}^{\mathcal{S}\rm eff}(\kappa) = \frac{1}{2} \int\limits_{-\infty}^{\infty} dz'\, \rho^*_{\rm wAv}(z') e^{\kappa z'} \tag{12.248}$$

and

$$\rho^*_{\rm wAv}(z') \equiv \lim_{\mathsf{A}\to\infty} \mathsf{A}^{-1} \int\limits_{\mathsf{A}} d\mathbf{r}'_{xy}\, \rho^*_{\rm w}(\mathbf{r}').$$

Together, the last two equations give the same $\sigma_{\rm w}^{\mathcal{S}\rm eff}(\kappa)$ as in Equation 12.234.

The x and y-dependent part of $\psi_{\rm w}(\mathbf{r})$ from the κ mode is

$$\psi_{\rm wR}(\mathbf{r},\kappa) = \frac{\mathscr{S}_{\rm wR}(\mathbf{r},\kappa)}{\mathcal{E}_r^{\rm eff}(\kappa)\varepsilon_0} \cdot \frac{e^{-\kappa z}}{\kappa},$$

where (cf. Equation 12.247)

$$\mathscr{S}_{\rm wR}(\mathbf{r},\kappa) = \frac{1}{8\pi^2} \int\limits_{-\infty}^{\infty} dz'\, e^{\kappa z'} \int d\mathbf{k}_{xy}\, \underset{\sim}{\rho}^*_{\rm wR}(\mathbf{k}_{xy}, z') \mathcal{F}(z - z', k_{xy}) e^{\mathrm{i}\mathbf{k}_{xy}\cdot\mathbf{r}_{xy}} \tag{12.249}$$

and (cf. Equation 12.243)

$$\underset{\sim}{\rho}^*_{\rm wR}(\mathbf{k}_{xy}, z') \equiv \underset{\sim}{\rho}^*_{\rm w}(\mathbf{k}_{xy}, z') - \rho^*_{\rm wAv}(z') 4\pi^2 \delta^{(2)}(\mathbf{k}_{xy}).$$

The function $\psi_{\rm wR}(\mathbf{r},\kappa)$ decays to zero faster than $e^{-\kappa z}$ when $z \to \infty$, so the x and y dependence of $\psi_{\rm wR}(\mathbf{r})$ fades out faster than the leading term, just like in the previous cases. The entire contribution to $\psi_{\rm w}(\mathbf{r})$ from the κ mode is given by Equation 12.235.

In the case of a periodic structure of $\sigma_{\rm w}(\mathbf{r})$, for instance when the wall consists of a crystal, the x and y dependence of $\mathscr{S}_{\rm wR}(\mathbf{r},\kappa)$ and hence of $\psi_{\rm w}(\mathbf{r})$ can be obtained analogously to that of $\psi_{\mathcal{S}_{\rm R}}(\mathbf{r})$ (defined in Equation 12.241). The $\mathbf{k}_{xy}$ integral in Equation 12.249 is then evaluated as shown in Equation 12.242, but with $\underset{\sim}{\sigma}^{\mathcal{S}}_{\rm R}(\mathbf{k}_{xy})$ replaced by $\underset{\sim}{\rho}^*_{\rm wR}(\mathbf{k}_{xy}, z')dz'$ and z by $z - z'$. Furthermore, $a(\boldsymbol{\tau}_{xy})$ is replaced by $a(\boldsymbol{\tau}_{xy}, z')$ because the magnitudes of the contributions at the lattice points in reciprocal space is different for each plane with coordinate z'. In case there are different kinds of periodic structure for various z', we have $\tau_{xy} = \tau_{xy}(z')$. The z' integration in Equation 12.249 provides the total contributions from all such planes with the appropriate weighing factor $e^{\kappa z'}$. The discussion of $\psi_{S_R}(\mathbf{r})$ above for periodic structures can be applied to the current case too. Also the discussion for a non-periodic distributions of charge in such planes can be applied here.

12.5 OVERSCREENING AND UNDERSCREENING IN ELECTROLYTES

The concepts of **overscreening** and **underscreening** in electrolytes are used to describe situations where the leading decay length of the charge-charge correlations and the screened interactions is smaller or larger, respectively, than the Debye length, so we have

$$\frac{\kappa}{\kappa_D} > 1 \quad \text{overscreening,} \qquad \frac{\kappa}{\kappa_D} < 1 \quad \text{underscreening.}$$

As we have seen, the decay length $1/\kappa$ is almost always different from the Debye length, so the normal situation is either one or the other, apart from very dilute electrolytes where $\kappa = \kappa_D$ to a very good approximation. For example, for monovalent electrolytes in aqueous solution at room temperature we have seen in Figure 12.2 that κ is *larger* than κ_D apart from very dilute solutions, so there is overscreening. On the other hand, for divalent (2:–2) electrolytes in aqueous solution at room temperature κ is *smaller* than κ_D, so there is underscreening as we saw in Figure 6.7 in Volume I. These two cases are also shown in Figure 12.8 as open symbols taken from MC simulation results[76] for primitive model electrolytes. The figure also includes MD simulation results[77] for other cases with about equal or higher electrostatic coupling as measured by the dimensionless quantity $\beta_R = \beta q^2/(4\pi\varepsilon_r^{\text{cont}}\varepsilon_0 d)$. In the figure we see that the underscreening becomes quite large when β_R is increased; for $\beta_R = 16.7$ the lowest value of κ/κ_D shown in the figure corresponds to a reduction in κ by 80 % compared to κ_D. There is a substantial underscreening also in dilute electrolytes when β_R is large, for example for a 0.5 mM aqueous solution of trivalent ions with $d = 3.8$ Å (a 3:–3 electrolyte with $\beta_R = 16.7$), we have $\kappa/\kappa_D \approx 0.4$ according to these results. Experimental measurements

[76] J. Ulander, and R. Kjellander, J. Chem. Phys. **114** (2001) 4893 (https://doi.org/10.1063/1.1350449).

[77] A. Härtel, M. Bültmann, and F. Coupette, Phys. Rev. Lett. **130** (2023) 108202 (https://doi.org/10.1103/PhysRevLett.130.108202) and Supplemental Material at http://link.aps.org/supplemental/10.1103/PhysRevLett.130.108202.

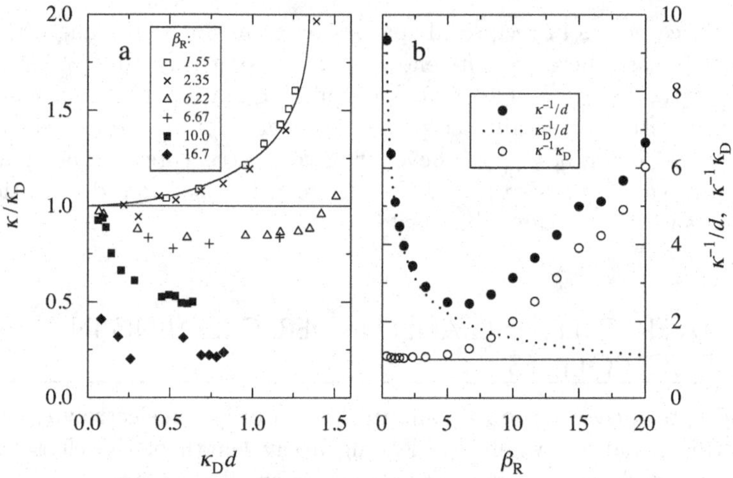

Figure 12.8 (a) The ratio κ/κ_D as function of $\kappa_D d$, where d is the ion diameter, for symmetric binary electrolytes with various values of $\beta_R = \beta q^2/(4\pi\varepsilon_r^{\text{cont}}\varepsilon_0 d)$. The open symbols (italics in the inserted caption) show MC simulation results[76] for ions in the restricted primitive model (with hard ionic cores) and the remaining symbols show MD simulation results[77] for ions modeled as charged soft spheres with a core potential that is strongly repulsive for $r < d$. The full curve is obtained from Equation 12.250. (b) The full circles show the decay length κ^{-1} in units of d and the open circles show $\kappa^{-1}\kappa_D$ as functions of β_R obtained from MD simulations[77] for an electrolyte of soft sphere ions with reduced density $n_R^{\text{b}} = 0.0016$ (the calculations were done for a 0.1 M electrolyte with $d = 3$ Å). The dotted curve shows κ_D^{-1}/d, that is, the Debye length in units of d.

of surface forces in electrolytes have also found large underscreening for very dilute solutions of multivalent ions in water.[78]

The MD results in Figure 12.8 have been obtained for a model system where the ionic core is modeled as a strongly repulsive soft-sphere potential of equal radii for anions and cations.[79] Recall that in the restricted primitive model with hard core ions the system is fully characterized by the two dimensionless quantities β_R and $\tau = \kappa_D d$ or any two combinations thereof, see footnote 42. This is not the case for soft-core ions, but when the core potential $u^s(r)$ is sufficiently repulsive (as in the present case), an electrolyte with a given $u^s(r)$ is almost fully characterized by two such parameters. The results

[78] A. M. Smith, Pl. Maroni, G. Trefalt, and M. Borkovec, J. Phys. Chem. B **123** (2019) 1733 (https://doi.org/10.1021/acs.jpcb.8b12246).

[79] The core repulsion for $r < d$ used in Ref. 77 is set equal to the short-range part $u^s(r)$ of the Weeks-Chandler-Anderson potential in Equations 9.85 and 9.86. Its parameters are $d^{\text{LJ}} = 2^{-1/6}d$, $\varepsilon^{\text{LJ}} = 5 \cdot 10^3 k_B T$ and r_0 is set at the minimum of the LJ potential. Thereby the core potential is strongly repulsive and has a continuous derivative at $r = d$.

for the hard core and soft core electrolytes with the same parameters differ, but not so much. As we can see in the figure, the results for $\beta_R = 6.22$ (hard cores) and $\beta_R = 6.67$ (soft cores) are nearly the same. The cases $\beta_R = 1.55$ and $\beta_R = 2.35$ agree well with the low-β_R approximation described in Example 12:A, that is, the leading solution κ of

$$\left(\frac{\kappa}{\kappa_{\mathrm{D}}}\right)^2 \approx \frac{e^{\kappa d}}{1+\kappa d} \qquad (\text{low } \beta_R) \tag{12.250}$$

(Equation 12.177). This approximation gives the full curve shown in the figure and as mentioned earlier it is valid for all β_R values lower than some threshold value of the order 2.

In Figure 12.8b the decay length $1/\kappa$ for an electrolyte with reduced density $n_R^{\mathrm{b}} \equiv n^{\mathrm{b}} d^3 = 0.0016$ is plotted in two different ways, κ^{-1}/d and $\kappa^{-1}\kappa_{\mathrm{D}}$, as functions of β_R, which is inversely proportional to temperature. The Debye length $1/\kappa_{\mathrm{D}}$ is plotted in the figure as $\kappa_{\mathrm{D}}^{-1}/d$ (the dotted curve). For low β_R (high temperatures, large $\varepsilon_r^{\mathrm{cont}}$ and/or small q) we can see that $\kappa^{-1}/d \approx \kappa_{\mathrm{D}}^{-1}/d$, so the decay length κ^{-1} is approximately equal to the Debye length. However, when β_R increases κ^{-1}/d starts to deviate from $\kappa_{\mathrm{D}}^{-1}/d$, passes a minimum and then κ^{-1} increases to become many times larger than the ionic diameter. At the same time the Debye length κ_{D}^{-1} decreases monotonically and becomes much smaller than d. As seen in the plot of $\kappa^{-1}\kappa_{\mathrm{D}}$, the decay length becomes several times larger than the Debye length for large β_R. This kind of behavior has be called **anomalous underscreening**, because $1/\kappa$ is *much* larger than what had been expected from the traditional Poisson-Boltzmann approximation. It must, however, be emphasized that underscreening or overscreening is normal in electrolytes and that the PB approximation is definitely not expected to be appropriate for the conditions where $1/\kappa \gg 1/\kappa_{\mathrm{D}}$. Therefore, this designation will be avoided and instead we will refer to such cases as "**large underscreening**," which means "**underscreening with $1/\kappa \gg 1/\kappa_{\mathrm{D}}$**."

It was a great surprise when this kind of underscreening was found experimentally in 2013 by Gebbie, Israelachvili and coworkers[80] in surface force measurements of room-temperature ionic liquids[81] with monovalent ions, where the screening was expected to be strong giving very short decay lengths since the density is very high, but where the decay lengths are very large, even orders of magnitude larger than expected. These results were soon verified by other labs and underscreening with $1/\kappa \gg 1/\kappa_{\mathrm{D}}$ has also been found experimentally

[80] M. A. Gebbie, M. Valtiner, X. Banquy, E. T. Fox, W. A. Henderson, and J. N. Israelachvili, Proc. Nat. Acad. Sci. **110** (2013) 9674 (https://doi.org/10.1073/pnas.1307871110); M. A. Gebbie, H. A. Dobbs, M. Valtiner, and J. N. Israelachvili, Proc. Nat. Acad. Sci. **112** (2015) 7432 (https://doi.org/10.1073/pnas.1508366112).

[81] Room-temperature ionic liquids are organic salts with a melting temperature below 100 °C.

in electrolyte solutions at high concentrations;[82] a recent review has been written by G. R. Elliot et al.[83] As we have seen in Figure 12.8, such underscreening occurs also at relatively low ionic densities when β_R is sufficiently large, like multivalent ions in water or classical plasmas of monovalent ions at sufficiently low temperatures. We will return to room-temperature ionic liquids in Chapter 13; here we will analyze the conditions under which underscreening occurs for electrolyte solutions in the primitive model and classical plasmas of spherical ions when κ is real. Thereafter we will investigate overscreening for such electrolytes.

For binary systems of spherical ions we have the exact relationship

$$\left(\frac{\kappa}{\kappa_{\mathrm{D}}}\right)^2 = \frac{q_+^{\mathrm{eff}}(\kappa) - q_-^{\mathrm{eff}}(\kappa)}{q_+ + |q_-|} \quad \text{(spherical ions)}$$

(Equation 12.169), which means that underscreening and overscreening are consequences of the magnitudes of the effective charges of the ions relative to the actual, bare charges. For symmetric electrolytes of equally sized ions, like the systems in Figure 12.8, this relationship implies that

$$\left(\frac{\kappa}{\kappa_{\mathrm{D}}}\right)^2 = \frac{q^{\mathrm{eff}}(\kappa)}{q} \quad \text{(RPM)} \tag{12.251}$$

(Equation 12.170). This means that systems that show underscreening with $1/\kappa \gg 1/\kappa_{\mathrm{D}}$ (i.e., $\kappa/\kappa_{\mathrm{D}} \ll 1$) have $q^{\mathrm{eff}}(\kappa) \ll q$. Since

$$q^{\mathrm{eff}}(\kappa) = \tilde{\rho}^{\star}(\mathrm{i}\kappa) = \int d\mathbf{r}'\, \rho^{\star}(r')\frac{\sinh(\kappa r')}{\kappa r'}$$

this occurs when

$$0 < q + \int d\mathbf{r}'\, \rho^{\mathrm{dress}}(r')\frac{\sinh(\kappa r')}{\kappa r'} \ll q, \tag{12.252}$$

where $\rho^{\mathrm{dress}} = \rho_+^{\mathrm{dress}} = -\rho_-^{\mathrm{dress}}$ and we have inserted $\rho_i^{\star}(r) = q_i\delta^{(3)}(r) + \rho_i^{\mathrm{dress}}(r)$. This means that the contribution to the effective charge from the dress *nearly neutralizes* the central charge q when there is underscreening with $1/\kappa \gg 1/\kappa_{\mathrm{D}}$. Let us now analyze the meaning of the condition 12.252.

In the cases that we are discussing here, the integral in Equation 12.252 is, in fact, not very different from the total charge $\int d\mathbf{r}'\, \rho^{\mathrm{dress}}(r')$ of the dress. In order to see this, we may first note that the factor $\sinh(\kappa r)/(\kappa r)$ in the integrand lies between 1 and 1.2 for $\kappa r \lesssim 1$ and between 1 and 1.8 for $\kappa r \lesssim 2$.

[82] M. A. Gebbie, A. M. Smith, H. A. Dobbs, A. A. Lee, G. G. Warr, X. Banquy, M. Valtiner, M. W. Rutland, J. N. Israelachvili, S. Perkin, and R. Atkin, Chem. Commun. **53** (2017) 1214 (https://doi.org/10.1039/C6CC08820A).

[83] G. R. Elliot, K. P. Gregory, H. Robertson, V. S. J. Craig, G. B. Webber, E.J. Wanless, and A. J. Page , Chem. Phys. Letters **843** (2024) 141190 (https://doi.org/10.1016/j.cplett.2024.141190).

For example, when $\kappa d \approx 1/3$ the latter applies for $r \lesssim 6d$ (a concrete example is the case $\beta_R \approx 10$ in Figure 12.8b). Since the main part of the total charge in the distribution $\rho^{\mathrm{dress}}(r)$ lies within a few ionic diameters from the ion surface it follows that the integral not very different from $\int d\mathbf{r}'\, \rho^{\mathrm{dress}}(r')$.

Recall that $q_i^{\mathrm{eff}}(\kappa)$ is given by the projection of $\rho_i^\star(r')$ on an exponentially decaying function $e^{-\kappa z'}$ in the z'-direction (or any other direction in space)

$$\int d\mathbf{r}'\, \rho_i^\star(r')e^{-\kappa z'} = \int d\mathbf{r}'\, \rho_i^\star(r')\frac{\sinh(\kappa r')}{\kappa r'} = q_i^{\mathrm{eff}}(\kappa)$$

(Equation 12.162). If the decay length $1/\kappa$ is so large that $e^{-\kappa z'}$ in the lhs does not deviate much from 1 near the origin ($\mathbf{r}' = \mathbf{0}$) where the dominating part of the charge distribution $\rho_i^\star(r')$ lies, we have $q_i^{\mathrm{eff}}(\kappa) \approx \int d\mathbf{r}'\, \rho_i^\star(r') = q_i^\star$, which does not depend on κ. If, on the other hand, $1/\kappa$ is much smaller, the value of $q_i^{\mathrm{eff}}(\kappa)$ will deviate a lot from $q_i^\star$ and depend strongly on κ. The cases that we are discussing here are closer to the former situation than the latter.

In order to continue the analysis of the effective charges, it is important to distinguish between the roles of the ion cloud of an ion $\rho_i^{\mathrm{cloud}}(r) = \sum_j q_j n_j^{\mathrm{b}} h_{ij}^{(2)}(r)$ and its dress $\rho_i^{\mathrm{dress}}(r_{12}) = \sum_j q_j n_j^{\mathrm{b}} h_{ij}^\star(r_{12})$ (for clarity have explicitly written "cloud" on $\rho_i^{\mathrm{cloud}} = \rho_i$). One may think that q_i^{eff} is a measure of the charge of an ion together with the charge of the cloud in its neighborhood, but this is not the case. It is the ionic charge together with the charge distribution *of the dress* that matters and ρ_i^{dress} is obtained from ρ_i^{cloud} by removing the linear electrostatic polarization response due to the i-ion

$$\rho_i^{\mathrm{dress}}(r) = \rho_i^{\mathrm{cloud}}(r) - \int d\mathbf{r}_3\, \psi_i(r')\chi^\star(|\mathbf{r} - \mathbf{r}'|)$$

(cf. Equation 12.87). Thereby we have removed all Yukawa function contributions from $\rho_i^{\mathrm{cloud}}(r)$ like[84]

$$-\frac{q_i^{\mathrm{eff}}(\kappa)\kappa^2}{\mathcal{E}_r^{\mathrm{eff}}(\kappa)} \cdot \frac{e^{-\kappa r}}{4\pi r} \quad \text{and} \quad -\frac{q_i^{\mathrm{eff}}(\kappa')(\kappa')^2}{\mathcal{E}_r^{\mathrm{eff}}(\kappa')} \cdot \frac{e^{-\kappa' r}}{4\pi r} \quad \text{etc.,}$$

that is, the Yukawa function contributions in Equation 12.172. What remains of $\rho_i^{\mathrm{cloud}}(r)$ in $\rho_i^{\mathrm{dress}}(r)$ are the *nonlinear* contributions of the electrostatic polarization response due to the ion and terms that arise from non-electrostatic interactions. As discussed in connection with Equation 12.203, the nonlinear

[84]This follows because from the Fourier transform of the integral $\int d\mathbf{r}_3\, \psi_i(r')\chi^\star(|\mathbf{r} - \mathbf{r}'|)$ we can deduce that

$$\tilde{\psi}_i(k)\tilde{\chi}^\star(k) = \tilde{\rho}_i^\star(k)\tilde{\phi}_{\mathrm{Coul}}^\star(k)\left[-\beta\sum_j n_j^{\mathrm{b}} q_j \tilde{\rho}_j^\star(k)\right] \sim -\frac{q_i^{\mathrm{eff}}(\kappa)\kappa^2}{(\kappa^2 + k^2)\mathcal{E}_r^{\mathrm{eff}}(\kappa)}$$

when $k \to i\kappa$, where we have used Equations 12.80, 12.83, 12.164 and 12.175. The rhs is the Fourier transform of the Yukawa function contribution to $\rho_i^{\mathrm{cloud}}(r)$ with decay parameter κ. The analogous result is obtained when $k \to i\kappa'$ and likewise for all other decay modes.

contributions have other functional forms than Yukawa functions and they have decay parameters 2κ, 3κ etc. and $2\kappa'$, $3\kappa'$ etc. as well as cross-terms. We can compare this with the Poisson-Boltzmann approximation, where we found that $\rho^{\star}(r)$ for $r \geq d^{\mathrm{h}}$ only contains nonlinear terms proportional to $\psi_i^2(r)$, $\psi_i^3(r)$ etc. as shown in Equation 12.135. These terms have decay parameters $2\kappa_{\mathrm{D}}$, $3\kappa_{\mathrm{D}}$ etc. because $\psi_i(r)$ decays like $e^{-\kappa_{\mathrm{D}} r}/r$.

The charge density of the cloud around a positive or negative ion is

$$\rho_{\pm}^{\mathrm{cloud}}(r) = \frac{\pm q n_{\mathrm{tot}}^{\mathrm{b}}}{2}[h_{++}^{(2)}(r) - h_{+-}^{(2)}(r)] = \pm q n_{\mathrm{tot}}^{\mathrm{b}} h_{\mathrm{D}}^{(2)}(r).$$

For large β_R, i.e., high values of the electrostatic coupling, the anion-cation correlations are very strong and the probability is high for anions to be near each cation (and vice versa). For example, when the reduced density is $n_R^{\mathrm{b}} = 0.0016$ (e.g. a 0.1 M electrolyte with $d = 3$ Å as in Figure 12.8b) $h_{\mathrm{D}}^{(2)}(r)$ has a minimum at $r \approx d$ of about -28, -83 and -220 when $\beta_R = 6.67$, 10.0 and 16.7, respectively. Such huge attractions mean that the electrostatic polarization response in the surroundings of each ion is large, so the nonlinear contributions will dominate in $\rho_i^{\mathrm{cloud}}(r)$ in the immediate neighborhood of the i-ion and these contributions are retained in $\rho_i^{\mathrm{dress}}(r)$. Thus, the anion-cation correlations contribute to a very large part of the dressed-ion charge density due to the strong attraction between such ions.

At least for low ionic densities, the large probability for anions and cations to be near each other can be described as a kind of transient aggregate formation, where the aggregates contain two or more ions each. A part of the aggregates can be ion pairs, where each pair consists of an anion and a cation, but the aggregation can also be a more general, transient association of ions involving several ions of opposite charges. This kind of aggregation for low ionic densities has been confirmed in the MD simulations cited in footnote 77. Aggregates that contain an equal number of anions and cations are electroneutral, while other aggregates as well as single ions will carry a charge. Aggregate formation due to strong attractions is clearly a nonlinear phenomenon. On average this means that from the point of view of each ion, the dress consists of a charge distribution of opposite charge that neutralizes the central charge to a large degree, which makes the effective charge small. As a consequence the decay length is large, as we have seen.

These matters can be illustrated in a concrete manner by investigating $h_{\mathrm{D}}^{(2)}(r)$ in some detail. This function, obtained by MD simulations, is plotted for various values of β_R in Figure 12.9.[85] For small β_R the function $h_{\mathrm{D}}^{(2)}(r)$ is monotonically decaying, as we have seen earlier and for large r it decays like a Yukawa function

$$h_{\mathrm{D}}^{(2)}(r) \sim -\frac{\beta[q^{\mathrm{eff}}(\kappa)]^2}{\mathcal{E}_r^{\mathrm{eff}}(\kappa)\varepsilon_0} \cdot \frac{e^{-\kappa r}}{4\pi r} \qquad \text{when } r \to \infty.$$

[85] Data are obtained from the Reference cited in footnote 77. Dr. Andreas Härtel is acknowledged for kindly providing the original data.

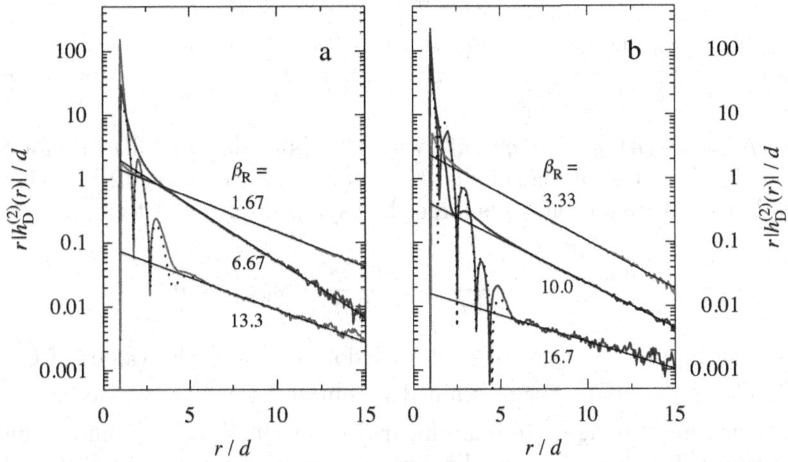

Figure 12.9 **(See color insert).** The difference-correlation function $h_D^{(2)}(r)$, which is proportional to the charge density $\rho_+^{\text{cloud}}(r)$ around a positive ion, plotted as $r|h_D^{(2)}(r)|$ for various values of β_R for symmetric electrolytes of soft sphere ions with diameter d and reduced density $n_R^{\text{b}} = 0.0016$. Frames (a) and (b) show the function for increasing β_R in an alternating manner: from top to bottom $\beta_R = 1.67$ (light blue in (a)), 3.33 (orange in (b)), 6.67 (green in (a)), 10.0 (dark blue in (b)), 13.3 (light red in (a)) and 16.7 (dark red in (b)). The colored full curves show data that have been obtained by MD simulations[85] and the straight black lines have been obtained by fitting a Yukawa function to the long-range tail of each function. The dotted black curves for the two highest values of β_R show the sum of a Yukawa function and an oscillatory Yukawa function that are both fitted, see text. For large r, where $h_D^{(2)}(r)$ is very small, there is numerical simulation noise in several curves.

The product $r|h_D^{(2)}(r)|$ plotted in a semi-logarithmic plot, as done in the figure, is a straight line for large r. The asymptote

$$h_D^{(2)}(r) \sim \mathsf{C}\frac{e^{-\kappa r}}{4\pi r} \tag{12.253}$$

with fitted values of C and κ is shown as straight black lines in the figure (note that $\mathsf{C} < 0$ and that the absolute value of $h_D^{(2)}$ is plotted).[86]

For large β_R the function $h_D^{(2)}(r)$ is oscillatory for small r and monotonic for large r, as seen in the figure. Then, the leading decay modes consist of one mode with decay parameter κ and one oscillatory short-range mode with

[86]The parameter values for the functions fitted to $h_D^{(2)}(r)$ that used in the present section are listed in the Reference cited in footnote 77.

decay parameters $\kappa''_{\Re} \pm i\kappa''_{\Im}$ where $\kappa''_{\Re} > \kappa$, so we have

$$h_{\rm D}^{(2)}(r) \sim -\frac{\beta[q^{\rm eff}]^2}{\mathcal{E}_r^{\rm eff}\varepsilon_0}\cdot\frac{e^{-\kappa r}}{4\pi r} - \frac{\beta|q''^{\,\rm eff}|^2}{|\mathcal{E}_r''^{\,\rm eff}|\varepsilon_0}\cdot\frac{e^{-\kappa''_{\Re}r}}{2\pi r}\cos(\kappa''_{\Im}r+\vartheta'') \quad \text{when } r\to\infty,$$

where $q^{\rm eff} = q^{\rm eff}(\kappa)$, $q''^{\,\rm eff} = q^{\rm eff}(\kappa'')$ with $\kappa'' = \kappa''_{\Re} + i\kappa''_{\Im}$ and likewise for $\mathcal{E}_r^{\rm eff}$, $\mathcal{E}_r''^{\,\rm eff}$ and where the phase shift is $\vartheta'' = \vartheta(\kappa'')$ (cf. Equation 12.193 with κ'' instead of κ). Thus, when β_R is large the asymptote has the form

$$h_{\rm D}^{(2)}(r) \sim \mathrm{C}\,\frac{e^{-\kappa r}}{4\pi r} + \mathrm{C}''\frac{e^{-\kappa''_{\Re}r}}{2\pi r}\cos(\kappa''_{\Im}r+\vartheta'')$$

and in the figure the rhs is plotted as a dotted curve (the values of C, C$''$, κ, $\kappa''_{\Re}$, $\kappa''_{\Im}$ and ϑ'' have been obtained by curve fitting[87]). Whenever $h_{\rm D}^{(2)}(r)$ passes zero and changes sign the logarithm $\ln(r|h_{\rm D}^{(2)}(r)|)$ becomes infinite, which causes the cusps seen in the figure (they are truncated at finite values in the plot).

In the plots we see the huge values of $|h_{\rm D}^{(2)}(r)|$ for $r \approx d$ when β_R is large, which is due to the very large attraction of ions of opposite charge to the central ion ($h_{\rm D}^{(2)}(r)$ and $\rho_+^{\rm cloud}(r)$ are negative). For the highest β_R values this leads to a layering of ions near the central ion and oscillations with alternating signs in $h_{\rm D}^{(2)}(r)$ and $\rho_+^{\rm cloud}(r)$ when r is increased, as apparent in the plot. Such oscillations are due to a sequential layering of anions and cations that give rise to out-of-phase oscillations[88] in $g_{++}^{(2)}(r)$ and $g_{+-}^{(2)}(r)$. The wave length of the oscillations in Figure 12.9 is close to $2d$, which is the sum of an anion and a cation diameter, and the decay length is about $0.4d$.

When r is further increased, the monotonic mode eventually takes over since it has the longest decay length, which is $4.2d$ and $5.1d$ for $\beta_R = 13.3$ and 16.7, respectively. The amplitude of the monotonic mode is much smaller than that of the oscillatory one: $|\mathrm{C}/\mathrm{C}''|$ is $8\cdot 10^{-4}$ and $8\cdot 10^{-5}$ for these two β_R values. A large part of this difference in amplitude of the modes is due to the difference in the decay parameters κ and κ'' because

$$\frac{\mathrm{C}}{\mathrm{C}''} = \frac{[q^{\rm eff}]^2}{|q''^{\,\rm eff}|^2}\cdot\frac{|\mathcal{E}_r''^{\,\rm eff}|}{\mathcal{E}_r^{\rm eff}} = \left[\frac{\kappa}{|\kappa''|}\right]^4\cdot\frac{|\mathcal{E}_r''^{\,\rm eff}|}{\mathcal{E}_r^{\rm eff}}, \tag{12.254}$$

where we have used the fact that $q^{\rm eff}(\kappa)$ is proportional to κ^2 (Equation 12.251). We have $[\kappa/|\kappa''|]^4 \approx 10^{-5}$ for the same β_R values. The rest of the amplitude difference is due to different values of $\mathcal{E}_r^{\rm eff}$ and $|\mathcal{E}_r''^{\,\rm eff}|$. The small magnitude of the Yukawa function contribution with large decay length is a notable feature of quite general significance as we will see.

Figure 12.10 shows $h_{\rm S}^{(2)}(r) = \frac{1}{2}[h_{++}^{(2)}(r) + h_{+-}^{(2)}(r)]$ for the same systems as in Figure 12.9. For the largest β_R values there are oscillations for small r

[87] See reference cited in footnote 77.

[88] Examples of out-of-phase and in-phase oscillations are given in Figure 12.7b.

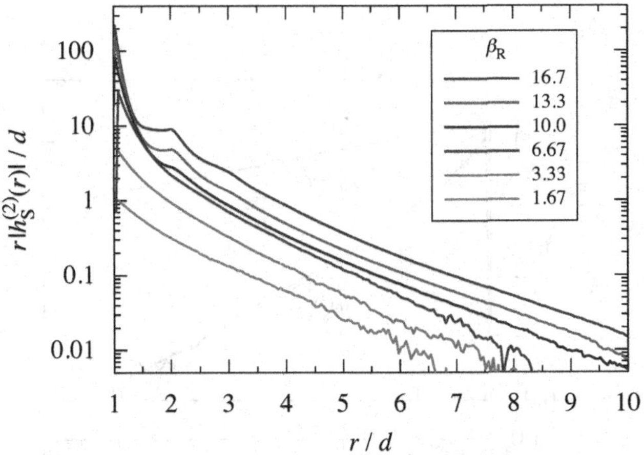

Figure 12.10 **(See color insert).** The sum-correlation function $h_S^{(2)}(r)$ for the same systems as in Figure 12.9 (the data are taken from the same MD simulations). From top to bottom $\beta_R = 16.7$ (dark red), 13.3 (light red), 10.0 (dark blue), 6.67 (green), 3.33 (orange) and 1.67 (light blue).

with a wave length $\approx d$. They are superimposed on a monotonically decaying function, which has a decay length that is considerably shorter than that of $h_D^{(2)}(r)$. These oscillations are due to in-phase oscillations of $g_{++}^{(2)}(r)$ and $g_{+-}^{(2)}(r)$ and are of a different type than those of $h_D^{(2)}(r)$ in Figure 12.9. The present ones are like the in-phase oscillations of $g_{++}^{(2)}$ and $g_{+-}^{(2)}$ in Figure 12.6a that are due to hard core layering and show up in $h_S^{(2)}(r)$.

Both the oscillations with wave length $\approx 2d$ in $h_D^{(2)}(r)$ and those with wave length $\approx d$ in $h_S^{(2)}(r)$ are due to a layering of ions near the central one, but the former are of electrostatic origin combined with excluded volume effects and the latter mainly due to layering of ionic cores irrespectively of the ionic charges. Recall that for these low ionic concentrations there is transient aggregate formation for large β_R and the "layering" occurs within such aggregates.

For the lowest electrostatic coupling, $\beta_R = 1.67$, the functions $h_S^{(2)}(r)$ and $h_D^{(2)}(r)$ are particularly simple because in this case there is only one decay mode that dominates for all r, namely

$$w^{(2)\mathrm{el}}(r) \approx \frac{\left[q^{\mathrm{eff}}\right]^2}{\mathcal{E}_r^{\mathrm{eff}}\varepsilon_0} \cdot \frac{e^{-\kappa r}}{4\pi r} = -\frac{\mathrm{C}}{\beta}\frac{e^{-\kappa r}}{4\pi r}$$

(the first term in Equation 12.216) and it is a very good approximation to set

$$g_S^{(2)}(r) = \cosh\left[\beta w^{(2)\mathrm{el}}(r)\right] \quad \text{and} \quad g_D^{(2)}(r) = -\sinh\left[\beta w^{(2)\mathrm{el}}(r)\right] \qquad (12.255)$$

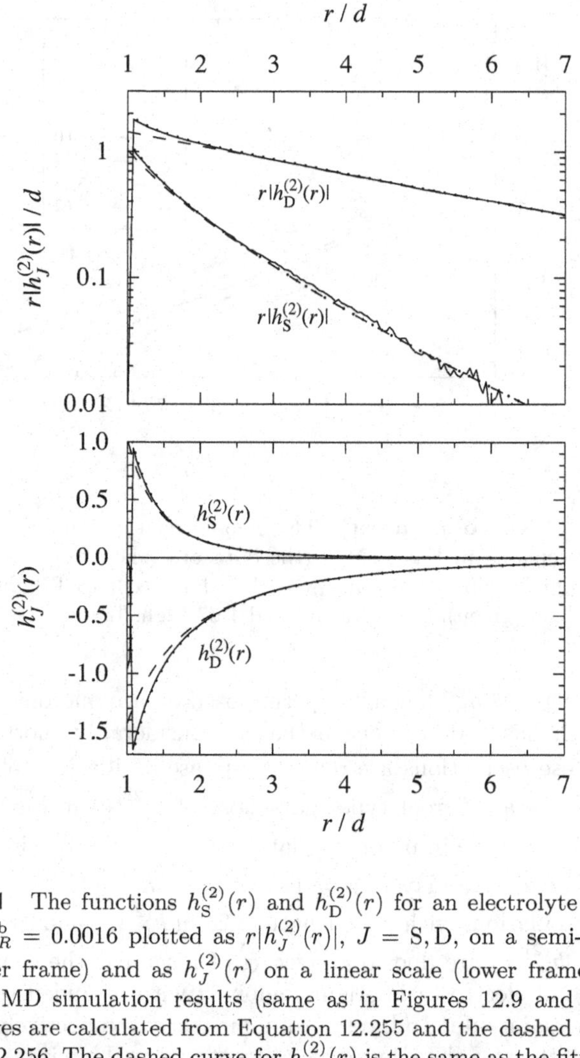

Figure 12.11 The functions $h_{\mathrm{S}}^{(2)}(r)$ and $h_{\mathrm{D}}^{(2)}(r)$ for an electrolyte with $\beta_R = 1.67$ and $n_R^{\mathrm{b}} = 0.0016$ plotted as $r|h_J^{(2)}(r)|$, $J = \mathrm{S}, \mathrm{D}$, on a semi-logarithmic scale (upper frame) and as $h_J^{(2)}(r)$ on a linear scale (lower frame). The full curves are MD simulation results (same as in Figures 12.9 and 12.10), the dotted curves are calculated from Equation 12.255 and the dashed curves from Equation 12.256. The dashed curve for $h_{\mathrm{D}}^{(2)}(r)$ is the same as the fitted Yukawa function plotted in Figure 12.9 for $\beta_R = 1.67$ as a light blue straight line.

outside the core region (cf. Equation 12.217 with $g^{(2)\mathrm{hs}}(r) = 1$ for $r > d$). This fact is demonstrated in Figure 12.11, where the simulated $h_{\mathrm{D}}^{(2)}(r)$ and $h_{\mathrm{S}}^{(2)}(r)$ from Figures 12.9 and 12.10 are plotted together with the results from Equation 12.255, where $g_{\mathrm{S}}^{(2)} = h_{\mathrm{S}}^{(2)} + 1$ and $g_{\mathrm{D}}^{(2)} = h_{\mathrm{D}}^{(2)}$. We see that the full and dotted curves are in excellent agreement for all r outside the core region. In the figure we have also plotted the lowest order power expansions of Equation

12.255, namely

$$\begin{aligned} h_{\rm S}^{(2)}(r) &\approx \tfrac{1}{2}\left[\beta w^{(2)\rm el}(r)\right]^2 \approx \frac{\rm C^2}{2}\cdot\frac{e^{-2\kappa r}}{(4\pi r)^2} \\ h_{\rm D}^{(2)}(r) &\approx -\beta w^{(2)\rm el}(r) \approx {\rm C}\frac{e^{-\kappa r}}{4\pi r} \end{aligned} \tag{12.256}$$

(cf. Equation 12.218). Note that the expression for $h_{\rm D}^{(2)}$ is the same as in Equation 12.253 and that $h_{\rm S}^{(2)}$ contains a squared Yukawa function.[89] This low-order approximation for $h_{\rm D}^{(2)}$ and $h_{\rm S}^{(2)}$ is appropriate for large r (see the dashed curves in Figure 12.11), but for small r the higher-order terms in the power expansion are needed: power 3 and upwards for $h_{\rm D}^{(2)}$ and power 4 and upwards for $h_{\rm S}^{(2)}$ that are given by the sinh and cosh functions in Equation 12.255. All contributions that are included here are solely due to the κ decay mode of $w^{(2)\rm el}(r)$. Accordingly, this mode gives both $h_{\rm D}^{(2)}(r)$ and $h_{\rm S}^{(2)}(r)$ and hence $h_{ij}^{(2)}(r)$ to a very good approximation for all r when $\beta_R = 1.67$.

For the case $\beta_R = 3.33$ the κ decay mode inserted in Equation 12.255 gives nearly the entire $h_{\rm D}^{(2)}(r)$ for all r, but only a rather small part of $h_{\rm S}^{(2)}(r)$ (these results are not shown). For the other β_R values shown in Figures 12.9 and 12.10, the contribution to $h_{\rm D}^{(2)}(r)$ from the κ decay mode gives, as we have seen, the dominating term for large r. However, the contributions to $h_{\rm S}^{(2)}(r)$ from this mode are negligible so other terms dominate in this function for all r in these cases.

Let us now turn to **overscreening**, that is, cases where $\kappa/\kappa_{\rm D} > 1$, and we will start with symmetrical electrolytes of equally sized ions. From Equation 12.251 it follows that we have $q^{\rm eff}(\kappa) > q$ in such cases and we therefore have to find a reason for an elevated effective charge. In Figure 12.8 the systems with $\beta_R = 1.67$ and 2.35 show overscreening. In both cases we can see in the figure that $\kappa/\kappa_{\rm D}$ for $\kappa_{\rm D}d \lesssim 1.3$ is given as a good approximation by Equation 12.250 (full curve in the figure), which is valid for sufficiently low β_R. This equation was derived in Example 12:A from

$$q_i^{\rm eff}(\kappa) = \frac{q_i e^{\kappa d}}{1+\kappa d} \tag{12.257}$$

(Equation 12.179), which is the same as the expression for $q_i^{\rm eff}(\kappa)$ in the Debye-Hückel approximation, that is, Equation 12.123 where $\kappa = \kappa_{\rm D}$. The DH approximation does, of course, not show overscreening, which is due to the fact that $q_i^{\rm eff}(\kappa_{\rm D})$ applies only to the central ion while the ions in the ion cloud are treated as bare point charges in this case. As we saw in Example 12:A, when *all* ions have their effective charges given by Equation 12.257 one obtains $\kappa/\kappa_{\rm D}$ according to Equation 12.250.

---

[89] The squared Yukawa function gives rise to a logarithmic branch point of $\tilde{h}_{ij}^{(2)}(k)$ in complex k space, cf. the shaded text box on page 351.

In order to see what Equation 12.257 stands for, let us consider the leading term in $\rho_i^{\text{cloud}}(r) \equiv \rho_i(r)$ for large r in

$$\rho_i(r) \sim -\frac{\kappa^2 q_i^{\text{eff}}(\kappa)}{\mathcal{E}_r^{\text{eff}}(\kappa)} \cdot \frac{e^{-\kappa r}}{4\pi r} = -\frac{\kappa^2 q_i}{(1+\kappa d)\mathcal{E}_r^{\text{eff}}(\kappa)} \cdot \frac{e^{-\kappa(r-d)}}{4\pi r}$$

(Equation 12.172), where the expression 12.257 for $q_i^{\text{eff}}(\kappa)$ has been inserted in order to obtain the last equality. When a single decay mode dominates and the rhs is a good approximation for $\rho_i(r)$ for all $r \geq d$, the condition of local electroneutrality implies that

$$-q_i = \int_d^{\infty} d\mathbf{r}\, \rho_i(r) \approx -\frac{\kappa^2 q_i}{(1+\kappa d)\mathcal{E}_r^{\text{eff}}(\kappa)} \int_d^{\infty} d\mathbf{r}\, \frac{e^{-\kappa(r-d)}}{4\pi r} = -\frac{q_i}{\mathcal{E}_r^{\text{eff}}(\kappa)}$$

since $\rho_i(r) = 0$ for $r < d$. Hence $\mathcal{E}_r^{\text{eff}}(\kappa) \approx 1$, which is a good approximation for $\mathcal{E}_r^{\text{eff}}(\kappa)$ when the ionic density is not too high, see Figure 12.3. (For an electrolyte in a dielectric continuum we have $\mathcal{E}_r^{\text{eff}}(\kappa)/\varepsilon_r^{\text{cont}} \approx 1$ instead.) Thus Equation 12.257 is a direct consequence of electroneutrality when a single decay mode dominates for all r outside the core region. An analogous interpretation also holds when two decay modes dominate in $\rho_i(r)$ for all r like in Example 12:A and then $\mathcal{E}_r^{\text{eff}}(\kappa)$ may deviate a lot from 1 as we saw there.[90]

It follows from Equation 12.257 that q_i^{eff}/q_i and hence κ/κ_{D} increases when d is increased at constant ion density (constant κ_{D}), so the overscreening becomes larger. This is solely a consequence of the fact that the countercharge distribution around the central ion is forced outwards when d is increased. The total charge $-q_i$ of the entire cloud is constant and these countercharges are not allowed to be inside the core.

Overscreening also occurs in asymmetric electrolytes that have $z_+ \neq -z_-$. An example is the 1:–2 electrolyte solution that we investigated in Figure 12.5, where $\kappa/\kappa_{\text{D}} > 1$ up to the Kirkwood cross-over point. In fact, for low electrolyte concentrations overscreening *always* occurs for asymmetric electrolytes because one can show that in the limit of zero ion density, we have[91]

$$\frac{\kappa^2}{\kappa_{\text{D}}^2} \sim 1 + \frac{\Lambda_{\text{e}} \ln 3}{4}(z_+ - |z_-|)^2 \quad \text{when } n_{\text{tot}}^{\text{b}} \to 0,$$

---

[90]When $q_i^{\text{eff}}(\kappa)$ is given by Equation 12.257 we have $\int_{r=d}^{r=\infty} d\mathbf{r}\, \kappa^2 q_i^{\text{eff}}(\kappa) e^{-\kappa r} / [4\pi r \mathcal{E}_r^{\text{eff}}(\kappa)] = q_i/\mathcal{E}_r^{\text{eff}}(\kappa)$, which was used when we applied the electroneutrality condition to the single-mode case and we concluded that $\mathcal{E}_r^{\text{eff}}(\kappa) \approx 1$. For the case of two decay modes as in Example 12:A, where both q_i^{eff} and $q_i'^{\text{eff}}$ are given by Equation 12.257, the same integral arises when the electroneutrality condition is applied and then Equation 12.181 for $\mathcal{E}_r^{\text{eff}}$ and $\mathcal{E}_r'^{\text{eff}}$ follows (see footnote 43 on page 339). Thus Equation 12.257 is connected to electroneutrality in this case too. Recall that a second equation for $\mathcal{E}_r^{\text{eff}}$ and $\mathcal{E}_r'^{\text{eff}}$ was obtained in the Example by applying the Stillinger-Lovett condition, whereby the results in Figure 12.3 were obtained.

[91]D. J. Mitchell and B. W. Ninham, Chem. Phys. Lett., **53** (1978) 397 (https://doi.org/10.1016/0009-2614(78)85426-8).

where $\Lambda_e = \ell_B \kappa_D$ is a coupling constant and $\ell_B = \beta q_e^2/(4\pi\varepsilon_r^{\text{cont}}\varepsilon_0)$ is the Bjerrum length (distinguish Λ_e from the de Broglie wave length Λ). This equation implies that $\kappa/\kappa_D > 1$ for low densities. The condition $n_{\text{tot}}^{\text{b}} \to 0$ in the equation can, of course, be replaced by $\kappa_D \to 0$ or $\Lambda_e \to 0$.

For symmetric electrolytes, where $z_+ = -z_-$, the second term vanishes identically and one must instead use the more complete expression[92]

$$\frac{\kappa^2}{\kappa_D^2} \sim 1 + \frac{\Lambda_e \ln 3}{4}(z_+ - |z_-|)^2 + \frac{\Lambda_e^2 \ln(\mathcal{L}\kappa_D)}{6}\left(z_+^2 - z_+|z_-| + z_-^2\right)^2 \quad \text{when } n_{\text{tot}}^{\text{b}} \to 0, \tag{12.258}$$

where $\mathcal{L}$ is a positive constant with the unit of length, for example $\mathcal{L} = \ell_B$ whereby we have $\ln(\mathcal{L}\kappa_D) = \ln \Lambda_e$. The choice of the value of $\mathcal{L}$ does not matter at very low densities because if one instead selects, say, $\mathcal{L}'$, the difference $\Lambda_e^2 \ln(\ell_B\kappa_D) - \Lambda_e^2 \ln(\mathcal{L}'\kappa_D) = \Lambda_e^2 \ln(\ell_B/\mathcal{L}')$ decays to zero faster than the terms shown in Equation 12.258 when $n_{\text{tot}}^{\text{b}} \to 0$ (a discussion about possible choices for $\mathcal{L}$ follows later). Any term of the form const $\times\, \Lambda_e^2$, like this difference, belong to the next order term of Equation 12.258 that is proportional to Λ_e^2 (and hence proportional to $n_{\text{tot}}^{\text{b}}$).

For symmetric electrolytes we have from Equation 12.258

$$\frac{\kappa^2}{\kappa_D^2} \sim 1 + \frac{\Lambda_e^2 \ln(\mathcal{L}\kappa_D)}{6} z^4 \quad \text{when } n_{\text{tot}}^{\text{b}} \to 0 \quad \text{(symmetric electr.)}, \tag{12.259}$$

where $z = z_+ = -z_-$. Since $\ln(\mathcal{L}\kappa_D) < 0$ when κ_D is small, this implies that there is *always underscreening* in symmetric electrolytes at sufficiently low

---

[92]The corresponding expression valid for any number of ionic species is

$$\frac{\kappa^2}{\kappa_D^2} = 1 + \frac{\Lambda_e \ln 3}{4}\left[\frac{\sum_j z_j^3 l_j}{\sum_j z_j^2 l_j}\right]^2 + \frac{\Lambda_e^2 \ln(\mathcal{L}\kappa_D)}{6}\left[\frac{\sum_j z_j^4 l_j}{\sum_j z_j^2 l_j}\right]^2 + \mathcal{O}(\Lambda_e^2)$$

where l_j is the stoichiometric number for species j in the salt (for example the salt $C_{l_+}^{z_+} A_{l_-}^{z_-}$ with anions A^{z_+} and cations C^{z_-}) and we likewise have

$$\frac{q_i^{\text{eff}}}{q_i} = 1 + z_i \frac{\Lambda_e \ln 3}{4} \cdot \frac{\sum_j z_j^3 l_j}{\sum_j z_j^2 l_j} + z_i^2 \frac{\Lambda_e^2 \ln(\mathcal{L}\kappa_D)}{6} \cdot \frac{\sum_j z_j^4 l_j}{\sum_j z_j^2 l_j} + \mathcal{O}(\Lambda_e^2),$$

see R. Kjellander and D. J. Mitchell, J. Chem. Phys. **101** (1994) 603 (https://doi.org/10.1063/1.468116), where $\mathcal{L} = \ell_B$ so $\ln(\mathcal{L}\kappa_D) = \ln \Lambda_e$.

These asymptotic formulas can be derived from $q_i^{\text{eff}}/q_i = 1 + \sum_j n_j^{\text{b}} q_j \tilde{h}_{ij}^{\star}(i\kappa)/q_i$ and the two terms originate from contributions to $h_{ij}^{\star}(r)$ that decay for large r like $[h_{ij}^{(2)}(r)]^\nu/\nu!$ with $\nu = 2$ and 3, respectively, that is, like $e^{-2\kappa r}/r^2$ and $e^{-3\kappa r}/r^3$ (they give terms in $\tilde{h}_{ij}^{\star}(i\kappa)$ that diverge like $1/\kappa$ and $\ln(\kappa)$, respectively, when $\kappa \sim \kappa_D \to 0$ and dominate in $\sum_j n_j^{\text{b}} q_j \tilde{h}_{ij}^{\star}(i\kappa)$ for small densities, see the reference previously cited). In fact, in the limit $\kappa \to 0$, solely the tails of $h_{ij}^{\star}(r)$ for $r > R$ matter, where R is an arbitrary distance (these tails cause the divergences). All other contributions to q_i^{eff} give rise to the higher order terms that are contained in $\mathcal{O}(\Lambda_e^2)$.

ionic densities. The underscreening increases a lot with increasing valency due to the factor z^4. These asymptotic formulas are exact and do not depend on any other ionic properties than the ionic charges. Ion sizes do not enter because these leading terms for low ionic densities originate solely from the long-range electrostatic correlations (cf. footnote 92). The higher order terms that are not shown in Equations 12.258 and 12.259, including the term proportional to Λ_e^2, depend on the charges as well as other ionic properties like the ion sizes.

We have seen in Figure 12.8 that there indeed is a large underscreening in symmetric electrolytes when β_R is large, but from Equation 12.258 it follows that underscreening also occurs for small β_R, at least for low ionic densities. This may seem like a contradiction to the results presented in this figure. We see from the full curve in the figure that Equation 12.250 gives a good prediction of κ/κ_D for systems with small β_R values, but this approximation does not show any underscreening because Equation 12.250 implies that κ/κ_D is always > 1 and that[93]

$$\frac{\kappa^2}{\kappa_\mathrm{D}^2} \sim 1 + \frac{(\kappa d)^2}{2} \sim 1 + \frac{d^2}{2}\kappa_\mathrm{D}^2 \quad \text{when } \kappa_\mathrm{D} \to 0. \tag{12.260}$$

However, in accurate calculations of κ/κ_D as function of $\kappa_\mathrm{D} d^\mathrm{h}$ there is a slight dip below 1 for very low densities[94] before κ/κ_D turns > 1 for slightly higher $n_\mathrm{tot}^\mathrm{b}$. The approximation 12.250 gives good results compared to simulations because the dip is really tiny and is therefore virtually negligible. Note that the κ_D^2 term in Equation 12.260 is proportional to Λ_e^2 and belongs to the higher order terms that are not shown in Equation 12.258.

As we concluded above, the value of the constant $\mathcal{L}$ does not matter for κ/κ_D when the ionic density is very low and terms that are proportional to Λ_e^2 are negligible. This means that $\mathcal{L}$ can be set to any value and the asymptotic formula is still valid, for example if we take $\mathcal{L} = C\ell_B$ for any positive constant C, we have

$$\Lambda_\mathrm{e}^2 \ln(\mathcal{L}\kappa_\mathrm{D}) = \Lambda_\mathrm{e}^2 \ln(C\Lambda_\mathrm{e}) = \Lambda_\mathrm{e}^2 \ln \Lambda_\mathrm{e} + \Lambda_\mathrm{e}^2 \ln C \sim \Lambda_\mathrm{e}^2 \ln \Lambda_\mathrm{e} \quad \text{when } \Lambda_\mathrm{e} \to 0,$$

so the limiting law 12.258 with $\mathcal{L}$ replaced by ℓ_B is valid irrespectively of C. The contribution $\Lambda_\mathrm{e}^2 \ln C$ belongs to the next order asymptotic term that is proportional to Λ_e^2.

A choice of $\mathcal{L}$ that is suitable for ions of diameter d is to set $\mathcal{L} = d$, so we have $\Lambda_\mathrm{e}^2 \ln(\mathcal{L}\kappa_\mathrm{D}) = \Lambda_\mathrm{e}^2 \ln(d\kappa_\mathrm{D})$. Thereby we have included an ion-size dependent contribution that belongs to the next order asymptotic term that is proportional to Λ_e^2. No exact simple expression for that term is presently known[95] so one needs to use an approximation for it in order to apply the

[93]Equation 12.260 is valid also in, for example, the Mean Spherical approximation.

[94]The minimum of κ/κ_D for a system with $\beta_\mathrm{R} = 1.55$ is about 0.99985 and occurs at $\kappa_\mathrm{D} d \approx 0.04$. It would be completely impossible to see this minimum on the scale of Figure 12.8.

[95]The term includes, for example, contributions from the pair bridge function.

asymptotic formulas outside the low-density regime. The choice $\mathcal{L} = d$ is, for example, suitable in the RPM because in this case the system is entirely determined by the parameters $\tau = \kappa_{\mathrm{D}} d^{\mathrm{h}}$ and β_{R}. Note that $\Lambda_{\mathrm{e}} = \beta_{\mathrm{R}}\tau/z^2$ so equation 12.259 becomes $(\kappa/\kappa_{\mathrm{D}})^2 \sim \beta_{\mathrm{R}}^2\tau^2 \ln(\tau)/6$ for the RPM.

APPENDIX

Appendix 12A

Extraction of the dressed-ion charge density $\rho_i^\star(r)$ and $q_i^\star$ from $h_{ij}^{(2)}(r)$

In this Appendix we will show how $\rho_i^\star(r)$ and $q_i^\star$ can be extracted numerically from the pair correlations $h_{ij}^{(2)}(r)$ of a bulk electrolyte consisting of spherical ions. We will thereby obtain $\tilde{\rho}_i^\star(k)$ and $q_i^\star = \tilde{\rho}_i^\star(0)$ in Fourier space, so $\rho_i^\star(r)$ will have to be calculated by doing an inverse Fourier transform (Equation 9.244).

We start with the charge distribution $\rho_i(r) = \sum_j q_j n_j^{\mathrm{b}} h_{ij}^{(2)}(r)$ around an i-ion, which can be calculated in any accurate approximate theory or by simulation. The total charge density of the ion is $\rho_i^{\mathrm{tot}}(r) = q_i\delta^{(3)}(r) + \rho_i(r)$. In Fourier space we have $\tilde{\rho}_i^\star(k) = \tilde{\rho}_i^{\mathrm{tot}}(k)\tilde{\varepsilon}(k)$ (Equation 12.108) and by inserting $\tilde{\varepsilon}(k)$ from Equation 12.101 we obtain

$$\tilde{\rho}_i^\star(k) = \frac{\tilde{\rho}_i^{\mathrm{tot}}(k)}{1 - \dfrac{\beta}{\varepsilon_0 k^2}\sum_j q_j n_j^{\mathrm{b}} \tilde{\rho}_j^{\mathrm{tot}}(k)}. \tag{12.261}$$

The numerator is zero for $k = 0$ due to local electroneutrality $\int d\mathbf{r}\rho_i^{\mathrm{tot}}(r) = 0$ and the denominator is zero there due to the Stillinger-Lovett condition (this follows from Equation 12.105 with $\tilde{\psi}_i(k) = \tilde{\rho}_i^{\mathrm{tot}}(k)/[\varepsilon_0 k^2]$). These zeros cancel but this must be handled carefully in order to avoid numerical problems for small k. In the numerator we therefore write

$$\tilde{\rho}_i^{\mathrm{tot}}(k) = \tilde{\rho}_i^{\mathrm{tot}}(k) - \int d\mathbf{r}\rho_i^{\mathrm{tot}}(r) = \int d\mathbf{r}\left[\frac{\sin(kr)}{kr} - 1\right]\rho_i^{\mathrm{tot}}(r),$$

where the square bracket is zero for $k = 0$. We use the series expansion of $\tilde{\rho}_i^{\mathrm{tot}}(k)$ in Equation 12.102 to rewrite the denominator, whereby we set

$$\begin{aligned}
\tilde{\rho}_j^{\mathrm{tot}}(k) &= \left[\tilde{\rho}_j^{\mathrm{tot}}(k) - \int d\mathbf{r}\rho_j^{\mathrm{tot}}(r) + \frac{k^2}{3!}\int d\mathbf{r}\, r^2\rho_j^{\mathrm{tot}}(r)\right] - \frac{k^2}{3!}\int d\mathbf{r}\, r^2\rho_j^{\mathrm{tot}}(r) \\
&= \int d\mathbf{r}\left[\frac{\sin(kr) - kr + (kr)^3/6}{kr}\right]\rho_j^{\mathrm{tot}}(r) - \frac{k^2}{6}\int d\mathbf{r}\, r^2\rho_j^{\mathrm{tot}}(r).
\end{aligned}$$

The last term gives the contribution -1 to the denominator because of the Stillinger-Lovett condition 12.103, so it cancels the first term 1.

By inserting these expressions into Equation 12.261 we obtain

$$\tilde{\rho}_i^*(k) = -\frac{\int d\mathbf{r}\left[\dfrac{\sin(kr)-kr}{kr}\right]\rho_i(r)}{\dfrac{\beta}{\varepsilon_0 k^2}\displaystyle\int d\mathbf{r}\left[\dfrac{\sin(kr)-kr+(kr)^3/6}{kr}\right]\sum_j q_j n_j^{\mathrm{b}}\rho_j(r)}$$

where we have entered ρ_i instead of ρ_i^{tot} because the contributions from the delta function vanish in the integrals (the square brackets are zero for $r = 0$). The zeros of the denominator and numerator for $k = 0$ can now be dealt with by dividing both by k^2 and we finally get

$$\tilde{\rho}_i^*(k) = -\frac{\varepsilon_0 \int d\mathbf{r}\, r^2\left[\dfrac{\sin(kr)-kr}{(kr)^3}\right]\rho_i(r)}{\beta\displaystyle\int d\mathbf{r}\, r^4\left[\dfrac{\sin(kr)-kr+(kr)^3/6}{(kr)^5}\right]\sum_j q_j n_j^{\mathrm{b}}\rho_j(r)}, \tag{12.262}$$

which can be used to extract $\tilde{\rho}_i^*(k)$ for all real-valued k from $\rho_j(r)$ for all j without the cancellation problems for small k. It may be important to have very accurate tails of $\rho_j(r)$ for large r because of the factors r^4 and r^6 in the integrals (there is an additional factor of r^2 in the integrals because $d\mathbf{r} = 4\pi r^2 dr$). This can be obtained by extending the tails.[96] For $k = 0$ the square bracket in the numerator is equal to $-1/3!$ and that in the denominator is $1/5!$, so we obtain

$$q_i^* = \tilde{\rho}_i^*(0) = \frac{20\varepsilon_0 \int d\mathbf{r}\, r^2 \rho_i(r)}{\beta \sum_j q_j n_j^{\mathrm{b}} \int d\mathbf{r}\, r^4 \rho_j(r)}. \tag{12.263}$$

[96]The tail extension can be done as described in J. Ulander and R. Kjellander, J. Chem. Phys. **114** (2001) 4893, where $\rho_i^\star(r)$, $h_{ij}^\star(r)$ and q_i^{eff} are denoted $\rho_i^0(r)$, $h_{ij}^0(r)$ and q_i^* (thus q_i^* has a different meaning there than it has here). This paper describes also a manner to extract $h_{ij}^\star(r)$ and $\rho_i^\star(r)$ from $h_{ij}^{(2)}(r)$ by utilizing the Ornstein-Zernike equation.

CHAPTER 13

Ionic Liquids and Electrolyte Solutions with Molecular Solvent

In this chapter we generalize the general, exact treatment of spherical-ion electrolytes presented in the previous chapter to the corresponding theory for electrolytes consisting of nonspherical ions and other particles. Thereby the treatment includes, for instance, ionic liquids, dispersions of macroparticles and electrolyte solutions with discrete, molecular solvent. We will find that the theory for both inhomogeneous and homogeneous fluids has mainly the same form as for the spherical case despite the added complication of rotational degree of freedom for the particles, which make some formulas to have a more complicated structure.

The formalism for the dielectric properties, the nonlocal electrostatics, the general screened Coulomb potential and the existence of multiple decay modes are essentially the same as before. A difference is, of course, that the interaction potential between nonspherical ions in an electrolyte depends on the particle orientations and this makes it necessary to introduce "effective charges" that depend on these orientations, entities that we call "multipolar effective charges." They describe the strength and orientation dependence of the ion-ion interaction for each decay mode, whereby each mode has its own value of these entities. In Volume I of this treatise we saw that multipolar effective charges also have an important role in the Poisson-Boltzmann theory for the description of interactions between nonspherical particles immersed in electrolytes, but in that approximation there is only one decay mode with a decay length equal to the Debye length.

An important difference between electrolytes and nonelectrolyte fluids is that all types of multipoles contribute to the interaction between charged nonspherical particles at all distances in electrolytes. For nonelectrolytes, on

DOI: 10.1201/9781003286882-13

the other hand, the quadrupolar contributions, for example, decay faster with distance than the dipolar ones, which in turn decays faster than the monopolar ones. This is also the case for multipolar electrostatic interactions in vacuum.

A concrete application of the general theory is the interpretation of experimental results of surface forces in room-temperature ionic liquids, for example the phenomenon of underscreening, and another is the analysis of the nature of hydration forces (more generally solvation forces). For homogeneous electrolyte solutions we also treat the case of electrolyte solutions with a nematic solvent at low electrolyte concentrations.

13.1 BASIC FORMULAS AND RELATIONSHIPS

In this chapter we are going to treat electrolytes consisting of a mixture of rigid, unpolarizable molecules/particles of arbitrary sizes, shapes and internal charge distributions (for simplicity we will call all constituents "particles"). For a particle of species i with orientation $\boldsymbol{\omega}$, the internal charge distribution is denoted $\sigma_i(\mathbf{r}, \boldsymbol{\omega})$, where $\mathbf{r}$ gives the coordinates *relative to the center of mass of the particle.* Like in Chapter 11, the orientational variable is normalized so that $\int d\boldsymbol{\omega} = 1$ (cf. Appendix 11A). The density may consist of discrete charges at certain sites or be a continuous charge distribution inside the particle. In the former case

$$\sigma_i(\mathbf{r}, \boldsymbol{\omega}) = \sum_\alpha q_{\alpha,i}\delta(\mathbf{r} - \mathbf{r}_{\alpha,i}(\boldsymbol{\omega})),$$

where the sum is over all sites, $q_{\alpha,i}$ is the partial charge of site α in species i and $\mathbf{r}_{\alpha,i}(\boldsymbol{\omega})$ specifies the coordinates of the site relative to the center of mass. For a particle placed at $\mathbf{r}_1$ with orientation $\boldsymbol{\omega}_1$, that is, a particle with coordinates $\mathbf{R}_1 \equiv (\mathbf{r}_1, \boldsymbol{\omega}_1)$, the internal charge density at $\mathbf{r}_2$ *in the laboratory frame* is given by $\sigma_i(\mathbf{r}_2|\mathbf{R}_1) \equiv \sigma_i(\mathbf{r}_{12}, \boldsymbol{\omega}_1)$, where $\mathbf{r}_{12} = \mathbf{r}_2 - \mathbf{r}_1$. The net charge q_i of a particle is

$$q_i = \int d\mathbf{r}\, \sigma_i(\mathbf{r}, \boldsymbol{\omega}), \tag{13.1}$$

which, of course, is independent of the orientation. The current chapter deals with electrolytes, so we will assume that there are *always* some charged species present, $q_i \neq 0$, with a non-zero number density. However, some molecular species, say s, may be uncharged, $q_s = 0$. The latter will be designated as being a solvent species in an **electrolyte solution with molecular solvent** (the solvent can, of course, consist of more that one species). For pure **ionic liquids** all particles are charged. In principle, ionic liquids can be any liquid consisting of ions, including molten salts of any kind, but often the term is used mainly for *room-temperature ionic liquids*, which are organic salts with a melting point below 100 °C. Here, we will treat any kind of ionic fluids, including classical plasmas and molten inorganic salts with high melting points.

The dipole moment vector $\mathbf{m}_i^{\mathrm{d}}(\boldsymbol{\omega})$ of a particle with orientation $\boldsymbol{\omega}$ is given by[1]

$$\mathbf{m}_i^{\mathrm{d}}(\boldsymbol{\omega}) = \int d\mathbf{r}\, \mathbf{r}\sigma_i(\mathbf{r}, \boldsymbol{\omega}) \tag{13.2}$$

and the quadrupole moment tensor $\mathbb{M}_i^{\mathrm{Q}}(\boldsymbol{\omega})$ is[2]

$$\mathbb{M}_i^{\mathrm{Q}}(\boldsymbol{\omega}) = \tfrac{1}{2} \int d\mathbf{r}\, \left[3\mathbf{r}\mathbf{r}^{\mathsf{T}} - r^2\mathbb{I}\right] \sigma_i(\mathbf{r}, \boldsymbol{\omega}), \tag{13.3}$$

where

$$\mathbf{r}\mathbf{r}^{\mathsf{T}} = \begin{bmatrix} x \\ y \\ z \end{bmatrix} [x,\ y,\ z] = \begin{bmatrix} xx & xy & xz \\ yx & yy & yz \\ zx & zy & zz \end{bmatrix},$$

superscript T denotes transpose, and $\mathbb{I}$ is the identity matrix (with diagonal elements equal to 1 and the rest equal to 0).

The particles interact with a pair interaction potential $u_{ij} = u_{ij}^{\mathrm{sh}} + u_{ij}^{\mathrm{el}}$, where u_{ij}^{sh} is a short-range, nonelectrostatic part and u_{ij}^{el} is the electrostatic part given by

$$\begin{aligned} u_{ij}^{\mathrm{el}}(\mathbf{R}_1, \mathbf{R}_2) &= \int d\mathbf{r}_3 d\mathbf{r}_4\, \sigma_i(\mathbf{r}_3|\mathbf{R}_1)\phi_{\mathrm{Coul}}(r_{34})\sigma_j(\mathbf{r}_4|\mathbf{R}_2) \\ &= \int d\mathbf{r}_3 d\mathbf{r}_4\, \sigma_i(\mathbf{r}_{13}, \boldsymbol{\omega}_1)\phi_{\mathrm{Coul}}(r_{34})\sigma_j(\mathbf{r}_{24}, \boldsymbol{\omega}_2), \end{aligned} \tag{13.4}$$

where $\phi_{\mathrm{Coul}}(r) = 1/(4\pi\varepsilon_0 r)$. This expression for u_{ij}^{el} is illustrated in Figure 13.1. The particles may also interact with an external potential $v_i = v_i^{\mathrm{sh}} + v_i^{\mathrm{el}}$, where v_i^{sh} is the short-range, nonelectrostatic part of the external potential and v_i^{el} is the interaction with an ***external electrostatic potential*** Ψ^{Ext}, that is, the potential from a charge distribution that is external to the system. We have

$$\begin{aligned} v_i^{\mathrm{el}}(\mathbf{R}_1) &= \int d\mathbf{r}_4\, \Psi^{\mathrm{Ext}}(\mathbf{r}_4)\sigma_i(\mathbf{r}_4|\mathbf{R}_1) \\ &= \int d\mathbf{r}_4\, \Psi^{\mathrm{Ext}}(\mathbf{r}_4)\sigma_i(\mathbf{r}_{14}, \boldsymbol{\omega}_1). \end{aligned} \tag{13.5}$$

The nonelectrostatic interactions decay more quickly to zero with increasing distance than the electrostatic ones, so we have $u_{ij}(\mathbf{R}_1, \mathbf{R}_2) \sim u_{ij}^{\mathrm{el}}(\mathbf{R}_1, \mathbf{R}_2)$ when $r_{12} \to \infty$ and like in Chapter 12 we will assume that u_{ij}^{sh} decays to zero

---

[1]The dipole moment is defined relative to the center of the particle. In this treatise we use the notation $\mathbf{m}^{\mathrm{d}}$ and $m^{\mathrm{d}} = |\mathbf{m}^{\mathrm{d}}|$ for the dipole moment rather than the more common $\boldsymbol{\mu}$ and $\mu = |\boldsymbol{\mu}|$ in order not to confuse with chemical potential μ.

[2]Note that this definition of quadrupole moment differs by a factor of 1/2 from the definition used, for example, in J. D. Jackson, 1999, *Classical Electrodynamics,* 3rd ed., New York: Wiley.

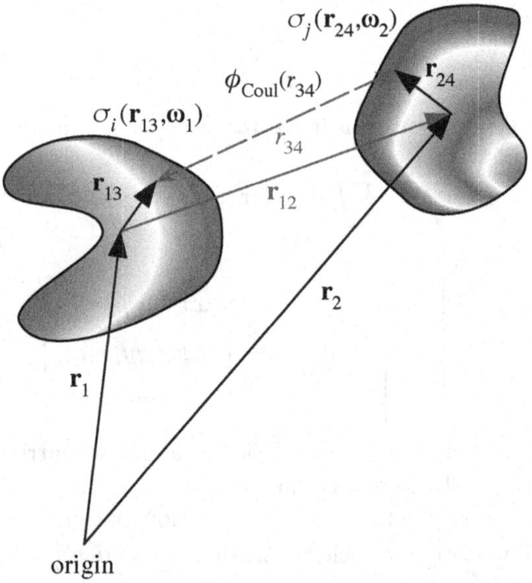

Figure 13.1 **(See color insert).** Sketch of how the electrostatic interaction between two particles is calculated via $u_{ij}^{\text{el}}(\mathbf{r}_1, \omega_1, \mathbf{r}_2, \omega_2) = \int d\mathbf{r}_3 d\mathbf{r}_4\, \sigma_i(\mathbf{r}_{13}, \omega_1)\phi_{\text{Coul}}(r_{34})\sigma_j(\mathbf{r}_{24}, \omega_2)$ (Equation 13.4). The charge distributions $\sigma_i(\mathbf{r}_{13}, \omega_1)$ and $\sigma_j(\mathbf{r}_{24}, \omega_2)$ of the particles are illustrated by color shading (red = positive, blue = negative).

at least exponentially fast with increasing r_{12}, unless we say otherwise, and the same applies to v_i^{sh} as function of distance from the body that gives rise to it. Otherwise u_{ij}^{sh} and v_i^{sh} are arbitrary.

It is worthwhile to see how Fourier transform techniques can be used to obtain and investigate the electrostatic potential from a dipole and the interaction potential between two dipoles, both for the case of dipoles that consist of two oppositely charged point charges separated by a finite distance and the limiting case of ideal dipoles, where the distance is zero. The potential for the latter contains an interesting technicality, namely a delta-function contribution to the interaction potential. We will see how the latter can be eliminated and then the Fourier transform of the potential must be changed in order to be consistent. The derivation of the expressions for these potentials are illustrative examples of the use of the Fourier techniques. In the shaded text box below we will give a summary; the details are given in Appendix 13A.

We consider a point charge q_1 located at the position $\mathbf{r}_1 + \mathbf{a}_1$ and a charge $-q_1$ located at $\mathbf{r}_1 - \mathbf{a}_1$. Together they form a dipole with center at $\mathbf{r}_1$ and dipole moment vector $\mathbf{m}_1^{\text{d}} = 2q_1\mathbf{a}_1$. In the limit $a_1 \to 0$ with $2q_1a_1 = m_1^{\text{d}} =$

constant, where $a_1 = |\mathbf{a}_1|$ and $m_1^{\mathrm{d}} = |\mathbf{m}_1^{\mathrm{d}}|$, they form an **ideal dipole** located at $\mathbf{r}_1$. An ideal dipole is a theoretical construct which is often used in simple models of polar fluids. We will denote a particle with an ideal dipole as a *purely dipolar particle* (such a particle only have a dipole moment but no multipole moments of higher orders). However, we will not pay much attention to such models in this treatise and mainly deal with particles with internal charges consisting of charged sites or a continuous distribution of charge, which are more realistic. A particle with a **finite dipole** (with nonzero a_1) belong to the latter category.

The "internal" charge density σ_1 of a finite dipole placed at $\mathbf{r}_1$ and with orientation ω_1 is

$$\begin{aligned}\sigma_1(\mathbf{r}_3|\mathbf{R}_1) &= \sigma_1(\mathbf{r}_{13}, \omega_1)\\ &= q_1\delta^{(3)}(\mathbf{r}_{13} - \mathbf{a}_1(\omega_1)) - q_1\delta^{(3)}(\mathbf{r}_{13} + \mathbf{a}_1(\omega_1)),\end{aligned}$$

where we have explicitly shown that $\mathbf{a}_1$ depends on the orientation ω_1 of the particle. The electrostatic potential ψ_1 from the dipole decays for large distances from its center as

$$\psi_1(\mathbf{r}, \omega_1) \sim \frac{\mathbf{m}_1^{\mathrm{d}}(\omega_1) \cdot \hat{\mathbf{r}}}{4\pi\varepsilon_0 r^2}, \quad \text{when } r \to \infty \qquad \text{(finite dipole)}$$

(Equation 13.122). In the limiting case of an ideal dipole we have for all r

$$\psi_1(\mathbf{r}, \omega_1) = \frac{\mathbf{m}_1^{\mathrm{d}}(\omega_1) \cdot \hat{\mathbf{r}}}{4\pi\varepsilon_0 r^2} \qquad \text{(ideal dipole)}$$

(Equation 13.121), which has the Fourier transform

$$\tilde{\psi}_1(\mathbf{k}, \omega_1) = -\frac{i\mathbf{k} \cdot \mathbf{m}_1^{\mathrm{d}}(\omega_1)}{\varepsilon_0 k^2} \qquad \text{(ideal dipole).}$$

For a finite dipole, $\tilde{\psi}_1(\mathbf{k}, \omega_1)$ contains a term that is equal to the rhs of this equation. That term determines the asymptotic decay of ψ_1.

The pair interaction potential $u_{12}^{\mathrm{el}}(\mathbf{r}_{12}, \omega_1, \omega_2)$ between two finite dipoles, dipole 1 at $\mathbf{r}_1$ and with orientation ω_1 and dipole 2 at $\mathbf{r}_2$ and with orientation ω_2, decays like

$$u_{12}^{\mathrm{el}}(\mathbf{r}, \omega_1, \omega_2) \sim -\frac{m_1^{\mathrm{d}} m_2^{\mathrm{d}} \mathsf{D}_{12}(\hat{\mathbf{r}}, \omega_1, \omega_2)}{4\pi\varepsilon_0 r^3} \quad \text{when } r \to \infty \qquad \text{(finite dipoles)} \tag{13.6}$$

(Equation 13.125), where

$$\mathsf{D}_{12}(\hat{\mathbf{r}}, \omega_1, \omega_2) = 3\left[\hat{\mathbf{r}} \cdot \hat{\mathbf{m}}_1^{\mathrm{d}}(\omega_1)\right]\left[\hat{\mathbf{r}} \cdot \hat{\mathbf{m}}_2^{\mathrm{d}}(\omega_2)\right] - \hat{\mathbf{m}}_1^{\mathrm{d}}(\omega_1) \cdot \hat{\mathbf{m}}_2^{\mathrm{d}}(\omega_2). \tag{13.7}$$

For two ideal dipoles we have for all r

$$\begin{aligned} u_{12}^{\mathrm{el}}(\mathbf{r},\omega_1,\omega_2) &= -\frac{m_1^{\mathrm{d}} m_2^{\mathrm{d}} \mathrm{D}_{12}(\hat{\mathbf{r}},\omega_1,\omega_2)}{4\pi\varepsilon_0 r^3} \\ &+ \delta^{(3)}(\mathbf{r}) \frac{\mathbf{m}_1^{\mathrm{d}}(\omega_1)\cdot \mathbf{m}_2^{\mathrm{d}}(\omega_2)}{3\varepsilon_0} \quad \text{(ideal dipoles)} \end{aligned} \tag{13.8}$$

(Equation 13.127), which has the Fourier transform

$$\tilde{u}_{12}^{\mathrm{el}}(\mathbf{k},\omega_1,\omega_2) = \frac{\mathbf{k}\cdot \mathbf{m}_1^{\mathrm{d}}(\omega_1)\, \mathbf{k}\cdot \mathbf{m}_2^{\mathrm{d}}(\omega_2)}{\varepsilon_0 k^2} \quad \text{(ideal dipoles).} \tag{13.9}$$

The rhs can be written as $\hat{\mathbf{k}}\cdot \mathbf{m}_1^{\mathrm{d}}(\omega_1)\, \hat{\mathbf{k}}\cdot \mathbf{m}_2^{\mathrm{d}}(\omega_2)/\varepsilon_0$ and does not depend on the magnitude of k. For finite dipoles, $\tilde{u}_{12}^{\mathrm{el}}$ contains this term and it determines the asymptotic decay in ordinary space. Note that

$$\frac{m_1^{\mathrm{d}} m_2^{\mathrm{d}} \mathrm{D}_{12}(\hat{\mathbf{r}},\omega_1,\omega_2)}{4\pi\varepsilon_0 r^3} = \left[\mathbf{m}_1^{\mathrm{d}}(\omega_1)\cdot\nabla\right]\left[\mathbf{m}_2^{\mathrm{d}}(\omega_1)\cdot\nabla\right]\frac{1}{4\pi\varepsilon_0 r}.$$

The delta function contribution for ideal dipoles has no consequence for the interaction potential between particles with finite sizes, but the term is needed for a mathematical consistent treatment of the singularity at $r = 0$ when performing Fourier transforms.

One can define an ideal dipole potential without the delta function contribution, but then the corresponding contribution has to be contained in the Fourier transform $\tilde{u}_{12}^{\mathrm{el}}$ instead. Then one has the alternative expressions

$$u_{12}^{\mathrm{el,\,alt.}}(\mathbf{r},\omega_1,\omega_2) = -\frac{m_1^{\mathrm{d}} m_2^{\mathrm{d}} \mathrm{D}_{12}(\hat{\mathbf{r}},\omega_1,\omega_2)}{4\pi\varepsilon_0 r^3} \quad \text{(ideal dipoles)}$$

and its Fourier transform

$$\begin{aligned} \tilde{u}_{12}^{\mathrm{el,\,alt.}}(\mathbf{k},\omega_1,\omega_2) &= \frac{\mathbf{k}\cdot \mathbf{m}_1^{\mathrm{d}}(\omega_1)\, \mathbf{k}\cdot \mathbf{m}_2^{\mathrm{d}}(\omega_2)}{\varepsilon_0 k^2} - \frac{\mathbf{m}_1^{\mathrm{d}}(\omega_1)\cdot \mathbf{m}_2^{\mathrm{d}}(\omega_2)}{3\varepsilon_0} \\ &= \frac{m_1^{\mathrm{d}} m_2^{\mathrm{d}} \mathrm{D}_{12}(\hat{\mathbf{k}},\omega_1,\omega_2)}{3\varepsilon_0} \quad \text{(ideal dipoles)} \end{aligned}$$

(Equation 13.129), where D_{12} is the same function as in Equation 13.7 but with $\hat{\mathbf{k}}$ inserted instead of $\hat{\mathbf{r}}$. Thus, the latter way of writing the potential has the appealing feature that D_{12} occurs in both the ordinary and the Fourier spaces.

In the current treatise we will, as stated above, mainly be concerned with particles that have internal distributions of charge. In the ideal dipole limit of such a distribution, Equations 13.8 and 13.9 give the appropriate potential, so we will use them in this treatise. When dealing with simple models of purely dipolar particles, the alternative expressions are, however, common in the scientific literature. Since realistic particles have finite size, the choice of behavior at $r = 0$ for the ideal dipole potential is often not a big issue, but one must be consistent in the theoretical treatment and use one or the other convention for u_{12}^{el} and $\tilde{u}_{12}^{\text{el}}$ when dealing with ideal dipoles.

For an inhomogeneous fluid with density distribution $\mathfrak{n}_j^{(1)}(\mathbf{r}_2, \omega_2) = \mathfrak{n}_j^{(1)}(\mathbf{R}_2)$ for species j, the mean charge density at position $\mathbf{r}_3$ from all species is[3]

$$\begin{aligned} \rho(\mathbf{r}_3) &= \sum_j \int d\mathbf{r}_2 d\omega_2\, \mathfrak{n}_j^{(1)}(\mathbf{r}_2, \omega_2)\sigma_j(\mathbf{r}_{23}, \omega_2) \\ &= \sum_j \int d\mathbf{R}_2\, \mathfrak{n}_j^{(1)}(\mathbf{R}_2)\sigma_j(\mathbf{r}_3|\mathbf{R}_2). \end{aligned}$$

This charge density originates both from the inhomogeneity in number density and from the anisotropy of the orientations of the particles. The ***total electrostatic potential*** Ψ (the mean potential) is the sum of Ψ^{Ext} and the potential due to the mean charge density ρ

$$\begin{aligned} \Psi(\mathbf{r}_4) &= \Psi^{\text{Ext}}(\mathbf{r}_4) + \int d\mathbf{r}_3\, \rho(\mathbf{r}_3)\phi_{\text{Coul}}(r_{34}) \\ &= \Psi^{\text{Ext}}(\mathbf{r}_4) + \sum_j \int d\mathbf{r}_3 d\mathbf{R}_2\, \mathfrak{n}_j^{(1)}(\mathbf{R}_2)\sigma_j(\mathbf{r}_3|\mathbf{R}_2)\phi_{\text{Coul}}(r_{34}). \quad (13.10) \end{aligned}$$

If we place a particle of species i with orientation ω_1 at position $\mathbf{r}_1$, the density distribution for j-particles changes from $\mathfrak{n}_j^{(1)}(\mathbf{R}_2)$ to

$$\mathfrak{n}_{j;i}(\mathbf{R}_2|\mathbf{R}_1) \equiv g_{ij}^{(2)}(\mathbf{R}_1, \mathbf{R}_2)\mathfrak{n}_j^{(1)}(\mathbf{R}_2)$$

and we have the charge distribution

$$\rho_i(\mathbf{r}_3|\mathbf{R}_1) = \sum_j \int d\mathbf{R}_2\, g_{ij}^{(2)}(\mathbf{R}_1, \mathbf{R}_2)\mathfrak{n}_j^{(1)}(\mathbf{R}_2)\sigma_j(\mathbf{r}_3|\mathbf{R}_2) \quad (13.11)$$

in the surroundings of the i-particle. The charge distribution that arises *because* we place the i-particle with orientation ω_1 at position $\mathbf{r}_1$ is accordingly

$$\begin{aligned} \Delta\rho_i(\mathbf{r}_3|\mathbf{R}_1) &= \rho_i(\mathbf{r}_3|\mathbf{R}_1) - \rho(\mathbf{r}_3) \\ &= \sum_j \int d\mathbf{R}_2\, h_{ij}^{(2)}(\mathbf{R}_1, \mathbf{R}_2)\mathfrak{n}_j^{(1)}(\mathbf{R}_2)\sigma_j(\mathbf{r}_3|\mathbf{R}_2). \quad (13.12) \end{aligned}$$

[3]A hint of how to read this kind of integral: We want to find out the local charge density at $\mathbf{r}_3$. Any particle that overlap with $\mathbf{r}_3$ can contribute to this. Say that the center of such a particle is located at point $\mathbf{r}_2$ in the neighborhood. The internal charge distribution $\sigma_j(\mathbf{r}_{23}, \omega_2)$ gives the charge density at $\mathbf{r}_3$ from a particle at $\mathbf{r}_2$ with orientation ω_2. The number of such particles is $\mathfrak{n}_j^{(1)}(\mathbf{r}_2, \omega_2)d\mathbf{r}_2 d\omega_2$ and the integral give the sum of all contributions.

Thus the surroundings of the particle is polarized by the interactions with it and $\Delta\rho_i$ is the induced charge density, which is caused by all kinds of interactions and not only the electrostatic ones. Analogously to the case of spherical ions in Chapter 12, we will denote $\Delta\rho_i(\mathbf{r}_2|\mathbf{r}_1)$ as the charge density of the **particle cloud** that surrounds the i-particle and we therefore write $\rho_i^{\mathrm{cloud}} = \Delta\rho_i$, that is,

$$\rho_i^{\mathrm{cloud}}(\mathbf{r}_3|\mathbf{R}_1) = \sum_j \int d\mathbf{R}_2\, h_{ij}^{(2)}(\mathbf{R}_1,\mathbf{R}_2)\mathfrak{n}_j^{(1)}(\mathbf{R}_2)\sigma_j(\mathbf{r}_3|\mathbf{R}_2). \tag{13.13}$$

A particle cloud is analogous an ion cloud for a system where all particles are ions. In electrolyte solutions with molecular solvent the cloud does not only consists of ions but also of solvent molecules with orientations and locations that are influenced by the interactions with the i-particle.

The *total* charge density of the particle and the induced charge in its surroundings is

$$\rho_i^{\mathrm{tot}}(\mathbf{r}_3|\mathbf{R}_1) = \sigma_i(\mathbf{r}_3|\mathbf{R}_1) + \rho_i^{\mathrm{cloud}}(\mathbf{r}_3|\mathbf{R}_1) \tag{13.14}$$

and is the entire charge density that is associated with an i-particle at $\mathbf{R}_1$. The mean electrostatic potential ψ_i from this charge density is

$$\psi_i(\mathbf{r}_4|\mathbf{R}_1) = \int d\mathbf{r}_3\, \rho_i^{\mathrm{tot}}(\mathbf{r}_3|\mathbf{R}_1)\, \phi_{\mathrm{Coul}}(r_{34}) \tag{13.15}$$

and it is the *mean electrostatic potential* due to the i-particle that satisfies the Poisson equation

$$-\varepsilon_0 \nabla_4^2 \psi_i(\mathbf{r}_4|\mathbf{R}_1) = \rho_i^{\mathrm{tot}}(\mathbf{r}_4|\mathbf{R}_1), \tag{13.16}$$

with boundary condition $\psi_i(\mathbf{r}_4|\mathbf{R}_1) \to 0$ when r_{14} becomes very large (for a finite system we here assume that $\mathbf{r}_4$ remains inside the electrolyte).

The formalism for fluids of nonspherical charged particles with or without solvent species is very similar to the formalism for spherical ions that we developed in Chapter 12. The orientation degrees of freedom add complexity to the formulas, but apart from this complication the formalism is very similar. Below we will consider electrolyte solutions and ionic liquids where all particles are free to move, that is, we do not have any particle placed fixed at some point. The external charge density that gives rise to $\Psi^{\mathrm{ext}}(\mathbf{r})$ is always stationary and like before we will only consider equilibrium systems so there is no time dependence of the various properties of the fluid.

13.2 DIELECTRIC PROPERTIES OF INHOMOGENEOUS ELECTROLYTE SOLUTIONS AND IONIC LIQUIDS

The theory of electric polarization and dielectric properties of inhomogeneous electrolytes with nonspherical particles can be developed analogously to the corresponding theory in Sections 12.2.1–12.2.3 for spherical ions and the formalism is very similar. Therefore, we will only give a brief overview of this topic here in order to establish the most important formulas. Several of the corresponding formulas for the spherical case can be found in Tables 12.1 and 12.2 on pages 290 and 303, respectively.

When the external potential $\Psi^{\rm ext}(\mathbf{r}_2)$ is changed by an infinitesimal amount $\delta\Psi^{\rm ext}(\mathbf{r}_2)$, a charge density $\delta\rho^{\rm pol}(\mathbf{r}_1)$ is induced in the electrolyte whereby $\rho(\mathbf{r}_1)$ becomes $\rho(\mathbf{r}_1) + \delta\rho^{\rm pol}(\mathbf{r}_1)$ and the total potential becomes $\Psi(\mathbf{r}_2) + \delta\Psi(\mathbf{r}_2)$, where

$$\delta\Psi(\mathbf{r}_2) = \delta\Psi^{\rm ext}(\mathbf{r}_2) + \delta\Psi^{\rm pol}(\mathbf{r}_2)$$

and

$$\delta\Psi^{\rm pol}(\mathbf{r}_2) = \int d\mathbf{r}_1\, \delta\rho^{\rm pol}(\mathbf{r}_1)\phi_{\rm Coul}(r_{12}).$$

We use the superscript "pol" on $\rho^{\rm pol}$ to mark that it is a polarization caused by the change in external electrostatic potential. As in the case of spherical ions, the polarization charge density is given by

$$\delta\rho^{\rm pol}(\mathbf{r}_1) = \int d\mathbf{r}_3\, \chi^{\rho}(\mathbf{r}_1,\mathbf{r}_3)\delta\Psi^{\rm ext}(\mathbf{r}_3), \tag{13.17}$$

where χ^{ρ} is the **polarization response function**

$$\chi^{\rho}(\mathbf{r}_1,\mathbf{r}_3) = -\beta q_{\rm e}^2 H_{\rm QQ}(\mathbf{r}_1,\mathbf{r}_3) \tag{13.18}$$

and $H_{\rm QQ}$ is the charge-charge correlation function, which in the present case is defined as

$$H_{\rm QQ}(\mathbf{r}_1,\mathbf{r}_3) \equiv \frac{1}{q_{\rm e}^2}\sum_{ij}\int d\mathbf{R}_4 d\mathbf{R}_2\, H_{ij}^{(2)}(\mathbf{R}_4,\mathbf{R}_2)\sigma_i(\mathbf{r}_1|\mathbf{R}_4)\sigma_j(\mathbf{r}_3|\mathbf{R}_2). \tag{13.19}$$

Equations 13.17 and 13.18 are derived in the shaded text box below and it is also shown there that

$$\chi^{\rho}(\mathbf{r}_1,\mathbf{r}_3) = -\beta\sum_{i}\int d\mathbf{R}_4\, \sigma_i(\mathbf{r}_1|\mathbf{R}_4)\mathrm{n}_i^{(1)}(\mathbf{R}_4)\rho_i^{\rm tot}(\mathbf{r}_3|\mathbf{R}_4). \tag{13.20}$$

(cf. Equation 12.20).

The expression 13.19 for $H_{\rm QQ}$ can be interpreted as follows. $H_{ij}^{(2)}(\mathbf{R}_4,\mathbf{R}_2)$ describes how fluctuations in density at $\mathbf{r}_4$ for i-particles with orientation ω_4 correlate with the corresponding fluctuations at $\mathbf{r}_2$ for j-particles with orientation ω_2 (cf. Equation 11.121 for the one-component case). The product $H_{ij}^{(2)}(\mathbf{R}_4,\mathbf{R}_2)\sigma_i(\mathbf{r}_1|\mathbf{R}_4)\sigma_j(\mathbf{r}_3|\mathbf{R}_2)$ contains the charge density at $\mathbf{r}_1$ of a particle with coordinates $\mathbf{R}_4$ and the charge density at $\mathbf{r}_3$ of a particle with coordinates $\mathbf{R}_2$, that is, particles that participate in the density fluctuations just described. The integrand hence describes the correlation in charge densities at $\mathbf{r}_1$ and $\mathbf{r}_3$

due to these density fluctuations. By integrating over $\mathbf{R}_4$ and $\mathbf{R}_2$ and summing over all species one obtains the correlation in charge densities fluctuations at $\mathbf{r}_1$ and at $\mathbf{r}_3$ due to all particles, that is, from all locations $\mathbf{r}_4$ and $\mathbf{r}_2$ for particles with all orientations $\boldsymbol{\omega}_4$ and $\boldsymbol{\omega}_2$.

From the definition of $H_{ij}^{(2)}$ in Equation 11.140 we see that

$$\begin{aligned} &H_{\mathrm{QQ}}(\mathbf{r}_1, \mathbf{r}_3) \\ &= \frac{1}{q_{\mathrm{e}}^2}\sum_i \int d\mathbf{R}_2\, \mathfrak{n}_i^{(1)}(\mathbf{R}_2)\sigma_i(\mathbf{r}_1|\mathbf{R}_2)\sigma_i(\mathbf{r}_3|\mathbf{R}_2) \\ &+ \frac{1}{q_{\mathrm{e}}^2}\sum_{ij} \int d\mathbf{R}_4 d\mathbf{R}_2\, h_{ij}^{(2)}(\mathbf{R}_4, \mathbf{R}_2)\mathfrak{n}_i^{(1)}(\mathbf{R}_4)\sigma_i(\mathbf{r}_1|\mathbf{R}_4)\mathfrak{n}_j^{(1)}(\mathbf{R}_2)\sigma_j(\mathbf{r}_3|\mathbf{R}_2) \end{aligned} \tag{13.21}$$

(cf. Equation 12.21), where the first term on the rhs is a self-term that describes the perfect correlations between the charges within each particle.

To show Equations 13.17-13.20 we start from

$$\delta\rho^{\mathrm{pol}}(\mathbf{r}_1) = \sum_i \int d\mathbf{R}_4\, \delta\mathfrak{n}_i^{(1)}(\mathbf{R}_4)\sigma_i(\mathbf{r}_1|\mathbf{R}_4),$$

where $\delta\mathfrak{n}_i^{(1)}$ is the variation in number density due to the change $\delta\Psi^{\mathrm{Ext}}$. This variation is given by the first Yvon equation 11.144 with $\delta v_j(\mathbf{R}_2) = \int d\mathbf{r}_3\, \sigma_j(\mathbf{r}_3|\mathbf{R}_2)\delta\Psi^{\mathrm{Ext}}(\mathbf{r}_3)$, which implies that

$$\begin{aligned} \delta\mathfrak{n}_i^{(1)}(\mathbf{R}_4) &= -\beta\sum_j \int d\mathbf{R}_2\, H_{ij}^{(2)}(\mathbf{R}_4, \mathbf{R}_2)\delta v_j(\mathbf{R}_2) \\ &= -\beta \int d\mathbf{r}_3 \sum_j \int d\mathbf{R}_2\, H_{ij}^{(2)}(\mathbf{R}_4, \mathbf{R}_2)\sigma_j(\mathbf{r}_3|\mathbf{R}_2)\delta\Psi^{\mathrm{Ext}}(\mathbf{r}_3). \end{aligned}$$

Equation 13.17 with 13.18 inserted follow from these expressions. To derive Equation 13.20, the definitions 11.140 of $H_{ij}^{(2)}$ and 13.14 of ρ_i^{tot} are also used.

It follows from the results in the shaded text box that

$$\delta\mathfrak{n}_i^{(1)}(\mathbf{R}_4) = -\beta\mathfrak{n}_i^{(1)}(\mathbf{R}_4) \int d\mathbf{r}_3\, \rho_i^{\mathrm{tot}}(\mathbf{r}_3|\mathbf{R}_4)\delta\Psi^{\mathrm{Ext}}(\mathbf{r}_3)$$

which implies that

$$\delta w_i(\mathbf{R}_4) = \int d\mathbf{r}_3\, \rho_i^{\mathrm{tot}}(\mathbf{r}_3|\mathbf{R}_4)\delta\Psi^{\mathrm{ext}}(\mathbf{r}_3)$$

(cf. Equation 12.14), that is,

$$\frac{\delta w_i(\mathbf{R}_4)}{\delta\Psi^{\mathrm{ext}}(\mathbf{r}_3)} = \rho_i^{\mathrm{tot}}(\mathbf{r}_3|\mathbf{R}_4). \tag{13.22}$$

Thus, the change in potential of mean force felt by an i-particle at $\mathbf{r}_4$ with orientation ω_4 is equal to the interaction energy between $\delta\Psi^{\text{ext}}$ and total charge density ρ_i^{tot} associated with the particle.

Like in the case of spherical ions, it is important to express these changes in terms of the potential Ψ that exist in the *presence* of the fluid instead of in terms of Ψ^{ext}, which is the potential due to the external charges in the *absence* of the fluid. Therefore, in addition to the polarization response function $\chi^{\rho}(\mathbf{r}_1, \mathbf{r}_3)$ that is given by

$$\chi^{\rho}(\mathbf{r}_1, \mathbf{r}_3) = \frac{\delta\rho^{\text{pol}}(\mathbf{r}_1)}{\delta\Psi^{\text{ext}}(\mathbf{r}_3)} = \frac{\delta\rho(\mathbf{r}_1)}{\delta\Psi^{\text{ext}}(\mathbf{r}_3)}$$

(Equation 12.22), we will consider the polarization response function $\chi^{\star}(\mathbf{r}_1, \mathbf{r}_3)$ that is given by

$$\chi^{\star}(\mathbf{r}_1, \mathbf{r}_3) = \frac{\delta\rho^{\text{pol}}(\mathbf{r}_1)}{\delta\Psi(\mathbf{r}_3)} = \frac{\delta\rho(\mathbf{r}_1)}{\delta\Psi(\mathbf{r}_3)}$$

(Equation 12.25). Exactly like before, these functions are related to each other via Equation 12.24, so when we know χ^{ρ} we also know $\chi^{\star}$, at least in principle. As we found earlier, the functions $\delta^{(3)}(\mathbf{r}_{23}) + \int d\mathbf{r}_4\, \chi^{\rho}(\mathbf{r}_4, \mathbf{r}_2)\phi_{\text{Coul}}(r_{43}) = \epsilon^{\text{R}}(\mathbf{r}_2; \mathbf{r}_3)$ and $\delta^{(3)}(\mathbf{r}_{12}) - \int d\mathbf{r}_5\, \chi^{\star}(\mathbf{r}_5, \mathbf{r}_1)\phi_{\text{Coul}}(r_{52}) = \epsilon(\mathbf{r}_1; \mathbf{r}_2)$ (defined in Equation 12.56 and 12.60) are functional inverses of each other.

In particular, we are interested to find the dressed charge distribution $\rho_i^{\star}(\mathbf{r}_3|\mathbf{R}_4)$ associated with the i-particle because it gives $\delta w_i(\mathbf{R}_4)$ in terms of $\delta\Psi(\mathbf{r}_3)$

$$\delta w_i(\mathbf{R}_4) = \int d\mathbf{r}_3\, \rho_i^{\star}(\mathbf{r}_3|\mathbf{R}_4)\delta\Psi(\mathbf{r}_3), \tag{13.23}$$

which is the same as

$$\frac{\delta w_i(\mathbf{R}_4)}{\delta\Psi(\mathbf{r}_3)} = \rho_i^{\star}(\mathbf{r}_3|\mathbf{R}_4).$$

This is the analogue of Equation 13.22. Like Equation Equation 12.26 in the spherical case, Equation 13.23 expresses the fact that ***electrostatics is non-local***. It is straightforward to show in the same way as for spherical ions that

$$\rho_i^{\star}(\mathbf{r}_3|\mathbf{R}_4) = \rho_i^{\text{tot}}(\mathbf{r}_3|\mathbf{R}_4) - \int d\mathbf{r}_2\, \psi_i(\mathbf{r}_2|\mathbf{R}_4)\chi^{\star}(\mathbf{r}_2, \mathbf{r}_3)$$

(cf. Equation 12.29), where the last term removes the *linear part* of the polarization response from the electrostatic field due to the i-particle. As before, this linear part, the integral, is denoted ρ_i^{lin}. When the particle is an ion, $\rho_i^{\star}$ is the **dressed-ion charge density**, while in general it may be called the **dressed-particle charge density**, and we have

$$\rho_i^{\star}(\mathbf{r}_3|\mathbf{R}_4) = \sigma(\mathbf{r}_3|\mathbf{R}_4) + \rho_i^{\text{dress}}(\mathbf{r}_3|\mathbf{R}_4)$$

where $\rho_i^{\text{dress}} = \rho_i^{\text{cloud}} - \rho_i^{\text{lin}}$, that is,

$$\rho_i^{\text{dress}}(\mathbf{r}_3|\mathbf{R}_4) = \rho_i^{\text{cloud}}(\mathbf{r}_3|\mathbf{R}_4) - \int d\mathbf{r}_2\, \psi_i(\mathbf{r}_2|\mathbf{R}_4)\chi^{\star}(\mathbf{r}_2, \mathbf{r}_3).$$

Like ρ_i^{cloud}, the charge distribution ρ_i^{dress} in electrolyte solutions is not only due to the surrounding ions but also solvent molecules with orientations and locations that are influenced by the interactions with the i-particle. Thus, the **dress of a particle** consists of both ions and other molecules. It is straightforward to show that

$$\chi^\star(\mathbf{r}_1,\mathbf{r}_3) = -\beta\sum_i\int d\mathbf{R}_4\,\sigma_i(\mathbf{r}_1|\mathbf{R}_4)\mathrm{n}_i^{(1)}(\mathbf{R}_4)\rho_i^\star(\mathbf{r}_3|\mathbf{R}_4) \tag{13.24}$$

(cf. the derivation of Equation 12.32), which has the same form as Equation 13.20 for χ^ρ.

As we did in the spherical case we now introduce the pair correlation functions $h_{ij}^\star$ that give ρ_i^{dress} in exactly the same way as ρ_i^{cloud} in Equation 13.13 is given in terms of $h_{ij}^{(2)}$. To this end we define the pair correlation function of the dress as

$$h_{ij}^\star(\mathbf{R}_1,\mathbf{R}_2) \equiv h_{ij}^{(2)}(\mathbf{R}_1,\mathbf{R}_2) + \beta\int d\mathbf{r}_3\,\psi_i(\mathbf{r}_3|\mathbf{R}_1)\rho_j^\star(\mathbf{r}_3|\mathbf{R}_2). \tag{13.25}$$

(cf. Equation 12.33) and it is straightforward to show that

$$\rho_i^\star(\mathbf{r}_3|\mathbf{R}_1) = \sigma_i(\mathbf{r}_3|\mathbf{R}_1) + \sum_j\int d\mathbf{R}_2\,h_{ij}^\star(\mathbf{R}_1,\mathbf{R}_2)\mathrm{n}_j^{(1)}(\mathbf{R}_2)\sigma_j(\mathbf{r}_3|\mathbf{R}_2) \tag{13.26}$$

(cf Equation 12.34), so we have the charge density of the dress

$$\rho_i^{\text{dress}}(\mathbf{r}_3|\mathbf{R}_1) = \sum_j\int d\mathbf{R}_2\,h_{ij}^\star(\mathbf{R}_1,\mathbf{R}_2)\mathrm{n}_j^{(1)}(\mathbf{R}_2)\sigma_j(\mathbf{r}_3|\mathbf{R}_2),$$

which has the same form as ρ_i^{cloud} in Equation 13.13. As in the spherical case, it is easy to show that $h_{ij}^\star$ is symmetric, $h_{ij}^\star(\mathbf{R}_1,\mathbf{R}_2) = h_{ji}^\star(\mathbf{R}_2,\mathbf{R}_1)$.

Like before we write $h_{ij}^\star$ as

$$h_{ij}^\star(\mathbf{R}_1,\mathbf{R}_2) \equiv h_{ij}^{(2)}(\mathbf{R}_1,\mathbf{R}_2) + \beta w_{ij}^{(2)\text{el}}(\mathbf{R}_1,\mathbf{R}_2), \tag{13.27}$$

where $\beta w_{ij}^{(2)\text{el}}$ is the integral in Equation 13.25 so that

$$\begin{aligned} w_{ij}^{(2)\text{el}}(\mathbf{R}_1,\mathbf{R}_2) &\equiv \int d\mathbf{r}_3\,\psi_i(\mathbf{r}_3|\mathbf{R}_1)\rho_j^\star(\mathbf{r}_3|\mathbf{R}_2) \\ &= \int d\mathbf{r}_4\,\rho_i^\star(\mathbf{r}_4|\mathbf{R}_1)\psi_j(\mathbf{r}_4|\mathbf{R}_2) \end{aligned} \tag{13.28}$$

(cf. Equation 12.37), where the last equality follows from symmetry. The function $w_{ij}^{(2)\text{el}}$ is the *linear electrostatic part* of $w_{ij}^{(2)}$ irrespectively of the magnitude of ψ_i and ψ_j.

To summarize the main points so far, we have seen that:

The dressed charge density $\rho_i^\star$ has the same role vis-à-vis the total potential $\delta\Psi$ as the total charge density ρ_i^{tot} has vis-à-vis the external potential $\delta\Psi^{\mathrm{ext}}$,
$\rho_i^\star$ has the same role vis-à-vis $\chi^\star$ as ρ_i^{tot} has vis-à-vis χ^ρ and
$h_{ij}^\star$ has the same role vis-à-vis $\rho_i^\star$ as $h_{ij}^{(2)}$ has vis-à-vis ρ_i^{tot}.

This is the same as we found for the spherical case (cf. Table 12.2 on page 303) and like in that case $\rho_i^\star$ and $h_{ij}^\star$ are essential ingredients in the analysis of properties of electrolytes.

In analogy to the definitions of $H_{ij}^{(2)}$ and H_{QQ} we introduce

$$H_{ij}^\star(\mathbf{R}_1,\mathbf{R}_2) \equiv \mathfrak{n}_i^{(1)}(\mathbf{R}_1)\delta_{ij}^{(3)}(\mathbf{R}_1-\mathbf{R}_2)+\mathfrak{n}_i^{(1)}(\mathbf{R}_1)\mathfrak{n}_j^{(1)}(\mathbf{R}_2)h_{ij}^\star(\mathbf{R}_1,\mathbf{R}_2) \quad (13.29)$$

and

$$H_{\mathrm{QQ}}^\star(\mathbf{r}_3,\mathbf{r}_4) \equiv \frac{1}{q_\mathrm{e}^2}\sum_{ij}\int d\mathbf{R}_1 d\mathbf{R}_2\, \sigma_i(\mathbf{r}_3|\mathbf{R}_1)\sigma_j(\mathbf{r}_4|\mathbf{R}_2)H_{ij}^\star(\mathbf{R}_1,\mathbf{R}_2),$$

whereby we can write

$$\chi^\star(\mathbf{r}_1,\mathbf{r}_2) = -\beta q_\mathrm{e}^2 H_{\mathrm{QQ}}^\star(\mathbf{r}_1,\mathbf{r}_2) \quad (13.30)$$

in analogy to χ^ρ in Equation 13.18. $H_{ij}^\star$ is the functional inverse of $C_{ij}^\star$, which is defined as

$$C_{ij}^\star(\mathbf{R}_1,\mathbf{R}_2) \equiv \frac{1}{\mathfrak{n}_i^{(1)}(\mathbf{R}_2)}\delta_{ij}^{(3)}(\mathbf{R}_1-\mathbf{R}_2) - c_{ij}^\star(\mathbf{R}_1,\mathbf{R}_2), \quad (13.31)$$

where

$$c_{ij}^\star(\mathbf{R}_1,\mathbf{R}_2) \equiv c_{ij}^{(2)}(\mathbf{R}_1,\mathbf{R}_2) + \beta u_{ij}^{\mathrm{el}}(\mathbf{R}_1,\mathbf{R}_2), \quad (13.32)$$

so we have

$$\sum_l \int d\mathbf{R}_3\, C_{il}^\star(\mathbf{R}_1,\mathbf{R}_3)H_{lj}^\star(\mathbf{R}_3,\mathbf{R}_2) = \delta_{ij}^{(3)}(\mathbf{R}_1-\mathbf{R}_2)$$

(cf. Equations 12.44 - 12.45). Since $c_{ij}^{(2)}(\mathbf{R}_1,\mathbf{R}_2) \sim -\beta u_{ij}(\mathbf{R}_1,\mathbf{R}_2) \sim -u_{ij}^{\mathrm{el}}(\mathbf{R}_1,\mathbf{R}_2)$ when $r_{12}\to\infty$ we can identify $c_{ij}^\star$ as the short-range part of $c_{ij}^{(2)}$ in the electrolyte.

The ***electric susceptibility function***

$$\chi^{\mathrm{el}}(\mathbf{r}_1;\mathbf{r}_2) = -\frac{\delta\Psi^{\mathrm{pol}}(\mathbf{r}_2)}{\delta\Psi(\mathbf{r}_1)},$$

the ***dielectric permittivity function***

$$\epsilon(\mathbf{r}_1;\mathbf{r}_2) \equiv \frac{\delta\Psi^{\mathrm{Ext}}(\mathbf{r}_2)}{\delta\Psi(\mathbf{r}_1)} = \delta^{(3)}(\mathbf{r}_{12}) + \chi^{\mathrm{el}}(\mathbf{r}_1;\mathbf{r}_2)$$

and the ***reciprocal dielectric-permittivity function***

$$\epsilon^{\mathrm{R}}(\mathbf{r}_2;\mathbf{r}_3) \equiv \frac{\delta\Psi(\mathbf{r}_3)}{\delta\Psi^{\mathrm{ext}}(\mathbf{r}_2)}$$

are introduced in exactly the same manner as in Section 12.2.3 and we have

$$\begin{aligned}
\chi^{\mathrm{el}}(\mathbf{r}_1;\mathbf{r}_2) &= -\int d\mathbf{r}_3\, \chi^{\star}(\mathbf{r}_1,\mathbf{r}_3)\phi_{\mathrm{Coul}}(r_{32}) \\
\epsilon(\mathbf{r}_1;\mathbf{r}_2) &= \delta^{(3)}(\mathbf{r}_{12}) - \int d\mathbf{r}_3\, \chi^{\star}(\mathbf{r}_1,\mathbf{r}_3)\phi_{\mathrm{Coul}}(r_{32}) \qquad (13.33)\\
\epsilon^{\mathrm{R}}(\mathbf{r}_2;\mathbf{r}_3) &= \delta^{(3)}(\mathbf{r}_{23}) + \int d\mathbf{r}_4\, \chi^{\rho}(\mathbf{r}_2,\mathbf{r}_4)\phi_{\mathrm{Coul}}(r_{43})
\end{aligned}$$

(Equations 12.54, 12.56 and 12.60). Using Equations 13.24 and 13.30 we can, for example, write $\epsilon(\mathbf{r}_1;\mathbf{r}_2)$ explicitly in terms of $\rho_i^{\star}$, $H_{ij}^{\star}$ or $h_{ij}^{\star}$.

The functions $\epsilon^{\mathrm{R}}(\mathbf{r}_2;\mathbf{r}_3)$ and $\epsilon(\mathbf{r}_1;\mathbf{r}_2)$ are functional inverses of each other and we have the useful relationships

$$\rho_i^{\mathrm{tot}}(\mathbf{r}_2|\mathbf{R}_1) = \int d\mathbf{r}_4\, \epsilon^{\mathrm{R}}(\mathbf{r}_2;\mathbf{r}_4)\rho_i^{\star}(\mathbf{r}_4|\mathbf{R}_1) \qquad (13.34)$$

$$\rho_j^{\star}(\mathbf{r}_3|\mathbf{R}_1) = \int d\mathbf{r}_2\, \epsilon(\mathbf{r}_3;\mathbf{r}_2)\rho_j^{\mathrm{tot}}(\mathbf{r}_2|\mathbf{R}_1) \qquad (13.35)$$

(cf. Equations 12.62 and 12.63). The ***unit screened Coulomb potential*** $\phi_{\mathrm{Coul}}^{\star}$ for inhomogeneous electrolytes is defined like in Equation 12.65, that is,

$$\phi_{\mathrm{Coul}}^{\star}(\mathbf{r}_4,\mathbf{r}_3) \equiv \int d\mathbf{r}_2\, \epsilon^{\mathrm{R}}(\mathbf{r}_2;\mathbf{r}_4)\phi_{\mathrm{Coul}}(r_{23}) \qquad (13.36)$$

and we have the ordinary and the screened versions of the Coulomb potential formula

$$\begin{aligned}
\psi_i(\mathbf{r}_3|\mathbf{R}_1) &= \int d\mathbf{r}_2\, \rho_i^{\mathrm{tot}}(\mathbf{r}_2|\mathbf{R}_1)\phi_{\mathrm{Coul}}(r_{23}) \\
&= \int d\mathbf{r}_4\, \rho_i^{\star}(\mathbf{r}_4|\mathbf{R}_1)\phi_{\mathrm{Coul}}^{\star}(\mathbf{r}_4,\mathbf{r}_3) \qquad (13.37)
\end{aligned}$$

(cf. Equation 12.64). We can easily derive the latter from the former using Equation 13.34. Alternatively, both versions follow from the facts that ϕ_{Coul} is the Green's function of the Poisson equation 13.16 and $\phi_{\mathrm{Coul}}^{\star}$ is the Green's function of the modified Poisson equation

$$-\varepsilon_0 \nabla_3^2 \psi_i(\mathbf{r}_3|\mathbf{R}_1) - \int d\mathbf{r}_4\, \psi_i(\mathbf{r}_4|\mathbf{R}_1)\chi^\star(\mathbf{r}_4,\mathbf{r}_3) = \rho_i^\star(\mathbf{r}_3|\mathbf{R}_1), \tag{13.38}$$

(cf. Equation 12.70). Thus, $\phi^\star_{\mathrm{Coul}}$ satisfies the Green's function equation

$$-\varepsilon_0 \nabla_3^2 \phi^\star_{\mathrm{Coul}}(\mathbf{r}_2,\mathbf{r}_3) - \int d\mathbf{r}_4\, \phi^\star_{\mathrm{Coul}}(\mathbf{r}_2,\mathbf{r}_4)\chi^\star(\mathbf{r}_4,\mathbf{r}_3) = \delta^{(3)}(\mathbf{r}_{23}), \tag{13.39}$$

(Equation 12.71).

From Equation 13.37 it follows that $w_{ij}^{(2)\mathrm{el}}$ in Equation 13.28 is given by

$$w_{ij}^{(2)\mathrm{el}}(\mathbf{R}_1,\mathbf{R}_2) = \int d\mathbf{r}_3\, d\mathbf{r}_4\, \rho_i^\star(\mathbf{r}_4|\mathbf{R}_1)\phi^\star_{\mathrm{Coul}}(\mathbf{r}_4,\mathbf{r}_3)\rho_j^\star(\mathbf{r}_3|\mathbf{R}_2),$$

which has the form of an electrostatic interactions energy between the charge distributions $\rho_i^\star$ and $\rho_j^\star$ as mediated by $\phi^\star_{\mathrm{Coul}}$. We have from Equation 13.27

$$h_{ij}^{(2)}(\mathbf{R}_1,\mathbf{R}_2) = h_{ij}^\star(\mathbf{R}_1,\mathbf{R}_2) - \beta \int d\mathbf{r}_3\, d\mathbf{r}_4\, \rho_i^\star(\mathbf{r}_4|\mathbf{R}_1)\phi^\star_{\mathrm{Coul}}(\mathbf{r}_4,\mathbf{r}_3)\rho_j^\star(\mathbf{r}_3|\mathbf{R}_2) \tag{13.40}$$

and from Equations 13.4 and 13.32 it follows that

$$c_{ij}^{(2)}(\mathbf{R}_1,\mathbf{R}_2) = c_{ij}^\star(\mathbf{R}_1,\mathbf{R}_2) - \beta \int d\mathbf{r}_3 d\mathbf{r}_4\, \sigma_i(\mathbf{r}_4|\mathbf{R}_1)\phi_{\mathrm{Coul}}(r_{43})\sigma_j(\mathbf{r}_3|\mathbf{R}_2), \tag{13.41}$$

which has a similar structure as $h_{ij}^{(2)}$ but uses the unscreened entities in the integral. Like for spherical ions, this expression for $h_{ij}^{(2)}$ constitute a starting point for a general analysis of correlations in bulk electrolytes in the exact case. This is the topic of the next section.

13.3 INTERACTIONS AND SCREENING IN ELECTROLYTE SOLUTIONS AND IONIC FLUIDS IN BULK, GENERAL ANALYSIS

13.3.1 General formalism for bulk electrolytes of nonspherical particles

In this section we will treat homogeneous bulk electrolytes consisting of rigid particles of arbitrary sizes, shapes and internal charge distributions. Thereby we will include **anisotropic bulk fluids** where the bulk density $\mathfrak{n}^{\mathrm{b}}(\boldsymbol{\omega})$ depends on the orientation variable $\boldsymbol{\omega}$ but not on position. Thus we include nematic liquid crystals but not smectic ones. Smectic liquid crystals can be regarded as a special case of inhomogeneous systems (in the limit of zero external potential) treated in the previous section.

The charge densities of the particle cloud and the dress of an i-ion are

$$\begin{aligned}
\rho_i^{\text{cloud}}(\mathbf{r}_3|\mathbf{R}_1) &= \rho_i^{\text{cloud}}(\mathbf{r}_{13},\omega_1) \\
&\equiv \sum_j \int d\mathbf{r}_2 \left\langle h_{ij}^{(2)}(\mathbf{r}_{12},\omega_1,\omega_2)\mathfrak{n}_j^{(1)}(\omega_2)\sigma_j(\mathbf{r}_{23},\omega_2)\right\rangle_{\omega_2} \\
\rho_i^{\text{dress}}(\mathbf{r}_3|\mathbf{R}_1) &= \rho_i^{\text{dress}}(\mathbf{r}_{13},\omega_1) \\
&\equiv \sum_j \int d\mathbf{r}_2 \left\langle h_{ij}^{\star}(\mathbf{r}_{12},\omega_1,\omega_2)\mathfrak{n}_j^{(1)}(\omega_2)\sigma_j(\mathbf{r}_{23},\omega_2)\right\rangle_{\omega_2},
\end{aligned}$$

respectively, since $h_{ij}^{(2)}(\mathbf{R}_1,\mathbf{R}_2) = h_{ij}^{(2)}(\mathbf{r}_2-\mathbf{r}_1,\omega_1,\omega_2)$ due to the homogeneity, whereby the pair functions depend on the vector $\mathbf{r}_{12}$ in an anisotropic medium and not only on the distance r_{12}. We have written the integral over ω_2 as an average because $\int d\omega f(\omega) = \int d\omega f(\omega)/\int d\omega' = \langle f(\omega)\rangle_\omega$ since ω is normalized. As before we have

$$\begin{aligned}
\rho_i^{\text{tot}}(\mathbf{r}_3|\mathbf{R}_1) &= \rho_i^{\text{tot}}(\mathbf{r}_{13},\omega_1) = \sigma_i(\mathbf{r}_{13},\omega_1) + \rho_i^{\text{cloud}}(\mathbf{r}_{13},\omega_1) \\
\rho_i^{\star}(\mathbf{r}_3|\mathbf{R}_1) &= \rho_i^{\star}(\mathbf{r}_{13},\omega_1) = \sigma_i(\mathbf{r}_{13},\omega_1) + \rho_i^{\text{dress}}(\mathbf{r}_{13},\omega_1).
\end{aligned}$$

For the isotropic case the only difference is that $\mathfrak{n}_j^{(1)}(\omega_2) = n_j^{\text{b}}$ in the formulas.[4]

The condition of local electroneutrality implies that

$$\int d\mathbf{r}_3\, \rho_i^{\text{tot}}(\mathbf{r}_{13},\omega_1) = 0$$

while the total charge of a dressed particle with orientation ω_1,

$$q_i^{\star}(\omega_1) \equiv \int d\mathbf{r}_3\, \rho_i^{\star}(\mathbf{r}_{13},\omega_1) = q_i + \int d\mathbf{r}_3\, \rho_i^{\text{dress}}(\mathbf{r}_{13},\omega_1),$$

is in general nonzero, $q_i^{\star}(\omega_1) \neq 0$, since ρ_i^{dress} does not contain the linear part of the polarization response in ρ_i^{cloud} due to the i-particle. The average charge of a dressed particle independently of orientation is

$$\overline{q}_i^{\star} \equiv \frac{1}{n_i^{\text{b}}}\left\langle \mathfrak{n}_i^{(1)}(\omega_1)q_i^{\star}(\omega_1)\right\rangle_{\omega_1} = \frac{1}{n_i^{\text{b}}}\int d\mathbf{r}_3 \left\langle \mathfrak{n}_i^{(1)}(\omega_1)\rho_i^{\star}(\mathbf{r}_{13},\omega_1)\right\rangle_{\omega_1}, \quad (13.42)$$

where we have used $\left\langle \mathfrak{n}_i^{(1)}(\omega)\right\rangle_\omega = n_i^{\text{b}}$. Note that in electrolyte solutions with polar solvents, only the ionic species contribute to $q_i^{\star}$ because the solvent molecules are uncharged. When the ion concentration goes to zero $q_i^{\star} \to q_i$ and $\overline{q}_i^{\star} \to q_i$ because in the definition of in ρ_i^{dress} the factor $\mathfrak{n}_j^{(1)}$ for $j \in \{$ionic species$\}$ goes to zero.

[4]The correlation function $h_{ij}^{(2)}(\mathbf{r}_{12},\omega_1,\omega_2)$ is invariant under any simultaneous rotation of the geometrical variables $\mathbf{r}_{12}$, ω_1 and ω_2 in an isotropic fluid and under any simultaneous rotation around the director vector of a nematic liquid crystal.

The charge density of the homogeneous bulk electrolyte is zero. When the electrolyte is exposed to a weak external electrostatic potential $\delta\Psi^{\text{ext}}(\mathbf{r})$, which is small variation from the value 0, the electrolyte is polarized and a polarization charge density $\delta\rho^{\text{pol}}(\mathbf{r}')$ arises. The resulting total electrostatic potential is $\delta\Psi = \delta\Psi^{\text{ext}} + \delta\Psi^{\text{pol}}$, where $\delta\Psi^{\text{pol}}$ arises from the induced $\delta\rho^{\text{pol}}$. When $\delta\Psi^{\text{ext}}$ is turned on, an i-particle at $\mathbf{r}_1$ with orientation ω_1 experiences a potential of mean force

$$\delta w_i(\mathbf{R}_1) = \int d\mathbf{r}_3\, \rho_i^{\text{tot}}(\mathbf{r}_3|\mathbf{R}_1)\delta\Psi^{\text{ext}}(\mathbf{r}_3) = \int d\mathbf{r}_3\, \rho_i^{\star}(\mathbf{r}_3|\mathbf{R}_1)\,\delta\Psi(\mathbf{r}_3),$$

which causes the fluid to become inhomogeneous. Both integrals have the form of an interaction energy between an electrostatic potential and a charge density associated with the i-particle: $\delta\Psi^{\text{ext}}$ and the charge density of the particle together with its entire cloud and the weak total potential $\delta\Psi$ and the charge density of the dressed particle. This is valid in the linear response domain, that is, for infinitesimally small $\delta\Psi^{\text{ext}}$ and $\delta\Psi$.

As before, the polarization response functions χ^{ρ} and $\chi^{\star}$ give the polarization charge density in terms of $\delta\Psi^{\text{ext}}$ and $\delta\Psi$, respectively. In this case we have

$$\delta\rho^{\text{pol}}(\mathbf{r}_1) = \int d\mathbf{r}_2\, \chi^{\rho}(\mathbf{r}_{12})\delta\Psi^{\text{ext}}(\mathbf{r}_2) = \int d\mathbf{r}_2\, \chi^{\star}(\mathbf{r}_{12})\delta\Psi(\mathbf{r}_2),$$

because $\chi^{\rho}(\mathbf{r}_1, \mathbf{r}_2) = \chi^{\rho}(\mathbf{r}_2 - \mathbf{r}_1)$ due to the homogeneity and likewise for $\chi^{\star}$. We have

$$\begin{aligned} \chi^{\rho}(\mathbf{r}_{12}) &= -\beta q_e^2 H_{QQ}(\mathbf{r}_{12}) && (13.43)\\ &= -\beta \sum_i \int d\mathbf{r}_3 \left\langle \sigma_i(\mathbf{r}_{31}, \omega_3)\mathrm{n}_i^{(1)}(\omega_3)\rho_i^{\text{tot}}(\mathbf{r}_{32}, \omega_3)\right\rangle_{\omega_3} \\ \chi^{\star}(\mathbf{r}_{12}) &= -\beta q_e^2 H_{QQ}^{\star}(\mathbf{r}_{12}) && (13.44)\\ &= -\beta \sum_i \int d\mathbf{r}_3 \left\langle \sigma_i(\mathbf{r}_{31}, \omega_3)\mathrm{n}_i^{(1)}(\omega_3)\rho_i^{\star}(\mathbf{r}_{32}, \omega_3)\right\rangle_{\omega_3}. \end{aligned}$$

The functions $\chi^{\rho}(\mathbf{r})$ and $\chi^{\star}(\mathbf{r})$ are related to each other via Equation 12.24, which becomes

$$\chi^{\rho}(\mathbf{r}_{14}) = \chi^{\star}(\mathbf{r}_{14}) + \int d\mathbf{r}_2 d\mathbf{r}_3\, \chi^{\star}(\mathbf{r}_{12})\phi_{\text{Coul}}(r_{23})\chi^{\rho}(\mathbf{r}_{34})$$

in the present case. Since the integrals are convolutions we have in Fourier space $\tilde{\chi}^{\rho}(\mathbf{k}) = \tilde{\chi}^{\star}(\mathbf{k}) + \tilde{\chi}^{\star}(\mathbf{k})\tilde{\phi}_{\text{Coul}}(k)\tilde{\chi}^{\rho}(\mathbf{k})$ and it follows that

$$1 - \tilde{\chi}^{\star}(\mathbf{k})\tilde{\phi}_{\text{Coul}}(k) = 1/[1 + \tilde{\chi}^{\rho}(\mathbf{k})\tilde{\phi}_{\text{Coul}}(k)].$$

Equation 13.33 becomes

$$\begin{aligned} \chi^{\mathrm{el}}(\mathbf{r}_{12}) &= -\int d\mathbf{r}_3\, \chi^{\star}(\mathbf{r}_{13})\phi_{\mathrm{Coul}}(r_{32}) \\ \epsilon(\mathbf{r}_{12}) &= \delta^{(3)}(\mathbf{r}_{12}) - \int d\mathbf{r}_3\, \chi^{\star}(\mathbf{r}_{13})\phi_{\mathrm{Coul}}(r_{32}) \\ \epsilon^{\mathrm{R}}(\mathbf{r}_{23}) &= \delta^{(3)}(\mathbf{r}_{23}) + \int d\mathbf{r}_4\, \chi^{\rho}(\mathbf{r}_{24})\phi_{\mathrm{Coul}}(r_{43}) \end{aligned}$$

and in Fourier space we have

$$\begin{aligned} \tilde{\chi}^{\mathrm{el}}(\mathbf{k}) &= -\tilde{\chi}^{\star}(\mathbf{k})\tilde{\phi}_{\mathrm{Coul}}(k) \\ \tilde{\epsilon}(\mathbf{k}) &= 1 - \tilde{\chi}^{\star}(\mathbf{k})\tilde{\phi}_{\mathrm{Coul}}(k) \;=\; 1 + \tilde{\chi}^{\mathrm{el}}(\mathbf{k}) \\ \tilde{\epsilon}^{\mathrm{R}}(\mathbf{k}) &= 1 + \tilde{\chi}^{\rho}(\mathbf{k})\tilde{\phi}_{\mathrm{Coul}}(k) \;=\; 1/\tilde{\epsilon}(\mathbf{k}). \end{aligned} \tag{13.45}$$

Like before, the function $\tilde{\epsilon}(\mathbf{k})$ is called the ***dielectric function*** (more precisely the *static* dielectric function), which should not be confused with the (spatial) dielectric permittivity function $\epsilon(\mathbf{r})$. We have

$$\delta\tilde{\Psi}^{\mathrm{pol}}(\mathbf{k}) = -\delta\tilde{\Psi}(\mathbf{k})\chi^{\mathrm{el}}(\mathbf{k}), \qquad \delta\tilde{\Psi}(\mathbf{k}) = \frac{\delta\tilde{\Psi}^{\mathrm{Ext}}(\mathbf{k})}{\tilde{\epsilon}(\mathbf{k})} \tag{13.46}$$

and the relationships between the corresponding weak electrostatic fields $\delta\mathbf{E}^{\mathrm{pol}}(\mathbf{r}) = -\nabla\Psi^{\mathrm{pol}}(\mathbf{r})$ etc. are

$$\delta\tilde{\mathbf{E}}^{\mathrm{pol}}(\mathbf{k}) = -\delta\tilde{\mathbf{E}}(\mathbf{k})\chi^{\mathrm{el}}(\mathbf{k}), \qquad \delta\mathbf{E}(\mathbf{k}) = \frac{\delta\mathbf{E}^{\mathrm{ext}}(\mathbf{k})}{\tilde{\epsilon}(\mathbf{k})}$$

because $\delta\tilde{\mathbf{E}}^{\mathrm{pol}}(\mathbf{k}) = -i\mathbf{k}\delta\tilde{\Psi}^{\mathrm{pol}}(\mathbf{k})$ and likewise for $\tilde{\mathbf{E}}$ and $\tilde{\mathbf{E}}^{\mathrm{ext}}$ (cf. the shaded text box on page 311).

Before we proceed, we will give the connection between $\tilde{\epsilon}(\mathbf{k})$ and the dielectric permittivity tensor $\tilde{\varepsilon}(\mathbf{k})$ that is commonly used in the theory of dielectric response. In fact, the dielectric function and the electric susceptibility are in general tensorial entities, but when we are dealing with equilibrium systems and do not consider time-dependent phenomena, it is sufficient to work with the scalar entities $\tilde{\epsilon}(\mathbf{k})$ and $\tilde{\chi}^{\mathrm{el}}(\mathbf{k})$. As we have seen in Section 13.2, the entities $\chi^{\mathrm{el}}(\mathbf{r}_1;\mathbf{r}_2)$, $\epsilon(\mathbf{r}_1;\mathbf{r}_2)$ and $\epsilon^{\mathrm{R}}(\mathbf{r}_1;\mathbf{r}_2)$ in inhomogeneous systems are also scalar functions.

As mentioned earlier, the total electrostatic field, $\mathbf{E}$, is often called the **Maxwell field**. The external field $\mathbf{E}^{\mathrm{Ext}}$ is related to the **displacement field** $\mathbf{D}$ by $\mathbf{D} = \varepsilon_0\mathbf{E}^{\mathrm{Ext}}$. In this treatise we prefer to use $\mathbf{E}^{\mathrm{Ext}}$ since it is simply equal to the electrostatic field from the external charge distribution in the absence of the fluid. Another common concept in the theory of dielectric response is the **polarization field P**, which is proportional to $\mathbf{E}^{\mathrm{pol}}(\mathbf{r})$ and is given by

$$\mathbf{P}(\mathbf{r}) = -\mathbf{E}^{\mathrm{pol}}(\mathbf{r})\varepsilon_0 = \nabla\Psi^{\mathrm{pol}}(\mathbf{r})\varepsilon_0 \tag{13.47}$$

in equilibrium systems. $\mathbf{P}(\mathbf{r})$ is related to the polarization charge density $\rho^{\mathrm{pol}}(\mathbf{r})$ via

$$-\boldsymbol{\nabla}\cdot\mathbf{P}(\mathbf{r}) = -\nabla^2\Psi^{\mathrm{pol}}(\mathbf{r})\varepsilon_0 = \rho^{\mathrm{pol}}(\mathbf{r}),$$

as follows from Poisson's equation, and in Fourier space $\tilde{\rho}^{\mathrm{pol}}(\mathbf{k}) = -i\mathbf{k}\cdot\mathbf{P}(\mathbf{k})$. In this treatise we will use Ψ^{pol} throughout rather than $\mathbf{P}$, because this is sufficient when one deals with equilibrium systems and is simpler in many respects. If $\mathbf{P}$ is required it can easily be obtained from Equation 13.47.

In the presence of a weak external field $\delta\mathbf{E}^{\mathrm{Ext}}$, the polarization $\delta\mathbf{P}$ is given in Fourier space in terms of the total field $\delta\tilde{\mathbf{E}}(\mathbf{k})$ by

$$\delta\tilde{\mathbf{P}}(\mathbf{k}) = \tilde{\chi}^{\mathrm{el}}(\mathbf{k})\cdot\delta\tilde{\mathbf{E}}(\mathbf{k})\epsilon_0,$$

which *defines* the **electric susceptibility tensor** $\tilde{\chi}^{\mathrm{el}}(\mathbf{k})$. The **dielectric permittivity tensor** is *defined* as

$$\tilde{\epsilon}(\mathbf{k}) \equiv \mathbb{I} + \tilde{\chi}^{\mathrm{el}}(\mathbf{k}).$$

We accordingly have

$$\delta\tilde{\rho}^{\mathrm{pol}}(\mathbf{k}) = -i\mathbf{k}\cdot\tilde{\chi}^{\mathrm{el}}(\mathbf{k})\cdot\delta\tilde{\mathbf{E}}(\mathbf{k})\epsilon_0 = -\mathbf{k}\cdot\tilde{\chi}^{\mathrm{el}}(\mathbf{k})\cdot\mathbf{k}\,\delta\tilde{\Psi}(\mathbf{k})\epsilon_0$$

and the potential due to the polarization charge density is

$$\delta\tilde{\Psi}^{\mathrm{Pol}}(\mathbf{k}) = \tilde{\phi}_{\mathrm{Coul}}(k)\delta\tilde{\rho}^{\mathrm{pol}}(\mathbf{k}) = -\hat{\mathbf{k}}\cdot\tilde{\chi}^{\mathrm{el}}(\mathbf{k})\cdot\hat{\mathbf{k}}\,\delta\tilde{\Psi}(\mathbf{k}),$$

where we have inserted $\tilde{\phi}_{\mathrm{Coul}}(k) = 1/(\varepsilon_0 k^2)$. By comparing with Equation 13.46 we see that $\chi^{\mathrm{el}}(\mathbf{k}) = \hat{\mathbf{k}}\cdot\tilde{\chi}^{\mathrm{el}}(\mathbf{k})\cdot\hat{\mathbf{k}}$ and hence, by using Equation 13.45, we obtain

$$\tilde{\epsilon}(\mathbf{k}) = 1 + \hat{\mathbf{k}}\cdot\tilde{\chi}^{\mathrm{el}}(\mathbf{k})\cdot\hat{\mathbf{k}} = \hat{\mathbf{k}}\cdot\tilde{\epsilon}(\mathbf{k})\cdot\hat{\mathbf{k}}.$$

The rhs shows that $\tilde{\epsilon}(\mathbf{k})$ is equal to the longitudinal projection of the tensor $\tilde{\epsilon}$. The transversal part is uninteresting for the electrostatics of equilibrium systems, but it is very important for dynamical phenomena. We are not dealing with the latter in the current treatise.

We can express $\tilde{\epsilon}(\mathbf{k})$ in its singular and regular parts

$$\tilde{\epsilon}(\mathbf{k}) = \tilde{\epsilon}_{\mathrm{sing}}(k) + \tilde{\epsilon}_{\mathrm{reg}}(\mathbf{k}) = -\frac{\tilde{\chi}^{\star}(0)}{\varepsilon_0 k^2} + \left(1 - \frac{\tilde{\chi}^{\star}(\mathbf{k}) - \tilde{\chi}^{\star}(0)}{\varepsilon_0 k^2}\right)$$

and we have from Equation 13.44

$$\begin{aligned}
\tilde{\epsilon}_{\text{sing}}(k) &\equiv -\frac{1}{\varepsilon_0 k^2}\int d\mathbf{r}_{12}\,\chi^\star(\mathbf{r}_{12}) = \{\text{select } \mathbf{r}_2 \text{ as the origin}\} = \\
&= \frac{\beta}{\varepsilon_0 k^2}\sum_i \int d\mathbf{r}_1 d\mathbf{r}_3 \left\langle \sigma_i(\mathbf{r}_{31},\omega_3)\mathfrak{n}_i^{(1)}(\omega_3)\rho_i^\star(\mathbf{r}_3,\omega_3)\right\rangle_{\omega_3} \\
&= \frac{\beta}{\varepsilon_0 k^2}\sum_i \int d\mathbf{r}_3 \left\langle q_i\mathfrak{n}_i^{(1)}(\omega_3)\rho_i^\star(\mathbf{r}_3,\omega_3)\right\rangle_{\omega_3}.
\end{aligned}$$

Using the definition 13.42 of the average $\bar{q}_i^\star$ we can conclude that

$$\tilde{\epsilon}_{\text{sing}}(k) = \frac{\beta}{\varepsilon_0 k^2}\sum_i n_i^{\text{b}} q_i \bar{q}_i^\star, \tag{13.48}$$

which has the same form as the expression 12.97 for $\tilde{\epsilon}_{\text{sing}}(k)$ for spherical ions, and we obtain

$$\tilde{\epsilon}(\mathbf{k}) = \frac{\beta}{\varepsilon_0 k^2}\sum_i n_i^{\text{b}} q_i \bar{q}_i^\star + \tilde{\epsilon}_{\text{reg}}(\mathbf{k}) \tag{13.49}$$

that corresponds to Equation 12.100. Note that only the charged species contribute to the sum $\sum_i n_i^{\text{b}} q_i \bar{q}_i^\star$ due to the factor q_i.

Equations 13.34 and 13.35 yield

$$\rho_i^{\text{tot}}(\mathbf{r}_2|\mathbf{R}_1) = \rho_i^{\text{tot}}(\mathbf{r}_{12},\omega_1) = \int d\mathbf{r}_4\,\epsilon^{\text{R}}(\mathbf{r}_{24})\rho_i^\star(\mathbf{r}_{14},\omega_1) \tag{13.50}$$

$$\rho_j^\star(\mathbf{r}_3|\mathbf{R}_1) = \rho_i^\star(\mathbf{r}_{13},\omega_1) = \int d\mathbf{r}_2\,\epsilon(\mathbf{r}_{32})\rho_i^{\text{tot}}(\mathbf{r}_{12},\omega_1). \tag{13.51}$$

Both become

$$\tilde{\rho}_i^{\text{tot}}(\mathbf{k},\omega_1) = \frac{\tilde{\rho}_i^\star(\mathbf{k},\omega_1)}{\tilde{\epsilon}(\mathbf{k})}$$

in Fourier space.

From Equation 13.36 it follows that the ***(unit) screened Coulomb potential*** is given by

$$\phi_{\text{Coul}}^\star(\mathbf{r}_{43}) \equiv \int d\mathbf{r}_2\,\epsilon^{\text{R}}(\mathbf{r}_{24})\phi_{\text{Coul}}(r_{23})$$

so we have

$$\tilde{\phi}_{\text{Coul}}^\star(\mathbf{k}) = \frac{\tilde{\phi}_{\text{Coul}}(k)}{\tilde{\epsilon}(\mathbf{k})} = \frac{1}{\varepsilon_0 k^2 \tilde{\epsilon}(\mathbf{k})} = \frac{1}{\varepsilon_0 k^2 - \chi^\star(\mathbf{k})}. \tag{13.52}$$

Note that $\phi_{\text{Coul}}^\star(\mathbf{r})$ is direction-dependent when the bulk electrolyte is anisotropic. Using Equation 13.49 we obtain

$$\tilde{\phi}_{\text{Coul}}^\star(\mathbf{k}) = \frac{1}{\beta\sum_i n_i^{\text{b}} q_i \bar{q}_i^\star + \varepsilon_0 k^2 \tilde{\epsilon}_{\text{reg}}(\mathbf{k})}. \tag{13.53}$$

The mean electrostatic potential ψ_i from an i-particle at $\mathbf{r}_1$ with orientation $\boldsymbol{\omega}_1$ is given by the Coulomb potential formula

$$\psi_i(\mathbf{r}_{13}, \boldsymbol{\omega}_1) = \int d\mathbf{r}_2\, \rho_i^{\text{tot}}(\mathbf{r}_{12}, \boldsymbol{\omega}_1) \phi_{\text{Coul}}(r_{23}) = \int d\mathbf{r}_4\, \rho_i^\star(\mathbf{r}_{14}, \boldsymbol{\omega}_1) \phi_{\text{Coul}}^\star(r_{43}),$$
$$(13.54)$$

which give the ordinary and the screened version of this formula. In Fourier space

$$\tilde{\psi}_i(\mathbf{k}, \boldsymbol{\omega}_1) = \frac{\tilde{\rho}_i^{\text{tot}}(\mathbf{k}, \boldsymbol{\omega}_1)}{\varepsilon_0 k^2} = \frac{\tilde{\rho}_i^\star(\mathbf{k}, \boldsymbol{\omega}_1)}{\varepsilon_0 k^2 \tilde{\epsilon}(\mathbf{k})}. \qquad (13.55)$$

{**Exercise 13.1:** Starting from Equations 13.52 and 13.55 show that ψ_i satisfies the modified Poisson equation

$$-\varepsilon_0 \nabla_3^2 \psi_i(\mathbf{r}_{13}, \boldsymbol{\omega}_1) - \int d\mathbf{r}_4\, \psi_i(\mathbf{r}_{14}, \boldsymbol{\omega}_1) \chi^\star(r_{43}) = \rho_i^\star(\mathbf{r}_{13}, \boldsymbol{\omega}_1)$$

and that $\phi_{\text{Coul}}^\star$ is the corresponding Green's function that satisfies

$$-\varepsilon_0 \nabla_3^2 \phi_{\text{Coul}}^\star(\mathbf{r}_{23}) - \int d\mathbf{r}_4\, \phi_{\text{Coul}}^\star(\mathbf{r}_{24}) \chi^\star(r_{43}) = \delta^{(3)}(\mathbf{r}_{23})$$

(cf. Equations 12.70 and 12.71 for inhomogeneous electrolytes consisting of spherical ions).}

The pair correlation function is

$$\begin{aligned} h_{ij}^{(2)}(\mathbf{r}_{12}, \boldsymbol{\omega}_1, \boldsymbol{\omega}_2) &= h_{ij}^\star(\mathbf{r}_{12}, \boldsymbol{\omega}_1, \boldsymbol{\omega}_2) \\ &- \beta \int d\mathbf{r}_3\, d\mathbf{r}_4\, \rho_i^\star(\mathbf{r}_{14}, \boldsymbol{\omega}_1) \phi_{\text{Coul}}^\star(r_{43}) \rho_j^\star(\mathbf{r}_{23}, \boldsymbol{\omega}_2) \end{aligned}$$

(from Equation 13.40), which is

$$\tilde{h}_{ij}^{(2)}(\mathbf{k}, \boldsymbol{\omega}_1, \boldsymbol{\omega}_2) = \tilde{h}_{ij}^\star(\mathbf{k}, \boldsymbol{\omega}_1, \boldsymbol{\omega}_2) - \beta \tilde{\rho}_i^\star(\mathbf{k}, \boldsymbol{\omega}_1) \tilde{\phi}_{\text{Coul}}^\star(\mathbf{k}) \tilde{\rho}_j^\star(-\mathbf{k}, \boldsymbol{\omega}_2) \qquad (13.56)$$

in Fourier space. The minus sign in the last factor appears because we have to change $\mathbf{r}_{23} \to -\mathbf{r}_{32}$ in order to make the $\mathbf{r}_3$ integral to be a convolution. The direct correlation function is

$$\begin{aligned} c_{ij}^{(2)}(\mathbf{r}_{12}, \boldsymbol{\omega}_1, \boldsymbol{\omega}_2) &= c_{ij}^\star(\mathbf{r}_{12}, \boldsymbol{\omega}_1, \boldsymbol{\omega}_2) \\ &- \beta \int d\mathbf{r}_3 d\mathbf{r}_4\, \sigma_i(\mathbf{r}_{14}, \boldsymbol{\omega}_1) \phi_{\text{Coul}}(r_{43}) \sigma_j(\mathbf{r}_{23}, \boldsymbol{\omega}_2), \end{aligned}$$

(from Equation 13.41) and in Fourier space

$$\tilde{c}_{ij}^{(2)}(\mathbf{k}, \boldsymbol{\omega}_1, \boldsymbol{\omega}_2) = \tilde{c}_{ij}^\star(\mathbf{k}, \boldsymbol{\omega}_1, \boldsymbol{\omega}_2) - \beta \tilde{\sigma}_i(\mathbf{k}, \boldsymbol{\omega}_1) \tilde{\phi}_{\text{Coul}}(k) \tilde{\sigma}_j(-\mathbf{k}, \boldsymbol{\omega}_2). \qquad (13.57)$$

13.3.2 Nematic electrolyte solutions at low concentrations

In order to evaluate $\tilde{\epsilon}_{\text{reg}}(\mathbf{k})$ for a nematic phase, let us select the z axis along the director vector[5] of the phase, whereby $\chi^\star(\mathbf{r})$ has cylindrical symmetry around this axis and is also invariant when the sign of z is changed. Setting $\mathbf{s} = (x, y)$ and $s = |\mathbf{s}|$ we can therefore write $\chi^\star(\mathbf{r}) = \chi^\star(s, z)$ and we have

$$\begin{aligned}\tilde{\chi}^\star(\mathbf{k}) &= \int d\mathbf{r}\, \chi^\star(\mathbf{r}) e^{-i\mathbf{k}\cdot\mathbf{r}} = \int d\mathbf{s}\, dz\, \chi^\star(s, z) e^{-i\mathbf{k}_\perp\cdot\mathbf{s}} e^{-ik_\parallel z} \\ &= \int d\mathbf{s}\, dz\, \chi^\star(s, z) J_0(k_\perp s)\cos(k_\parallel z) \equiv \tilde{\chi}^\star(k_\perp, k_\parallel), \end{aligned} \quad (13.58)$$

where $k_\perp = |\mathbf{k}_\perp|$ is the wave vector component perpendicular to the director, $k_\parallel$ is the component along the director and we have used the fact that $\chi^\star(s, z) = \chi^\star(s, -z)$ and Equation 9.256 in Appendix 9B to simplify the Fourier integral. We obtain

$$\begin{aligned}\tilde{\epsilon}_{\text{reg}}(\mathbf{k}) &= \tilde{\epsilon}_{\text{reg}}(k_\perp, k_\parallel) = 1 - \frac{\tilde{\chi}^\star(k_\perp, k_\parallel) - \tilde{\chi}^\star(0)}{\varepsilon_0 k^2} \\ &= 1 - \frac{1}{\varepsilon_0}\int d\mathbf{s}\, dz\, \frac{J_0(k_\perp s)\cos(k_\parallel z) - 1}{k_\perp^2 + k_\parallel^2}\chi^\star(s, z). \end{aligned} \quad (13.59)$$

When $k \to 0$ we have[6]

$$\begin{aligned}\tilde{\epsilon}_{\text{reg}}(k_\perp, k_\parallel) &\to 1 + \frac{1}{\varepsilon_0}\int d\mathbf{s}\, dz\, \frac{(k_\perp s)^2 + 2(k_\parallel z)^2}{4\left(k_\perp^2 + k_\parallel^2\right)}\chi^\star(s, z) \\ &= \mathcal{E}_\perp^{\star 0}\left[\frac{k_\perp}{k}\right]^2 + \mathcal{E}_\parallel^{\star 0}\left[\frac{k_\parallel}{k}\right]^2 = \mathcal{E}_\perp^{\star 0}\sin^2(\theta_{\mathbf{k}}) + \mathcal{E}_\parallel^{\star 0}\cos^2(\theta_{\mathbf{k}}) \end{aligned} \quad (13.60)$$

where $\theta_{\mathbf{k}}$ is the angle between $\hat{\mathbf{k}}$ and the director (the z axis) and

$$\begin{aligned}\mathcal{E}_\perp^{\star 0} &= 1 + \frac{1}{4\varepsilon_0}\int d\mathbf{s}\, dz\, s^2 \chi^\star(s, z) \\ \mathcal{E}_\parallel^{\star 0} &= 1 + \frac{1}{2\varepsilon_0}\int d\mathbf{s}\, dz\, z^2 \chi^\star(s, z). \end{aligned} \quad (13.61)$$

Thus, $\tilde{\epsilon}_{\text{reg}}(k_\perp, k_\parallel)$ in the limit $k \to 0$ is a finite number that is independent of k but depends on the direction of $\hat{\mathbf{k}}$, whereby $\mathcal{E}_\parallel^{\star 0}$ is the value when $\hat{\mathbf{k}}$ is directed along the director and $\mathcal{E}_\perp^{\star 0}$ is the value when $\hat{\mathbf{k}}$ is perpendicular to it. Note that $\mathcal{E}_\perp^{\star 0}$ and $\mathcal{E}_\parallel^{\star 0}$ contain contributions from the both the charged and uncharged species. When the electrolyte concentration is nonzero they are different from the dielectric constants $\varepsilon_\perp$ and $\varepsilon_\parallel$ of the pure, anisotropic solvent phase.

[5] The **director vector** gives the common direction for the average orientation of particles in a nematic liquid crystal. For prolate particles it is the direction along which their long-axes are aligned on average.

[6] The MacLaurin expansion of the Bessel function is $J_0(x) = 1 - x^2/4 + x^4/64 + \mathcal{O}(x^6)$.

The direction dependence of $\tilde{\epsilon}_{\text{reg}}(\mathbf{k})|_{k\to 0}$ can alternatively be expressed in terms of a dielectric permittivity tensor $\tilde{\mathbb{\epsilon}}_{\text{reg}}(\mathbf{k})$ that is independent of *both* the direction and magnitude of $\mathbf{k}$ in the limit $k \to 0$

$$\tilde{\mathbb{\epsilon}}_{\text{reg}}(\mathbf{0}) = \begin{bmatrix} \mathcal{E}_{\perp}^{\star 0} & 0 & 0 \\ 0 & \mathcal{E}_{\perp}^{\star 0} & 0 \\ 0 & 0 & \mathcal{E}_{\parallel}^{\star 0} \end{bmatrix} = \mathcal{E}_{\perp}^{\star 0}(\hat{\mathbf{x}}\hat{\mathbf{x}}^{\intercal} + \hat{\mathbf{y}}\hat{\mathbf{y}}^{\intercal}) + \mathcal{E}_{\parallel}^{\star 0}\hat{\mathbf{z}}\hat{\mathbf{z}}^{\intercal},$$

where the form of the matrix is suggested by the symmetry of the system. The longitudinal projection is $\hat{\mathbf{k}} \cdot \tilde{\mathbb{\epsilon}}_{\text{reg}}(\mathbf{0}) \cdot \hat{\mathbf{k}} = \mathcal{E}_{\perp}^{\star 0} \sin^2(\theta_{\mathbf{k}}) + \mathcal{E}_{\parallel}^{\star 0} \cos^2(\theta_{\mathbf{k}}) = \tilde{\epsilon}_{\text{reg}}(\mathbf{k})|_{k\to 0}$ in accordance with the result we obtained above. The direction dependence here is solely a consequence of the tensorial nature of $\tilde{\mathbb{\epsilon}}_{\text{reg}}(\mathbf{0})$ while it is due to the $\hat{\mathbf{k}}$-dependence of the scalar $\tilde{\epsilon}_{\text{reg}}(\mathbf{k})|_{k\to 0}$.

{**Exercise 13.2:** For isotropic bulk fluids $\chi^\star$ does not depend on the direction of $\mathbf{r}$ and we have

$$\mathcal{E}_r^{\star 0} \equiv \tilde{\epsilon}_{\text{reg}}(0) = 1 + \frac{1}{6\varepsilon_0} \int d\mathbf{r}\, r^2 \chi^\star(r), \tag{13.62}$$

which is the same as for the case of spherical particles, Equation 12.99. Show that $\mathcal{E}_{\parallel}^{\star 0}$ and $\mathcal{E}_{\perp}^{\star 0}$ defined in Equation 13.61 becomes equal to $\mathcal{E}_r^{\star 0}$ when $\chi^\star(s, z) = \chi^\star(\mathbf{r}) = \chi^\star(r)$ as in an isotropic system.}

As an example of electrostatic interactions in a nematic bulk phase we will investigate a dilute solution of an electrolyte in a solvent that form a nematic liquid crystal. Thereby we will find out how the anisotropy affects the electrostatic screening, in particular how $\phi^\star_{\text{Coul}}(\mathbf{r})$ behaves for large r. As we have seen, the asymptotic behavior of electrostatic interactions in dilute electrolytes is determined by $\tilde{\phi}^\star_{\text{Coul}}(\mathbf{k})$ for small k values, so it follows from Equation 13.53 that it is determined by $\tilde{\epsilon}_{\text{reg}}(\mathbf{k})$ for small k. Using Equation 13.60 we see that

$$\tilde{\phi}^\star_{\text{Coul}}(\mathbf{k}) \sim \frac{1}{\beta \sum_i n_i^{\text{b}} q_i \bar{q}_i^\star + \varepsilon_0 \left[\mathcal{E}_{\perp}^{\star 0} k_{\perp}^2 + \mathcal{E}_{\parallel}^{\star 0} k_{\parallel}^2\right]}, \quad k \to 0. \tag{13.63}$$

If we take this as an approximation for all $\mathbf{k}$ we obtain $\phi^\star_{\text{Coul}}(\mathbf{r})$ for all $\mathbf{r}$, but then we have neglected terms in $\tilde{\phi}^\star_{\text{Coul}}(\mathbf{k})$ that give important contributions for small $\mathbf{r}$. To use Equation 13.63 for all $\mathbf{k}$ corresponds to a theory on the level of the Poisson-Boltzmann approximation for the ions. For large r, the treatment of both the ions and the solvent is, however, essentially exact here for sufficiently low concentrations.

Thus, we will investigate the approximation where $\tilde{\phi}^{\star}_{\text{Coul}}(k_\perp, k_\parallel)$ is equal to the rhs of Equation 13.63, so we have after rearrangement

$$\left(\frac{k_\perp^2}{\kappa_{0\perp}^2} + \frac{k_\parallel^2}{\kappa_{0\parallel}^2}\right) \tilde{\phi}^{\star}_{\text{Coul}}(k_\perp, k_\parallel) + \tilde{\phi}^{\star}_{\text{Coul}}(k_\perp, k_\parallel) = \frac{1}{\beta \sum_i n_i^{\text{b}} q_i \bar{q}_i^{\star}}, \tag{13.64}$$

where we have introduced

$$\kappa_{0\perp}^2 = \frac{\beta \sum_i n_i^{\text{b}} q_i \bar{q}_i^{\star}}{\varepsilon_0 \mathcal{E}_\perp^{\star 0}}, \qquad \kappa_{0\parallel}^2 = \frac{\beta \sum_i n_i^{\text{b}} q_i \bar{q}_i^{\star}}{\varepsilon_0 \mathcal{E}_\parallel^{\star 0}}. \tag{13.65}$$

We will now use the facts that $k_\perp^2 = -\mathrm{i}\mathbf{k}_\perp \cdot \mathrm{i}\mathbf{k}_\perp$ and that $\mathrm{i}\mathbf{k}_\perp = \mathrm{i}(k_x, k_y)$ corresponds to the gradient $\nabla_{\mathbf{s}} = (\partial/\partial x, \partial/\partial y)$ in the ordinary two-dimensional space, see Equation 9.257. Hence $k_\perp^2 \tilde{\phi}^{\star}_{\text{Coul}}$ becomes $-\nabla_{\mathbf{s}}^2 \phi^{\star}_{\text{Coul}} \equiv -[\partial^2/\partial z^2 + \partial^2/\partial y^2]\phi^{\star}_{\text{Coul}}$ in $\mathbf{r}$-space and likewise $k_\parallel^2 \tilde{\phi}^{\star}_{\text{Coul}}$ becomes $-\partial^2 \phi^{\star}_{\text{Coul}}/\partial z^2$. This means that Equation 13.64 implies that

$$-\frac{1}{\kappa_{0\perp}^2} \nabla_{\mathbf{s}}^2 \phi^{\star}_{\text{Coul}}(\mathbf{s}, z) - \frac{1}{\kappa_{0\parallel}^2} \frac{\partial^2 \phi^{\star}_{\text{Coul}}(\mathbf{s}, z)}{\partial z^2} + \phi^{\star}_{\text{Coul}}(\mathbf{s}, z) = \frac{\delta^{(2)}(\mathbf{s})\delta(z)}{\beta \sum_i n_i^{\text{b}} q_i \bar{q}_i^{\star}},$$

where $\delta^{(2)}(\mathbf{s})$ is the two-dimensional delta function. This equation can be solved analytically[7] after variable substitution $(\kappa_{0\perp}\mathbf{s}, \kappa_{0\parallel} z) \to (\mathbf{s}^\circ, z^\circ) = \mathbf{r}^\circ$ and the final solution is

$$\begin{aligned}
&\phi^{\star}_{\text{Coul}}(s, z) \\
&\quad = K_0 \frac{\exp(-r^\circ)}{4\pi r^\circ} = K_0 \frac{\exp\left(-\left\{[\kappa_{0\perp} s]^2 + [\kappa_{0\parallel} z]^2\right\}^{1/2}\right)}{4\pi \left\{[\kappa_{0\perp} s]^2 + [\kappa_{0\parallel} z]^2\right\}^{1/2}} \\
&\quad = \frac{1}{\varepsilon_0 \mathcal{E}_\perp^{\star 0} \left[\mathcal{E}_\parallel^{\star 0}\right]^{1/2}} \cdot \frac{\exp\left(-\left\{[\kappa_{0\perp} s]^2 + [\kappa_{0\parallel} z]^2\right\}^{1/2}\right)}{4\pi \left\{s^2/\mathcal{E}_\perp^{\star 0} + z^2/\mathcal{E}_\parallel^{\star 0}\right\}^{1/2}} \quad \text{(dilute soln.)},
\end{aligned}$$

where $K_0 = \kappa_{0\perp}^2 \kappa_{0\parallel}/[\beta \sum_i n_i^{\text{b}} q_i \bar{q}_i^{\star}]$ and $r^{\circ 2} = |(\mathbf{s}^\circ, z^\circ)|^2 = [\kappa_{0\perp} s]^2 + [\kappa_{0\parallel} z]^2$.

We conclude that there are two decay parameters $\kappa_{0\perp}$ and $\kappa_{0\parallel}$ and that the anisotropic solution, as expected, screens the electrostatic interactions differently in different directions. For a sufficiently dilute electrolyte solution

---

[7]We make the variable substitution $(\kappa_{0\perp}\mathbf{s}, \kappa_{0\parallel} z) \to (\mathbf{s}^\circ, z^\circ) = \mathbf{r}^\circ$ and in Fourier space $(\mathbf{k}_\perp/\kappa_{0\perp}, k_\parallel/\kappa_{0\parallel}) \to (\mathbf{k}_{\mathbf{s}}^\circ, k_z^\circ) = (k_x^\circ, k_y^\circ, k_z^\circ) = \mathbf{k}^\circ$, which implies that

$$\tilde{\phi}^{\star}_{\text{Coul}}(|\mathbf{k}_{\mathbf{s}}^\circ|, k_z^\circ) = \frac{1}{[|\mathbf{k}_{\mathbf{s}}^\circ|^2 + (k_z^\circ)^2 + 1]\, C_0} = \frac{1}{[(k^\circ)^2 + 1]\, C_0},$$

where $C_0 = \beta \sum_i n_i^{\text{b}} q_i \bar{q}_i^{\star}$. The rhs is the Fourier transform of a Yukawa function and we obtain $\phi^{\star}_{\text{Coul}}(s, z) = K_0 e^{-r^\circ}/(4\pi r^\circ)$, where $K_0 = \kappa_{0\perp}^2 \kappa_{0\parallel}/C_0$. The factor $\kappa_{0\perp}^2 \kappa_{0\parallel}$ arises from the differential $d\mathbf{k}_\perp dk_\parallel = \kappa_{0\perp}^2 \kappa_{0\parallel}\, d\mathbf{k}_{\mathbf{s}}^\circ dk_z^\circ$ in the inverse transform integration.

we accordingly have

$$\phi^{\star}_{\text{Coul}}(s,z) \sim \begin{cases} \dfrac{1}{\varepsilon_0 \mathcal{E}^{\star 0}_{\perp}} \cdot \dfrac{e^{-\kappa_{0\|}|z|}}{4\pi|z|}, & \text{along the director} \\ \dfrac{1}{\varepsilon_0 [\mathcal{E}^{\star 0}_{\perp}\mathcal{E}^{\star 0}_{\|}]^{1/2}} \cdot \dfrac{e^{-\kappa_{0\perp} s}}{4\pi s}, & \text{perpendicularly} \end{cases}$$

when $z \to \pm\infty$ and $s \to \infty$, respectively. Note that the magnitude depends on the value of $\mathcal{E}^{\star 0}_{\perp}$ or $\mathcal{E}^{\star 0}_{\|}$ in the *other* direction. In isotropic systems where $\mathcal{E}^{\star 0}_{\perp} = \mathcal{E}^{\star 0}_{\|} = \mathcal{E}^{\star 0}_{r}$, both expression become the usual Yukawa function decay, that is, $\phi^{\star}_{\text{Coul}}(r) \sim e^{-\kappa_0 r}/(4\pi\varepsilon_0\mathcal{E}^{\star 0}_r r)$, where $\kappa_0^2 = \beta \sum_i n_i^{\text{b}} q_i \bar{q}_i^{\star}/[\varepsilon_0 \mathcal{E}^{\star 0}_r]$.

For the pure solvent consisting of electroneutral molecules we have $\tilde{\epsilon}_{\text{sing}}(k) \equiv 0$ (in Equation 13.48 all $q_j = 0$), so

$$\tilde{\epsilon}(\mathbf{k}) = \tilde{\epsilon}_{\text{reg}}(\mathbf{k}) = 1 - \frac{\tilde{\chi}^{\star}(\mathbf{k}) - \tilde{\chi}^{\star}(0)}{\varepsilon_0 k^2} \qquad \text{(pure solvent).} \tag{13.66}$$

As we have seen, the rhs stays finite when $k \to 0$. For an isotropic system $\tilde{\epsilon}(\mathbf{k}) = \tilde{\epsilon}(k)$ and we have

$$\varepsilon_r = \lim_{k\to 0} \tilde{\epsilon}(k) \qquad \text{(isotropic pure solvent),} \tag{13.67}$$

which can be regarded as the definition of the **dielectric constant** for non-electrolytes (this topic will be treated in more detail in Section 14). For an isotropic electrolyte solution in the limit of *infinite dilution* $\mathcal{E}^{\star 0}_r \to \varepsilon_r$. It follows that ε_r is given by the rhs of Equation 13.62 evaluated for the pure solvent phase. Note that ε_r is obtained here from a molecular description of the solvent; its value not assumed as in the dielectric continuum model, where we used the notation $\varepsilon_r^{\text{cont}}$ in order to make a distinction from the actual ε_r.

For a nematic electrolyte solution $\mathcal{E}^{\star 0}_{\|} \to \varepsilon_{\|}$ and $\mathcal{E}^{\star 0}_{\perp} \to \varepsilon_{\perp}$ in the infinite dilution limit, where $\varepsilon_{\perp}$ and $\varepsilon_{\|}$ are the dielectric constants for the nematic pure solvent along the director and in the perpendicular direction, respectively. In this case $\varepsilon_{\perp}$ and $\varepsilon_{\|}$ are given by the rhs of Equation 13.61 evaluated for the pure solvent. In all cases $q_i^{\star} \to q_i$ at infinite dilution, so the expressions for $\kappa_{0\perp}$ and $\kappa_{0\|}$ in Equation 13.65 are replaced by

$$\kappa^2_{\text{D}\perp} = \frac{\beta \sum_i n_i^{\text{b}} q_i^2}{\varepsilon_0 \varepsilon_{\perp}}, \qquad \kappa^2_{\text{D}\|} = \frac{\beta \sum_i n_i^{\text{b}} q_i^2}{\varepsilon_0 \varepsilon_{\|}},$$

that is, they take on Debye-Hückel-like forms.[8]

[8] A Poisson-Boltzmann treatment of anisotropic screening in nematic solvents is presented in J. C. Everts, B. Senyuk, H. Mundoor, M. Ravnik, and I. I. Smalyukh, Sci. Adv. 7 (2021) eabd0662 (https://doi.org/10.1126/sciadv.abd0662). Supplementary material: https://www.science.org/doi/suppl/10.1126/sciadv.abd0662/suppl_file/abd0662_sm.pdf.

13.3.3 Isotropic bulk electrolytes, general analysis

13.3.3.1 *Electrostatic potentials, decay modes and effective charges for nonspherical particles*

In isotropic phases the dielectric function $\tilde{\epsilon}(k) = \beta \sum_i n_i^{\mathrm{b}} q_i q_i^{\star}/[\varepsilon_0 k^2] + \tilde{\epsilon}_{\mathrm{reg}}(k)$ is spherically symmetric and so are the response functions $\chi^{\rho}(r) = -\beta q_{\mathrm{e}}^2 H_{\mathrm{QQ}}^{(2)}(r)$, $\chi^{\star}(r) = -\beta q_{\mathrm{e}}^2 H_{\mathrm{QQ}}^{\star}(r)$ and the screened Coulomb potential $\phi_{\mathrm{Coul}}^{\star}(r)$. As before we have

$$\tilde{\phi}_{\mathrm{Coul}}^{\star}(k) = \frac{1}{\varepsilon_0 k^2 \tilde{\epsilon}(k)} = \frac{1}{\beta \sum_i n_i^{\mathrm{b}} q_i q_i^{\star} + \varepsilon_0 k^2 \tilde{\epsilon}_{\mathrm{reg}}(k)}. \tag{13.68}$$

Several exact results for the case of spherical ions in Section 12.3.3 carry over unchanged. For instance, we have for sufficiently large r values

$$\phi_{\mathrm{Coul}}^{\star}(r) = \frac{1}{\mathcal{E}_r^{\mathrm{eff}}(\kappa)\varepsilon_0} \cdot \frac{e^{-\kappa r}}{4\pi r} + \frac{1}{\mathcal{E}_r^{\mathrm{eff}}(\kappa')\varepsilon_0} \cdot \frac{e^{-\kappa' r}}{4\pi r} + \frac{1}{\mathcal{E}_r^{\mathrm{eff}}(\kappa'')\varepsilon_0} \cdot \frac{e^{-\kappa'' r}}{4\pi r} + \ldots \tag{13.69}$$

(cf. Equation 12.171), where we have explicitly shown only the Yukawa function terms of the first three decay modes, and the decay parameters κ, κ', κ'', ... are solutions to $\tilde{\epsilon}(k) = 0$, that is, they are solutions of

$$\kappa^2 = \frac{\beta}{\mathcal{E}_r^{\star}(\kappa)\varepsilon_0} \sum_i n_i^{\mathrm{b}} q_i q_i^{\star}, \tag{13.70}$$

where, as before, the dielectric factor is $\mathcal{E}_r^{\star}(\kappa) \equiv \tilde{\epsilon}_{\mathrm{reg}}(\mathrm{i}\kappa)$ and the effective dielectric permittivity is

$$\mathcal{E}_r^{\mathrm{eff}}(\kappa) = \mathcal{E}_r^{\star}(\kappa) + \frac{\mathrm{i}\kappa}{2}\tilde{\epsilon}_{\mathrm{reg}}'(\mathrm{i}\kappa) = \frac{\mathrm{i}\kappa}{2}\tilde{\epsilon}'(\mathrm{i}\kappa) = 1 - \tilde{\chi}^{\star\prime}(\mathrm{i}\kappa)/[2\varepsilon_0 \mathrm{i}\kappa]. \tag{13.71}$$

They have different values for the various modes. Explicitly in terms of $\chi^{\star}(r)$ we have, also as before,

$$\begin{aligned}
\mathcal{E}_r^{\star}(\kappa) &= 1 + \frac{1}{\varepsilon_0} \int d\mathbf{r}\, r^2 \left[\frac{\sinh(\kappa r) - \kappa r}{(\kappa r)^3}\right] \chi^{\star}(r), \\
\mathcal{E}_r^{\mathrm{eff}}(\kappa) &= 1 + \frac{1}{2\varepsilon_0} \int d\mathbf{r}\, r^2 \left[\frac{\kappa r \cosh(\kappa r) - \sinh(\kappa r)}{(\kappa r)^3}\right] \chi^{\star}(r).
\end{aligned}$$

These relationships can be proven in the same way as in the spherical case. The expressions for $\chi^{\rho}(r)$ in terms of ρ_i^{tot} or $h_{ij}^{(2)}$ and $\chi^{\star}(r)$ in terms of $\rho_i^{\star}$ or $h_{ij}^{\star}$ are, of course, different for nonspherical particles, because they involve integration over particle orientations as we have seen in, for example, Equation 13.44, where the integration is expressed as an average. Like in the spherical case there are additional contributions to $\phi_{\mathrm{Coul}}^{\star}(r)$ with other functional form, which have shorter range than the first term.

The mean electrostatic potential due to a particle of species i placed at $\mathbf{r}_1$ with orientation $\boldsymbol{\omega}_1$ is given by the rhs of Equations 13.54. When κ of the leading decay mode is real, ψ_i decays when $r_{12} \to 0$

$$\psi_i(\mathbf{r}_{12}, \boldsymbol{\omega}_1) \sim \frac{1}{\mathcal{E}_r^{\mathrm{eff}}(\kappa)\varepsilon_0} \int d\mathbf{r}_4\, \rho_i^\star(\mathbf{r}_{14}, \boldsymbol{\omega}_1) \frac{e^{-\kappa r_{42}}}{4\pi r_{42}}, \tag{13.72}$$

where we have inserted the first term of Equation 13.69. As shown in Chapter 6 of Volume I and in the shaded text box below on page 429, this implies that

$$\psi_i(\mathbf{r}_{12}, \boldsymbol{\omega}_1) \sim \frac{\mathrm{Q}_i^{\mathrm{eff}}(\hat{\mathbf{r}}_{12}, \boldsymbol{\omega}_1, \kappa)}{\mathcal{E}_r^{\mathrm{eff}}(\kappa)\varepsilon_0} \cdot \frac{e^{-\kappa r_{12}}}{4\pi r_{12}}, \tag{13.73}$$

where

$$\mathrm{Q}_i^{\mathrm{eff}}(\hat{\mathbf{r}}_{12}, \boldsymbol{\omega}_1, \kappa) = \int d\mathbf{r}_4\, \rho_i^\star(\mathbf{r}_{14}, \boldsymbol{\omega}_1) e^{\kappa \mathbf{r}_{14}\cdot\hat{\mathbf{r}}_{12}} \tag{13.74}$$

is a direction-dependent entity that has the unit of charge. Note that

$$\mathrm{Q}_i^{\mathrm{eff}}(\hat{\mathbf{r}}_{12}, \boldsymbol{\omega}_1, \kappa) = \tilde{\rho}_i^\star(i\kappa\hat{\mathbf{r}}_{12}, \boldsymbol{\omega}_1) \tag{13.75}$$

as follows from the definition of the Fourier transform of $\rho_i^\star$. $\mathrm{Q}_i^{\mathrm{eff}}$ is the "**multipolar effective charge**" of the i-particle; a concept that we introduced for the Poisson-Boltzmann approximation in Equation I:6.85 and for the general case in Equation I:6.154. This expression shows that the Yukawa function decay of $\psi_i(\mathbf{r}_{12}, \boldsymbol{\omega}_1)$ has different magnitudes in different directions as expressed by the $\hat{\mathbf{r}}_{12}$ dependence of the prefactor $\mathrm{Q}_i^{\mathrm{eff}}(\hat{\mathbf{r}}_{12}, \boldsymbol{\omega}_1, \kappa)$. In Chapter 6 we saw that Equation 13.73 is valid for the potential from a single nonspherical particle immersed in an electrolyte consisting of spherical ions. Here, we have generalized this result to electrolytes constituted by nonspherical particles. In the present case, the i-particle can be any of the particles in the electrolyte.

The direction-independent part of $\mathrm{Q}_i^{\mathrm{eff}}$ is

$$q_i^{\mathrm{eff}}(\kappa) \equiv \left\langle \mathrm{Q}_i^{\mathrm{eff}}(\hat{\mathbf{r}}_{12}, \boldsymbol{\omega}_1, \kappa) \right\rangle_{\hat{\mathbf{r}}_{12}} = \left\langle \mathrm{Q}_i^{\mathrm{eff}}(\hat{\mathbf{r}}_{12}, \boldsymbol{\omega}_1, \kappa) \right\rangle_{\boldsymbol{\omega}_1}, \tag{13.76}$$

which is the coefficient for the asymptotic decay of ψ_i averaged over the directions of $\hat{\mathbf{r}}_{12}$

$$\langle \psi_i(\mathbf{r}_{12}, \boldsymbol{\omega}_1) \rangle_{\hat{\mathbf{r}}_{12}} \sim \frac{q_i^{\mathrm{eff}}(\kappa)}{\mathcal{E}_r^{\mathrm{eff}}(\kappa)\varepsilon_0} \cdot \frac{e^{-\kappa r_{12}}}{4\pi r_{12}}.$$

This means that q_i^{eff} can be described as the **effective charge** of the nonspherical particle. This decay is the same as for a spherically symmetric particle, for which $\mathrm{Q}_i^{\mathrm{eff}}(\hat{\mathbf{r}}_{12}, \boldsymbol{\omega}_1, \kappa) = q_i^{\mathrm{eff}}(\kappa)$ is independent of $\hat{\mathbf{r}}_{12}$ and $\boldsymbol{\omega}_1$.

For spherical particles the Yukawa function term $e^{-\kappa r_{12}}/r_{12}$ is the only contribution to ψ_i that decays as a function that contains the factor $e^{-\kappa r_{12}}$. This is not the case for nonspherical particles, where there are contributions that decay like $e^{-\kappa r_{12}}/r_{12}^2$, $e^{-\kappa r_{12}}/r_{12}^3$ etc. in addition to the Yukawa term. In the shaded text box on the next page it is shown that when $r_{12} \to \infty$

$$\begin{aligned}\psi_i(\mathbf{r}_{12},\boldsymbol{\omega}_1) &\sim \sum_{m=0}^{\infty} \frac{\mathrm{Q}_i^{[m]\mathrm{eff}}(\hat{\mathbf{r}}_{12},\boldsymbol{\omega}_1,\kappa)}{\mathcal{E}_r^{\mathrm{eff}}(\kappa)\varepsilon_0}\cdot\frac{e^{-\kappa r_{12}}}{4\pi r_{12}^{m+1}} \\ &= \frac{\mathrm{Q}_i^{\mathrm{eff}}(\hat{\mathbf{r}}_{12},\boldsymbol{\omega}_1,\kappa)}{\mathcal{E}_r^{\mathrm{eff}}(\kappa)\varepsilon_0}\cdot\frac{e^{-\kappa r_{12}}}{4\pi r_{12}} + \frac{\mathrm{Q}_i^{[1]\mathrm{eff}}(\hat{\mathbf{r}}_{12},\boldsymbol{\omega}_1,\kappa)}{\mathcal{E}_r^{\mathrm{eff}}(\kappa)\varepsilon_0}\cdot\frac{e^{-\kappa r_{12}}}{4\pi r_{12}^2} + \dots,\end{aligned} \quad (13.77)$$

where

$$\begin{aligned}\mathrm{Q}_i^{[1]\mathrm{eff}}(\hat{\mathbf{r}}_{12},\boldsymbol{\omega}_1,\kappa) &= \frac{1}{2}\int d\mathbf{r}_4\,\rho_i^\star(\mathbf{r}_{14},\boldsymbol{\omega}_1)e^{\kappa\mathbf{r}_{14}\cdot\hat{\mathbf{r}}_{12}} \\ &\times \left(\kappa\left[(\mathbf{r}_{14}\cdot\hat{\mathbf{r}}_{12})^2 - r_{14}^2\right] + 2\mathbf{r}_{14}\cdot\hat{\mathbf{r}}_{12}\right)\end{aligned} \quad (13.78)$$

is another direction-dependent entity associated with the i-particle. It has the unit charge×length, that is, the same as a dipole moment. The next term (not shown) contains the factor

$$\begin{aligned}\mathrm{Q}_i^{[2]\mathrm{eff}}(\hat{\mathbf{r}}_{12},\boldsymbol{\omega}_1,\kappa) &= \frac{1}{8}\int d\mathbf{r}_4\,\rho_i^\star(\mathbf{r}_{14},\boldsymbol{\omega}_1)e^{\kappa\mathbf{r}_{14}\cdot\hat{\mathbf{r}}_{12}} \\ &\times \Big(\kappa\left[(\mathbf{r}_{14}\cdot\hat{\mathbf{r}}_{12})^2 - r_{14}^2\right]8\mathbf{r}_{14}\cdot\hat{\mathbf{r}}_{12} \\ &+ \kappa^2\left[(\mathbf{r}_{14}\cdot\hat{\mathbf{r}}_{12})^2 - r_{14}^2\right]^2 + 4\left[3(\mathbf{r}_{14}\cdot\hat{\mathbf{r}}_{12})^2 - r_{14}^2\right]\Big)\end{aligned} \quad (13.79)$$

which has the unit charge×length^2, i.e., the same unit as a quadrupole moment.

For an electrolyte solution with molecular solvent in the limit of zero ionic concentration, $\kappa \to 0$, it follows from Equations 13.74, 13.78 and 13.79 that

$$\begin{aligned}&\mathrm{Q}_i^{\mathrm{eff}}(\hat{\mathbf{r}}_{12},\boldsymbol{\omega}_1,\kappa) \to q_i, \qquad \mathrm{Q}_i^{[1]\mathrm{eff}}(\hat{\mathbf{r}}_{12},\boldsymbol{\omega}_1,\kappa) \to \mathbf{m}_i^{\mathrm{d}\star}(\boldsymbol{\omega}_1)\cdot\hat{\mathbf{r}}_{12}, \\ &\mathrm{Q}_i^{[2]\mathrm{eff}}(\hat{\mathbf{r}}_{12},\boldsymbol{\omega}_1,\kappa) \to \hat{\mathbf{r}}_{12}\cdot\mathbb{M}_i^{\mathrm{Q}\star}(\boldsymbol{\omega})\cdot\hat{\mathbf{r}}_{12},\end{aligned} \quad (13.80)$$

where

$$\mathbf{m}_i^{\mathrm{d}\star}(\boldsymbol{\omega}) = \int d\mathbf{r}\,\mathbf{r}\rho_i^\star(\mathbf{r},\boldsymbol{\omega}), \quad \mathbb{M}_i^{\mathrm{Q}\star}(\boldsymbol{\omega}) = \tfrac{1}{2}\int d\mathbf{r}\,\left[3\mathbf{r}\mathbf{r}^{\mathsf{T}} - r^2\mathbb{I}\right]\rho_i^\star(\mathbf{r},\boldsymbol{\omega}) \quad (13.81)$$

are the dipole and quadrupole moments of the dressed i-particle.[9] These quantities are the effective dipole and quadrupole moments of the particles in dilute solutions in the limit of zero electrolyte concentration (see the shaded text box on the facing page for the effective quantities in the general case). Since in this limit $\mathcal{E}_r^{\mathrm{eff}}(\kappa) \to \mathcal{E}_r^{\star 0} \to \varepsilon_r$, i.e., the dielectric constant of the pure solvent,

[9]Alternatively, Equation 13.80 can be derived from Equation 13.83 because we have $\mathrm{Q}_i^{[0]\mathrm{eff}}(\hat{\mathbf{r}},\boldsymbol{\omega},\kappa) \to q_i^{\mathrm{eff}}(\kappa) \to q_i$ and $\mathrm{Q}_i^{[1]\mathrm{eff}}(\hat{\mathbf{r}},\boldsymbol{\omega},\kappa) \to \mathbf{m}_i^{\mathrm{d,eff}}(\boldsymbol{\omega},\kappa)\cdot\hat{\mathbf{r}} \to \mathbf{m}_i^{\mathrm{d}\star}(\boldsymbol{\omega})\cdot\hat{\mathbf{r}}$ when $\kappa \to 0$.

Equation 13.77 becomes

$$\psi_i(\mathbf{r}_{12},\boldsymbol{\omega}_1) \sim \frac{1}{4\pi\varepsilon_r\varepsilon_0}\left[\frac{q_i}{r_{12}} + \frac{\mathbf{m}_i^{\mathrm{d}\star}(\boldsymbol{\omega}_1)\cdot\hat{\mathbf{r}}_{12}}{r_{12}^2} + \frac{\hat{\mathbf{r}}_{12}\cdot\mathrm{M}_i^{\mathrm{Q}\star}(\boldsymbol{\omega}_1)\cdot\hat{\mathbf{r}}_{12}}{r_{12}^3}\ldots\right] \quad \begin{array}{l}\text{for } r_{12}\to\infty\\ \text{when } \kappa=0.\end{array} \tag{13.82}$$

It has the same form as an ordinary multipole expansion[10] but contains dressed particle dipoles, quadrupoles etc. and it is the appropriate asymptotic formula for the electrostatic potential from a charged particle immersed in a pure polar liquid. Thus, Equation 13.77 is the *generalization to electrolytes of the multipole expansion.* For charges surrounded by vacuum or placed in a polar fluid, the quadrupolar term decays faster with increasing distance than the dipolar one and finally only the spherically symmetric contribution from the total particle charge remains far from the particle. An important difference between electrolytes and polar liquids is that for electrolytes the leading coefficient $\mathrm{Q}_i^{\mathrm{eff}}$ has a multipolar nature; it contains contributions of monopolar, dipolar, quadrupolar etc. character (see the shaded text box below). Thus, the electrostatic potential from a nonspherical charged particle in an electrolyte maintains its direction dependence all the way to infinity. $\mathrm{Q}_i^{[1]\mathrm{eff}}$ in the second term in the rhs of Equation 13.77 contains contributions of dipolar, quadrupolar etc. character. $\mathrm{Q}_i^{[2]\mathrm{eff}}$ contains contributions of quadrupolar and higher multipole moment character.

Equations 13.73 and 13.77 can be derived from Equation 13.72 as follows. Using the law of cosine in trigonometry[11] we obtain when $r_{12}\to\infty$

$$\begin{aligned} r_{42} &= |\mathbf{r}_{12}-\mathbf{r}_{14}| = \left[r_{12}^2+r_{14}^2-2\mathbf{r}_{14}\cdot\mathbf{r}_{12}\right]^{1/2}\\ &= r_{12}\left[1+r_{14}^2/r_{12}^2-2\mathbf{r}_{14}\cdot\hat{\mathbf{r}}_{12}/r_{12}\right]^{1/2}\\ &\sim r_{12}-\mathbf{r}_{14}\cdot\hat{\mathbf{r}}_{12}-\frac{1}{2r_{12}}\left[(\mathbf{r}_{14}\cdot\hat{\mathbf{r}}_{12})^2-r_{14}^2\right]+\mathcal{O}(1/r_{12}^2)\end{aligned}$$

provided that $r_{14}\ll r_{12}$. This means that

$$\frac{e^{-\kappa r_{42}}}{r_{42}} \sim \frac{e^{-\kappa r_{12}}}{r_{12}}\times e^{\kappa\mathbf{r}_{14}\cdot\hat{\mathbf{r}}_{12}}\times\frac{\exp\left(\kappa\left[(\mathbf{r}_{14}\cdot\hat{\mathbf{r}}_{12})^2-r_{14}^2\right]/[2r_{12}]\right)}{1-\mathbf{r}_{14}\cdot\hat{\mathbf{r}}_{12}/r_{12}}$$

[10]See Chapter 4 in: J. D. Jackson. 1999. *Classical Electrodynamics* 3rd ed. New York: Wiley, where the definition of the quadrupole moment differs by a factor of 1/2 from the one used in this treatise.

[11]The law of cosine says that $r_{ab}^2=r_a^2+r_b^2-2r_ar_b\cos\theta_{ab}$, where θ_{ab} is the angle between the vectors $\mathbf{r}_a$ and $\mathbf{r}_b$. The last term can be written as $-2\mathbf{r}_a\cdot\mathbf{r}_b$.

when $r_{12} \to \infty$. In the last factor we have omitted the terms that decays like $\mathcal{O}(1/r_{12}^2)$ in the numerator and the denominator. This factor decays like

$$\frac{\exp\left(\kappa\left[(\mathbf{r}_{14}\cdot\hat{\mathbf{r}}_{12})^2 - r_{14}^2\right]/[2r_{12}]\right)}{1-\mathbf{r}_{14}\cdot\hat{\mathbf{r}}_{12}/r_{12}} \sim 1 + \frac{\kappa\left[(\mathbf{r}_{14}\cdot\hat{\mathbf{r}}_{12})^2 - r_{14}^2\right] + 2\mathbf{r}_{14}\cdot\hat{\mathbf{r}}_{12}}{2r_{12}} + \mathcal{O}(1/r_{12}^2).$$

It follows that when $r_{12} \to \infty$

$$\int d\mathbf{r}_{14}\,\rho_i^\star(\mathbf{r}_{14},\omega_1)\frac{e^{-\kappa r_{42}}}{r_{42}} \sim \frac{e^{-\kappa r_{12}}}{r_{12}}\left[\mathsf{Q}_i^{\mathrm{eff}}(\hat{\mathbf{r}}_{12},\omega_1,\kappa) + \frac{\mathsf{Q}_i^{[1]\mathrm{eff}}(\hat{\mathbf{r}}_{12},\omega_1,\kappa)}{r_{12}} + \mathcal{O}(1/r_{12}^2)\right],$$

which gives Equation 13.77 and hence also Equation 13.73. In the evaluation of the integral the condition $r_{14} \ll r_{12}$ is fulfilled in practice when $r_{12} \to \infty$ because the dominating contributions from the integrand comes from relatively small r_{14} since the decay length of $\rho_i^\star(\mathbf{r}_{14},\omega_1)$ is shorter than $1/\kappa$ in the cases we are considering here. One can obtain the expression for $\mathsf{Q}_i^{[2]\mathrm{eff}}$ in Equation 13.79 by continuing the expansions above one further step.

★ Each $\mathsf{Q}_i^{[m]\mathrm{eff}}$ in Equation 13.77 can be written as a sum of effective multipolar contributions from the i-particle, for instance[12]

$$\begin{aligned}
\mathsf{Q}_i^{[0]\mathrm{eff}}(\hat{\mathbf{r}},\omega,\kappa) &= q_i^{\mathrm{eff}}(\kappa) + \kappa\mathbf{m}_i^{\mathrm{d,eff}}(\omega,\kappa)\cdot\hat{\mathbf{r}} + \kappa^2\hat{\mathbf{r}}\cdot\mathbb{M}_i^{\mathrm{Q,eff}}(\omega,\kappa)\cdot\hat{\mathbf{r}}/3 + \dots \\
\mathsf{Q}_i^{[1]\mathrm{eff}}(\hat{\mathbf{r}},\omega,\kappa) &= \mathbf{m}_i^{\mathrm{d,eff}}(\omega,\kappa)\cdot\hat{\mathbf{r}} + \kappa\hat{\mathbf{r}}\cdot\mathbb{M}_i^{\mathrm{Q,eff}}(\omega,\kappa)\cdot\hat{\mathbf{r}} + \dots \qquad (13.83)\\
\mathsf{Q}_i^{[2]\mathrm{eff}}(\hat{\mathbf{r}},\omega,\kappa) &= \hat{\mathbf{r}}\cdot\mathbb{M}_i^{\mathrm{Q,eff}}(\omega,\kappa)\cdot\hat{\mathbf{r}} + \dots,
\end{aligned}$$

where $\mathbf{m}_i^{\mathrm{d,eff}}(\omega,\kappa) = \mathtt{M}_i^{[1]}(\omega,\kappa)$ is the **effective dipole moment**, $\mathbb{M}_i^{\mathrm{Q,eff}}(\omega,\kappa) = \mathtt{M}_i^{[2]}(\omega,\kappa)$ is the **effective quadrupole moment** defined from

$$\mathtt{M}_i^{[m]}(\omega,\kappa) = (2m+1)!!\int d\mathbf{r}\,\rho_i^\star(\mathbf{r},\omega)\mathtt{P}_m(\mathbf{r})j_m(\mathrm{i}\kappa r)/(\mathrm{i}\kappa r)^m,$$

where $(2m+1)!! = 1\cdot 3\cdot 5\cdot\ldots\cdot 2m+1$, $j_m(x)$ is a spherical Bessel function, and $\mathtt{P}_1(\mathbf{r}) = \mathbf{r}$, $\mathtt{P}_2(\mathbf{r}) = [3\mathbf{r}\mathbf{r}^{\intercal} - r^2\mathbb{I}]/2$ are solid Legendre polynomial tensors. The effective charge is given by $q_i^{\mathrm{eff}}(\kappa) = \mathtt{M}_i^{[0]}(\kappa)$ and we have $\mathtt{P}_0(\mathbf{r}) = 1$ and $j_0(\mathrm{i}x) = \sinh(x)/x$.

[12]For derivations of the formulas in the shaded text box see R. Ramirez and R. Kjellander, J. Chem. Phys. **125** (2006) 144110 (https://doi.org/10.1063/1.2355486).

The general $\mathrm{Q}_i^{[m]\mathrm{eff}}(\hat{\mathbf{r}},\omega,\kappa)$ is given by the recurrence relationship

$$\mathrm{Q}_i^{[m]\mathrm{eff}}(\hat{\mathbf{r}},\omega,\kappa)=\left[L^2-m(m-1)\right]\mathrm{Q}_i^{[m-1]\mathrm{eff}}(\hat{\mathbf{r}},\omega,\kappa)/(2\kappa m),$$

that starts from $m=1$ with $\mathrm{Q}_i^{[0]\mathrm{eff}}(\hat{\mathbf{r}},\omega,\kappa)\equiv\mathrm{Q}_i^{\mathrm{eff}}(\hat{\mathbf{r}},\omega,\kappa)$ in the rhs. $L^2=\mathbf{L}^2$ is the squared angular momentum operator $\mathbf{L}$ that only operates on the angular variables in $\hat{\mathbf{r}}$ (in polar coordinates). It is related to the Laplace operator via $\nabla^2=\nabla_r^2-r^{-2}L^2$, where $\nabla_r^2=r^{-1}[\partial^2/\partial r^2]r$. The recurrence relationship implies that

$$\mathrm{Q}_i^{[m]\mathrm{eff}}(\hat{\mathbf{r}},\omega,\kappa)=\mathcal{M}_m\mathrm{Q}_i^{[0]\mathrm{eff}}(\hat{\mathbf{r}},\omega,\kappa)=\int d\mathbf{r}'\,\rho_i^\star(\mathbf{r}',\omega)\mathcal{M}_m e^{\kappa\mathbf{r}'\cdot\hat{\mathbf{r}}}$$

where $\mathcal{M}_m$ is an operator defined as $\mathcal{M}_m\equiv\prod_{\nu=1}^m\left[L^2-\nu(\nu-1)\right]/(2\kappa\nu)$.

Each one of the decay modes of $\phi^\star_{\mathrm{Coul}}(r)$ in Equation 13.69 gives rise to terms analogous to those in Equation 13.77, each one having its own values of $\mathcal{E}_r^{\mathrm{eff}}(\kappa')$, $\mathrm{Q}_i^{\mathrm{eff}}(\hat{\mathbf{r}}_{12},\omega_1,\kappa')$, $\mathrm{Q}_i^{[1]\mathrm{eff}}(\hat{\mathbf{r}}_{12},\omega_1,\kappa')$ etc.[13] As in the spherical case, some decay modes can have complex κ and therefore they give rise to oscillatory contributions to $\phi^\star_{\mathrm{Coul}}(r)$ that decay like in Equation 12.173. Then, there are pairs of such modes with, say, $\kappa=\kappa_\Re+i\kappa_\Im$ and $\kappa'=\kappa_\Re-i\kappa_\Im$, whereby $\psi_i(\mathbf{r},\omega)$ has a sequence of contributions with coefficients $\mathrm{Q}_i^{\mathrm{eff}}$, $\mathrm{Q}_i^{[1]\mathrm{eff}}$, $\mathrm{Q}_i^{[2]\mathrm{eff}}$, ... obtained from Equations 13.74, 13.78 and 13.79 for the two modes. The sum of these two modes is real-valued. The leading term when $r_{12}\to\infty$ is

$$\psi_i(\mathbf{r}_{12},\omega_1)\;\sim\;\frac{|\mathrm{Q}_i^{\mathrm{eff}}(\hat{\mathbf{r}}_{12},\omega_1,\kappa)|}{|\mathcal{E}_r^{\mathrm{eff}}(\kappa)|\varepsilon_0}\cdot\frac{e^{-\kappa_\Re r}}{2\pi r}\cos(\kappa_\Im r+\vartheta_i(\hat{\mathbf{r}}_{12},\omega_1,\kappa)-\vartheta_{\mathcal{E}}(\kappa))\tag{13.84}$$

where $\mathcal{E}_r^{\mathrm{eff}}(\kappa)=|\mathcal{E}_r^{\mathrm{eff}}(\kappa)|e^{-i\vartheta_{\mathcal{E}}(\kappa)}$ and $\mathrm{Q}_i^{\mathrm{eff}}(\hat{\mathbf{r}},\omega,\kappa)=|\mathrm{Q}_i^{\mathrm{eff}}(\hat{\mathbf{r}},\omega,\kappa)|\,e^{-i\vartheta_i(\hat{\mathbf{r}},\omega,\kappa)}$. Note that both the magnitude and the phase depends on the direction of $\hat{\mathbf{r}}_{12}$. After the leading term follow analogous terms that decay like $e^{-\kappa_\Re r_{12}}/r_{12}^2$, $e^{-\kappa\kappa_\Re}/r_{12}^3$ etc. times a cosine function.

For the present case of isotropic electrolytes we have

$$\chi^\star(r_{12})=-\beta\sum_i n_i^{\mathrm{b}}\int d\mathbf{r}_3\,\langle\sigma_i(\mathbf{r}_{31},\omega_3)\rho_i^\star(\mathbf{r}_{32},\omega_3)\rangle_{\omega_3}$$

(from Equation 13.44) and we can write the dielectric function $\tilde{\epsilon}(k)=1-\tilde{\phi}_{\mathrm{Coul}}(k)\tilde{\chi}^\star(k)$ as

$$\tilde{\epsilon}(k)=1+\frac{\beta}{\varepsilon_0 k^2}\sum_i n_i^{\mathrm{b}}\,\langle\tilde{\sigma}_i(-\mathbf{k},\omega)\tilde{\rho}_i^\star(\mathbf{k},\omega)\rangle_\omega\,.$$

[13] In addition there are terms that decay like $e^{-2\kappa r_{12}}/r_{12}^2$, $e^{-3\kappa r_{12}}/r_{12}^3$ etc. times slowly varying functions as discussed for $h_{ij}^{(2)}$ in Section 12.3.4 for spherical ions.

Since $\tilde{\epsilon}(\mathrm{i}\kappa) = 0$ and $\mathbf{k} = k\hat{\mathbf{k}}$ we obtain

$$\kappa^2 = \frac{\beta}{\varepsilon_0} \sum_i n_i^{\mathrm{b}} \left\langle \tilde{\sigma}_i(-\mathrm{i}\kappa\hat{\mathbf{k}}, \boldsymbol{\omega}) \tilde{\rho}_i^{\star}(\mathrm{i}\kappa\hat{\mathbf{k}}, \boldsymbol{\omega}) \right\rangle_{\boldsymbol{\omega}},$$

which is independent of the direction of $\hat{\mathbf{k}}$ due to the averaging over $\boldsymbol{\omega}$. Introducing the notation

$$\mathrm{Q}_i(\hat{\mathbf{r}}, \boldsymbol{\omega}, \kappa) \equiv \tilde{\sigma}_i(\mathrm{i}\kappa\hat{\mathbf{r}}, \boldsymbol{\omega}) \tag{13.85}$$

in analogy to $\mathrm{Q}_i^{\mathrm{eff}}(\hat{\mathbf{r}}, \boldsymbol{\omega}, \kappa) = \tilde{\rho}_i^{\star}(\mathrm{i}\kappa\hat{\mathbf{r}}, \boldsymbol{\omega})$ in Equation 13.75 we can write this relationship as

$$\kappa^2 = \frac{\beta}{\varepsilon_0} \sum_i n_i^{\mathrm{b}} \left\langle \mathrm{Q}_i(-\hat{\mathbf{r}}, \boldsymbol{\omega}, \kappa) \mathrm{Q}_i^{\mathrm{eff}}(\hat{\mathbf{r}}, \boldsymbol{\omega}, \kappa) \right\rangle_{\boldsymbol{\omega}}, \tag{13.86}$$

where we have written $\hat{\mathbf{r}}$ instead of $\hat{\mathbf{k}}$. This equation for the decay parameter κ is a mathematically equivalent alternative to $\kappa^2 = \beta \sum_i n_i^{\mathrm{b}} q_i q_i^{\star} / [\mathcal{E}_r^{\star}(\kappa)\varepsilon_0]$ (Equation 13.70) but has a much more complicated structure. For spherically symmetric particles it reduces to the much simpler equation $\kappa^2 = \beta \sum_j n_j^{\mathrm{b}} q_j q_j^{\mathrm{eff}}(\kappa)/\varepsilon_0$ (Equation 12.164). Equation 13.86 is, however, useful because it relates κ to the prefactor of the leading term for ψ_i in Equation 13.73 and hence to the magnitude of the electrostatic potential (and other entities as we will see). Table 12.3 on page 329 contains a summary of the various equivalent equations for κ.

13.3.3.2 Pair correlation functions for nonspherical ions and solvent particles

The pair correlation functions $h_{ij}^{(2)}(\mathbf{r}_{12}, \boldsymbol{\omega}_1, \boldsymbol{\omega}_2)$ can be expressed in Fourier space as

$$\begin{aligned} \tilde{h}_{ij}^{(2)}(\mathbf{k}, \boldsymbol{\omega}_1, \boldsymbol{\omega}_2) &= \tilde{h}_{ij}^{\star}(\mathbf{k}, \boldsymbol{\omega}_1, \boldsymbol{\omega}_2) - \beta \tilde{\rho}_i^{\star}(\mathbf{k}, \boldsymbol{\omega}_1) \tilde{\phi}_{\mathrm{Coul}}^{\star}(k) \tilde{\rho}_j^{\star}(-\mathbf{k}, \boldsymbol{\omega}_2) \\ &= \tilde{h}_{ij}^{\star}(\mathbf{k}, \boldsymbol{\omega}_1, \boldsymbol{\omega}_2) - \frac{\beta \tilde{\rho}_i^{\star}(\mathbf{k}, \boldsymbol{\omega}_1) \tilde{\rho}_j^{\star}(-\mathbf{k}, \boldsymbol{\omega}_2)}{\varepsilon_0 k^2 \tilde{\epsilon}(k)} \end{aligned} \tag{13.87}$$

(Equation 13.56). As we will see, like in the spherical case the decay modes of $h_{ij}^{(2)}$ are in the vast majority of cases the same as those of $\phi_{\mathrm{Coul}}^{\star}(r)$ and $\psi_i(\mathbf{r}_{12}, \boldsymbol{\omega}_1)$, so they are given by the zeros of $\tilde{\epsilon}(k)$. The exceptions are rare cases of charge-inversion invariant model systems, for instance the simplest model of an electrolyte dissolved in a polar fluid: a $\mathrm{z} : -\mathrm{z}$ electrolyte solutions of spherical ions with the same diameter where the solvent molecules are modeled as spheres with a centrally located ideal dipole. If the dipole is displaced from the center in almost any direction, the charge-inversion invariance is broken. A realistic water molecule is, of course, not charge-inversion invariant so aqueous solutions of simple ions are not invariant even if the anions and cations differ only in the sign of their charges. Since anions and cations differ by much more than the sign of charge, both electrolyte solutions and pure ionic liquids are not charge-inversion invariant.

Like in the spherical case, the second term in Equation 13.87 normally gives the leading decay of $h_{ij}^{(2)}$ (see the box on page 344) and we will show that for sufficiently large r_{12}

$$\begin{aligned} & h_{ij}^{(2)}(\mathbf{r}_{12},\boldsymbol{\omega}_1,\boldsymbol{\omega}_2) \\ & = -\frac{\beta}{\mathcal{E}_r^{\text{eff}}(\kappa)\varepsilon_0}\left[\mathsf{Q}_i^{\text{eff}}(\hat{\mathbf{r}}_{12},\boldsymbol{\omega}_1,\kappa)\mathsf{Q}_j^{\text{eff}}(-\hat{\mathbf{r}}_{12},\boldsymbol{\omega}_2,\kappa)\frac{e^{-\kappa r_{12}}}{4\pi r_{12}}+\ldots\right] \\ & - \frac{\beta}{\mathcal{E}_r^{\text{eff}}(\kappa')\varepsilon_0}\left[\mathsf{Q}_i^{\text{eff}}(\hat{\mathbf{r}}_{12},\boldsymbol{\omega}_1,\kappa')\mathsf{Q}_j^{\text{eff}}(-\hat{\mathbf{r}}_{12},\boldsymbol{\omega}_2,\kappa')\frac{e^{-\kappa' r_{12}}}{4\pi r_{12}}+\ldots\right]+\ \ldots, \end{aligned} \tag{13.88}$$

where we have only shown the leading Yukawa terms of two modes. The orientation dependence of these terms is *described by the multipolar effective charges* of the two particles. The missing terms in the square brackets are the terms that decay like $e^{-\kappa r_{12}}/r_{12}^2$, $e^{-\kappa r_{12}}/r_{12}^3$ etc.[14] Like in the spherical case treated in Section 12.3.4 there are also terms that decay like $e^{-2\kappa r_{12}}/r_{12}^2$, $e^{-3\kappa r_{12}}/r_{12}^3$ etc. times slowly varying functions and cross-terms of the various modes. The Yukawa function terms are, however, the most interesting ones and in many cases they are dominant. When κ and κ' are complex conjugates of each other we have for $r_{12} \to \infty$

$$\begin{aligned} & h_{ij}^{(2)}(\mathbf{r}_{12},\boldsymbol{\omega}_1,\boldsymbol{\omega}_2) \\ & \sim -\frac{\beta}{|\mathcal{E}_r^{\text{eff}}(\kappa)|\varepsilon_0}\left[|\mathsf{Q}_i^{\text{eff}}\mathsf{Q}_j^{\text{eff}}|\frac{e^{-\kappa_\Re r_{12}}}{2\pi r_{12}}\right. \\ & \left.\times\ \cos\Big(\kappa_\Im r_{12}-\vartheta_{\mathcal{E}}(\kappa)+\vartheta_i(\hat{\mathbf{r}}_{12},\boldsymbol{\omega}_1,\kappa)+\vartheta_j(-\hat{\mathbf{r}}_{12},\boldsymbol{\omega}_2,\kappa)\Big)+\ldots\right], \end{aligned} \tag{13.89}$$

where we have suppressed the arguments of $\mathsf{Q}_i^{\text{eff}}$ and $\mathsf{Q}_j^{\text{eff}}$ (they are the same as in Equation 13.88).

In order to show that in the vast majority of cases all decay modes of $h_{ij}^{(2)}(\mathbf{r}_{12},\boldsymbol{\omega}_1,\boldsymbol{\omega}_2)$ are given by the zeros of $\tilde{\varepsilon}(k)$ we need to prove that each pole of $\tilde{h}_{ij}^{\star}$ in Equation 13.87 is cancelled by a corresponding pole of the last term of this equation. This implies that $\tilde{h}_{ij}^{(2)}$ and $\tilde{h}_{ij}^{\star}$ cannot have poles for the same k values, so the poles of $\tilde{h}_{ij}^{(2)}$ are given by the second term in Equation 13.87 as anticipated earlier. The proof is very similar to the corresponding one for spherical particles in the shaded text box on page 345, where the details were shown only for binary systems. The present case is a bit more

[14] The coefficients of these terms satisfy the same recurrence relationship as in the shaded text box on page 429, whereby the prefactor of the term that decays like $e^{-\kappa r}/r^{m+1}$ is given by $\mathcal{M}_m\left[\mathsf{Q}_i^{[0]\text{eff}}(\hat{\mathbf{r}},\omega_1,\kappa)\mathsf{Q}_j^{[0]\text{eff}}(-\hat{\mathbf{r}},\omega_2,\kappa)\right]$.

complicated because it involves several components and pair correlations that have orientational variables.

The Ornstein-Zernike equation for $H_{ij}^{(2)}$ and $C_{ij}^{(2)}$ is

$$\sum_l \int d\mathbf{r}_3\, d\omega_3\, C_{il}^{(2)}(\mathbf{r}_{13},\omega_1,\omega_3) H_{lj}^{(2)}(\mathbf{r}_{32},\omega_3,\omega_2) = \delta_{ij}^{(3)}(\mathbf{r}_{12})\delta^{\omega}(\omega_1-\omega_2)$$

(cf. Equation 11.142). In Fourier space we can write this and the OZ equation for $H_{ij}^{\star}$ and $C_{ij}^{\star}$ in the form

$$\mathbb{C}^{(2)}(\mathbf{k}) \star \mathbb{H}^{(2)}(\mathbf{k}) = \mathbb{1} \quad \text{and} \quad \mathbb{C}^{\star}(\mathbf{k}) \star \mathbb{H}^{\star}(\mathbf{k}) = \mathbb{1},$$

where we have defined the square matrices $\mathbb{1} = \{\delta_{ij}\delta^{\omega}(\omega_1-\omega_2)\}$, $\mathbb{H}^{(2)} = \{\tilde{H}_{ij}(\mathbf{k},\omega_1,\omega_2)\}$, $\mathbb{C}^{(2)} = \{\tilde{C}_{ij}(\mathbf{k},\omega_1,\omega_2)\}$ and likewise for $\mathbb{H}^{\star}$ and $\mathbb{C}^{\star}$ and where the operation $\star$ denotes combined matrix multiplication for species indices and integration for orientation variables. By discretizing the orientational variable ω on a very dense uniform grid and thereby replacing integrals of the kind $\int d\omega$ by discrete sums $\sum_{\omega} d\omega$, we can express the OZ equation by the same expression in matrix form also for the orientational variables, whereby $\star$ is solely an ordinary matrix multiplication. This is the convention we will use for the matrix definitions henceforth.

$H_{ij}^{(2)}(k\hat{\mathbf{k}},\omega_1,\omega_2)$ and $\tilde{h}_{ij}^{(2)}(k\hat{\mathbf{k}},\omega_1,\omega_2)$ have poles for the same k values because

$$H_{ij}^{(2)}(\mathbf{k},\omega_1,\omega_2) = n_i^{\mathrm{b}}\delta_{ij}\delta^{\omega}(\omega_1-\omega_2) + n_i^{\mathrm{b}}n_j^{\mathrm{b}}\tilde{h}_{ij}^{(2)}(\mathbf{k},\omega_1,\omega_2) \tag{13.90}$$

and the corresponding applies to $\tilde{H}_{ij}^{\star}$ and $\tilde{h}_{ij}^{\star}$. We start be investigating $\tilde{H}_{ij}^{\star}$ and we have

$$\mathbb{H}^{\star}(\mathbf{k}) = [\mathbb{C}^{\star}(\mathbf{k})]^{-1} = \frac{\mathrm{Adj}\,\mathbb{C}^{\star}(\mathbf{k})}{D^{\star}(k)}, \tag{13.91}$$

where $D^{\star}(k) = \det \mathbb{C}^{\star}(\mathbf{k})$ is the determinant[15] and $\mathrm{Adj}\,\mathbb{C}^{\star}(\mathbf{k})$ is the adjugate matrix of $\mathbb{C}^{\star}(\mathbf{k})$. $\tilde{H}_{ij}^{\star}$ has a (simple) pole when $D^{\star}(k)$ has a (simple) zero provided that the corresponding elements in $\mathrm{Adj}\,\mathbb{C}^{\star}(\mathbf{k})$ is nonzero. Say that the determinant is zero for $k = ib$, where b may be a real or a complex number,

[15] The determinant $\det \mathbb{C}^{\star}(\mathbf{k})$ is independent of $\hat{\mathbf{k}}$ due to integrations over orientations, which are represented by a summations over the dense uniform grid. This can be understood from the following. Each term in the determinant is a product of $\tilde{C}_{ij}(\mathbf{k},\omega_1,\omega_2)$ over combinations of the two composite indices i,ω_1 and j,ω_2 according to the rules to calculate a determinant. Let $\hat{\mathbf{k}}$ change from $\hat{\mathbf{k}}'$ to $\hat{\mathbf{k}}''$ by means of a rotation. For each term in the determinant for $\hat{\mathbf{k}} = \hat{\mathbf{k}}'$ there is always a term for $\hat{\mathbf{k}} = \hat{\mathbf{k}}''$ with the same value. The combinations of orientations ω_1,ω_2 in the factors $\tilde{C}_{ij}(\mathbf{k},\omega_1,\omega_2)$ in the term for $\hat{\mathbf{k}} = \hat{\mathbf{k}}''$ are simply those that appear in the term for $\hat{\mathbf{k}} = \hat{\mathbf{k}}'$ before the rotation. i.e., the orientations for $\hat{\mathbf{k}} = \hat{\mathbf{k}}''$ are obtained from those for $\hat{\mathbf{k}} = \hat{\mathbf{k}}'$ by the rotation.

so $\tilde{H}^{\star}_{ij}$ has a pole for $k = ib$. Thus, $\tilde{H}^{\star}_{ij}$ is dominated for k values near ib by the diverging contribution

$$\tilde{H}^{\star}_{ij}(\mathbf{k}, \omega_1, \omega_2) \sim \frac{\tilde{F}_{ij}(ib\hat{\mathbf{k}}, \omega_1, \omega_2)}{k^2 + b^2} \quad \text{when } k \to ib,$$

where $\tilde{F}_{ij}(k\hat{\mathbf{k}}, \omega_1, \omega_2)$ is a function that is finite at $k = \pm ib$.

We now use the mathematical fact that the adjugate matrix $\mathbb{A}$ of a square matrix with zero determinant has elements $A_{\nu\nu'}$ that factorize such that $A_{\nu\nu'} = \mathcal{C}\,\tau_\nu \tau_{\nu'}$, where $\mathcal{C}$ is some number and τ_ν are components of some vector. In our case this means that $\mathbb{C}^{\star}(k\hat{\mathbf{k}})$ has this property for $k = ib$ since $\det \mathbb{C}^{\star}(ib\hat{\mathbf{k}}) = 0$. This implies that $\tilde{F}_{ij}$ also factorizes in this manner, so we have

$$\tilde{H}^{\star}_{ij}(\mathbf{k}, \omega_1, \omega_2) = \frac{\alpha_i(ib\hat{\mathbf{k}}, \omega_1)\alpha_j(-ib\hat{\mathbf{k}}, \omega_2)}{(k^2 + b^2)\mathcal{K}^{\star}(b)} + \tilde{H}^{\star\mathrm{rest}}_{ij}(\mathbf{k}, \omega_1, \omega_2) \qquad (13.92)$$

where $\mathcal{K}^{\star}(b)$ is a constant and $\tilde{H}^{\star\mathrm{rest}}_{ij}$ is the part of $\tilde{H}^{\star}_{ij}$ that stays finite at $k = ib$. Note that $n^{\mathrm{b}}_i n^{\mathrm{b}}_j \tilde{h}^{\star}_{ij}$ and $n^{\mathrm{b}}_i n^{\mathrm{b}}_j \tilde{h}^{\star\mathrm{rest}}_{ij}$, analogously defined, satisfy the same relationship taking the places of $\tilde{H}^{\star}_{ij}$ and $\tilde{H}^{\star\mathrm{rest}}_{ij}$, respectively.

Furthermore, we obtain using Equation 13.92

$$\begin{aligned} n^{\mathrm{b}}_i \tilde{\rho}^{\star}_i(\mathbf{k}, \omega_1) &= \sum_j \int d\omega_2\, \tilde{H}^{\star}_{ij}(\mathbf{k}, \omega_1, \omega_2)\tilde{\sigma}_j(\mathbf{k}, \omega_2) \\ &= \frac{\alpha_i(ib\hat{\mathbf{k}}, \omega_1)\mathcal{F}(b)}{(k^2 + b^2)\mathcal{K}^{\star}(b)} + n^{\mathrm{b}}_i \tilde{\rho}^{\star\mathrm{rest}}_i(\mathbf{k}, \omega_1), \end{aligned} \qquad (13.93)$$

where

$$\mathcal{F}(b) = \sum_l \int d\omega\, \alpha_l(-ib\hat{\mathbf{k}}, \omega)\tilde{\sigma}_l(ib\hat{\mathbf{k}}, \omega) \qquad (13.94)$$

and $n^{\mathrm{b}}_i \tilde{\rho}^{\star\mathrm{rest}}_i$ is the remaining part of $n^{\mathrm{b}}_i \tilde{\rho}^{\star}_i$ that stays finite at $k = ib$. $\mathcal{F}(b)$ does not depend on the direction of $\hat{\mathbf{k}}$ due to the averaging over ω.

The dielectric function $\tilde{\epsilon}(k) = 1 - \tilde{\phi}_{\mathrm{Coul}}(k)\tilde{\chi}^{\star}(k)$ can be written as the first line in the following equation and by inserting Equation 13.92 we obtain the second line

$$\begin{aligned} \tilde{\epsilon}(k) &= 1 + \beta\tilde{\phi}_{\mathrm{Coul}}(k)\sum_{ij}\int d\omega_1 d\omega_2\, \sigma_i(-\mathbf{k}, \omega_1)\sigma_j(\mathbf{k}, \omega_2)\tilde{H}^{\star}_{ij}(\mathbf{k}, \omega_1, \omega_2) \\ &= 1 + \beta\tilde{\phi}_{\mathrm{Coul}}(k)\frac{\mathcal{F}^2(b)}{(k^2 + b^2)\mathcal{K}^{\star}(b)} + \tilde{\epsilon}^{\mathrm{rest}}(k), \end{aligned} \qquad (13.95)$$

where $\tilde{\epsilon}^{\mathrm{rest}}$ stays finite at $k = ib$. Provided that $\mathcal{F}(b)$ is nonzero, the functions $\tilde{\rho}^{\star}_i$ and $\tilde{\epsilon}(k)$ diverge when $k \to ib$ due to $k^2 + b^2$ in the denominator.

We can now investigate the second term in Equation 13.87 when $k \to ib$ and we have

$$n_i^{\mathrm{b}} n_j^{\mathrm{b}} \frac{\beta \rho_i^\star(\mathbf{k}, \boldsymbol{\omega}_1) \rho_j^\star(-\mathbf{k}, \boldsymbol{\omega}_2)}{\varepsilon_0 k^2 \tilde{\epsilon}(k)} \sim \frac{\beta \dfrac{\alpha_i(\hat{\mathbf{k}}, \boldsymbol{\omega}_1, b)\mathcal{F}(b)}{(k^2+b^2)\mathcal{K}^\star(b)} \cdot \dfrac{\alpha_j(-\hat{\mathbf{k}}, \boldsymbol{\omega}_2, b)\mathcal{F}(b)}{(k^2+b^2)\mathcal{K}^\star(b)}}{\beta \dfrac{\mathcal{F}^2(b)}{(k^2+b^2)\mathcal{K}^\star(b)}}$$
$$= \frac{\alpha_i(\hat{\mathbf{k}}, \boldsymbol{\omega}_1, b)\alpha_j(-\hat{\mathbf{k}}, \boldsymbol{\omega}_1, b)}{(k^2+b^2)\mathcal{K}^\star(b)}.$$

This is the same as the contribution from the pole of $n_i^{\mathrm{b}} n_j^{\mathrm{b}} h_{ij}^\star(\mathbf{k}, \boldsymbol{\omega}_1, \boldsymbol{\omega}_2)$, which means that the rhs of Equation 13.87 stays finite for $k = ib$, that is, $h_{ij}^{(2)}(\mathbf{k}, \boldsymbol{\omega}_1, \boldsymbol{\omega}_2)$ does not have any pole for $k = ib$. Thus, the pole of $h_{ij}^\star(\mathbf{k}, \boldsymbol{\omega}_1, \boldsymbol{\omega}_2)$ is cancelled by the pole of the second term in Equation 13.87 as anticipated. This holds true provided that $\mathcal{F}(b)$ is nonzero, which is normally the case. Exceptions occur for charge-inversion invariant systems where $\mathcal{F}(b) = 0$ for some poles $k = ib$ of $h_{ij}^\star$, as we saw in Section 12.3.4 for the spherical case, whereby this pole is not cancelled by the second term and $h_{ij}^{(2)}$ *does* have a pole for $k = ib$. Such systems will be investigated later. For other systems, this can happen at particular points in the system's parameter space where $\mathcal{F}(b)$ fortuitously happen to be equal to zero, which means that the cancellation may not take place at such a point. Otherwise the cancellation always occurs and the poles of $\tilde{h}_{ij}^{(2)}$ are given by the zeros of $\tilde{\epsilon}(k)$.

The various poles of $\tilde{h}_{ij}^\star$ at $k = ib$, ib' ... give rise to Yukawa function terms

$$\begin{aligned} & h_{ij}^\star(\mathbf{r}_{12}, \boldsymbol{\omega}_1, \boldsymbol{\omega}_2) \\ &= \frac{a_i(\hat{\mathbf{r}}_{12}, \boldsymbol{\omega}_1, b) a_j(-\hat{\mathbf{r}}_{12}, \boldsymbol{\omega}_2, b)}{\mathcal{K}^\star(b)} \cdot \frac{e^{-br_{12}}}{4\pi r_{12}} \\ &+ \frac{a_i(\hat{\mathbf{r}}_{12}, \boldsymbol{\omega}_1, b') a_j(-\hat{\mathbf{r}}_{12}, \boldsymbol{\omega}_2, b')}{\mathcal{K}^\star(b')} \cdot \frac{e^{-b'r_{12}}}{4\pi r_{12}} + \dots, \end{aligned}$$

where $a_i(\hat{\mathbf{r}}_{12}, \boldsymbol{\omega}_1, b) = \alpha_i(ib\hat{\mathbf{r}}_{12}, \boldsymbol{\omega}_1)/n_i$, and there are also contributions of other functional forms as before. When ib is the singularity of $\tilde{h}_{ij}^\star$ closest to the real k axis in complex k-space, the first term is leading when $r_{12} \to \infty$. For a real-valued decay parameter b the factor $a_i(\hat{\mathbf{r}}_{12}, \boldsymbol{\omega}_1, b)$ is real and when there is a pair of complex-valued of decay parameters that are complex conjugates of each other, for example b and b', we obtain an oscillatory Yukawa function contribution as usual.

Let us now investigate the poles of $\tilde{H}_{ij}^\star$ and $\tilde{H}_{ij}^{(2)}$ further. We have

$$\mathbb{H}^{(2)}(\mathbf{k}) = [\mathbb{C}^{(2)}(\mathbf{k})]^{-1} = \frac{\operatorname{Adj}\mathbb{C}^{(2)}(\mathbf{k})}{D^{(2)}(k)},$$

where $D^{(2)}(k) = \det \mathbb{C}^{(2)}(\mathbf{k})$ is the determinant. Equation 13.57 can be written as the matrix equation

$$\mathbb{C}^{(2)}(\mathbf{k}) = \mathbb{C}^{\star}(\mathbf{k}) + \beta\tilde{\phi}_{\mathrm{Coul}}(k)\boldsymbol{\sigma}(\mathbf{k})\boldsymbol{\sigma}^{\dagger}(\mathbf{k}),$$

where $\boldsymbol{\sigma}(\mathbf{k}) = \{\tilde{\sigma}_i(\mathbf{k},\boldsymbol{\omega})\}$ is a column vector and $\dagger$ denotes the conjugate transpose operation (Hermitian conjugate). We now make use of the matrix identity $\det(\mathbb{A}+\mathbf{a}\,\mathbf{b}^{\dagger}) = \det(\mathbb{A}) + \mathbf{b}^{\dagger}\,\mathrm{Adj}(\mathbb{A})\mathbf{a}$, where $\mathbb{A}$ is a square matrix and $\mathbf{a}$ and $\mathbf{b}$ are column vectors. By setting $\mathbb{A} = \mathbb{C}^{\star}$, $\mathbf{a} = \boldsymbol{\sigma}$ and $\mathbf{b} = \beta\tilde{\phi}_{\mathrm{Coul}}\boldsymbol{\sigma}$ we get

$$D^{(2)}(k) = D^{\star}(k) + \beta\tilde{\phi}_{\mathrm{Coul}}(k)\boldsymbol{\sigma}^{\dagger}(\mathbf{k})\mathrm{Adj}\,\mathbb{C}^{\star}(\mathbf{k})\boldsymbol{\sigma}(\mathbf{k}). \tag{13.96}$$

This implies that $D^{(2)}(k)$ and $D^{\star}(k)$ cannot simultaneously be equal to zero for some k value unless the last term is equal to zero for this k, which only occurs in exceptional cases.

This is, in fact, in line with our findings in the shaded text box above and we will see that the exceptions are the same. In order to see this, let us consider a single zero of $D^{\star}(k)$, say, $k = ib$ as investigated there ($k = -ib$ must also be a zero). Then we can write $D^{\star}$ as $D^{\star}(k) = (k^2+b^2)f(k,b)$, where the function $f(k,b) \neq 0$ when $k = \pm ib$. Since $\mathrm{Adj}\,\mathbb{C}^{\star}(\mathbf{k}) = \mathbb{H}^{\star}(\mathbf{k})D^{\star}(k) = \mathbb{H}^{\star}(\mathbf{k})(k^2+b^2)f(k,b)$ we can write

$$\boldsymbol{\sigma}^{\dagger}(\mathbf{k})\mathrm{Adj}\,\mathbb{C}^{\star}(\mathbf{k})\boldsymbol{\sigma}(\mathbf{k}) = \boldsymbol{\sigma}^{\dagger}(\mathbf{k})\mathbb{H}^{\star}(\mathbf{k})\boldsymbol{\sigma}(\mathbf{k})(k^2+b^2)f(k,b).$$

Equation 13.92 implies that

$$\mathbb{H}^{\star}(\mathbf{k}) \sim \frac{\boldsymbol{\alpha}(ib\hat{\mathbf{k}})\boldsymbol{\alpha}^{\dagger}(ib\hat{\mathbf{k}})}{(k^2+b^2)\mathcal{K}^{\star}(b)} \quad \text{when } k \to ib,$$

where we have introduced the column vector $\boldsymbol{\alpha}(ib\hat{\mathbf{k}}) = \{\alpha_i(ib\hat{\mathbf{k}},\boldsymbol{\omega}_1)\}$, and therefore when $k \to ib$ we have

$$\boldsymbol{\sigma}^{\dagger}(\mathbf{k})\mathrm{Adj}\,\mathbb{C}^{\star}(\mathbf{k})\boldsymbol{\sigma}(\mathbf{k}) \to \left.\frac{\boldsymbol{\sigma}^{\dagger}(\mathbf{k})\boldsymbol{\alpha}(ib\hat{\mathbf{k}})\boldsymbol{\alpha}^{\dagger}(ib\hat{\mathbf{k}},b)\boldsymbol{\sigma}(\mathbf{k})f(k,b)}{\mathcal{K}^{\star}(b)}\right|_{k=ib}.$$

Now, $\boldsymbol{\alpha}^{\dagger}(ib\hat{\mathbf{k}})\boldsymbol{\sigma}(\mathbf{k})$ for $k = ib$ is the same as $\mathcal{F}(b)$ defined in Equation 13.94, because the matrix multiplication $\boldsymbol{\alpha}^{\dagger}\boldsymbol{\sigma}$ involves both sum over species index and integration over orientations. Hence,

$$\boldsymbol{\sigma}^{\dagger}(\mathbf{k})\mathrm{Adj}\,\mathbb{C}^{\star}(\mathbf{k})\boldsymbol{\sigma}(\mathbf{k}) \to \left.\frac{\mathcal{F}^2(b)f(k,b)}{\mathcal{K}^{\star}(b)}\right|_{k=ib} \quad \text{when } k \to ib$$

and Equation 13.96 implies that

$$D^{(2)}(k) \to \beta\left.\frac{\tilde{\phi}_{\mathrm{Coul}}(k)\mathcal{F}^2(b)f(k,b)}{\mathcal{K}^{\star}(b)}\right|_{k=ib}$$

since $D^{\star}(k) = 0$ there. Thus, we can conclude that $D^{(2)}(k)$ is nonzero when $k = ib$ unless $\mathcal{F}(b) = 0$, which is in agreement with what we found earlier.

In the case of spherical particles we found in Section 12.3.4 that $D^{(2)}(k) = D^{\star}(k)\tilde{\epsilon}(k)$ (Equation 12.202) and we concluded that for charge-inversion invariant systems it is possible to have $D^{(2)}(k) = 0$ when either $D^{*}(k) = 0$ or $\tilde{\epsilon}(k) = 0$, but that in general, $D^{(2)}(k)$ is zero *only* when $\tilde{\epsilon}(k)$ is zero because $D^{(2)}(k)$ and $D^{\star}(k)$ are not simultaneously equal to zero. This is true also in the present case because Equation 13.96 can be written as

$$D^{(2)}(k) = D^{\star}(k)\left[1 + \beta\tilde{\phi}_{\mathrm{Coul}}(k)\boldsymbol{\sigma}^{\dagger}(\mathbf{k})\mathbb{H}^{\star}(\mathbf{k})\boldsymbol{\sigma}(\mathbf{k})\right] = D^{\star}(k)\tilde{\epsilon}(k), \qquad (13.97)$$

where we have used Equation 13.91 and the fact that

$$\tilde{\epsilon}(k) = 1 + \beta\tilde{\phi}_{\mathrm{Coul}}(k)\boldsymbol{\sigma}^{\dagger}(\mathbf{k})\mathbb{H}^{\star}(\mathbf{k})\boldsymbol{\sigma}(\mathbf{k})$$

(from the first line of Equation 13.95).

13.3.3.3 The density-density, density-charge and charge-charge correlation functions

Let us now turn to the density-density, density-charge and charge-charge correlation functions, $H_{\mathrm{NN}}(r)$, $H_{\mathrm{NQ}}(r) \equiv H_{\mathrm{QN}}(r)$ and $H_{\mathrm{QQ}}(r)$. They depend only on the distance r in isotropic systems and we have

$$\begin{aligned}
H_{\mathrm{NN}}(r_{12}) &= \sum_{ij}\left\langle H^{(2)}_{ij}(\mathbf{r}_{12},\boldsymbol{\omega}_1,\boldsymbol{\omega}_2)\right\rangle_{\boldsymbol{\omega}_1,\boldsymbol{\omega}_2} \\
H_{\mathrm{NQ}}(r_{12}) &= \frac{1}{q_{\mathrm{e}}}\sum_{ij}\int d\mathbf{r}_4 \left\langle H^{(2)}_{ij}(\mathbf{r}_{14},\boldsymbol{\omega}_1,\boldsymbol{\omega}_4)\sigma_j(\mathbf{r}_{42},\boldsymbol{\omega}_4)\right\rangle_{\boldsymbol{\omega}_1,\boldsymbol{\omega}_4} \\
H_{\mathrm{QQ}}(r_{12}) &= \frac{1}{q_{\mathrm{e}}^2}\sum_{ij}\int d\mathbf{r}_3 d\mathbf{r}_4 \qquad\qquad (13.98)\\
&\quad\times \left\langle H^{(2)}_{ij}(\mathbf{r}_{34},\boldsymbol{\omega}_3,\boldsymbol{\omega}_4)\sigma_i(\mathbf{r}_{31},\boldsymbol{\omega}_3)\sigma_j(\mathbf{r}_{42},\boldsymbol{\omega}_4)\right\rangle_{\boldsymbol{\omega}_3,\boldsymbol{\omega}_4}
\end{aligned}$$

(cf. the definition of H_{QQ} in Equation 13.19, where the expression is interpreted in terms of charge density fluctuations). The starred functions $H^{\star}_{IJ}(r_{12})$ for $I, J = \mathrm{Q}$ and N are given in terms of $H^{\star}_{ij}$ by the analogous equations.

We will now use Equation 13.87 to obtain the analogous expression for $H^{(2)}_{IJ}$ and $H^{\star}_{IJ}$. Using the Fourier transform of Equation 13.98 and the definitions of $H^{(2)}_{ij}$ and $H^{\star}_{ij}$ we get

$$\begin{aligned}
\tilde{H}_{\mathrm{NN}}(k) &= \tilde{H}^{\star}_{\mathrm{NN}}(k) - \beta(n^{\mathrm{b}}_{\mathrm{tot}})^2\frac{\left[\tilde{\rho}^{\star}_{\mathrm{n}}(k)\right]^2}{\varepsilon_0 k^2\tilde{\epsilon}(k)} \\
\tilde{H}_{\mathrm{NQ}}(k) &= \tilde{H}^{\star}_{\mathrm{NQ}}(k) - \beta(n^{\mathrm{b}}_{\mathrm{tot}})^2\frac{\tilde{\rho}^{\star}_{\mathrm{n}}(k)\tilde{\rho}^{\star}_{\mathrm{q}}(k)}{\varepsilon_0 k^2\tilde{\epsilon}(k)} \qquad (13.99)\\
\tilde{H}_{\mathrm{QQ}}(k) &= \tilde{H}^{\star}_{\mathrm{QQ}}(k) - \beta(n^{\mathrm{b}}_{\mathrm{tot}})^2\frac{\left[\tilde{\rho}^{\star}_{\mathrm{q}}(k)\right]^2}{\varepsilon_0 k^2\tilde{\epsilon}(k)},
\end{aligned}$$

where

$$\begin{aligned}
\tilde{\rho}_{\rm n}^{\star}(k) &= \frac{1}{n_{\rm tot}^{\rm b}} \sum_i n_i^{\rm b} \langle \tilde{\rho}_i^{\star}(\mathbf{k},\omega) \rangle_\omega \\
\tilde{\rho}_{\rm q}^{\star}(k) &= \frac{1}{n_{\rm tot}^{\rm b} q_{\rm e}} \sum_i n_i^{\rm b} \langle \tilde{\sigma}_i(-\mathbf{k},\omega) \tilde{\rho}_i^{\star}(\mathbf{k},\omega) \rangle_\omega
\end{aligned}$$

are real-valued functions because they do not depend on the direction of $\mathbf{k}$ due to the averaging over orientations. Apart from the definitions of $\tilde{\rho}_{\rm n}^{\star}(k)$ and $\tilde{\rho}_{\rm q}^{\star}(k)$, this is the same as Equation 12.205 for the spherical case.

We will continue with an analysis of the main case where all decay modes of $h_{ij}^{(2)}$ are the same as those of $\phi_{\rm Coul}^{\star}(r)$ and are given the zeros of $\tilde{\epsilon}(k)$. The decay modes of $H_{\rm NN}(r)$, $H_{\rm NQ}(r)$ and $H_{\rm QQ}(r)$ are then determined by the last term in each line of Equation 13.99. We will treat the exceptional case of charge-inversion invariant systems later.

From Equation 13.99 it follows that the first two Yukawa function terms are given by

$$\begin{aligned}
H_{\rm NN}(r) &= -\frac{\left[q_{\rm n}^{\rm eff}(\kappa)\right]^2}{\mathcal{B}(\kappa)} \cdot \frac{e^{-\kappa r}}{4\pi r} - \frac{\left[q_{\rm n}^{\rm eff}(\kappa')\right]^2}{\mathcal{B}(\kappa')} \cdot \frac{e^{-\kappa' r}}{4\pi r} - \dots \\
H_{\rm NQ}(r) &= -\frac{q_{\rm n}^{\rm eff}(\kappa) q_{\rm q}^{\rm eff}(\kappa)}{\mathcal{B}(\kappa)} \cdot \frac{e^{-\kappa r}}{4\pi r} - \frac{q_{\rm n}^{\rm eff}(\kappa') q_{\rm q}^{\rm eff}(\kappa')}{\mathcal{B}(\kappa')} \cdot \frac{e^{-\kappa' r}}{4\pi r} - \dots \\
H_{\rm QQ}(r) &= -\frac{\left[q_{\rm q}^{\rm eff}(\kappa)\right]^2}{\mathcal{B}(\kappa)} \cdot \frac{e^{-\kappa r}}{4\pi r} - \frac{\left[q_{\rm q}^{\rm eff}(\kappa')\right]^2}{\mathcal{B}(\kappa')} \cdot \frac{e^{-\kappa' r}}{4\pi r} - \dots, \qquad (13.100)
\end{aligned}$$

where

$$\mathcal{B}(\kappa) = \frac{\mathcal{E}_r^{\rm eff}(\kappa)\varepsilon_0}{\beta (n_{\rm tot}^{\rm b})^2}$$

and

$$\begin{aligned}
q_{\rm n}^{\rm eff}(\kappa) \equiv \tilde{\rho}_{\rm n}^{\star}(i\kappa) &= \sum_i \mathrm{x}_i^{\rm b} \left\langle \tilde{\rho}_i^{\star}(i\kappa\hat{\mathbf{k}},\omega) \right\rangle_\omega = \sum_i \mathrm{x}_i^{\rm b} q_i^{\rm eff}(\kappa) \\
q_{\rm q}^{\rm eff}(\kappa) \equiv \tilde{\rho}_{\rm q}^{\star}(i\kappa) &= q_{\rm e}^{-1} \sum_i \mathrm{x}_i^{\rm b} \left\langle \tilde{\sigma}_i(-i\kappa\hat{\mathbf{k}},\omega) \tilde{\rho}_i^{\star}(i\kappa\hat{\mathbf{k}},\omega) \right\rangle_\omega \\
&= q_{\rm e}^{-1} \sum_i \mathrm{x}_i^{\rm b} \left\langle \mathrm{Q}_i(-\hat{\mathbf{k}},\omega,\kappa) \mathrm{Q}_i^{\rm eff}(\hat{\mathbf{k}},\omega,\kappa) \right\rangle_\omega,
\end{aligned}$$

where we have used the definitions of $\mathrm{Q}_i^{\rm eff}$, $q_i^{\rm eff}$ and Q_i (Equations 13.75, 13.76 and 13.85) in order to obtain the rhs of each equation. Due to the averaging over orientations, these expressions are independent of the direction of $\hat{\mathbf{k}}$. The equation for $q_{\rm q}^{\rm eff}$ is much more complicated than that for $q_{\rm n}^{\rm eff}$, but by using Equation 13.86 for κ^2 we can simplify it and thereby we obtain

$$q_{\rm q}^{\rm eff}(\kappa) = \frac{\kappa^2 \varepsilon_0}{\beta q_{\rm e} n_{\rm tot}^{\rm b}}, \qquad q_{\rm n}^{\rm eff}(\kappa) = \sum_i \mathrm{x}_i^{\rm b} q_i^{\rm eff}(\kappa) \qquad (13.101)$$

which is the same as for the spherical case. Thus, Equation 13.100 together with the latter expressions for $q_{\rm q}^{\rm eff}(\kappa)$ and $q_{\rm n}^{\rm eff}(\kappa)$ are *exactly the same for electrolytes consisting of either spherical or nonspherical particles.* Note that we can write the $q_{\rm q}^{\rm eff}(\kappa)$-expression as

$$\kappa^2 = \frac{\beta n_{\rm tot}^{\rm b} q_{\rm e} q_{\rm q}^{\rm eff}(\kappa)}{\varepsilon_0},$$

which is yet another exact equation for κ that is equivalent to the other ones.

A summary of the decay modes for $H_{\rm NN}(r)$, $H_{\rm NQ}(r)$ and $H_{\rm QQ}(r)$ is given Table 12.4 on page 356 and *applies in general, both for spherical electrolytes and nonspherical particles.*

$H_{\rm NN}(r)$, $H_{\rm NQ}(r)$ and $H_{\rm QQ}(r)$ have the same decay length $1/\kappa$, but as we saw for the spherical case, the leading term of $H_{\rm QQ}(r)$ can be larger than that of $H_{\rm NN}(r)$ or vice versa depending on the values of the system parameters. Due to Equation 13.101 the magnitudes of $H_{\rm QQ}(r)$ and $H_{\rm NQ}(r)$ are directly linked to the decay length $1/\kappa$, while the magnitude of $H_{\rm NN}(r)$ does not have such a direct link. The leading term of $H_{\rm QQ}(r)$ contains the factor κ^4 and is small when the decay length is large. We have

$$H_{\rm QQ}(r) = -\frac{\kappa^4 \varepsilon_0}{\mathcal{E}_r^{\rm eff}(\kappa)\beta q_{\rm e}^2} \cdot \frac{e^{-\kappa r}}{4\pi r} - \frac{(\kappa')^4 \varepsilon_0}{\mathcal{E}_r^{\rm eff}(\kappa')\beta q_{\rm e}^2} \cdot \frac{e^{-\kappa' r}}{4\pi r} - \ldots .$$

This implies that the amplitudes of the decay modes of

$$H_{\rm QQ}(r) = \mathsf{C}\frac{e^{-\kappa r}}{4\pi r} + \mathsf{C}'\frac{e^{-\kappa' r}}{4\pi r} + \ldots,$$

in all cases satisfy

$$\frac{\mathsf{C}}{\mathsf{C}'} = \left[\frac{\kappa}{\kappa'}\right]^4 \cdot \frac{\mathcal{E}_r^{\rm eff}(\kappa')}{\mathcal{E}_r^{\rm eff}(\kappa)} \tag{13.102}$$

when κ and κ' are real.

When there is an oscillatory mode, for example when the leading decay modes consist of one mode with decay parameter κ and there is an oscillatory mode with decay parameters $\kappa''_{\Re} \pm {\rm i}\kappa''_{\Im}$ with $\kappa''_{\Re} > \kappa$, we have from Equations 13.100 and 13.101

$$\begin{aligned} H_{\rm QQ}(r) \sim & -\frac{\varepsilon_0}{\beta q_{\rm e}^2}\left[\frac{\kappa^4}{\mathcal{E}_r^{\rm eff}(\kappa)} \cdot \frac{e^{-\kappa r}}{4\pi r}\right. \qquad (13.103) \\ & \left. + \frac{|\kappa''|^4}{|\mathcal{E}_r^{\rm eff}(\kappa'')|} \cdot \frac{e^{-\kappa''_{\Re} r}}{2\pi r} \cos(\kappa''_{\Im} r + 4\vartheta_{\kappa''} - \vartheta_{\mathcal{E}}(\kappa'')) + \ldots\right], \end{aligned}$$

where $\kappa''_{\Re} + {\rm i}\kappa''_{\Im} = |\kappa''| \exp(-{\rm i}\vartheta_{\kappa''})$. This has the functional form

$$H_{\rm QQ}(r) \sim \mathsf{C}\frac{e^{-\kappa r}}{4\pi r} + \mathsf{C}''\frac{e^{-\kappa''_{\Re} r}}{2\pi r} \cos(\kappa''_{\Im} r + \vartheta'') + \ldots \tag{13.104}$$

and it follows that

$$\frac{\mathsf{C}}{\mathsf{C}''} = \left[\frac{\kappa}{|\kappa''|}\right]^4 \cdot \frac{|\mathcal{E}_r^{\text{eff}}(\kappa'')|}{\mathcal{E}_r^{\text{eff}}(\kappa)}. \tag{13.105}$$

This is the same as we obtained for electrolytes in the restricted primitive model in Equation 12.254, where we investigated $h_{\text{D}}^{(2)}(r)$ that is proportional to $H_{\text{QQ}}(r)$. These results have general validity and the amplitudes of the various modes of $H_{\text{QQ}}(r)$ scales as the forth power of the decay parameter divided by $\mathcal{E}_r^{\text{eff}}$.

Let us now treat ***charge-inversion invariant systems***, for which $H_{\text{NQ}}(r)$ and $\tilde{\rho}_{\text{n}}^{\star}(k)$ are identically equal to zero for the same reasons as in the spherical case. Therefore

$$\begin{aligned} \tilde{H}_{\text{NN}}(k) &= \tilde{H}_{\text{NN}}^{\star}(k) \\ \tilde{H}_{\text{QQ}}(k) &= \tilde{H}_{\text{QQ}}^{\star}(k) - \beta(n_{\text{tot}}^{\text{b}})^2 \cdot \frac{\left[\tilde{\rho}_{\text{q}}^{\star}(k)\right]^2}{\varepsilon_0 k^2 \tilde{\epsilon}(k)} \\ H_{\text{NQ}}(k) &\equiv 0 \qquad \text{(charge-inv. invar.)} \end{aligned} \tag{13.106}$$

(obtained from Equation 13.99), where $H_{\text{NN}}(r)$ and $H_{\text{QQ}}(r)$ have different decay lengths. As mentioned earlier, since $D^{(2)}(k) = D^{\star}(k)\tilde{\epsilon}(k)$ we have $D^{(2)}(k) = 0$ when either $D^{*}(k) = 0$ or $\tilde{\epsilon}(k) = 0$ for charge-inversion invariant systems, so $H_{ij}^{(2)}$ has poles due to both $D^{*}(k) = 0$, i.e., $k = ib'$, ib'', ... , and $\tilde{\epsilon}(k) = 0$, i.e., $k = i\kappa'$, $i\kappa''$, ... The latter give the decay modes of $H_{\text{QQ}}(r)$. The decay parameters b', b'', ... of $H_{\text{NN}}(r) = H_{\text{NN}}^{\star}(r)$ satisfy the condition $\mathcal{F}(b) = 0$ because this is required in order for $D^{(2)}(k)$ and $D^{*}(k)$ to be zero for the same k. $\mathcal{F}(b)$ is defined in Equation 13.94.

Like in the spherical case the decay parameters fall into *three distinct sets*: the solutions of

(i) $\tilde{\epsilon}(i\kappa) = 0$ (Yukawa terms of H_{QQ}),
(ii) $D^{\star}(ib) = 0$ with $\mathcal{F}(b) = 0$ (Yukawa terms of $H_{\text{NN}} = H_{\text{NN}}^{\star}$),
(iii) $D^{\star}(ib) = 0$ with $\mathcal{F}(b) \neq 0$ (Yukawa terms of $H_{\text{QQ}}^{\star}$).

Thus, H_{QQ} has the same decay modes as before

$$H_{\text{QQ}}(r) = -\frac{\beta\left[n_{\text{tot}}^{\text{b}} q_{\text{q}}^{\text{eff}}(\kappa)\right]^2}{\mathcal{E}_r^{\text{eff}}(\kappa)\varepsilon_0} \cdot \frac{e^{-\kappa r}}{4\pi r} - \frac{\beta\left[n_{\text{tot}}^{\text{b}} q_{\text{q}}^{\text{eff}}(\kappa')\right]^2}{\mathcal{E}_r^{\text{eff}}(\kappa')\varepsilon_0} \cdot \frac{e^{-\kappa' r}}{4\pi r} - \dots. \tag{13.107}$$

Cases (ii) and (iii) are treated in Exercise 13.3.

{**Exercise 13.3:** Starting from

$$\tilde{H}_{ij}^{\star}(\mathbf{k}, \omega_1, \omega_2) \sim \frac{\alpha_i(ib\hat{\mathbf{k}}, \omega_1)\alpha_j(-ib\hat{\mathbf{k}}, \omega_2)}{(k^2 + b^2)\mathcal{K}^{\star}(b)} \quad \text{when } k \to ib,$$

show that (ii) and (iii) apply for $H_{NN}^\star$ and $H_{QQ}^\star$. Show also that

$$H_{NN}(r) \sim \frac{[\mathcal{C}_n(b)]^2}{\mathcal{K}^\star(b)} \cdot \frac{e^{-br}}{4\pi r},\qquad (13.108)$$

when $r \to \infty$, where

$$\mathcal{C}_n(b) = \sum_l \left\langle \alpha_l(ib\hat{\mathbf{k}}, \boldsymbol{\omega}) \right\rangle_{\boldsymbol{\omega}}$$

and the leading zero $k = ib$ of $D^\star(k)$ has a real-valued b.}

For binary electrolytes we have $\tilde{H}_{++}^\star \equiv \tilde{H}_{--}^\star$ and $\tilde{\sigma}_+ \equiv -\tilde{\sigma}_-$ due to the charge inversion invariance and it follows that

$$\mathcal{F}(b) = \int d\boldsymbol{\omega}\, \left[\alpha_+(-ib\hat{\mathbf{k}}, \boldsymbol{\omega}) - \alpha_-(-ib\hat{\mathbf{k}}, \boldsymbol{\omega})\right]\tilde{\sigma}_+(ib\hat{\mathbf{k}}, \boldsymbol{\omega})$$

(from Equation 13.94). There are two possibilities $\alpha_+(-ib\hat{\mathbf{k}}, \boldsymbol{\omega}) = \pm\alpha_-(-ib\hat{\mathbf{k}}, \boldsymbol{\omega})$ because[16] $\alpha_+^2(-ib\hat{\mathbf{k}}, \boldsymbol{\omega}) = \alpha_-^2(-ib\hat{\mathbf{k}}, \boldsymbol{\omega})$. We have $\mathcal{F}(b) = 0$ when $\alpha_+ = \alpha_-$ and $\mathcal{F}(b) \neq 0$ when $\alpha_+ = -\alpha_-$.

The decay parameters of the Yukawa function contributions of the pair correlation function $h_{ij}^{(2)}(\mathbf{r}_{12}, \boldsymbol{\omega}_1, \boldsymbol{\omega}_2)$ belong to the sets (i) and (ii) above and we have for sufficiently large r values

$$h_{ij}^{(2)}(\mathbf{r}_{12}, \boldsymbol{\omega}_1, \boldsymbol{\omega}_2)$$
$$= -\frac{\beta Q_i^{\text{eff}}(\hat{\mathbf{r}}_{12}, \boldsymbol{\omega}_1, \kappa) Q_j^{\text{eff}}(-\hat{\mathbf{r}}_{12}, \boldsymbol{\omega}_2, \kappa)}{\mathcal{E}_r^{\text{eff}}(\kappa)\varepsilon_0} \cdot \frac{e^{-\kappa r_{12}}}{4\pi r_{12}} + \dots$$
$$+ \frac{a_i(\hat{\mathbf{r}}_{12}, \boldsymbol{\omega}_1, b) a_j(-\hat{\mathbf{r}}_{12}, \boldsymbol{\omega}_2, b)}{\mathcal{K}^\star(b)} \cdot \frac{e^{-br_{12}}}{4\pi r_{12}} + \dots \quad \text{(charge-inv. invar.)},$$
$$(13.109)$$

where each line continues with the corresponding Yukawa function terms for κ', κ'', $\dots$ and b', b'', $\dots$, respectively. Like before there are also terms with other functional forms. A pair of complex decay parameters gives an oscillatory Yukawa function term as usual. Note that the a_i coefficients that appear in Equation 13.109 fulfill

$$\sum_i \int d\boldsymbol{\omega}\, n_i^{\text{b}} a_i(-\hat{\mathbf{r}}, \boldsymbol{\omega}, b)\tilde{\sigma}_i(ib\hat{\mathbf{r}}, \boldsymbol{\omega}) = 0$$

due to the condition $\mathcal{F}(b) = 0$. The charge distribution $\rho_i(\mathbf{r}_{12}, \boldsymbol{\omega}_1)$ and the potential $\psi_i(\mathbf{r}_{12}, \boldsymbol{\omega}_1)$ contain only the decay modes in the second line of Equation

---

[16]Due to the charge inversion invariance we have $\alpha_+(ib\hat{\mathbf{k}}, \boldsymbol{\omega}_1)\alpha_+(-ib\hat{\mathbf{k}}, \boldsymbol{\omega}_2) = \alpha_-(ib\hat{\mathbf{k}}, \boldsymbol{\omega}_1)\alpha_-(-ib\hat{\mathbf{k}}, \boldsymbol{\omega}_2)$ for all $\hat{\mathbf{k}}$, $\boldsymbol{\omega}_1$ and $\boldsymbol{\omega}_2$. Since $\alpha_l(-ib\hat{\mathbf{k}}, \boldsymbol{\omega}_2)$ for $l = +$ and $-$ is unchanged if one changes the direction of $\hat{\mathbf{k}}$ to $-\hat{\mathbf{k}}$ and simultaneously rotates the particle half a turn, it is always possible to select $\boldsymbol{\omega}_1$ such that $\alpha_l(ib\hat{\mathbf{k}}, \boldsymbol{\omega}_1) = \alpha_l(-ib\hat{\mathbf{k}}, \boldsymbol{\omega}_2)$. Therefore, it follows that $\alpha_+^2(-ib\hat{\mathbf{k}}, \boldsymbol{\omega}_2) = \alpha_-^2(-ib\hat{\mathbf{k}}, \boldsymbol{\omega}_2)$.

13.109, while the contributions from the third line cancel in these two functions.

As we saw for the spherical case in Section 12.3.5, the charge-inversion invariance, which implies that H_{NQ} is identically equal to zero, makes the charge density and the number density distributions are decoupled on the linear level, but there remains a coupling on the non-linear level. We also saw that *borderline cases*, which are close to being charge-inversion invariant, have properties that are very similar to the corresponding charge-inversion invariant system. For such systems the density-charge correlations make $H_{\mathrm{NN}}(r)$ and $H_{\mathrm{QQ}}(r)$ to be coupled also on the linear level, whereby they have the same decay modes instead of distinct ones. They belong to the main case where all decay modes $h_{ij}^{(2)}$ are the same as those of ρ_i, ψ_i and $\phi^{\star}_{\mathrm{Coul}}(r)$. This applies also to the present case of nonspherical particles.

The analysis in Section 12.3.5 of the behavior of the poles of $\tilde{H}_{\mathrm{NN}}(k)$ and $\tilde{H}_{\mathrm{QQ}}(k)$ in the transition from borderline cases to charge-inversion invariance also applies here for the same reasons. In summary: The zeros of $\tilde{\epsilon}(k)$ belong to either of two sets, A and B. The former consists of the zeros that lie close to some zero of $D^{\star}(k)$ and will coincide with the latter when the system becomes charge-inversion invariant. Set B consists of all the other zeros. For borderline cases both A and B give rise to contributions to $H_{\mathrm{QQ}}(r)$ and $H_{\mathrm{NN}}(r)$, but not for the invariant systems. For the latter, set A contributes to $H_{\mathrm{NN}}(r)$ and set B to $H_{\mathrm{QQ}}(r)$, but there is no contribution from A in $H_{\mathrm{QQ}}(r)$ and none from B in $H_{\mathrm{NN}}(r)$. Therefore, in borderline cases the terms in $H_{\mathrm{QQ}}(r)$ from set A are much smaller the corresponding terms in $H_{\mathrm{NN}}(r)$ and vice versa for the terms from set B.

13.4 SURFACE INTERACTIONS IN IONIC LIQUIDS AND ELECTROLYTE SOLUTIONS

13.4.1 Ionic liquids and concentrated solutions

In Section 12.4 we treated electrolytes with spherical ions in contact with surfaces. We investigated the electrostatic potential outside a planar wall, ion-wall interactions for spherical ions and surface interactions between two walls. As we will see, the final results apply also to electrolytes with nonspherical ions, but in order to show this we need to generalize the treatment so it includes distributions of nonspherical particles outside a wall and the interactions between such particles and the wall. Furthermore, we will obtain the results for surface forces in a different, more general manner than before. In this section we select the coordinate system with the origin at a wall surface and with the z axis perpendicular to it. We will *only treat our main case* of the previous section that includes all electrolyte systems except the charge-inversion invariant ones. Thus, the realistic electrolytes are included.

The theory in the previous sections for electrolytes consisting of nonspherical particles applies to any rigid particles independently of their sizes, shapes and internal charge distributions and therefore it applies also when one particle is a wall. This case can be regarded as the limit where one particle with a flat surface has grown infinitely large in the lateral directions along that surface.

The inhomogeneous electrolyte outside a single wall w has the number density $\mathfrak{n}_j(\mathbf{r},\omega)$ for species j and $\mathfrak{n}_j(\mathbf{r},\omega) \to n_j^{\mathrm{b}}$ when $z \to \infty$. The charge density distribution $\rho_{\mathtt{w}}$ of the electrolyte outside the wall and the associated total charge density $\rho_{\mathtt{w}}^{\mathrm{tot}}$, which includes the charge density $\sigma_{\mathtt{w}}$ inside the wall, are

$$\rho_{\mathtt{w}}(\mathbf{r}_1) = \sum_j \int d\mathbf{r}_2 d\omega_2 \mathfrak{n}_j(\mathbf{r}_2,\omega_2)\sigma_j(\mathbf{r}_{21},\omega_2) \quad \text{and} \quad \rho_{\mathtt{w}}^{\mathrm{tot}}(\mathbf{r}) = \sigma_{\mathtt{w}}(\mathbf{r}) + \rho_{\mathtt{w}}(\mathbf{r}).$$

As before, the dressed charge distribution $\rho_{\mathtt{w}}^{\star}(\mathbf{r})$ of the wall is given by

$$\rho_{\mathtt{w}}^{\star}(\mathbf{r}) = \rho_{\mathtt{w}}^{\mathrm{tot}}(\mathbf{r}) - \int d\mathbf{r}'\, \psi_{\mathtt{w}}(\mathbf{r}')\chi^{\star}(|\mathbf{r}-\mathbf{r}'|),$$

where $\chi^{\star}(r)$ is the polarization response function of the bulk electrolyte and

$$\psi_{\mathtt{w}}(\mathbf{r}) = \int d\mathbf{r}'\, \rho_{\mathtt{w}}^{\mathrm{tot}}(\mathbf{r}')\phi_{\mathrm{Coul}}(|\mathbf{r}-\mathbf{r}'|) = \int d\mathbf{r}'\, \rho_{\mathtt{w}}^{\star}(\mathbf{r}')\phi_{\mathrm{Coul}}^{\star}(|\mathbf{r}-\mathbf{r}'|),$$

where $\phi_{\mathrm{Coul}}^{\star}(r)$ is the unit screened Coulomb potential for the bulk phase. As shown in Section 12.4, $\psi_{\mathtt{w}}(\mathbf{r})$ sufficiently far away from the surface $\psi_{\mathtt{w}}$ depends only on z, so we have $\psi_{\mathtt{w}}(\mathbf{r}) \sim \psi_{\mathtt{w}}(z) \to 0$ when $z \to \infty$, but closer in it depends in general also on x and y, see Equation 12.235. When κ is real and $k = i\kappa$ is the singularity of $\tilde{\phi}_{\mathrm{Coul}}^{\star}(k)$ closest to the real axis in complex k space, we have as before

$$\psi_{\mathtt{w}}(z) \sim \frac{\sigma_{\mathtt{w}}^{\mathcal{S}\mathrm{eff}}(\kappa)}{\mathcal{E}_r^{\mathrm{eff}}(\kappa)\varepsilon_0} \cdot \frac{e^{-\kappa z}}{\kappa} \qquad \text{when } z \to \infty, \tag{13.110}$$

where

$$\sigma_{\mathtt{w}}^{\mathcal{S}\mathrm{eff}}(\kappa) = \lim_{\mathsf{A}\to\infty} \frac{1}{2\mathsf{A}} \int d\mathbf{r}' \rho_{\mathtt{w}}^{\star}(\mathbf{r}')e^{\kappa z'}. \tag{13.111}$$

is the effective surface charge density of the wall in the limit of infinite surface area A.

Since the electrolyte theory is applicable also when one particle is a wall, the distribution functions for small particles outside the wall surface can be regarded as a special case of pair distributions between a large flat particle and the small ones. In Section 13.3.3.2 we saw that all decay modes of the pair correlation function $h_{ij}^{(2)}$ in bulk (for the main case that we are treating here) are given by the zeros of $\tilde{\epsilon}(k)$ and that their contributions originate from the term

$$-\beta\tilde{\rho}_i^{\star}(\mathbf{k},\omega_1)\tilde{\phi}_{\mathrm{Coul}}^{\star}(k)\tilde{\rho}_j^{\star}(-\mathbf{k},\omega_2) \;=\; -\beta\psi_i(\mathbf{k},\omega_1)\tilde{\rho}_j^{\star}(-\mathbf{k},\omega_2)$$

in the rhs of Equation 13.87. When $h_{ij}^{(2)}$ is small we have $h_{ij}^{(2)}(r) = e^{-\beta w_{ij}^{(2)}} - 1 \sim -\beta w_{ij}^{(2)}$, so the decay modes of $\tilde{w}_{ij}^{(2)}$ are included in $\psi_i(\mathbf{k}, \omega_1)\tilde{\rho}_j^{\star}(-\mathbf{k}, \omega_2)$. In real space this means that

$$w_{ij}^{(2)}(\mathbf{r}_{12}, \omega_1, \omega_2) \sim \int d\mathbf{r}_3 \psi_i(\mathbf{r}_{13}, \omega_1)\rho_j^*(\mathbf{r}_{23}, \omega_2) \qquad \text{when } r_{12} \to \infty$$

since the leading decay mode of ψ_i dominates in the rhs.

If we apply this to *surface-particle interactions* where the i-particle is a planar wall with infinite surface area and $\mathbf{r}_1$ is placed at the origin, i.e., the wall surface, the potential of mean force $w_{j;\mathtt{w}}$ of the j-particle outside the wall $\mathtt{w}$ decays like

$$w_{j;\mathtt{w}}(\mathbf{r}_2, \omega_2) \sim \int d\mathbf{r}_3 \psi_{\mathtt{w}}(\mathbf{r}_3)\rho_j^*(\mathbf{r}_{23}, \omega_2) \qquad \text{when } z_2 \to \infty.$$

If we insert Equation 13.110 this implies that

$$\begin{aligned} w_{j;\mathtt{w}}(\mathbf{r}_2, \omega_2) &\sim \frac{\sigma_{\mathtt{w}}^{\mathcal{S}\mathrm{eff}}(\kappa)}{\mathcal{E}_r^{\mathrm{eff}}(\kappa)\varepsilon_0\kappa} \int d\mathbf{r}_3 e^{-\kappa z_3} \rho_j^*(\mathbf{r}_{23}, \omega_2) \\ &= \frac{\sigma_{\mathtt{w}}^{\mathcal{S}\mathrm{eff}}(\kappa) e^{-\kappa z_2}}{\mathcal{E}_r^{\mathrm{eff}}(\kappa)\varepsilon_0\kappa} \int d\mathbf{r}_3 e^{-\kappa(z_3 - z_2)} \rho_j^*(\mathbf{r}_3 - \mathbf{r}_2, \omega_2) \end{aligned}$$

when $z_2 \to \infty$. Since $z_3 - z_2 = \mathbf{r}_{23} \cdot \hat{\mathbf{z}}$, where $\hat{\mathbf{z}}$ is perpendicular to the wall, this result can be written as

$$w_{j;\mathtt{w}}(\mathbf{r}_2, \omega_2) \sim \frac{\sigma_{\mathtt{w}}^{\mathcal{S}\mathrm{eff}}(\kappa)\mathrm{Q}_j^{\mathrm{eff}}(-\hat{\mathbf{z}}_2, \omega_2, \kappa)}{\mathcal{E}_r^{\mathrm{eff}}(\kappa)\varepsilon_0} \cdot \frac{e^{-\kappa z_2}}{\kappa} \qquad \text{when } z_2 \to \infty, \qquad (13.112)$$

where $\mathrm{Q}_j^{\mathrm{eff}}$ is the multipolar effective charge defined in Equation 13.74. The vector $-\hat{\mathbf{z}}_2$ points from the center of the j-particle toward the wall and is perpendicular to the latter. Compare this with the factor $\mathrm{Q}_j^{\mathrm{eff}}(-\hat{\mathbf{r}}_{12}, \omega_2, \kappa)$ in Equation 13.88 where $-\hat{\mathbf{r}}_{12}$ points from the center of the j-particle toward the center of the i-particle. The leading term from each decay mode $\kappa, \kappa', \ldots$ has the same form so for sufficiently large z_2, we have

$$\begin{aligned} w_{j;\mathtt{w}}(\mathbf{r}_2, \omega_2) &= \frac{\sigma_{\mathtt{w}}^{\mathcal{S}\mathrm{eff}}(\kappa)\mathrm{Q}_j^{\mathrm{eff}}(-\hat{\mathbf{z}}_2, \omega_2, \kappa)}{\mathcal{E}_r^{\mathrm{eff}}(\kappa)\varepsilon_0} \cdot \frac{e^{-\kappa z_2}}{\kappa} \\ &+ \frac{\sigma_{\mathtt{w}}^{\mathcal{S}\mathrm{eff}}(\kappa')\mathrm{Q}_j^{\mathrm{eff}}(-\hat{\mathbf{z}}_2, \omega_2, \kappa')}{\mathcal{E}_r^{\mathrm{eff}}(\kappa')\varepsilon_0} \cdot \frac{e^{-\kappa' z_2}}{\kappa'} + \ldots \end{aligned}$$

and as before there are also terms of other functional form that decay to zero faster with increasing z_2 than the first term.

Next, we treat *surface interactions* between two planar, parallel walls $\mathtt{w}_1$ and $\mathtt{w}_2$ that have an electrolyte in the slit between the surfaces. The electrolyte is in equilibrium with a bulk electrolyte with density n_i^{b} for each species i. Wall w_1 is infinitely large laterally, but initially w_2 has a finite, albeit very large, surface area A_2. As before we will increase the area to infinity, whereby the end result will be two infinite planar walls with separation $D^{\mathcal{S}}$. We assume that the walls are very thick.

Since our result in Equation 13.112 is of general validity for any rigid particle placed at coordinate z_2, it is also valid for the wall w_2 placed with its planar surface located at z_2 and gives the potential of mean force $w_{\mathrm{w}_1,\mathrm{w}_2}$ that acts on w_2 in the presence of w_1. For w_2 we have

$$\mathrm{Q}_{\mathrm{w}_2}^{\mathrm{eff}}(-\hat{\mathbf{z}}_2,\boldsymbol{\omega}_2,\kappa) = \mathrm{Q}_{\mathrm{w}_2}^{\mathrm{eff}}(\hat{\mathbf{z}}_{\mathrm{w}_2},\boldsymbol{\omega}_2,\kappa) = \int d\mathbf{r}' e^{\kappa \mathbf{r}'\cdot\hat{\mathbf{z}}_{\mathrm{w}_2}} \rho_{\mathrm{w}_2}^*(\mathbf{r}',\boldsymbol{\omega}_2),$$

where $\hat{\mathbf{z}}_{\mathrm{w}_2}$ is the local normal of the surface of w_2 that faces wall w_1. The orientation $\boldsymbol{\omega}_2$, which is written explicitly here, is redundant because $\boldsymbol{\omega}_2$ is always parallel to $\boldsymbol{\omega}_1$ and this information is implicit in $\rho_{\mathrm{w}_2}^*(\mathbf{r}')$. By setting

$$\sigma_{\mathrm{w}_2}^{\mathcal{S}\mathrm{eff}} = \mathrm{Q}_{\mathrm{w}_2}^{\mathrm{eff}}/[2\mathrm{A}_2] = \int d\mathbf{r}' e^{\kappa \mathbf{r}'\cdot\hat{\mathbf{z}}_{\mathrm{w}_2}} \rho_{\mathrm{w}_2}^*(\mathbf{r}')/[2\mathrm{A}_2]$$

and letting $\mathrm{A}_2 \to \infty$ we obtain the same $\sigma_{\mathrm{w}_2}^{\mathcal{S}\mathrm{eff}}$ as from Equation 13.111 because $\mathbf{r}'\cdot\hat{\mathbf{z}}_{\mathrm{w}_2} = z'$ in the local coordinate system for w_2, where the z' axis points the $\hat{\mathbf{z}}_{\mathrm{w}_2}$ direction.[17] By using this result and defining the **free energy of interaction** (potential of mean force) **per unit area** between w_1 and w_2 as $\mathscr{W}_{\mathrm{w}_1,\mathrm{w}_2}^{\mathcal{S}} = w_{\mathrm{w}_1,\mathrm{w}_2}/\mathrm{A}_2$, we obtain from Equation 13.112 in the limit $\mathrm{A}_2 \to \infty$

$$\mathscr{W}_{\mathrm{w}_1,\mathrm{w}_2}^{\mathcal{S}}(D^{\mathcal{S}}) \sim \frac{2\sigma_{\mathrm{w}_1}^{\mathcal{S}\mathrm{eff}}(\kappa)\sigma_{\mathrm{w}_2}^{\mathcal{S}\mathrm{eff}}(\kappa)}{\mathcal{E}_r^{\mathrm{eff}}(\kappa)\varepsilon_0}\cdot\frac{e^{-\kappa D^{\mathcal{S}}}}{\kappa} \quad \text{when } D^{\mathcal{S}} \to \infty, \tag{13.113}$$

where we have replaced the distance z_2 from the surface of w_1 in Equation 13.112 by $D^{\mathcal{S}}$, which is the distance between the surfaces. This result is the same as that we obtained for spherical ions in Section 12.4, which hence is generalized to electrolytes consisting nonspherical particles. The surface force per unit area between the two walls, that is, the disjoining pressure $P^{\mathcal{S}}$ is given by $P_{\mathrm{w}_1,\mathrm{w}_2}^{\mathcal{S}} = -\mathscr{W}_{\mathrm{w}_1,\mathrm{w}_2}^{\mathcal{S}}(D^{\mathcal{S}})/dD^{\mathcal{S}}$.

The leading term in $\mathscr{W}_{\mathrm{w}_1,\mathrm{w}_2}^{\mathcal{S}}$ from each decay mode $\kappa, \kappa', \kappa'', \ldots$ of the bulk electrolyte has the same form as in Equation 13.113, so Equation 12.230 in Section 12.4 is valid also in the nonspherical case. As we noted there, $\mathscr{W}_{\mathrm{w}_1,\mathrm{w}_2}^{\mathcal{S}}$ also contain more short-range terms with other functional forms and there may be some additional short-range contribution induced by the surfaces. Only contributions that are commensurate with the decay modes of the bulk survive with increasing $D^{\mathcal{S}}$.

The decay parameters κ, κ', κ'' etc. are those of the bulk electrolyte that is in equilibrium with the inhomogeneous electrolyte between the surfaces,

[17] This coordinate system is used for $\mathbf{r}'$ in the integral for $\sigma_{\mathrm{w}_2}^{\mathcal{S}\mathrm{eff}}$, which is in congruence with the choice of coordinate system in Equation 13.111 when applied to wall w_1.

so the decay lengths and, for the oscillatory modes, the wave lengths *are the same as in bulk.* The parameters κ, κ', κ'', ... are the solutions of

$$\kappa^2 = \frac{\beta}{\mathcal{E}_r^\star(\kappa)\varepsilon_0} \sum_i n_i^{\rm b} q_i q_i^\star \tag{13.114}$$

(Equation 13.70). The walls affect only the amplitude of each decay mode in $\mathscr{W}^S_{\rm w_1,w_2}$ and, for the oscillatory modes, also the phase of the oscillations. The influence occurs via the prefactors $\sigma^{S\rm eff}_{\rm w_\nu}(\kappa)$. For some surfaces, one decay mode can give much larger contribution than another mode, while for other surfaces immersed in the same electrolyte the reverse may be be true.

When the leading decay mode has a real-valued decay parameter κ and two other decay modes have complex κ'' and κ''' such that $\kappa'' = \kappa''_{\Re} + i\kappa''_{\Im}$ and $\kappa''' = \kappa''_{\Re} - i\kappa''_{\Im}$, we have for sufficiently large D^S

$$\begin{aligned} \mathscr{W}^S_{\rm w_1,w_2}(D^S) &= \frac{2\sigma^{S\rm eff}_{\rm w_1}(\kappa)\sigma^{S\rm eff}_{\rm w_2}(\kappa)}{\mathcal{E}_r^{\rm eff}(\kappa)\varepsilon_0} \cdot \frac{e^{-\kappa D^S}}{\kappa} \\ &+ \frac{4\left|\sigma^{S\rm eff}_{\rm w_1}(\kappa'')\sigma^{S\rm eff}_{\rm w_2}(\kappa'')\right|}{\left|\mathcal{E}_r^{\rm eff}(\kappa'')\right|\varepsilon_0} \\ &\times \frac{e^{-\kappa''_{\Re} D^S}}{|\kappa''|} \cos\left(\kappa''_{\Im} D^S + \vartheta_{\rm w_1,w_2}(\kappa'')\right) + \ldots, \end{aligned} \tag{13.115}$$

where $\kappa''_{\Re} + i\kappa''_{\Im} = |\kappa''|\exp(-i\vartheta_{\kappa''})$ and $\sigma^{S\rm eff}_{\rm w_\nu} = |\sigma^{S\rm eff}_{\rm w_\nu}|e^{-i\vartheta_{\rm w_\nu}}$. We assume here that $1/\kappa''_{\Re} < 1/\kappa$ and that any other decay mode has an even smaller decay length. The combined phase shift is $\vartheta_{\rm w_1,w_2}(\kappa'') = \vartheta_{\rm w_1}(\kappa'') + \vartheta_{\rm w_2}(\kappa'') - \vartheta_{\mathcal{E}}(\kappa'') - \vartheta_{\kappa''}$. The mathematical form of this expression is

$$\mathscr{W}^S(D^S) \sim \mathsf{B}\, e^{-\kappa D^S} + \mathsf{B}'' e^{-\kappa''_{\Re} D^S} \cos(\kappa''_{\Im} D^S + \vartheta''_{\rm w,w}) + \ldots \tag{13.116}$$

for large D^S, where the parameters can be identified with the corresponding ones in Equation 13.115.

As an example, let us consider surface forces in a room-temperature ionic liquid. In Figure 13.2 the measured force[18] between two curved, atomically smooth mica surfaces immersed in $[C_2C_1Im]^+[NTf_2]^-$ (see Figure 13.3 for structural formulas of the ions) is plotted as a function of the surface separation D^S. The force F between curved surfaces with radius $\mathcal{R}$ is related to the free energy of interaction per unit area $\mathscr{W}^S$ between flat plates via the Derjaguin approximation[19] $F/\mathcal{R} = 2\pi\mathscr{W}^S$ and this quantity is plotted in the figure. The experimental results are obtained using the surface force apparatus, SFA (also called the surface force balance),[20] which measures the force

[18] T. S. Groves and S. Perkin, 2024, *Wave mechanics in an ionic liquid mixture.* To appear in Faraday Discuss. (https://doi.org/10.1039/D4FD00040D). Timothy Groves and Susan Perkin are acknowledged for kindly providing the original data.

[19] See Section 8.1.1 in Volume I.

[20] A recent presentation of the surface force apparatus technique can be found in: H. J. Hayler, T. S. Groves, A. Guerrini, A. Southam, W. Zheng, and S. Perkin, Rep. Prog. Phys. 87 (2024) 046601 (https://doi.org/10.1088/1361-6633/ad2b9b).

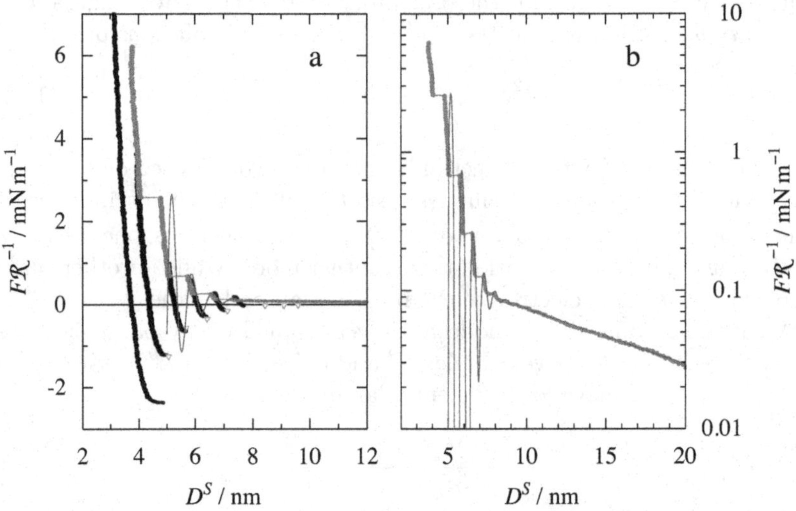

Figure 13.2 **(See color insert).** The force F as function of the surface separation D^S between two curved mica surfaces of radius $\mathcal{R} = 6.5$ mm immersed in a room-temperature ionic liquid, $[C_2C_1Im]^+[NTf_2]^-$, at $T = 295$ K measured using a surface force apparatus (SFA). $F/\mathcal{R}$ between curved surfaces is equivalent to $2\pi\mathscr{W}^S$, where $\mathscr{W}^S$ is the free energy of interaction per unit area between flat plates. The results are plotted on linear scales (a) and a semilogarithmic scale (b). The blue data are measured on approach of the two surfaces toward each other and the black data on retractions (see footnote 21). For the former the surfaces jump toward each other at the horizontal sections of the plotted data and for the latter they jump apart at the green symbols. The black data together with the blue ones (apart from the horizontal parts) show the parts of the oscillations with negative derivative. The red curve is obtained from the theoretical expression in Equation 13.116 by adjusting its parameters; in the right frame the positive parts of this curve are plotted. This curve is an aid to the eye rather than an actual fitting to the experimental data. The experimental data are taken from the work by Groves and Perkin.[18]

between the two curved surfaces immersed in a liquid with 0.1 nm resolution of the surface separation.

The measured force is oscillatory for small D^S and monotonically decaying for large D^S. The *entire* oscillatory force curve can often not be measured with the SFA technique for technical reasons,[21] but the parts of the curve

[21] In the surface force apparatus the cylindrical bodies with the two surfaces are mounted on a spring mechanism, whereby the force is measured mechanically by the amount of spring deflection and the surface separation is measured by an optical interference technique. For oscillatory forces, the spring mechanism often makes two approaching surfaces to jump together close to the points of maximum force (due to a mechanical instability of the mechanism), whereby the surfaces reach the next oscillation of the force curve for smaller D^S.

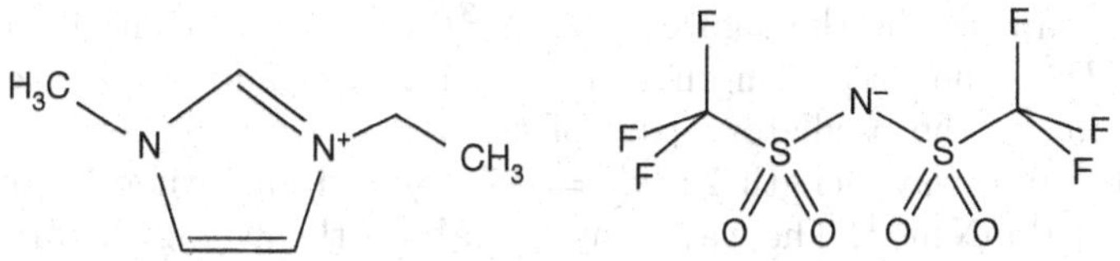

$[C_2C_1Im]^+$ $[NTf_2]^-$

Figure 13.3 The ions in the room-temperature ionic liquid $[C_2C_1Im]^+[NTf_2]^-$ (1-ethyl-3-methylimidazolium bis(trifluoromethylsulfonyl)imide).

with negative derivative can be obtained. In the figure, these parts are shown as blue and black data points, see the figure caption. The green symbols lie close to the minima of the oscillatory force curve.

There are also other technical limitations in the experimental measurements, like surface flattening at large forces (small D^S), imperfections in the surface separation measurements and estimation of the point of zero D^S (an accuracy of about 0.5 nm). The blue data in the figure and the green symbols are not fully consistent with each other and neither are the inner branches of the measured curves. For these reasons, a comparison between the experimental data and theoretical predictions have some limitations. Force curves predicted by Equation 13.116 are asymptotic results for the main decay modes that can be valid for sufficiently large D^S and as we have seen there are also other types of contributions to the interactions that may be important particularly for small D^S. Nevertheless, by adjusting the parameters in Equation 13.116 one can obtain a theoretical curve that contains major features of the experimental data for large and intermediate D^S. This is the red curve in the figure and its role is mainly to help in the interpretation of the experimental data rather than to give an optimal curve fit.[22]

On retraction, the surfaces instead jump apart close to the points of minimum force and reach some very large separation. The points where the jumps occur depend on the stiffness of the spring; the stiffer the spring, the larger part of the curves can be measured close to the extrema.

[22]The estimate of the amplitude of $\mathcal{W}^S(D^S)$ is, in fact, affected by the convention used to measure the distance D^S between two macroscopic bodies. It is, for example, not clear exactly where the surface plane is located for an atomic surface. In surface force measurements one estimates the location of zero surface separation by pressing the surface firmly together and the location of $D^S = 0$ always carries un uncertainty. Let us change the definition of separation so that it becomes $D^S_a = D^S + a$ instead of D^S without moving the surfaces. $\mathcal{W}^S$ as function of surface separation will be denoted $\mathcal{W}^S(D^S)$ for the first definition and $\mathcal{W}^S_a(D^S_a)$ for the second. We have $\mathcal{W}^S(D^S) = \mathcal{W}^S_a(D^S_a) = \mathcal{W}^S_a(D^S + a)$ because we have only changed the definition of D^S without moving the surfaces. Consider, for example, a monotonic decay mode, so we have $\mathcal{W}^S(D^S) \sim \mathsf{B}\, e^{-\kappa D^S}$ and $\mathcal{W}^S_a(D^S_a) \sim \mathsf{B}_a\, e^{-\kappa D^S_a}$. Since $\mathcal{W}^S(D^S) = \mathcal{W}^S_a(D^S_a)$ if follows that $\mathsf{B}\, e^{-\kappa D^S} = \mathsf{B}_a\, e^{-\kappa D^S_a}$ and therefore $\mathsf{B}\, e^{-\kappa D^S} = \mathsf{B}_a\, e^{-\kappa(D^S+a)}$. Hence $\mathsf{B}_a = \mathsf{B}e^{\kappa a}$ so the value of the amplitude depends on the choice of definition for D^S. For an oscillatory mode the values of both the amplitude and the phase depend on this definition.

As we can see in the figure 13.2, $\mathscr{W}^S(D^S)$ is exponentially decaying for large D^S. The decay length is $1/\kappa = 10.7$ nm and amplitude is $\mathsf{B} = 0.030$ mN m^{-1}. The oscillatory part of the red curve has the decay length $1/\kappa''_{\Re} = 0.5$ nm, wave length $2\pi/\kappa''_{\Im} = 0.7$ nm and amplitude B'' of the order[23] $1.3 \cdot 10^4$ mN m^{-1}. The wave length is about the average diameter of an anion-cation pair (≈ 0.75 nm). The same decay modes exist for the correlation functions in bulk and the features of the oscillatory mode in the surface forces correspond to pair correlations in bulk with out-of-phase oscillations characteristic of anion/cation layering (cf. Figure 12.7b). This kind of modes give large contributions to the charge-charge correlations $H_{\mathrm{QQ}}(r)$ in bulk. There are other contributions to the experimental surface forces, like van der Waals interactions and the influence of higher order electrostatic terms, but the terms in Equation 13.116 are the most important contributions to the surface interactions in the present case except for small surface separations.

In the scientific literature it common to compare $1/\kappa$ with the Debye length $1/\kappa_{\mathrm{D}}$, which is defined by

$$\kappa_{\mathrm{D}}^2 = \frac{\beta}{\varepsilon_r^{\mathrm{M}}\varepsilon_0}\sum_j n_j^{\mathrm{b}} q_j^2$$

where $\varepsilon_r^{\mathrm{M}}$ is a "macroscopic dielectric constant." In the dielectric-continuum model we have $\varepsilon_r^{\mathrm{M}} = \varepsilon_r^{\mathrm{cont}}$ (Equation 12.66). For a dilute electrolyte it is reasonable to set $\varepsilon_r^{\mathrm{M}}$ equal to the dielectric constant of the pure solvent, but it is not self-evident what is a suitable value of $\varepsilon_r^{\mathrm{M}}$ to be used for an ionic liquid or a concentrated electrolyte solution. In such systems $\varepsilon_r^{\mathrm{M}}$ depends on the ion density and it is common to use a value of $\varepsilon_r^{\mathrm{M}}$ that has been obtained from the *macroscopic* dielectric response of the electrolyte when it is exposed to a time dependent electric field.[24] However, as we have seen, screening is *in general* not given by the macroscopic properties of the electrolyte.

In the exact expression 13.114 for κ the dielectric factor $\mathcal{E}_r^\star(\kappa) = \tilde{\epsilon}_{\mathrm{reg}}(i\kappa)$ occurs instead of $\varepsilon_r^{\mathrm{M}}$. This factor is a microscopic property, but when the decay length $1/\kappa$ is very large it is approximately equal to $\mathcal{E}_r^{\star 0} \equiv \tilde{\epsilon}_{\mathrm{reg}}(0)$, that is, the macroscopic ($k = 0$) value of $\tilde{\epsilon}_{\mathrm{reg}}(k)$. Recall that $\tilde{\epsilon}_{\mathrm{reg}}(k)$ is the part of $\tilde{\epsilon}(k)$ that remains when the divergence at $k = 0$ has been removed. For ionic liquids and electrolyte solutions, $\mathcal{E}_r^{\star 0}$ is the macroscopic entity that

[23] A small change in $1/\kappa''_{\Re}$ can lead to a rather large variation of the fitted value of B''. Furthermore, the experimental estimate of zero surface separation affects the values of B'' and B as discussed in footnote 22.

[24] One common way to estimate $\varepsilon_r^{\mathrm{M}}$ is to do experimental measurements of the frequency-dependent dielectric response in the limit of zero frequency. For the time-dependent response, the dielectric function $\tilde{\epsilon}(k,\omega)$, where ω is angular frequency, is relevant. The static dielectric function $\tilde{\epsilon}(k)$ is $\tilde{\epsilon}(k) = \lim_{\omega\to 0}\tilde{\epsilon}(k,\omega)$ and the macroscopic, frequency-dependent function is $\tilde{\epsilon}(k,\omega)$ at $k = 0$, that is, $\tilde{\epsilon}(0,\omega)$, which is a complex-valued entity. Since the ionic liquid is a conductor, $\tilde{\epsilon}(0,\omega)$ diverges when $\omega \to 0$ due to conduction, but one can remove the diverging contribution using a value of the direct-current conductivity (at $\omega = 0$) that is measured separately. Recall that the static dielectric function $\tilde{\epsilon}(k)$ diverges at $k = 0$ because electrolytes are conductors and screen the macroscopic electrostatic field perfectly.

may be designated as a concentration dependent, static "dielectric constant." Likewise, the effective dielectric permittivity $\mathcal{E}_r^{\text{eff}}(\kappa)$ is approximately equal to $\mathcal{E}_r^{\star 0}$ when $1/\kappa$ is large. In dilute electrolyte solutions $\mathcal{E}_r^{\star 0}$ is approximately equal to the dielectric constant of the solvent. In general, however, $\mathcal{E}_r^*(\kappa)$ and $\mathcal{E}_r^{\text{eff}}(\kappa)$ should be used and, of course, the correct expression for the decay parameter has $q_i q_i^\star$ instead of q_i^2.

For $[C_2C_1Im]^+[NTf_2]^-$ the value $\varepsilon_r^{\text{M}} = 12$ has been obtained experimentally,[25] whereby $1/\kappa_{\text{D}} \approx 0.060$ nm at $T = 295$ K. Thus the decay length $1/\kappa \approx 10.7$ nm of the monotonic decay is huge compared to the Debye length; we have $\kappa/\kappa_{\text{D}} \approx 0.0056$. This ionic liquid is a typical example of *large underscreening with* $1/\kappa \gg 1/\kappa_{\text{D}}$ that was discussed in Section 12.5 and that is characterized by $\kappa/\kappa_{\text{D}} \ll 1$. Such underscreening has been seen for many (but not for all) room-temperature ionic liquids and also for concentrated electrolyte solutions.[26] It has been called "*anomalous underscreening.*"

There is a remarkable qualitative similarity between the surface force curves for ionic liquids and the charge-charge correlation function for dilute bulk electrolytes at high electrostatic coupling; the latter is proportional to the difference-correlation function $h_{\text{D}}^{(2)}(r)$ plotted in Figure 12.9 for simple electrolytes in the restricted primitive model. For the highest couplings $\beta_R = 13.3$ and 16.7 in this figure, the correlation function is oscillatory with a large amplitude for small distances r and exponentially decaying with a large decay-length and small amplitude for large distances. This is similar to the results in Figure 13.2. Recall that the decay length for large β_R in the dilute electrolyte is many times larger than the Debye length, see Figure 12.8b where the decay length $1/\kappa \approx 6.7/\kappa_{\text{D}}$ and $1/\kappa \approx 6.0d$ at the highest coupling $\beta_R = 20$. In the analysis of these results we found that the long decay length is due to a small effective charge of the ions, which arises because of the strong anion-cation attraction. These attractions lead to a large nonlinear contribution to the charge density of the dress around each ion – a charge distribution of opposite sign than that of the bare ionic charge. As we will see, similar considerations apply for the ionic liquids.

In any electrolyte, the decay parameters in bulk are the solutions of Equation 13.114 and they are also the decay parameters for the surface forces in systems that are in equilibrium with the bulk phase. For a binary ionic fluid with anions and cations of equal absolute valency (usually monovalent), we have $q_+ = -q_- = q$ and $n_+^{\text{b}} = n_-^{\text{b}} = n^{\text{b}}$. Then this equation becomes

$$\kappa^2 = \frac{\beta n^{\text{b}}(q_+^* - q_-^*)q}{\mathcal{E}_r^*(\kappa)\varepsilon_0}. \tag{13.117}$$

Note that $q_+^* \neq -q_-^*$ because anions and cations are quite different in many respects, so ions in their surroundings correlate with them in different manners.

---

[25] H. Weingärtner, J. Molec. Liquids **192** (2014) 185 (https://doi.org/10.1016/j.molliq.2013.07.020).

[26] See references in the footnotes 80 on page 387 and 82 on page 388.

The corresponding equation for the Debye parameter is

$$\kappa_{\rm D}^2 = \frac{\beta n^{\rm b}(q_+ - q_-)q}{\varepsilon_r^{\rm M}\varepsilon_0} = \frac{\beta n^{\rm b}2q^2}{\varepsilon_r^{\rm M}\varepsilon_0}$$

so we have

$$\left(\frac{\kappa}{\kappa_{\rm D}}\right)^2 = \frac{q_+^* - q_-^*}{2q} \cdot \frac{\varepsilon_r^{\rm M}}{\mathcal{E}_r^*(\kappa)}. \tag{13.118}$$

Apart from the leading solution κ to this equation, there are other solutions like $\kappa'' = \kappa''_{\Re} + \mathrm{i}\kappa''_{\Im}$ and $\kappa''' = \kappa''_{\Re} - \mathrm{i}\kappa''_{\Im}$ that together form the oscillatory decay mode mentioned earlier.

For $[C_2C_1Im]^+[NTf_2]^-$ the monotonic decay mode with very long decay length has $\kappa/\kappa_{\rm D} \approx 0.0056$, which means that the lhs of Equation 13.118 is $3 \cdot 10^{-5}$. Thus, either $|q_+^* - q_-^*|$ is much smaller than $2q$ or $|\mathcal{E}_r^*(\kappa)|$ is much larger than $\varepsilon_r^{\rm M} \approx 12$ or both. At the time of writing this book, this issue has not been fully resolved, but $|q_+^* - q_-^*|/2q$ is most likely very small since a sufficiently large value of $|\mathcal{E}_r^*(\kappa)|$ needed to explain the huge difference in magnitude of κ^2 and $\kappa_{\rm D}^2$ is not likely. In the present case $\kappa \approx 0.09$ nm^{-1}, which is a small number, and when $\kappa \approx 0$ we have $\mathcal{E}_r^*(\kappa) \approx \mathcal{E}_r^{*0}$, that is, the macroscopic value of $\mathcal{E}_r^*$ mentioned earlier.

Let us compare this with our findings for spherical ions in Figure 12.8b. In the restricted primitive model, we have $q_+^{\rm eff}(\kappa) = -q_-^{\rm eff}(\kappa) \equiv q^{\rm eff}(\kappa)$ and $q_+^\star = -q_-^\star \equiv q^\star$. In this case it follows from Equations 12.251 and 13.118 that[27]

$$\left(\frac{\kappa}{\kappa_{\rm D}}\right)^2 = \frac{q^{\rm eff}(\kappa)}{q} = \frac{q^\star}{q} \cdot \frac{\varepsilon_r^{\rm M}}{\mathcal{E}_r^\star(\kappa)} \qquad \text{(RPM)}$$

and *large underscreening with* $1/\kappa \gg 1/\kappa_{\rm D}$ takes place when $q^{\rm eff}(\kappa)$ is small compared to q. The reasons for the small effective charge in the dilute electrolytes at high coupling was mentioned above and the charge $q^\star$ of the dressed ion is small for the same reasons, i.e., the bare charge q of each ion is nearly neutralized by charge in the dress, which has opposite sign to the ion itself.

The common features of the ionic liquid and the dilute electrolyte at high coupling are the very strong anion-cation attractions and the consequent large nonlinear contribution to the charge distribution of the dress of each ion, which leads to a small $q_i^\star$. Otherwise, they are of course very different. As we have seen, the strong electrostatic attractions in the dilute electrolytes cause transient ion clustering, where each cluster contains a few ions. In the ionic liquid, on the other hand, the strong attractions and the resulting strong correlations concern many ions in the surroundings of each ion. This gives transient "associations" where the ions have high probability to be in the immediate neighborhood of ions of opposite charge due to Coulomb attractions and effects of many-body correlations in the fluid. This causes very small q_i^* values

[27] Note that $\sum_i n_i^{\rm b} q_i q_i^\star/\mathcal{E}_r^\star(\kappa) \neq \sum_i n_i^{\rm b} q_i q_i^{\rm eff}(\kappa)/\varepsilon_r^{\rm cont}$ in the general case. RPM is an exception.

and hence a very large decay-length $1/\kappa$ of the leading decay mode. These effects of strong anion-cation attractions can in some cases be augmented by short-range intermolecular interactions, for instance hydrogen bonding.

It has been hypothesized that the long decay length in room-temperature ionic liquids is due to extensive ion-pair formation. The number of free ions is then lowered and ion pairs do not contribute to screening because they have no net charge. In other words, the effective charge of the ions becomes lower on average. However, a low effective charge q_i^{eff} and a low $q_i^\star$ do not mean that the ions necessarily form ion pairs. Instead the low q_i^{eff} and $q_i^\star$ are due to a more general, *transient association involving many ions.* In an ionic fluid where each ion is densely surrounded by many other ions at close range, it would be virtually impossible to say which ion forms a pair with which ion. Even in the case of dilute electrolyte at high coupling, we have seen that there is an ion clustering of both two and more ions.

Next, we investigate the other decay modes like the oscillatory one in Figure 13.2. In order to do this it is useful to temporarily consider a binary ionic liquid with $q_+ = -q_- = q$ that is mixed with a polar fluid, which consists of uncharged molecules. For this system Equation 13.117 still applies,[28] but $\mathcal{E}_r^*(\kappa)$ contains contributions from the polar fluid. Our case of a pure ionic liquid is, of course, obtained when the density of the polar fluid is set to zero and then $\mathcal{E}_r^*(\kappa)$ is the same as before.

Let us write Equation 13.117 as

$$\mathcal{E}_r^*(\kappa)\kappa^2 = \frac{q_+^* - q_-^*}{2q}\,\kappa_{\text{D,pl}}^2,$$

where $\kappa_{\text{D,pl}}^2 = \beta n^{\text{b}} 2q^2/\varepsilon_0$ is the Debye parameter for a classical plasma with the same ionic density n^{b} (subscript "pl" stands for plasma). When $|q_+^* - q_-^*|/2q$ is very small, the solutions to this equation encompass $\kappa \approx 0$ and the solutions κ'' etc. to $\mathcal{E}_r^*(\kappa'') \approx 0$. In order to understand what this means it is illuminating to consider another system where the rhs is small, namely an electrolyte dissolved in a polar fluid at very low concentration, whereby $\kappa_{\text{D,pl}}^2 \approx 0$. For such a system we also have $\kappa \approx 0$ and the solutions κ'' etc. to $\mathcal{E}_r^*(\kappa'') \approx 0$. This case is treated in the next section, where we will see that $\kappa \approx 0$ gives the usual monotonic decay mode with a long decay length (like in the Debye-Hückel approximation for dilute electrolytes). In addition there are oscillatory contributions from the κ''-mode, which originate from the liquid structure of the solvent, and we will see that they play an important role.[29] For a pure polar fluid with no ions $\kappa_{\text{D,pl}}^2 = 0$. Then the modes encompass the leading one with $\kappa = 0$ (giving a Coulombic $1/r$ interaction) and the solutions κ'' etc. to $\mathcal{E}_r^*(\kappa'') = 0$ that include the oscillatory mode (see the end of the

[28] Note that the sum in the general Equation 13.114 only have contribution from charged particles.

[29] We will, however, find that the wave length of the oscillations in dilute solutions are close to the diameter of the solvent molecules, while it is about the average diameter of an anion-cation pair for the ionic liquid.

next section and Chapter 14). In all cases the oscillatory contributions are important.

If we now go back to the pure room-temperature ionic liquid, we can give the following likely interpretation of the results. The solutions to $\mathcal{E}_r^*(\kappa'') \approx 0$ give the decay modes due to the structure of a liquid that is dominated by strong anion-cation attractions due to the *short-range* effects of the electrostatic interactions (in some cases also due to short-range intermolecular interactions of other kinds). This causes transient associations of ions that result in small values of the charges of the dressed ions. It is as if the ionic fluid with such dressed ions correspond to a fluid of actual particles with just a fraction of their real charges. The long-range Coulomb interactions then give only a weak tail in the electrostatic interactions over large distances.

The different magnitudes of the modes in the bulk can also be understood as follows. The amplitudes of the various modes of $H_{QQ}(r)$ scales as the forth power of the decay parameter divided by $\mathcal{E}_r^{\text{eff}}$ (Equations 13.103 and 13.105). For $[C_2C_1Im]^+[NTf_2]^-$ we have $[\kappa/|\kappa''|] \approx 10^{-2}$, which means the ratio C/C'' between the amplitudes of the leading monotonic mode and the short-range oscillatory mode in $H_{QQ}(r)$ contain a factor 10^{-8}. For the dilute RPM electrolyte in Figure 12.9 for $\beta_R = 13.3$ and 16.7 we have $[\kappa/|\kappa''|]^4 \approx 10^{-5}$ and we found that $|C/C''|$ is $8 \cdot 10^{-4}$ and $8 \cdot 10^{-5}$ in these two cases. The monotonic tail with very large decay length $1/\kappa$ is accordingly expected to have a very small amplitude compared with the oscillatory mode with much shorter decay length.

13.4.2 Dilute electrolyte solutions with molecular solvent

For dilute electrolyte solutions in the bulk we have seen that the screened Coulomb potential $\phi^\star_{\text{Coul}}(r)$ is dominated for large r by the term

$$\phi^\star_{\text{Coul}}(r) \sim \frac{1}{\mathcal{E}_r^{\text{eff}}(\kappa)\varepsilon_0} \cdot \frac{e^{-\kappa r}}{4\pi r},$$

where in the limit of infinite dilution $\kappa/\kappa_D \to 1$ and $\mathcal{E}_r^{\text{eff}}(\kappa) \to \mathcal{E}_r^{\star 0} \to \varepsilon_r$, i.e., the dielectric constant of the pure solvent. In the dielectric-continuum solvent model described in Section 12.3.2, the dielectric effects due to the solvent were approximated by the use of an imposed dielectric constant $\varepsilon_r^{\text{cont}}$. This is reasonable for the leading term at large r only for dilute solutions. When we deal with electrolyte solutions with molecular solvent, the dielectric effects due to the solvent are dependent on the electrolyte concentration. Both ions and solvent molecules contribute to $\mathcal{E}_r^\star(\kappa) \equiv \tilde{\varepsilon}_{\text{reg}}(\mathrm{i}\kappa)$ and $\mathcal{E}_r^{\text{eff}}(\kappa) \equiv \tilde{\varepsilon}_{\text{reg}}(\mathrm{i}\kappa) + \mathrm{i}\kappa\, \tilde{\varepsilon}'_{\text{reg}}(\mathrm{i}\kappa)/2$.

The intermolecular interactions in solutions have substantial contributions from the liquid structure of all species present. In dilute solutions such effects from the solvent molecules are additional to the dielectric effects. Therefore, in addition to the leading κ-mode, there are other decay modes like the solutions $k = \mathrm{i}\kappa''$ and $k = \mathrm{i}\kappa'''$ of the equation $\epsilon(k) = 0$. If, for example, the decay parameters κ'' and κ''' of these modes are complex-valued $\kappa'' = \kappa''_{\Re} + i\kappa''_{\Im}$ and $\kappa''' = \kappa''_{\Re} - i\kappa''_{\Im}$ we have when $r \to \infty$

$$\phi^{\star}_{\mathrm{Coul}}(r) \sim \frac{1}{\mathcal{E}_r^{\mathrm{eff}}(\kappa)\varepsilon_0} \cdot \frac{e^{-\kappa r}}{4\pi r} + \frac{1}{|\mathcal{E}_r^{\mathrm{eff}}(\kappa'')|\,\varepsilon_0} \cdot \frac{e^{-\kappa''_{\Re} r}}{2\pi r} \cos(\kappa''_{\Im} r - \vartheta_{\mathcal{E}}(\kappa'')). \tag{13.119}$$

For low electrolyte concentrations, where κ is small, the first term dominates for large r, while the oscillatory term can contribute significantly for smaller r. The corresponding is true for intermolecular interactions, pair distributions and surface forces in dilute electrolytes because all of these entities have the same decay modes.[30] From the results in the previous section it follows that $\mathscr{W}^S(D^S)$ between two planar walls immersed in an electrolyte solution decays like in Equations 13.115 and 13.116 when $\phi^{\star}_{\mathrm{Coul}}(r)$ decays as in Equation 13.119.

As an example, we consider interactions in dilute solutions of simple electrolytes. Figure 13.4 shows the surface forces measured[31] by SFA between two mica surface immersed in propylene carbonate containing ions at very low concentration, of the order 10^{-5} M. Some of these ions have dissociated from the mica surface[32] and some have leached from other surfaces (e.g. glass). The force F is plotted as $F/\mathcal{R}$ in the figure and as before the free energy of interaction per unit area $\mathscr{W}^S$ between flat plates is equal to $\mathscr{W}^S = F/[2\pi\mathcal{R}]$. For small and intermediate surface separations $\mathscr{W}^S$ is an oscillatory function of D^S as shown by the blue and black data points, which are measured on approach of the two surfaces toward each other, and by the green symbols that are located close to the minima measured during retractions. These data sets have been obtained in different measurements. When the surface separation is increased $\mathscr{W}^S$ goes over to a monotonic exponentially decaying function at large D^S as seen in the semilogarithmic plot (in the right frame of the figure).

The red curve is obtained from the theoretical expression in Equation 13.116; in the right frame the positive parts of this curve are plotted. The parameters of this equation are selected such that the resulting curve represents the experimental data in an approximate manner. Like in Figure 13.2, its role is mainly to help in the interpretation of the experimental data for large and intermediate surface separations D^S rather than to give an optimal curve fit. As always, there are other contributions to the experimental force, like the influence of higher order terms and van der Waals interactions, but apparently the decay modes in Equations 13.116 and 13.119 are the important

[30] We are here only dealing with realistic models of electrolyte solutions so we do not include charge-inversion invariant systems.

[31] Unpublished data obtained by A. Smith at Prof. Perkin's lab at Physical and Theoretical Chemistry Laboratory, University of Oxford. I want to thank Alex Smith and Susan Perkin for kindly providing me with the these data.

[32] Mica is a silica mineral that can be split into extremely thin, negatively charged plates with adsorbed cation. Some of these cations dissociate from the surface when mica is placed in contact with a polar liquid.

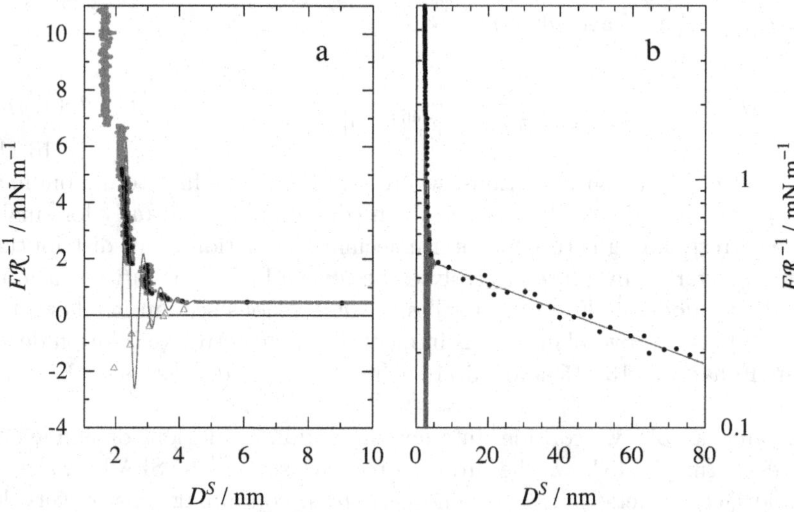

Figure 13.4 **(See color insert).** The normalized force $F/\mathcal{R}$ between two curved mica surfaces of radius $\mathcal{R}$ immersed in propylene carbonate at $T = 294$ K measured[31] by SFA and plotted in the same manner as the force in Figure 13.2 on linear scales (a) and a semi-logarithmic scale (b). The liquid contains a very low concentration of electrolyte (see text). The blue and black data points are measured on approach of the two surfaces toward each other and the green symbols show the places where the surfaces jump apart upon retraction. The red curve is obtained from Equation 13.116 by adjusting its parameters and is an aid to the eye rather than an actual fitting to the experimental data.

ones for most separations, so these decay modes are relevant for the surface forces.

The electrolyte concentration is very low so the decay length of $\mathscr{W}^S$ for large D^S is virtually the same as the Debye length given by the Debye-Hückel approximation ($\varepsilon_r = 64$ for propylene carbonate), so in the first term of Equation 13.116 we have $\kappa = \kappa_D$ in practice. This term gives the straight part of the red curve in the semilogarithmic plot, which has decay length 80 nm and an amplitude $C = 0.078$ mN m^{-1}. The oscillatory part of the red curve for intermediate D^S has the decay length $1/\kappa''_{\Re} = 0.45$ nm, wave length $2\pi/\kappa''_{\Im} = 0.54$ nm and amplitude C'' of the order 160 mN m^{-1}, which is much larger than for the monotonic part. The wave length is approximately equal to the molecular diameter of propylene carbonate and the oscillations are due to the packing of the solvent molecules. Note the difference from the results for the ionic liquid in Figure 13.2, where the wave length is equal to mean diameter of anion-cation pairs and not the individual ions. In the present case the density-density correlations for the uncharged solvent molecules are dominant for small and

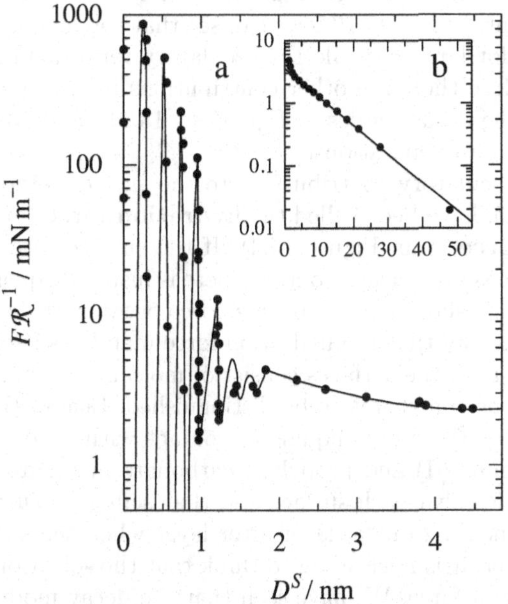

Figure 13.5 (a) The experimentally measured force F (symbols) between two curved mica surfaces of radius $\mathcal{R} = 2$ cm immersed in 10^{-3} M aqueous solution of KCl plotted as the normalized force $F/\mathcal{R}$ as function of the surface separation D^S. The curve constitutes an aid to the eye. The inset (b) shows $F/\mathcal{R}$ (symbols) for large D^S on a logarithmic scale and the full curve is the Poisson-Boltzmann prediction for the double layer interaction between two planar surfaces with surface potential −78 mV. The data are taken from the work by J. N. Israelachvili and R. M. Pashley.[33]

intermediate D^S, while the charge-charge correlations between the ions are dominant in the ionic-liquid.

The next example is a dilute aqueous solution. Figure 13.5 shows the normalized surface force $F/\mathcal{R}$ between two curved, charged surfaces immersed in 10^{-3} M aqueous solution of KCl obtained from SFA measurements by Israelachvili and Pashley.[33] As in the case of propylene carbonate, $\mathscr{W}^S(D^S)$ is oscillatory for small D^S and exponentially decaying for large D^S. The oscillations have a mean periodicity of 0.25 nm which is close to the diameter of a water molecule and the decay length of the monotonic part is about 9.6 nm, which agrees with the Debye length $1/\kappa_D$ of the bulk electrolyte solution.

[33] J. N. Israelachvili and R. M. Pashley, Nature **306** (1983) 249 (https://doi.org/10.1038/306249a0). See also Section 15.8 in: J. N. Israelachvili. 2011. *Intermolecular and Surface Forces*, 3rd edition. Amsterdam: Elsevier.

Also for this system, the three decay modes κ, κ'' and κ''' of Equation 13.119 are relevant. In Figure 13.5 we can see that the monotonic exponential term dominates for large r, while the oscillatory term dominates greatly for smaller r. As before there are other contributions to the experimental force, but apparently the decay modes in Equations 13.116 and 13.119 are the important ones for most separations.

Short-range oscillatory contributions to the surface interactions, like the one in Figure 13.5, have been called the **hydration force** (or more generally, the **solvation force** as in Figure 13.4). If the surfaces are either rough or soft, the oscillations are smeared out, so there is instead a quasi-exponentially decaying monotonic short-range repulsive force between the surfaces, which is the most common way that the hydration force manifests itself. In the experiments of Figure 13.5, the surfaces consist of molecularly smooth mica sheets and the oscillations can therefore be distinguished. Like in the case of structural surface forces for simple liquids in contact with hard walls (treated in Section 8.4 of Volume I) and propylene carbonate in Figure 13.4, there is a layering of particles near each surface and the oscillatory surface force corresponds to the squeezing out of layer after layer when the surface separation decreases. However, it is misleading to think that the solvation of the surfaces *causes* the solvation force. We have seen that the decay modes of the surface forces are the same as for the bulk fluid, so *the oscillations are rather an intrinsic property of the bulk fluid and are not caused by the interactions with the surfaces.* As pointed out earlier, there *can* be contributions to the surface interactions that are induced by the surfaces, but they die out quite quickly with increasing surface separations unless they are commensurate with some decay mode(s) of the bulk.

In order to verify that the oscillatory force is an intrinsic property of the bulk fluid, let us consider pure bulk water where the interactions between the vast majority of water molecules are very similar (or identical) to those in very dilute electrolyte solutions. In pure water we have $\kappa = 0$ so Equation 13.119 becomes

$$\phi^{\star}_{\text{Coul}}(r) \sim \frac{1}{4\pi\varepsilon_r\varepsilon_0 r} + \frac{1}{|\mathcal{E}^{\text{eff}}_r(\kappa'')|\,\varepsilon_0} \cdot \frac{e^{-\kappa''_{\Re} r}}{2\pi r} \cos(\kappa''_{\Im} r - \vartheta_{\mathcal{E}}(\kappa'')) \qquad (13.120)$$

when $r \to \infty$, where the first term is the usual Coulomb potential divided by the dielectric constant ε_r of water. The screened Coulomb potential in pure water obtained by Monte Carlo simulations[34] is plotted in Figure 13.6a, which

[34]Unpublished data from Luc Belloni, personal communication. I want to thank Dr. Belloni for providing me with the data and giving me permission to use them here. The calculations from which $\phi^{\star}_{\text{Coul}}(r)$ has been extracted are described in L. Belloni, J. Chem. Phys. **147** (2017) 164121 (https://doi.org/10.1063/1.5001684). The Monte Carlo simulations were enhanced by advanced integral equation techniques in this work, which is important in order to achieve high accuracy.

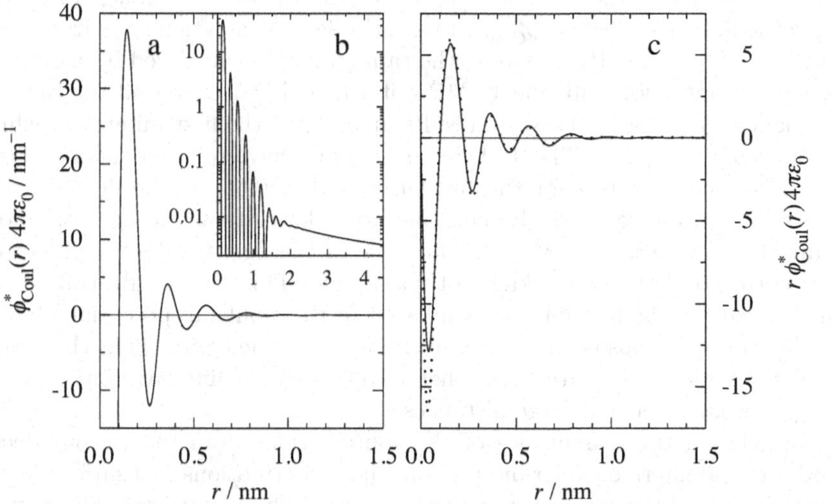

Figure 13.6 (a) The screened Coulomb potential $\phi^\star_{\mathrm{Coul}}(r)$ times $4\pi\varepsilon_0$ for pure bulk water calculated by MC simulation[34] for the SPC/E water model. (b) The positive part of $\phi^\star_{\mathrm{Coul}}(r)4\pi\varepsilon_0$ plotted on a logarithmic scale. (c) The function $r\phi^\star_{\mathrm{Coul}}(r)4\pi\varepsilon_0$ from the same simulations for SPC/E water (full curve) and the same function from MD simulations[37] for the TIP4P water model (dotted curve). When $r \to \infty$ the function $r\phi^\star_{\mathrm{Coul}}(r)4\pi\varepsilon_0 \to 1/\varepsilon_r$, where ε_r is the dielectric constant for pure water.

shows $\phi^\star_{\mathrm{Coul}}(r)$ multiplied by $4\pi\varepsilon_0$ for pure water modeled as SPC/E water.[35] The inset shows the positive part of the same function plotted on a logarithmic scale. The simulated $\phi^\star_{\mathrm{Coul}}(r)$ agrees with the behavior predicted in Equation 13.120. For small r the function $\phi^\star_{\mathrm{Coul}}(r)$ is oscillatory with a wave length about 0.21 nm. The magnitude of the oscillations is much larger than the leading $1/r$ term for $r \lesssim 1.3$ nm, while the latter dominates thereafter.[36] The resemblance to Figure 13.5 is striking including the range of the oscillations, although different functions are plotted.

Thus, the water structure gives rise to a very large oscillatory contribution to the electrostatic interactions in aqueous systems, which causes the hydration force both in bulk and for surface interactions. This contribution is *much larger* than the usual Coulomb $1/r$ potential up to distance of about 1.3 nm corresponding to about six full oscillations. The difference in magnitudes of

[35]In the SPC/E (extended simple point charge) water model, the water molecule is modeled as a Lennard-Jones sphere with three internal point charges located on the vertices of a rigid isosceles triangle. One of the charges represents the partial charge of the oxygen atom and two represent those of the hydrogen atoms. The LJ sphere is centered at the oxygen.

[36]The oscillations continue as ripples on the $1/r$ curve, but beyond $r \approx 2.5$ nm all contributions apart from the leading $1/r$ term are completely negligible.

the oscillatory and the $1/r$ parts can also be illustrated as follows. In Figure 13.6c the full curve shows $r\phi^{\star}_{\text{Coul}}(r)$ from the same simulation as in frames (a) and (b), while the dotted curve is the same function calculated by molecular dynamics simulation[37] for the TIP4P water model.[38] We see that the two water models give nearly the same results apart from the first minimum, which is somewhat deeper for TIP4P. When $r \to \infty$ we have $r\phi^{\star}_{\text{Coul}}(r)4\pi\varepsilon_0 \to 1/\varepsilon_r$, which is equal to 0.014 for the full curve and ≈ 0.02 for the dotted curve ($\varepsilon_r = 71$ and 50, respectively, i.e., somewhat lower than the dielectric constant of real water, $\varepsilon_r \approx 80$). On the scale of the figure, values of the order 0.02 correspond to the thickness of r-axis line. This vividly illustrates how much larger the oscillations are compared to the Coulomb potential $1/r$ tail of $\phi^{\star}_{\text{Coul}}(r)$, which also can be seen in frame (b) of the figure. Thus the actual Coulomb potential in water and other solvents is very different from a simple $1/r$ potential except for large distances.

In general, the magnitudes of the contributions from the various decay modes to the intermolecular interactions, pair distributions and surface forces depend, as we have seen, on the particles and surfaces involved. These magnitudes depend on the effective charges, which are different for the various modes. As mentioned earlier, for a certain particle/surface one decay mode can give much larger contribution than another mode, while for other particles/surfaces immersed in the same electrolyte the reverse may be be true. This is universally true and for the case of dilute electrolyte solutions this implies that the magnitude of the hydration force for various surfaces depends on $|\sigma_{\text{w}}^{S\text{eff}}|$ and $|\mathcal{E}_r^{\text{eff}}|$ evaluated for the oscillatory mode with decay parameter κ' of the bulk solution. For interactions with particles it is $|\mathrm{Q}_i^{\text{eff}}|$ and $|\mathcal{E}_r^{\text{eff}}|$ that determine the magnitude.

To conclude we have seen that apart from the usual screened Coulomb interaction with decay length $1/\kappa$ in dilute electrolyte solutions, there is an oscillatory short-range contribution to the electrostatic interactions. The latter is due to the structure of the solvent and *can be identified with the solvation force* (hydration force in aqueous systems), which is oscillatory for hard surfaces. The decay length and the wave length of this force are determined by the structure of the bulk solution. The magnitude and the phase of the oscillations are determined by the surfaces and the electrolyte. For rough or soft surfaces the oscillations are smoothed out and the solvation force appears instead as a monotonic, quasi-exponentially decaying force, but it has the same molecular origin and its magnitude depends on $|\sigma_{\text{w}}^{S\text{eff}}|$ and $|\mathcal{E}_r^{\text{eff}}|$.

[37] I. P. Omelyan, Molec. Phys. **93** (1998) 123.

[38] The TIP4P model is a rigid, four-site interaction potential model for water, where the point charge representing the oxygen is located away from the oxygen nucleus toward the hydrogens that are represented by two point charges. The fourth site is a Lennard-Jones sphere centered at the oxygen, which lies in the same plane as the other three sites.

APPENDIX

Appendix 13A
Dipolar electrostatic potentials

Let us consider a point charge q_1 located at the position $\mathbf{r}_1 + \mathbf{a}_1$ and a charge $-q_1$ located at $\mathbf{r}_1 - \mathbf{a}_1$. Together they form a dipole with center at $\mathbf{r}_1$ and dipole moment vector $\mathbf{m}_1^{\mathrm{d}} = 2q_1\mathbf{a}_1$. In the limit $a_1 \to 0$ with $2q_1 a_1 = m_1^{\mathrm{d}} =$ constant, where $a_1 = |\mathbf{a}_1|$ and $m_1^{\mathrm{d}} = |\mathbf{m}_1^{\mathrm{d}}|$, they form an ideal dipole located at $\mathbf{r}_1$.

For a dipole placed at $\mathbf{r}_1$ and with orientation $\boldsymbol{\omega}_1$, we have the "internal" charge density σ_1

$$\begin{aligned}\sigma_1(\mathbf{r}_3|\mathbf{R}_1) &= \sigma_1(\mathbf{r}_{13}, \boldsymbol{\omega}_1)\\ &= q_1\delta^{(3)}(\mathbf{r}_{13} - \mathbf{a}_1(\boldsymbol{\omega}_1)) - q_1\delta^{(3)}(\mathbf{r}_{13} + \mathbf{a}_1(\boldsymbol{\omega}_1)).\end{aligned}$$

The electrostatic potential from the dipole is

$$\psi_1(\mathbf{r}_{14}, \boldsymbol{\omega}_1) = \int d\mathbf{r}_3 \sigma_1(\mathbf{r}_{13}, \boldsymbol{\omega}_1)\phi_{\mathrm{Coul}}(r_{34}).$$

The integral is a convolution, so in Fourier space, we have

$$\tilde{\psi}_1(\mathbf{k}, \boldsymbol{\omega}_1) = \tilde{\sigma}_1(\mathbf{k}, \boldsymbol{\omega}_1)\tilde{\phi}_{\mathrm{Coul}}(k) = \frac{\tilde{\sigma}_1(\mathbf{k}, \boldsymbol{\omega}_1)}{\varepsilon_0 k^2}$$

since $\tilde{\phi}_{\mathrm{Coul}}(k) = 1/(\varepsilon_0 k^2)$. Now, the Fourier transform of $\delta^{(3)}(\mathbf{r}_{13} + \mathbf{a})$ is $e^{\mathrm{i}\mathbf{k}\cdot\mathbf{a}}$, as follows from Equations 9.246 and 9.247, so we have

$$\tilde{\sigma}_1(\mathbf{k}, \boldsymbol{\omega}_1) = q_1[e^{-\mathrm{i}\mathbf{k}\cdot\mathbf{a}_1(\boldsymbol{\omega}_1)} - e^{\mathrm{i}\mathbf{k}\cdot\mathbf{a}_1(\boldsymbol{\omega}_1)}] = -\mathrm{i}2q_1 \sin(\mathbf{k}\cdot\mathbf{a}_1(\boldsymbol{\omega}_1))$$

and therefore

$$\tilde{\psi}_1(\mathbf{k}, \boldsymbol{\omega}_1) = -\frac{\mathrm{i}2q_1 \sin(\mathbf{k}\cdot\mathbf{a}_1(\boldsymbol{\omega}_1))}{\varepsilon_0 k^2}.$$

When $k \to 0$ or $a_1 \to 0$ (keeping m_1^{d} constant), we have $\sin(\mathbf{k}\cdot\mathbf{a}) \sim \mathbf{k}\cdot\mathbf{a}$ and

$$\tilde{\psi}_1(\mathbf{k}, \boldsymbol{\omega}_1) \sim -\frac{\mathrm{i}2q_1\mathbf{k}\cdot\mathbf{a}_1(\boldsymbol{\omega}_1)}{\varepsilon_0 k^2} = -\frac{\mathrm{i}\mathbf{k}\cdot\mathbf{m}_1^{\mathrm{d}}(\boldsymbol{\omega}_1)}{\varepsilon_0 k^2}.$$

Since a multiplication by $\mathrm{i}\mathbf{k}$ in Fourier space corresponds to ∇ in real space, we have

$$\psi_1(\mathbf{r}, \boldsymbol{\omega}_1) \sim -\mathbf{m}_1^{\mathrm{d}}(\boldsymbol{\omega}_1)\cdot\nabla\frac{1}{4\pi\varepsilon_0 r} = \frac{\mathbf{m}_1^{\mathrm{d}}(\boldsymbol{\omega}_1)\cdot\mathbf{r}}{4\pi\varepsilon_0 r^3} = \frac{\mathbf{m}_1^{\mathrm{d}}(\boldsymbol{\omega}_1)\cdot\hat{\mathbf{r}}}{4\pi\varepsilon_0 r^2},$$

where the unit vector $\hat{\mathbf{r}} = \mathbf{r}/r$. The rhs is the potential from an ideal dipole ($a_1 \to 0$) and it is also the potential from a finite dipole, $a_1 > 0$, for large r. The latter corresponds to small k since the behavior of the Fourier transform for small k (long wave lengths) describe the features in real space that changes slowly with varying $\mathbf{r}$, in our case for large r. To summarize, we have

$$\psi_1(\mathbf{r}, \boldsymbol{\omega}_1) = \frac{\mathbf{m}_1^{\mathrm{d}}(\boldsymbol{\omega}_1) \cdot \hat{\mathbf{r}}}{4\pi\varepsilon_0 r^2}, \quad \text{for } r > 0 \qquad \text{(ideal dipole)} \tag{13.121}$$

and

$$\psi_1(\mathbf{r}, \boldsymbol{\omega}_1) \sim \frac{\mathbf{m}_1^{\mathrm{d}}(\boldsymbol{\omega}_1) \cdot \hat{\mathbf{r}}}{4\pi\varepsilon_0 r^2}, \quad \text{when } r \to \infty \qquad \text{(finite dipole).} \tag{13.122}$$

Both correspond to the term $-i\mathbf{k} \cdot \mathbf{m}_1^{\mathrm{d}}(\boldsymbol{\omega}_1)/(\varepsilon_0 k^2)$ in the Fourier-transformed potential, $\tilde{\psi}_1(\mathbf{k}, \boldsymbol{\omega}_1)$.

Next, we investigate the pair interaction potential $u_{12}^{\mathrm{el}}(\mathbf{r}_{12}, \boldsymbol{\omega}_1, \boldsymbol{\omega}_2)$ between two finite dipoles: dipole 1 at $\mathbf{r}_1$ and with orientation $\boldsymbol{\omega}_1$ and dipole 2 at $\mathbf{r}_2$ and with orientation $\boldsymbol{\omega}_2$. The electrostatic potential energy for dipole 2 when it interacts with the electrostatic potential $\psi_1(\mathbf{r}_4)$ from dipole 1 is

$$\begin{aligned} u_{12}^{\mathrm{el}}(\mathbf{r}_{12}, \boldsymbol{\omega}_1, \boldsymbol{\omega}_2) &= \int d\mathbf{r}_4 \psi_1(\mathbf{r}_4, \boldsymbol{\omega}_1)\sigma_2(\mathbf{r}_{24}, \boldsymbol{\omega}_2) \\ &= \int d\mathbf{r}_3 d\mathbf{r}_4 \sigma_1(\mathbf{r}_{13}, \boldsymbol{\omega}_1)\phi_{\mathrm{Coul}}(r_{34})\sigma_2(\mathbf{r}_{24}, \boldsymbol{\omega}_2), \end{aligned}$$

where $\sigma_2(\mathbf{r}_{24}, \boldsymbol{\omega}_2) = q_2\delta^{(3)}(\mathbf{r}_{24} - \mathbf{a}_2(\boldsymbol{\omega}_2)) - q_2\delta^{(3)}(\mathbf{r}_{24} + \mathbf{a}_2(\boldsymbol{\omega}_2))$. In Fourier space we have

$$\begin{aligned} \tilde{u}_{12}^{\mathrm{el}}(\mathbf{k}, \boldsymbol{\omega}_1, \boldsymbol{\omega}_2) &= \tilde{\sigma}_1(\mathbf{k}, \boldsymbol{\omega}_1)\tilde{\phi}_{\mathrm{Coul}}(k)\tilde{\sigma}_2(-\mathbf{k}, \boldsymbol{\omega}_2) \\ &= \frac{4q_1q_2 \sin(\mathbf{k} \cdot \mathbf{a}_1(\boldsymbol{\omega}_1)) \sin(\mathbf{k} \cdot \mathbf{a}_2(\boldsymbol{\omega}_2))}{\varepsilon_0 k^2} \end{aligned}$$

where $-\mathbf{k}$ occurs in $\tilde{\sigma}_2(-\mathbf{k}, \boldsymbol{\omega}_2)$ because the integral over $\mathbf{r}_4$ is not a convolution unless $\sigma_2(\mathbf{r}_{24}, \boldsymbol{\omega}_2)$ is changed to $\sigma_2(\mathbf{r}_{42}, \boldsymbol{\omega}_2)$ with $\mathbf{r}_{42} = -\mathbf{r}_{24}$.

When $k \to 0$ or when both a_1 and $a_2 \to 0$ (keeping m_1^{d} and m_2^{d} constant) we have

$$\tilde{u}_{12}^{\mathrm{el}}(\mathbf{k}, \boldsymbol{\omega}_1, \boldsymbol{\omega}_2) \sim \frac{4q_1q_2\mathbf{k} \cdot \mathbf{a}_1(\boldsymbol{\omega}_1)\, \mathbf{k} \cdot \mathbf{a}_2(\boldsymbol{\omega}_2)}{\varepsilon_0 k^2} = \frac{\mathbf{k} \cdot \mathbf{m}_1^{\mathrm{d}}(\boldsymbol{\omega}_1)\, \mathbf{k} \cdot \mathbf{m}_2^{\mathrm{d}}(\boldsymbol{\omega}_2)}{\varepsilon_0 k^2}. \tag{13.123}$$

In real space this implies that

$$u_{12}^{\mathrm{el}}(\mathbf{r}, \boldsymbol{\omega}_1, \boldsymbol{\omega}_2) \sim -\left[\mathbf{m}_1^{\mathrm{d}}(\boldsymbol{\omega}_1) \cdot \nabla\right] \left[\mathbf{m}_2^{\mathrm{d}}(\boldsymbol{\omega}_1) \cdot \nabla\right] \frac{1}{4\pi\varepsilon_0 r}.$$

By performing the derivatives and simplifying we obtain

$$u_{12}^{\mathrm{el}}(\mathbf{r}, \boldsymbol{\omega}_1, \boldsymbol{\omega}_2) \sim \frac{\mathbf{m}_1^{\mathrm{d}}(\boldsymbol{\omega}_1) \cdot \mathbf{m}_2^{\mathrm{d}}(\boldsymbol{\omega}_2) - 3\left[\mathbf{m}_1^{\mathrm{d}}(\boldsymbol{\omega}_1) \cdot \hat{\mathbf{r}}\right] \left[\mathbf{m}_2^{\mathrm{d}}(\boldsymbol{\omega}_2) \cdot \hat{\mathbf{r}}\right]}{4\pi\varepsilon_0 r^3}.$$

The rhs is the interaction potential for two ideal dipoles (a_1 and $a_2 \to 0$) and it is also the interaction potential for two finite dipoles, $a_1 > 0$ and $a_2 > 0$, for large r. To summarize, we have

$$u_{12}^{\mathrm{el}}(\mathbf{r}, \boldsymbol{\omega}_1, \boldsymbol{\omega}_2) = -\frac{m_1^{\mathrm{d}} m_2^{\mathrm{d}} \mathsf{D}_{12}(\hat{\mathbf{r}}, \boldsymbol{\omega}_1, \boldsymbol{\omega}_2)}{4\pi\varepsilon_0 r^3} \quad \text{for } r > 0 \qquad \text{(ideal dipoles)} \tag{13.124}$$

and

$$u_{12}^{\mathrm{el}}(\mathbf{r}, \boldsymbol{\omega}_1, \boldsymbol{\omega}_2) \sim -\frac{m_1^{\mathrm{d}} m_2^{\mathrm{d}} \mathsf{D}_{12}(\hat{\mathbf{r}}, \boldsymbol{\omega}_1, \boldsymbol{\omega}_2)}{4\pi\varepsilon_0 r^3} \quad \text{when } r \to \infty \qquad \text{(finite dipoles).} \tag{13.125}$$

where

$$\mathsf{D}_{12}(\hat{\mathbf{r}}, \boldsymbol{\omega}_1, \boldsymbol{\omega}_2) = 3\left[\hat{\mathbf{r}} \cdot \hat{\mathbf{m}}_1^{\mathrm{d}}(\boldsymbol{\omega}_1)\right]\left[\hat{\mathbf{r}} \cdot \hat{\mathbf{m}}_2^{\mathrm{d}}(\boldsymbol{\omega}_2)\right] - \hat{\mathbf{m}}_1^{\mathrm{d}}(\boldsymbol{\omega}_1) \cdot \hat{\mathbf{m}}_2^{\mathrm{d}}(\boldsymbol{\omega}_2). \tag{13.126}$$

The dipolar potential term in both Equations 13.124 and 13.125 corresponds, as we have seen, to the term in the rhs of Equation 13.123. This term can be written as $\hat{\mathbf{k}} \cdot \mathbf{m}_1^{\mathrm{d}}(\boldsymbol{\omega}_1)\, \hat{\mathbf{k}} \cdot \mathbf{m}_2^{\mathrm{d}}(\boldsymbol{\omega}_2)/\varepsilon_0$ so it does not depend on the magnitude of k.

In fact, for ideal dipoles there is also a term that contributes to u_{12}^{el} when $r = 0$. The complete expression for u_{12}^{el} in this case is

$$u_{12}^{\mathrm{el}}(\mathbf{r}, \boldsymbol{\omega}_1, \boldsymbol{\omega}_2) = -\frac{m_1^{\mathrm{d}} m_2^{\mathrm{d}} \mathsf{D}_{12}(\hat{\mathbf{r}}, \boldsymbol{\omega}_1, \boldsymbol{\omega}_2)}{4\pi\varepsilon_0 r^3} + \delta^{(3)}(\mathbf{r}) \frac{\mathbf{m}_1^{\mathrm{d}}(\boldsymbol{\omega}_1) \cdot \mathbf{m}_2^{\mathrm{d}}(\boldsymbol{\omega}_2)}{3\varepsilon_0} \quad \text{for all } r \qquad \text{(ideal dipoles).} \tag{13.127}$$

In practice the delta function contribution has no consequence for the interaction potential (the two ideal dipoles are normally not allowed to coincide), but the term is needed for a mathematical consistent treatment of the singularity at $r = 0$ when performing Fourier transforms. It is shown in the shaded text box below how to obtain this contribution from the expression

$$\tilde{u}_{12}^{\mathrm{el}}(\mathbf{k}, \boldsymbol{\omega}_1, \boldsymbol{\omega}_2) = \frac{\mathbf{k} \cdot \mathbf{m}_1^{\mathrm{d}}(\boldsymbol{\omega}_1)\, \mathbf{k} \cdot \mathbf{m}_2^{\mathrm{d}}(\boldsymbol{\omega}_2)}{\varepsilon_0 k^2} \qquad \text{(ideal dipoles)} \tag{13.128}$$

in Fourier space.

One can define an ideal dipole potential without the delta function contribution, but then the corresponding contribution has to be contained in the Fourier transform $\tilde{u}_{12}^{\mathrm{el}}$ instead. We then have the alternative expression

$$u_{12}^{\mathrm{el,\,alt.}}(\mathbf{r}, \boldsymbol{\omega}_1, \boldsymbol{\omega}_2) = -\frac{m_1^{\mathrm{d}} m_2^{\mathrm{d}} \mathsf{D}_{12}(\hat{\mathbf{r}}, \boldsymbol{\omega}_1, \boldsymbol{\omega}_2)}{4\pi\varepsilon_0 r^3} \qquad \text{(ideal dipoles)}$$

and in Fourier space, where we have subtracted the Fourier transform of the delta function term, we have

$$\begin{aligned}\tilde{u}_{12}^{\text{el, alt.}}(\mathbf{k},\omega_1,\omega_2) &= \frac{\mathbf{k}\cdot\mathbf{m}_1^{\mathrm{d}}(\omega_1)\,\mathbf{k}\cdot\mathbf{m}_2^{\mathrm{d}}(\omega_2)}{\varepsilon_0 k^2} - \frac{\mathbf{m}_1^{\mathrm{d}}(\omega_1)\cdot\mathbf{m}_2^{\mathrm{d}}(\omega_2)}{3\varepsilon_0} \\ &= \frac{m_1^{\mathrm{d}} m_2^{\mathrm{d}} \mathrm{D}_{12}(\hat{\mathbf{k}},\omega_1,\omega_2)}{3\varepsilon_0} \quad \text{(ideal dipoles)}, \qquad (13.129)\end{aligned}$$

where D_{12} is the same function as in Equation 13.126 but with $\hat{\mathbf{k}}$ inserted instead of $\hat{\mathbf{r}}$.

Here, we will show how to obtain the delta function term in $u_{12}^{\mathrm{el}}(\mathbf{r},\omega_1,\omega_2)$ of Equation 13.127 for ideal dipoles starting from Equation 13.128. Let us consider a well-behaved arbitrary function $f(r)$, a so-called test-function, and investigate

$$\begin{aligned}\int d\mathbf{r}\, u_{12}^{\mathrm{el}}(\mathbf{r},\omega_1,\omega_2) f(r) &= \frac{1}{(2\pi)^3}\int d\mathbf{r}\, \tilde{u}_{12}^{\mathrm{el}}(\mathbf{k},\omega_1,\omega_2)\tilde{f}(k) \\ &= \frac{1}{\varepsilon_0(2\pi)^3}\int d\mathbf{k}\, \frac{\mathbf{k}\cdot\mathbf{m}_1^{\mathrm{d}}(\omega_1)\,\mathbf{k}\cdot\mathbf{m}_2^{\mathrm{d}}(\omega_2)}{k^2}\tilde{f}(k),\end{aligned}$$

where we have used Parseval's formula 9.253 to obtain the first equality. The integral can be written as

$$\begin{aligned}&\int d\mathbf{k}\, \frac{\mathbf{m}_1^{\mathrm{d}}(\omega_1)\cdot\mathbf{k}\,\mathbf{k}\cdot\mathbf{m}_2^{\mathrm{d}}(\omega_2)}{k^2}\tilde{f}(k) \\ &= \mathbf{m}_1^{\mathrm{d}}(\omega_1)\cdot\int d\mathbf{k}\, \frac{1}{k^2}\begin{pmatrix} (k_x)^2 & k_x k_y & k_x k_z \\ k_y k_x & (k_y)^2 & k_y k_z \\ k_z k_x & k_z k_y & (k_z)^2 \end{pmatrix}\tilde{f}(k)\cdot\mathbf{m}_2^{\mathrm{d}}(\omega_2).\end{aligned}$$

Now, we have

$$\int d\mathbf{k}\, \frac{k_\nu k_{\nu'}}{k^2}\tilde{f}(k) = \begin{cases} 0, & \nu \neq \nu' \\ \frac{1}{3}\int d\mathbf{k}\, \tilde{f}(k), & \nu = \nu', \end{cases}$$

where the result for $\nu \neq \nu'$ follows since the integrand is an odd function and the other result can be deduced the fact that

$$\int d\mathbf{k}\, \frac{(k_x)^2}{k^2}\tilde{f}(k) = \int d\mathbf{k}\, \frac{(k_y)^2}{k^2}\tilde{f}(k) = \int d\mathbf{k}\, \frac{(k_z)^2}{k^2}\tilde{f}(k),$$

so, for example,

$$\int d\mathbf{k}\, \frac{(k_x)^2}{k^2}\tilde{f}(k) = \frac{1}{3}\int d\mathbf{k}\, \frac{(k_x)^2+(k_y)^2+(k_z)^2}{k^2}\tilde{f}(k) = \frac{1}{3}\int d\mathbf{k}\, \tilde{f}(k).$$

Thus,

$$\int d\mathbf{k}\,\frac{\mathbf{m}_1^{\mathrm{d}}(\omega_1)\cdot\mathbf{k}\,\mathbf{k}\cdot\mathbf{m}_2^{\mathrm{d}}(\omega_2)}{k^2}\tilde{f}(k)=\tfrac{1}{3}\int d\mathbf{k}\,\mathbf{m}_1^{\mathrm{d}}(\omega_1)\cdot\mathbb{I}\cdot\mathbf{m}_2^{\mathrm{d}}(\omega_2)\tilde{f}(k)$$

and we conclude that

$$\begin{aligned}\int d\mathbf{r}\,u_{12}^{\mathrm{el}}(\mathbf{r},\omega_1,\omega_2)f(r) &= \frac{\mathbf{m}_1^{\mathrm{d}}(\omega_1)\cdot\mathbf{m}_2^{\mathrm{d}}(\omega_2)}{3\varepsilon_0(2\pi)^3}\int d\mathbf{k}\,\tilde{f}(k)\\ &= \frac{\mathbf{m}_1^{\mathrm{d}}(\omega_1)\cdot\mathbf{m}_2^{\mathrm{d}}(\omega_2)}{3\varepsilon_0}f(0),\end{aligned}$$

where we have used Equation 9.243 with $\mathbf{r}=0$ to obtain the last equality. Since $f(r)$ is arbitrary, this implies that $u_{12}^{\mathrm{el}}(\mathbf{r},\omega_1,\omega_2)$ contains the delta function term in Equation 13.127 (the term with D_{12} gives the contribution zero to $\int d\mathbf{r}\,u_{12}^{\mathrm{el}}(\mathbf{r},\omega_1,\omega_2)f(r)$).

CHAPTER 14

Polar Fluids

Polar fluids consist of electroneutral molecules/particles with an internal charge distribution that is not spherically symmetric. In this chapter we build on the theory presented in the previous chapters, whereby we present the last part of the unified exact treatment of electrolytes and polar fluids. An important task is to find out how the theory must be modified in order to accommodate the consequences of an intermolecular pair potential that have a dipolar or multipolar character for large separations. We will treat both nematic and isotropic polar fluids, with the main focus on the latter. One important topic is the dielectric properties of polar fluids, especially the static dielectric function and the dielectric constant. Another topic is interparticle interactions and correlations. We will find that apart from the leading multipolar Coulomb potential contributions, there are multiple decay modes of the same kinds as for electrolytes with exponentially decaying contributions. The latter have exactly the same relationship to the dielectric function as in electrolytes. The oscillatory contributions dominate for short separations up to a distance of several molecular diameters because they have very much larger amplitude than the Coulombic terms. They originate from the structure of the fluid and give rise to a kind of structural force (often called "hydration force" in aqueous systems), as we saw in the previous chapter.

14.1 PAIR CORRELATIONS, DIELECTRIC RESPONSE AND ELECTROSTATIC INTERACTIONS

In this chapter we are going to treat liquids consisting of polar rigid molecules and all molecules in the fluid are electroneutral, $q_i = 0$. However, the most important polar fluid, water, is in reality a very dilute electrolyte due to the self-ionization into H_3O^+ and OH^-, whereby the total ionic concentration is about $2 \cdot 10^{-7}$ M at pH $= 7$. Therefore, the statistical mechanics of water actually belongs to the previous chapter and the screened Coulomb potential of water decays like a Yukawa function with a decay length of about 960 nm. However, for virtually all practical purposes, $\phi^{\star}_{\mathrm{Coul}}(r)$ for bulk water decays

DOI: 10.1201/9781003286882-14

for large r as $1/r$ like in Equation 13.120 and there is also a large, short-range oscillatory contribution due to the structure of water that gives rise to the oscillatory term in this equation as discussed in the previous section. In the general case of polar fluids without any ions, there is a $1/r$ tail of $\phi^{\star}_{\text{Coul}}(r)$ for large r, but for small distances the interaction is affected by the molecular structure of the fluid.

The $1/r$ potential has an enormous range and leads, as we will see, to a pair correlation function $h^{(2)}_{ij}$ between the particles in a pure polar fluid that decays like a dipole-dipole interaction potential at large distances,[1] that is, like $1/r^3$. Such long-range correlations implies, in fact, that some properties of a macroscopically large sample of a polar liquid depend on the size and the shape of the sample due to boundary effects. The correlations with any particle at the center of the sample have not died down to zero for particles at the boundary of the sample despite the huge distances involved.[2] This leads to some mathematical subtleties in the statistical mechanical treatment of polar fluids that are absent for electrolyte systems, where the screening leads to an exponential decay of the electrostatic interactions irrespectively of how dilute the electrolyte is. In the present treatment we will assume that the system is infinitely large.

The theory of polar fluids has a long involved history, where an early landmark was the famous work by Kirkwood,[3] and it would lead too far to review it. In the previous chapter we obtained some results for pure pure polar fluids by taking the limit of zero ionic concentration of the corresponding results for electrolytes. This is the strategy we will follow in the present chapter. In fact, by formulating the statistical mechanics for pure polar fluids in terms of the starred pair correlation functions $h^{\star}_{ij}$ and $c^{\star}_{ij}$ that have shorter ranges than $h^{(2)}_{ij}$ and $c^{(2)}_{ij}$, one can avoid most mathematical subtleties due to the long range of the interactions, which is an approach that has been used in the past. In the general formalism that we utilized for electrolytes and that is valid here as well, the relationships between the two sets of functions are

$$h^{(2)}_{ij}(\mathbf{R}_1,\mathbf{R}_2) = h^{\star}_{ij}(\mathbf{R}_1,\mathbf{R}_2) - \beta \int d\mathbf{r}_3\, d\mathbf{r}_4\, \rho^{\star}_i(\mathbf{r}_4|\mathbf{R}_1)\phi^{\star}_{\text{Coul}}(\mathbf{r}_4,\mathbf{r}_3)\rho^{\star}_j(\mathbf{r}_3|\mathbf{R}_2) \tag{14.1}$$

[1]The dipole moment vectors of the particles in the asymptotic decay of $h^{(2)}_{ij}$ thereby assume effective values due to the correlations with the surrounding particles, G. Nienhuis and J. M. Deutch, J. Chem. Phys. **55** (1971) 4213 (https://doi.org/10.1063/1.1676739). The explicit formula is given below in Equation 14.12.

[2]Let us consider particles that are located at distances between r and $r + dr$ from the central particle and in directions within the differential solid angle $d\phi \sin\theta\, d\theta$ in polar coordinates. The total magnitude of their correlations with the central particle is given by $r^2 dr\, d\phi \sin\theta\, d\theta \times 1/r^3$, which decays like $1/r$. This means that the correlations between the central particle and the particles at the boundary of a macroscopically large sample cannot be neglected.

[3]J. G. Kirkwood, J. Chem. Phys. **7** (1939) 911 (https://doi.org/10.1063/1.1750343).

and

$$c_{ij}^{(2)}(\mathbf{R}_1,\mathbf{R}_2) = c_{ij}^{\star}(\mathbf{R}_1,\mathbf{R}_2) - \beta \int d\mathbf{r}_3 d\mathbf{r}_4 \, \sigma_i(\mathbf{r}_4|\mathbf{R}_1)\phi_{\mathrm{Coul}}(r_{43})\sigma_j(\mathbf{r}_3|\mathbf{R}_2) \tag{14.2}$$

(Equations 13.40 and 13.41). Furthermore, $h_{ij}^{(2)}$ and $c_{ij}^{(2)}$ satisfy the Ornstein-Zernike equation in both cases and the same is true for $h_{ij}^{\star}$ and $c_{ij}^{\star}$.

The long-range nature of $h_{ij}^{(2)}$ and $c_{ij}^{(2)}$ for polar fluids arises from the slow decays of $\phi_{\mathrm{Coul}}^{\star}$ and ϕ_{Coul}, respectively, in the last term in each of these two equations; for $c_{ij}^{(2)}$ the last term is equal to $-\beta u_{ij}^{\mathrm{el}}(\mathbf{R}_1,\mathbf{R}_2)$, which decays to zero like a dipole potential when $r_{12} \to \infty$ because the molecules are uncharged (u_{ij}^{el} decays faster to zero if any of the particles lack dipole moment). Note that $\rho_i^{\star}$ in the equation for $h_{ij}^{(2)}$ cannot have longer range than the functions $h_{ij}^{\star}$ because $\rho_i^{\star}$ is essentially a sum of $h_{il}^{\star}$ functions convoluted with σ_l, which has a finite range (cf. Equation 13.26). Let us recapitulate that since $c_{ij}^{(2)}(\mathbf{R}_1,\mathbf{R}_2) \sim -\beta u_{ij}^{\mathrm{el}}(\mathbf{R}_1,\mathbf{R}_2)$ when $r_{12} \to \infty$, the function $c_{ij}^{\star} \equiv c_{ij}^{(2)} + \beta u_{ij}^{\mathrm{el}}$ has a shorter range than $c_{ij}^{(2)}$ and since $g_{ij}^{(2)} = h_{ij}^{(2)} + 1 = \exp(-\beta[u_{ij}^{\mathrm{el}} + u_{ij}^{\mathrm{sh}}] + h_{ij}^{(2)} - c_{ij}^{(2)} + b_{ij}^{(2)})$ it follows that $c_{ij}^{\star}$ decays as $\frac{1}{2}[h_{ij}^{(2)}]^2 + b_{ij}^{(2)}$ (see Equation 12.52). The functions $[h_{ij}^{(2)}]^2$ and $b_{ij}^{(2)}$ decay about equally fast to zero when r_{12} goes to infinity and the former decays like $1/r_{12}^6$.

For a bulk fluid, the polarization response function $\chi^{\star}$ is

$$\begin{aligned} \chi^{\star}(\mathbf{r}_{12}) &= -\beta q_{\mathrm{e}}^2 H_{\mathrm{QQ}}^{\star}(\mathbf{r}_{12}) \\ &= -\beta \sum_i \int d\mathbf{r}_3 \left\langle \sigma_i(\mathbf{r}_{31},\boldsymbol{\omega}_3)\mathfrak{n}_i^{(1)}(\boldsymbol{\omega}_3)\rho_i^{\star}(\mathbf{r}_{32},\boldsymbol{\omega}_3)\right\rangle_{\boldsymbol{\omega}_3} \end{aligned} \tag{14.3}$$

(Equation 13.44) and

$$\begin{aligned} \tilde{\chi}^{\mathrm{el}}(\mathbf{k}) &= -\tilde{\chi}^{\star}(\mathbf{k})\tilde{\phi}_{\mathrm{Coul}}(k) \\ \tilde{\epsilon}(\mathbf{k}) &= 1 - \tilde{\chi}^{\star}(\mathbf{k})\tilde{\phi}_{\mathrm{Coul}}(k) = 1 + \tilde{\chi}^{\mathrm{el}}(\mathbf{k}) \\ \tilde{\epsilon}^{\mathrm{R}}(\mathbf{k}) &= 1/\tilde{\epsilon}(\mathbf{k}). \end{aligned} \tag{14.4}$$

(Equation 13.45). Explicitly, in terms of the pair correlations we have

$$\tilde{\epsilon}(\mathbf{k}) = 1 + \frac{\beta}{\varepsilon_0 k^2} \sum_{i,j} \langle \tilde{H}_{ij}^{\star}(\mathbf{k},\boldsymbol{\omega}_1,\boldsymbol{\omega}_2)\sigma_i(\mathbf{k},\boldsymbol{\omega}_1)\sigma_j(-\mathbf{k},\boldsymbol{\omega}_2)\rangle_{\boldsymbol{\omega}_1,\boldsymbol{\omega}_2}. \tag{14.5}$$

Since $q_i = 0$ for all i we have $\tilde{\epsilon}_{\mathrm{sing}}(k) = 0$, so $\tilde{\epsilon}(\mathbf{k}) = \tilde{\epsilon}_{\mathrm{reg}}(\mathbf{k})$, which remains finite when $k \to 0$ (cf. Equation 13.49).

For a nematic, liquid crystalline phase we have $\tilde{\epsilon}(\mathbf{k}) = \tilde{\epsilon}_{\mathrm{reg}}(k_\perp, k_\parallel)$, where $k_\perp = |\mathbf{k}_\perp|$ and $k_\parallel$ are the wave vector components perpendicular to and parallel with the director of the system, respectively, and $\tilde{\epsilon}_{\mathrm{reg}}(k_\perp, k_\parallel)$ is given by Equation 13.59. In the limit $k \to 0$, we have

$$\tilde{\varepsilon}_{\text{reg}}(k_\perp, k_\parallel) \to \varepsilon_\perp \left[\frac{k_\perp}{k}\right]^2 + \varepsilon_\parallel \left[\frac{k_\parallel}{k}\right]^2 \quad \text{(nematic polar fluid)},$$

(cf. Equation 13.60), where $\varepsilon_\perp$ and $\varepsilon_\parallel$ are the **dielectric constants of the anisotropic phase** in the perpendicular and parallel directions, respectively. They are given by

$$\begin{aligned} \varepsilon_\perp &= 1 + \frac{1}{4\varepsilon_0} \int ds\, dz\, s^2 \chi^\star(s,z) \\ \varepsilon_\parallel &= 1 + \frac{1}{2\varepsilon_0} \int ds\, dz\, z^2 \chi^\star(s,z). \end{aligned} \tag{14.6}$$

(cf. Equation 13.61).

For a homogeneous and isotropic polar fluid $\mathrm{n}_i^{(1)}(\boldsymbol{\omega}_3) = n_i^{\mathrm{b}}$ and $\tilde{\chi}^{\mathrm{el}}(k)$ and $\tilde{\varepsilon}(k)$ depend only on k. The ***dielectric constant*** ε_r is given by $\tilde{\varepsilon}(k)$ for $k = 0$, that is,

$$\varepsilon_r \equiv \tilde{\varepsilon}(0) = \tilde{\varepsilon}_{\text{reg}}(0) \equiv \mathcal{E}^{\star 0} \quad \text{(isotropic polar fluid)}. \tag{14.7}$$

Such fluids constitute the main case in what follows. The fact that $\tilde{\varepsilon}(k)$ of a polar medium depends on the wave number k is a consequence of ***nonlocal electrostatics*** that we discussed for electrolytes with spherical ions on page 287. In the present case

$$\delta w_i(\mathbf{r}_4, \boldsymbol{\omega}_4) = \int d\mathbf{r}_3\, \rho_i^\star(\mathbf{r}_{43}, \boldsymbol{\omega}_4) \delta\Psi(\mathbf{r}_3)$$

(Equation 13.23), which shows that the potential of mean force for the i-particle at $\mathbf{r}_4$ is affected by the total electrostatic potential in an entire neighborhood of that point. The nonlocality is described by the charge density $\rho_i^\star$, which gives $\tilde{\varepsilon}(k) = \tilde{\varepsilon}_{\text{reg}}(k)$ via Equations 14.3 and 14.4. In approximations where the solvent is treated as a dielectric continuum one disregards the k dependence of $\tilde{\varepsilon}(k)$, whereby the electrostatics of the solvent is assumed to be local. In effect, one then treats the solvent as in classical electrostatics for macroscopic systems.

As mentioned earlier, in macroscopic electrostatics and also the theory of dielectric response it is common to use the polarization field $\mathbf{P}(\mathbf{r}) = -\mathbf{E}^{\text{pol}}(\mathbf{r})\varepsilon_0 = \nabla\Psi^{\text{pol}}(\mathbf{r})\varepsilon_0$, where Ψ^{pol} is the potential from the polarization charge density $\rho^{\text{pol}}(\mathbf{r})$ that is induced by the external potential $\Psi^{\text{ext}}(\mathbf{r})$. We have $\rho^{\text{pol}}(\mathbf{r}) = -\boldsymbol{\nabla} \cdot \mathbf{P}(\mathbf{r}) = -\nabla^2\Psi^{\text{pol}}(\mathbf{r})\varepsilon_0$. For bulk fluid exposed to a weak external potential $\delta\Psi^{\text{ext}}(\mathbf{r})$ there is a linear polarization response and we have

$$\delta\tilde{\mathbf{P}}(\mathbf{k}) = -\delta\tilde{\mathbf{E}}^{\text{pol}}(\mathbf{k})\varepsilon_0 = \tilde{\chi}^{\text{el}}(\mathbf{k})\delta\tilde{\mathbf{E}}(\mathbf{k})\varepsilon_0,$$

where $\delta\tilde{\mathbf{E}}(\mathbf{k}) = -\boldsymbol{\nabla}\delta\tilde{\Psi}(\mathbf{k})$ is the total electrostatic field and we have used $\delta\Psi^{\text{pol}}(\mathbf{k}) = -\tilde{\chi}^{\text{el}}(\mathbf{k})\delta\tilde{\Psi}(\mathbf{k})$.

When the fluid consists of purely dipolar particles (each particle contains solely an ideal dipole) one can obtain the polarization vector from

$$\mathbf{P}(\mathbf{r}) = \left\langle \sum_{i,\nu} \mathbf{m}_i^{\mathrm{d}}(\boldsymbol{\omega}_\nu)\delta^{(3)}(\mathbf{r}-\mathbf{r}_\nu) \right\rangle_{\mathbf{E}^{\mathrm{Ext}}(\mathbf{r})} \qquad \text{(ideal dipoles)}, \tag{14.8}$$

where $\langle \cdot \rangle_{\mathbf{E}^{\mathrm{Ext}}(\mathbf{r})}$ denotes the ensemble average for the particles (labelled ν) taken in the presence of the external field $\mathbf{E}^{\mathrm{Ext}}(\mathbf{r})$. This expression is sometimes used as a definition of $\mathbf{P}$. However, when the particles have internal charge distribution $\sigma_i(\mathbf{r}, \boldsymbol{\omega})$, the polarization charge density $\rho^{\mathrm{pol}}(\mathbf{r})$ and hence $\Psi^{\mathrm{pol}}(\mathbf{r})$ are not given by the distribution of molecular dipoles. Then the polarization vector $\mathbf{P}$ includes additional contributions from, for example, the electric multipole moments of the particles. As we will see, the multipole moments contribute to $\tilde{\epsilon}(\mathbf{k})$ for $\mathbf{k} \neq 0$. In order to obtain $\tilde{\epsilon}(\mathbf{k})$ it is therefore *not sufficient* in the general case to consider a polarization $\mathbf{P}$ and the corresponding potential Ψ^{pol} based solely on the distribution of dipole moment vectors of the particles (as in Equation 14.8).

Let us investigate electrostatic interactions in an *isotropic* polar fluid. We have

$$\tilde{\phi}^{\star}_{\mathrm{Coul}}(k) = \frac{1}{\tilde{\epsilon}(k)\varepsilon_0 k^2} \sim \frac{1}{\tilde{\epsilon}(0)\varepsilon_0 k^2} = \frac{1}{\varepsilon_r \varepsilon_0 k^2} \qquad \text{when } k \to 0,$$

so there is a $1/k^2$ divergence of $\tilde{\phi}^{\star}_{\mathrm{Coul}}(k)$ at $k = 0$ in Fourier space. This is typical for a function that decays like $1/r$ for large r in real space. To see this, let us consider the inverse Fourier transform (Equation 9.244) of $\tilde{\phi}^{\star}_{\mathrm{Coul}}(k)$

$$\begin{aligned} \phi^{\star}_{\mathrm{Coul}}(r) &= \frac{1}{2\pi^2}\int_0^\infty dk\, k^2 \frac{\sin(kr)/(kr)}{\tilde{\epsilon}(k)\varepsilon_0 k^2} = \frac{1}{2\pi^2 \varepsilon_0 r}\int_0^\infty ds\, \frac{\sin(s)/s}{\tilde{\epsilon}\left(\frac{s}{r}\right)} \\ &\sim \frac{1}{2\pi^2 \varepsilon_0 r}\int_0^\infty ds\, \frac{\sin(s)/s}{\tilde{\epsilon}(0)} \qquad \text{when } r \to \infty, \end{aligned}$$

where we have made the substitution $s = kr$. It follows that

$$\phi^{\star}_{\mathrm{Coul}}(r) \sim \frac{1}{4\pi\varepsilon_r\varepsilon_0 r} \qquad \text{when } r \to \infty \qquad \text{(polar bulk fluid)}, \tag{14.9}$$

since $\int_0^\infty ds\, \sin(s)/s = \pi/2$. This is in agreement with what we found earlier. The important difference between a pure polar fluid and an electrolyte is that for the latter we have $\tilde{\epsilon}(k) = \tilde{\epsilon}_{\mathrm{sing}}(k) + \tilde{\epsilon}_{\mathrm{reg}}(k)$, where $\tilde{\epsilon}_{\mathrm{sing}}(k)$ diverges like $1/k^2$ and $\tilde{\epsilon}_{\mathrm{reg}}(k) \to \mathcal{E}^{\star 0}$ when $k \to 0$, so $\tilde{\phi}^{\star}_{\mathrm{Coul}}(k)$ stays finite when $k \to 0$. The $1/r$ decay is then replaced by a Yukawa function decay as we have seen.

The mean electrostatic potential from an i-particle in the polar fluid can be obtained from the screened version of the Coulomb potential formula

$$\psi_i(\mathbf{r}_{12}, \boldsymbol{\omega}_1) = \int d\mathbf{r}_3\, \rho_i^{\star}(\mathbf{r}_{13}, \boldsymbol{\omega}_1)\phi^{\star}_{\mathrm{Coul}}(\mathbf{r}_{32})$$

(Equation 13.54) and the leading terms for large r_{12} are[4]

$$
\begin{aligned}
\psi_i(\mathbf{r}_{12},\boldsymbol{\omega}_1) &\sim \int d\mathbf{r}_3\, \rho_i^\star(\mathbf{r}_{13},\boldsymbol{\omega}_1)\frac{1}{4\pi\varepsilon_r\varepsilon_0 r_{32}} \\
&\sim \frac{1}{4\pi\varepsilon_r\varepsilon_0}\left[\frac{q_i}{r_{12}} + \frac{\mathbf{m}_i^{\mathrm{d}\star}(\boldsymbol{\omega}_1)\cdot\hat{\mathbf{r}}_{12}}{r_{12}^2}\right. \\
&+ \left.\frac{\hat{\mathbf{r}}_{12}\cdot\mathrm{M}_i^{\mathrm{Q}\star}(\boldsymbol{\omega}_1)\cdot\hat{\mathbf{r}}_{12}}{r_{12}^3}\ldots\right] \quad \text{when } r_{12}\to\infty, \qquad (14.10)
\end{aligned}
$$

where we have used the fact that $\int d\mathbf{r}\,\rho_i^\star(\mathbf{r},\boldsymbol{\omega}) = \int d\mathbf{r}\,\sigma_i(\mathbf{r},\boldsymbol{\omega}) = q_i$ because the dress of the i-particle consists of uncharged molecules and where $\mathbf{m}_i^{\mathrm{d}\star}$ is the dipole moment and $\mathrm{M}_i^{\mathrm{Q}\star}$ is the quadrupole moment of the dressed i-particle defined in Equation 13.81, that is, the moments of the charge density $\rho_i^\star$. When the i-particle is one of the particles in the polar fluid we have $q_i = 0$, but Equation 14.10 with $q_i \neq 0$ is valid for a single charged i-particle immersed in a polar fluid (whereby $n_i^{\mathrm{b}} = 0$). The rhs of this equation was obtained in Section 13.3.3 from the expression for ψ_i of an electrolyte solution in the limit of $\kappa \to 0$, see Equations 13.77 and 13.82. Alternatively, we can obtain these results from $\tilde{\psi}_i(\mathbf{k},\boldsymbol{\omega}_1) = \tilde{\rho}_i^\star(\mathbf{k},\boldsymbol{\omega}_1)\tilde{\phi}_{\mathrm{Coul}}^\star(k) \sim \tilde{\rho}_i^\star(\mathbf{k},\boldsymbol{\omega}_1)/[\varepsilon_r\varepsilon_0 k^2]$ by using the Fourier transform techniques of Appendix 13A.

The dipole moment $\mathbf{m}_i^{\mathrm{d}\star}$ of the dressed particle can be written as

$$
\mathbf{m}_i^{\mathrm{d}\star}(\boldsymbol{\omega}) = \int d\mathbf{r}\,\mathbf{r}\rho_i^\star(\mathbf{r},\boldsymbol{\omega}) = \mathbf{m}_i^{\mathrm{d}}(\boldsymbol{\omega}) + \mathbf{m}_i^{\mathrm{dress}}(\boldsymbol{\omega}),
$$

where $\mathbf{m}_i^{\mathrm{d}}$ is the dipole moment of the particle itself (defined as in Equation 13.2) and $\mathbf{m}_i^{\mathrm{dress}}$ is the dipole moment of its dress $\rho_i^{\mathrm{dress}}(\mathbf{r},\boldsymbol{\omega})$. In general, the vectors $\mathbf{m}_i^{\mathrm{d}\star}(\boldsymbol{\omega})$ and $\mathbf{m}_i^{\mathrm{d}}(\boldsymbol{\omega})$ have different directions unless the symmetry of the particle dictates them to be aligned. Furthermore, $\mathbf{m}_i^{\mathrm{dress}}$ can be non-zero even for a particle without dipole moment since the non-electrostatic interactions can make $\rho_i^\star$ to have a dipole moment even when σ_i does not.

Next, we will investigate the leading decay of the pair correlation function in Equation 14.1, which becomes

$$
\begin{aligned}
h_{ij}^{(2)}(\mathbf{r}_{12},\boldsymbol{\omega}_1,\boldsymbol{\omega}_2) &= h_{ij}^\star(\mathbf{r}_{12},\boldsymbol{\omega}_1,\boldsymbol{\omega}_2) \\
&- \beta\int d\mathbf{r}_3\,d\mathbf{r}_4\,\rho_i^\star(\mathbf{r}_{13},\boldsymbol{\omega}_1)\phi_{\mathrm{Coul}}^\star(r_{34})\rho_j^\star(\mathbf{r}_{24},\boldsymbol{\omega}_2) \qquad (14.11)
\end{aligned}
$$

in a homogeneous, isotropic phase. The leading contribution arises from the last term because $\phi_{\mathrm{Coul}}^\star(r) \sim 1/[4\pi\varepsilon_r\varepsilon_0 r]$ for large r and it can be obtained

[4]The integral involving $1/[4\pi\varepsilon_r\varepsilon_0 r_{32}]$ can be evaluated exactly like a potential from a charge distribution in ordinary electrostatics provided that the dressed particle charge distribution $\rho_i^\star$ is used as the source. The result is the multipole expansion of ψ_i in terms of multipole moments of $\rho_i^\star$, Equation 14.10.

in the same manner as in ordinary electrostatics. Alternatively, this can be worked out from $\tilde{h}_{ij}^{(2)}(\mathbf{k}, \omega_1, \omega_2) \sim -\beta\tilde{\rho}_i^\star(\mathbf{k}, \omega_1)\tilde{\rho}_j^\star(-\mathbf{k}, \omega_2)/[\varepsilon_r\varepsilon_0 k^2]$ by using the Fourier transform techniques of Appendix 13A. The end result is

$$h_{ij}^{(2)}(\mathbf{r}_{12}, \omega_1, \omega_2) \sim \beta \frac{m_i^{\mathrm{d}\star} m_j^{\mathrm{d}\star} \mathsf{D}_{ij}^\star(\hat{\mathbf{r}}_{12}, \omega_1, \omega_2)}{4\pi\varepsilon_r\varepsilon_0 r_{12}^3} \quad \text{when } r_{12} \to \infty, \tag{14.12}$$

where $m_i^{\mathrm{d}\star} = |\mathbf{m}_i^{\mathrm{d}\star}|$ and

$$\mathsf{D}_{ij}^\star(\hat{\mathbf{r}}, \omega_1, \omega_2) = 3\left[\hat{\mathbf{r}} \cdot \hat{\mathbf{m}}_i^{\mathrm{d}\star}(\omega_1)\right]\left[\hat{\mathbf{r}} \cdot \hat{\mathbf{m}}_j^{\mathrm{d}\star}(\omega_2)\right] - \hat{\mathbf{m}}_i^{\mathrm{d}\star}(\omega_1) \cdot \hat{\mathbf{m}}_j^{\mathrm{d}\star}(\omega_2). \tag{14.13}$$

$\mathsf{D}_{ij}^\star$ is analogous to the function D_{12} in Equations 13.6 and 13.7 for the dipole-dipole interaction potential, but it contains the unit vector $\hat{\mathbf{m}}_\nu^{\mathrm{d}\star}$ for the dipole moment of the dressed particle instead of $\hat{\mathbf{m}}_\nu^{\mathrm{d}}$. Thus, we have verified that the pair correlation function between the molecules in a pure polar fluid decays like a dipole-dipole potential at large distances. In Fourier space Equation 14.12 corresponds to

$$\tilde{h}_{ij}^{(2)}(\mathbf{k}, \omega_1, \omega_2) \to -\frac{\beta}{\varepsilon_r\varepsilon_0}\left[\hat{\mathbf{k}} \cdot \mathbf{m}_i^{\mathrm{d}\star}(\omega_1)\right]\left[\hat{\mathbf{k}} \cdot \mathbf{m}_j^{\mathrm{d}\star}(\omega_2)\right] \quad \text{when } k \to 0, \tag{14.14}$$

that is, the same as relationship as between $\tilde{u}_{12}^{\mathrm{el}}$ in Equation 13.123 and u_{12}^{el} in Equation 13.125. The following terms in the decay of $h_{ij}^{(2)}$ are from quadrupole/dipole and quadrupole/quadrupole interactions involving $\mathbb{M}_l^{\mathrm{Q}\star}(\omega)$ for $l = i, j$ and thereafter follow higher order multipolar terms.

The special case of a one-component system of species s where the dressed-particle dipole moment vector $\mathbf{m}_s^{\mathrm{d}\star}(\omega)$ of a molecule is parallel with its internal dipole moment vector $\mathbf{m}_s^{\mathrm{d}}(\omega)$ has an interesting feature. The main case is when the two vectors are parallel for symmetry reasons, for example for a dipolar particle with $\mathbf{m}_s^{\mathrm{d}}$ lying along an axis of rotational symmetry or for a molecule like water, but the two vectors can just fortuitously happen to be parallel. In all such cases we have $\mathbf{m}_s^{\mathrm{d}\star} = \pm[m_s^{\mathrm{d}\star}/m_s^{\mathrm{d}}]\mathbf{m}_s^{\mathrm{d}}$, where the minus sign applies when the vectors are oppositely directed, and $\mathsf{D}_{ss}^\star(\hat{\mathbf{r}}, \omega_1, \omega_2) = \mathsf{D}_{ss}(\hat{\mathbf{r}}, \omega_1, \omega_2)$ with D_{ss} defined in 13.126. To proceed we must use an expression for ε_r that we will derive later, Equation 14.21, and it implies that $\mathbf{m}_s^{\mathrm{d}\star} \cdot \mathbf{m}_s^{\mathrm{d}} = (\varepsilon_r - 1)3\varepsilon_0/(\beta n^{\mathrm{b}})$. Since $\mathbf{m}_s^{\mathrm{d}\star} \cdot \mathbf{m}_s^{\mathrm{d}} = \pm m_s^{\mathrm{d}\star} m_s^{\mathrm{d}}$ in the present case, we obtain from Equation 14.12 when $r_{12} \to \infty$

$$h_{ss}^{(2)}(\mathbf{r}_{12}, \omega_1, \omega_2) \sim \frac{(\varepsilon_r - 1)^2}{4\pi \mathsf{y} n^{\mathrm{b}} \varepsilon_r} \frac{\mathsf{D}_{ss}(\hat{\mathbf{r}}_{12}, \omega_1, \omega_2)}{r_{12}^3} \quad \text{if } \mathbf{m}_s^{\mathrm{d}\star} \parallel \mathbf{m}_s^{\mathrm{d}},$$

where $\mathsf{y} = \beta n^{\mathrm{b}}(m_s^{\mathrm{d}})^2/9\varepsilon_0$. The magnitude of the pair correlation function for large separations is in this case solely *given by the dielectric constant*, apart from the system parameters y and n^{b}, and the direction dependence is the same as for the electrostatic pair interaction potential u_{ss}^{el} (since $\mathbf{m}_s^{\mathrm{d}\star}$ and $\mathbf{m}_s^{\mathrm{d}}$ are parallel).

Next, we will consider the interaction $w_{ij}^{(2)}$ between two single particles (or molecules) of species i and j placed at $\mathbf{r}_1$ and $\mathbf{r}_2$, respectively, in a pure polar fluid and we have $n_i^{\mathrm{b}} = 0$ and $n_j^{\mathrm{b}} = 0$. The leading term in the interaction between these particles decays like $-h_{ij}^{(2)}/\beta$, which is dominated for large r_{12} by the integral in the last term of Equation 14.11. If both particles have a net charge their interaction potential goes like $q_i q_j/(4\pi\varepsilon_r\varepsilon_0 r_{12})$ at large distances and if only one has a net charge the interaction goes like a charge-dipole interaction potential with the dipole moment $\mathbf{m}^{\mathrm{d}\star}$ of the dressed uncharged species inserted. For two uncharged particles it follows that their interaction goes like a dipole-dipole interaction potential with the dipole moments $\mathbf{m}_i^{\mathrm{d}\star}$ and $\mathbf{m}_j^{\mathrm{d}\star}$ of the dressed particles inserted. Let us now return to pure polar fluids.

Apart from the leading Coulombic terms in the pair correlation function, we have *contributions of the same kind as in electrolyte systems*, in particular those that are given by the poles of $\tilde{\phi}^{\star}_{\mathrm{Coul}}(k)$ for $k \neq 0$, that is, the zeros of $\tilde{\varepsilon}(k)$. Thus, there are decay modes with decay parameters κ', κ", etc. that are the solutions of $\tilde{\varepsilon}(\mathrm{i}\kappa') = 0$, just as in electrolytes. For a pure polar fluid this equation can be written as $\mathcal{E}_r^{\star}(\kappa') = 0$. Recall that the leading κ-mode in electrolytes turns into the Coulombic $1/r$ interaction when the ion concentration goes to zero and $\kappa \to 0$, which leads to the dipolar and multipolar contributions to the intermolecular interactions and pair correlations in pure polar fluids.

As we have seen, the screened Coulomb potential in water decays like

$$\phi^{\star}_{\mathrm{Coul}}(r) \sim \frac{1}{4\pi\varepsilon_r\varepsilon_0 r} + \frac{1}{|\mathcal{E}_r^{\mathrm{eff}}(\kappa'')|\,\varepsilon_0} \cdot \frac{e^{-\kappa''_{\Re} r}}{2\pi r}\cos(\kappa''_{\Im} r - \vartheta_{\mathcal{E}}(\kappa'')) \qquad (14.15)$$

when $r \to \infty$ (Equation 13.120), where the second term gives the oscillations of $\tilde{\phi}^{\star}_{\mathrm{Coul}}(k)$ in Figure 13.6. The oscillatory decay mode of $\phi^{\star}_{\mathrm{Coul}}(r)$ gives rise to oscillatory contributions to the pair correlation function and we obtain

$$\begin{aligned}
&h_{ij}^{(2)}(\mathbf{r}_{12}, \boldsymbol{\omega}_1, \boldsymbol{\omega}_2) \\
&\quad\sim \frac{\beta m_i^{\mathrm{d}\star} m_j^{\mathrm{d}\star} \mathrm{D}^{\star}_{ij}(\hat{\mathbf{r}}_{12}, \boldsymbol{\omega}_1, \boldsymbol{\omega}_2)}{4\pi\varepsilon_r\varepsilon_0 r_{12}^3} + \text{multipolar Coulombic terms} \\
&\quad - \frac{\beta}{|\mathcal{E}_r^{\mathrm{eff}}(\kappa'')|\,\varepsilon_0}\left[\left|\mathrm{Q}_i^{\mathrm{eff}}(\hat{\mathbf{r}}_{12}, \boldsymbol{\omega}_1, \kappa'')\mathrm{Q}_j^{\mathrm{eff}}(-\hat{\mathbf{r}}_{12}, \boldsymbol{\omega}_2, \kappa'')\right| \frac{e^{-\kappa''_{\Re} r_{12}}}{2\pi r_{12}}\right. \\
&\quad \left.\times \cos\Big(\kappa''_{\Im} r_{12} - \vartheta_{\mathcal{E}}(\kappa'') + \vartheta_i(\hat{\mathbf{r}}_{12}, \boldsymbol{\omega}_1, \kappa'') + \vartheta_j(-\hat{\mathbf{r}}_{12}, \boldsymbol{\omega}_2, \kappa'')\Big) + \ldots\right]
\end{aligned}$$

when $r_{12} \to \infty$ (cf. the rhs of Equation 13.89). The ellipsis in the square bracket indicates higher order oscillatory terms with an exponential factor $e^{-\kappa''_{\Re} r_{12}}/r^{\nu}$ with $\nu \geq 2$ that are not explicitly shown (cf. Section 13.3.3.1).

Each decay mode of $\phi^{\star}_{\mathrm{Coul}}(r)$ gives rise to a Yukawa function decay in $h_{ij}^{(2)}$ like in the case of electrolytes, see Equation 13.88. Likewise, the forces between two surfaces in a polar fluid has contributions with the same decay parameters as $\phi^{\star}_{\text{Coul}}(r)$ apart from the leading electrostatic term for large surface separations. If the surfaces are charged there are counterions present in the system.

As we concluded in the discussion of Figures 13.5 and 13.6, the oscillatory contribution *can be identified with the solvation force* (hydration force in aqueous systems) between the surfaces. Its decay length and wave length are determined by the structure of the bulk phase of the polar fluid, while the surface properties only affect the magnitude and the phase of the interaction. For rough or soft surfaces the oscillations in the surface forces are smoothed out. In that case, the solvation force instead appears as a monotonic, quasi-exponentially decaying force. However, it has the same molecular origin.

14.2 THE STATIC DIELECTRIC CONSTANT AND DIELECTRIC FUNCTION

14.2.1 The dielectric constant calculated from $h^\star$ or $\rho^\star$

In this section we will limit ourselves to a homogeneous and isotropic one-component polar fluid of density n^{b}. The multicomponent case can be treated similarly. In order to obtain $\varepsilon_r = \tilde{\epsilon}(0)$ we will investigate $\tilde{\epsilon}(k)$ for small k. Equation 14.5 is in the current case[5]

$$
\begin{aligned}
\tilde{\epsilon}(k) &= 1 + \frac{\beta}{\varepsilon_0 k^2} \left\langle \tilde{H}^\star(\mathbf{k}, \omega_1, \omega_2) \sigma(\mathbf{k}, \omega_1) \sigma(-\mathbf{k}, \omega_2) \right\rangle_{\omega_1, \omega_2} \\
&= 1 + \frac{\beta n^{\text{b}}}{\varepsilon_0 k^2} \Big[\left\langle \sigma(\mathbf{k}, \omega_1) \sigma(-\mathbf{k}, \omega_1) \right\rangle_{\omega_1} \\
&+ \; n^{\text{b}} \left\langle \tilde{h}^\star(\mathbf{k}, \omega_1, \omega_2) \sigma(\mathbf{k}, \omega_1) \sigma(-\mathbf{k}, \omega_2) \right\rangle_{\omega_1, \omega_2} \Big]
\end{aligned}
\tag{14.16}
$$

(we have dropped the species index because there is only one component). In order to extract $\tilde{\epsilon}(0)$ we expand $\tilde{\sigma}(\mathbf{k}, \omega)$ in a Taylor series around $\mathbf{k} = 0$

$$
\begin{aligned}
\tilde{\sigma}(\mathbf{k}, \omega) &= \int d\mathbf{r}\, \sigma(\mathbf{r}, \omega) e^{-i\mathbf{k}\cdot\mathbf{r}} \\
&= \int d\mathbf{r}\, \sigma(\mathbf{r}, \omega) \left[1 - i\mathbf{k} \cdot \mathbf{r} + O(k^2) \right] \\
&= \int d\mathbf{r}\, \sigma(\mathbf{r}, \omega) - i\mathbf{k} \cdot \left[\int d\mathbf{r}\, \sigma(\mathbf{r}, \omega) \mathbf{r} \right] + \mathcal{O}(k^2)
\end{aligned}
$$

and hence we have

$$
\sigma(\mathbf{k}, \omega) = -i\mathbf{k} \cdot \mathbf{m}^{\text{d}}(\omega) + \mathcal{O}(k^2) \tag{14.17}
$$

[5]This equation implies that $\lim_{k\to\infty} \tilde{\epsilon}(k) = 1$ because the particles are not polarizable. When effects of the electronic degrees of freedom are included $\lim_{k\to\infty} \tilde{\epsilon}(k) > 1$.

because $q = 0$. We insert this into Equation 14.16, whereby we note that the product $\sigma(\mathbf{k}, \omega_2)\sigma(-\mathbf{k}, \omega_1)/k^2 \to [\hat{\mathbf{k}} \cdot \mathbf{m}^{\mathrm{d}}(\omega_2)]\,[\hat{\mathbf{k}} \cdot \mathbf{m}^{\mathrm{d}}(\omega_1)]/\varepsilon_0$ when $k \to 0$. Using the fact that[6] $\int d\omega_1\, [\hat{\mathbf{k}} \cdot \mathbf{m}^{\mathrm{d}}(\omega_1)]^2 = (m^{\mathrm{d}})^2/3$ we obtain

$$\begin{aligned}\tilde{\epsilon}(k) &= 1 + \frac{\beta n^{\mathrm{b}} (m^{\mathrm{d}})^2}{\varepsilon_0} \\ &\times \left[\tfrac{1}{3} + n^{\mathrm{b}} \left\langle \tilde{h}^{\star}(\mathbf{k}, \omega_1, \omega_2)\, \hat{\mathbf{k}} \cdot \hat{\mathbf{m}}^{\mathrm{d}}(\omega_1)\, \hat{\mathbf{k}} \cdot \hat{\mathbf{m}}^{\mathrm{d}}(\omega_2) \right\rangle_{\omega_1, \omega_2} \right] + \mathcal{O}(k^2).\end{aligned} \tag{14.18}$$

In order to obtain ε_r we take the limit $k \to 0$ of this expression and by using the fact that

$$\begin{aligned}&\left\langle \tilde{h}^{\star}(0, \omega_1, \omega_2)\, \hat{\mathbf{k}} \cdot \hat{\mathbf{m}}^{\mathrm{d}}(\omega_2)\, \hat{\mathbf{k}} \cdot \hat{\mathbf{m}}^{\mathrm{d}}(\omega_1) \right\rangle_{\omega_1, \omega_2} \\ &= \tfrac{1}{3} \left\langle \tilde{h}^{\star}(0, \omega_1, \omega_2)\, \hat{\mathbf{m}}^{\mathrm{d}}(\omega_2) \cdot \hat{\mathbf{m}}^{\mathrm{d}}(\omega_1) \right\rangle_{\omega_1, \omega_2}\end{aligned} \tag{14.19}$$

(see Exercise 14.1) we obtain

$$\varepsilon_r = 1 + \frac{\beta n^{\mathrm{b}} (m^{\mathrm{d}})^2}{3\varepsilon_0} \left[1 + n^{\mathrm{b}} \left\langle \tilde{h}^{\star}(0, \omega_1, \omega_2)\, \mathit{\Delta}(\omega_1, \omega_2) \right\rangle_{\omega_1, \omega_2} \right], \tag{14.20}$$

where we have introduced the function

$$\mathit{\Delta}(\omega_1, \omega_2) = \hat{\mathbf{m}}^{\mathrm{d}}(\omega_1) \cdot \hat{\mathbf{m}}^{\mathrm{d}}(\omega_2)$$

(distinguish $\mathit{\Delta}$ from the difference symbol Δ). Equation 14.20 gives ε_r in terms of distributions of dipole moment vectors of the molecules. The higher multipole moments of the molecules contribute only indirectly via their influence on $\tilde{h}^{\star}$. Note, however, that when $\tilde{\epsilon}(k)$ is going to be calculated for $k > 0$, it is in general is not sufficient to only consider the distributions of dipole moment vectors since the terms that decay like $\mathcal{O}(k^2)$ for small k then cannot be neglected.

{**Exercise 14.1:** Show Equation 14.19 by using an argument like that in footnote 6.}

Another expression for ε_r can be obtained from

$$\tilde{\epsilon}(k) = 1 + \frac{\beta n^{\mathrm{b}}}{\varepsilon_0 k^2} \left\langle \tilde{\rho}^{\star}(\mathbf{k}, \omega_1) \tilde{\sigma}(-\mathbf{k}, \omega_1) \right\rangle_{\omega_1},$$

[6]The integral $\int d\omega_1\, [\hat{\mathbf{k}} \cdot \mathbf{m}^{\mathrm{d}}(\omega_1)]^2$ can be evaluated as follows. Since result does not depend on the direction of $\hat{\mathbf{k}}$ we may, for example, let $\hat{\mathbf{k}}$ coincide with the x axis and then we have $[\hat{\mathbf{k}} \cdot \mathbf{m}^{\mathrm{d}}(\omega_1)]^2 = [m^{\mathrm{d}}_x(\omega_1)]^2$. If we instead select the y or z axes, we obtain $[m^{\mathrm{d}}_y(\omega_1)]^2$ and $[m^{\mathrm{d}}_z(\omega_1)]^2$, respectively. Since the integral is independent of the choice of axis, we obtain the same value of the integral in all three cases, so if we sum the three results we obtain a value that is three times too large. Therefore, $\int d\omega_1\, [\hat{\mathbf{k}} \cdot \mathbf{m}^{\mathrm{d}}(\omega_1)]^2 = \frac{1}{3} \int d\omega_1\, \{[m^{\mathrm{d}}_x(\omega_1)]^2 + [m^{\mathrm{d}}_y(\omega_1)]^2 + [m^{\mathrm{d}}_z(\omega_1)]^2\}$. Since the sum in braces equals $(m^{\mathrm{d}})^2$, which is independent of ω_1, we finally obtain the result $(m^{\mathrm{d}})^2/3$.

which can be derived by inserting the Fourier transform of Equation 14.3 (applied to a bulk phase) into $\tilde{\epsilon}(k) = 1 - \tilde{\chi}^\star(k)\tilde{\phi}_{\mathrm{Coul}}(k)$. Since $\rho^\star$ is built up by electroneutral molecules in the present case it has zero total charge (i.e., $q^\star = 0$), so we have

$$\tilde{\rho}^\star(\mathbf{k}, \omega_1) = \int d\mathbf{r}\, \rho^\star(\mathbf{r}, \omega_1) e^{-i\mathbf{k}\cdot\mathbf{r}} = -i\mathbf{k} \cdot \mathbf{m}^{\mathrm{d}\star}(\omega_1) + \mathcal{O}(k^2).$$

In the limit $k \to 0$ we obtain[7]

$$\varepsilon_r = 1 + \frac{\beta n^{\mathrm{b}}}{\varepsilon_0} \left\langle \hat{\mathbf{k}} \cdot \mathbf{m}^{\mathrm{d}\star}(\omega_1)\, \hat{\mathbf{k}} \cdot \mathbf{m}^{\mathrm{d}}(\omega_1) \right\rangle_{\omega_1} = 1 + \frac{\beta n^{\mathrm{b}}}{3\varepsilon_0} \mathbf{m}^{\mathrm{d}\star} \cdot \mathbf{m}^{\mathrm{d}}. \quad (14.21)$$

This expression can alternatively be derived from Equation 14.20.

14.2.2 The dielectric function and ε_r calculated from H_{QQ} or $h^{(2)}$

There exists also an alternative way to obtain $\tilde{\epsilon}(\mathbf{k})$ from pair correlation functions, namely via polarization response function χ^ρ and the charge-charge correlation function H_{QQ}

$$\begin{aligned} \chi^\rho(\mathbf{r}_{34}) &= -\beta q_{\mathrm{e}}^2 H_{\mathrm{QQ}}(\mathbf{r}_{12}) \quad (14.22) \\ &= -\beta \int d\mathbf{r}_1 d\mathbf{r}_2 \left\langle H^{(2)}(\mathbf{r}_{12}, \omega_1, \omega_2)\sigma(\mathbf{r}_{13}, \omega_1)\sigma(\mathbf{r}_{24}, \omega_2) \right\rangle_{\omega_1, \omega_2}. \end{aligned}$$

(Equation 13.43) and we have

$$\tilde{\epsilon}(\mathbf{k}) = \frac{1}{1 + \tilde{\chi}^\rho(\mathbf{k})\tilde{\phi}_{\mathrm{Coul}}(k)}. \quad (14.23)$$

(Equations 13.43 and 13.45), where we still have limited ourselves to a one-component polar fluid in the bulk. Hence we have the alternative expression for the dielectric function

$$\frac{1}{\tilde{\epsilon}(\mathbf{k})} = 1 - \frac{\beta}{\varepsilon_0 k^2} \left\langle \tilde{H}^{(2)}(\mathbf{k}, \omega_1, \omega_2)\sigma(\mathbf{k}, \omega_1)\sigma(-\mathbf{k}, \omega_2) \right\rangle_{\omega_1, \omega_2}. \quad (14.24)$$

Since the response function χ^ρ directly involves the long-ranged function $h^{(2)}$ (expressed as $H^{(2)}$), it is more difficult to handle than $\chi^\star$ in particular for a finite sample of the polar fluid (our treatment here is for an infinite system). If the calculation of $\tilde{\epsilon}(\mathbf{k})$ is done via a path that needs $h^{(2)}$ one can, however, not avoid dealing with the long-range nature of the latter. Numerically it is a demanding task to calculate $\tilde{\epsilon}(\mathbf{k})$ accurately for small k values and hence to calculate the dielectric constant since this requires that the long range tails of the correlation functions are treated accurately. The tails contribute a lot to the Fourier transforms for small k since they are slowly varying in $\mathbf{r}$.

[7]The last equality follows by using an argument like that in footnote 6 and from the fact that the scalar product $\mathbf{m}^{\mathrm{d}*}(\omega_1) \cdot \mathbf{m}^{\mathrm{d}}(\omega_1)$ is independent of the orientation.

Let us from now on consider an isotropic bulk liquid. $\tilde{H}^{(2)}$ is given in terms of $\tilde{h}^{(2)}$ by Equation 13.90 and if we insert this into Equation 14.24 we obtain in an analogous manner as we obtained Equation 14.18

$$\frac{1}{\tilde{\varepsilon}(k)} = 1 - \frac{\beta n^{\mathrm{b}}(m^{\mathrm{d}})^2}{\varepsilon_0} \times \left[\frac{1}{3} + n^{\mathrm{b}}\left\langle \tilde{h}^{(2)}(\mathbf{k},\omega_1,\omega_2)\,\hat{\mathbf{k}}\cdot\hat{\mathbf{m}}^{\mathrm{d}}(\omega_1)\,\hat{\mathbf{k}}\cdot\hat{\mathbf{m}}^{\mathrm{d}}(\omega_2)\right\rangle_{\omega_1,\omega_2}\right] + \mathcal{O}(k^2). \tag{14.25}$$

Again we see that solely the dipole moment distribution in the fluid contributes to $\tilde{\varepsilon}(k)$ in the limit $k \to 0$ and therefore to ε_r. The multipole moments enter in the contributions that vanish like $\mathcal{O}(k^2)$ when $k \to 0$. In the general case, when $\tilde{\varepsilon}(k)$ is going to be calculated for $k > 0$ it is, as we concluded earlier, not sufficient to only consider the distributions of dipole moment vectors.

In contrast to the arguments that lead from Equation 14.18 to Equation 14.20 when $k \to 0$, we cannot simply make the corresponding operation for Equation 14.25 to obtain an expression for $1/\varepsilon_r$ analogous to Equation 14.20. The reason is that $\tilde{h}^{(2)}(\mathbf{k},\omega_1,\omega_2)$ depends on $\hat{\mathbf{k}}$ in the limit $k \to 0$ as seen in Equation 14.14 and this part has to be treated differently than the rest of $\tilde{h}^{(2)}(\mathbf{k},\omega_1,\omega_2)$.

For reasons that will be apparent soon we write $\tilde{h}^{(2)}$ as

$$\tilde{h}^{(2)}(\mathbf{k},\omega_1,\omega_2) = -\frac{\beta(m^{\mathrm{d}\star})^2}{3\varepsilon_r\varepsilon_0}\mathsf{D}^\star(\hat{\mathbf{k}},\omega_1,\omega_2) + \tilde{h}^{(2)\mathrm{rest}}(\mathbf{k},\omega_1,\omega_2), \tag{14.26}$$

where

$$\mathsf{D}^\star(\hat{\mathbf{k}},\omega_1,\omega_2) = 3\left[\hat{\mathbf{k}}\cdot\hat{\mathbf{m}}^{\mathrm{d}\star}(\omega_1)\right]\left[\hat{\mathbf{k}}\cdot\hat{\mathbf{m}}^{\mathrm{d}\star}(\omega_2)\right] - \hat{\mathbf{m}}^{\mathrm{d}\star}(\omega_1)\cdot\hat{\mathbf{m}}^{\mathrm{d}\star}(\omega_2). \tag{14.27}$$

(cf. Equation 14.13 but with $\hat{\mathbf{k}}$ inserted instead of $\hat{\mathbf{r}}$). Note that the entire $\hat{\mathbf{k}}$ dependence of $\tilde{h}^{(2)}$ in the limit $k \to 0$ is included in the first term of $\mathsf{D}^\star$. Therefore, the function $\tilde{h}^{(2)\mathrm{rest}}$ is independent of $\hat{\mathbf{k}}$ in this limit and the contribution from $\tilde{h}^{(2)\mathrm{rest}}$ to the average in Equation 14.25 at $k = 0$ can be obtained in the same manner as in Equation 14.19

$$\begin{aligned}
&\left\langle \tilde{h}^{(2)\mathrm{rest}}(0,\omega_1,\omega_2)\,\hat{\mathbf{k}}\cdot\hat{\mathbf{m}}^{\mathrm{d}}(\omega_2)\,\hat{\mathbf{k}}\cdot\hat{\mathbf{m}}^{\mathrm{d}}(\omega_1)\right\rangle_{\omega_1,\omega_2} \\
&= \tfrac{1}{3}\left\langle \tilde{h}^{(2)\mathrm{rest}}(0,\omega_1,\omega_2)\,\hat{\mathbf{m}}^{\mathrm{d}}(\omega_2)\cdot\hat{\mathbf{m}}^{\mathrm{d}}(\omega_1)\right\rangle_{\omega_1,\omega_2} \\
&= \tfrac{1}{3}\left\langle \tilde{h}^{(2)\mathrm{rest}}(0,\omega_1,\omega_2)\,\Delta(\omega_1,\omega_2)\right\rangle_{\omega_1,\omega_2}.
\end{aligned}$$

By inserting Equation 14.26 into Equation 14.25 and taking limit $k \to 0$, we therefore obtain

$$\begin{aligned}\frac{1}{\varepsilon_r} &= 1 - \frac{\beta n^{\mathrm{b}}(m^{\mathrm{d}})^2}{3\varepsilon_0}\left[1 + n^{\mathrm{b}}\left\langle \tilde{h}^{(2)\mathrm{rest}}(0,\omega_1,\omega_2)\,\Delta(\omega_1,\omega_2)\right\rangle_{\omega_1,\omega_2}\right] \\ &+ \frac{1}{3\varepsilon_r}\left(\frac{\beta n^{\mathrm{b}} m^{\mathrm{d}} m^{\mathrm{d}\star}}{\varepsilon_0}\right)^2 \\ &\times \left\langle \mathsf{D}^\star(\hat{\mathbf{k}},\omega_1,\omega_2)\,\hat{\mathbf{k}}\cdot\hat{\mathbf{m}}^{\mathrm{d}}(\omega_1)\,\hat{\mathbf{k}}\cdot\hat{\mathbf{m}}^{\mathrm{d}}(\omega_2)\right\rangle_{\omega_1,\omega_2}. \end{aligned} \tag{14.28}$$

This expression can be considerably simplified. First, we have

$$\left\langle \tilde{h}^{(2)\mathrm{rest}}(0,\omega_1,\omega_2)\,\Delta(\omega_1,\omega_2)\right\rangle_{\omega_1,\omega_2} = \left\langle \tilde{h}^{(2)}(0,\omega_1,\omega_2)\,\Delta(\omega_1,\omega_2)\right\rangle_{\omega_1,\omega_2},$$

which is a consequence of the fact that (see the shaded text box below)

$$\left\langle \mathsf{D}^\star(\hat{\mathbf{k}},\omega_1,\omega_2)\,\Delta(\omega_1,\omega_2)\right\rangle_{\omega_1,\omega_2} = 0, \tag{14.29}$$

that is, the functions $\mathsf{D}^\star(\hat{\mathbf{k}},\omega_1,\omega_2)$ and $\Delta(\omega_1,\omega_2)$ are orthogonal to each other in orientation space. This is the reason why we used $\mathsf{D}^\star(\hat{\mathbf{k}},\omega_1,\omega_2)$ in the splitting of $\tilde{h}^{(2)}$ into two terms in Equation 14.26. Secondly, as also shown in the shaded text box below, the average in the last term of Equation 14.28 is given by

$$\left\langle \mathsf{D}^\star(\hat{\mathbf{k}},\omega_1,\omega_2)\,\hat{\mathbf{k}}\cdot\hat{\mathbf{m}}^{\mathrm{d}}(\omega_2)\,\hat{\mathbf{k}}\cdot\hat{\mathbf{m}}^{\mathrm{d}}(\omega_1)\right\rangle_{\omega_1,\omega_2} = 2\left(\frac{\varepsilon_0}{\beta n^{\mathrm{b}} m^{\mathrm{d}} m^{\mathrm{d}\star}}\right)^2(\varepsilon_r - 1)^2. \tag{14.30}$$

These facts imply that can write Equation 14.28 as

$$\frac{(\varepsilon_r - 1)(2\varepsilon_r + 1)}{\varepsilon_r} = \frac{\beta n^{\mathrm{b}}(m^{\mathrm{d}})^2}{\varepsilon_0}\left[1 + n^{\mathrm{b}}\left\langle \tilde{h}^{(2)}(0,\omega_1,\omega_2)\,\Delta(\omega_1,\omega_2)\right\rangle_{\omega_1,\omega_2}\right].$$

This relationship, which gives a way to determine ε_r from the pair correlation function, is often written

$$\frac{(\varepsilon_r - 1)(2\varepsilon_r + 1)}{9\varepsilon_r} = \mathsf{y}\mathsf{g}_{\mathrm{K}}, \tag{14.31}$$

where the constant $\mathsf{y} = \beta n^{\mathrm{b}}(m^{\mathrm{d}})^2/9\varepsilon_0$ only contains system parameters and

$$\begin{aligned}\mathsf{g}_{\mathrm{K}} &= 1 + n^{\mathrm{b}}\left\langle \tilde{h}^{(2)}(0,\omega_1,\omega_2)\,\Delta(\omega_1,\omega_2)\right\rangle_{\omega_1,\omega_2} \\ &= 1 + n^{\mathrm{b}}\int d\mathbf{r}\left\langle h^{(2)}(\mathbf{r},\omega_1,\omega_2)\,\Delta(\omega_1,\omega_2)\right\rangle_{\omega_1,\omega_2}\end{aligned}$$

is the so-called **Kirkwood g-factor**. Equation 14.31 is known as the **Kirkwood equation** for ε_r.

By taking the average $\langle h^{(2)}(\mathbf{r},\omega_1,\omega_2)\,\Delta(\omega_1,\omega_2)\rangle_{\omega_1,\omega_2}$ one has extracted out short-ranged correlations between the polar molecules; contributions like the long-ranged dipole-dipole correlations are removed from the average since Δ is orthogonal to $D^\star$ in orientation space. If one neglects these short range correlations and set $\mathsf{g}_\mathrm{K}=1$, one obtains the approximation $(\varepsilon_r-1)(2\varepsilon_r+1)/(9\varepsilon_r)=\mathsf{y}$, which is known as the **Onsager equation** for ε_r.

The orthogonality of $\mathsf{D}^\star(\hat{\mathbf{k}},\omega_1,\omega_2)$ and $\Delta(\omega_1,\omega_2)$ can be proven as follows. By using an argument like that in footnote 6 and selecting $\hat{\mathbf{k}}$ along the x-axis, we can see that

$$\begin{aligned}
&3\left\langle\left[\hat{\mathbf{k}}\cdot\hat{\mathbf{m}}^{\mathrm{d}\star}(\omega_1)\right]\left[\hat{\mathbf{k}}\cdot\hat{\mathbf{m}}^{\mathrm{d}\star}(\omega_2)\right]\,\hat{\mathbf{m}}^{\mathrm{d}}(\omega_2)\cdot\hat{\mathbf{m}}^{\mathrm{d}}(\omega_1)\right\rangle_{\omega_1,\omega_2}\\
&=3\left\langle\left[\hat{m}_x^{\mathrm{d}\star}(\omega_1)\,\hat{m}_x^{\mathrm{d}\star}(\omega_2)\right]\,\hat{\mathbf{m}}^{\mathrm{d}}(\omega_2)\cdot\hat{\mathbf{m}}^{\mathrm{d}}(\omega_1)\right\rangle_{\omega_1,\omega_2}\\
&=\left\langle\left[\hat{\mathbf{m}}^{\mathrm{d}\star}(\omega_1)\cdot\hat{\mathbf{m}}^{\mathrm{d}\star}(\omega_2)\right]\,\hat{\mathbf{m}}^{\mathrm{d}}(\omega_2)\cdot\hat{\mathbf{m}}^{\mathrm{d}}(\omega_1)\right\rangle_{\omega_1,\omega_2}
\end{aligned}$$

because the average in the middle line has the same value if $\hat{\mathbf{k}}$ lies along the y or z axes. This implies that in Equation 14.29, the average of $\Delta(\omega_1,\omega_2)$ times the first term of $\mathsf{D}^\star(\hat{\mathbf{k}},\omega_1,\omega_2)$ in Equation 14.27 cancels the corresponding average involving the second term in $\mathsf{D}^\star(\hat{\mathbf{k}},\omega_1,\omega_2)$ and the orthogonality follows.

The average $\left\langle\mathsf{D}^\star(\hat{\mathbf{k}},\omega_1,\omega_2)\,\hat{\mathbf{k}}\cdot\hat{\mathbf{m}}^{\mathrm{d}}(\omega_2)\,\hat{\mathbf{k}}\cdot\hat{\mathbf{m}}^{\mathrm{d}}(\omega_1)\right\rangle_{\omega_1,\omega_2}$ in Equation 14.30 can be evaluated as follows. We first note that

$$\begin{aligned}
&\left\langle\left[\hat{\mathbf{k}}\cdot\hat{\mathbf{m}}^{\mathrm{d}\star}(\omega_1)\right]\left[\hat{\mathbf{k}}\cdot\hat{\mathbf{m}}^{\mathrm{d}\star}(\omega_2)\right]\,\hat{\mathbf{k}}\cdot\hat{\mathbf{m}}^{\mathrm{d}}(\omega_1)\,\hat{\mathbf{k}}\cdot\hat{\mathbf{m}}^{\mathrm{d}}(\omega_2)\right\rangle_{\omega_1,\omega_2}\\
&=\frac{1}{(m^{\mathrm{d}}m^{\mathrm{d}\star})^2}\left\langle\hat{\mathbf{k}}\cdot\mathbf{m}^{\mathrm{d}\star}(\omega_1)\,\hat{\mathbf{k}}\cdot\mathbf{m}^{\mathrm{d}}(\omega_1)\right\rangle_{\omega_1}\left\langle\hat{\mathbf{k}}\cdot\mathbf{m}^{\mathrm{d}\star}(\omega_2)\,\hat{\mathbf{k}}\cdot\mathbf{m}^{\mathrm{d}}(\omega_2)\right\rangle_{\omega_2}\\
&=\frac{1}{9(m^{\mathrm{d}}m^{\mathrm{d}\star})^2}\left[\mathbf{m}^{\mathrm{d}\star}\cdot\mathbf{m}^{\mathrm{d}}\right]^2=\left(\frac{\varepsilon_0}{\beta n^{\mathrm{b}}m^{\mathrm{d}}m^{\mathrm{d}\star}}\right)^2(\varepsilon_r-1)^2,
\end{aligned}\tag{14.32}$$

where we have used Equation 14.21 to obtain the last equality. This means that first term of $\mathsf{D}^\star(\hat{\mathbf{k}},\omega_1,\omega_2)$ in Equation 14.27 contributes to $\left\langle\mathsf{D}^\star(\hat{\mathbf{k}},\omega_1,\omega_2)\,\hat{\mathbf{k}}\cdot\hat{\mathbf{m}}^{\mathrm{d}}(\omega_2)\,\hat{\mathbf{k}}\cdot\hat{\mathbf{m}}^{\mathrm{d}}(\omega_1)\right\rangle_{\omega_1,\omega_2}$ with three times the value obtained in Equation 14.32. In fact, the second term in $\mathsf{D}^\star(\hat{\mathbf{k}},\omega_1,\omega_2)$ contributes with minus the same value, so in total we obtain the result Equation 14.30. This follows because for the second term, we have[8]

$$\begin{aligned}
&\left\langle\left[\hat{\mathbf{m}}^{\mathrm{d}\star}(\omega_1)\cdot\hat{\mathbf{m}}^{\mathrm{d}\star}(\omega_2)\right]\,\hat{\mathbf{k}}\cdot\hat{\mathbf{m}}^{\mathrm{d}}(\omega_1)\,\hat{\mathbf{k}}\cdot\hat{\mathbf{m}}^{\mathrm{d}}(\omega_2)\right\rangle_{\omega_1,\omega_2}\\
&=\left\langle\left[\hat{\mathbf{k}}\cdot\hat{\mathbf{m}}^{\mathrm{d}\star}(\omega_1)\,\hat{\mathbf{k}}\cdot\hat{\mathbf{m}}^{\mathrm{d}\star}(\omega_2)\right]\,\hat{\mathbf{k}}\cdot\hat{\mathbf{m}}^{\mathrm{d}}(\omega_1)\,\hat{\mathbf{k}}\cdot\hat{\mathbf{m}}^{\mathrm{d}}(\omega_2)\right\rangle_{\omega_1,\omega_2},
\end{aligned}\tag{14.33}$$

which is the same as in Equation 14.32.

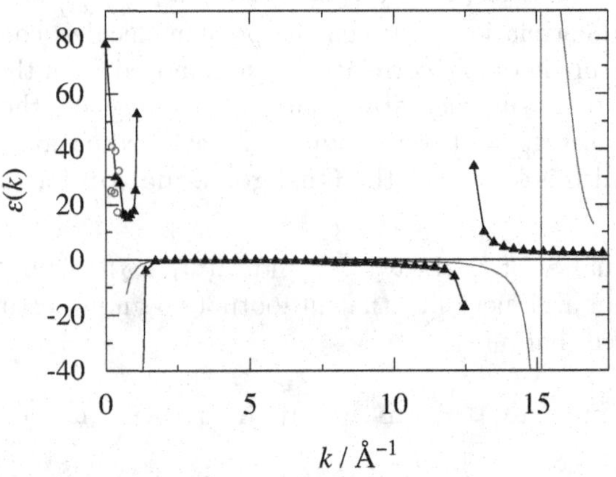

Figure 14.1 **(See color insert).** The dielectric function $\tilde{\varepsilon}(k)$ for water as obtained by Molecular Dynamics simulation for two different models of water, the BJH model (black curves and triangles)[9] and SPC/E water (red curves and open circles).[10]

As a concrete example we show the dielectric function $\tilde{\varepsilon}(k)$ for water in Figure 14.1 calculated by Molecular Dynamics simulations[9,10] for two different water models: the SPC/E model[11] and the BJH central force model.[12] At $k = 0$ the function $\tilde{\varepsilon}(k)$ assumes a value $\varepsilon_r \approx 78$ for the BJH model and 71 for the SPC/E model. Then it initially decreases quickly with increasing k to reach a minimum, after which it increases. For BJH water it goes to infinity at $k \approx 1.1$ Å^{-1} and jumps to minus infinity, while for SPC/E water this occurs for somewhat smaller k. This is a singularity that occurs because $1/\tilde{\varepsilon}(k)$ is zero there and changes sign. For larger k values $\tilde{\varepsilon}(k)$ is negative and

[8] Equation 14.33 is independent of the direction of $\hat{\mathbf{k}}$ and if we select $\hat{\mathbf{k}}$ along the x-axes it becomes

$$\left\langle \left[\hat{m}_x^{\mathrm{d}\star}(\omega_1)\hat{m}_x^{\mathrm{d}\star}(\omega_2) + \hat{m}_y^{\mathrm{d}\star}(\omega_1)\hat{m}_y^{\mathrm{d}\star}(\omega_2) + \hat{m}_z^{\mathrm{d}\star}(\omega_1)\hat{m}_z^{\mathrm{d}\star}(\omega_2)\right] \hat{m}_x^{\mathrm{d}}(\omega_1)\hat{m}_x^{\mathrm{d}}(\omega_2)\right\rangle_{\omega_1,\omega_2}$$
$$= \left\langle \left[\hat{m}_x^{\mathrm{d}\star}(\omega_1)\hat{m}_x^{\mathrm{d}\star}(\omega_2)\right] \hat{m}_x^{\mathrm{d}}(\omega_1)\hat{m}_x^{\mathrm{d}}(\omega_2)\right\rangle_{\omega_1,\omega_2}$$

which is true because $\left\langle \hat{m}_y^{\mathrm{d}\star}(\omega)\hat{m}_x^{\mathrm{d}}(\omega)\right\rangle_\omega = \left\langle \hat{m}_z^{\mathrm{d}\star}(\omega)\hat{m}_x^{\mathrm{d}}(\omega)\right\rangle_\omega = 0$. The latter result follows because when we rotate around the x-axes the vector component $\hat{m}_x^{\mathrm{d}}$ is unchanged while the average of $\hat{m}_\nu^{\mathrm{d}\star}$ for $\nu = y, z$ is zero.

[9] P. A. Bopp, A. A. Kornyshev, and G. Sutmann, J. Chem. Phys. **109** (1998) 1939 (https://doi.org/10.1063/1.476884). Prof. Sutmann is acknowledged for kindly providing the original data.

[10] J. G. Hedley, H. Berthoumieux, and A. A. Kornyshev, J. Phys. Chem. C **127** (2023) 8429 (https://doi.org/10.1021/acs.jpcc.3c00262). Dr. Berthoumieux is acknowledged for kindly providing the original data.

[11] See footnote 35 on page 459.

[12] P. Bopp, G. Jancsó, and K. Heinzinger, Chem. Phys. Letters **98** (1983) 129 (https://doi.org/10.1016/0009-2614(83)87112-7); G. Jancsó, P. Bopp, and K. Heinzinger, Chem. Phys. **85** (1984) 377 (https://doi.org/10.1016/0301-0104(84)85264-7).

has another singularity at $k \approx 13$ Å^{-1} for BJH water where $1/\tilde{\epsilon}(k) = 0$. This second singularity occurs for SPC/E water at $k \approx 15$ Å^{-1}. Thereafter $\tilde{\epsilon}(k)$ is positive and for large k it decreases to the limit $\lim_{k\to\infty} \tilde{\epsilon}(k) = 1$, which is appropriate for a model without electronic polarization.

A singularity may appear to signal something dramatic, but this is not the case. Since the potential $\delta\Psi$ due to a weak external potential $\delta\Psi^{\mathrm{ext}}$ is given by $\delta\tilde{\Psi}(\mathbf{k}) = \delta\tilde{\Psi}^{\mathrm{Ext}}(\mathbf{k})/\tilde{\epsilon}(k)$ (Equation 13.46), a singularity of $\tilde{\epsilon}(k)$ for a certain k value (where $1/\tilde{\epsilon}(k) = 0$) simply means that $\delta\tilde{\Psi}(\mathbf{k})$ is zero there and changes its sign as function of k, while $\delta\tilde{\Psi}^{\mathrm{Ext}}(\mathbf{k})$ does not. The external electrostatic field is given by $\delta\tilde{\mathbf{E}}^{\mathrm{Ext}}(\mathbf{k}) = -i\mathbf{k}\delta\tilde{\Psi}^{\mathrm{Ext}}(\mathbf{k})$ and the total field is $\delta\tilde{\mathbf{E}}(\mathbf{k}) = -i\mathbf{k}\delta\tilde{\Psi}(\mathbf{k})$, which equals $-i\mathbf{k}\delta\tilde{\Psi}^{\mathrm{Ext}}(\mathbf{k})/\tilde{\epsilon}(k)$. For the k values between the singularities, where $\tilde{\epsilon}(k)$ is negative, $\delta\tilde{\mathbf{E}}^{\mathrm{Ext}}(\mathbf{k})$ and $\delta\tilde{\mathbf{E}}(\mathbf{k})$ accordingly have opposite signs. Thus, the total field $\delta\mathbf{E}(\mathbf{r})$ at each point is *directed in the opposite direction* to the sinusoidally varying external field $\delta\mathbf{E}^{\mathrm{Ext}}(\mathbf{r})$ (in form of a planar wave with wave number k).[13]

From $\delta\mathbf{E} = \delta\mathbf{E}^{\mathrm{Ext}} + \delta\mathbf{E}^{\mathrm{pol}}$, where $\delta\mathbf{E}^{\mathrm{pol}}$ is the field from the induced polarization charges in the bulk fluid, it follows that $\delta\mathbf{E}^{\mathrm{pol}}$ for such k values also must be directed in the opposite direction to $\delta\mathbf{E}^{\mathrm{Ext}}$. This situation can in general occur due to interactions between the field and the molecular structure of the polar fluid; a kind of resonance effect due to a commensurability between the short-range periodicity of the liquid structure and the wave length of the imposed field. For such k values the $\delta\mathbf{E}^{\mathrm{pol}}$ strongly opposes $\delta\tilde{\mathbf{E}}^{\mathrm{Ext}}(\mathbf{k})$ leading to the sign change, while for other k the polarization may instead enhance the applied field.

As we have seen, the dielectric function is related to the polarization response function $\chi^{\rho}(r)$, which gives the polarization charge density due to changes in the external potential according to $\delta\rho^{\mathrm{pol}}(\mathbf{k}) = \tilde{\chi}^{\rho}(k)\delta\tilde{\Psi}^{\mathrm{Ext}}(\mathbf{k})$. We have $1 + \tilde{\chi}^{\rho}(k)\tilde{\phi}_{\mathrm{Coul}}(k) = 1/\tilde{\epsilon}(k)$ (Equation 13.45) and since $\tilde{\chi}^{\rho}(k) = -\beta q_{\mathrm{e}}^2 \tilde{H}_{\mathrm{QQ}}(k)$ (from Equation 13.43) we have the following relationship between $\tilde{\epsilon}(k)$ and the charge-charge correlation function

$$\frac{\beta q_{\mathrm{e}}^2}{\varepsilon_0 k^2}\tilde{H}_{\mathrm{QQ}}(k) = 1 - \frac{1}{\tilde{\epsilon}(k)}.$$

This entity must be positive so the only possibilities are that $\tilde{\epsilon}(k) > 1$ or $\tilde{\epsilon}(k) < 0$ because when $0 < \tilde{\epsilon}(k) < 1$ the rhs is negative.[14] Since values of $\tilde{\epsilon}(k)$ between 0 and 1 are forbidden, $\tilde{\epsilon}(k)$ can change sign only via a singularity as

[13]Note that $\delta\tilde{\Psi}(\mathbf{k})$ is a complex-valued function and that both its real and imaginary parts change sign at the singularity (recall that the real and the imaginary parts are given by the cosine and sine transforms of $\delta\Psi(\mathbf{r})$, respectively). Each part of $\delta\tilde{\Psi}(\mathbf{k})$ gives the total electrostatic potential when the bulk fluid is exposed to a weak external potential $\delta\tilde{\Psi}^{\mathrm{Ext}}(\mathbf{k})$ in the form of a sinusoidal plane wave with wave vector $\mathbf{k}$, i.e., a real-valued sine or cosine wave, respectively (cf. the discussion of the plane wave interpretation of Fourier transforms in Appendix 9B). For external fields with other coordinate dependences than plane waves, the fields and the polarization response are superpositions of such planar waves.

[14]O. V. Dolgov, D. A. Kirzhnits, and E. G. Maksimov, Rev. Mod. Phys. **53** (1981) 81 (https://doi.org/10.1103/RevModPhys.53.81).

in Figure 14.1. The k interval where $\tilde{\epsilon}(k) < 0$ lies where $\beta q_e^2 \tilde{H}_{QQ}(k)/[\varepsilon_0 k^2]$ is larger than 1.[15]

The knowledge of $\tilde{\epsilon}(k)$ for a pure polar solvent allows for an approximate treatment of nonlocal electrostatics in solutions. As a simple example we can consider a dilute aqueous electrolyte solution and set $\tilde{\epsilon}_{\mathrm{reg}}(k) \approx \tilde{\epsilon}_{\mathrm{aq}}(k)$, where $\tilde{\epsilon}_{\mathrm{aq}}(k)$ is the dielectric function for pure water. This is reasonable because the vast majority of the water molecules have very few ions in their vicinity. Thereby, we have $\tilde{\epsilon}(k) \approx \tilde{\epsilon}_{\mathrm{aq}}(k) + \beta \sum_i n_i^{\mathrm{b}} q_i q_i^\star/[\varepsilon_0 k^2]$ and hence[16]

$$\tilde{\phi}_{\mathrm{Coul}}^\star(k) \approx \frac{1}{\beta \sum_i n_i^{\mathrm{b}} q_i q_i^\star + \varepsilon_0 k^2 \tilde{\epsilon}_{\mathrm{aq}}(k)}$$

(cf. Equation 13.68). The zeros of the denominator in this approximate $\tilde{\phi}_{\mathrm{Coul}}^\star(k)$ give the decay modes. For the zero $k = i\kappa$ we have

$$\kappa^2 \approx \frac{\beta}{\tilde{\epsilon}_{\mathrm{aq}}(i\kappa)\varepsilon_0} \sum_i n_i^{\mathrm{b}} q_i q_i^\star$$

and likewise for the other zeros. When the leading solutions to this equation are κ and $\kappa''_{\Re} \pm i\kappa''_{\Im}$ we have for $r \to \infty$

$$\phi_{\mathrm{Coul}}^\star(r) \sim \frac{1}{\mathcal{E}_r^{\mathrm{eff}}(\kappa)\varepsilon_0} \cdot \frac{e^{-\kappa r}}{4\pi r} + \frac{1}{|\mathcal{E}_r^{\mathrm{eff}}(\kappa'')|\,\varepsilon_0} \cdot \frac{e^{-\kappa''_{\Re} r}}{2\pi r} \cos(\kappa''_{\Im} r - \vartheta_{\mathcal{E}}(\kappa''))$$

as in Equation 13.119, which we used for a dilute electrolyte solution in Figure 13.4. For zero electrolyte concentration one obtains $\phi_{\mathrm{Coul}}^\star(r)$ for pure water that decays like in Equation 14.15, where $\varepsilon_r = \tilde{\epsilon}_{\mathrm{aq}}(0)$. This case is plotted in Figure 13.6.

For very dilute electrolytes we can set $q_i^\star \approx q_i$ as a further approximation because $q_i^\star \to q_i$ when the concentrations decreases to zero. The leading decay parameter κ then is approximately equal to κ_{D} because $\tilde{\epsilon}_{\mathrm{aq}}(i\kappa) \approx \tilde{\epsilon}_{\mathrm{aq}}(0)$ and the decay parameters $\kappa''_{\Re} \pm i\kappa''_{\Im}$ are approximately the same as in pure water, that is, the oscillatory modes are virtually the same in the dilute solution and in water.

[15]The function $1 - 1/\tilde{\epsilon}(k)$ is plotted in the original papers cited in footnotes 9 and 10. It has a sharp maximum at around $k \approx 3$ Å^{-1}, that is, in the region where $\tilde{\epsilon}(k) < 0$. The SPC/E model gives a higher maximum than the BJH model, but they agree qualitatively with each other.

[16]This approximation has the same form for $\tilde{\phi}_{\mathrm{Coul}}^\star(k)$ as in A. A. Kornyshev, J. Chem. Soc. Faraday Trans. 2, **79** (1983) 651 (https://doi.org/10.1039/F29837900651) if κ_0^2 in that paper is replaced by $\beta \sum_i n_i^{\mathrm{b}} q_i q_i^\star$.

Index

Printed in the United States
by Baker & Taylor Publisher Services